Relativistic Cosmology

Cosmology has been transformed by dramatic progress in high-precision observations and theoretical modelling. This book surveys key developments and open issues for graduate students and researchers. Using a relativistic geometric approach, it focuses on the general concepts and relations that underpin the standard model of the Universe.

Part 1 covers foundations of relativistic cosmology, whilst Part 2 develops the dynamical and observational relations for all models of the Universe based on general relativity. Part 3 focuses on the standard model of cosmology, including inflation, dark matter, dark energy, perturbation theory, the cosmic microwave background, structure formation and gravitational lensing. It also examines modified gravity and inhomogeneity as possible alternatives to dark energy. Anisotropic and inhomogeneous models are described in Part 4, and Part 5 reviews deeper issues, such as quantum cosmology, the start of the universe and the multiverse proposal. Colour versions of some figures are available at www.cambridge.org/9780521381154.

George F. R. Ellis FRS is Professor Emeritus at the University of Cape Town, South Africa. He is co-author with Stephen Hawking of *The Large Scale Structure of Space-Time*.

Roy Maartens holds an SKA Research Chair at the University of the Western Cape, South Africa, and is Professor of Cosmology at the University of Portsmouth, UK.

Malcolm A. H. MacCallum is Director of the Heilbronn Institute at Bristol, and is President of the International Society on General Relativity and Gravitation.

Relativistic Cosmology

GEORGE F. R. ELLIS
University of Cape Town

ROY MAARTENS
University of Portsmouth and University of the Western Cape

MALCOLM A. H. MACCALLUM
University of Bristol

CAMBRIDGE
UNIVERSITY PRESS

CAMBRIDGE
UNIVERSITY PRESS

University Printing House, Cambridge CB2 8BS, United Kingdom

One Liberty Plaza, 20th Floor, New York, NY 10006, USA

477 Williamstown Road, Port Melbourne, VIC 3207, Australia

314-321, 3rd Floor, Plot 3, Splendor Forum, Jasola District Centre, New Delhi - 110025, India

79 Anson Road, #06-04/06, Singapore 079906

Cambridge University Press is part of the University of Cambridge.

It furthers the University's mission by disseminating knowledge in the pursuit of education, learning and research at the highest international levels of excellence.

www.cambridge.org
Information on this title: www.cambridge.org/9781108812764

First published 2012
4th printing 2013
First paperback edition 2021

A catalogue record for this publication is available from the British Library

Library of Congress Cataloging in Publication data
Ellis, George F. R. (George Francis Rayner)
Relativistic cosmology / George Ellis, Roy Maartens, Malcolm MacCallum.
p. cm.
Includes bibliographical references and index.
ISBN 978-0-521-38115-4
1. Cosmology.   2. Relativistic astrophysics.   3. Relativistic quantum theory.
I. Maartens, R. (Roy)   II. MacCallum, M. A. H.   III. Title.
QB981.E4654 2012
523.1–dc23   2011040518

ISBN 978-0-521-38115-4 Hardback
ISBN 978-1-108-81276-4 Paperback

Additional resources for this publication at www.cambridge.org/9781108812764.

Contents

Part 3   The standard model and extensions

Part 4   Anisotropic and inhomogeneous models

Part 5   Broader perspectives

Preface

This book provides a survey of modern cosmology emphasizing the relativistic approach. It is shaped by a number of guiding principles.

- **Adopt a geometric approach** Cosmology is crucially based in spacetime geometry, because the dominant force shaping the universe is gravity; and the best classical theory of gravity we have is Einstein's general theory of relativity, which is at heart a geometric theory. One should therefore explore the spacetime geometry of cosmological models as a key feature of cosmology.
- **Move from general to special** One can best understand the rather special models most used in cosmology by understanding relationships which hold in general, in all space-times, rather than by only considering special high symmetry cases. The properties of these solutions are then seen as specialized cases of general relations.
- **Explore geometric as well as matter degrees of freedom** As well as exploring matter degrees of freedom in cosmology, one should examine the geometric degrees of free-dom. This applies in particular in examining the possible explanations of the apparent acceleration of the expansion of the universe in recent times.
- **Determine exact properties and solutions where possible** Because of the nonlinearity of the Einstein field equations, approximate solutions may omit important aspects of what occurs in the full theory. Realistic solutions will necessarily involve approximation methods, but we aim where possible to develop exact relations that are true generically, on the one hand, and exact solutions of the field equations that are of cosmological interest, on the other.
- **Explore the degree of generality or speciality of models** A key theme in recent cos-mological writing is the idea of 'fine tuning', and it is typically taken to be bad if a universe model is rather special. One can, however, only explore the degree of speciality of specific models by embedding them in a larger context of geometrically and physically more general models.
- **Clearly relate theory to testability** Because of the special nature of cosmology, theory runs into the limits of the possibility of observational testing. One should therefore pur-sue all possible observational consistency checks, and be wary of claiming theories as scientific when they may not in principle be testable observationally.
- **Focus on physical and cosmological relevance** The physics proposed should be plau-sibly integrated into the rest of physics, where it is not directly testable; and the cosmological models proposed should be observationally testable, and be relevant to the astronomical situation we see around us.

• **Search for enduring rather than ephemeral aspects** We have attempted to focus on issues that appear to be of more fundamental importance, and therefore will not fade away, but will continue to be of importance in cosmological studies in the long term, as opposed to ephemeral topics that come and go.

Part 1 presents the foundations of relativistic cosmology. Part 2 is a comprehensive discussion of the dynamical and observational relations that are valid in all models of the universe based on general relativity. In particular, we analyse to what extent the geometry of spacetime can be determined from observations on the past light-cone. The standard Friedmann–Lemaître–Robertson–Walker (FLRW) universes are discussed in depth in Part 3, covering both the background and perturbed models. We present the theory of perturbations in both the standard coordinate-based and the 1+3 covariant approaches, and then apply the theory to inflation, the cosmic microwave background, structure formation and gravitational lensing. We review the key unsolved issue of the apparent acceleration of the expansion of the universe, covering dark energy models and modified gravity models. Then we look at alternative explanations in terms of large scale inhomogeneity or small scale inhomogeneity.

Anisotropic homogeneous (Gödel, Kantowski-Sachs and Bianchi) and inhomogeneous universes (including the Szekeres models) are the focus of Part 4, giving the larger context of the family of possible models that contains the standard FLRW models as a special case. In all cases the relation of the models to astronomical observations is a central feature of the presentation.

The text concludes in Part 5 with a brief review of some of the deeper issues underlying all cosmological models. This includes quantum gravity and the start of the universe, the relation between local physics and cosmology, why the universe is so special that it allows intelligent life to exist, and the issue of testability of proposals such as the multiverse.

The text is at an advanced level; it presumes a basic knowledge of general relativity (e.g. as in the recent introductory texts of Carroll (2004), Stephani (2004), Hobson, Efstathiou and Lasenby (2006) and Schutz (2009)) and of the broad nature of cosmology and cosmological observations (e.g. as in the recent introductory books of Harrison (2000), Ferreira (2007) and Silk (2008)). However, we provide a self-contained, although brief, survey of Riemannian geometry, general relativity and observations.

Our approach is similar to that of our previous reviews, Ellis (1971a, 1973), MacCallum (1973, 1979), Ellis and van Elst (1999a) and Tsagas, Challinor and Maartens (2008), and it builds on foundations laid by Eisenhart (1924), Synge (1937), Heckmann and Schucking (1962), Ehlers (1961), Trümper (1962, and unpublished), Hawking (1966) and Kristian and Sachs (1966). This approach differs from the approach in the excellent recent texts by Peacock (1999), Dodelson (2003), Mukhanov (2005), Weinberg (2008), Durrer (2008), Lyth and Liddle (2009) and Peter and Uzan (2009), in that we emphasize a covariant and geometrical approach to curved spacetimes and where possible consider general geometries instead of restricting considerations to the FLRW geometries that underlie the standard models of cosmology.

A further feature of our presentation is that although it is solidly grounded in relativity theory, we recognize the usefulness of Newtonian cosmological models and calculations. We detail how the Newtonian limit follows from the relativistic theory in situations of cosmological interest, and make clear when Newtonian calculations give a good approximation to the results of the relativistic theory and when they do not.

It is not possible to cover all of modern cosmology in depth in one book. We present a summary of present cosmological observations and of modern astrophysical understanding of cosmology, drawing out their implications for the theoretical models of the universe, but we often refer to other texts for in-depth coverage of particular topics. We are relatively complete in the theory of relativistic cosmological models, but even here the literature is so vast that we are obliged to refer to other texts for fuller details. In particular, the very extensive discussions of spatially homogeneous cosmologies and of inhomogeneous cosmologies in the books by Wainwright and Ellis (1997), Krasiński (1997), and Bolejko et al. (2010) complement and extend our much shorter summaries of those topics in Part 4. Our guiding aim is to present a coherent core of theory that is not too ephemeral, i.e. that in our opinion will remain significant even when some present theories and observations have fallen away. Only the passage of time will tell how good our judgement has been.

We have given numerical values for the key cosmological parameters, but these should be interpreted only as indicative approximations. The values and their error bars change as observations develop, so that no book can give definitive values. Furthermore, there are inherent limitations to parameter values and error bars – which depend on the particular observations used, on the assumptions made in reducing the observational data, on the chosen theoretical model needed to interpret the observations, and on the type of statistical analysis used.

In the text we have two kinds of interventions apart from the usual apparatus of footnotes and references: namely, exercises and problems. The *Exercises* enable the reader to develop and test his or her understanding of the main material; we believe we know the answers to all the exercises, or at least where the answer is given in the literature (in which case an appropriate reference is provided). By contrast, the *Problems* are unsolved questions whose solution would be of some interest, or in some cases would be a major contribution to our understanding.

We are grateful to numerous people who have played an important role in developing our understanding of cosmology: we cannot name them all (though most of their names will be found in the reference list at the end), but we would particularly like to thank John Barrow, Bruce Bassett, Hermann Bondi,[1] Marco Bruni, Anthony Challinor, Chris Clarkson, Peter Coles, Rob Crittenden, Peter Dunsby, Ruth Durrer, Jürgen Ehlers,[1] Henk van Elst, Pedro Ferreira, Stephen Hawking, Charles Hellaby, Kazuya Koyama, Julien Larena, David Matravers, Charles Misner, Jeff Murugan, Bob Nichol, Roger Penrose, Felix Pirani, Alan Rendall, Wolfgang Rindler, Tony Rothman, Rainer Sachs, Varun Sahni, Misao Sasaki, Bernd Schmidt, Engelbert Schucking, Dennis Sciama,[1] Stephen Siklos, John Stewart, Bill Stoeger,

---

[1] deceased

Reza Tavakol, Manfred Trümper, Christos Tsagas, Jean-Philippe Uzan, John Wainwright and David Wands for insights that have helped shape much of what is presented here. We thank the FRD and NRF (South Africa), the STFC and Royal Society (UK), and our departments, for financial support that has contributed to this work.

George F. R. Ellis
Roy Maartens
Malcolm A. H. MacCallum

PART 1

FOUNDATIONS

1 The nature of cosmology

1.1 The aims of cosmology

The physical universe is the maximal set of physical objects which are locally causally connected to each other and to the region of spacetime that is accessible to us by astronomical observation. The scientific theory of cosmology is concerned with the study of the large-scale structure of the observable region of the universe, and its relation to local physics on the one hand and to the rest of the universe on the other.

Thus cosmology deals with the distribution and motion of radiation and of galaxies, clusters of galaxies, radio sources, quasi-stellar objects, and other astronomical objects observable at large distances, and so – in response to the astronomical observations – contemplates the nature and history of the expanding universe. Following the evolution of matter back into the past, this inevitably leads to consideration of physical processes in the hot early universe (the 'Hot Big Bang', or HBB), and even contemplation of the origin of the universe itself. Such studies underlie our current – still incomplete – understanding of the origin of galaxies, and in particular of our own Galaxy, which is the environment in which the Solar System and the Earth developed. Hence, as well as providing an observationally based analysis of what we can see in distant regions and how it got to be as it is, cosmology provides important information on the environment in which life – including ourselves — could come to exist in the universe, and so sets the background against which any philosophy of life in the universe must be set.

Thus, when understood in the widest sense, cosmology has both narrow and broad aims. It has aspects similar to normal physics, at least in its role as an explanatory theory for astrophysical objects (even if laboratory experiments are impossible in this context); aspects peculiar to scientific theories dealing with unique observable objects (and in particular the universe itself, regarded as a physical object); and one can use it as a starting point when considering aspects that stretch beyond science to metaphysics and philosophy.

Sciences vary in their mix of explanatory power, verifiability and links with the rest of science. The relative value one puts on those different qualities of scientific theories affects one's view of the nature of cosmology as a science, and hence one's approach to cosmology. The importance of considering such issues arises from one of the fundamental limitations of cosmology: there is only one universe. We cannot compare it with similar objects, so neither repeatable nor statistical experiments are possible. Thus a prime problem in cosmology is *the uniqueness of the universe* (Bondi, 1960, Harré, 1962, North, 1965, McCrea, 1970, Ellis, 2007). This means we have to pay even more care and attention than in other sciences to extracting as much as possible from data and theory; and we have to be very aware of the limitations of what we can state with reasonable certainty. These issues will be developed in the analysis that follows.

1.1.1 Scientific cosmology

The starting point of cosmology is a *description* of what there is in the universe and how it is distributed and moving – the geography of matter in the large; this is sometimes called 'cosmography'. It inevitably involves a filter of theory through which the raw data has been passed. At this level, the main aim is descriptive and work of this type provides the most accurate representation of the actual universe. We can refer to it as *observational cosmology*. It often leads to unexpected discoveries: for example, the expansion of the universe – and its acceleration, the existence of dark matter, massive walls and voids in the large-scale distribution of matter and large-scale motions of matter.

However, the cosmologist also seeks to *explain* the observations, to give an understanding of what processes are occurring and how they have led to the structures we see – an explanation of the nature and operation of the universe in physical terms. This explores the dynamics of the expansion of the universe in the large, but can also be at the level of the structure and evolution of large-scale objects, e.g. the physics of galaxy formation, the evolution of radio sources and the clustering of galaxies, as well as considering micro-processes in the HBB epoch, such as nucleosynthesis and the decoupling of matter and radiation. These studies can be called *physical cosmology*. It is usual here to take as the background model of the universe on the largest scales one of the Friedmann–Lemaître–Robertson–Walker (FLRW) class, and study the inhomogeneities by considering perturbed FLRW models: the 'standard model' is such a perturbed FLRW model.

The great potential significance of quantum and particle physics for the evolution of the early universe in the big-bang picture has come to the fore in recent years; this field may be called *particle cosmology*. As with physical cosmology, the background model is usually assumed to be an FLRW universe. Aspects of particle cosmology, such as the concept of inflation – an extremely brief era of extraordinarily rapid expansion in the very early universe – are regarded by most cosmologists as part of the standard model of cosmology. This approach is extended by some to *quantum cosmology*, which attempts to describe the very origin of spacetime and of physics. That attempt is still speculative and controversial, inter alia because it involves an engagement with quantum gravity, an as yet speculative theory, and also necessarily raises profound questions about the nature of quantum theory itself.

Finally, this all takes place in the context of gravitational theories based on Einstein's General Relativity (GR) theory. Spacetime curvature – and hence the evolution of the universe – is determined by the matter present via the Einstein Field Equations (EFE). Both the motion of matter in the universe, and the paths of light rays by which we observe it, are determined by this curvature. Therefore an exploration of these features ultimately underlies understanding of the others. *Relativistic cosmology* puts emphasis on the curved spaces demanded by GR and related theories, and focuses on the spacetime geometry of the universe and its consequences for observational and physical cosmology. In order to situate our understanding fully, it considers wider families of universe models than the FLRW models. This is our main approach, and its importance has become apparent from subtleties in applying the standard framework to such issues as horizons, lensing, gauge invariance, chaotic inflation and the supernovae data. A further key issue, which we also explore, is whether GR itself is an adequate theory of gravity for explaining the universe on cosmological scales, or whether some generalization is required.

These approaches have to some extent developed as a historical sequence of new 'paradigms' for cosmology, each offering new depth in our understanding (Ellis, 1993). We believe each of them offers important insights, and that a full understanding of the universe can only come about from the interaction of these approaches, to their mutual enrichment. Thus while our own expertise and emphasis is on the relativistic approach, which is perhaps the most neglected at the present time, we shall endeavour to link this fully to the other views. The full depth of the subject of cosmology involves all of them.

1.1.2 Cosmology's wider implications

An investigation of the universe as a whole inevitably has implications for philosophy and the humanities. For example, we may seek some view on how the cosmos relates to humanity in general and our own individual lives in particular – some conceptualization of how cosmology relates to meaning. This necessarily takes one beyond purely scientific concerns to broader philosophical issues, constrained by the scientific data and theories but not encompassed by them. Science itself cannot resolve the metaphysical issues posed by seeking reasons for existence of the universe, the existence of any physical laws at all, or the nature of the specific physical laws that actually hold, because we cannot devise experimental tests that will answer such questions; they are inevitably philosophical and metaphysical. However such issues lie at the foundation of cosmology.

This book is concerned with the scientific and technical aspects of cosmology. It will not specifically deal with the wider concerns, except for some brief comments towards the end. However, it will contribute to these wider concerns by attempting to delineate carefully the boundaries of what can be reliably achieved in cosmology by use of the scientific method. This involves in particular a careful review of which aspects of cosmological theory are testable by presently possible observations, or by observations that will conceivably be possible some day. These limitations are not always taken seriously in writings on cosmology.

1.2 Observational evidence and its limitations

There are three broad ways in which we obtain the evidence used in cosmology (all of them discussed in more depth later).

1.2.1 Evidence from astronomical observations

By observing the sky with telescopes and other instruments – detecting electromagnetic radiation (infrared, radio, optical, ultraviolet, X-ray and γ-ray), neutrinos, and gravitational waves – we aim to determine the distribution of matter around us. We observe discrete objects and hydrogen clouds up to very large distances, and indirectly observe the total matter (dark plus baryonic) via weak lensing. We also observe background radiation of various kinds that does not come from identifiable discrete sources. The most important such radiation is the blackbody Cosmic Microwave Background (CMB) that we identify as being relic radiation from the HBB. Its study is a central part of present day cosmology. It has propagated freely through space since its emission by hot matter on the Last Scattering Surface (LSS) in the early universe at the time of decoupling of matter and radiation, as the universe cooled through its ionization temperature. The universe was opaque at earlier times.

All electromagnetic radiation travels to us at the speed of light, so, *via electromagnetic phenomena, we can only observe the universe on our past light cone*; hence, as we observe to greater distances, we also observe to earlier times: each object is seen when it emitted the radiation, at a 'look back time' determined by the speed of light. In addition, we can observe massive high-energy particles ('cosmic rays'), but because they are charged they are strongly affected by local magnetic fields, so only very high-energy cosmic rays could carry information across cosmological distances.

Although we have strong evidence for our estimates of distances to the nearer galaxies, determining the *distance* of objects further away is difficult and often controversial. The basic problem is that we have direct observational access only to a two-dimensional projection of a three-dimensional spacetime region: we have to de-convolve these data to recover a three-dimensional picture of what is out there. However, this problem is ameliorated because we can observe at many wavelengths, and so can obtain spectral information about the objects we observe. We can also separate out different polarizations of the radiation received.

Experimentally there are problems in measuring faint signals and excluding effects of intervening matter, theoretically we have to make assumptions about the physical laws and conditions at the sources, and from both together we have to try to establish the intrinsic properties of the sources. The essential idea is to determine some class of 'standard candles' whose intrinsic luminosity is known and whose measured luminosity therefore gives a well-defined distance (relationships such as the Tully–Fisher relation between luminosity and rotational velocity for spiral galaxies are used, as well as classes of objects, like Cepheid variable stars or brightest cluster galaxies).

In spite of these difficulties, we understand quite a lot about the broad nature of what lies around us, as we describe in the next section.

Size of the universe

Astronomical length scales are determined by a variety of methods. Perhaps the most important thing we learn from these scales is that *the universe is extremely large relative to our own size*; even the immensities of our own Galaxy are insignificant compared with the scale

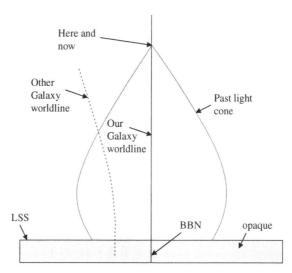

Here and
now

Other
Galaxy
worldline

Past light
cone

Our
Galaxy
worldline

LSS   BBN   opaque

Fig. 1.1   Regions from which we have astronomical and 'geological' evidence, following Hoyle (1962).

of the observed region of the universe, which is of the order of 10^{10} light years (whereas the diameter of our Galaxy is of the order of $50,000$ light years, and the distance to the nearest other galaxy is about 10^6 light years).

This is the primary reason for our major observational problems in cosmology: in effect we can only observe the universe from one spacetime event, dubbed 'here and now', with all our direct observational information coming to us on a single light-cone (see Figure 1.1), supplemented by 'geological' data relating to the early history of our part of the universe (see below). Even a long-term astronomical data collection and analysis programme (say, collating data obtained over the next $10,000$ years by all available means including rocket probes able to travel at the speed of light) would not enable us to evade this restriction by observing the universe from an essentially different spatial or temporal vantage point, as, on cosmological scales, it would not move us from the point labelled 'here and now' in that spacetime diagram. Such a time scale is far too small to be detected relative to 10^{10} years, the scale of the universe itself.

1.2.2 Evidence of a geological nature

Additionally we obtain much useful information from evidence of a 'geological' nature, i.e. by careful study of the history of locally occurring objects as implied by their present-day structure and abundances. Particularly useful are measures of the abundances of elements, together with studies of the nature and hence the inferred ages of local astronomical objects such as star clusters. These observations test features of the early universe at times well before the earliest times accessible with telescopes (though only at points near our world line), thus enabling us to probe the physical evolution of matter in our vicinity at very early times (see Figure 1.1), for example testing Big Bang Nucleosynthesis (BBN) near our world line long before the LSS.

1.2.3 Evidence from local physics

Thirdly, a line of argument due to Mach, Olbers and others (see e.g. Bondi (1960)), argues that local physical conditions and even physical laws would be different if the universe were different; thus we can in principle use the nature of local physical conditions as evidence of the nature of the distant universe.

Mach raised this issue as regards the origin of inertia, and his proposal that inertia depends on the most distant matter in the universe had a profound influence on Einstein's cosmological thinking. Because the strength of the gravitational coupling constant G might be related to inertial properties, and so could depend on the state of the universe, this suggests there might be a *time-varying gravitational 'constant', $G = G(t)$* (Dirac, 1938). Two other examples are,

(a) The *dark night sky* ('Olbers' paradox') – the simplest static universe models suggest the entire sky at night (and, indeed, also during the day) should be as bright as the surface of the Sun. So why is the night sky dark?

(b) The *'arrow of time'* – the effects of the macroscopic laws of physics are dominated by irreversible processes with a unique arrow of time, despite the time reversibility of the fundamental local physical laws.

Plausibly, both may result from boundary conditions in the distant universe at very early times (Ellis and Sciama, 1972, Ellis, 2002), but they certainly have a profound effect on local physics. The essential point is that boundary conditions at the edge of the universe strongly affect the experienced nature of local physical laws, and conceivably affect the nature of the laws themselves – the distinction becomes blurred in the case of cosmology, where the boundary conditions are given and not open to change. We return to these issues in Sections 21.1 and 21.2.

1.2.4 Existence of horizons

Not only do signals fade with distance: if we live in an almost FLRW universe, as is commonly assumed, there is a series of horizons that limit what we can ever observationally or experimentally test in the cosmological context.

Firstly, the HBB era ends when the universe cools so much that matter and radiation, tightly coupled at earlier times, decouple from each other at the LSS in the early universe, which is the source of the CMB we detect today. The universe suddenly becomes transparent at this time: it was opaque to all electromagnetic radiation before then. Hence the earliest times we can access by electromagnetic experiments of all kinds are limited by a *visual horizon*: we can in principle have seen anything this side of the visual horizon, and cannot possibly have seen anything further out – and this will remain true, no matter how technology develops in the future.

There are two important provisos here. Firstly, there is no visual horizon if we live in a *small universe*, that is a universe spatially closed on such small scales that we have already seen around the entire universe more than once. This is a possibility we shall discuss below. Secondly, neutrino and gravitational wave detectors can in principle see to greater distances and earlier times. But they too will have their own horizons, limiting what they can ever see.

Because causal communication is limited by the speed of light, unless we live in a small universe, there exists outside the visual horizon a *particle horizon* limiting causality in the universe. We can have some kind of causal connection to any matter inside the particle horizon, but none whatever with matter outside it. This is a fundamental limitation on physical possibilities in the early universe. The proviso is that geometry at very early times may have been quite unlike that of an FLRW universe, and the situation may be different in FLRW models that collapse to a minimum radius, and then bounce to start a new expansion era. These possibilities also need investigation.

Because the energies we can attain in particle accelerators are limited by practical considerations (e.g. we cannot build a particle accelerator larger than the Solar System), there is a limit to our ability to experimentally determine the nature of the physical interactions that dominate what occurs at extremely early times, and in particular in the quantum gravity era. Hence there is a *physics horizon* preventing us from experimentally testing the relevant physics when we try to apply physical reasoning to earlier times (Section 20.5). Known, or at least potentially testable, physics applies at more recent times; what occurs at earlier times involves physics that cannot be directly observed or confirmed.

Unlike the other two horizons, this is technologically dependent, and the energies determining its location may change with time; nevertheless we may be certain that such a horizon exists. The ability of physical investigations to determine the nature of processes relevant to the very early universe is limited by technological and economic practicalities.

1.3 A summary of current observations

The current state of observations is discussed in detail in Chapter 13. The huge increase in available data and in accuracy of observations is a result of numerous technical developments such as space and balloon-borne telescopes, multi-mirror telescopes, interferometer techniques, adaptive optics, fibre optics, photon multipliers, CCDs, massive computing capabilities and so on, all coming together in an ability to do precision multi-wavelength observations (from radio through optical and infrared to gamma ray) across the entire sky. We shall not describe these developments in this book, but acknowledge that it is only through them that the era of data-based 'precision cosmology' has become possible (Bothun, 1998, Lena, Lebrun and Mignard, 2010). It is this solid grounding in observations and data that makes cosmology the exciting science that it is.

1.3.1 Expansion of the universe – and its acceleration

After Hubble determined the distance of other galaxies (and hence their nature) by observing Cepheid variables in them, the earliest observational result of modern cosmology was Hubble's 1929 law relating the magnitude and redshift of galaxies. The redshift z can be interpreted as due to the Doppler effect of a velocity of recession (since the measurements are made by comparing spectra and using known spectra of different elements, the interpretation depends on assuming that these were the same in the past). The flux received from a distant source depends on its distance, and may also be given as the source's apparent magnitude *m*. For 'standard candles', such as supernovae, the flux is related to distance by the inverse square law; so magnitude is a proxy for distance. For relatively nearby sources believed to be intrinsically alike, the magnitude is related linearly to the redshift, as seen in Figure 1.2. This can be interpreted as a linear relation $v = H_0 d$ between velocity v and distance d, which is then in turn interpreted as due to *expansion of the universe*. The Hubble constant is $H_0 = 100h$ km/s/Mpc. For a long time there were uncertainties in its value of up to a factor 2, but recent observations have given much more accurately determined values. For example the Hubble Space Telescope Key Project gave $h = 0.73 \pm 0.06$ km/s/Mpc (Freedman and Madore, 2010). The constant H_0 gives a time scale $1/H_0$ for the present day expansion: using a linear extrapolation, this would be the time since a moment when all galaxies were in the same place, which gives an estimate of the age of the universe.

The fact that the universe is expanding does not necessarily imply it is evolving: it could conceivably be in a steady state, with the expansion rate always the same and a steady creation of matter keeping the density constant (Hoyle, 1948, Bondi, 1960). However there is a greater number density of radio sources at some distance than there is nearby, which disagrees with a 'steady state' picture. The initial rise and later fall in numbers as we go to fainter fluxes is consistent with a picture of an HBB universe in which radio sources form after the big bang, and their numbers rise to a peak and then start to decrease as their energy sources become exhausted. This is evidence of the crucial feature that *the universe*

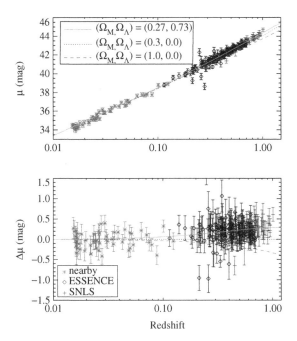

Fig. 1.2 *Top*: Magnitude-redshift diagram for SNIa from three surveys. The solid (green) curve gives the best fit. *Bottom*: Deviation from a model with no dark energy. (From Krisciunas (2008), courtesy of Kevin Krisciunas and the ESSENCE Supernova Search Team. Note that these are the preliminary results as of 2008.) A colour version of this figure is available online.

is evolving as it expands, confirmed by QSO source counts, and also by the discovery of the CMB, as discussed below.

Extension of the magnitude–redshift relation to higher z has depended on using observations of supernovae of type Ia (SNIa), which are believed to behave as good standard candles. The first results were announced in 1998–9. A recent compilation is shown in Figure 1.2. This shows that if the universe has an FLRW geometry, it is expanding more slowly at larger redshift than nearby: the expansion is accelerating. This requires a *'dark energy'*, additional to and with a very different equation of state from the dark matter that we discuss shortly. Accounting for this dark energy is both a central issue for present day cosmology, and a major problem for theoretical physics (see Chapters 14–16).

1.3.2 Nucleosynthesis and the hot big bang

The observed relative abundances of the chemical elements are well accounted for by the HBB and subsequent processing in stars, and poorly accounted for by other hypotheses (Wagoner, Fowler and Hoyle, 1967, Smith, Kawano and Malaney, 1993). Comparing the results (see Figure 1.3) of cosmological nucleosynthesis theory (based in our understanding of nuclear physics), with element abundance data determined from stellar spectra, the observed detailed abundances of the light elements determine the density of baryonic matter in the universe, $\Omega_b h^2 \approx 0.01$ (Steigman, 2006). Fitting all four observed primordial element abundances in this way is a triumph for cosmological theory, because it confirms the application of nuclear physics to the early universe, with the outcome determined via the

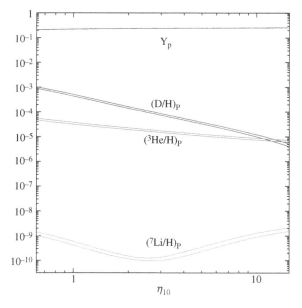

Fig. 1.3   Production of elements in the early universe versus entropy per baryon. (From Steigman (2006). © World Scientific (2006). Reproduced with permission from World Scientific Publishing Co. Pte. Ltd.)

EFE (which govern the expansion rate of the universe, and hence control the rate of change of temperature with time).

1.3.3 Cosmic microwave background and the hot big bang

As well as radiation which we attribute to distinct sources, we also measure background radiation not attributable to such sources. In particular, we detect radiation in the microwave region that is an excellent approximation to black body radiation. The CMB has a temperature $T \approx 3\,\mathrm{K}$. In the HBB picture, this was emitted by the primeval matter at $z \approx 1100$, when the universe became transparent as the temperature dropped below 4000 K, and combination of electrons and nuclei to form atoms took place. At earlier times the universe was opaque because the mean free path for radiation was very small and so matter and radiation were tightly coupled before then. The time of *decoupling of matter and radiation*, when photons last scatter, defines the LSS.

The CMB has been shown to have a very precise black body spectrum (Mather *et al.*, 1990, Fixsen *et al.*, 1996), as shown in Figure 1.4. This is a very important result, with two major consequences. Firstly, this shows that quantum theory as tested in laboratories today held in precisely the same form 13 billion years ago – since a precise black body spectrum for thermal radiation is the unique outcome of quantum theory, as shown by Max Planck over a century ago. Thus this confirms one of the most basic underlying assumptions of cosmology: that physical laws hold unchanged throughout the history of the universe. Secondly, it shows the radiation is well thermalized, fitting the HBB model well. Although the energy density in this radiation is not very high, producing such a spectrum in any model other than an HBB is very difficult.

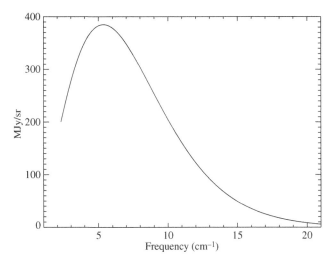

Fig. 1.4   Black body spectrum of the CMB. The RMS errors are 50 parts per million of peak brightness, a small fraction of the line thickness. (From Fixsen *et al.* (1996). Data from FIRAS/COBE. Reproduced by permission of the AAS.)

Fig. 1.5 CMB temperature anisotropy over the sky (WMAP 7-year data). Dark (blue) regions are cold and light (red) are hot: the magnitude of the variation is of the order of 10^{-5}. (From http://map.gsfc.nasa.gov/, reproduced courtesy of NASA/WMAP Science Team.) A colour version of this figure is available online.

Isotropy and homogeneity

The CMB has an extraordinarily high observed degree of *isotropy* about us: after allowance for the motion of the Earth, Sun and Galaxy through the universe (which combine to give a dipole variation), the temperature variations around the sky in the CMB are of order $|\Delta T/T| \lesssim 10^{-5}$, as illustrated in Figure 1.5. These fluctuations mark the presence of density perturbations on the LSS, which will later form the observed galaxies and clusters. Currently the best model for the origin of the fluctuations is inflation, as described below.

This high level of isotropy is the primary evidence supporting our belief in *the homogeneity of the universe on the largest length scales*, because most forms of large-scale inhomogeneity would lead to temperature anisotropies. The distribution of galaxies on large scales shows no evidence of significant anisotropy, but definite conclusions are hindered by the lack of all-sky coverage. Radio source numbers are isotropic to below 5%, and the diffuse X-ray background to below 3%. The CMB data are clearly the best we have.

It should be noted that more direct evidence of homogeneity from discrete sources is hard to obtain. We see distant objects as they were a long time ago, so to compare them to closer objects we would need a deeper understanding of the evolution of galaxies than we have. In testing whether the distribution of galaxies becomes truly homogeneous at large distances, one has to be very sure one has measured beyond the radius at which statistical fluctuations would dominate. There is still controversy about the true situation (Joyce *et al.*, 2005).

The study of the small-scale anisotropies encoded in these temperature variations, and their relation to galaxy formation processes, is a key area of modern cosmology: it will be discussed in detail in Chapter 13. Polarization studies of the CMB, for which first large-angle results have been given by WMAP (Hinshaw *et al.*, 2007), are likely to be another key feature in the future.

Other background radiation

In addition to the CMB, electromagnetic background radiation has been observed in detail in the radio, microwave, X-ray and γ-ray bands. Study of its detailed spectrum and relation to the observed matter density and thermal history is an important part of astrophysical cosmology. There may also be other forms of background radiation, in particular cosmological fluxes of neutrinos and gravitational waves. These are at levels unobservable today, but their effects are indirectly measurable via the CMB (e.g. gravity waves induce B-mode polarization) and the large-scale distribution of matter (e.g. neutrinos affect the matter power spectrum).

1.3.4 Structure formation and the very early universe

Matter is distributed into great voids, filaments and walls populated with clusters of galaxies, apparently arising through a process of structure formation based on a remarkable confluence of particle physics processes and large-scale properties of the universe. This is based on the idea of inflation, mentioned above. This very rapid period of expansion in the very early universe amplifies quantum fluctuations to macroscopic scales, where they become very small density fluctuations that are then the seeds of large-scale structure growth later on. These fluctuations on the LSS are visible to us as fluctuations in the CMB power spectrum, with a particular large peak that has been observed on an angular scale of about 1 minute of arc – and these CMB fluctuations are related to corresponding peaks in the matter angular power spectrum, which have also been observed. Theory and observation fit extremely well.

Also, some observable features of the universe might be due to the properties of quantum gravity, dominant in the times preceding inflation. Salam (1990) has referred to such epochs as the 'speculative era', since we have so little chance of directly testing our theories of the behaviour then. However, the consequences of the physics of such epochs could be very far-reaching, so in recent years much effort has gone into proposed models about what happens then (see Chapter 20). Indeed one potential use of cosmology is to be a laboratory for probing quantum gravity – it is difficult to test theories of quantum gravity in any other context.

1.3.5 Baryonic matter: galaxy distribution and acoustic peak

Baryonic matter occurs in luminous (stars) and non-luminous (gas) forms. On the cosmological scale, we observe an essentially hierarchical structure of stars, star clusters, galaxies and clusters of galaxies; these in turn form even larger structures (voids, walls and filaments) and massive concentrations of matter such as the 'Great Attractor', with associated large-scale motions of matter. There are about 10^{11} galaxies in the observable region of the universe, each containing on the order of 10^{11} stars. It is likely that other kinds of object such as radio sources, quasi-stellar objects (QSOs or quasars), X-ray sources, and active galactic nuclei (AGNs) are related to galaxies, so that understanding their nature is probably a part of the study of galactic structure and evolution.

Massive redshift surveys – the 2dF and the SDSS – have mapped the distribution of galaxies in exquisite detail, over significant fractions of the sky. These surveys allow us to determine the power spectrum of the galaxies observed on various scales, and the related two-point correlation function of sources seen in the sky. See Figure 1.6 for an illustration from the SDSS galaxy redshift survey. Future massive radio surveys, such as the planned

Distribution of galaxies. *Left:* a section through the SDSS survey. *Right:* SDSS optical image. (From
http://cmb.as.arizona.edu/~eisenste/acousticpeak/figs/pie_lrg.eps.gz . Reproduced courtesy of Daniel Eisenstein
and the Sloan Digital Sky Survey.)

SKA, will map the hydrogen on cosmological scales, opening up an important new frontier
in our map of the large-scale distribution of matter.

Inferring densities requires a distance scale such as the Hubble scale, and densities are
usually expressed in dimensionless Ω quantities, or, where the relevant distance scale is
uncertain, Ωh^2. Luminous matter provides $\Omega \approx 0.01$. Intra- and inter-galactic clouds of
gas (e.g. detected by the HI line or by absorption in the Lyman-α forest in quasar spectra),
account for $\sim 45\%$ of the baryonic matter inferred from nucleosynthesis (Nicastro *et al.*,
2005).

A critical test of our theory of structure formation is to trace the evolution of the acoustic
scale in the primordial plasma from the moment of decoupling onwards. This scale arises
from the acoustic waves in the tightly coupled baryon–photon plasma before decoupling,
and is frozen into the radiation and the matter at the time of decoupling. In radiation, it plays
a crucial role in linking the CMB anisotropies to properties of the universe (see Chapter 11).
In baryonic matter, the scale is imprinted as a slight rise in the 2-point correlator of galaxies
(see Chapter 12). Confirmation of this Baryon Acoustic Oscillation (BAO) peak came in
2005 from the 2dF and the SDSS surveys; see Figure 1.7.

1.3.6 Dark matter

We can directly observe the luminous baryonic stars since they radiate, and indirectly we can
detect the non-luminous gas via absorption and emission. Indirect evidence now strongly
indicates that the visible galaxies and surrounding gas are only a very small part of all that
there is.

From various sources of evidence – rotation curves of galaxies (Figure 1.8), dynamics
of galaxy clusters, X-ray emitting gas in clusters, gravitational lensing – interpreted using
GR, it has been known for some time that there is much more non-luminous 'dark' matter
than visible matter (we consider the evidence further in Section 12.3). Dark matter provides

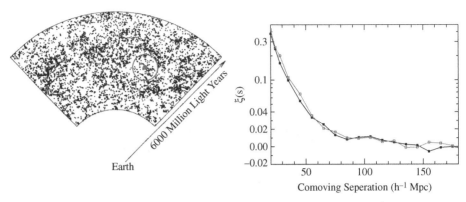

Fig. 1.7 *Left:* Section of a sphere centred on a galaxy in the SDSS survey, with comoving radius $\sim 100h^{-1}$ Mpc, the BAO scale at which an excess of neighbours should be found. (From http://space.mit.edu/home/tegmark/sdss/. Reproduced courtesy of Max Tegmark/SDSS Collaboration.) *Right:* Confirmation of the BAO in the correlation function, in two redshift slices, $0.16 < z < 0.36$ (filled squares, black) and $0.36 < z < 0.47$ (open squares, red). (From Eisenstein *et al.* (2005). Reproduced by permission of the AAS.) A colour version of the right-hand figure is available online.

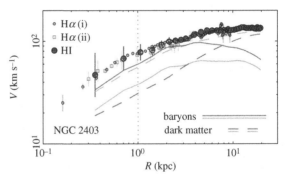

Fig. 1.8 An example of the evidence for dark matter from rotation velocities of stars and gas in spiral galaxies. The measured rotation curve for the spiral galaxy NGC 2403 is compared with the baryonic (solid) and dark matter (dashed) contributions, for two different models of the stellar mass-to-light ratio. (From McGaugh *et al.* (2007). Reproduced by permission of the AAS.)

$\Omega \approx 0.3$, but uncertainty about its physical nature and detailed distribution is one of the major uncertainties of modern cosmology.

There is a 'bias' between the clustering of baryonic and dark matter, but its nature is unknown and phenomenological models for the bias are necessary, supplemented by empirical relations based on N-body simulations. If we know the bias, in principle we can deduce the distribution of dark matter from that of baryonic matter. A major new development in cosmology is an independent probe of the total matter distribution, and hence a handle on the distribution of dark matter. This probe is in the form of gravitational lensing (see Section 12.4), which is now being used (Massey *et al.*, 2007) to map the dark matter density; see Figure 1.9.

Fig. 1.9   Distribution of dark and baryonic matter, mapped via weak gravitational lensing using the COSMOS survey. (From http://www.spacetelescope.org/images/ . Reproduced courtesy of NASA, ESA/Hubble and R. Massey.) A colour version of this figure is available online.

The cosmological effects of dark matter, and the particle physics models of candidate dark matter particles, are consistent with a pressure-free equation of state, and hence it is known as Cold Dark Matter (CDM). A key feature is the *non-baryonic* nature of CDM. The point is that the amount of dark matter detected by astrophysical methods is much larger than the amount of baryons compatible with the nucleosynthesis results mentioned above. Major theoretical and experimental efforts are underway to determine what non-baryonic forms of CDM are plausible. With the advent of gauge theories of physics and the various forms of grand unified theory (including supersymmetric and superstring theories), a wide variety of possible exotic particle species can have cosmological significance, and could indeed constitute the CDM. Many astronomical and laboratory experiments are presently underway to try to detect these candidates.

1.4 Cosmological concepts

Having outlined the definite and possible constituents of the cosmos, we have to decide how to theoretically represent the matter present, the physical theories governing its behaviour, and the spacetime geometries relevant to cosmology.

1.4.1 Matter description

As regards the matter in the universe, in practice a fluid description, associated with a continuum approximation, is very widely employed. This may or may not be justified; alternative possibilities are that the universe could be chaotic or hierarchical. The situation will be clearer if we contrast these cases.

Consider a plot of measured average density ρ against the averaging scale L used in the measurement of matter in some domain. When L is very small, there will be considerable fluctuations in density as L is changed; at these scales, individual particles affect the measured density appreciably. There may be a significant intermediate range of scales where the value obtained is essentially independent of smoothing scale chosen: fluctuations are smoothed out by averaging, giving values insensitive to details of the averaging volume chosen (Batchelor, 1967). These are the scales where a *fluid description* is appropriate. At very large scales, macroscopic gradients become important, and the fluid representation is then again inadequate for an average at those scales.

In the *chaotic case*, the curve obtained is not smooth on any scale, so no well-defined value is defined by any suitable averaging procedure (the value always depends crucially on averaging size, time and position).

In the *hierarchical case*, like the fluid situation, the curve becomes smooth at some averaging size, but it never levels out at a finite radius to become approximately constant: the average determined at a given scale changes as the averaging volume changes size, and in hierarchical cosmologies it is usually assumed that the density goes to zero on the largest scales (though other behaviours might be possible). Thus no good density function is defined because the result obtained is never approximately independent of the size of the averaging volume used.

Which of these possible descriptions applies to the real universe can only be determined by observation. Much of what follows depends on assumption of the existence of averaging scales where the fluid approximation (smooth representation) is applicable. We believe this description is valid because we have evidence for smoothness of the matter flow on a macroscopic scale; such evidence is given by the smooth magnitude–redshift relation (see Figure 1.2) on scales of 50 Mpc to 200 Mpc. Thus from the present viewpoint, an essential achievement of those measurements is to validate the smooth (fluid) picture on these scales, with perhaps some caveats arising from the sponginess of the structures revealed by the deep redshift surveys.

Hence a fluid description of the matter present is the core of many dynamical studies in modern cosmology. However, it should be noted (Heller, 1974) that there is an important point usually glossed over, namely that the fluid picture for the early universe demands a very different averaging scale: the consequently required transitions between different averaging scales at different cosmological epochs are rarely considered at all.

The fluid picture essentially arises from averaging over a particle distribution, and one very useful way to describe that at a microscopic level is *kinetic theory*, which we outline in Section 5.4. This description plays a key role in studies of the CMB anisotropies.

Magnetic fields with micro-gauss strength affect the dynamics and evolution of galaxies and clusters of galaxies. Cosmic magnetic fields on larger scales are generated (at a very weak level) by second-order effects during recombination; stronger large-scale fields could be generated by other mechanisms, and could leave a detectable imprint on the CMB. Thus as well as fluids, some questions involve investigation of magnetic effects.

In the early universe quantum fields are important, and it is conventional to model these effects by a scalar field. Such a scalar field, with its potential energy dominating its kinetic energy (and hence with negative pressure), can behave like a positive cosmological constant, and drive inflation. It is perhaps problematic here that we have not yet physically detected a single scalar field in a laboratory or particle accelerator experiment; nevertheless it is plausible that such fields will give reasonable descriptions of effective theories for various quantum fields. In any case they are widely used in discussions of the dynamics of the early universe, so we shall consider them too.

Finally, what description should we use for cold dark matter and dark energy? The former is conventionally modelled as pressure-free matter ('dust'), and the latter as either a cosmological constant, or a scalar field ('quintessence').

1.4.2 Dynamics

It is now generally agreed that local physical laws (applied everywhere) can be used to describe the evolution of the universe; what is controversial is the issue (mentioned above) of the extent to which the nature of the universe affects the nature of local physical laws, perhaps causing an evolution of those laws, as for example in Dirac's proposal (see Section 1.2.3) of a time-varying gravitational constant as a way of taking cognizance of Mach's principle.

Our present day understanding of physics is based on there being four fundamental forces of nature effective at the present time (that were probably unified at early times): the strong and weak nuclear forces, electromagnetism and gravity. Of these, only gravity and electromagnetism are long-range and are therefore candidates for determining the spacetime curvature in cosmology.

Electromagnetic dynamics

If there were an overall electric charge on galaxies, e.g. if there were different numbers of protons and electrons in astronomical bodies, or if the proton charge were infinitesimally different from the electron charge (Lyttleton and Bondi, 1959), then electromagnetism would be the dominant force on astronomical scales. However, it would be hard to develop a scheme consistent with observation in which some astronomical bodies had overall positive charges and some negative, as they seem so similar in structure. We do not have evidence of intergalactic lightning, nor is it easy to develop a scheme in which close bodies are matter–antimatter pairs with opposite charges and associated matter–antimatter annihilation. But if all bodies had like kinds of charges, and therefore repelled each other, then forces on the cosmological scale would be repulsive, while astronomical objects, such as our Local Group of galaxies, form bound systems from terrestrial scales up to the scales of clusters of galaxies and thus clearly must be governed by the attractive force we call gravity. Hence dominance of electromagnetism on the largest scales does not seem very plausible.

Gravitational dynamics

Thus it is now believed that gravity is the dominant force on large scales. Since GR is the best available classical theory of gravity, it is clear that it is the most appropriate theory to use to describe cosmological models, representing the geometry of the universe through a curved Riemannian spacetime model, and using the EFE to determine the evolution of spacetime curvature.

In much cosmological work of an astrophysical nature the fundamental role of GR can be hidden by the use of only a few time-evolution equations for the FLRW models and quasi-Newtonian equations for structure formation studies. This can be problematic, and it is one of the aims of this book to display the importance of relativistic concepts in cosmology. Among them are the principle of equivalence, and the absence of any independently fixed background within which the gravitational effects can be calculated, both of which have, as we shall discuss, implications for the handling of calculations of the evolution of structure.

The nonlinearity of GR implies the possibility of unique features such as existence of black holes and associated spacetime singularities (on small scales), and existence of spacetime singularities at the start of the universe (on large scales). Indeed the expanding relativistic models and associated singularity theorems raise the possibility of a *beginning to the universe*, an edge to spacetime, where there is an origin of matter and radiation in the universe, of space and time, and even of physical laws. If this in fact occurs, it is one of the most important physical features of the cosmos, for the predictability of science breaks down here; the nature of causation comes into question at such an event.

Close to such a boundary, the matter density may be very high, and a quantum gravity theory would appear to be necessary to describe gravity at the very high densities of the earliest stages of a HBB. Quantum gravity effects may very well allow singularity avoidance at the start of the universe, and traces of the quantum gravity epoch may still remain in the details of the CMB anisotropy patterns. These are important features to consider; however, this volume will only briefly consider issues in quantum cosmology, which is a major topic in its own right.

1.5 Cosmological models

Taking all this into account, key ingredients of relativistic cosmology are:

(1) a *spacetime,* with a Lorentzian metric and associated torsion-free connection (these terms are defined in Chapter 2), determined through the EFE from the matter and radiation present;
(2) a *description of matter and radiation*, with appropriate thermodynamic, kinetic or field-theoretic models that determine their local physical properties;
(3) a uniquely defined family of *fundamental observers*, whose motion represents the average motion of matter in the universe. This matter should be expanding during some epoch which plausibly corresponds to the universe domain we see around us at present – so these motions should correspond well with astronomical observations.

Observationally, the 4-velocity of such a family can be determined either (a) by measuring the motion of matter in an averaging volume (e.g. a local cluster of galaxies) and determining a suitable average of those motions, or (b) from the CMB anisotropy measurements. There is a preferred frame of motion in the real universe such that the radiation background is (approximately) isotropic; this is a classic case of a broken symmetry (the solution breaks the symmetry of the equations).[1] We move with almost that preferred velocity, which can be dynamically related to that of the matter present in the universe (the 'Great Attractor' is thought to be responsible for our own peculiar velocity relative to the cosmological background). Our usual assumption is that the matter and CMB velocities agree. If not, we can model this situation too, but with more complex models involving relative motion of matter and radiation.

With suitable astrophysical assumptions, we also require

(4) A set of *observational relations* that follow from the geometry and the interactions between the matter and radiation in it.

Putting this all together,

> a COSMOLOGICAL MODEL consists of a spacetime with well-defined, physically realistic, matter and radiation content plus a uniquely defined family of fundamental observers whose world lines are expanding away from each other in some universe domain, resulting in a well-defined set of observational predictions for that domain.

In constructing cosmological models, one attempts to fit them with the observations, but they are also used to provide explanatory frameworks. There is a tension between these two roles of the models which is the source of the different approaches to cosmological modelling that we discuss in the sequel.

1.5.1 Averaging scales

It is fundamentally important that each attempt at modelling is based on an implied *averaging scale*, determined by the description used, and a *range of applicability* for both the physical and geometric descriptions used (together, these determine the physical effects taken into account). The model will be related to observations which also have an implied averaging scale (determined by the resolution) and a range of applicability (determined either by limitations of the method of observation, or by imposing a cutoff in the data). Clearly the observational technique should involve an averaging scale suitable to the model.

As we shall discuss in Chapter 16, the crucial problem of averaging remains unresolved in GR, due to the complexities following from the nonlinearity of the theory. Attempts to solve this problem are of fundamental importance.

When we speak of cosmological models, we imply that the models do not describe small-scale structure; they are valid as a description only above some scale of averaging, which should be made explicit. In general we choose an averaging volume large enough that – in the observed universe – it has a positive matter density.

The models we use will also be restricted in the epoch of their applicability: different matter models and levels of description may be described at different epochs. Thus there will usually be some time in our past before which the model is considered inapplicable. An overall cosmological model is in fact usually a patchwork of models applicable to different

---

[1] One cannot observe this velocity from within an isolated box, e.g. if closed off in a laboratory with no windows; thus this does not violate the principle of special relativity.

epochs (inflation, nucleosynthesis, decoupling, galaxy formation, etc.), which are loosely linked to each other with appropriate junction conditions, to form the overall model.

1.5.2 Specific models

Although we shall cover the full range of available specific models in detail in Parts 3 and 4, it is useful during the general discussion in Part 2 to have some simple models to refer to as examples.

FLRW models

The most important and basic of these are the FLRW models, i.e. the 'standard models' of cosmology. These universes are exactly spatially homogeneous (they have no centre, and no feature distinguishing any one spatial point from any other) and isotropic (all spatial directions are equivalent at every point). They are discussed in detail in Part 3.

The FLRW models are extraordinarily effective in their power to explain the broad features of cosmology – particularly the expansion and evolution of the universe. However, they are inadequate as realistic universe models in that they are *exactly* homogeneous and isotropic: the real universe is clearly neither. Thus we need more elaborate models.

We can obtain a great deal of useful understanding by considering perturbations of the FLRW universes, enabling us to construct reasonably realistic model universes that are like FLRW models on average, and to investigate issues such as the growth of structure in an almost-FLRW universe and observational properties of such models. These important models will be discussed at length in Chapters 10–13. However they will not enable us to investigate all the theoretical issues of interest that arise, so we shall also consider universe models that are intrinsically anisotropic or inhomogeneous. This will lead to useful new insights on the possible nature of the universe.

We shall consider whether models quite different from the FLRW models could explain present observations. This can also throw light on the crucial unresolved problem of averaging. Another major unsolved issue in cosmology is whether the universe was initially very smooth, and then developed inhomogeneity by physical processes, or was initially very chaotic, and was then smoothed by physical processes. The first possibility can be adequately studied by using the standard FLRW models and their perturbations; the second requires the study of alternative (inhomogeneous and anisotropic) models.

Spherically symmetric inhomogeneous models

Lemaître–Tolman–Bondi (LTB) models (Section 15.1) are the simplest inhomogeneous expanding models, spherically symmetric about a centre. They have been used to give exact nonlinear models of inhomogeneous cosmologies where no dark energy is needed – the apparent acceleration of the universe seen in the supernova data is not a consequence of dark energy (as in an FLRW model), but is due to spatial inhomogeneity. This is an important alternative to the standard interpretation, and is discussed in detail in Chapter 15. Further cosmological uses of these models are discussed in Chapter 19.

Lumpy inhomogeneous models

'Swiss cheese' models (Section 16.4.1) insert spherical regions representing virialized structures into a smooth universe (such as FLRW). In the general case they provide simple exact models which are anisotropic and inhomogeneous. The Einstein–Strauss solution consists of an interior Schwarzschild region, representing a black hole or the vacuum exterior to a spherical body, joined across a spherical boundary to an exterior FLRW solution. One can alternatively take a collapsing FLRW region as the interior spherical body. It is possible to have any number of such non-overlapping Schwarzschild regions in a surrounding FLRW solution, with centres distributed in an arbitrary manner, and this makes a model suitable for nonlinear modelling of a distribution of galaxies or clusters – e.g. for a simplified study of observational effects due to multiple gravitational lensing. This is discussed in Chapter 16; their further uses are discussed in Chapter 19.

Spatially homogeneous models

Bianchi universe models (Chapter 18) are anisotropic expanding models that are spatially homogeneous. They are useful for investigations aimed at bounding the anisotropy of the real universe, or concerned with such matters as the nature of the HBB singularity, the onset of inflation, a lack of particle horizons in at least some directions, variations in nucleosynthesis outcomes and complex CMB angular variations and polarization patterns. The simplest such models that have been investigated in this regard are Bianchi I models, with flat spatial sections. More general Bianchi models exhibit much more complex behaviours, for example chaotic dynamics and hesitation dynamics.

1.6 Overview

An outline of the book is as follows:

In **Part 1** ('Foundations'), we consider the nature of cosmology in Chapter 1, and the geometric and physical foundations of cosmological studies successively in Chapters 2 and 3.

In **Part 2** ('Relativistic cosmological models'), we look at the description of generic cosmological models. Their geometry and kinematics are described by covariant variables in Chapter 4. Appropriate matter descriptions and the consequences of their dynamical equations and conservation equations are covered in Chapter 5. The dynamic consequences of the gravitational equations (the EFE) are considered in Chapter 6. The nature of observations in generic cosmological models is considered in Chapter 7, and the light-cone approach to observations is discussed in Chapter 8.

In **Part 3** ('The standard model and extensions'), we examine in depth the FLRW models (Chapter 9) and their perturbations (Chapter 10). We consider the CMB anisotropies in these models (Chapter 11), and then structure formation and gravitational lensing in

Chapter 12. On this basis, we confront these models with current astronomical observations (Chapter 13). This leads on to a consideration of the crucial issue of how to explain the supernova, CMB and galaxy distribution data that indicate a speeding up of the expansion of the universe in recent times. In Chapter 14 we consider dark energy or modified theories of gravity causing acceleration of the expansion in a FLRW universe. Chapter 15 considers the possibility that the observations can be explained geometrically, with no dark energy needed, through large-scale spatial inhomogeneity (the Copernican Principle is not satisfied). Chapter 16 considers the more radical alternative that the apparent acceleration may be at least in part due to dynamical back-reaction effects and/or averaging effects on cosmological observations due to local inhomogeneities.

In **Part 4** ('Anisotropic and inhomogeneous models'), we first look at the space of possible anisotropic and inhomogeneous models in Chapter 17, handling their geometry and dynamics in an exact covariant manner. Chapter 18 looks at the geometry and dynamics of Bianchi spatially homogeneous but anisotropic universe models, which can be nicely described through a dynamical systems approach. Chapter 19 looks at how exact inhomogeneous models may be used to illuminate aspects of structure formation in cosmology.

In **Part 5** ('Broader perspectives'), we discuss the issue of quantum cosmology and a start to the universe in Chapter 20, emphasizing problems due to difficulties in testing the relevant physics for this era. Chapter 21 looks at the relation of cosmology to local physics and to the existence of life in the universe, including the vexed issues of the Anthropic Principle and the possible existence of a multiverse. A concluding overview and perspective are given in Chapter 22.

Appendix and References: An appendix summarizes issues to do with notation, units chosen and common abbreviations. Finally an extensive list of references (arranged alphabetically by first author, with full titles of all articles referred to) is a hopefully helpful resource in its own right.

Problem 1.1 Consider whether cosmic rays may give us useful cosmological information. (A key issue is how far away from us they originate.)

Problem 1.2 Give a complete and rigorous account of the transitions between and validity of the differing bases for a fluid picture during the evolution of the universe.

Problem 1.3 Consider the physical conditions needed for charge neutrality of astronomical objects like stars and galaxies. Are there any fundamental reasons that they should be fulfilled, or is it just an environmental quirk that has led to this situation?

Problem 1.4 Sometimes when new evidence contradicts a model, it can be adapted to accommodate this new data; sometimes it has to be abandoned, and replaced by something quite different. What kind of data would be sufficient to cause abandonment of a cosmological model, in general; and of the present standard model of cosmology, in particular?

2 Geometry

Physics usually begins with some concept of a space of events, or of positions at which objects or fields can be present. While this could be a discrete space (e.g. in lattice models) or a topological space without extra structure, it is usually assumed to be a continuum on which one can carry out the operations of calculus, i.e. a differential manifold.

Most of modern theoretical physics can be written in the language of differential geometry and topology, though it has only become common to do so since gauge theories assumed their present prominent role. Many advanced notions in these areas find a place in physics (see e.g. Nakahara (1990)). We shall be careful below to distinguish between concepts dependent on and independent of the presence of a metric, since gauge theories usually do not assume a metric.

While it is not true that every geometric object is of physical significance, it will be true of the geometric quantities we discuss in subsequent chapters. So when we consider geometric questions, it is important to recognize that these can also be understood as physical. Indeed, one of our aims is to show how powerful geometric methods can be in discussing cosmological questions, once one has mastered the necessary tools.

Dirac in his classic book (Dirac, 1981) emphasizes that quantum mechanics rests on a number of principles. The first two are that observables are operators, which can be expressed in different bases, and that the core of the dynamics lies in non-zero commutators for these operators. The same is true of gravity realized as geometry, in particular in GR. (However, GR does not display Dirac's two other main principles, namely that superposition holds, implying linearity whereas GR is essentially nonlinear, and that only probabilities are predicted, not definite outcomes.)

Firstly, an operator is an entity that exists in its own right, and can be dealt with as such (e.g. in defining a commutator), but can be represented in many different bases, leading to various different representations: in QM Dirac showed this and uncovered the profound equivalence between apparently completely different representations. We shall describe how, in GR, vectors are differential operators; 1-forms and, more generally, tensors are multilinear algebraic operators on them; the process of 'dragging along' (Section 2.4) is an operator, generated by its differential version, the Lie derivative; parallel transport is an operator, generated by its differential form – the covariant derivative (Section 2.5); and there is a Hamiltonian operator in GR (Section 3.3.2), which is important in the link to quantum gravity.

Vectors and tensors have many different representations related by the usual tensor transformation laws, as well as both coordinate and tetrad representations (Section 2.8). This illustrates how different representations, which can look very different, may refer to the same entity. By referring to the operator itself rather than just to some specific representation one can make manifest the unity behind a diversity of appearances.

Secondly, we observe that the essential gravitational physics in GR lies in the curvature, the commutator of covariant derivatives. Curvature can be thought of as the operator generating holonomy (the basic feature underlying gauge theories). Consequently geodesic deviation is also an operator acting on connecting vectors, which is what leads to tidal gravitational effects, including null geodesic focusing and gravitational lensing. Moreover, the commutator of two Lie derivatives is another Lie derivative, the basic feature of Lie algebras and hence of spacetime symmetries.

We therefore emphasize these two aspects of geometry in this chapter. We believe that doing so leads to conceptual clarity that is useful in understanding GR.

As discussed in the preface, we assume the reader has studied special relativity and an introduction to general relativity such as the texts named there.[1] Such courses have to introduce manifolds and tensors, including tangent (contravariant) vectors and 1-forms (covariant vectors), and, at least for the Riemannian case, covariant derivatives, the Levi-Civita connection, and curvature. Nevertheless, for reference, to fix notation, and to make our treatment self-contained, we introduce these concepts briefly in Sections 2.1–2.7. For those to whom differential geometry is new, a suitable relativity text, or works such as Schutz (1980), might be more digestible, while those to whom it is very familiar can sensibly skip to what is new to them. Stephani *et al.* (2003) provides a more formal but still concise summary with a somewhat different selection of topics.

We also assume the reader is familiar with linear algebra (including scalar products and duals), and with continuity and differentiability and their basic applications including three-dimensional vector calculus.

This chapter also covers a number of ideas needed in relativistic cosmology but not routinely treated in introductory courses. To help readers decide whether to skip this chapter, the main ones are maps between spacetimes, aspects of tensor algebra, Lie derivatives, holonomy, geodesic deviation, symmetries, the Levi-Civita 4-form, the Weyl tensor, sectional curvature, use of general bases, and geometry of hypersurfaces.

2.1 Manifolds

Essentially, a differential manifold is a space that can be given coordinates locally, though there may be no coordinate system covering the whole space. Coordinates can be chosen arbitrarily, provided that the relations between any two coordinate systems are differentiable.

We take a set of points M with a well-defined topology.[2] If p is a point in M, an open set U containing p is called a *neighbourhood* of p. Note that an open set in $\mathbb{R}^n$ does *not*

---

[1] More advanced material can be found in, for example, Wald (1984), Stewart (1994) and Penrose and Rindler (1984, 1985).

[2] A topology defines and is defined by which sets are open sets: these must include the empty set and M itself, and have the properties that any union of open sets is open, and any intersection of a finite number of open sets is open. In spacetime, we want to ensure that distinct events can be taken to lie in distinct open sets, and that it is possible to define distance and integration without problems. Mathematically these aims are achieved by assuming that the topology is paracompact and Hausdorff, terms defined in introductory texts, e.g. Brickell and Clark (1970).

contain any of its boundary points. As an example, open sets in $\mathbb{R}$ are just (combinations of) open intervals (a, b).

In a manifold we can assign coordinates $\mathbf{x} = (x^1, x^2, \ldots, x^m)$ on a neighbourhood around any point p in M. A neighbourhood U together with the coordinates on it is called a *coordinate chart*. We often write $\mathbf{x}$ as (x^a), where $a = 1 \ldots m$.

For differentiability we require that if we have two distinct charts which overlap in an open set W and assign coordinates $(x^1, x^2, \ldots, x^m)$ and $(y^1, y^2, \ldots, y^m)$ on them, then on W the functions $y^a(x^b)$ for $a, b = 1 \ldots m$ are differentiable, and so are their inverses $x^b(y^a)$. The manifold is said to be of class C^k if the coordinate transformations are continuously differentiable k times. Charts obeying these conditions on the coordinate transformation $y^a(x^b)$ are called *compatible*. Assuming M is connected (i.e. cannot be split into disjoint non-empty open sets) compatibility implies that m is the same everywhere in M: it is the *dimension* of M. For simplicity and physical relevance we consider only connected manifolds.

A set of charts covering all of M is called an *atlas*, and one can add to it any other compatible chart, i.e. one has effectively an arbitrary choice of coordinates. Note that in general an atlas requires more than one chart: for instance, the circle requires at least two charts. (The usual description, that the circle is labelled by an angle $0 \leq \theta \leq 2\pi$ and $0 \equiv 2\pi$, is not strictly permitted because it becomes 1-2 at the point labelled by 0. However, the properties of angular coordinates are so well understood that we usually ignore this, recalling it only when it becomes important.)

Although coordinates are essential to the structure of a manifold, it is important to characterize physical objects in a coordinate-independent way. This can be done by providing a coordinate-free definition, or by giving a definition in coordinate terms and proving it gives the same object whatever coordinates are used. Once we know an object is coordinate-independent, we can safely use its coordinate representation (or its form in a general basis, as in Section 2.8) for calculations.

Continuous (or *differentiable*) *functions* f *on* M are those which are continuous (or differentiable) when expressed as functions of the coordinates. A map $h : M \rightarrow N$ between two differential manifolds M and N, of dimensions m and n respectively, is a C^k *differentiable map* if for coordinates x^a in M and $y^{b'}$ in N, $y^{b'}(x^a)$ is C^k differentiable.

While for some purposes, such as proving singularity theorems (Hawking and Ellis, 1973), physicists specify k carefully, for most purposes it is good enough to assume k is infinite, or even assume the stronger restriction that the coordinate transformations are analytic, i.e. have Taylor series which converge to the actual map in some neighbourhood. We shall usually assume without comment that our manifolds (and the various objects on them that we introduce) are as smooth as necessary for the differentiations we make. To check this physically would imply arbitrarily accurate measurement, so the assumption is unlikely to restrict the physical phenomena we can discuss.

The manifolds we use as cosmological models will be four-dimensional spacetimes. We shall often want to consider lower-dimensional manifolds contained in a spacetime, for instance a spatial surface at a given time, or more generally consider manifolds containing others of lower dimension, so we need to introduce submanifolds. A p-dimensional differential *submanifold* N of M is a subspace of M such that at any point in N there are coordinates $(x^1, \ldots, x^p, x^{p+1}, \ldots, x^m)$ for M in which points in N have coordinates $(x^1, \ldots, x^p, 0, 0, \ldots, 0)$.

In a differential manifold of dimension m, submanifolds of dimension $m - 1$ are often of interest, e.g. the spatial sections $t = $ const in the FLRW spacetimes (2.65). They are called *hypersurfaces* and are discussed further in Section 2.9. In cosmology, it is a useful convention, which we shall follow, to denote the three-dimensional components in a (usually spacelike) hypersurface Σ by indices from the middle of the alphabet, e.g. i, j, k, reserving the early alphabet e.g. a, b, c for the fully four-dimensional objects.

Modelling the interface between two regions as a sharp boundary, e.g. to discuss a shock wave, the surface of a star or a brane-world cosmology, requires the concept of a *manifold M with boundary* ∂M, which is similar to a manifold except that coordinates take values in the region of $\mathbb{R}^m$ with $x^m \geq 0$ (note that such sets need not be open in $\mathbb{R}^m$). The boundary ∂M, whose intersection with each coordinate patch is $x^m = 0$, is like a hypersurface (but may have several separate pieces), except that the manifold lies on one side of it only. When joining regions in spacetime with different matter content or other properties, we can consider two manifolds with boundary joined at a hypersurface (the 'junction surface'), at which junction conditions for physical quantities can be derived by an appropriate matching procedure (see Section 16.4.2).

Note that the Cartesian product $M = Q \times N$ of two manifolds Q and N (the set of pairs (q, n) for q in Q and n in N) is also a manifold in an obvious way. In this case, for any q in Q the points (q, n) for n in N form a submanifold of M.

A manifold is called *orientable* if one can cover it with systems of coordinates such that all the Jacobian matrices of the coordinate transformations, which have entries of the form $\partial y^{b'} / \partial x^a$, have positive determinants. There is then a second set of coordinate systems whose Jacobian determinants relative to the first are all negative, and these two sets give the two possible choices of orientation (in three dimensions these correspond to the choice between a left-hand rule and the usual right-hand rule). An orientation of the manifold with boundary implies an orientation in the boundary. If a manifold is not orientable, problems arise in globally defining quantities such as spinors.

2.2 Tangent vectors and 1-forms

Physics makes extensive use of vectors for many entities, such as forces and fields. They can be introduced in a number of ways, for example as differential operators, as tangents to (equivalence classes of) curves or via a fibre bundle. These ways can then be proved to be equivalent, see Auslander and MacKenzie (1963), Hicks (1965) and Lang (1962).

Following our earlier remarks on the significance of operators, we shall take a tangent vector to be a directional derivative operator, obeying the usual rules for differentiation of sums and products (the 'chain rule' can then be derived from the rules for differentiable maps between manifolds and the chain rule for spaces $\mathbb{R}^m$ and $\mathbb{R}^n$). The space of all tangent vectors at p, $T_p(M)$, is a vector space, with the usual rules for linear combinations.

Writing a differentiable function f in terms of coordinates as $f(x^a)$, and using the chain rule, we obtain for the directional derivative $\mathbf{V}(f)$ of f in the direction $\mathbf{V}$,

$$\mathbf{V}(f) = \frac{\partial f}{\partial x^a} \mathbf{V}(x^a) = V^a \frac{\partial f}{\partial x^a}, \tag{2.1}$$

where $V^a = \mathbf{V}(x^a)$. Here we use, as we shall from now on, the *Einstein summation convention* that an index repeated exactly twice in a product, once as a subscript and once as a superscript, is to be summed over all its possible values; such indices are called *dummy* indices. Indices appearing more than twice, or at the same level, are either mistyped or require explicit summation. Indices appearing just once are called *free* indices, should appear in the same way in all terms in an equation and imply that the equation is true for each of their possible values. The a in $\partial f/\partial x^a$ is considered to be a subscript. Occasionally we mildly break this convention when denoting dependences on indexed variables: e.g. $y^b = y^b(x^a)$. From (2.1), the coordinate form of $\mathbf{V}$ is

$$\mathbf{V} = V^a \frac{\partial}{\partial x^a}. \tag{2.2}$$

We often refer to the vector just as V^a. For brevity we often denote $\partial f/\partial x^a$ by $f_{,a}$ and $\partial/\partial x^a$ by ∂_a. Equation (2.2) implies that $T_p(M)$ is m-dimensional and hence isomorphic with $\mathbb{R}^m$.

Now let us consider a change of coordinates, to new coordinates $y^{b'}$. By applying the vector to an arbitrary function, we find that the new coordinate components of $\mathbf{V}$ are given by

$$V^{b'} = \mathbf{V}(y^{b'}) = V^a \frac{\partial y^{b'}}{\partial x^a}. \tag{2.3}$$

Any set of quantities obeying (2.3), for sets of coordinates, form the components of a *vector*.

We may need to relate vectors on two distinct manifolds, not necessarily of the same dimension: for example, perturbed and background cosmological models, or a spatial surface and the spacetime containing it, or a higher-dimensional model and four dimensions within it. If we have a differentiable map of manifolds $h : M \to N$, with dimensions m and n, a vector $\mathbf{V}$ at p in M is mapped to a vector $h_* \mathbf{V}$ at $h(p)$ in N by the rule that for any f on N,

$$h_* \mathbf{V}(f) = \mathbf{V}(f(h(x))). \tag{2.4}$$

Here, h_* is called *the push-forward map*. Evaluation of this in coordinates leads again to the formula (2.3) but the indices a and b' now have different ranges $(1,\ldots,m$ and $1,\ldots,n$ respectively); $y^{b'}$ now refers to the coordinates in N of the point $h(p)$ for p in M, and $V^{b'}$ to the corresponding components of $h_* \mathbf{V}$ in N. Regarding the previous interpretation of (2.3) as the special case where $M = N$ and h is the identity map $p \mapsto p$ we see why this should be so.

A *curve* can be expressed in coordinates by m functions $x^a(v)$ where v in $\mathbb{R}$ is the curve parameter, which could be taken to lie in the interval $\mathcal{I} = [0, 1]$. The tangent vector at any point of a differentiable curve is then $V^a = dx^a/dv$. This suggests a way of relating vectors in differential geometry to the elementary idea of vectors as displacements, by identifying the vector with the displacement from p to a point at unit parameter distance along a curve α with tangent $\mathbf{V}$. However, since such curves α are not unique, making this concept precise requires a way to fix α.

Many of the entities in physical theories, such as an electric field, are vector fields rather than vectors, i.e. they require a specification of a vector at each point of spacetime. We shall then need to be able to compare vectors at neighbouring points, so we need to make the set of all tangent spaces $T_p(M)$ at all points p in M (the *tangent bundle* $T(M)$ of M) a differential manifold. To do so, the coordinates on a neighbourhood U in M can be extended to a neighbourhood $U \times \mathbb{R}^m$ in $T(M)$ by using coordinates (x^a, V^b), V^b as in (2.2). The tangent bundle thus has dimension $2m$. *Vector fields* are then (differentiable) maps from M to $T(M)$ determining a particular vector $\mathbf{V}_p$ in $T_p(M)$ at each point p in M. Since we so frequently use vector fields, we often say 'vector' when strictly we mean 'vector field'.

Any set of m linearly independent vector fields forms a basis for $T_p(M)$ at each point. Usually introductory texts discuss only coordinate bases $\{\partial_a\}$, but in spacetimes non-coordinate bases, in particular bases of vectors whose scalar products are constants, are often useful (see Section 2.8). Hence where possible we obtain relations in an arbitrary basis: the specialization to constant scalar product or coordinate bases is usually easy.

We can compute the *commutator* of (differentiable) vector fields $\mathbf{V}$ and $\mathbf{W}$, provided that the functions f are at least C^2. In that case, $\mathbf{V}$ gives a new function $\mathbf{V}(f)$ to which $\mathbf{W}$ can be applied. The commutator

$$[\mathbf{V}, \mathbf{W}](f) := \mathbf{V}(\mathbf{W}(f)) - \mathbf{W}(\mathbf{V}(f)) \tag{2.5}$$

is a vector field which has coordinate components

$$V^b \frac{\partial W^a}{\partial x^b} - W^b \frac{\partial V^a}{\partial x^b} . \tag{2.6}$$

Direct evaluation shows that such commutators always obey the *Jacobi identity*

$$[\mathbf{U}, [\mathbf{V}, \mathbf{W}]] + [\mathbf{V}, [\mathbf{W}, \mathbf{U}]] + [\mathbf{W}, [\mathbf{U}, \mathbf{V}]] = \mathbf{0} \tag{2.7}$$

for any three (sufficiently differentiable) vector fields $\mathbf{U}$, $\mathbf{V}$ and $\mathbf{W}$. Commutators are preserved by a map (2.4), i.e.

$$[h_*\mathbf{V}, h_*\mathbf{W}] = h_*[\mathbf{V}, \mathbf{W}] . \tag{2.8}$$

A *covariant vector* is characterized by the analogue

$$\omega_{b'} = \omega_a \frac{\partial x^a}{\partial y^{b'}} . \tag{2.9}$$

of (2.3); examples are provided by differentials of functions df. In this language the tangent vectors above are 'contravariant', and these names continue to be used for the upper and lower index positions on tensors – see Section 2.3. The more modern name for a covariant vector is a 1-form. Given a vector $\mathbf{V}$, $\boldsymbol{\omega}$ can operate on it to give a scalar $\boldsymbol{\omega}(\mathbf{V}) = \omega_a V^a$, so covariant vectors are elements of the algebraic dual of the space of tangent vectors. Although this means that the spaces of tangent vectors and of 1-forms at a point are each vector spaces of dimension m, and hence isomorphic, there is no unique map between them.

Following the way we built the tangent bundle from tangent vectors, we can build the 1-form bundle $T^*(M)$ of M, also called the *cotangent bundle* of M, and 1-form fields (usually called just 1-forms, for brevity). From a map of manifolds $h : M \to N$ we obtain a map $h^* : T^*(N) \to T^*(M)$, the *pullback*, by the rule that for any ω in $T^*(N)$ and $\mathbf{v}$ in $T(M)$,

$$(h^*\omega)(\mathbf{v}) = \omega(h_*\mathbf{v}) , \tag{2.10}$$

using (2.4). Using (2.3) this gives

$$(h_*\omega)_a = \frac{\partial y^{b'}}{\partial x^a}\omega_{b'} \tag{2.11}$$

for the coordinate components of 1-forms, and (2.9) can be seen to be the case when $M = N$. Thus (2.9) can be re-interpreted in the general case as (2.3) was, so that the coordinates now relate to two different manifolds. Note that if h maps M with coordinates x^a to N with coordinates $y^{b'}$, in general only the derivatives $\partial y^{b'}/\partial x^a$ are defined, because $h(M)$ will be only a submanifold of N, so one has only the maps h_* of vectors on M to vectors on N and h^* of 1-forms on N to those on M, and not maps of vectors on N to vectors on M or 1-forms on M to those on N.

2.3 Tensors

Physics uses tensors widely, physicists usually encountering first some of rank two. Among them are: in the mechanics of rigid bodies, the tensor composed of the moments and products of inertia, mapping angular velocity (a vector) linearly to angular momentum (another vector); in fluid dynamics, or elastic media, the shear, strain and stress tensors; and in electromagnetic theory in special relativity, the Maxwell field tensor and its associated stress tensor.

In linear algebra, a tensor T is an operator acting linearly on each of a number of copies of a vector space V and its dual V^*, giving a numerical value $T(v_1,\ldots, v_q, w^1,\ldots, w^p)$ for each choice of a set of vectors v_b, $b = 1,\ldots, q$, in V and 1-forms w^a, $a = 1,\ldots, p$, in V^*. Such a tensor is said to have rank $p + q$ and type (p, q). The tensors of type $(1, 0)$ constitute V itself, and those of type $(0, 1)$ constitute V^*. Scalars, i.e. numbers, are regarded as tensors of type $(0, 0)$. In a basis $\{e_a\}$ of V with dual basis $\{\omega^b\}$ of V^*, the components of a tensor T are the values $T(e_f,\ldots,e_h,\omega^a,\ldots,\omega^c)$. From linearity in each argument we then get

$$T(v_1,\ldots, v_q, w^1,\ldots, w^p) = T^{ab\cdots c}{}_{fg\cdots h} w_a^1 w_b^2 \cdots w_c^p v_1^f v_2^g \cdots v_q^h , \tag{2.12}$$

where we must remember that the superscripts on w_a^q and subscripts on v_b^c are not component indices, just labels. So far the order of upper indices relative to lower indices is not significant.

The space of type (p, q) tensors is itself a vector space. A basis of this space is provided by the product of basis elements of V and V^* so that we could write, for instance, $\mathbf{T} = T^a{}_b e_a \omega^b$.

In the context of differential geometry, we are interested in the tensors constructed from the tangent vectors. These are given by the following formula for the change of coordinate components of tensors:

$$T^{a'b'\cdots}{}_{i'j'\cdots} = \frac{\partial y^{a'}}{\partial x^c} \frac{\partial y^{b'}}{\partial x^d} \cdots \frac{\partial x^k}{\partial y^{i'}} \frac{\partial x^m}{\partial y^{j'}} \cdots T^{cd\cdots}{}_{km\cdots} \,, \tag{2.13}$$

which contains as special cases the rules (2.3) and (2.9). The same formula of course results from expansion of (2.12) in terms of coordinates, using the fact that the Jacobian matrix gives the relevant linear transformation of bases. (As already mentioned, the superscript indices are referred to as contravariant and the subscripts as covariant.) From this it is easy to show that the *Kronecker delta* symbol $\delta^a{}_b$, which has the value 1 if $a = b$ and zero otherwise, is a tensor.[3]

From the requirement that for all choices of vectors and 1-forms the formula (2.12) gives a scalar, one can deduce that the components of T form a tensor obeying (2.13), a result known as the 'tensor detection theorem' (sometimes called the 'quotient theorem' although no division is involved).

Many of the tensors used in general relativity, and physics generally, are either symmetric or skew-symmetric (also called antisymmetric). Taking a rank two covariant tensor, its *symmetric part* is the tensor whose components are

$$T_{(ab)} := \tfrac{1}{2}(T_{ab} + T_{ba}) \,. \tag{2.14}$$

Symmetrization over n indices similarly implies taking all possible permutations of those indices, and dividing by the number of such permutations;

$$T_{(abc)} := \tfrac{1}{6}(T_{abc} + T_{cab} + T_{bca} + T_{acb} + T_{bac} + T_{cba}) \,, \tag{2.15}$$

for example. Antisymmetrization is analogous, except that each term involving an odd permutation (i.e. one obtainable by an odd number of exchanges of pairs of indices) is multiplied by a factor (-1); thus the antisymmetric, or skew, part of a rank two tensor is

$$T_{[ab]} := \tfrac{1}{2}(T_{ab} - T_{ba}) \,, \tag{2.16}$$

and for a rank three tensor,

$$T_{[abc]} := \tfrac{1}{6}(T_{abc} + T_{cab} + T_{bca} - T_{acb} - T_{bac} - T_{cba}) \,. \tag{2.17}$$

Tensors of rank greater than two can be symmetrized or skewed on any set of indices (two or more) of the same type. The notation is extended, if the indices (anti)symmetrized over are not adjacent, by using a vertical bar to mark the limits of the (anti)symmetrization; thus, for example,

$$T_{(a|cd|b)} := \tfrac{1}{2}(T_{acdb} + T_{bcda}) \,. \tag{2.18}$$

It is a simple exercise to check that (anti)symmetrization is a property invariant under basis changes.

Covariant tensors skew on all their p indices are referred to as p-forms. One can introduce a derivative d on the space of p-forms (extending the concept of df above), giving the

---

[3] Later we shall use δ_{ab}, with the same values for given a and b, but which, unlike δ^b_a, is not a tensor. It is used, for example, to write the metric of a three-dimensional spacelike surface in an orthonormal basis.

exterior calculus. For information on this see e.g. Schutz (1980) or Stephani *et al.* (2003), Chapter 2. The *p*-form fields and their exterior calculus play an important role in gauge theories and string theory (see Section 20.3.1).

For a rank two tensor, one easily sees that it can be uniquely split into symmetric and skew parts:

$$T_{ab} = T_{(ab)} + T_{[ab]}, \tag{2.19}$$

and similar, but more complicated, combinations hold for higher ranks (Young tableaux, familiar in quantum mechanics, are a means of working out the combinations required: see Agacy (1997)).

In simplifying complicated tensor expressions it is often useful to remember the easily proved rule that 'symmetric contracted with skew gives zero', i.e. if $A^{ab\cdots}{}_{mn\cdots}$ is a tensor symmetric in ab and $B^{pq\cdots}{}_{abk\cdots}$ is a tensor skew in ab, then $A^{ab\cdots}\cdots B^{\cdots}{}_{ab\cdots}$ is 0. In particular if we have a symmetric metric g^{ab} (see Section 2.7.1) the trace of a skew tensor is zero: if $A_{ab} = A_{[ab]}$, then $A := A^a{}_a = g^{ab}A_{ab} = 0$.

Now we consider various algebraic operations on tensors. Addition and subtraction, and multiplication by scalars, are simple. The *(outer) product* of two tensors, T_1 of type (p_1, q_1) and T_2 of type (p_2, q_2), is the tensor acting on $(p_1 + p_2)$ copies of V^* and $(q_1 + q_2)$ copies of V whose action uses T_1 on the first p_1 copies of V^* and q_1 copies of V, and T_2 on the rest. In terms of components we can write the new tensor T as

$$T^{ab\cdots cd\cdots e}{}_{fg\cdots hi\cdots m} = T_1{}^{ab\cdots c}{}_{fg\cdots h}T_2{}^{d\cdots e}{}_{i\cdots m} . \tag{2.20}$$

The other algebraic operation we need on tensors is *contraction*. The contraction on the jth contravariant index and kth covariant index is the tensor whose components are

$$T^{ab\cdots n\cdots c}{}_{de\cdots m\cdots f}\delta^m{}_n = T^{ab\cdots m\cdots c}{}_{de\cdots m\cdots f}, \tag{2.21}$$

where on the left side m is the kth covariant index and n is the jth contravariant index on T. Calculations involving substantial numbers of contractions are hard to notate in an index-free way: indices become almost essential (compare Schouten (1954), Section I.11). It should also be noted that the (dummy) indices on which contraction takes place can be renamed without changing the value of the resulting expression, so any name other than those of the free indices or of other dummy indices can be used. Useful scalars can often be obtained from tensors by suitable contractions.

We can now construct tensor bundles (as we constructed the tangent vector bundle and cotangent bundle), tensor fields (and scalar fields) and maps h_* for tensors of type $(p, 0)$ on M and h^* for tensors of type $(0, q)$ on N from a map $h : M \to N$ (by the obvious generalizations of (2.4) and (2.10)). Maps of more general tensors are only possible when h is invertible, so that we can use $(h^{-1})^*$ to map forms from M to N and $(h^{-1})_*$ to map vectors from N to M; in this case the formula (2.13), appropriately re-interpreted, gives the relation between the components.

One of the most important characteristics of tensors, and the one which led to their use as the means of formulating the Maxwell equations in special relativity and the Einstein field equations for gravity in general relativity, is that *an equation between tensors of the same type will take the same form in all bases or coordinate systems*, so it models the physical situation in a way which is completely coordinate or basis invariant: note that such equations are only meaningful if every term has the same type. Moreover, if such an equation is satisfied in one basis or coordinate system, it is true in all, which sometimes provides simple proofs.

Some applications, such as integrals, need tensor densities, which, for manifolds with a metric g_{ab} as defined in Section 2.7, transform like a tensor multiplied by a power of $\sqrt{|g|}$, so that when a coordinate transformation is applied they pick up a factor of the determinant of the transformation matrix, as well as the factors in (2.13).

Having set up the algebra of tensors, we want to be able to differentiate tensor fields, but in general the partial derivative of a tensor is not a tensor, as direct calculation shows. In order to take derivatives we need to say which tensors at neighbouring points are to be regarded as equal: then, given a point $q \in M$ and a neighbouring point p, we can compare the actual tensor at p with the tensor at p 'equal' to the actual tensor at q, and thus find the change in the tensor due to the displacement from q to p. Next we discuss the two main ways to specify such an equality, the Lie derivative and the covariant derivative. The Lie derivative requires a given vector field $\mathbf{v}$ (not just a vector at a point, because, except for scalar arguments, the value of the Lie derivative depends on the derivatives of $\mathbf{v}$). The covariant derivative uses an additional structure, the connection.

2.4 Lie derivatives

The Lie derivative with respect to a vector field $\mathbf{v}$ is given for scalars, vectors and 1-forms respectively by the relations

$$\mathcal{L}_{\mathbf{v}} f = f_{,a} v^a \, , \tag{2.22}$$

$$\mathcal{L}_{\mathbf{v}} \mathbf{w} = [\mathbf{v}, \mathbf{w}] \Leftrightarrow (\mathcal{L}_{\mathbf{v}} \mathbf{w})^a = w^a{}_{,b} v^b - v^a{}_{,b} w^b \, , \tag{2.23}$$

$$(\mathcal{L}_{\mathbf{v}} \boldsymbol{\sigma})_a = \sigma_{a,m} v^m + \sigma_m v^m{}_{,a} \, , \tag{2.24}$$

from which the Lie derivative of a tensor of arbitrary type can be deduced using the Leibniz rule.

A coordinate-free definition of the Lie derivative can be obtained by dragging objects along a congruence of curves. For each point p in M, a vector field $\mathbf{v}$ fixes a unique curve $\gamma_p(t)$ such that $\gamma_p(0) = p$ and $\mathbf{v}$ is tangent to the curve at all points. The family of such curves – one through each point – is called the *congruence* associated with the vector field, and is said to be generated by $\mathbf{v}$. Conversely, a congruence of (differentiable) curves, one through each point in a neighbourhood, induces a vector field on that neighbourhood.

'Dragging along' (Schouten, 1954) through a parameter distance t means taking the map of manifolds which maps p to $q = \gamma_p(t)$ and using the associated push-forward and pullback maps. The Lie derivative then takes the limit of the difference of the value at a point and a dragged-along value, divided by the change in t. An object with zero Lie derivative is said to be Lie dragged. Using dragging, Lie derivatives can be calculated for non-tensorial geometric objects such as the connection introduced in the next section.

If we have a vector field **v**, and a corresponding congruence of curves, a vector **p** dragged along a curve is called a *Jacobi field* along the curve. Assuming **v** and **p** are not parallel, then **p**, considered as defining an infinitesimal displacement between curves, always connects the same two curves of the congruence, and is therefore called a *connecting vector* for the congruence. Thus connecting vectors obey

$$\mathcal{L}_{\mathbf{v}}\mathbf{p} = [\mathbf{v}, \mathbf{p}] = -\mathcal{L}_{\mathbf{p}}\mathbf{v} = p^a{}_{,b}v^b - v^a{}_{,b}p^b = 0 \ . \tag{2.25}$$

Another important application of the Lie derivative is to spacetime symmetries (see Section 2.7.1).

2.5 Connections and covariant derivatives

We now describe a second way of specifying equality of tensors at neighbouring points, introducing the connection. From that we can obtain a derivative, the *covariant derivative*, by the usual limiting processes.

Take the case of vectors first. For every small displacement $\delta\mathbf{x}$, from p to q, say, and for each vector **v** at p, we need the 'equal' vector at q. To preserve the vector space structure, the map chosen must be linear in **v** and tend to the identity as $\delta\mathbf{x} \to 0$. To introduce a derivative, we also want linearity in $\delta\mathbf{x}$, in lowest order approximation. Hence for small $\delta\mathbf{x}$, the transformation can be approximated as the identity minus[4] a small transformation **Γ** where **Γ** depends linearly on $\delta\mathbf{x}$. Thinking of $\delta\mathbf{x}$ as being in $T_p(M)$, the resulting map $T_p(M) \to L$, where L is the space of linear maps of vectors, is called a *connection* because it connects vectors at neighbouring points of the manifold. With coordinates $\{x^a\}$, we can write the transformation for $\delta\mathbf{x}$ as $\delta^a{}_b - \Gamma^a{}_{bc}\delta x^c$: the $\Gamma^a{}_{bc}$ are the components of the connection. The covariant derivative ∇_c in the x^c direction of a vector with values v^a is now

$$\nabla_c v^a = v^a{}_{,c} + \Gamma^a{}_{bc}v^b \ , \tag{2.26}$$

which is usually denoted $v^a{}_{;c}$.

We can introduce a connection in the same way for vector spaces other than tangent vectors. If at each point there is a vector space, with vectors v^I, $I = 1, \ldots, k$ say, the connection will have the form $\Gamma^I{}_{Jc}$; in physics these are gauge potentials, often denoted A with indices suppressed. Note that for electromagnetism $k = 1$ so the indices I, J are dropped and the resulting $A = \Gamma$ is the usual vector potential: the corresponding covariant derivative appears, for example, in calculating the Zeeman effect.

We assume that for scalar functions $\nabla_c f = \partial_c f = f_{,c}$. Then in order for covariant differentiation of tensors to obey the Leibniz rule we have to take specific associated connections on the cotangent bundle, and the various tensor bundles. These are given, relative to the

---

[4] Minus by convention: plus could have been used.

same basis, by

$$\nabla_a \omega_b = \omega_{b,a} - \Gamma^c{}_{ba} \omega_c, \tag{2.27}$$

$$\nabla_q T^{ab\cdots c}{}_{\ell m\cdots p} = T^{ab\cdots c}{}_{\ell m\cdots p,q} + \Gamma^a{}_{rq} T^{rb\cdots c}{}_{\ell m\cdots p} \cdots + \Gamma^c{}_{rq} T^{ab\cdots r}{}_{\ell m\cdots p}$$
$$- \Gamma^r{}_{\ell q} T^{ab\cdots c}{}_{rm\cdots p} \cdots - \Gamma^r{}_{pq} T^{ab\cdots c}{}_{\ell m\cdots r}, \tag{2.28}$$

where in the last formula there is one term with a product of $\mathbf{\Gamma}$ and $\mathbf{T}$ for each index on $\mathbf{T}$ and the indices in these terms are arranged so that each index of $\mathbf{T}$ is contracted in turn with one on $\mathbf{\Gamma}$. For brevity we write the components of the covariant derivative of a tensor, $\nabla_q T^{ab\cdots c}{}_{\ell m\cdots p}$ say, as $T^{ab\cdots c}{}_{\ell m\cdots p;q}$. The covariant derivative itself, being a tensor, can of course be covariantly differentiated, and we notate this as, e.g. $T^{ab\cdots c}{}_{\ell m\cdots p;qr}$.

We can combine the derivatives ∇_a into a single operator ∇ which maps vectors to differential operators, such that $\nabla_{\mathbf{u}} = u^a \nabla_a$; we can regard ∇_a as the result of applying ∇ to the basis vector ∂_a. The components of $\nabla_{\mathbf{u}} \mathbf{T}$ for the tensor above would be

$$\nabla_{\mathbf{u}} T^{ab\cdots c}{}_{\ell m\cdots p} = T^{ab\cdots c}{}_{\ell m\cdots p;q} u^q , \tag{2.29}$$

so ∇ maps tensors of type (p, q) to tensors of type $(p, q + 1)$. For a vector $\mathbf{v}$,

$$\nabla \mathbf{v} = v^a{}_{;b} \boldsymbol{e}_a \boldsymbol{\omega}^b , \tag{2.30}$$

using bases $\{\boldsymbol{e}_a\}$ of V and $\{\boldsymbol{\omega}^b\}$ of V^*. If we have a curve with parameter λ and tangent vector $\mathbf{u}$ we can define, for any quantity Q,

$$\frac{DQ}{D\lambda} = Q_{;b} u^b . \tag{2.31}$$

The usual differentiation of (tangent) vectors in $\mathbb{R}^n$ does involve a connection, but all of its components are zero in Cartesian coordinates: however, its components are not zero in general curvilinear coordinates, and they appear in the formulae for three-dimensional vector calculus in, for example, spherical polar coordinates. Because the 'equal' vectors in $\mathbb{R}^n$ are the parallel vectors in the usual sense, the connection used in (2.26) is regarded as a generalization of the concept of parallellism. A vector or tensor field with zero covariant derivative along a curve is therefore said to be *parallelly propagated* (or transferred or transported). In curved space (see Section 2.6), parallel propagation is only meaningful along a particular curve, and not globally.

A curve whose tangent vector v^a is itself parallelly propagated along the curve is called an *autoparallel* and obeys the equation

$$v^a{}_{;b} v^b = 0 , \tag{2.32}$$

or in component form with $v^a = \mathrm{d}x^a / \mathrm{d}\lambda$,

$$\frac{\mathrm{d}^2 x^a}{\mathrm{d}\lambda^2} + \Gamma^a{}_{bc} \frac{\mathrm{d}x^b}{\mathrm{d}\lambda} \frac{\mathrm{d}x^c}{\mathrm{d}\lambda} = 0. \tag{2.33}$$

This is often called the geodesic equation, although when we introduce geodesics (see Section 2.7.2) we shall find that strictly this is only correct for Riemannian spaces. Even in that case, where the parameter w along the autoparallel curve such that $v^a = \mathrm{d}x^a / \mathrm{d}w$

is called an *affine parameter*, we find that the same curve expressed with a non-affine parameter would obey a more complicated expression but would still be geodesic.

The connection provides a specific link between vectors and displacements. For each vector $\mathbf{V}$ at p take an autoparallel curve whose tangent vector at p is $\mathbf{V}$, and move a unit affine parameter distance along it, to q say: this is called *the exponential map*, mapping $\mathbf{V} \rightarrow q$, and exists for a maximal region U around p called a *normal neighbourhood*. Choosing a basis of $T_p(M) \cong \mathbb{R}^m$ at p, the components of $\mathbf{V}$ then give coordinates for q, these being the *Riemannian normal coordinates* in U.

Since for any pair of vectors $\mathbf{v}$ and $\mathbf{w}$, and any basis, the combination

$$\nabla_{\mathbf{v}}\mathbf{w} - \nabla_{\mathbf{w}}\mathbf{v} - [\mathbf{v}, \mathbf{w}] = (\Gamma^a{}_{bc} - \Gamma^a{}_{cb})v^b w^c \partial_a \tag{2.34}$$

is a vector, we see that the components

$$\Gamma^a{}_{bc} - \Gamma^a{}_{cb} =: T^a{}_{bc} \tag{2.35}$$

specify a tensor, the *torsion $T^a{}_{bc}$*.

It should be noted that the connection is *not* a tensor. This is natural since the partial derivative of a tensor is also not a tensor, while the combination of partial derivative and connection terms is. In fact a direct calculation gives

$$\Gamma^{a'}{}_{b'c'} = \left(\frac{\partial y^{a'}}{\partial x^a} \Gamma^a{}_{bc} - \frac{\partial^2 y^{a'}}{\partial x^b \partial x^c} \right) \frac{\partial x^b}{\partial y^{b'}} \frac{\partial x^c}{\partial y^{c'}} . \tag{2.36}$$

However, the difference between two connections (on the tangent bundle) is a tensor; the torsion is an example, being (twice) the difference between the connection and its symmetric part which is also a connection.

We can understand the physical meaning of symmetry of the connection ($T^a{}_{bc} = 0$) as follows. If we take a point with coordinates x^a and construct a 'parallellogram' by taking two infinitesimal displacements δx_1^a and δx_2^a and parallelly transporting the displacement δx_1^a along δx_2^c and vice versa, the figure closes for all choices of δx_1^a and δx_2^a if and only if

$$\Gamma^a{}_{bc} = \Gamma^a{}_{cb} \Leftrightarrow T^a{}_{bc} = 0. \tag{2.37}$$

Thus the name torsion refers to the idea that some parallelograms will no longer close due to some kind of twisting of the space. From now on, unless otherwise stated, we shall assume that the torsion is zero.

2.6 The curvature tensor

2.6.1 Curvature and covariant differentiation

Curvature arises from the noncommutation of covariant derivatives. In components the *curvature tensor* or *Riemann tensor $R^d{}_{abc}$* is given by

$$w^d{}_{;bc} - w^d{}_{;cb} = -R^d{}_{abc} w^a \tag{2.38}$$

for arbitrary w^a (some authors use the opposite sign convention here). Substitution from (2.26) gives

$$R^d{}_{abc} := \Gamma^d{}_{ac,b} - \Gamma^d{}_{ab,c} + \Gamma^d{}_{ec}\Gamma^e{}_{ab} - \Gamma^d{}_{eb}\Gamma^e{}_{ac}. \tag{2.39}$$

To show that this is actually a tensor, we can directly apply a coordinate transformation, or introduce the type $(1, 3)$ tensor $\mathbf{R}$ such that

$$\mathbf{R}(\boldsymbol{\sigma}, \mathbf{u}, \mathbf{v}, \mathbf{w}) = \boldsymbol{\sigma}((\nabla_{\mathbf{u}}\nabla_{\mathbf{v}} - \nabla_{\mathbf{v}}\nabla_{\mathbf{u}} - \nabla_{[\mathbf{u},\mathbf{v}]})\mathbf{w}) \tag{2.40}$$

for arbitrary 1-form and vector fields, $\boldsymbol{\sigma}$, $\mathbf{u}$, $\mathbf{v}$ and $\mathbf{w}$ and show that this has the component form (2.38).

From these forms we see that $\mathbf{R}$ is always skew in its last two indices:

$$R^d{}_{abc} = R^d{}_{acb} \Leftrightarrow R^d{}_{a(bc)} = 0 . \tag{2.41}$$

Direct calculation shows that in addition

$$R^a{}_{[bcd]} = 0; \tag{2.42}$$

this is called the *first Bianchi identity*. It can be viewed as a version of the Jacobi identity (2.7) applied to the basis vectors ∂_a. We also obtain the *second Bianchi identities*, often called simply *the* Bianchi identities, which can be written

$$R^a{}_{b[cd;e]} = 0 . \tag{2.43}$$

The contraction

$$R_{ab} := R^c{}_{acb} = -R^c{}_{abc} \tag{2.44}$$

of the curvature tensor yields the *Ricci tensor* (some authors use the opposite sign convention here). This in general has no symmetry, but for Riemannian spaces it is symmetric (see Section 2.7). The Ricci tensor satisfies the (once) contracted Bianchi identities

$$R_{bd;e} - R_{be;d} + R^a{}_{bde;a} = 0 . \tag{2.45}$$

An alternative interpretation (and definition) of curvature is provided by considering parallel transport around a closed curve. In general this does not bring a vector back to its original value. The effect is called *holonomy*, and the group of all transformations of the tangent space obtained from parallel transport around different closed curves is called *the holonomy group*. Curvature arises from considering holonomy around infinitesimally small closed curves: direct calculation shows that the change between the original and final vectors is given by the curvature contracted with a skew two-index tensor giving the area (and the plane in which it lies). We can understand how this is related to the (non)commutation of covariant derivatives in (2.38) by remembering that parallel transport corresponds to a zero covariant derivative.

One can compute curvatures by direct application of the formula (2.39) but this is cumbersome (even if using the methods of Section 2.8). The exterior calculus provides very neat and efficient methods if one is calculating by hand (see e.g. Schutz (1980)). Nowadays the most effective way is to use one of several available computer programs, which can exploit the symmetries of the objects involved.

In a gauge theory, the commutator of covariant derivatives gives a curvature $R^I{}_{Jcd}$ which is the gauge field, usually denoted $\textbf{F}$ with indices suppressed. The key difference is that in GR the tangent vectors appear both as the spacetime and the internal gauge vectors. Correspondingly, relativity can and does have different dynamical equations, which use the possibility of contraction between 'internal' and spacetime indices (see Section 3.3).

2.6.2 Curvature and geodesic deviation

A third way of arriving at curvature is to consider 'geodesic deviation' (strictly, since we do not yet have a metric, autoparallel deviation), which is geometrically illuminating and of direct physical importance.

Consider a pair of neighbouring curves taken from a congruence of autoparallels; let them be given in terms of affine parameters λ as $x^a(\lambda)$, and have tangent vectors $\textbf{u}$, so $u^a{}_{;b}u^b = 0$. Let $p^b(\lambda) = x_1^b(\lambda) - x_2^b(\lambda)$, where $x_1^b(\lambda)$ and $x_2^b(\lambda)$ are the nearby autoparallels. The connecting vectors are dragged along by the tangent vector field to the autoparallels, and we have from (2.25), for a symmetric connection,

$$u^a{}_{;b}p^b = p^a{}_{;b}u^b. \tag{2.46}$$

Hence we find that (Synge and Schild, 1949)

$$\frac{D^2 p^a}{D\lambda^2} = -R^a{}_{dbc}u^d p^b u^c. \tag{2.47}$$

Thus we see how to describe the relative acceleration between autoparallels (and, in Riemannian spaces, geodesics) in terms of the curvature. This is fundamental in seeing how curvature focuses light rays (see e.g. Ellis and van Elst (1999b)) and causes tidal forces when acting on freely moving matter (Pirani, 1956).

By a similar calculation we see that connecting vectors between curves which are not autoparallel obey

$$\frac{D^2 p^a}{D\lambda^2} = -R^a{}_{dbc}u^d p^b u^c + (u^a{}_{;b}u^b)_{;c}p^c, \tag{2.48}$$

which will show how matter experiencing other forces responds to the gravitational field.

2.7 Riemannian geometry

2.7.1 The metric tensor

The spacetimes of general relativity have an important additional structure, the *Riemannian metric* g_{ab} which is a (differentiable) symmetric type $(0, 2)$ tensor field and gives a specific map between vectors and 1-forms, assumed to be $1-1$. The metric, operating on a pair of vectors, gives a scalar product

$$\textbf{v} \cdot \textbf{w} = g_{ab}v^a w^b. \tag{2.49}$$

Sometimes the term Riemannian is reserved for the case where the quadratic form (2.49) is positive definite, in which case the indefinite metric used in relativity would be called pseudo-Riemannian or semi-Riemannian, but we do not generally make this distinction. (A metric which is not $1-1$, and hence not Riemannian, is said to be degenerate.)

In a basis (coordinate or general) the metric gives us a quadratic form,

$$\mathbf{g}(d\mathbf{x}, d\mathbf{x}) = ds^2 = g_{ab}dx^a dx^b, \tag{2.50}$$

on infinitesimal displacements, sometimes called the *line element*. The notation ds^2 comes from the fact that in the positive definite case one can, for a displacement, consider ds to be its length.

Two vectors $\mathbf{v}$, $\mathbf{w}$, are said to be orthogonal if $\mathbf{v}.\mathbf{w} = 0$. In relativity, spacetime has dimension 4 and the metric has signature ± 2 where the sign depends on choice of conventions; we choose $+2$ so there are three positive eigenvalues and one negative. A non-zero vector is then said to be *spacelike*, *timelike* or *null* (or lightlike) depending on whether $\mathbf{v}.\mathbf{v}$ is positive, negative or zero, respectively.

At any point in spacetime, the set of timelike vectors and the set of non-zero null vectors each have two disjoint subsets, which we can call the future and past. The null vectors form a cone, the *null cone* or *light-cone*, which divides the timelike vectors from the spacelike ones. It has two parts, the future and past light-cones, with the zero vector as the common apex. We usually assume that spacetime is *time orientable*, i.e. that the choice of the future direction can be made consistently over the whole manifold.

If one makes a change of the metric by

$$g'_{ab} = \Omega^2 g_{ab}, \tag{2.51}$$

lengths are changed but angles between vectors are not. For this reason the change is called a *conformal transformation*. Under conformal transformations null vectors remain null and the light-cones are unaltered. Hence the causal structure of the spacetime is unaffected, where causal structure is determined by whether pairs of points can be joined by everywhere timelike or null future-pointing curves; note that this assumes time orientability. The field of light-cones can therefore be regarded as specifying the conformal structure of the spacetime.

For positive ds^2 (spacelike displacements), the *arc length* along a curve $x^a(\lambda)$, where λ is the curve parameter, is

$$s = \int \sqrt{\left(g_{ab} \frac{dx^a}{d\lambda} \frac{dx^b}{d\lambda} \right)} \, d\lambda . \tag{2.52}$$

For negative ds^2 (timelike displacements), $d\tau = \sqrt{-ds^2}$ defines the proper time τ along the curve, with the same physical meaning in spacetime as in special relativity: for such curves τ is the arc length. (Only if we considered curves which were partly timelike and partly spacelike would this inconsistency in the definition of arc length be a problem.)

A non-degenerate metric has an inverse, which is a symmetric $(2, 0)$ tensor denoted g^{ab} that obeys

$$g^{ab} g_{bc} = \delta^a{}_c . \tag{2.53}$$

The metric and its inverse then enable us to raise or lower any index on a tensor. Consequently we can no longer collect all indices of the same type together, as we have up to now, because we need to keep track of the raising and lowering of indices; it is not in general true that $g^{ab}T_{bcd} = g^{ab}T_{cbd}$, so we cannot write both as $T^a{}_{cd}$. The convention is to maintain a fixed horizontal order of indices whether they are raised or lowered, so we would write $g^{ab}T_{cbd} = T_c{}^a{}_d$ and $g^{ab}T_{cdb} = T_{cd}{}^a$.

2.7.2 Geodesics and the Levi-Civita connection

Using arc length (2.52), we can characterize *geodesics* as extremal curves, i.e. those whose total arc length is stationary under small variations of the curve. For spacelike or timelike curves we can easily carry out the variational calculation (things are not so simple for the null case, but we can treat this as a limit of the nonnull cases). Taking the spacelike case, we obtain a variational problem with Lagrangian

$$\mathcal{L} = \sqrt{g_{ab}\frac{dx^a}{d\lambda}\frac{dx^b}{d\lambda}} = \frac{ds}{d\lambda} . \tag{2.54}$$

The resulting Euler–Lagrange equation is

$$\frac{d}{d\lambda}\left(g_{ab}\frac{dx^b}{d\lambda}\right) - \frac{1}{2}g_{bc,a}\frac{dx^b}{d\lambda}\frac{dx^c}{d\lambda} = \frac{d^2s/d\lambda^2}{ds/d\lambda}g_{ab}\frac{dx^b}{d\lambda} . \tag{2.55}$$

We can always choose λ so that $d^2s/d\lambda^2 = 0$. Such a λ is called an *affine parameter*. In spacetime, s itself is an affine parameter on spacelike curves, and τ is an affine parameter on timelike curves (and is the one which is almost always used). From the equation satisfied by an affine parameter it is easy to show that all affine parameters for a given curve are related by linear equations with constant coefficients. With an affine parameter we have

$$\frac{d}{d\lambda}\left(g_{ab}\frac{dx^b}{d\lambda}\right) - \frac{1}{2}g_{bc,a}\frac{dx^b}{d\lambda}\frac{dx^c}{d\lambda} = 0,$$

which can be rewritten as

$$\frac{d^2x^d}{d\lambda^2} + g^{da}\left(g_{ab,c} - \frac{1}{2}g_{bc,a}\right)\frac{dx^b}{d\lambda}\frac{dx^c}{d\lambda} = 0 . \tag{2.56}$$

As a consequence of the symmetry of $\dfrac{dx^b}{d\lambda}\dfrac{dx^c}{d\lambda}$ in bc and the rule 'symmetric contracted with skew is zero' we need only take the part of the first term in the bracket symmetric in bc, so obtaining

$$\frac{d^2x^d}{d\lambda^2} + \tfrac{1}{2}g^{da}(g_{ab,c} + g_{ac,b} - g_{bc,a})\frac{dx^b}{d\lambda}\frac{dx^c}{d\lambda} = 0 . \tag{2.57}$$

We thus find that using an affine parameter casts (2.55) into the form of the autoparallel equation (2.32), so that the coefficients of the terms $\dfrac{dx^b}{d\lambda}\dfrac{dx^c}{d\lambda}$ for varying b and c in (2.57) must form a connection, called *the Levi-Civita connection*. This connection has as its coordinate components the *Christoffel symbols* (of the second kind) $\left\{{d \atop bc}\right\}$, i.e. in a coordinate basis,

$$\Gamma^d{}_{bc} = \left\{{d \atop bc}\right\} := \tfrac{1}{2}g^{da}(g_{ab,c} + g_{ac,b} - g_{bc,a}) \, . \tag{2.58}$$

For hand calculations of the Levi-Civita connection it is useful to note that the Lagrangian $\mathcal{L} = g_{ab}(\mathrm{d}x^a/\mathrm{d}\lambda)(\mathrm{d}x^b/\mathrm{d}\lambda)$ gives (2.57) as its Euler–Lagrange equations.

It is now obvious that in a space with a Riemannian metric, if there is no torsion and the geodesics are also to be autoparallels, i.e. if these curves are to satisfy the generalizations of both of the natural definitions of 'straight line' in Euclidean space, the connection must be the Levi-Civita one. Substitution from (2.58) gives

$$g_{ab;c} := g_{ab,c} - \Gamma_{abc} - \Gamma_{bac} = 0 \, . \tag{2.59}$$

Since this is a tensorial equation it must be true in all bases, so parallel transport preserves lengths and angles given by the scalar product (2.49). Conversely, (2.59) and absence of torsion lead to the Levi-Civita connection and so ensure that geodesics are the same as autoparallels. Moreover (2.59) implies that indices can be raised or lowered freely inside a covariant derivative with this connection, and (2.38) can then easily be extended to tensors of all ranks.

A space with Riemannian metric and Levi-Civita connection is called a *Riemannian manifold*.[5] Many generalizations of general relativity also use a space with a metric (which, as Dicke (1963) has argued, is necessary if particle motion under gravity is to be obtainable from a Lagrangian formulation) but may use a different connection on the tangent bundle: such manifolds are called *metric-affine*, with more specific names depending on the nature of the assumed connection.

2.7.3 Spacetimes with symmetry

Considering the metric as a sum of products of one-forms, we see from (2.24) that

$$(\mathcal{L}_\mathbf{v}\mathbf{g})_{ab} = g_{ab,m}v^m + g_{mb}v^m{}_{,a} + g_{am}v^m{}_{,b} \, .$$

Using (2.59) and (2.26) we find that the metric tensor is Lie dragged into itself by the transformations generated by a vector field $\mathbf{v}$ if and only if

$$(\mathcal{L}_\mathbf{v}\mathbf{g})_{ab} = v_{a;b} + v_{b;a} = 0 \, , \tag{2.60}$$

this being *Killing's equation*. This implies that if a Killing vector field $\boldsymbol{\xi}$ has $\xi^a = 0$ and $\xi_{a;b} = 0$ at a point p, then $\boldsymbol{\xi} \equiv \mathbf{0}$.

The weaker symmetries in which

$$(\mathcal{L}_\mathbf{v}\mathbf{g})_{ab} = 2\phi g_{ab} \tag{2.61}$$

---

[5] Or, if one needs to distinguish the positive definite and indefinite cases, a pseudo-Riemannian or semi-Riemannian manifold in the indefinite case.

for some function ϕ are *conformal motions*: the sub-cases where $\phi = \text{const} \neq 0$, called *homotheties* or self-similarities, are of some importance in cosmology. The vectors $\mathbf{v}$ in (2.60) and (2.61) are called respectively *Killing vectors* and *conformal Killing vectors*.

It is easy to see that linear combinations of Killing vectors with constant coefficients are also Killing. The set of all Killing vectors on a manifold forms a Lie algebra, i.e. a vector space with a bilinear commutator operation obeying (2.7), since the rules for Lie derivatives imply that the commutators are Killing. If $\{\boldsymbol{\xi}_A : A = 1, \ldots, r\}$ is a basis of the Lie algebra, we must have

$$[\boldsymbol{\xi}_A, \boldsymbol{\xi}_B] = C^C{}_{AB} \boldsymbol{\xi}_C, \quad C^C{}_{AB} = -C^C{}_{BA} . \tag{2.62}$$

The coefficients $C^A{}_{BC}$ are called the *structure constants* of the algebra. The Jacobi identity for $\{\boldsymbol{\xi}_A : A = 1, \ldots, r\}$ expressed in terms of these constants reads

$$C^E{}_{[AB} C^F{}_{C]E} = 0 . \tag{2.63}$$

The set of finite transformations generated by $\{\boldsymbol{\xi}_A\}$ forms the corresponding Lie group G of symmetries of the spacetime. Its dimension is often shown by the notation G_r, and the group is said to have r parameters.

The *orbit* (or *trajectory*, or *minimum invariant variety*) $\mathcal{O}_p$ of G through a fixed p is the set of points to which elements of G map p. It is a submanifold of $\mathcal{M}$. The group G is said to be *transitive* on its orbits, and to be either *transitive* (when $\mathcal{O}_p = M$) or *intransitive* ($\mathcal{O}_p \neq M$) on M. It is *simply transitive* on an orbit if for any q in $\mathcal{O}_p$ the transformation from p is unique; otherwise it is *multiply transitive*.

The set of g in G which maps p to itself forms a subgroup of G called the *isotropy group* $H(p)$ of p for groups of motions (or, more generally, the *stability group*): its generators have $\boldsymbol{v} = \mathbf{0}$ at p. For any q in $\mathcal{O}_p$, $H(p)$ and $H(q)$ are conjugate subgroups of G, and have the same dimension, s say; one can thus for brevity refer to the isotropy subgroup H_s of an orbit. Denoting the dimension of $\mathcal{O}_p$ by d we thus have

$$r = d + s . \tag{2.64}$$

For more information on Lie groups, their related Lie algebras and their application as transformations of manifolds and spacetimes see Chapter 8 of Stephani *et al.* (2003) and references therein.

The Riemannian structure is invariant under the transformations in G_r, called *motions* or *isometries* and discussed further in Chapter 17. Consequently the connection, curvature tensor and all quantities defined uniquely by them in a covariant way will also be invariant. Kerr (1963) proved that in a four-dimensional Einstein space, the number of functionally independent scalar invariants is $4 - d$, where d is the dimension of the orbits of the maximal group of motions.

A universe model is *spatially homogeneous* if there are spacelike surfaces $\{t = \text{const}\}$ in which any point can be moved to any other point by an isometry. This will be the case if and only if there are at least three independent Killing vector fields everywhere in these surfaces. A model is *spherically symmetric* if there exist spacelike 2-spheres S^2 everywhere (except possibly at the centres of the spheres) in which the rotation group $O(3)$ acts as an isometry group.

Example: The Robertson–Walker (RW) metric We can choose coordinates $\{t, r, \theta, \phi\}$ for a RW model such that the metric is

$$\mathrm{d}s^2 = -\mathrm{d}t^2 + a^2(t)[\mathrm{d}r^2 + f^2(r)(\mathrm{d}\theta^2 + \sin^2\theta\,\mathrm{d}\phi^2)], \tag{2.65}$$

where $a(t)$ is the 'scale factor' showing how the size of the universe changes with time[6] and $f(r) = \sin r, r$ or $\sinh r$ if the universe has spatial sections of positive, zero or negative curvature respectively (but compare Section 9.1.3). Fundamental observers move on the lines $\{r, \theta, \phi\} = \text{const}$, so their 4-velocity is $u^a = \mathrm{d}x^a/\mathrm{d}t = \delta_0^a$. The simplest case is the Einstein–de Sitter model, with flat space sections and $a(t) = t^{2/3}$.

This is the standard metric used for the universe in cosmology. It is both spatially homogeneous and isotropic about every point. The geometry and dynamics of RW spacetimes will be explored in detail in Part 3.

2.7.4 Riemannian curvature

When the manifold is Riemannian, then in addition to the curvature symmetries given above, i.e. (2.41) and (2.42), one can show by evaluation in coordinates using (2.58) that the curvature obeys

$$R_{(ab)cd} = 0, \quad R_{abcd} = R_{cdab}\,. \tag{2.66}$$

The first of these results can also be obtained by applying the Ricci identity (2.38) to g_{ab} and using (2.59), and the second by using the first, (2.41), and the cyclic identity (2.42).

For (four-dimensional) spacetime, these symmetries imply that of the 256 components of R_{abcd} only 20 are independent. This follows by first considering only the 36 possible components obtained from distinct skew pairs ab and cd, then noting that the 6×6 matrix thus obtained is symmetric (with 21 components), and finally evaluating the first Bianchi identity (2.42) which gives one more equation.

From the second of (2.66) it follows that the Ricci tensor now obeys

$$R_{ab} = R_{ba}\,. \tag{2.67}$$

Moreover, with a metric we can now obtain the *Ricci scalar*,

$$R := g^{ab}R_{ab}\,. \tag{2.68}$$

Using the metric we can contract the Bianchi identities (2.45) a second time and obtain

$$(R^a{}_b - \tfrac{1}{2}R\delta^a{}_b)_{;a} = 0; \tag{2.69}$$

$G^a{}_b = R^a{}_b - \tfrac{1}{2}R\delta^a{}_b$ is called the *Einstein tensor*.

In a Riemannian space, whatever the initial coordinates, since the matrix g_{ab} of components of the metric at a point $x^c = X^c$ (for some constants X^c) is symmetric, a linear transformation $y^{a'} = L^{a'}{}_a x^a$, where the components of $L^{a'}{}_a$ are constants, can be used to

---

[6] This notation is the most common one now: in older literature the corresponding quantity may be denoted ℓ for 'length', R for 'radius' or S for 'scale factor'.

transform it into diagonal form. Thus there are coordinates, in any spacetime, in which at a given point X^c,

$$ds^2 = A^2 dx^2 + B^2 dy^2 + C^2 dz^2 - D^2 dt^2. \tag{2.70}$$

provided that the metric has the correct 'signature', i.e. that three of the eigenvalues of the matrix of components are positive and one is negative. This is an invariant property of tensors, and independent of the choice of initial coordinates. At X^a, a simple further transformation, re-scaling each coordinate, sets $A = B = C = D = 1$ in (2.70) as required by the condition that the metric is locally that of special relativity. We can also shift the origin to X^a.

Using the coordinates with these properties, we can additionally require that the connection vanishes at the origin, so that geodesics are approximately straight coordinate lines there. Making a transformation to new special coordinates $y^{a'}$ we find from (2.36) that to achieve this we need

$$\Gamma^a{}_{bc} = \frac{\partial^2 y^{a'}}{\partial x^b \partial x^c} \frac{\partial x^a}{\partial y^{a'}} \quad \text{at the origin,}$$

which is satisfied for any symmetric connection if

$$\frac{\partial x^a}{\partial y^{a'}} = \delta^a{}_{a'} \quad \text{and} \quad \frac{\partial^2 y^e}{\partial x^f \partial x^g} = \Gamma^e{}_{fg} \quad \text{at the origin}$$

and this in turn is satisfied if

$$y^{a'} = \delta^{a'}{}_e (x^e + \tfrac{1}{2} \Gamma^e{}_{fg} x^f x^g). \tag{2.71}$$

Coordinates which obey the restrictions so far imposed are called *locally orthogonal geodesic coordinates* (for brevity, LOGC). In LOGC the metric can be written as

$$ds^2 = -dt^2 + dx^2 + dy^2 + dz^2 + k_{ab} dx^a dx^b, \tag{2.72}$$

where k_{ab} is of order 2 in the x^a. Riemannian normal coordinates based on an orthonormal basis of $T_p(M)$ are LOGC and for them a power series for k_{ab} can be computed whose first term is $\tfrac{1}{2} R_{acbd} x^c x^d$. Various other refinements of LOGC are used for special purposes. LOGC are often helpful in checking a tensor equation by evaluating it in a special coordinate system.

The extreme case is that of zero curvature. It is easy to show (by evaluating (2.39) in Minkowski coordinates) that Minkowski space is flat (i.e. has zero curvature). In fact Minkowski space (or a space derived from it by topological identifications) is the only flat space. To show this, take LOGC at a point X^a. Take the unit vector $\overset{(a)}{v}$ along the x^a axis at X^a and let the corresponding vector at any other point y^a be given by parallel transport along any curve connecting y^a and X^a. That the result is independent of the curve chosen follows from the holonomy interpretation of curvature. Thus we have obtained a vector field $\overset{(a)}{v}$ which has (from its method of definition) a zero covariant derivative. Hence,

$$\overset{(a)}{v}{}_{c;b} - \overset{(a)}{v}{}_{b;c} = 0 \Rightarrow \overset{(a)}{v}{}_{c,b} - \overset{(a)}{v}{}_{b,a} = 0,$$

since the connection is symmetric. This implies $\overset{(a)}{v}$ is a gradient, i.e. there is a ξ^a such that

$$\overset{(a)}{v}{}_d = \xi^a{}_{,d}.$$

Using the ξ^a as new coordinates, the metric in these coordinates must obey

$$g^{ab}{}_{,c} = (g^{db}\,\overset{(a)}{v}{}_d\,\overset{(b)}{v}{}_b)_{;c} = \overset{(a)}{v}{}^d{}_{;c}\,\overset{(b)}{v}{}_d + \overset{(a)}{v}{}^d\,\overset{(b)}{v}{}_{d;c} = 0,$$

so its components are constants, and by evaluation at X^a, it must be the Minkowski metric. We have thus proved that $R_{abcd} = 0$ in a region if and only if it is a region of Minkowski space.

2.7.5 The Levi-Civita volume form in spacetime

Given a metric in spacetime, *the Levi-Civita 4-form* has components

$$\eta_{abcd} = -\sqrt{|g|}\epsilon_{abcd}, \tag{2.73}$$

where g is the determinant of the matrix g_{ab} of components of the metric, and ϵ_{abcd} is the totally skew tensor density whose components are fixed by $\epsilon_{0123} = 1$. Note that the value depends on the orientation of the coordinates (if one interchanged the labels 2 and 3 on coordinates, the sign of η_{abcd} would reverse). The Levi-Civita form gives an infinitesimal volume element,

$$dV = \eta_{abcd}\,dx^a dx^b dx^c dx^d, \tag{2.74}$$

which enables one to integrate functions f in a coordinate-independent way: the eventual integrand in given coordinates is the scalar density $\sqrt{|g|}f$. (In fact such integrals are the only invariantly defined ones.)

The Levi-Civita 4-form obeys some very useful contraction relations, namely,

$$\eta_{abcd}\,\eta^{efgh} = -24\delta^e{}_{[a}\delta^f{}_b\delta^g{}_c\delta^h{}_{d]}, \tag{2.75}$$

$$\eta_{abcd}\,\eta^{afgh} = -6\delta^f{}_{[b}\delta^g{}_c\delta^h{}_{d]}, \quad \eta_{abcd}\,\eta^{abgh} = -4\delta^g{}_{[c}\delta^h{}_{d]}, \tag{2.76}$$

$$\eta_{abcd}\,\eta^{abch} = -6\delta^h{}_d, \quad \eta_{abcd}\,\eta^{abcd} = -24. \tag{2.77}$$

The last four are easily deduced from the first one, which itself follows because both sides are zero unless $abcd$ and $efgh$ are each permutations of 0123, while, if they are such permutations, the two sides will each be minus (for spacetimes) the sign of the permutation turning $abcd$ into $efgh$ (the minus arising from the signature of g_{ab}). In a Riemannian space, $\eta_{abcd;e} = 0$, so we can freely move η inside a covariant differentiation.

Our characterization of η_{abcd} above is local: in practice we assume that η_{abcd} exists globally, which means that spacetime must be orientable.

In three dimensions, the analogous object is denoted η_{abc}. It has identities similar to, and deducible from, those above. For an observer moving with 4-velocity u^a, the covariant relation between the four-dimensional Levi-Civita form and a three-dimensional one for the orthogonal tangent planes is simply

$$\eta_{abc} = \eta_{abcd}u^d : \tag{2.78}$$

when the three-dimensional planes knit together to form 3-surfaces, this gives a volume form for the space sections.

The Levi-Civita form and metric enable us to compute, for any tensor skew in $0 < m < 4$ indices, a corresponding dual object with $(4 - m)$ skew indices, by contracting the skew indices with the last m indices of $\eta_{abcd}/m!$, raised as necessary: in spacetime, carrying the operation out again on the $(4 - m)$ skew indices arising gives $(-1)^{m-1}$ times the original tensor. This operation (which can be stated similarly for any dimension) is called *Hodge duality*. It is denoted by prepending or appending $*$ to the symbol for the tensor. In spacetime, such dualization is especially commonly applied to pairs cd, say, of free indices, by contracting with $\frac{1}{2}\eta_{abcd}$: for example, for the electromagnetic field tensor F_{ab} we have a dual $F^*_{ab} = \frac{1}{2}\eta_{ab}{}^{cd}F_{cd}$, while for the curvature we have two duals $*R_{abcd} = \frac{1}{2}\eta_{ab}{}^{ef}R_{efcd}$ and $R^*_{abcd} = \frac{1}{2}\eta_{cd}{}^{ef}R_{abef}$, in which the position of the $*$ shows where the duality has been applied. (If the metric is not available, the duality is between m-forms and multi-vectors, elements of the space generated by skewed products of vectors (Schutz, 1980); for an n-dimensional space, a double use of $*$, with the Levi-Civita form, gives $(\text{sgn} \det g)(-1)^{m(n-m)}$ times the original tensor.)

2.7.6 The Weyl tensor

For (four-dimensional) spacetime the Ricci tensor will have 10 independent components; its trace forms the Ricci scalar. The remaining 10 independent components of the Riemann tensor are contained in the Weyl tensor,

$$C^{ab}{}_{cd} = R^{ab}{}_{cd} + \frac{1}{2}(\delta^a{}_d R^b{}_c - \delta^a{}_c R^b{}_d + \delta^b{}_c R^a{}_d - \delta^b{}_d R^a{}_c) + \frac{1}{6}R(\delta^a{}_c\delta^b{}_d - \delta^a{}_d\delta^b{}_c).$$

(2.79)

This has all the symmetries of the curvature tensor, but in addition is trace-free on all indices:

$$C_{abcd} = C_{[ab][cd]} = C_{cdab}, \quad C_{a[bcd]} = 0, \quad C^a{}_{bad} = 0.$$

(2.80)

Thus it can be thought of as the trace-free part of $R^{ab}{}_{cd}$. It is in many ways similar to the electromagnetic field tensor F_{ab} (see Section 5.5). Physically, we can regard it as the 'free gravitational field', i.e. the part of the spacetime curvature not determined locally by the matter at a point, but rather determined by conditions elsewhere. Thus it represents both a Coulomb-type part of the field and gravitational radiation.

The Weyl tensor is also known as the conformal curvature tensor, because (a) the tensor $C^a{}_{bcd}$ is unaltered by conformal transformation and (b) if it is zero, the spacetime can be locally conformally transformed to flat space (or any other conformally flat manifold, such as the Einstein static universe (Hawking and Ellis, 1973)). These remarks also apply to dimensions > 4, provided the coefficients in (2.79) are suitably amended. All two-dimensional Riemannian manifolds are conformally flat. Three-dimensional Riemannian manifolds are conformally flat if and only if the Cotton tensor,[7]

$$C^a{}_{bc} := (R^a{}_b - \frac{1}{4}R\delta^a{}_b)_{;c} - (R^a{}_c - \frac{1}{4}R\delta^a{}_c)_{;b},$$

(2.81)

---

[7] Also associated with Schouten, Weyl and York.

vanishes, see Schouten (1954). It may be noted that in three dimensions the Riemann tensor is completely specified by the Ricci tensor,

$$R^{ab}{}_{cd} = (\delta^a{}_c R^b{}_d - \delta^a{}_d R^b{}_c + \delta^b{}_d R^a{}_c - \delta^b{}_c R^a{}_d) - \tfrac{1}{2} R(\delta^a{}_c \delta^b{}_d - \delta^a{}_d \delta^b{}_c). \tag{2.82}$$

In two dimensions it is specified by the Ricci scalar,

$$R_{abcd} = \tfrac{1}{2} R(g_{ac} g_{bd} - g_{ad} g_{bc}). \tag{2.83}$$

Hence four dimensions is the minimum in which the curvature tensor carries 'free' information not specified by the Ricci tensor, and so makes gravitational radiation and tidal forces possible.

The algebraic structure of the Weyl tensor can be characterized by its Petrov type. This uses the principal null directions (PNDs) $\boldsymbol{k}$ obeying

$$k_{[e} C_{a]bc[d} k_{f]} k^b k^c = 0. \tag{2.84}$$

There are at most four such null vectors. If they are all distinct, the spacetime is said to be algebraically general (Petrov type I): otherwise it is algebraically special. It is of Petrov type II if two PNDs coincide and the others are distinct; type D if two distinct pairs of PNDs coincide; type III if three PNDs coincide and the other is distinct; and type N if all four coincide. The conformally flat case $C_{abcd} = 0$ is sometimes called Petrov type O. The Petrov types feature prominently in discussions of gravitational radiation and in obtaining exact solutions. For example, the Kerr solution for rotating black holes was discovered via its Petrov type (which is D). For more on Petrov types and applications see e.g. Stewart (1994) and Stephani *et al.* (2003).

The fundamental integrability conditions for the curvature tensor, ensuring that it does come from a connection as in (2.39), are the Bianchi identities (2.43). There is a useful form of these identities (for spacetime) in terms of the Weyl tensor, which follows from the properties of η_{abcd} (in four dimensions). They can be written as

$$R^{*abcd}{}_{;d} = 0, \tag{2.85}$$

which in turn implies $*R^{*abcd}{}_{;d} = 0$, where $*R^{*abcd} = \tfrac{1}{4}\eta^{abst} R_{stuv} \eta^{cduv}$ (the double-dual) of the Riemann tensor). After some algebra, this shows that

$$C^{abcd}{}_{;d} = R^{c[a;b]} - \tfrac{1}{6} g^{c[a} R^{;b]} =: J^{abc}, \tag{2.86}$$

where the 'current' J^{abc} necessarily has a vanishing divergence,

$$C^{abcd}{}_{;dc} = 0 \;\Rightarrow\; J^{abc}{}_{;c} = 0. \tag{2.87}$$

In the vacuum case,

$$T_{ab} = 0 \;\Rightarrow\; R_{ab} = 0 \;\Rightarrow\; J^{abc} = 0 \;\Rightarrow\; C^{abcd}{}_{;d} = 0. \tag{2.88}$$

Equation (2.86) has a striking similarity to the Maxwell equations (Section 5.5) and (2.87) to their consequent current conservation equation $J^a{}_{;a} = 0$, as we shall see in detail in Section 6.4.

2.7.7 Sectional curvature and constant curvature

In a plane specified by two unit vectors $\mathbf{u}$ and $\mathbf{w}$, we can evaluate the effect of curvature in that plane by taking $R_{abcd}u^a w^b u^c w^d$, which is called the *sectional curvature*. If all sectional curvatures are equal, the space is said to be of constant curvature. In this case the curvature has the form

$$R_{abcd} = \kappa (g_{ac}g_{bd} - g_{ad}g_{bc}) , \tag{2.89}$$

where κ is a constant (note that in (2.83) R need not be constant). Spaces of constant curvature are highly symmetric and we can find special coordinate systems in which they take a simple form.

As an example we consider the three-dimensional spaces of constant curvature with positive definite metric which are used for the spatial sections of the 'standard model' of cosmology with metric (2.65); they are isotropic at every point. We have the freedom to rescale $a \to \lambda a$ with constant λ. When $\kappa \neq 0$ we can use this freedom to set $K = \kappa a^2$ to ± 1, so we now assume $K = 1, 0$ or -1. (Later we will use a different normalization of a.) This uniquely determines the scale factor a except when $K = 0$, when there is no intrinsic length scale, and we retain this scaling freedom.

Now choose any point p in the surface $\{\Sigma_1 : t = \text{const} = t_1\}$ and draw the radial geodesics γ of Σ_1 through p, with curve parameter the radial distance r as measured by the induced three-dimensional metric $f_{ij} = g_{ij}/a^2$. The actual distance will then be $d = a(t_1)r$, so since $a(t_1)$ is constant along each of these curves r is an affine parameter on each of them. Isotropy about every world line implies the 3-metrics are spherically symmetric about p so the surface $\{S_d : r = d/a(t_1)\}$ in Σ_1 is a 2-sphere orthogonal to the geodesics γ, with metric proportional to that of a unit 2-sphere. Putting this together, the 3-space metric form is

$$d\sigma^2 := h_{ij}(x^c)dx^i dx^b = a^2(t)[dr^2 + f^2(r)(d\theta^2 + \sin^2\theta \, d\phi^2)], \tag{2.90}$$

where the function of proportionality $f(r)$ is independent of θ and ϕ because of isotropy, and must obey the limit $f(r) \sim r$ as $r \to 0$ because the origin of coordinates is a regular spacetime point.

To determine the function $f(r)$, we use the geodesic deviation equation (2.47) for the radial geodesics γ with tangent vector $X^b = dx^b/dr = \delta_r^b$ and connecting vector $\eta^c = dx^c/d\theta = \delta_\theta^c$. These vectors (each orthogonal to u^a) must commute: $X^b{}_{,c}\eta^c = \eta^b{}_{,c}X^c$, and are orthogonal to each other: $X^b g_{bc}\eta^c = X^b\eta_b = 0$. They have magnitudes $X^2 = X^b g_{bc}X^c = a^2(t), \eta^2 = \eta^b g_{bc}\eta^c = f^2(r)a^2(t)$. Thus the geodesic deviation equation $d^2\eta^b/dv^2 = -R^b{}_{cde}X^c\eta^d X^e$ becomes

$$\frac{d^2\eta^b}{dv^2} = -K(h^b{}_d h_{ce} - h^b{}_e h_{cd})X^c\eta^d X^e = -\frac{K}{a^2}\eta^b X^2 = -K\eta^b. \tag{2.91}$$

To turn the covariant derivatives into ordinary derivatives, we use orthonormal basis vectors $\{e_a\}$ parallelly propagated along the geodesics γ,

$$e_1{}^b = a^{-1}(t)\delta_r^b, \; e_2{}^b = (a(t)f(r))^{-1}\delta_\theta^b, \; e_3{}^b = (a(t)f(r)\sin\theta)^{-1}\delta_\phi^b. \tag{2.92}$$

(That these basis vectors are parallelly propagated can be seen from the fact that they have constant magnitude and angles, and the rotational symmetry implies they must have parallelly propagated directions.) On using this basis, the covariant derivative along the radial curves becomes an ordinary derivative, v can be taken to be r, and the components of η are $\eta^b = a(t) f(r) \delta_\theta^b$, so the equation becomes

$$\frac{\mathrm{d}^2 f(r)}{\mathrm{d}r^2} + K f(r) = 0. \tag{2.93}$$

The solutions with the appropriate limit behaviour $f(r) \to 0$ as $r \to 0$ are

$$f(r) = \sin r \text{ if } K = 1, \quad r \text{ if } K = 0, \quad \sinh r \text{ if } K = -1. \tag{2.94}$$

Note that if we abandon the link with spherical polars at $r = 0$ then $\cosh r$ and $\cos r$ are also possible cases, among others; indeed they show how the Euclidean parallel postulate breaks down in these spaces.

Exercise 2.7.1 Calculate the equations governing radial null geodesics in the RW metric (2.65). Show that the fundamental world lines with tangent vector $u^a = \delta_0^a$ are timelike geodesics.

Exercise 2.7.2 Calculate the curvature tensor for the RW metric (2.65) and hence show that $C_{abcd} = 0$ for these metrics.
Comment: you may find it useful to compare doing this calculation in terms of a coordinate basis (as used above) and a tetrad basis (see Exercise 2.8.1).

Exercise 2.7.3 Determine the induced metric of the surfaces $\{r = \text{const}\}$ in the metric (2.90). Show that they are 2-spheres of constant curvature.

Exercise 2.7.4 Derive (2.66) by applying the Ricci identity (2.38) to g_{ab}, using (2.59), to get the first result, and then using this, (2.41) and the cyclic identity (2.42) [Hint: several times] to get the second result.

Exercise 2.7.5 The metric for Bianchi I models can be written

$$\mathrm{d}s^2 = -\mathrm{d}t^2 + X(t)^2 \mathrm{d}x^2 + Y(t)^2 \mathrm{d}y^2 + Z(t)^2 \mathrm{d}z^2, \tag{2.95}$$

with fundamental observers moving on lines $\{x, y, z\} = \text{const}$ with 4-velocity $u^a = \mathrm{d}x^a/\mathrm{d}t = \delta_0^a$. Instead of the single scale factor $a(t)$ of the FLRW models, there are now three scale factors $X(t)$, $Y(t)$ and $Z(t)$ corresponding to the expansions along three orthogonal directions.

Show that these models are spatially homogeneous, and in general not isotropic. (These models are explored further in Chapter 18.)

Exercise 2.7.6 The Lemaître–Tolman–Bondi (LTB) models have the metric

$$\mathrm{d}s^2 = -\mathrm{d}t^2 + R^2(r,t)[\mathrm{d}\vartheta^2 + \sin^2(\vartheta)\mathrm{d}\varphi^2] + \frac{R'^2 \mathrm{d}r^2}{1 - \varepsilon f^2(r)}, \tag{2.96}$$

where the matter world lines have tangent vector $u^a = \delta_0^a$.
Show that these models are spherically symmetric.

Hint: the 2-spheres with metric $d\vartheta^2 + \sin^2(\vartheta) d\varphi^2$ are spherically symmetric. (These models are explored further in Sections 15.1 and 19.1.)

2.8 General bases and tetrads

General bases of $T_p(M)$ are widely used nowadays. They have a number of advantages both for practical calculations and, when the basis vectors can be chosen in some way relevant to the problem, in physical interpretation. In a basis of m independent vectors $\{e_a\}$ at p, we write a vector V as

$$V = V^a e_a . \tag{2.97}$$

In spacetime such a basis is often called a *tetrad*. It is conventional to write $e_a(f)$ as $\partial_a f$ or $f_{,a}$. We continue to use Latin indices when we write covariant expressions which could be in any basis, but change to Greek indices for expressions which are specifically in a coordinate basis (especially when a particular coordinate basis, or, as for the FLRW perturbations discussed in Chapter 10, one of a restricted set of coordinate systems, is in use). Each of the e_a can itself be written, in a particular coordinate system, in the form

$$e_a = e_a{}^\mu \frac{\partial}{\partial x^\mu} . \tag{2.98}$$

If a change of basis is made, then the components of V in the old and new bases are related by

$$V^{b'} = M^{b'}{}_a V^a \tag{2.99}$$

if the basis transformation is

$$e_{b'} = M^{-1}{}_{b'}{}^a e_a, \tag{2.100}$$

where $M^{b'}{}_a$ and $M^{-1}{}_{b'}{}^a$ are a pair of mutually inverse matrices, since $V^a e_a = V^{b'} e_{b'}$. Clearly this is a generalization of (2.3).

In such a basis a metric tensor has components

$$g_{ab} = g_{\mu\nu} e_a{}^\mu e_b{}^\nu = e_a \cdot e_b \tag{2.101}$$

which give the scalar products between the basis vectors. The term tetrad is often taken to imply that these scalar products are constants. The most frequently used such special type of tetrad in cosmology is an orthonormal basis, where $g_{ab} = \mathrm{diag}(-1,1,1,1)$.

The commutators of the basis vectors play an important role in tetrad methods. In the case of a coordinate basis $\{\partial/\partial x^\mu\}$ any two basis vectors commute, but in a general basis one will have non-zero commutators,

$$[e_a, e_b] = \gamma^c{}_{ab} e_c; \tag{2.102}$$

the coefficients $\gamma^c{}_{ab}$ are called the *commutation coefficients* of the basis. They characterize the extent to which the vector fields do not commute with each other, and are the components of the Lie derivatives of the basis vector fields relative to each other (see Section 2.4). They vanish if and only if the basis vector fields commute, which is true if and only if these vectors form a coordinate basis for some set of coordinates $\{x^\mu\}$.

The Jacobi identity, or first Bianchi identity, (2.42), reads

$$\partial_a \gamma^d{}_{bc} + \partial_b \gamma^d{}_{ca} + \partial_c \gamma^d{}_{ab} + \gamma^e{}_{ab}\gamma^d{}_{ce} + \gamma^e{}_{bc}\gamma^d{}_{ae} + \gamma^e{}_{ca}\gamma^d{}_{be} = 0, \qquad (2.103)$$

for every triplet of basis vectors $\{e_a, e_b, e_c\}$. This is the integrability condition which a set of functions $\gamma^c{}_{ab}(x^\mu)$ must satisfy if it is to specify the commutators of a set of vector fields $\{e_a\}$ according to (2.102).

Covariant derivatives can be written in general bases as follows. If we choose a basis $\{e_a : a = 1, \ldots, m\}$, the covariant derivative ∇_c in the e_c direction of a vector with components v^a is

$$\nabla_c v^a = v^a{}_{,c} + \Gamma^a{}_{bc} v^b \,, \qquad (2.104)$$

and the corresponding formulae for other tensors take the forms (2.27) and (2.28) with the coordinate indices replaced by general basis indices. Here the connection components are just the covariant derivatives of the basis vectors,

$$\nabla_c e_b = \Gamma^a{}_{bc} e_a \Leftrightarrow \Gamma^a{}_{bc} = e^a{}_\nu e_b{}^\nu{}_{;\mu} e_c{}^\mu, \qquad (2.105)$$

these quantities being known as the *Ricci rotation coefficients*. In a Riemannian space, if a tetrad with constant scalar products between the basis vectors (e.g. an orthonormal tetrad) is chosen, then

$$g_{ab;c} = 0 \Leftrightarrow \Gamma_{abc} + \Gamma_{bac} = 0 \qquad (2.106)$$

(Γ_{bac} is skew in its first pair of indices), indices being raised and lowered with g_{ab} and its inverse. Together with (2.102) this leads to,

$$\gamma_{abc} = (\Gamma_{acb} - \Gamma_{abc}) \Leftrightarrow \Gamma_{cab} = \tfrac{1}{2}(\gamma_{bca} + \gamma_{acb} - \gamma_{cab}), \qquad (2.107)$$

where $\gamma_{abc} = g_{ad}\gamma^d{}_{bc}$, relating the connection components to the commutator coefficients (so Γ_{bac} is not skew in its last pair of indices unless we have a coordinate basis; but a coordinate basis with constant scalar products is only possible in flat spacetime). The corresponding formulae for the general case can be found in Stephani *et al.* (2003), Section 3.2.

General tensorial equations take the same form in coordinate and general bases. Expanding covariant derivatives in terms of the commutator coefficients is straightforward using the above equations (commuted derivatives giving rise to terms involving $\gamma^c{}_{ab}$): e.g. (2.40) leads to

$$R^a{}_{bcd} = \Gamma^a{}_{bd,c} - \Gamma^a{}_{bc,d} + \Gamma^a{}_{ec}\Gamma^e{}_{bd} - \Gamma^a{}_{ed}\Gamma^e{}_{bc} - \Gamma^a{}_{be}\gamma^e{}_{cd}. \qquad (2.108)$$

Exercise 2.8.1 Derive the following results.

For the FLRW metric (2.65) we take the basis

$$e_1 = (1/a)\,\partial_r, \quad e_2 = (1/(fa))\,\partial_\theta, \quad e_3 = (1/(fa\sin\theta))\,\partial_\phi, \quad e_4 = \partial_t, \qquad (2.109)$$

so we have as (2.102),

$$[e_1, e_2] = -\frac{f'}{fa}e_2, \quad [e_2, e_3] = -\frac{\cot\theta}{fa}e_3, \quad [e_3, e_1] = \frac{f'}{fa}e_3,$$

$$[e_1, e_4] = -\frac{\dot{a}}{a}e_1, \quad [e_2, e_4] = -\frac{\dot{a}}{a}e_2, \quad [e_3, e_4] = -\frac{\dot{a}}{a}e_3,$$

where the prime and dot are the derivatives with respect to r and t respectively. The condition (2.106) applies so the non-zero connection coefficients are

$$\Gamma_{141} = -\Gamma_{411} = \Gamma_{242} = -\Gamma_{422} = \Gamma_{343} = -\Gamma_{433} = -\dot{a}/a,$$

$$\Gamma_{212} = -\Gamma_{122} = -f'/fa, \qquad \Gamma_{313} = -\Gamma_{133} = -f'/fa,$$

$$\Gamma_{322} = -\Gamma_{232} = -\cot\theta/fa.$$

Hence the non-zero curvature components computed from (2.108) are given by

$$R_{1414} = R_{2424} = R_{3434} = \ddot{a}/a$$

$$R_{1212} = R_{1313} = f''/fa^2 - (\dot{a})^2/a^2$$

$$R_{2323} = (f'^2 - 1)/f^2a^2 - (\dot{a})^2/a^2,$$

and the corresponding components obtained by exchanging indices. Note that since f satisfies $ff'' = f'^2 - 1$, we have $R_{1212} = R_{2323}$.

Now obtain the corresponding results for the Bianchi I metric (2.95).

2.9 Hypersurfaces

As already mentioned in Section 2.1, we often want to consider hypersurfaces, i.e. submanifolds of dimension $m-1$ in an m-dimensional manifold. One can always choose coordinates $(u_1, u_2, u_3, \ldots y)$ such that the hypersurface (or each hypersurface in a non-intersecting family) is $y = $ const.

There is, as for any submanifold, an injection map $i : \Sigma \to M$ from the hypersurface Σ to the manifold M, fixed by mapping each point of Σ to itself considered as a point of M. In particular, the four-dimensional metric g of spacetime gives a three-dimensional metric $h = i^*g$ on Σ, often called the *first fundamental form* of Σ, that will determine distances and angles in Σ. In the special coordinates (u_1, u_2, u_3), this has the components

$$h_{\mu\nu} = g_{ab}e_\mu^a e_\nu^b, \quad (\mu, \nu = 1, 2, 3), \tag{2.110}$$

where e_μ^a are coordinate components of the basis vectors $\{e^a\}$. The metric $h_{\mu\nu}$ completely characterizes the intrinsic geometry of the surfaces. If there is a nonnull unit vector $\boldsymbol{n}$ orthogonal to Σ, the first fundamental form gives a four-dimensional projection tensor $h_{ab} = g_{ab} - n_a n_b/(n_c n^c)$ (beware that in Section 4.4 a similar formula is used with a vector field $\boldsymbol{n} = \boldsymbol{u}$ which is not necessarily hypersurface-orthogonal). Then one can choose *Gaussian normal coordinates* in which y is an affine parameter along the geodesics with the unit normal n as the initial tangent vector, and the u^i are constant along those geodesics; such coordinates exist in a neighbourhood of Σ. In the case of timelike n, Gaussian normal coordinates are called *synchronous* and the metric, with y renamed t, is then

$$ds^2 = -dt^2 + h_{\mu\nu}(t, u^\rho)\,du^\mu du^\nu, \quad (\mu, \nu, \rho = 1, 2, 3). \tag{2.111}$$

Since i is only defined on Σ, not on some open set in M, it does not define an inverse and so we cannot map vectors from M to Σ. When there is a nonnull unit normal n to Σ, the projection h_{ab} defined by n resolves this difficulty. When the normal n to Σ is null, we can obtain a projection into Σ by using a non-zero vector field l not lying in Σ (called a *rigging*), which may be chosen so that $n(l) = 1$. Then the tensor $P^a{}_b = \delta^a{}_b - l^a n_b$ is a projection which maps a vector v^b to $v^a - l^a(n_b v^b)$ in Σ.

The vector field n also gives the *second fundamental form* or *extrinsic curvature* on Σ as follows. Take any extension of n off Σ; then $K = i^*(\nabla n)$. Using coordinates (u^1, u^2, u^3) in Σ, this has coordinate components

$$K_{\mu\nu} = e^a_\mu e^b_\nu n_{a;b} = -n_a e^a{}_{\mu;b} e^b_\nu, \quad (\mu, \nu = 1, 2, 3), \tag{2.112}$$

and is symmetric (because the connection is); it can be calculated entirely on Σ. When n is a nonnull unit vector, K can be considered to be a four-dimensional tensor $K_{ab} = h_a{}^c h_b{}^d n_{c;d}$, using the first fundamental form, and can be written as the Lie derivative (as introduced in Section 2.4) $K_{ab} = \frac{1}{2}\mathcal{L}_n h_{ab}$; this follows from the formula for h_{ab} and equations (2.24) and (2.59). In Gaussian normal coordinates (u^1, u^2, u^3, y) where Σ is $y = 0$, the possibly non-zero components are

$$K_{\mu\nu} = \frac{1}{2} h_{\mu\nu,y}, \tag{2.113}$$

for $\mu, \nu = 1, 2, 3$, which is the same as $\frac{1}{2} h_{\mu\nu,a} n^a$. The second fundamental form determines the embedding of the surfaces in the spacetime (it characterizes how their normals diverge).

If $|$ denotes the covariant differentiation in the spacelike surface with unit normal n and first fundamental form h_{ab}, so the curvature tensor ${}^3R_{ijkl}$ of the 3-spaces is given by the Ricci identity $V_{j|kl} - V_{j|lk} = {}^3R_{ijkl}V^i$ for each vector V^a orthogonal to n^a ($V^b n_b = 0$), then

$$V_{j|lk} = h_k{}^m h_l{}^n h_j{}^p (V_{p|n})_{;m} = h_k{}^m h_l{}^n h_j{}^p (h_n{}^q h_p{}^t V_{t;q})_{;m}.$$

Substituting for h_{ij} and using (2.112) we obtain the Gauss equation,

$${}^3R_{ijkl} = R_{ijkl} - K_{ik}K_{jl} + K_{jk}K_{li}, \tag{2.114}$$

showing that the three-dimensional curvature is the projection of the four-dimensional curvature, corrected by terms (with the correct symmetries of a curvature tensor) involving the second fundamental form. This equation actually holds for any number n of dimensions, and, with appropriate sign changes (see Stephani *et al.* (2003)), any signature, if the superscript 3 is replaced by $(n - 1)$, e.g. 2 for a two-dimensional cylinder or a two-dimensional sphere in three-dimensional flat space (the embedding space is flat, the two-dimensional space being flat in the first case but curved in the second).

For such a spacelike surface one also finds

$$R^a{}_{jkm}n_a = K_{jm|k} - K_{jk|m},$$ (2.115)

$$R^a{}_{jbm}n_a n^b = K_{jk}K^k{}_m + \mathcal{L}_{\boldsymbol{n}}K_{jm} + \dot{n}_j\dot{n}_m + \dot{n}_{(j;m)} \,.$$ (2.116)

The first of these is the Codazzi equation.

Exercise 2.9.1 Determine the second fundamental form of the surfaces $\{r = \text{const}\}$ in the metric (2.90). Show that it is proportional to the induced metric tensor in those surfaces.

Classical physics and gravity

In standard cosmology, gravity is modelled by GR. In this chapter we review how, in GR, gravity is represented by a curved spacetime, with matter moving on timelike geodesics and photons on null geodesics. There is no definition of gravitational force or gravitational energy. Thus although GR has a good Newtonian limit, it has totally different conceptual foundations. It is only in restricted circumstances that gravity will be well represented by Newtonian theory. GR also has its limits: it can only be a good description if quantum gravity effects are negligible. Then it is very good: there are no data requiring us to alter it in such contexts, which include all of cosmology except the very earliest times.

This chapter discusses the Einstein field equations of GR, after a short discussion of physics in a curved spacetime and the energy–momentum tensor. We give a brief introduction to the physical foundations of GR such as the equivalence principle and the motivation of the form adopted for the field equations but do not cover the experimental tests (for which see Will (2006); note that except for the binary pulsar data, these tests are essentially tests of the weak field slow motion regime).

3.1 Equivalence principles, gravity and local physics

Using our understanding of spacetime geometry, we now consider how to describe local physics in a curved spacetime. Two principles underlie the way we do this: namely, use of tensor equations, and minimal coupling based on covariant differentiation. After motivating use of tensor equations to describe physics, we explain why gravity is such an exceptional phenomenon and how this leads to the curved spacetime view and the Einstein field equations.

3.1.1 Tensor equations

As explained in Section 2.3, the fundamental advantage of tensors is that:

> if a tensor equation is true in one coordinate or basis system, it is true in all such systems.

We would like this property to be true for physically meaningful equations: they should hold independently of the coordinate system used (otherwise we could change an effect, or even make it 'go away', by simply changing the coordinate system). Since if, for example, $T_{ab} = R_{ab}$ in an initial coordinate system, where T_{ab} and R_{ab} are components of tensors, then $T_{ab} - R_{ab}$ is a zero tensor, the above statement follows immediately from the simpler result that:

if a tensor vanishes in one coordinate or basis system, it vanishes in all such systems.

This is true for any tensor, S_{ab} say, because in any new system $S_{a'b'} = A_{a'}{}^a A_{b'}{}^b S_{ab} = 0$, using (2.13).

Thus from now on we assume that physically meaningful equations are tensor equations (although, as we shall explain in Section 3.1.4, other possibilities exist). They can be evaluated in any coordinate system or basis.

3.1.2 The weak equivalence principle

When one allows arbitrary choices of coordinates and reference frames, there is no vector G^a describing the gravitational force in a way analogous to the description of a Newtonian gravitational force by a 3-vector g^i. One can transform the gravitational force away by changing to a freely falling reference frame; and, conversely, one can generate an apparent gravitational field by changing from a freely falling frame to a relatively accelerating one. This is the burden of Einstein's famous 'thought experiments' concerning an observer in a lift. Because all objects accelerate at the same rate in the Earth's gravitational field (as shown by the legendary Tower of Pisa experiment of Galileo, and in more modern Eötvös experiments, see Will (2006)), an observer isolated in a lift cannot, by any experiments carried out wholly in the lift, distinguish between the Earth's gravitational field and a uniform acceleration. For example, if the observer drops a weight it will apparently float in the air alongside him or her – since it will accelerate at exactly the same rate as the observer. This is no longer just a 'thought experiment': it is now commonplace to see films of astronauts floating in their spacecraft as though gravity had been abolished.

Einstein formalized this as the (weak) *principle of equivalence* (WEP):

gravity and inertia are equivalent, as far as local physical experiments are concerned. They cannot be distinguished from each other experimentally.

Furthermore, gravity and inertia both depend on the frame of reference adopted, and their combined effect can be transformed to zero by appropriate choice of reference frame. Thus the gravitational force is not a tensor quantity, for such a quantity vanishes in all frames if it vanishes in one. Gravity does not have this property.

How then do we represent gravity? The key idea is that *a particle moves on a spacetime geodesic if in free fall, i.e. if it moves under gravity and inertia alone*. This is the interpretation of (2.32) or (2.33) and it satisfies another formulation of the WEP, namely that the motion of a test body subject to no non-gravitational force is determined only by its initial position and velocity (Universality of Free Fall). In special relativity we implicitly regard the geodesic equation as the equation of motion of a body moving under inertia alone; we now interpret it as representing any freely falling object, moving under the combined effects of gravity and inertia, but with no other forces acting. Motion under a non-gravitational force F^a per unit mass is described by

$$v^a{}_{;b} v^b = F^a \tag{3.1}$$

where, in coordinates, $v^\mu = \mathrm{d}x^\mu/\mathrm{d}\lambda$ and λ is an affine parameter. (Remember that in GR, geodesics are autoparallel when an affine parameter is used.) This is the curved space version of Newton's first and second laws.

The point then is that in flat spacetime, inertial forces are represented by the Γs in (2.32). We can see this for example by transforming from Minkowski coordinates to (a) rotating coordinates, and (b) uniformly accelerated coordinates. The equivalence principle, however, says that we cannot locally separate gravity from inertia. Thus we conclude that *both gravitational and inertial forces are incorporated in the $\Gamma^a{}_{bc}$* in (2.32) and (3.1); this is consistent because (like the gravitational–inertial force) they can locally be set to zero by a change of coordinates (compare the discussion of LOGC in Section 2.7.4).

We thus arrive at the astonishing conclusion that there is no need to define a special vector to describe the gravitational force: it is already incorporated in the force law (3.1) via the implicit $\Gamma^a{}_{bc}$. When we adopt (2.32) to describe free motion of a particle, this will (in a curved spacetime) automatically include a description of the effect of the gravitational field on its motion. However, it would be incorrect to say that these quantities describe gravity only; they describe 'gravity plus inertia' together, which cannot intrinsically be separated from each other by any local physical experiments. A theory of gravity based in this way on a manifold with a metric of the usual Minkowskian signature and with a symmetric affine connection which affects the other physical laws by a minimal coupling prescription will also meet the requirements of the strong equivalence principle discussed in Section 3.1.3 (by the arguments in Section 2.7.4).

What is measurable and cannot be transformed away is the relative motion of objects in free fall, physically caused by tidal forces and gravitational waves, and mathematically represented by the geodesic deviation equation (2.47). Consequently these gravitational effects must be represented by the source term in that equation, namely the spacetime curvature tensor. Thus the gravitational effect of matter is exerted by that matter causing a curvature of spacetime; this idea is made precise by the Einstein field equations (see Section 3.3 below). In turn this spacetime curvature determines how the matter moves in the spacetime, whose curvature is determined by the matter; this basic nonlinearity arises in any self-consistent modelling of gravitational effects.

It is fundamental that as a consequence of this, *there is no fixed background spacetime underlying the curved spacetimes of GR*. Whenever such a background spacetime is introduced, one has a two-metric theory, rather than true GR. It is this lack of a background spacetime that presents problems with quantizing gravity, as discussed in Section 20.2.1, and which underlies the gauge problem of perturbation theory in relativity, which we discuss in detail in Chapter 10.

3.1.3 The strong equivalence principle and minimal coupling

The way we describe local physics in a curved spacetime is usually by assuming *minimal coupling*, i.e. making the simplest possible transition from known physics in a flat spacetime to covariant equations in a curved spacetime. We aim for a spacetime (four-dimensional) tensor form of all physical equations.

The form taken is also chosen to satisfy the *strong equivalence principle* (SEP) that all non-gravitational phenomena locally take the same form in GR as in special relativity. Given an algebraic physical law (i.e. one that does not involve differentiation) in flat spacetime, we can satisfy minimal coupling and the SEP by assuming that the same form holds in an orthonormal local system in curved spacetime; but what about equations that involve derivatives? Given a known physical law in Cartesian/Minkowski coordinates in flat spacetime, we

(a) change all partial derivatives to covariant derivatives, obtaining the general tensor form of the equations in flat spacetime (valid in any coordinate system); and then

(b) assume the same form holds unchanged in curved spacetime.

In this way we obtain the simplest covariant form of the equations that reduces to the known flat-space form; in particular, we thereby assume that there is no explicit coupling to spacetime curvature in the equations.

This procedure will usually, but not always, give us a unique way of extending known physics from flat to curved spacetime. However,

(a) in some cases minimal coupling is not unique, and we have to make a choice between alternatives – for example, taking Maxwell's equations to be minimally coupled, as in (5.115), then the gauge potential equation (5.141) is not minimally coupled;

(b) sometimes theoretical reasons may suggest non-minimal coupling (see e.g. Balakin and Ni (2010)) – the challenge then is to give experimental evidence that this is physically correct;

(c) in some cases, the minimal coupling idea may be quite incorrect. The most striking example of this kind is gravity. The fundamental insight of Einstein was to realize that gravity is unlike every other force, so the minimal coupling idea completely fails in this case. Ultimately this is the reason why gravity is best described by a curved spacetime structure.

3.1.4 Remarks on the other 'principles' of general relativity

Before turning to the mathematical description of the matter content of spacetime and the gravitational field equations, we mention some more of the 'principles' which have played a part in relativity theory.

There is an even stronger form of the usual SEP, called the Einstein Equivalence Principle (Will, 2006), which reads: *all physics in freely falling systems is (locally) the same as in special relativity.* This principle, which is true in GR provided the physical laws concerned are minimally coupled, has the consequence that it is impossible to define a local gravitational energy–momentum tensor, since any such tensor would be zero in the special relativity approximation, and would thus be zero in all frames (see Section 3.1.1). Therefore any valid definition of gravitational energy, or of the total energy of a system, in GR must be non-local, e.g. be defined by an integral over a finite region.

This clash between the equivalence principle and the concept of a local gravitational energy is the source of the difficulties with the energy concept in GR which have been a continual topic of research. The problem has, in our view, so far only been completely solved in the context of defining a total energy for an isolated ('asymptotically flat') system (see e.g. Chruściel, Jezierski and MacCallum (1998)).

One should also note that in a situation where the actual metric is approximated by an averaged metric, there may be terms of the form of an energy–momentum representing the effect of, e.g., high-frequency gravitational waves on the averaged metric (Isaacson, 1968) (see Section 16.1): this is not in disagreement with the previous statement since such terms only arise from transferring part of the curvature of the actual metric to the other side of the equation after splitting the actual curvature into an averaged part and a high-frequency part.

Another principle, which certainly was a motivation for Einstein, is the 'principle of covariance'. Unfortunately, this principle is not always carefully formulated. Its simplest expression is the requirement that the laws of physics should be stated in a way which makes them independent of the choice of coordinates. (Although explicit calculations may require the use of coordinates, the physical results should be independent of that choice: for comparison of metrics in a coordinate-free way see Section 17.2.)[1] It is sometimes supposed that this implies the laws must be in tensor form; however, the use of tensors, though sufficient (see Section 3.1.1 above), is not necessary to achieve covariance, for the following reason.

Ricci and Levi-Civita (1901) and Kretschmann (1917) showed that virtually any theory describing spacetime as a manifold is expressible in coordinate-independent form. A coordinate is a scalar function, which could be regarded as a (zero rank) tensor, so even giving particular coordinates some physical significance need not violate covariance, and equations involving them could be rewritten in an arbitrary coordinate system. It is also possible to use geometric objects which are not tensors to construct equations that are expressed in the same way in all coordinate systems; in particular, the curvature tensor is defined from the (non-tensorial) connection. One essential requirement for covariance is that if all non-zero quantities in an equation are written on one side of the equality sign, they must transform in a way consistent with the rule that 0 transforms to 0 on the other side.

Thus the reason for adopting tensor equations, as we do, is that they provide an especially simple and manageable way to satisfy the covariance principle, and moreover one which works well in practice, but this is not a forced choice.

We shall later make extensive use of a 'covariant' 1+3 splitting of spacetime, based on a choice of a physically preferred four-velocity defined by the matter content of the universe (or sometimes in another invariant way). The covariance here refers to the fact that the choice is not arbitrary and does not involve a choice of coordinates: in those respects it is essentially different from approaches based on choosing three-dimensional spatial sections in spacetime in a non-unique manner (which we shall refer to as 3+1 rather than 1+3 approaches), such as are frequently used in discussions of cosmological perturbation theory (compare Chapter 10).

One can also argue that in cosmology and other contexts completely general covariance is not the correct approach, physically, and a number of aspects and examples of restricted

---

[1] One can consider general covariance to mean invariance under the group of coordinate transformations, and then similarly define covariance under other groups, e.g. Lorentz covariance under Lorentz transformations.

covariance have been examined (Ellis and Matravers, 1995, Zalaletdinov, Tavakol and Ellis, 1996).

It is sometimes said that Einstein 'geometrized' physics. In particular, attempts have been made, as in Kaluza–Klein theory (Kaluza, 1921, Klein, 1926) or Einstein's own non-symmetric theories (Einstein, 1956), to describe a wider range of physical fields by a generalization of the four-dimensional Riemannian metric and connection of GR. In the sense of differential geometry, any set of fields on a manifold is geometric, but authors following this idea sometimes seem to just pile up unrelated structures, so that it is not clear how helpful the concept of 'geometrization' is. It could be said to have clarified the structure of gauge theories when the general description of them in terms of connections (see Section 2.5) was recognized.

Another principle Einstein had in mind was Mach's principle. We discuss this in Section 21.1.3.

3.2 Conservation equations

Conservation is usually concerned with quantities integrated over a finite or infinite hyper-surface and asks: do the integrals remain constant as we vary the hypersurface in spacetime? The fundamental way to develop conservation laws is to integrate over a volume or surface. For an n-dimensional manifold, the integrand must involve an n-form (see Section 2.3): a volume form enables us to integrate scalars, and these are the physically interesting cases, enabling covariantly valid conservation laws (see Chapter 20 of Stephani (2004) for further discussion). Any attempt to define an integral of a vector over a volume by integrating its components, for example, will not be invariant under general changes of coordinates that are position dependent within the volume, and so will not be well defined.

Note, however, that when integrating over nonnull hypersurfaces the induced volume element can be considered as a multiple of the normal vector to the hypersurface, like a vectorial surface area element in three-dimensional vector calculus. For a spacelike surface S with unit timelike normal n^a, the 3-volume is $\mathrm{d}^3 V = \eta_{\mu\nu\kappa}\,\mathrm{d}x^\mu \mathrm{d}x^\nu \mathrm{d}x^\kappa /3! = \sqrt{|^3 g|}\mathrm{d}^3 x$ with coordinate volume $\mathrm{d}^3 x = \mathrm{d}x^1 \mathrm{d}x^2 \mathrm{d}x^3$, and the relevant vector is $\mathrm{d}S_a = -|\mathrm{d}^3 V|n_a$, which can be contracted with another vector to give a scalar on the hypersurface. (The minus in the definition of $\mathrm{d}S_a$ arises from our choice of signature; when there is no potential ambiguity with the four-dimensional or other volumes we will drop the ^{3} superscript.) Similar remarks apply to other submanifolds.

Scalar conservation

If we have a 4-vector with vanishing divergence, then this expresses conservation of some quantity (mass, charge, etc.). Consider a timelike vector J^a, and a 3-surface volume element $\mathrm{d}S_a$; contracting J^a with $\mathrm{d}S_a$ and integrating over the surface S gives the flux I of J^a through S: $I = \int_S J^a \mathrm{d}S_a$. In the case of a spacelike surface element $I = -\int_S J^a n_a \mathrm{d}^3 V$.

In a volume V which is a tube with timelike sides bounded top and bottom by spacelike surfaces labelled S_1 and S_2 respectively, and for which either (a) the sides are parallel to J^a, or (b) the sides are in a region where $J^a = 0$ (for example lying very far away in an asymptotically flat spacetime), there is no flux across the sides. Thus the integral over the surface of V becomes $I_1 - I_2 \equiv \int_{S_1} J^a \mathrm{d}S_a - \int_{S_2} J^a \mathrm{d}S_a$. Then the (four-dimensional) divergence theorem shows

$$I_1 = I_2 + \int_V J^a{}_{;a} \mathrm{d}V .$$

Consequently

$$J^a{}_{;a} = 0 \;\Rightarrow\; I_1 = I_2, \tag{3.2}$$

that is, vanishing divergence of J^a implies I is a conserved quantity for a 4-volume with sides parallel to J^a or with sides located where $J^a = 0$ (it is independent of the particular choices of S_1 and S_2).

To see what this means physically, split J^a into its spacetime direction u^a and its magnitude ρ by the equation

$$J^a = \rho u^a, \;\; u^a u_a = -1 . \tag{3.3}$$

This defines the average 4-velocity u^a of the conserved quantity represented by J^a, and its density ρ measured by an observer moving at that average velocity (rest mass density, electric charge density, etc.). Now for a spacelike surface S, by (3.3),

$$I = \int_S J^a \mathrm{d}S_a = \int \rho(-u^a n_a) \mathrm{d}^3 V . \tag{3.4}$$

Thus

$$I = \int_S \rho \cosh \beta \sqrt{|{}^3 g|} \mathrm{d}^3 x, \;\; \text{where } \cosh \beta := -u^a n_a, \tag{3.5}$$

is the total conserved quantity crossing the surface (rest mass, electric charge, number of particles, etc., depending on the nature of the conserved current J^a; in these different cases, ρ is the rest mass density, electric charge density, number density respectively, measured by an observer moving with velocity u^a). In terms of the quantities in (3.3), the divergence equation (3.2) is

$$(\rho u^a){}_{;a} = 0 \;\Leftrightarrow\; \dot{\rho} + \rho \Theta = 0, \tag{3.6}$$

where we have defined the expansion Θ of u^a by $\Theta = u^a{}_{;a}$.

It is convenient to define a representative length $\ell(x^a)$ by $\ell^3 = \sqrt{|{}^3 g|}$. If we consider a very narrow tube around a particular world line C that is an integral curve of u^a, with n^a chosen parallel to u^a in this tube in the limit as it shrinks to zero (so that $u^a n_a = -1$ there), we find that

$$I = \int_S \rho \ell^3 \mathrm{d}^3 v = \rho \ell^3 \epsilon$$

is constant, where $\epsilon := \int_S d^3 v$ is the (constant) comoving coordinate volume of this thin tube. ('Comoving' refers to propagation along the integral curves of u^a: comoving coordinates are discussed more fully in Section 4.1.) Thus (as in Newtonian theory) conservation is expressed by the differential relation

$$\rho\ell^3 =: M, \quad \dot{M} = 0 \iff \dot{\rho} + 3\rho\dot{\ell}/\ell = 0, \tag{3.7}$$

where the second form is obtained by taking the covariant derivative of the first along this tube (in the direction u^a), and we denote the (comoving) time derivative measured by an observer with 4-velocity u^a by a dot (compare Section 4.3): e.g. $\dot{\rho} = \rho_{;a} u^a$. Equations (3.6) and (3.7) together show that the expansion Θ gives the rate of change of volume,

$$\Theta = 3\dot{\ell}/\ell = (d^3 V)^{\cdot}/(d^3 V), \tag{3.8}$$

which also follows either from detailed analysis of the fluid flow characterized by u^a (see Ellis (1971a)) or from the useful identity

$$u^\mu{}_{;\mu} = \frac{1}{\sqrt{|g|}} \frac{\partial}{\partial x^\mu} (\sqrt{|g|} u^\mu), \tag{3.9}$$

expressing the divergence in terms of partial derivatives (to obtain (3.8) use comoving coordinates (see Section 4.1) such that $u^\mu = \delta_0^\mu = n^\mu$, noting that then $g = -{}^3g = -\ell^6$). This equation gives a quick way of calculating Θ.

Conservation of mass, charge, particle number, etc., can thus be expressed in a variety of forms. We shall usually come across them either in the form (3.6) or (3.7), which for example describe the conservation of rest-mass in cosmology (see Ellis (1971a) and Section 5.1.1).

Energy–momentum conservation

The relativistic energy, momentum and stresses of whatever matter fields are present are described by the symmetric *energy–momentum–stress tensor* $T_{ab} = T_{ba}$, which gives the energy and 4-momentum crossing a surface element dS_a by the relation $T^a = T^{ab} dS_b$. The symmetry of T_{ab} is a fundamental property of relativity theory expressing the equivalence of mass and relativistic energy ($T_{0i} = T_{i0}$) and the absence of macroscopic spin in the matter ($T_{ij} = T_{ji}$). (This result can be derived by considering the balance of the net fluxes of energy and momentum across all faces of an infinitesimally small volume.) Conservation of energy and momentum is given by the equation

$$T^{ab}{}_{;b} = 0, \tag{3.10}$$

generalizing the flat-space conservation laws to curved space in the standard way. However, we cannot integrate the quantities T^a over a finite surface to get a vector conserved quantity in general, because (as mentioned above) we cannot integrate a vector covariantly over a volume. Nevertheless (3.10) represents the local conservation of energy and momentum, as we see later in the case of various examples such as perfect fluids, electromagnetic fields and scalar fields.

Three points should be noted about the stress tensor.

Firstly, because of the principle of equivalence T^{ab} does not include a contribution from gravitational energy (unless it represents an averaging over high-frequency parts: see Section 16.1).

Secondly, if there is a Killing vector ξ^a in a spacetime, then we get a conserved vectorial quantity by contracting the Killing vector with the stress tensor:

$$\xi_{a;b} = \xi_{[a;b]}, \quad J^a := T^{ab}\xi_b \quad \Rightarrow \quad J^a{}_{;a} = 0 \,. \tag{3.11}$$

Thus there is an associated conserved quantity for each Killing vector (i.e. for each space-time symmetry). This is helpful in understanding specific exact solutions of the Einstein equations; however, a realistic spacetime has no Killing vectors.

Thirdly, in general there will be various matter components present in spacetime (baryons, photons, neutrinos, a scalar field, etc.) The total stress-tensor of such multi-component systems is obtained by adding the stress-tensors of the components: $T^{ab} = \sum_I T_I^{ab}$ where I labels the different components. Now while total energy and momentum are necessarily conserved, interchange of energy and momentum between the components is of course possible, so the energy and momentum of the individual components is not necessarily conserved. This is represented by interchange vectors Q_I^a showing how much energy and momentum has been gained or lost by each component, their total summing to zero to guarantee conservation of total energy–momentum:

$$T_I^{ab}{}_{;b} = Q_I^a \quad \Rightarrow \quad T^{ab}{}_{;b} = \sum_I Q_I^a = 0. \tag{3.12}$$

The quantities Q_I^a will be determined by the physics of interactions between the components.

3.3 The field equations in relativity and their structure

Using the Einstein tensor,

$$G_{ab} = R_{ab} - \tfrac{1}{2}Rg_{ab},$$

we can write Einstein's field equations (EFE) as

$$G_{ab} + \Lambda g_{ab} = R_{ab} - \tfrac{1}{2}Rg_{ab} + \Lambda g_{ab} = \kappa T_{ab}, \tag{3.13}$$

where Λ is the cosmological constant and κ is the gravitational constant ($\kappa = 8\pi G$ in our units: see Appendix A). The essential points about this formulation are that:

(1) these are non-linear, but quasi-linear, second-order partial differential equations for the gravitational potentials g_{ab}, and give the Newtonian limit in the weak-field and slow motion approximation, and

(2) they guarantee energy–momentum conservation: from these equations,[2]

$$G^{ab}{}_{;b} = 0 \quad \Leftrightarrow \quad T^{ab}{}_{;b} = 0 \tag{3.14}$$

as a consequence of the Bianchi identities (2.69).

---

[2] Provided Λ and κ are indeed constant! One would have to reconsider the formulation if variation of these quantities were allowed.

These are the reasons for the form chosen for the equations.

Defining

$$T := T^a{}_a,$$

we find that (3.13) gives

$$R - 2R + 4\Lambda = 8\pi GT \Rightarrow R_{ab} = 8\pi G(T_{ab} - \tfrac{1}{2}Tg_{ab}) + \Lambda g_{ab}, \qquad (3.15)$$

which is often more convenient for practical use than (3.13).

Matter present locally fixes the Ricci tensor completely through (3.15), but this is only a contraction of the full curvature. The remaining part, the Weyl tensor (see (2.79)), is not fixed by the local matter but is related to it by the Bianchi identities (2.86), which are differential equations with source terms given by the matter tensor through the EFE (3.15), implying that the Weyl tensor also requires boundary or initial conditions for its full specification.

It is in the choice of the equations (3.13) that GR differs essentially from standard gauge theories of physics. Such theories follow the example of the electromagnetic field, and have a free-field Lagrangian of the form $R^I{}_{Jab} R^J{}_I{}^{ab}$. This is an expression quadratic in first derivatives of the gauge potentials (the connection) and leads to differential equations of second order for those potentials. In the Maxwell case these will be the first of (5.115), the second being integrability conditions for the existence of the gauge potential. In GR we instead use the extra contraction possibilities made available by working on the tangent bundle (to reach R_{ab} from $R^c{}_{acb}$) and using a metric (to reach R), and write equations which are of second order not for the gauge potential (connection) but for the metric itself, which can be regarded as a kind of pre-potential in this context. If one were to take a theory of gravity with a purely quadratic Lagrangian such as $R^{abcd}R_{abcd}$, the equations would in general be fourth-order equations for the metric, and such theories are therefore not compatible with experiment: however, theories with a Lagrangian of some form such as $R + \alpha R^2$ are often considered, in particular because such terms arise naturally in attempts to perturbatively renormalize GR (see Section 20.2.1).

A cosmological constant is permitted by requirements (1) and (2) above, and could account for the accelerated expansion suggested by SNIa observations (see Section 13.2), but causes some difficulties. Even before the inference of a definite value from the SNIa data, Λ was known to be small, observationally, or it would have affected (e.g.) dynamics of galaxy clusters, but attempts to give it an origin in particle physics naturally produced very large values (Weinberg, 1989). It is then hard to see why the quantum effects should cancel leaving only a small residual value (rather than either a large value or zero). As a classical field, it has the drawback of acting on everything but being acted on by nothing, which makes it different from all other fields in not obeying Newton's law of action and reaction. The recent observations oblige us to include such 'dark energy' throughout (or find an alternative explanation for the SNIa observations, as discussed in Chapters 15–16), but it remains poorly motivated as a field to be expected in the universe.

3.3.1 Evolution from initial values

Now suppose the EFE are true on an initial spacelike surface S given by $t = t_0$, and consider under what conditions they will remain true off this surface. The characteristics of the equations are null rays; in physical terms, gravitational waves travel at the speed of light (i.e. the equations are hyperbolic if written in suitable coordinates; we discuss this in more detail in Section 6.6.2). Therefore we obtain a unique solution to the future of S from initial data on S only in the spacetime region $D^+(S)$ called the *future Cauchy development of* S (see Tipler, Clarke and Ellis (1980) for a discussion), which is the region such that all past-directed timelike and null curves from each point in $D^+(S)$ intersect S (in brief: it is the future region of spacetime such that all information arriving there at less than or equal to the speed of light has had to cross S, so conditions there are completely determined by data on S). The *past Cauchy development* $D^-(S)$ is similarly defined. The *Cauchy development* $D(S)$ of S is the union of these two regions and S and so is the complete region of spacetime that is determined by data on S.

Define the tensor $A^{ab} = G^{ab} - \kappa T^{ab}$, which is symmetric. Then the EFE (3.13) are equivalent to the equations $A^{ab} = 0$. From (2.69) and (3.10), A^{ab} has vanishing divergence,

$$A^{ab}{}_{;b} = 0 \quad \Leftrightarrow \quad A^{ab}{}_{,b} + A^{cb}\Gamma^a{}_{bc} + A^{ac}\Gamma^b{}_{bc} = 0. \tag{3.16}$$

Separating out the time and space summations, this can be written

$$A^{a0}{}_{,0} + A^{ai}{}_{,i} + A^{c0}\Gamma^a{}_{0c} + A^{0i}\Gamma^a{}_{i0} + A^{ji}\Gamma^a{}_{ij} + A^{a0}\Gamma^b{}_{b0} + A^{ai}\Gamma^b{}_{bi} = 0. \tag{3.17}$$

Now suppose that the equations $A^{ij} = 0$ are true in a spacetime region V containing S and lying in the Cauchy development of S while the equations $A^{0a} \equiv A^{a0} = 0$ are true on the initial surface S. Then the form of the first-order linear set of differential equations (3.17) (which give the time development of A^{a0} off S) guarantees a unique solution locally in V from given initial data for A^{a0} on S. However, there is a solution of these equations given by $A^{a0} = 0$ in V, which of course implies $A^{a0} = 0$ on S, and, by the uniqueness, is implied by those initial conditions. Hence we have shown the following: *if the four initial value equations $A^{a0} = 0$ are true on S and the six propagation equations $A^{ij} = 0$ are true in V, then $A^{a0} = 0$ will hold in V.* Thus the four Einstein equations $A^{a0} = 0$ are first integrals of the other six equations (provided the energy–momentum conservation equations (3.14) are satisfied). This structure is helpful in actually solving the equations.

To see what initial data for the field equations is like, it is convenient to use Gaussian normal coordinates, giving a metric of the form

$$ds^2 = -dt^2 + h_{ij}\,dx^i\,dx^j \tag{3.18}$$

(compare (2.111)). The propagation equations $A^{ij} = 0$ turn out to be equations for $d^2 h_{ij}/dt^2$ in terms of the metric h_{ij} and its first derivatives $dh_{ij}/dt, h_{ij,k}$, while the constraint equations $A^{a0} = 0$ turn out to be equations involving only the first time derivatives of h_{ij}, h_{ij} itself and its spatial derivatives. Thus the initial data for the EFE on S are,

(a) the *first fundamental form* h_{ij} (just the intrinsic metric of that surface), plus
(b) the *second fundamental form* $K_{ij} = dh_{ij}/dt$ (the derivative of the metric with respect to proper time measured along the orthogonal geodesics, which characterizes how the 3-space is imbedded in the 4-space), together with
(c) initial data for whatever matter fields may be present.

These data must be chosen to satisfy the constraint equations $A^{0a} = 0$; then the propagation equations $A^{ij} = 0$ together with the evolution equations for the matter will determine the solution off S. The constraint equations will remain true off S if they are true on S, as just proved above. The solution will be determined by this data within the Cauchy development $D(S)$ of S, but not outside this spacetime region.

One can apply similar considerations to equations (2.86) regarded as equations giving the divergence of the Weyl tensor in terms of matter variables (on using (3.15) to replace the Ricci terms by matter terms). The corresponding conservation equations are (2.87) which can be analysed similarly to (3.16), with the constraint equations in the set preserved by the dynamical evolution equations. We shall in effect be using this result in the 1+3 analyses of the Weyl tensor propagation in the following chapters.

3.3.2 Variational formulation

One can obtain the gravitational field equations from a variational principle with Einstein–Hilbert action,

$$S = \frac{1}{16\pi G} \int R \, dV, \qquad (3.19)$$

where R is the curvature scalar (see e.g. Section 22.4 of Stephani (2004)).

Although Lagrangian principles, and the related Hamiltonian principles, are of great importance, they may be of limited use in applications to specific metrics for the following reasons.

(a) Varying S above with respect to a general metric, we get the EFE (3.13); if we now choose a particular metric form, the equations (3.13) specialize to the EFE for that family of metrics. These are the equations to be solved for specific solutions of the EFE with a metric of the chosen form, but the variational principle may be of little help in obtaining them by this route, since the specialization may be a long calculation.
(b) On the other hand we can calculate the curvature scalar R directly from the specific metric form, and then put this into (3.19) to get a variational principle for those metrics. Performing the variation, we get another set of equations for metrics of the chosen form.

The latter may provide a quick derivation of equations, but now the problem is that *in general these two sets of equations are not the same*: the operations of performing the variation and specializing the metric do not commute. The reason has to do with the boundary terms derived in the variation, which we normally assume will be zero. When we use method (b), in general this will not be so. Hence method (b), the obvious way to use the variational principle to simplify derivation of the field equations, may give the wrong answer (MacCallum and Taub, 1972, Sneddon, 1975).

Thus in practice one should use method (b) with caution, checking whether the surface terms vanish or not. One could of course check the result from (b) with that from (a), but then one may as well just use the result obtained by (a), or calculate the curvature directly.

3.3.3 ADM Hamiltonian formulation

The Arnowitt, Deser and Misner (1962) (ADM) formalism (see Peter and Uzan (2009) for a recent account) foliates the spacetime into spatial hypersurfaces and then uses the splitting of the metric in these coordinates to rewrite the gravitational action in a way that produces a Hamiltonian. This Hamiltonian approach is particularly important for aspects of quantum gravity. However, the ADM form of the metric and field equations has wider applicability.

The ADM metric is

$$ds^2 = -N^2 dt^2 + \gamma_{ij}(N^i dt + dx^i)(N^j dt + dx^j), \tag{3.20}$$

where N is the 'lapse' function and N^i is the 'shift' vector. Geometry can now be expressed in terms of the curvatures of the Riemannian 3-spaces $t = \text{const}$. The Einstein–Hilbert action (3.19) can be written as (we set $8\pi G = 1$ for convenience in this subsection)

$$S = \int dt \int d^3x \sqrt{\gamma}\, N \left(^3R + K_{ij}K^{ij} - K^2\right), \tag{3.21}$$

where the extrinsic curvature is

$$K_{ij} = \frac{1}{2N}\left(\gamma_{ik}D_j N^k + \gamma_{jk}D_i N^k - \dot{\gamma}_{ij}\right), \tag{3.22}$$

D_i being the covariant derivative defined by γ_{ij}. The trace is $K = \gamma^{ij}K_{ij}$.

The action is considered as a function of the Lagrangian variable $q = (\gamma_{ij}, N, N^i)$ and $\dot{q}$, and the Lagrangian density is

$$\mathcal{L}[q, \dot{q}] = \sqrt{\gamma}\, N \left(^3R + K_{ij}K^{ij} - K^2\right). \tag{3.23}$$

There is no dependence on N, N^i, so these variables are not dynamical and are associated with constraints. The dynamical variable γ_{ij} defines a conjugate momentum via the variational derivative:

$$\pi^{ij} := \frac{\delta \mathcal{L}}{\delta \gamma_{ij}} = \sqrt{\gamma}\left(K\gamma^{ij} - K^{ij}\right). \tag{3.24}$$

Then the Hamiltonian density in vacuum is

$$\mathcal{H} = \pi^{ij}\dot{\pi}_{ij} - \mathcal{L}, \tag{3.25}$$

and the Hamiltonian becomes (after dropping a divergence term)

$$H = -\int d^3x \sqrt{\gamma}\left(N C_0 - 2N^i C_i\right), \tag{3.26}$$

where the constraints are

$$C_0 = {}^3R - K_{ij}K^{ij} + K^2, \tag{3.27}$$

$$C_i = -D_i K + D_j K_i^j. \tag{3.28}$$

$C_0 = 0$ is the 'energy constraint', and $C_i = 0$ is the 'momentum constraint' (these are respectively the trace of (2.114), and (2.115), for a vacuum). There is also an evolution equation from the Hamilton equation $\dot{q} = \delta H / \delta p$; in vacuum, it is

$$\dot{\gamma}_{ij} = -2N K_{ij} + D_i N_j + D_j N_i . \tag{3.29}$$

Exercise 3.3.1 Show in detail how the analysis of (2.86) mentioned at the end of Section 3.3.1 works.

3.4 Relation to Newtonian theory

An important issue is to formulate the Newtonian limit of GR in a cosmological context, when conditions relating to gravitational physics are significantly different from the more traditional quasi-stationary and asymptotically flat situations where the Newtonian and post-Newtonian limits are usually derived and where they have been subjected to precise experimental tests, enabling us to evaluate the κ of (3.13) in terms of the Newtonian gravitational constant G. The importance arises because many astrophysical calculations on the formation of large-scale structure in the universe are done in a Newtonian way and so depend on such a limit. But there are major differences between Newtonian theory and GR, particularly because in Newtonian theory (a) there is a preferred time coordinate, (b) spacetime is flat; neither is true in GR.

There are of course many derivations of Newtonian gravity in terms of the weak field limit of GR, but in the astrophysical context such linearized derivations cease to be useful just when the theories become important – namely in the context of nonlinear structure formation. One should therefore note the following:

(A) The appropriate classical theory of gravitation is GR; Newtonian theory is only a good theory of gravitation when it is a good approximation to the results obtained from GR.
(B) We have to extend standard Newtonian theory (which strictly can only deal with quasi-stationary isolated systems in an asymptotically flat spacetime) in some way or another to deal with a non-stationary cosmological context. This extension needs to be clearly spelt out (see Section 6.8).

We can give close Newtonian analogues to many of the major covariant relations presented in this book (see also Ellis (1971a)), but significant questions still remain with regard to both points. These include:

(1) Full GR involves 10 *gravitational potentials* (combined in a tensorial variable), subject to the 10 Einstein field equations ('EFE'), but Newtonian theory involves only one (a scalar variable – the acceleration potential), subject to the one Poisson field equation; how does it arise that the other nine potentials and equations can be ignored in the Newtonian limit? Given that nine equations of the full theory are not satisfied even in some limiting sense, how do we know when Newtonian cosmological solutions correspond to consistent relativistic solutions of the full set of equations? Part of the answer is

that the coordinate freedom of GR accounts for four of these potentials; however, this still leaves another five to account for, and we have examples of Newtonian solutions with *no* GR analogues (Ellis, 1997, van Elst and Ellis, 1998), as we shall discuss in Section 6.8.2. Consequently we need to be concerned about how well standard Newtonian theory represents the results of GR in the cosmological context we have in mind, when we take a 'Newtonian limit'.

(2) The issue of *boundary conditions* for Newtonian theory in the cosmological context is problematic even in the context of exactly spatially homogeneous (and spatially isotropic) cosmological models (Heckmann and Schucking, 1956); no fully adequate theory exists in the more realistic almost-homogeneous case. The essential difficulty lies in specifying the boundary conditions at spatial infinity. In normal Newtonian theory one assumes that the potential becomes constant there: that assumption is not compatible with an infinite universe of constant or near-constant density, since it would not satisfy the Poisson field equation (this is why Einstein so strongly supported the idea of closed spatial sections in cosmology – then no 'infinity' would exist where boundary conditions had to be specified). Numerical simulations, for example, usually rely on periodic boundary conditions, which correspond to the real universe only if we live in a 'small universe' (see Section 9.1.6) in which there is a long-wavelength cutoff in the spectrum of inhomogeneities of its large-scale structure. Analytic solutions usually rely on asymptotically flat conditions which are manifestly not true in a realistic, almost-FLRW situation (they implicitly or explicitly assume that inhomogeneities die off sufficiently far away from the region of interest).

(3) How do we obtain a *unique propagation equation* for the gravitational scalar potential in a Newtonian cosmology, when Newtonian theory proper has no such equation? In standard Newtonian theory this is related to the previous problem: since, unlike GR, the Newtonian gravitational field equation allows infinitely fast propagation of influences from infinity, boundary conditions have to be imposed at infinity at every instant to obtain unique propagation.

(4) How do we satisfactorily handle the *gauge dependence* that underlies most derivations of a Newtonian limit? Equivalently, most derivations of the Newtonian limit are highly coordinate dependent, basically because Newtonian theory depends fundamentally on its preferred time coordinate; there is no such unique coordinate in the perturbed FLRW models used for studies of structure formation (see discussion of the gauge problem in perturbation theory, Chapter 10).

We shall return to these issues in Section 6.8 and Chapter 21.

PART 2

RELATIVISTIC COSMOLOGICAL MODELS

4   Kinematics of cosmological models

In cosmology, the matter components allow us to make a physically motivated choice of preferred motion. For example, we could choose the CMB frame, in which the radiation dipole vanishes, or the frame in which the total momentum density of all components vanishes. Such a choice corresponds to a preferred 4-velocity field u^a that generates a family of preferred world lines. We can then make a 1+3 split relative to u^a, in order to relate the physics and geometry to the observations. In this chapter we discuss how to do this for the kinematics of cosmological models; the following chapter will consider the dynamics.

The (real or fictitious) observers are comoving with the matter-defined 4-velocity u^a, and we can call the observers and the 4-velocity 'fundamental'. If we change our choice of fundamental 4-velocity, the kinematics and dynamics transform in a well-defined way, as discussed in the following chapter.

4.1 Comoving coordinates

To describe the spacetime geometry it is convenient to use comoving (Lagrangian) coordinates, adapted to the fundamental world lines. These are locally defined as follows.[1]

(1) Choose a surface S that intersects each world line once only (note that no unique choice is available in general). Label each world line where it intersects this surface; as the surface is three-dimensional, three labels y^i, $(i = 1, 2, 3)$, are required to label all the world lines.

(2) Extend this labelling off the surface S by maintaining the same labelling for the world lines at later and earlier times. Thus the y^i are *comoving coordinates*: the value of the coordinate is maintained along each world line, and in fact the world lines (and so the fundamental particles) are labelled by these coordinates.

(3) Define a time coordinate t along the fluid flow lines (it must be a function that increases along each flow line).

Then (t, y^i) are *comoving coordinates* adapted to the flow lines. Note that the surfaces $t = $ const will in general not be orthogonal to the fundamental world lines; indeed, in general it is not possible to choose a time coordinate for which these surfaces are orthogonal (see Section 4.6 below).

---

[1] In this section we are not concerned about possible global problems with coordinates; these are considered in Section 6.7.

The coordinate freedoms available which preserve this form are (a) *time transformations* $t' = t'(t, y^i)$, $y^{i'} = y^i$, corresponding to a new choice of time surfaces, and (b) *relabelling of the world lines* by choosing new coordinates in the initial surface: $t' = t$, $y^{i'} = y^{i'}(y^i)$.

A particular choice for t which is often convenient is the *normalized time* $s = s_0 + \tau$, where τ is proper time measured along the fundamental world lines from S (positive to the future of S, negative to the past) and s_0 is an arbitrary constant. With this choice, $x^\mu = (s, y^i)$ are *normalized comoving coordinates*, s measuring proper time from the surface S (on S, $s = s_0$) along the world lines, which lie in the intersection in spacetime of the surfaces $y^i = $const. The remaining time freedom is then $s' = s + f(y^i)$, corresponding to choice of the initial surface S. For example, the standard coordinates (t, r, θ, ϕ) in an FLRW universe model (2.65) are such normalized comoving coooordinates.

General coordinates x^μ will be related to the normalized comoving coordinates by a coordinate transformation,

$$x^\mu = x^\mu(s, y^i), \qquad \mu = 0, \dots, 3, \qquad i = 1, \dots, 3, \qquad (4.1)$$

the spatial parts of which are similar to the transformation from Lagrangian to Eulerian coordinates in Newtonian theory. Indeed one can define Newtonian-like quasi-Eulerian ('fixed', non-comoving) coordinates in general relativity, defined by the physical (proper) distance and direction from some chosen world line in preferred space sections. They can provide a useful alternative to the more usual Lagrangian coordinates when taking the Newtonian limit (for the FLRW case, see Ellis and Rothman (1993)).

In Newtonian theory, because unique spatial sections exist in spacetime, there is a natural time coordinate t, uniquely defined (up to a constant), which measures proper time along all lines. In Newtonian cosmology, one can again choose comoving spatial ('Lagrangian') coordinates y^i as in the relativistic case, obtaining spacetime coordinates (t, y^i).

4.2 The fundamental 4-velocity

The preferred matter motion implies a preferred 4-velocity at each point. Geometrically, this can be depicted as an arrow pointing along the fundamental world lines. If the preferred world lines are given in terms of *local* coordinates x^μ by $x^\mu = x^\mu(\tau)$ where τ is proper time along the world lines, then the preferred 4-velocity is the unit timelike vector

$$u^\mu = \frac{\mathrm{d}x^\mu}{\mathrm{d}\tau} \quad \Rightarrow \quad u^\mu u_\mu = -1. \qquad (4.2)$$

The implication follows by considering the integral for proper time along a world line: $\tau = \int [-(\mathrm{d}x^\mu/\mathrm{d}\tau)(\mathrm{d}x^b/\mathrm{d}\tau)g_{ab}]^{1/2}\mathrm{d}\tau = \int (-u^\mu u_\mu)^{1/2}\mathrm{d}\tau$. In normalized comoving coordinates $x^\mu = (s, y^i)$ this becomes

$$u^\mu = \delta^\mu_0 \quad \Leftrightarrow \quad \frac{\mathrm{d}s}{\mathrm{d}\tau} = 1, \; \frac{\mathrm{d}y^i}{\mathrm{d}\tau} = 0. \qquad (4.3)$$

These are also sufficient conditions such that the coordinates are normalized comoving coordinates (the curve parameter is proper time, and the vector u^μ is tangent to the direction where all the y^i are constant).

In general coordinates x^μ, this vector will be given by

$$u^\mu = \left(\frac{\partial x^\mu}{\partial s}\right)_{y^i = \text{const}}. \tag{4.4}$$

This is obtained by applying the coordinate transformation (4.1) to (4.3); conversely, specializing the coordinates in (4.4) to normalized comoving coordinates, we recover (4.3).

The component of any vector X^a parallel to u^a is

$$X^a_\parallel = U^a{}_b X^b, \quad U^a{}_b := -u^a u_b, \tag{4.5}$$

where $U^a{}_b$ is a projection tensor ($U^a{}_b U^b{}_c = U^a{}_c$) into the one-dimensional tangent line ($U^a{}_a = 1$) parallel to u^a ($U^a{}_b u^b = u^a$). For example, the fundamental 4-velocity in an FLRW universe model in the standard coordinates (2.65) is given by $u^\mu = \delta^\mu_0$, $u_\mu = -\delta^0_\mu$. Thus in this case $U^\mu{}_\nu = \delta^\mu_0 \delta^0_\nu$.

In Newtonian theory, a 3-velocity v^i representing the average motion of matter at each point will be defined. Lagrangian coordinates (the Newtonian version of comoving coordinates) are characterized by the condition $v^i = 0$.

Exercise 4.2.1 Show that

(a) if we use general comoving coordinates (t, y^i), then $u^\mu = v^{-1} \delta^\mu_0$, where $v(x^\alpha) = ds/dt$, and that $u_\mu = g_{\mu 0}/v$, $g_{00} = -v^2$, $u_0 = -v$;
(b) under a time transformation, $t' = t'(t, y^i)$, $y^{i'} = y^i$, these relations are preserved with $v \to v' = v/(\partial t'/\partial t)$;
(c) (4.4) reduces to (4.3) on using normalized comoving coordinates;
(d) we have normalized comoving coordinates if and only if $u_\mu = g_{\mu 0}$.

4.3 Time derivatives and the acceleration vector

The time derivative of any tensor $S^{a\cdots}{}_{b\cdots}$ along the fluid flow lines is

$$\dot{S}^{a\cdots}{}_{b\cdots} = u^c \nabla_c S^{a\cdots}{}_{b\cdots}, \tag{4.6}$$

and by (2.28) this is of the form

$$\dot{S}^{a\cdots}{}_{b\cdots} = \frac{d}{d\tau} S^{a\cdots}{}_{b\cdots} + S^{c\cdots}{}_{b\cdots} \Gamma^a{}_{cd} u^d + \cdots - S^{a\cdots}{}_{c\cdots} \Gamma^c{}_{bd} u^d - \cdots. \tag{4.7}$$

The first term is the apparent derivative (i.e. the value obtained by only taking the directional derivative of the components) relative to the coordinate frame along the world lines, and the others correct this apparent derivative to give the covariant derivative along the world lines. When a frame which is parallelly propagated along the world lines is used, this reduces to $\dot{S}^{a\cdots}{}_{b\cdots} = \mathrm{d}S^{a\cdots}{}_{b\cdots}/\mathrm{d}\tau = \partial S^{a\cdots}{}_{b\cdots}/\partial\tau + S^{a\cdots}{}_{b\cdots,i}u^i$ (the last term vanishes if comoving coordinates are used).

A particular application of time differentiation is the derivative of the 4-velocity itself in its own direction: this determines the acceleration vector,

$$\dot{u}^a = u^b\nabla_b u^a \;\Rightarrow\; \dot{u}^a u_a = 0, \tag{4.8}$$

which vanishes if and only if the flow lines are geodesics. Physically, this is the case if they represent motion under gravity and inertia alone, i.e. no non-gravitational force acts (see Section 3.1.2). From this definition,

$$\nabla_b u_a = h_a{}^c h_b{}^d \nabla_d u_c - \dot{u}_a u_b, \tag{4.9}$$

where the first term on the right is orthogonal to u^a, with h_{ab} defined in (4.10).

The corresponding Newtonian derivative is the 'convective derivative' (Batchelor, 1967), $\dot{T}^{ij\cdots}{}_{\cdots k\ell} = \partial T^{ij\cdots}{}_{\cdots k\ell}/\partial t + T^{ij\cdots}{}_{\cdots k\ell,m}v^m$, determining the rate of change of $T^{ij\cdots}{}_{\cdots k\ell}$ relative to the fluid. As we have just seen, this is essentially what is obtained from the general relativity equations if a parallelly propagated frame is used.

Exercise 4.3.1 Show that in normalized comoving coordinates, $\dot{u}^\mu = \Gamma^\mu{}_{00}$, so that $\dot{u}_\mu = \frac{1}{2}(2\partial g_{0\mu}/\partial t - \partial g_{00}/\partial x^\mu)$.

Exercise 4.3.2

(a) Show that the Newtonian analogue of the 'acceleration vector' is $a_i = \dot{v}_i + \Phi_{,i}$ where Φ is the Newtonian gravitational potential. Deduce that even in Newtonian theory, we are unable to separate the gravitational and inertial parts of a_i invariantly if the matter density does not go to zero at infinity. (Hint: see Heckmann and Schucking (1955), Bondi (1960), Trautman (1965)). (Note: when comparing relativistic and Newtonian equations, $\dot{u}^\mu$ should be compared with a_i not $\dot{v}_i$; the difference between the two cases can be considered to arise because in relativity covariant differentiation already has the gravitational effects coded into it.)

(b) Show that when comoving coordinates are used, $a_i = \Phi_{,i}$.

(c) Consider when this form can be obtained as a limit of the relativistic equations in the previous exercise (i.e. when does there exist a scalar potential Φ for $\dot{u}_a$?).

4.4 Projection to give three-dimensional relations

The existence of a preferred velocity at each point implies the existence of preferred rest frames at each point; locally these define surfaces of simultaneity for the fundamental observers.

4.4.1 Orthogonal projection

The (induced) effective metric tensor for these surfaces is the tensor

$$h_{ab} = g_{ab} + u_a u_b. \tag{4.10}$$

We see this as follows: from the above definition and (4.2),

$$h^a{}_b h^b{}_c = h^a{}_c, \quad h^a{}_a = 3, \quad h^a{}_b u^b = 0, \tag{4.11}$$

that is, $h^a{}_b$ is a projection tensor projecting into the three-dimensional tangent plane orthogonal to u^a.

Any vector X^a can be projected, by means of $h^a{}_b$, to its part $X^a_\perp$ orthogonal to u^a (i.e. its component in the instantaneous rest-space of an observer moving with the 4-velocity u^a):

$$X^a_\perp = h^a{}_b X^b \quad \Rightarrow \quad X^a_\perp u_a = 0, \ (X^a_\perp)_\perp = X^a_\perp. \tag{4.12}$$

The projection tensor is the metric tensor for the rest space; for if X^a and Y^b are any vectors orthogonal to u^a ($X^a u_a = 0 = Y^b u_b$), then $X \cdot Y = X^a g_{ab} Y^b = X^a h_{ab} Y^b$, that is, h_{ab} determines scalar products and so angles and magnitudes for all vectors in the rest space of u^a. It corresponds precisely to the Newtonian metric tensor h_{ij} determining magnitudes and angles in Newtonian theory.

U_{ab} and h_{ab} enable projection of any tensor into parts parallel and perpendicular to u^a. A particular example is the metric tensor itself: $g_{\perp ab} = h_a{}^d h_b{}^e g_{de} = h_{ab}$, $g_{\parallel ab} = U_a{}^d U_b{}^e g_{de} = U_{ab}$, so its splitting into parallel and perpendicular parts is given by

$$g_{ab} = h_{ab} + U_{ab} = h_{ab} - u_a u_b. \tag{4.13}$$

This shows that the interval ds^2 associated with an arbitrary displacement $x^\mu \rightarrow x^\mu + dx^\mu$ can be decomposed by

$$ds^2 = g_{\mu\nu} dx^\mu dx^\nu = h_{\mu\nu} dx^\mu dx^\nu - (u_\mu dx^\mu)^2 = (\delta l)^2 - (\delta t)^2 \tag{4.14}$$

into a time difference $\delta t = (U_{\mu\nu} dx^\mu dx^\nu)^{1/2} = (-u_\mu dx^\mu) \ (= c \, \delta t$ in units in which $c \neq 1)$ and a spatial distance $\delta l = (h_{\mu\nu} dx^\mu dx^\nu)^{1/2}$ as measured by an observer moving with 4-velocity u^μ. This decomposition implies the usual special relativistic length contraction and time dilation formulae (as these quantities are related to the interval in the standard special relativistic way, and all the equations hold for the projection tensors associated with arbitrary 4-velocities; we will get a different decomposition of g_{ab} into U_{ab} and h_{ab} if we choose a different 4-velocity u_a).

For example, in comoving coordinates, $U^{ij} = 0$, $h_{\mu 0} = 0$ in any universe, while the components of U^{ab}, h_{ab} in an orthonormal tetrad $\{u, e_i\}$ are $U^{ab} = \text{diag}(-1, 0, 0, 0)$, $h_{ab} = \text{diag}(0, 1, 1, 1)$. The components of h_{ab} in an FLRW universe in the standard FLRW coordinates (2.65) are $h_{\mu\nu} = a^2(t) \, \text{diag}(0, 1, f^2(r), f^2(r) \sin^2 \theta)$. In Newtonian theory, we have $\delta l = (h_{ij} dx^i dx^j)^{1/2}$, $\delta t = (-U_{ij} dx^i dx^j)^{1/2}$ still, but now the flat 3-space metric h_{ij} is given and fixed, and the time metric U^{ij} is also fixed. Using the preferred time coordinate t, $h_{\mu 0} = 0$, $h_{ij} = \delta_{ij}$ (in Cartesian coordinates) and $U^{\mu\nu} = -\delta^\mu_0 \delta^\nu_0$. These relations also follow from (4.13) in the slow-motion limit.

In the sequel we shall very frequently have occasion to project vectors orthogonal to u^a, and to take the traceless parts, orthogonal to u^μ, of rank two symmetric tensors – i.e. the projected symmetric tracefree, or PSTF, parts. For brevity we use angled brackets on indices to denote the PSTF parts. For convenience we use the term PSTF to include projected rank one tensors (vectors). For any V_a and tensor S_{ab}, the PSTF parts are given by

$$V_{\langle a \rangle} = h_a{}^b V_b \,, \quad S_{\langle ab \rangle} = \left\{ h_{(a}{}^c h_{b)}{}^d - \tfrac{1}{3} h_{ab} h^{cd} \right\} S_{cd} \,. \tag{4.15}$$

We write equations so that all terms are manifestly PSTF. We continue to use $\perp$ to indicate the projection in other cases. One may note that a general rank two tensor can be written as

$$
\begin{aligned}
S_{ab} &= (h_a{}^c + U_a{}^c)(h_b{}^d + U_b{}^d) S_{cd} \\
&= \tfrac{1}{3} h_{ab} (h^{cd} S_{cd}) + S_{\langle ab \rangle} + h_a{}^c h_b{}^d S_{[cd]} \\
&\quad - (h_a{}^c S_{cd} u^d) u_b - u_a (u^c S_{cd} h_b{}^d) + u_a u_b (u^c u^d S_{cd}) \,.
\end{aligned} \tag{4.16}
$$

This decomposition splits the tensor into parts irreducible under the rotational freedom in the hyperplane orthogonal to u^a. For the projection of time derivatives we use the notation

$$\dot{V}^{\langle a \rangle} = h^a{}_b \dot{V}^b \,, \tag{4.17}$$

and similarly for $\dot{S}^{\langle ab \rangle}$.

Any skew 2-tensor orthogonal to u^a can be written as

$$A_{ab} = A_{[ab]} = \eta_{abc} A^c \Leftrightarrow A^c = \tfrac{1}{2} \eta^{abc} A_{bc}, \tag{4.18}$$

where the projected alternating tensor η_{abc} was defined in Section 2.7.5. Then in (4.16) all terms can be expressed using u^a, scalars, projected vectors obeying $V_a = V_{\langle a \rangle}$, and PSTF 2-tensors $S_{ab} = S_{\langle ab \rangle}$. Note that such a PSTF tensor has five independent components. By evaluating components in an orthonormal tetrad aligned with u^μ it is easily found that η_{abc} has the same components as the skew object in three-dimensional vector calculus used, for instance, in defining vector products and the curl of a vector.

4.4.2 Orthogonal spatial derivatives

One can project the four-dimensional covariant derivatives to give three-dimensional derivative operators:

$$\overline{\nabla}_c S^{a\cdots}{}_{b\cdots} = h_c{}^f h^a{}_d \cdots h_b{}^e \cdots \nabla_f S^{d\cdots}{}_{e\cdots} \,. \tag{4.19}$$

Other notations for this derivative in the literature are ${}^{(3)}\nabla_a, \widehat{\nabla}_a, \vec{\nabla}_a$ and D_a.

Following Maartens (1997), we define

$$\operatorname{div} V = \overline{\nabla}^a V_a \,, \quad (\operatorname{div} S)_a = \overline{\nabla}^b S_{ab} \,, \tag{4.20}$$

$$\operatorname{curl} V_a = \eta_{abc} \overline{\nabla}^b V^c \,, \quad \operatorname{curl} S_{ab} = \eta_{cd(a} \overline{\nabla}^c S_{b)}{}^d \,. \tag{4.21}$$

The covariant div and curl preserve the PSTF property. Note that $\overline{\nabla}$, div and curl are defined as operators in a 3-manifold only if the vorticity vanishes (see below). When vorticity is non-zero, they are only operators in the tangent hyperplane at each point and not on a manifold.

One should be cautious in using three-dimensional concepts and notation in the case with non-zero vorticity – since in that case there are no hypersurfaces to which h_{ab} is everywhere tangent (see Section 4.6). This means in particular that there is in general no three-dimensional manifold in which Poincaré's Lemma can be applied to obtain scalar or vector potentials. However, when vorticity vanishes, we have a powerful covariant spatial calculus of vectors and tensors. Even with vorticity, it dramatically shortens subsequent equations, makes derivations far easier and more transparent and facilitates new insights.

With these definitions we find that

$$\overline{\nabla}_c h_{ab} = 0, \ \overline{\nabla}_d \eta_{abc} = 0, \ \dot{h}_{ab} = 2u_{(a}\dot{u}_{b)}, \ \dot{\eta}_{abc} = 3u_{[a}\eta_{bc]d}\dot{u}^d. \tag{4.22}$$

Identities obeyed by div and curl are collected in Section 4.8.

Exercise 4.4.1

(a) Show from (4.14) that the proper time $d\tau$ experienced by a particle between events P and Q for which an observer O determines coordinates x^μ, $x^\mu + dx^\mu$, respectively, is $d\tau = \delta t/\gamma$ where $\gamma = (1 - v^2/c^2)^{-1/2}$ and $v = \delta l/\delta t$ is the velocity of the particle relative to O.

(b) Show that

$$\eta_{abcd} = 2u_{[a}\eta_{b]cd} - 2\eta_{ab[c}u_{d]}, \ \eta_{abc}\eta^{def} = 3!h_{[a}{}^d h_b{}^e h_{c]}{}^f, \tag{4.23}$$

$$(V_{\langle a \rangle})\dot{} = \dot{V}_{\langle a \rangle} + V_b \dot{u}^b u_a. \tag{4.24}$$

(d) Derive (4.22).

4.5 Relative position and velocity

4.5.1 Relative position vectors

Consider a curve $y^i = y^i(v)$ in a surface $S : s = s_0$, where (s, y^i) are comoving coordinates. This curve links a set of fundamental observer world lines (which we shall assume for now are also galaxy world lines) in that surface; at all later times, the same curve links the same set of world lines, that is, the curve is dragged along by the world lines from the surface S to any other surface $s = $ const.

Similarly the vector $\beta^\mu = (dx^\mu/dv)\delta v$ tangent to this curve, given in comoving coordinates by $\beta^\mu = (0, \delta y^i)$, where $\delta y^i = (dy^i/dv)\delta v$, links the same pair of world lines O: $y^i = c^i = $ const and G: $y^i = c^i + \delta y^i$, $\delta y^i = $ const, at all times, provided δv is small (so that the displacement represented by the vector is a good approximation to displacement along the curve). This is a *connecting vector* as described in Section 2.4, since it always joins the same pair of fundamental world lines. In general coordinates x^μ, this vector will be given by

$$\beta^\mu = \left(\frac{\partial x^\mu}{\partial y^i}\right)_{s=\text{const}} \delta y^i. \tag{4.25}$$

An observer on O would at all times find that the spacetime position defined by β^μ lies on the world line of the galaxy G.

There is, however, a catch. The vector β^a will not in general be orthogonal to the fluid flow lines; thus it will represent both a spatial displacement from O to G and a time increment, i.e. it will link O to G at an earlier or later time as measured by O. What we wish to obtain, however, is the analogue of the Newtonian relative position vector, which represents an instantaneous spatial displacement as measured in O's rest frame, and so is orthogonal to u^a. We obtain this by projecting β^a orthogonal to u^a; that is, the *relative position vector* of G as measured by O is $\beta^{\langle a \rangle}$. This projected vector will represent a spacetime displacement from G to O provided the relative velocity of G and O is not too large, which will be true in the limit of small δv. We shall find it useful to decompose this vector into a *relative distance* δl and a *direction* e^a, where e^a is a unit vector in the rest space of O:

$$\beta^{\langle a \rangle} = e^a \delta l, \;\; e^a e_a = 1, \;\; e^a u_a = 0 \;\; \Rightarrow \;\; \beta^{\langle a \rangle} \beta_{\langle a \rangle} = (\delta l)^2. \tag{4.26}$$

The set of all directions about an observer O can just be considered as the sphere at unit radius about O.

4.5.2 Relative velocity

Given the definition of relative position, the way to define relative velocity is clear: take the time derivative of the relative position vector, and then project orthogonal to u^a to produce a vector in the rest frame of u^a:

$$v^a = v^{\langle a \rangle} = h^a{}_b u^d \nabla_d (h^b{}_c \beta^c) = \dot{\beta}^{\langle a \rangle}. \tag{4.27}$$

Now by its definition as a connecting vector (dragged along by the 4-velocity u^a), the Lie derivative of β^a with respect to u^a vanishes:

$$[\boldsymbol{u}, \boldsymbol{\beta}]^a = u^a{}_{,b}\beta^b - \beta^a{}_{,b}u^b = \beta^b \nabla_b u^a - u^b \nabla_b \beta^a = 0. \tag{4.28}$$

This is a direct consequence of (4.4), (4.25). It follows that,

$$v^a = V^a{}_b \beta^{\langle b \rangle}, \;\; V_{ab} := h_a{}^c h_b{}^d \nabla_d u_c = \overline{\nabla}_b u_a \tag{4.29}$$

showing that the relative velocity of nearby particles is given from their relative position by a linear transformation, the transformation matrix being simply the orthogonal projection of the covariant derivative of the 4-velocity vector.

Exercise 4.5.1 Show that if the fluid flow lines are non-geodesic, β^a cannot remain orthogonal to u^a even if it is orthogonal initially.

4.6 The kinematic quantities

To examine this further, we substitute the decomposition of $\beta^{\langle a \rangle}$ in terms of relative distance and direction into (4.29), and split V_{ab} up into its irreducible parts:

$$V_{ab} = V_{(ab)} + V_{[ab]} = \Theta_{ab} + \omega_{ab}, \tag{4.30}$$

where $\Theta_{ab} = \Theta_{(ab)} = \overline{\nabla}_{(a}u_{b)}$, the *expansion tensor*, and $\omega_{ab} = \omega_{[ab]} = \overline{\nabla}_{[b}u_{a]}$, the *vorticity tensor*, are the symmetric and skew-symmetric parts of the projected tensor V_{ab} respectively. Further,

$$\Theta_{ab} = \Theta_{\langle ab \rangle} + \tfrac{1}{3}\Theta^c{}_c h_{ab} \equiv \sigma_{ab} + \tfrac{1}{3}\Theta h_{ab}, \qquad (4.31)$$

where σ_{ab}, the *shear tensor*, is the PSTF part of Θ_{ab} (so $\sigma_{ab} = \overline{\nabla}_{\langle a}u_{b \rangle}$) and Θ, the (volume) *expansion*, is the trace part ($\Theta = \overline{\nabla}_a u^a$). Note that this is an invariant splitting: because these are tensor equations, the splitting will be the same irrespective of what coordinates are used.

In terms of these quantities, we derive from (4.26, 4.27) firstly the relation

$$\frac{\delta l\,\dot{}}{\delta l} = \Theta_{ab}e^a e^b = \tfrac{1}{3}\Theta + \sigma_{ab}e^a e^b, \qquad (4.32)$$

the *generalized Hubble relation*, showing that the rate of change of distance of neighbouring galaxies is proportional to their distance, with a ratio of proportionality which is in general direction dependent. Secondly, we obtain

$$\dot{e}_{\langle a \rangle} = \omega_{ab}e^b + \sigma_{ab}e^b - (\sigma_{cd}e^c e^d)e_a, \qquad (4.33)$$

the *rate of change of direction* equation.

It is important to ask: relative to what frame is this rate of change of relative direction determined? The answer is: a frame for which each basis vector e^a obeys the Fermi equation $\dot{e}_{\langle a \rangle} = 0$. Physically, this corresponds to a non-rotating local inertial reference frame as determined by local dynamical experiments (by gyroscopes, Foucault pendula, etc.): see e.g. Trautman (1965). Relation (4.33) is an observable relation in the sense that if one can determine a local non-rotating reference frame from local dynamics, then one can measure the rate of change of direction of galaxies relative to this frame. Thus in principle the left-hand side of this relation is directly observable; the components on the right-hand side can then be determined from these observations. This is also true for (4.32), distance being estimated from apparent size and velocity from redshift; the way this is done will be discussed in detail later.

The results above are somewhat surprising: we have deduced the generalized Hubble relation (4.32) apparently out of nothing! Historically, it took many years of devoted observation to show such a relation in the real universe. How have we arrived at the result theoretically? What is the origin of these relations?

Basically, it is because we have used a linearized representation of properties of the fluid flow that follows from the existence and differentiability of the average velocity vector field u^a. For this reason the results are only a first approximation in the region close to the point of observation. The derivation will be correct provided the continuum (fluid) approximation is a good description of the matter distribution and velocities in the universe, as discussed in Section 1.4.

4.6.1 Kinematical effects

To understand these equations better, it is convenient to consider successively the effect on relative position of pure expansion, shear and vorticity in turn.

In the case of *pure expansion*, $\omega_{ab} = \sigma_{ab} = 0$, so the rate of change of relative distance becomes $\delta l'/\delta l = \Theta/3$, independent of direction, while the rate of change of relative position becomes $\dot{e}^{\langle a \rangle} = 0$. Thus if we consider a sphere of galaxies of radius δl around us at time t, at time $t + \delta t$ the distances to all of the galaxies have increased by $dl = \Theta \delta l \, \delta t/3$ and their directions have all remained unchanged, so the galaxies then form a larger sphere (assuming $\Theta > 0$) with each galaxy lying in the same direction as before. Hence we have a distortion-free expansion without any rotation.

In the case of *pure shear*, $\omega_{ab} = \Theta = 0$, so the rate of change of relative distance becomes $\delta l'/\delta l = \sigma_{ab}e^a e^b$, and the rate of change of relative position becomes $\dot{e}_{\langle a \rangle} = \sigma_{ab}e^b - (\sigma_{cd}e^c e^d)e_a$. Since the shear tensor is symmetric, we can choose an orthonormal basis of shear eigenvectors, so the components of σ_{ab} become $\sigma_{ab} = \text{diag}(0, \sigma_1, \sigma_2, \sigma_3)$, where $\sigma_1 + \sigma_2 + \sigma_3 = 0$ (because this tensor is trace-free). Then if there is an expansion in the 1-direction ($\sigma_1 > 0$), there must be a contraction in at least one other direction (say $\sigma_2 < 0$). If in this case we consider a sphere of galaxies around us at time t, at time $t + \delta t$ the distances to galaxies in the principal j-axis direction will have changed by $\dot{dl} = \sigma_j \, \delta l \, \delta t$ and their directions remain unchanged. Thus the galaxies then form an ellipsoid, expanded in the 1-direction but contracted in the 2-direction, with the same volume as before. Each galaxy lying in a shear eigendirection will be in the same direction as before; all others will appear to have moved in the sky, but the average change of direction, integrated over the whole sky, will be zero (for each galactic apparent motion there will be an equal and opposite apparent motion of another galaxy to compensate). Hence we have a pure distortion, without rotation or change of volume.

In the case of *pure vorticity*, $\sigma_{ab} = \Theta = 0$, so the rate of change of relative distance becomes $\delta l'/\delta l = 0$ (all relative distances are unchanged), and the rate of change of relative position becomes $\dot{e}_{\langle a \rangle} = \omega_{ab}e^b$. By definition, a rotation preserves all distances, so these relations show that the change is a pure rotation. To examine this further, it is convenient to define the *vorticity vector* ω^a by the relations

$$\omega_a = -\tfrac{1}{2}\text{curl } u_a = \tfrac{1}{2}\eta_{abc}\omega^{bc} \iff \omega_{ab} = \eta_{abc}\omega^c, \tag{4.34}$$

showing that ω^a is a vector orthogonal to u^b which is an eigenvector of ω_{ab} with eigenvalue zero ($\omega^a \omega_{ab} = 0$). This implies that it defines the axis of the rotation (which is simply the set of directions invariant under the rotation). We can choose an orthonormal basis with $e_0 = u$ and e_1 parallel to ω; the components of ω^a and ω_{ab} then become $\omega^1 = \omega_{23} = -\omega_{32} = \omega$, the rest being zero. The galaxies in the ω direction momentarily remain unchanged in direction, as time increases, and all other galaxies remain at the same distance but appear to revolve around this axis. Thus this represents a pure rotation, without distortion or expansion.

Note that in the Newtonian limit, $\omega^i = -\tfrac{1}{2}\omega_N^i$, where $\omega_N = \nabla \times v$.

In a *general fluid flow*, all these quantities will be non-zero, so a combination of effects (volume change, rotation, distortion) will occur. It is still true, however, that there will always be two fixed points in the sky, where (instantaneously) the galaxy directions for a given celestial sphere of galaxies remain constant; this follows from the fixed point theorem for vector fields on the 2-sphere, applied to the vector field representing apparent motions in the sky, together with the fact that equation (4.33) shows that if e^a is such a fixed direction, so is $-e^a$. The volume will still change by an amount proportionate to Θ: $V \to V(1 + \Theta\delta t)$.

It is convenient to define a representative *length scale* $\ell(\tau)$ (in agreement with the ℓ of Section 3.2) by the relation

$$\frac{\dot{\ell}}{\ell} = \tfrac{1}{3}\Theta \qquad\qquad (4.35)$$

determining ℓ up to a constant scale factor along each world line. Then we shall always have the change of volume along the fluid flow characterized by $V \propto \ell^3$. The quantity ℓ here corresponds to the FLRW scale-factor a in (2.65), but is defined for an arbitrary flow field (determining the average distance behaviour of that flow field). The *Hubble parameter* for the flow is $H := \dot{\ell}/\ell = \tfrac{1}{3}\Theta$. Its present-day value $H_0 = (\dot{\ell}/\ell)_0$ is the *Hubble constant*.

4.6.2 Non-zero vorticity and cosmic time

We have seen that $\omega_a \neq 0$ if and only if the local inertial frame rotates relative to the rest frame defined by distant galaxies. The second important characterization is that $\omega_a \neq 0$ implies u_a is *not* a gradient. In detail:

$$\omega_a = 0 \Leftrightarrow u_{[b}\nabla_c u_{d]} = 0 \Leftrightarrow u_{[b}u_{c,d]} = 0$$

$$\Leftrightarrow \text{locally there are functions } r,t \text{ such that } u_a = -r t_{,a}, \qquad (4.36)$$

that is, u_a is proportional to a gradient (the implication from left to right is trivial; the implication from right to left follows from Darboux's theorem). Analytically, t is a potential function for the direction of u_a. Geometrically, this means u^a is orthogonal to the surfaces $t = \text{const}$, for $X^a u_a = 0 \Leftrightarrow X^a t_{,a} = 0$, i.e. the derivative of t in the direction X^a is zero for every vector X^a orthogonal to u^a.

We can think of this geometrically as follows: at each point the tangent plane orthogonal to u^a is spanned by h_{ab}, but in general these surface elements do not mesh together to form a surface in spacetime. We can obtain a geometrical picture of this situation by thinking of a twisted rope where the central strands are nearly vertical and the outer ones lie in a flatter spiral. Starting at the centre and moving along a curve always orthogonal to the strands, one can arrive back at the central strand above or below where one departed from it, i.e. the set of curves orthogonal to the strands do not integrate together to form a 2-space orthogonal to all the strands. The tangent elements orthogonal to u^a mesh together to form a 3-surface in spacetime (with the 3-space defined by h_{ab} tangent to these surfaces at each point) precisely when $\omega_a = 0$, these surfaces being the surfaces $t = \text{const}$: the surfaces are unique because the vector field is unique. When $\omega_a \neq 0$, no such orthogonal surfaces exist.

What is the physical meaning? The orthogonal tangent planes are instantaneous rest spaces for observers moving with 4-velocity u^a; these fit together coherently if and only if $\omega_a = 0$, that is, vanishing vorticity is the condition for the existence of a *cosmic time* for the fundamental observers.[2] Such a function allows the fundamental observers to synchronize their clocks, determining the event q on a world line G simultaneous with an event p on a

---

[2] Time functions exist which consistently order events on all timelike lines, whenever the stable causality condition holds (Hawking and Ellis, 1973), and such a time may even measure proper time along all the fundamental world lines; but it will not locally determine simultaneity as measured by all the fundamental observers unless it satisfies (4.36).

world line O. When the vorticity is non-zero, this is not possible, for starting from p and moving on a path that is everywhere orthogonal to u^a, one can return to O at events earlier or later than p; thus one does not even obtain a unique result for the cosmic time on the world line O itself. When vorticity vanishes, synchronization is unique, for each such curve lies in a surface $t = t_p$ and returns to O at the unique event where this surface intersects the world line. Thus this time extends simultaneity from any one world line to its neighbours; and as the surfaces $t = $ const are spacelike it gives the same time ordering on all timelike and null world lines.

However, the function t does not necessarily measure *proper time* along the world lines. Indeed, the derivative of t along the world lines with respect to proper time is $\dot{t} = t_{,a} u^a = -r^{-1} u_a u^a = r^{-1}$. Thus t can be chosen to measure proper time along the world lines only if $r = r(t)$, for only then can we choose $r = 1$ (by rescaling $t \to t'(t)$); such a t is a *normalized cosmic time*, which both determines the rest space of each fundamental observer and measures proper time along all the fundamental world lines. The condition for this to be possible is that u_a is a gradient, which will be true if both the vorticity and the acceleration vanish:

$$\omega_a = 0 = \dot{u}^a \Leftrightarrow \nabla_{[a} u_{b]} = 0 \Leftrightarrow u_{[a,b]} = 0$$

$$\Leftrightarrow \text{locally there is a function } t \text{ such that } u_a = -t_{,a}. \tag{4.37}$$

When $\omega_a = 0$, $\dot{u}^a \neq 0$, one can normalize the cosmic time to measure proper time along one world line but then, even though it synchronizes instantaneous events on the different world lines, it will not measure proper time along other world lines. For example, the time coordinate t in the FLRW universes is a fundamental cosmic time that measures proper time along each world line. The standard time coordinate t in a Schwarzschild solution is a cosmic time for the static observers, but does not measure proper time along their world lines (because their acceleration is non-zero).

Another example (and a warning) is provided by the Gödel universe (Gödel, 1949, Hawking and Ellis, 1973). In this rotating universe there exist normalized comoving coordinates; the 'time' coordinate t exists globally and measures proper time along every fundamental world line. However, there is no good time function whatever: the surfaces $t = $ const cannot be chosen to be spacelike everywhere. Thus, of necessity, such surfaces become timelike somewhere; such a 'time' can order events along the fundamental world lines, but does not provide a unique time ordering along arbitrary timelike or null world lines. This feature can occur because causality is violated in this universe. If $\omega = 0$ and the topology is simply connected, there is a global time coordinate t. We may thus expect causality violations to be associated either with rotation, if the rotation occurs over a large enough part of the universe, or with closed topologies in the timelike direction (Tipler, Clarke and Ellis, 1980, Ellis, 1996).

4.6.3 Characterizing the fluid flow

The quantities we have now defined (the acceleration, expansion, shear and vorticity) are called the *kinematic quantities* because they characterize the kinematic features of the fluid flow. More precisely, on the one hand these quantities are all defined from the first covariant derivative of the 4-velocity vector field u^a; on the other, it follows from (4.8) and

(4.29)–(4.31) that

$$\nabla_b u_a = \omega_{ab} + \sigma_{ab} + \tfrac{1}{3}\Theta h_{ab} - \dot{u}_a u_b, \tag{4.38}$$

which shows that this derivative is completely determined by the kinematic quantities. Thus they contain precisely the same information as the first derivative $\nabla_b u_a$ of u_a (there are 12 independent components of $\nabla_b u_a$, which is orthogonal to u^a on the index a, while there are five independent components of σ_{ab}, three of ω_{ab}, one of Θ and three of $\dot{u}_a$). Their geometric meaning has been emphasized above. In principle, they are directly measurable from observations of nearby galaxies through equations (4.32) and (4.33). Their magnitudes are defined as follows:

$$\omega^2 = \tfrac{1}{2}\omega_{ab}\omega^{ab} = \omega^a \omega_a \text{ so that } \omega^2 = 0 \Leftrightarrow \omega^a = 0 \Leftrightarrow \omega_{ab} = 0, \tag{4.39}$$

$$\sigma^2 = \tfrac{1}{2}\sigma_{ab}\sigma^{ab} \text{ so that } \sigma^2 = 0 \Leftrightarrow \sigma_{ab} = 0, \tag{4.40}$$

the implication following because these are spacelike tensors, orthogonal to u^a. They can be used to characterize some simple universe models. For example, in an Einstein static universe $\omega = \sigma = \dot{u}^a = \Theta = 0$; in all other FLRW universes, $\omega = \sigma = \dot{u}^a = 0$; $\Theta \neq 0$. In a Gödel universe, $\Theta = \sigma = \dot{u}^a = 0$; $\omega \neq 0$. In a static star model, $\Theta = \sigma = \omega = 0$; $\dot{u}^a \neq 0$.

In the real universe, what are the current limits on the present-day values of these quantities? Direct observations show $\Theta_0 > 0$ (as the Hubble constant is positive), and put upper limits on σ_0 and ω_0: $\sigma_0 < \tfrac{1}{4}\Theta_0$, $\omega_0 < \tfrac{1}{3}\Theta_0$ (Kristian and Sachs, 1966). Indirect evidence (from nucleosynthesis and CMB isotropy) is much more stringent (see Chapter 13). However, even if these values are very low today, this does not imply that these quantities are unimportant; indeed they can dominate the early expansion of the universe even if very small today.

The Newtonian analogue of (4.38) is the pair of equations $v_{i,j} = \omega_{ij} + \sigma_{ij} + \tfrac{1}{3}\Theta h_{ij}$, $\partial v_j / \partial t = a_j - v_{j,i} v^i - \Phi_{,j}$.

Exercise 4.6.1 Show that relations essentially identical to (4.25)–(4.35) hold in Newtonian theory, with $V_{ij} = v_{i,j}$.

Exercise 4.6.2

(a) Show that (4.36) implies $\dot{u}_a = \overline{\nabla}_a(\ln r)$, i.e. that (irrespective of the fluid equation of state) if $\omega = 0$ there necessarily exists an acceleration potential r. [Note: this follows from (4.43) also.]

(b) Remembering that $\dot{t} = r^{-1}$, show that this implies $\overline{\nabla}_a \dot{t} = -\dot{t}\dot{u}_a$.

(c) Consider neighbouring world lines O and G intersecting the surfaces $t = t_0$ and $t = t_0 + \delta t$ where the corresponding proper times along the world lines O and G are $\delta\tau_O$ and $\delta\tau_G$ respectively and dx^a is a relative position vector from O to G. Show that then $\delta\tau_G = \delta\tau_0(1 + \dot{u}_a dx^a)$.

(d) In the case of a perfect fluid, r will be given (up to a multiplicative constant) by (5.41) where the integral is taken along the fluid flow lines from some initial surface S. Show that then $\delta\tau_G/\delta\tau_O = 1 - \delta p/(\rho + p)$ where $\delta p = p_G - p_O$ is the pressure difference between O and G. If further $p = w\rho$ where w is constant, this becomes $\delta\tau_G/\delta\tau_O = 1 - w\delta\rho/[\rho(1+w)]$.

4.7 Curvature and the Ricci identities for the 4-velocity

Here we gather the purely kinematic identities arising from the Ricci identity for the velocity field u^a, that is,

$$(\nabla_c \nabla_d - \nabla_d \nabla_c) u_a = R_{abcd} u^b. \tag{4.41}$$

Some of these equations become dynamic when the Ricci tensor has been related to the matter content through the Einstein field equations (3.13), and we therefore postpone detailed discussion of the consequences of these equations until Chapter 6, after we have discussed the possible matter content in Chapter 5. The Newtonian equivalents of the Ricci identities are the equations $\partial(v_{j,i})/\partial t = (\partial v_j/\partial t)_{,i}$, $v_{i,j,k} = v_{i,k,j}$ which follow because the space sections are locally flat and evenly spaced in the Newtonian limit.

4.7.1 Ricci tensor relations

We note first that on contracting (4.41) with u^c, we obtain a propagation equation for $\nabla_d u_a$ along the fluid flow lines:

$$(\nabla_d u_a)\dot{} - \nabla_d \dot{u}_a + (\nabla_d u^c)(\nabla_c u_a) = R_{abcd} u^b u^c . \tag{4.42}$$

If we now substitute for $u_{a;d}$ in terms of the kinematic quantities, we obtain propagation equations for expansion, shear and vorticity ((4.43), (4.46) and (4.51) below) but not for acceleration, due to the Riemann tensor symmetries. The Newtonian analogue of (4.42) is $\dot{v}_{ij} - a_{i,j} + v_{ik}v^k{}_j + \Phi_{,i,j} = 0$, which follows from the definitions of the 'acceleration vector' a_j and the convective derivative.

The Riemann tensor symmetries imply that of the 24 components arising from the four possible a and six distinct pairs cd in (4.41), six, the projections on u^a, vanish trivially by (2.66), three equations are given by contraction of (4.41) with u^c and one by contraction with η^{abc}. Respectively, these last four give (multiplying the first by η^{ade})

$$\dot{\omega}^{\langle e \rangle} = -\tfrac{2}{3}\Theta\omega^e + \sigma^{ed}\omega_d - \tfrac{1}{2}\mathrm{curl}\,\dot{u}^e , \tag{4.43}$$

$$\overline{\nabla}_a \omega^a = \omega^a \dot{u}_a . \tag{4.44}$$

Equation (4.43) gives the basis for the discussion of vorticity conservation in Section 6.2.

The first four of the remaining 14 components of (4.41) are obtained by contraction on ac, which we call the trace. This can be split into three spatial parts and one time part. The latter is the same as the contraction of (4.42) and gives

$$(\nabla_a u^a)\dot{} - \nabla_a \dot{u}^a + (\nabla^a u^c)(\nabla_c u_a) = R^a{}_{bca} u^b u^c = -R_{bc} u^b u^c . \tag{4.45}$$

In terms of the kinematic quantities, this is

$$\dot{\Theta} + \tfrac{1}{3}\Theta^2 + 2(\sigma^2 - \omega^2) - \overline{\nabla}_a \dot{u}^a + \dot{u}^a \dot{u}_a = -R_{bc} u^b u^c . \tag{4.46}$$

The spatial projection of the trace gives,

$$h_a{}^b \nabla_c \left(\sigma^c{}_b + \omega^c{}_b \right) - \tfrac{2}{3} \overline{\nabla}_a \Theta - (\omega_{ab} + \sigma_{ab}) \dot{u}^b = h_a{}^b R_b{}^c u_c \, , \tag{4.47}$$

which can be written

$$\overline{\nabla}^b \sigma_{ab} - \mathrm{curl}\, \omega_a - \tfrac{2}{3} \overline{\nabla}_a \Theta + 2\eta_{abc} \omega^b \dot{u}^c = R_{\langle a \rangle b} u^b \, . \tag{4.48}$$

4.7.2 Weyl tensor relations

To express the remaining results we shall use the decomposition of the Weyl curvature into its 'electric' and 'magnetic' parts E_{ab} and H_{ab}, discussed in more detail in Section 6.4 and defined by,

$$E_{ab} = C_{acbd} u^b u^d = E_{\langle ab \rangle}, \quad H_{ab} := \tfrac{1}{2} \eta_{acd} C^{cd}{}_{be} u^e := H_{\langle ab \rangle} \tag{4.49}$$

where we used the Weyl tensor symmetries. Then

$$C_{ab}{}^{cd} = 4 \left(h_{[a}{}^{[c} + u_{[a} u^{[c} \right) E_{b]}{}^{d]} + 2\eta_{abe} u^{[c} H^{d]e} + 2\eta^{cde} u_{[a} H_{b]e} \, . \tag{4.50}$$

The two PSTF tensors E_{ab} and H_{ab} match with the residual parts of (4.41) which are the five components arising from the PSTF part of (4.42), and five from the PSTF part of the contraction with η^{cde}. These are,

$$E_{ab} - \tfrac{1}{2} R_{\langle ab \rangle} = -\dot{\sigma}_{\langle ab \rangle} - \tfrac{2}{3} \Theta \sigma_{ab} + \overline{\nabla}_{\langle a} \dot{u}_{b \rangle} + \dot{u}_{\langle a} \dot{u}_{b \rangle} - \omega_{\langle a} \omega_{b \rangle} - \sigma_{c \langle a} \sigma_{b \rangle}{}^c \tag{4.51}$$

$$H_{ab} = \mathrm{curl}\, \sigma_{ab} + \overline{\nabla}_{\langle a} \omega_{b \rangle} + 2\dot{u}_{\langle a} \omega_{b \rangle} \, , \tag{4.52}$$

where we have removed the trace (4.44).

To summarize, the symmetries of the Riemann tensor give (4.43) and (4.44), and the remaining components give (4.46), (4.48), (4.51) and (4.52). Of these, (4.43), (4.46) and (4.51) come from (4.42), which is the contraction of (4.41) with u^c, and (4.46) and (4.48) come from the 'trace' of (4.41). All these equations are kinematic identities, i.e. they are true whatever the dynamics in action. They obtain a dynamical content when we join them with the EFE, which we do in the next chapter.

Exercise 4.7.1

(a) Verify the above equations, and find their Newtonian counterparts. (Hint: See Ellis (1971a)).
(b) Show that

$$C_{abcd} = (g_{abpq} g_{cdrs} - \eta_{abpq} \eta_{cdrs}) u^p u^r E^{qs}$$

$$- (\eta_{abpq} g_{cdrs} + g_{abpq} \eta_{cdrs}) u^p u^r H^{qs} \, , \tag{4.53}$$

where $g_{abcd} := g_{ac} g_{bd} - g_{ad} g_{bc}$. (This corrects a sign error in Ellis (1971a).)

4.8 Identities for the projected covariant derivatives

The projected spatial derivative is related to the covariant derivative by

$$\nabla_a f = -u_a \dot{f} + \overline{\nabla}_a f \,, \tag{4.54}$$

$$\nabla_b V_a = -u_b \left\{ \dot{V}_{\langle a \rangle} + \dot{u}_c V^c u_a \right\}$$
$$+ u_a \left\{ \tfrac{1}{3} \Theta V_b + \sigma_{bc} V^c + \eta_{bcd} \omega^c V^d \right\} + \overline{\nabla}_b V_a \,, \tag{4.55}$$

$$\nabla_c S_{ab} = -u_c \left\{ \dot{S}_{\langle ab \rangle} + 2 u_{(a} S_{b)d} \dot{u}^d \right\}$$
$$+ 2 u_{(a} \left\{ \tfrac{1}{3} \Theta S_{b)c} + S_{b)}{}^d \left(\sigma_{cd} - \eta_{cde} \omega^e \right) \right\} + \overline{\nabla}_c S_{ab} \,, \tag{4.56}$$

where

$$\overline{\nabla}_b V_a = \tfrac{1}{3} \overline{\nabla}^c V_c h_{ab} - \tfrac{1}{2} \eta_{abc} \operatorname{curl} V^c + \overline{\nabla}_{\langle a} V_{b \rangle} \,, \tag{4.57}$$

$$\overline{\nabla}_c S_{ab} = \tfrac{3}{5} \overline{\nabla}^d S_{d\langle a} h_{b \rangle c} - \tfrac{2}{3} \eta_{dc(a} \operatorname{curl} S_{b)}{}^d + \overline{\nabla}_{\langle a} S_{bc \rangle} \,. \tag{4.58}$$

Thus the spatial derivatives of projected vectors and rank-2 PSTF tensors are made up of a covariant divergence and curl, and a 'distortion' derivative (Maartens, Ellis and Siklos, 1997). In the rank-2 tensor case, the distortion is a rank-3 PSTF tensor, which can be written as

$$\overline{\nabla}_{\langle a} S_{bc \rangle} = \overline{\nabla}_{(a} S_{bc)} - \tfrac{2}{5} h_{(ab} \overline{\nabla}^d S_{c)d} \,. \tag{4.59}$$

The covariant spatial gradient, divergence and curl obey identities which are a covariant generalization of Newtonian vector calculus identities. A selection of important identities is the following (see Maartens (1997), van Elst (1996), Maartens, Ellis and Siklos (1997), Maartens (1998), Maartens and Bassett (1998)):

$$\operatorname{curl} \overline{\nabla}_a f = -2 \dot{f} \omega_a \iff \overline{\nabla}_{[a} \overline{\nabla}_{b]} f = -\dot{f} \omega_{ab} \,, \tag{4.60}$$

$$\overline{\nabla}_a \operatorname{curl} V^a = -\tfrac{2}{3} \omega_a \left(3 \dot{V}^{\langle a \rangle} + \Theta V^a + 3 \sigma^{ab} V_b \right) \,, \tag{4.61}$$

$$(\overline{\nabla}_a f)\dot{} = \overline{\nabla}_a \dot{f} + (\dot{u}^b \overline{\nabla}_b f) u_a + \dot{u}_a \dot{f} - \tfrac{1}{3} \Theta \overline{\nabla}_a f$$
$$- \sigma_{ab} \overline{\nabla}^b f + \eta_{abc} \omega^b \overline{\nabla}^c f \,. \tag{4.62}$$

The first two show that, when $\omega_a \neq 0$, the Newtonian identities, curl grad $= 0$ and div curl $= 0$, no longer hold. The third identity shows how the Newtonian identity $(\nabla f)\dot{} = \nabla \dot{f}$ is generalized. This identity is important for commuting time and space derivatives of gradient quantities.

The extensions of these identities to expressions for $\overline{\nabla}_{[a} \overline{\nabla}_{b]} V_c$, $\overline{\nabla}_{[a} \overline{\nabla}_{b]} S_{cd}$, $\overline{\nabla}_b \operatorname{curl} S^{ab}$, $(\overline{\nabla}_a V_b)\dot{}$, etc. are complicated, and may be found in the references cited above. In the case of an almost FLRW spacetime, the identities simplify – see Section 10.4.1.

5 Matter in the universe

Our current understanding of the contents of the universe is based on the Standard Model of particle physics and its extensions (see e.g. Mukhanov (2005), Peter and Uzan (2009)). The Standard Model incorporates the strong, weak and electromagnetic interactions. The hadrons, made of quarks and anti-quarks, feel the strong interaction (and the weak). They are fermionic baryons and bosonic mesons. In cosmology, the key hadrons are the baryonic proton and neutron, but many more hadrons have been detected. Fermionic leptons feel the weak interaction; these include the electron and the three neutrino species. All charged hadrons and leptons feel the electromagnetic interaction. See Table 9.3 for a summary.

This model is able to explain all particles so far observed in colliders and particle detectors, except that experiments have recently detected neutrino oscillations, so that at least two of the neutrinos must have mass. The candidate particles for cold dark matter also cannot be explained within the Standard Model.

This Standard Model allows us to understand the ultra-relativistic early universe, for times $t \gtrsim 10^{-10}$ s and energies $E \lesssim 1$ TeV. The Large Hadron Collider is beginning to probe $E \gtrsim 1$ TeV at the time of writing. One of the outstanding successes of the model is the prediction of light element nucleosynthesis. A brief overview of particle physics in the early universe is given in Section 9.6. Our focus in this chapter is on the universe after matter–radiation equality, when the relevant contents of the universe are:

- *Standard-model matter:* protons, electrons, atoms, molecules, photons and neutrinos, all of which are observed in non-gravitational experiments. (Massive neutrinos require a minimal extension of the Standard Model.) Baryonic matter aggregates under gravity into gas, stars, galaxies, clusters.
- *Cold dark matter:* indirectly required by astrophysical and cosmological dynamics. Non-Standard-Model candidates are proposed, but so far there is no non-gravitational detection.
- *Dark energy:* deduced purely from cosmological gravitational effects and dependent on the cosmological model.

These constituents may be modelled using fluids, gases and fields,[1] and in this chapter we discuss fluids and their thermodynamics, scalar fields, multiple fluids and fields, electromagnetic fields, kinetic theory and quantum field theory.

---

[1] Although solid, or partially solid, bodies occur, they are not important in cosmological discussions.

5.1 Conservation laws

There are two different kinds of conservation law which constrain the behaviour of matter.

5.1.1 Average 4-velocities and conserved quantities

How can we define the average 4-velocity of non-relativistic matter, assumed to be made up of particles of identical mass, whose number is conserved (each 'particle' of matter for the present-day universe may be a cluster of galaxies)? Consider an averaging volume of scale size L and volume dV, measured by an observer O, which is small relative to the curvature of spacetime. O can associate with each particle a 3-velocity v_* and a rest-mass m_*. From this O can define the rest-mass density current 4-vector,

$$J^a = \frac{1}{dV} \Sigma_*(m_*, m_* v_*), \tag{5.1}$$

where the sum is over all particles in the volume. If this vector is well defined for one observer, it is well defined for all, i.e. the same 4-vector will be found no matter which observer makes the measurements (this is far from obvious; it follows most easily from relativistic kinetic theory, see Section 5.4). We now split the vector into a magnitude ρ_N (the rest-mass density) and a unit timelike vector $u^a_{(b)}$:

$$J^a = \rho_N u^a_{(b)}, \quad u^a_{(b)} u_{(b)a} = -1. \tag{5.2}$$

This defines the *barycentric frame*.

To clarify its meaning, suppose we consider an observer O moving with this 4-velocity. From (5.2), for O the components will be $J^a = (\rho_N, 0, 0, 0)$. However, as the definition (5.1) holds in all frames, we see that in this frame

$$\rho_N = \frac{1}{dV} \Sigma_* m_*, \quad \Sigma_* m_* v_* = 0. \tag{5.3}$$

Thus ρ_N is the rest-mass density measured by $u^a_{(b)}$ and the frame is the centre of rest-mass frame (the total momentum measured relative to this frame is zero).

In the Newtonian case, the second of (5.3) is the definition of the centre-of-rest frame.

At late times in the universe, rest-mass of galaxies is conserved. Hence by the arguments in Section 3.2 we shall have

$$\rho_N = M\ell^{-3}, \ \dot{M} = 0 \ \Leftrightarrow \ \dot{\rho_N} + \rho_N \Theta = 0 \ \Leftrightarrow \ \nabla_a J^a = 0. \tag{5.4}$$

If one is considering matter at a time when rest-mass is not conserved (e.g. when nucleosynthesis is important), one must define the average velocity in terms of some other quantity that is conserved at that time: e.g. baryon or lepton number, or electric charge. Then we obtain the baryon or lepton density current vector, or the charge density current vector (compare Section 5.5) respectively.

If the particle number is conserved, we can use the particle 4-current density, defined in a general frame by

$$N^a = nu^a + n^a, \quad n^a u_a = 0, \tag{5.5}$$

where $n = -u_a N^a$ is the particle number density and $n^a = N^{\langle a \rangle}$ is the particle flux vector, as measured by u^a. For a non-relativistic fluid, $N^a = m^{-1} J^a$, where $m = \rho_N / n$ is the particle mass. If particle number is conserved, then

$$\nabla_a N^a = 0 \quad \Rightarrow \quad \dot{n} + \Theta n + \overline{\nabla}^a n_a + \dot{u}^a n_a = 0. \tag{5.6}$$

We can define a unique 4-velocity by requiring that there is no particle number flux relative to it:

$$N^a = n_{(\mathrm{p})} u^a_{(\mathrm{p})}. \tag{5.7}$$

The 4-velocity $u^a_{(\mathrm{p})}$ defines the *particle (or Eckart) frame*. It coincides with the barycentric frame for a non-relativistic fluid of identical mass particles, but it also applies to massless and varying-mass particles, provided that the total particle number is conserved.

An alternative frame is defined via energy flux. When the strong energy condition holds (see below), the energy–momentum tensor (5.9) of a fluid has a unique timelike 4-velocity eigenvector $u^a_{(\mathrm{e})}$, characterized by

$$q^a_{(\mathrm{e})} := T_{bc} h^{ab}_{(\mathrm{e})} u^c_{(\mathrm{e})} = 0. \tag{5.8}$$

This defines the *energy (or Landau–Lifshitz) frame*, in which there is no energy flux. The energy frame 4-velocity is uniquely characterized as the only 4-velocity that is an eigenvector of the stress tensor: $T^a_b u^b_{(\mathrm{e})} = -\rho_{(\mathrm{e})} u^a_{(\mathrm{e})}$.

For a perfect fluid, $u^a_{(\mathrm{e})} = u^a_{(\mathrm{p})}$, and they define a unique hydrodynamic 4-velocity vector. This is the only frame in which the energy–momentum tensor has perfect-fluid form.

What is clear is that if no quantity is conserved one cannot define an average velocity, for one cannot then identify the same quantity at the beginning and end of the time period used to measure the 4-velocities. Thus every such definition of a 4-velocity is based on a conserved quantity, leading to a conserved 4-current.

Determination of the average velocity of matter locally is a major issue in observational cosmology, leading to the concepts of large-scale streaming velocities and the 'Great Attractor' (Bertschinger *et al.*, 1990, Burstein, Faber and Dressler, 1990), which attempt to reconcile the average velocities estimated by observation of galaxies as discussed above with the average velocity defined by the microwave background radiation. As discussed later, we usually assume these velocities are the same. If this is not true at some cosmological scale, there are serious implications for cosmology: a 1-component fluid description must be replaced by at least a 2-component description at that scale.

5.1.2 Energy–momentum conservation

Energy and momentum conservation is of course a cornerstone of physical theory.

The energy–momentum tensor

As a result of (4.16) and its symmetry, T_{ab}, as measured by an observer moving with 4-velocity u^a, can be split up into its parts parallel and orthogonal to u^a as follows:

$$T_{ab} = \rho u_a u_b + q_a u_b + u_a q_b + p h_{ab} + \pi_{ab}, \quad q_a = q_{\langle a \rangle}, \quad \pi_{ab} = \pi_{\langle ab \rangle}. \tag{5.9}$$

The observer measures that

$\rho = T_{ab}u^a u^b$ is the relativistic energy density (the rest mass density plus the total internal energy due to heat, chemical energy, etc.);

$p = h^{ab}T_{ab}/3$ is the relativistic pressure;

$q_a = -T_{\langle a \rangle b}u^b$ is the relativistic momentum density (due to processes such as diffusion and heat conduction), which (because of the equivalence of mass and energy) is also the energy flux relative to u^a;

$\pi_{ab} = T_{\langle ab \rangle}$ is the relativistic anisotropic (trace-free) stress tensor due to effects such as viscosity or free-streaming or magnetic fields.

The 10 components of T_{ab} are thus represented by the two scalar quantities ρ and p, the three components of the vector q_a and the five components of the tensor π_{ab}.

Given a choice of u^a, this splitting into parallel and perpendicular parts can be applied to any stress–energy tensor whatever; the physics of the matter is then given by *equations of state* relating the quantities ρ, p, q_a, π_{ab} and possibly other thermodynamic variables such as the temperature and entropy. The special relativistic transformation laws between energy density, momentum density and stresses are contained in the decomposition above, since a different observer will have a different 4-velocity and obtain a different decomposition of the same stress tensor into such components. When the observer is moving with the physically defined average velocity, these components will embody the physical nature of the matter. When the observer is not moving with the average velocity, these interpretations would not be physically meaningful in a way intrinsic to the matter content itself.

In Newtonian theory, the mass density ρ_N and energy density ϵ are independent of each other, and q_a and π_{ab} are separately defined (unlike the GR case where, together with ρ, they are components of a single tensor).

The conservation laws

The energy–momentum conservation equations are given by the four-dimensional equation

$$\nabla_b T^{ab} = 0. \tag{5.10}$$

In terms of the 1+3 decomposition (5.9), the component of these equations parallel to u^a is the *energy conservation* equation,

$$\dot{\rho} + (\rho + p)\Theta + \pi^{ab}\sigma_{ab} + \overline{\nabla}_a q^a + 2\dot{u}_a q^a = 0, \tag{5.11}$$

which determines the rate of change of relativistic energy along the world lines. The projection orthogonal to u^a gives the *momentum conservation equation*

$$\dot{q}_{\langle a \rangle} + \tfrac{4}{3}\Theta q_a + (\rho + p)\dot{u}_a + \overline{\nabla}_a p + \overline{\nabla}^b \pi_{ab} + \dot{u}^b \pi_{ab} + \left(\sigma_{ab} + \eta_{abc}\omega^c\right)q^b = 0, \tag{5.12}$$

which determines the acceleration caused by various pressure contributions. This shows that the *inertial mass density* of matter is $\rho + p$, so any form of internal energy (e.g. heat or chemical energy) contributes to the effective inertial mass both directly (by increasing ρ) and indirectly (by contributing to p).

Unlike conservation laws of the general form given by the last expression in (5.4), the law (5.10) does not in general lead to integral forms of conservation law such as the first of (5.4). Thus total energy–momentum of an extended region cannot readily be defined: this is related (see Section 3.1) to the absence of a general definition of gravitational energy in a curved spacetime.

These conservation laws hold for the total matter stress tensor; if there are several matter components, the total energy and momentum is conserved, but energy and momentum conservation for each component may be modified when interactions between the components are taken into account, as discussed in Section 5.3.

The Newtonian analogue of (5.12) is the Navier–Stokes equation,

$$\dot{v}_i = -\Phi_{,i} - \rho_N^{-1}\left(p_{,i} + \pi_i{}^j{}_{,j}\right) \quad \Leftrightarrow \quad \rho_N a_i + p_{,i} + \pi_i{}^j{}_{,j} = 0. \tag{5.13}$$

The energy equation ((5.27) below) is deduced separately.

5.1.3 General physical constraints

There are many possible descriptions of the matter and radiation in the universe. However, irrespective of the detailed description, there are some general constraints that will normally be applied to classical matter.

Firstly, the speed of sound must be less than the speed of light, or else we can have violations of special relativistic causality (we can send a signal faster by sound than light). Furthermore, local mechanical stability demands that the speed of sound be real, for if it is imaginary an impulse applied to the fluid and causing a perturbation $\propto e^{-ic_s t}$, where c_s is the speed of sound, will cause a collapse of the matter instead of a wave. Now for a barotropic fluid, $p = p(\rho)$, the speed of sound is given by the adiabatic formula,

$$c_s^2 = \frac{\mathrm{d}p}{\mathrm{d}\rho} \quad \text{(adiabatic)}. \tag{5.14}$$

(For a proof, see Exercise 10.2.4.) Hence, in this case, the conditions are

$$0 \le \frac{\mathrm{d}p}{\mathrm{d}\rho} \le 1 \quad \text{for } p = p(\rho). \tag{5.15}$$

The top limit is a rigorous limit which cannot be violated unless we abandon relativity theory. The bottom limit will apply to a stable situation and certainly to ordinary barotropic matter. In the case of non-barotropic fluids, the limits would have to be re-evaluated – and for scalar fields the limits do not apply (see Section 5.6). But in these cases, the speed of sound is not given by (5.14), and the principle remains: no signals can be sent faster than light, and we usually demand stable matter.

Secondly, there is the condition that the inertial mass density of matter is positive (i.e. matter will tend to move in the direction of a pressure gradient applied to it, rather than in the opposite direction). By (5.12), this condition, known as the weak energy condition, is

$$\rho + p > 0 \quad \text{weak energy condition.} \tag{5.16}$$

By (5.11), this is also the condition that, when the matter expands, its density decreases rather than increases.

Thirdly, there is the condition that the gravitational mass density of matter is positive, known as the strong energy condition. We shall show in the next section that this is equivalent to

$$\rho + 3p > 0 \quad \text{strong energy condition.} \tag{5.17}$$

For ordinary fluids, we expect each component of matter present to obey both of these conditions, and so the total stress tensor will also do so. However, scalar fields can violate (5.17), and the vacuum energy of quantum fields reaches the limiting value in (5.16).

In Newtonian theory, we would normally expect $\rho_N \geq 0$, $p \geq 0$.

Exercise 5.1.1
The change of stress tensor splitting with change of 4-velocity.
Consider an observer with 4-velocity $\tilde{u}^a$, moving relative to the u^a frame:

$$\tilde{u}^a = \gamma(u^a + v^a), \quad \gamma = (1 - v^2)^{-1/2}, \quad v_a u^a = 0, \tag{5.18}$$

where v_a is the relative velocity measured by u^a. We can decompose the energy–momentum tensor of the matter relative to the $\tilde{u}^a$ frame: $T_{ab} = \tilde{\rho}\tilde{u}_a\tilde{u}_b + \tilde{p}\tilde{h}_{ab} + 2\tilde{q}_{(a}\tilde{u}_{b)} + \tilde{\pi}_{ab}$. Show that

$$\tilde{\rho} = \rho + \gamma^2 \left[v^2(\rho + p) - 2q_a v^a + \pi_{ab} v^a v^b \right], \tag{5.19}$$

$$\tilde{p} = p + \tfrac{1}{3}\gamma^2 \left[v^2(\rho + p) - 2q_a v^a + \pi_{ab} v^a v^b \right], \tag{5.20}$$

$$\tilde{q}_a = \gamma q_a - \gamma \pi_{ab} v^b - \gamma^3 \left[(\rho + p) - 2q_b v^b + \pi_{bc} v^b v^c \right] v_a$$
$$- \gamma^3 \left[v^2(\rho + p) - (1 + v^2)q_b v^b + \pi_{bc} v^b v^c \right] u_a, \tag{5.21}$$

$$\tilde{\pi}_{ab} = \pi_{ab} + 2\gamma^2 v^c \pi_{c(a} \left\{ u_{b)} + v_{b)} \right\} - 2v^2\gamma^2 q_{(a}u_{b)} - 2\gamma^2 q_{\langle a} v_{b\rangle}$$
$$- \tfrac{1}{3}\gamma^2 \left[v^2(\rho + p) + \pi_{cd} v^c v^d \right] h_{ab}$$
$$+ \tfrac{1}{3}\gamma^4 \left[2v^4(\rho + p) - 4v^2 q_c v^c + (3 - v^2)\pi_{cd} v^c v^d \right] u_a u_b$$
$$+ \tfrac{2}{3}\gamma^4 \left[2v^2(\rho + p) - (1 + 3v^2)q_c v^c + 2\pi_{cd} v^c v^d \right] u_{(a}v_{b)}$$
$$+ \tfrac{1}{3}\gamma^4 \left[(3 - v^2)(\rho + p) - 4q_c v^c + 2\pi_{cd} v^c v^d \right] v_a v_b. \tag{5.22}$$

Exercise 5.1.2 Use the expression for the Weyl tensor in terms of the gravito-electric/magnetic fields,

$$C_{ab}{}^{cd} = 4 \left\{ u_{[a}u^{[c} + h_{[a}{}^{[c} \right\} E_{b]}{}^{d]} + 2\eta_{abe}u^{[c}H^{d]e} + 2u_{[a}H_{b]e}\eta^{cde} \tag{5.23}$$

$$= 4 \left\{ \tilde{u}_{[a}\tilde{u}^{[c} + \tilde{h}_{[a}{}^{[c} \right\} \tilde{E}_{b]}{}^{d]} + 2\tilde{\eta}_{abe}\tilde{u}^{[c}\tilde{H}^{d]e} + 2\tilde{u}_{[a}\tilde{H}_{b]e}\tilde{\eta}^{cde}, \tag{5.24}$$

to show that the gravito-electric/magnetic fields transform under a velocity boost (5.18) as follows:

$$\tilde{E}_{ab} = \gamma^2 \left\{ (1+v^2)E_{ab} + v^c \left[2\eta_{cd(a}H_{b)}{}^d + 2E_{c(a}u_{b)} \right. \right.$$
$$\left. \left. + (u_a u_b + h_{ab})E_{cd}v^d - 2E_{c(a}v_{b)} + 2u_{(a}\eta_{b)cd}H^{de}v_e \right] \right\}, \tag{5.25}$$

$$\tilde{H}_{ab} = \gamma^2 \left\{ (1+v^2)H_{ab} + v^c \left[-2\eta_{cd(a}E_{b)}{}^d + 2H_{c(a}u_{b)} \right. \right.$$
$$\left. \left. + (u_a u_b + h_{ab})H_{cd}v^d - 2H_{c(a}v_{b)} - 2u_{(a}\eta_{b)cd}E^{de}v_e \right] \right\}. \tag{5.26}$$

Exercise 5.1.3 Show that the conserved quantity arising as in (3.11), which is an energy if ξ^a is timelike, and a momentum component if ξ^a is spacelike, generalizes to the case of a conformal Killing vector obeying (2.61) if $T^a{}_a = 0$. This is the case for isotropic radiation and for the electromagnetic field (see below).

5.2 Fluids

The general equation of state for fluids can be expressed in terms of thermodynamic quantities. Defining the specific internal energy ϵ by $\rho = (1+\epsilon)\rho_N$ and the specific volume v by $v = 1/\rho_N$, the temperature T and specific entropy S are determined by the first law of thermodynamics, i.e.

$$d\epsilon + p_t dv = T dS, \tag{5.27}$$

where p_t is the pressure in thermodynamic equilibrium. It follows that

$$\rho_N T \dot{S} + (p - p_t)\Theta = \dot{\rho} + (\rho + p)\Theta. \tag{5.28}$$

Combining this with the energy conservation equation (5.11) we find

$$\rho_N T \dot{S} + (p - p_t)\Theta = -(\pi^{ab}\sigma_{ab} + \overline{\nabla}_a q^a + 2\dot{u}_a q^a). \tag{5.29}$$

This enables us to calculate the divergence of the entropy flow density vector S^a, defined by the relation

$$S^a = \rho_N S u^a + \frac{q^a}{T}, \tag{5.30}$$

the first term being the convection term (entropy carried along with the fluid flow) and the second the conduction and diffusion term (entropy carried by energy flow in the rest frame of the fluid). We obtain

$$\nabla_a S^a = -\frac{1}{T}\left[\pi^{ab}\sigma_{ab} + q^a \left(\dot{u}_a + \overline{\nabla}_a \ln T \right) + (p - p_t)\Theta \right]. \tag{5.31}$$

Now if we consider an isolated fluid flow (a timelike tube of fluid $\mathcal{T}$ such that $\rho_N > 0$ in $\mathcal{T}$ but $\rho_N = q^a = 0$ outside $\mathcal{T}$), entropy production must always be positive by the second law of thermodynamics: the entropy density integrated across the tube at an initial time s_1

must be less than or equal to that at a final time s_2. By the divergence theorem, this is the requirement that

$$\nabla_a S^a \geq 0. \tag{5.32}$$

This will necessarily be true for arbitrary fluid flows if

$$\pi_{ab} = -\lambda \sigma_{ab}, \tag{5.33}$$

$$q_a = -\kappa (\overline{\nabla}_a T + T \dot{u}_a), \tag{5.34}$$

$$p - p_t = -\zeta \Theta, \tag{5.35}$$

where $\lambda \geq 0$ is the *viscosity* coefficient, $\kappa \geq 0$ is the *heat conduction* coefficient, and $\zeta \geq 0$ is the *bulk viscosity* coefficient. Then,

$$\nabla_a S^a = \frac{1}{T} \left(\lambda \sigma^{ab} \sigma_{ab} + \frac{T}{\kappa} q^a q_a + \zeta \Theta^2 \right), \tag{5.36}$$

which is clearly non-negative. Thus (5.33)–(5.35) are the simplest thermodynamically viable equations of state for the dissipative fluid properties. Note that they do not contain any explicit term for entropy of the gravitational field, just as there are no local gravitational energy effects (see Section 3.1.4). Attempts to define a gravitational entropy will be discussed in Chapter 21.

Equations (5.33)–(5.35) are the relativistic generalizations of the Newtonian equations. However, these equations can violate the causality condition that no influence can propagate faster than light. Strictly, they can only be used in non-relativistic conditions. A much more complex (14-coefficient) description is required to represent dissipative processes in a relativistically correct approximation based on kinetic theory (Israel and Stewart, 1979). Luckily, nine of these modes are strongly damped in the long-wave length limit (i.e. compared to the typical mean-free-path), two propagate at the adiabatic sound speed, two transverse shear modes decay at the classical viscous damping rate, and the final mode decays at the classical thermal diffusion rate (Hiscock and Lindblom, 1987). Thus this set reduces to the familiar dynamics of classical fluids in this limit, and the above equations will then be adequate.

It is important to realize that the form of the equation of state depends on the choice of the average 4-vector u^a relative to which the 1+3 decomposition of T_{ab} is taken. The forms of the equations of state given here correspond to the barycentric choice above (the 4-velocity represents the average motion of rest-mass). One can as an alternative choose the 4-velocity to represent the average motion of relativistic energy, i.e. as the timelike eigenvector of the stress–energy tensor (if there is such an eigenvector, which will normally be the case). With this choice, by definition we will find $q_a = 0$; however, if dissipative processes are taking place, there will then be an average mass-flux relative to this 4-velocity (the barycentric 4-velocity defined above will have non-zero spatial components relative to this frame), and thermodynamics will look more complex than in the description above.

In Newtonian theory, the same thermodynamic relations (5.27)–(5.35) hold, except that the term $\dot{u}^i$ does not occur in (5.34).

5.2.1 Perfect fluids

At most times in cosmology, we can assume that the anisotropic dissipative terms are negligible. The fluid stress tensor then takes the 'perfect fluid' form,

$$T_{ab} = \rho u_a u_b + p h_{ab} = (\rho + p)u_a u_b + p g_{ab} \Leftrightarrow q_a = 0, \ \pi_{ab} = 0, \tag{5.37}$$

where u^a is the unique 4-velocity for which the stress tensor has this form. The perfect fluid form is usually understood to imply no viscous processes – but it is compatible with bulk viscosity. In general, the energy and momentum conservation equations (5.11) and (5.12) for a perfect fluid become

$$\dot{\rho} + (\rho + p)\Theta = 0 \Leftrightarrow \rho_N \dot{S} = -\frac{(p - p_t)}{T}\Theta, \tag{5.38}$$

$$(\rho + p)\dot{u}_a + \overline{\nabla}_a p = 0, \tag{5.39}$$

respectively.

The stress tensor of an FLRW universe must always take the perfect fluid form relative to the preferred family of observers, because of the isotropy of those spacetimes as seen by those observers.

An observer O moving relative to a perfect fluid will *not* determine it to have the perfect fluid form, but will see an effective non-zero momentum density and anisotropic stress tensor. For example, by (5.21), O will measure a momentum density $\tilde{q}_a = -\gamma^3(\rho + p)(v_a + v^2 u_a)$. The dipole anisotropy of the microwave background radiation is interpreted as a peculiar velocity of the Galaxy relative to the CMB rest frame. Thus if we live in a universe that is represented to a good approximation by an FLRW model, we are moving relative to the fundamental velocity at this speed, and will experience an anisotropic stress-tensor of this form.

Reversible flows and barotropic fluids

In general the physics of a 'perfect fluid' is determined by giving equations of state such as $p = p(T, \rho)$. Note, however, that the name is misleading; the fluid flow is reversible (i.e. isentropic) only if there is a barotropic equation of state $p = p(\rho)$, or if the fluid moves in such a way that such a relation effectively holds; for only then does the general case of two thermodynamic variables reduce effectively to one. This distinction is of some importance, for it confirms that irreversible processes can indeed take place in an FLRW universe despite the perfect fluid form of the energy–momentum tensor, i.e. (5.37) is compatible with irreversible processes. Reversible flows occur when there is a *barotropic* equation of state:

$$p = p(\rho) \Leftrightarrow p = p_t \Leftrightarrow \dot{S} = 0. \tag{5.40}$$

Then we can define the enthalpy and acceleration potential,

$$W = \exp\left(\int_{\rho_0}^{\rho} \frac{\mathrm{d}\rho}{3(\rho + p)}\right), \quad r = \exp\left(\int_{p_0}^{p} \frac{\mathrm{d}p}{\rho + p}\right), \tag{5.41}$$

which in effect integrate the energy and momentum conservation equations (5.38) and (5.39) in the form

$$W = W_0 \frac{\ell_0}{\ell}, \quad \dot{W}_0 = 0, \quad \dot{u}_a = -\overline{\nabla}_a \ln r. \tag{5.42}$$

In the case of a barotropic fluid, using the results of Exercise 5.2.3 below, comoving coordinates can be found such that

$$ds^2 = h_{ij}\, dx^i dx^j - \frac{1}{r^2}\Big[dx^0 + a_i(x^j)dx^i\Big]^2, \tag{5.43}$$

$$u^\mu = r\delta_0^\mu, \quad u_\mu = -r^{-1}(1, a_i), \tag{5.44}$$

which imply

$$\dot{u}_\mu = r^{-1}(0, -r_{,i}), \quad \omega_{\mu 0} = 0, \quad \omega_{ij} = -r^{-1}a_{[i,j]}, \tag{5.45}$$

$$\Theta_{\mu 0} = 0, \quad \Theta_{ij} = \frac{r}{2}h_{ij,0}, \tag{5.46}$$

and the conservation equations (5.39) are identically fulfilled. The coordinate transformation $x^{0'} = x^0 + f(x^i)$, $x^{i'} = x^i$, preserves these conditions, as does the relabelling $x^{0'} = x^0$, $y^{i'} = y^{i'}(y^i)$.

With this coordinate choice, $\Theta = (r/2)g^{ij}h_{ij,0}$ and so (4.35) implies $\ell(x^\mu) = \exp \int g^{ij}h_{ij,0}\, dt$, with the integral taken along the integral curves of u^μ from $t = t_0$. If we define f_{ij} by

$$h_{ij} = \ell^2 f_{ij} \Rightarrow g^{ij} f_{ij,0} = 0, \tag{5.47}$$

then the expansion and shear are given by

$$\Theta = 3r\frac{\ell_{,0}}{\ell}, \quad \sigma_{\mu 0} = 0, \quad \sigma_{ij} = \frac{r}{2}\ell^2 f_{ij,0}. \tag{5.48}$$

5.2.2 Simple equations of state

The barotropic linear equation of state,

$$p = w\rho, \quad w = \text{const}, \tag{5.49}$$

covers the simple matter models in cosmology. From Exercise 5.2.2 (b), this equation of state is compatible with the perfect-gas form $p \propto \rho_N^{1+w}$ with $B = 0$ ($w = \gamma - 1$).

Pressure-free matter ('dust'):

$$w = 0 \Rightarrow \dot{u}^a = 0, \quad \rho = \rho_0 \left(\frac{\ell_0}{\ell}\right)^3, \tag{5.50}$$

from the momentum and energy conservation equations (5.39) and (5.38). This is a good description of baryonic and cold dark matter at late times in the universe (the random velocities of CDM particles, of atoms after recombination and of galaxies are small, so the corresponding kinetic pressures are negligible). The matter must move geodesically (there are no pressure gradients to make it deviate from free-fall), and the density evolves as 1/volume. The temperature of dust is strictly zero, but if we take into account the very small velocity dispersion, the monatomic gas property, $T \propto 1/\ell^2$, can be applied. (This is consistent with $p \simeq 0$ as the kinetic energies are small compared with rest-mass energy.)

Radiation (*incoherent*):

$$w = \tfrac{1}{3} \;\Rightarrow\; \rho = \rho_0 \left(\frac{\ell_0}{\ell}\right)^4, \; T = T_0 \frac{\ell_0}{\ell}. \tag{5.51}$$

Here the temperature T is given by $\rho = a_R T^4$, where a_R is the radiation constant. Any distribution of particles that move at the speed of light (photons, zero-mass neutrinos, etc.) will have a stress tensor of the form $T_{ab} = \sum_* k_{*a} k_{*b}$ where $k_{*a} k_*{}^a = 0$, which implies $T^a{}_a = 0$. Isotropy implies the perfect fluid form, and for a perfect fluid $T^a{}_a = -\rho + 3p$ so this is just the condition $\rho = 3p$. It will be a good approximation for relativistic particles in the early universe, until they become non-relativistic – such as protons or massive neutrinos.

Vacuum energy or cosmological constant:

$$w = -1 \;\Rightarrow\; \rho = \text{const}. \tag{5.52}$$

This exceptional equation of state is at the limit of violating (5.16). The stress tensor is Lorentz invariant, i.e. $T_{ab} \propto g_{ab}$, since by (5.37) this will occur if and only if $p + \rho = 0$. As a consequence, there is no unique 4-velocity defined by the medium; in particular, every 4-velocity is an eigenvector of T_{ab} and thus an energy-frame 4-velocity. Therefore $\dot{\rho} := \rho_{,a} \dot{u}^a$ vanishes by energy conservation (5.11) for any choice of u^a, and thus ρ is a constant: expansion does not affect the energy density. Furthermore, the acceleration is no longer determined by momentum conservation: since $\rho + p = 0 = \overline{\nabla}_a p$, (5.12) does not constrain $\dot{u}_a$. This equation of state, with $\rho \geq 0$, also violates (5.17), for in this case $\rho + 3p = -2\rho$. However, vacuum energy is well behaved: it has no speed of sound, since it does not support (classical) fluctuations. A slow-rolling scalar field – as in simple models of inflation (see Section 9.7) – obeys $p \approx -\rho$ and has a perfectly well-defined speed of sound $c_{s\,\text{eff}} = 1$ that determines the speed of pressure fluctuations. A scalar field is not barotropic (nor adiabatic), so that its effective speed of sound is not the adiabatic sound speed, (5.14); see Section 5.6.

Stiff matter:

$$w = 1 \;\Rightarrow\; \rho = \rho_0 \left(\frac{\ell_0}{\ell}\right)^6. \tag{5.53}$$

This is the stiffest equation of state one can have for a fluid – with higher pressures, the speed of sound will exceed the speed of light by (5.15), violating the consistency of special relativity. It was proposed by Zel'dovich for a very early era, but it is not clear whether there is a realistic fluid with this equation of state. It is also a limiting case for a scalar field – when the potential energy vanishes, $w = 1$ (see Section 5.6).

From (5.49) and (5.15) we may expect adiabatic perfect fluids to have an energy density that, as the fluid is compressed or expanded, lies between $\rho \propto \ell^{-3}$ and $\rho \propto \ell^{-6}$.

5.2.3 Unphysical exact solutions

In order that a cosmological model should be meaningful, it is crucial that the matter description used is physically realistic, which (in the cosmological context) means it corresponds to one of the models of matter described in the previous sections of this chapter, and also obeys suitable energy conditions. We make this remark because there are published solutions claiming physical relevance, but which do not obey this requirement. Most often this is because, in one form or another, authors rediscover the simple trick criticized by Synge (1971): one can run the EFE (3.13) from left to right, instead of from right to left, so determining the stress tensor required to give an exact solution of the EFE. But the resultant solution will in general not be physically meaningful.

Thus in this case, instead of specifying suitable matter content and then solving the EFE for that matter with appropriate initial/boundary conditions, one simply postulates some geometry or other, and then differentiates the assumed metric so as to determine the Riemann and Ricci tensors. The tensor T_{ab} required to exactly balance the EFE follows trivially (read the equation from left to right). But then, even if restrictions are put on assuring that the energy conditions are obeyed, *the resulting 'matter' will almost always be non-physical*: it will not correspond to any of the matter forms discussed above. It will simply be a formal solution of the field equations.

In particular, this method of 'solution' of the EFE often happens in the form of unacceptable 'imperfect fluids': some geometrical or mathematical assumption is used to provide an 'imperfect fluid' solution of the EFE, i.e. a matter tensor with anisotropic pressures. But then there is little chance this 'fluid' will satisfy (5.33), (5.34) or other physically motivated equations of state. Unless it does so, this is just an arbitrary mathematical solution of the EFE, with no physical content: calling it 'imperfect fluid' solution is a misnomer. It is not a fluid in any meaningful sense, and is not relevant to physical cosmology.

Exercise 5.2.1

(a) *Dominant energy condition.* Show that arbitrary observers in a perfect-fluid-filled spacetime, moving with 4-velocity $\tilde{u}^a$, will find $\tilde{\rho} := T_{ab}\tilde{u}^a\tilde{u}^b \geq 0$ if and only if $\rho \geq 0$, $\rho + p \geq 0$.

(b) *Strong energy condition.* Show that $R_{ab}\tilde{u}^a\tilde{u}^b = (T_{ab} - \frac{1}{2}T^c{}_c g_{ab})\tilde{u}^a\tilde{u}^b > 0$ for arbitrary observers if and only if $\rho + 3p > 0$, $\rho + p > 0$.

Exercise 5.2.2 Consider the case of a perfect fluid obeying the ideal gas equation of state $p = \alpha\rho_N^\gamma$ (α, γ constant).

(a) Show from the conservation equations for ρ_N and ρ that if $\gamma = 1$ then $\rho = Cp + p\ln(p/p_0)$, ($\dot{C} = 0 = \dot{p}_0$). Why is this solution physically unrealistic?

(b) Show that when $\gamma \neq 1$, $\rho = B\rho_N + p/(\gamma - 1)$ with $\dot{B} = 0$, and that the time evolution of such a fluid is given by $\rho = M/\ell^3 + N/\ell^{3\gamma}$, $\dot{M} = 0$, $\dot{N} = 0$. If $B = 1$, what is the internal energy density ϵ of the fluid?

(c) Show that when $\gamma \neq 1$, the effective relativistic coefficient $\tilde{\gamma}$, defined by $p = (\tilde{\gamma} - 1)\rho$, takes the form

$$\tilde{\gamma} = \frac{B(\gamma - 1) + \gamma A\ell^{1-3\gamma}}{B(\gamma - 1) + A\ell^{1-3\gamma}}, \quad \dot{A} = 0.$$

Plot a graph showing how $\tilde{\gamma}$ varies with ℓ.

Exercise 5.2.3 Consider the case of a barotropic fluid. Show that

(a) if we use comoving coordinates $x^\mu = (t, y^i)$, we can choose a coordinate transformation so that $u^\mu = r(x^\nu)\delta_0^\mu$, $g_{00} = -1/r^2$, $u_0 = -1/r$; and this form is preserved under the coordinate transformations, $x^{0'} = x^0 + f(x^i)$, $x^{i'} = x^i$.

(b) If we now write $u_i = -a_i(x^\mu)/r$, then $g_{0i} = -a_i/r^2$ and $\dot{u}_0 = 0$, $\dot{u}_i = -a_{i,0} + (r_{,0}/r)a_i - r_{,i}/r$.

(c) The conservation equations (5.39) are now equivalent to $a_{i,0} = 0$, and so may be integrated to give $a_i = a_i(x^j)$.

(d) Defining W by (5.41), show that $\ell = \lambda/W$ for some function $\lambda(x^i)$, so that we can relabel $\lambda^2 f_{ij} \to f_{ij}$, $\ell \to 1/W$ in (5.47) to obtain $h_{ij} = W^{-2}(x^\mu) f_{ij}(x^\nu)$, $\hat{f}^{ij} f_{ij,0} = 0$, where $\hat{f}^{ij} f_{jk} = \delta^i{}_k$. The conservation equations (5.38) are then identically satisfied.

Exercise 5.2.4 Show that for a barotropic perfect fluid with w constant, the enthalpy $W = (\rho/\rho_0)^{1/3(1+w)}$ and $r = (p/p_0)^{w/(1+w)}$. Deduce that then $\rho = M/\ell^{3(1+w)}$, $\dot{M} = 0$ and $r = r_0/\ell^{3w}$, $\dot{r}_0 = 0$.

5.3 Multiple fluids

The universe at different times contains a variety of matter components, and there may in addition be interactions (exchanges of energy and momentum) between some of these components. Here we shall treat the components as general fluids, without regard to the detailed properties of each fluid, so that the treatment applies to any form of matter that has an energy–momentum tensor.

Since the components will in general have different 4-velocities u_I^a, where I labels the components, we need to choose a reference 4-velocity u^a. This could be one of the u_I^a, e.g. the component that is dominating the universe at the time being studied, or it could be a combination of the velocities, e.g. the velocity which gives zero total momentum density, $q^a = 0$. Given a choice of u^a, the individual 4-velocities are related to this choice via (5.18):

$$u_I^a = \gamma_I(u^a + v_I^a), \quad \gamma_I = (1 - v_I^2)^{-1/2}, \quad v_I^a u_a = 0, \tag{5.54}$$

where v_I^a are the relative velocities as measured by a u^a observer. In cosmology, the components typically are photons ($I = \gamma$), baryonic matter ($I = $ b) modelled as a perfect fluid, cold dark matter ($I = $ c) modelled as dust, neutrinos ($I = $ v) modelled as a collisionless distribution, and a cosmological constant ($I = \Lambda$), or more generally a dynamical form of dark energy ($I = $ de).

The dynamical quantities in the field equations are the total quantities, with contributions from all dynamically significant species. Thus

$$T^{ab} = \sum_I T_I^{ab} = \rho u^a u^b + p h^{ab} + 2q^{(a} u^{b)} + \pi^{ab}, \tag{5.55}$$

$$T_I^{ab} = \rho_I^* u_I^a u_I^b + p_I^* h_I^{ab} + 2q_I^{*(a} u_I^{b)} + \pi_I^{*ab}, \tag{5.56}$$

where I labels the species. The dynamical quantities in (5.56) with an asterisk,[2] are the intrinsic quantities, i.e. as measured in the I-frame.

Then the following intrinsic relations hold:

$$p_c^* = 0 = q_c^{*a} = \pi_c^{*ab}, \quad q_b^{*a} = 0 = \pi_b^{*ab}, \tag{5.57}$$

$$p_\gamma^* = \tfrac{1}{3}\rho_\gamma^*, \quad p_\nu^* = \tfrac{1}{3}\rho_\nu^*, \tag{5.58}$$

where we have chosen the unique 4-velocity in the cold dark matter and baryonic cases which follows from modelling these fluids as perfect. After recombination, the baryonic pressure drops to zero, and eventually the neutrinos become non-relativistic. The cosmological constant is characterized by

$$p_\Lambda^* = -\rho_\Lambda^* = -\Lambda, \, q_\Lambda^{*a} = 0 = \pi_\Lambda^{*ab}, \, v_\Lambda^a = 0, \tag{5.59}$$

whereas dynamical dark energy will have an evolving equation of state depending on the particular model, and will not have zero relative velocity.

The conservation equations for the species are best given in the overall u^a-frame, in terms of the velocities v_I^a of species I relative to this frame. Furthermore, the evolution and constraint equations of Chapter 6 are all given in terms of the u^a-frame. Thus we need the expressions for the partial dynamic quantities as measured in the overall frame. Following Maartens, Gebbie and Ellis (1999), we find (Exercise 5.3.2) the exact (fully nonlinear) equations for the dynamical quantities of species I as measured in the overall u^a-frame:[3]

$$\rho_I = \rho_I^* + \left\{ \gamma_I^2 v_I^2 \left(\rho_I^* + p_I^* \right) + 2\gamma_I q_I^{*a} v_{Ia} + \pi_I^{*ab} v_{Ia} v_{Ib} \right\}, \tag{5.60}$$

$$p_I = p_I^* + \tfrac{1}{3} \left\{ \gamma_I^2 v_I^2 \left(\rho_I^* + p_I^* \right) + 2\gamma_I q_I^{*a} v_{Ia} + \pi_I^{*ab} v_{Ia} v_{Ib} \right\}, \tag{5.61}$$

$$\begin{aligned} q_I^a = q_I^{*a} + (\rho_I^* + p_I^*) v_I^a &+ \left\{ (\gamma_I - 1) q_I^{*a} + \gamma_I^2 v_I^2 \left(\rho_I^* + p_I^* \right) v_I^a \right. \\ &\left. + \gamma_I q_I^{*b} v_{Ib} v^a - \gamma_I q_I^{*b} v_{Ib} u^a + \pi_I^{*ab} v_{Ib} - \pi_I^{*bc} v_{Ib} v_{Ic} u^a \right\}, \end{aligned} \tag{5.62}$$

$$\begin{aligned} \pi_I^{ab} = \pi_I^{*ab} &+ \left\{ \gamma_I^2 \left(\rho_I^* + p_I^* \right) v_I^{\langle a} v_I^{b \rangle} - 2u^{\langle a} \pi_I^{*b \rangle c} v_{Ic} + \pi_I^{*bc} v_{Ib} v_{Ic} u^a u^b \right. \\ &\left. - \tfrac{1}{3} \pi_I^{*cd} v_{Ic} v_{Id} h^{ab} - 2\gamma_I q_I^{*c} v_I^c u^{\langle a} v_I^{b \rangle} + 2\gamma_I v_I^{\langle a} q_I^{b \rangle} \right\}. \end{aligned} \tag{5.63}$$

These equations have been written to make clear the linear parts, which will be applicable in an almost FLRW model; in that case, all terms in braces will be neglected. The total dynamical quantities are simply given by

$$\rho = \sum_I \rho_I, \, p = \sum_I p_I, \, q^a = \sum_I q_I^a, \, \pi^{ab} = \sum_I \pi_I^{ab}. \tag{5.64}$$

---

[2] This is the reverse of the asterisk notation used in Maartens, Gebbie and Ellis (1999).
[3] With minor corrections to Maartens, Gebbie and Ellis (1999), following Clarkson and Maartens (2010).

A convenient choice for each partial four-velocity u_I^a is the energy frame, i.e.

$$q_I^{*a} = 0, \qquad (5.65)$$

for each I (this is the obvious choice in the cases $I = $ b,c). As measured in the fundamental frame, the partial energy fluxes do not vanish, i.e. $q_I^a \neq 0$ – see (5.62). With this choice, using the above equations, we find the following expressions for the dynamic quantities of matter as measured in the fundamental frame. For cold dark matter:

$$\rho_c = \gamma_c^2 \rho_c^*, \quad p_c = \tfrac{1}{3}\gamma_c^2 v_c^2 \rho_c^*, \qquad (5.66)$$
$$q_c^a = \gamma_c^2 \rho_c^* v_c^a, \quad \pi_c^{ab} = \gamma_c^2 \rho_c^* v_c^{\langle a} v_c^{b\rangle}. \qquad (5.67)$$

For baryonic matter:

$$\rho_b = \gamma_b^2 \left(1 + w_b v_b^2\right) \rho_b^*, \quad p_b = \left[w_b + \tfrac{1}{3}\gamma_b^2 v_b^2 (1 + w_b)\right]\rho_b^*, \qquad (5.68)$$
$$q_b^a = \gamma_b^2 (1 + w_b)\rho_b^* v_b^a, \quad \pi_b^{ab} = \gamma_b^2 (1 + w_b)\rho_b^* v_b^{\langle a} v_b^{b\rangle}, \qquad (5.69)$$

where $w_b := p_b/\rho_b$. In the case of radiation and neutrinos, we shall evaluate the dynamic quantities relative to the u^a-frame directly via kinetic theory (see below).

The total energy–momentum tensor is conserved, i.e. $\nabla_b T^{ab} = 0$. The partial energy–momentum tensors obey

$$\nabla_b T_I^{ab} = Q_I^a = Q_I u^a + \mathcal{Q}_I^a, \qquad (5.70)$$

where Q_I is the rate of energy density transfer to species I as measured in the u^a-frame, and $\mathcal{Q}_I^a = M_I^{\langle a\rangle}$ is the rate of momentum density transfer to species I, as measured in the u^a-frame. Cold dark matter and neutrinos are decoupled during the period of relevance for CMB anisotropies, while radiation and baryons are coupled through Thomson (more generally, Compton) scattering. Thus,

$$Q_c^{*a} = 0 = Q_v^{*a}, \quad Q_\gamma^{*a} = -Q_b^{*a} \propto n_e \sigma_T, \qquad (5.71)$$

where n_e is the free electron number density, and σ_T is the Thomson cross-section.

Exercise 5.3.1 Suppose there is a mixture of two perfect fluids with different 4-velocities. Determine the unit timelike eigenvector U^a of T_{ab} and associated eigenvalue. Hint: the eigenvector will lie in the plane spanned by the two velocity vectors u_1^a, u_2^b; then find the effective equation of state for the fluid relative to the 4-velocity U^a (it will not be equivalent to a perfect fluid).

Exercise 5.3.2 Show that the velocity formula inverse to equation (5.54) is

$$u^a = \gamma_I \left(u_I^a + \hat{v}_I^a\right), \quad \hat{v}_I^a = -\gamma_I \left(v_I^a + v_I^2 u^a\right), \qquad (5.72)$$

where $\hat{v}_I^a u_{Ia} = 0$, and $\hat{v}_I^a \hat{v}_{Ia} = v_I^a v_{Ia}$. Show that $\hat{v}_I^a$ can also be written as

$$\hat{v}_I^a = -\gamma_I^{-1} \left(v_I^a + \gamma_I v_I^2 u_I^a\right). \qquad (5.73)$$

Now write h_{ab} in terms of $\hat{v}_I^a$ and u_I^a. Then derive (5.60)–(5.63) by using the expressions

$$\rho_I = T_I^{ab} u_a u_b, \quad p_I = \tfrac{1}{3} T_I^{ab} h_{ab}, \tag{5.74}$$

$$q_I^a = -T_I^{ab} u_b - \rho_I u^a, \quad \pi_I^{ab} = T_I^{cd} h_c^a h_d^b - p_I h^{ab}, \tag{5.75}$$

and expressing T_I^{ab} in terms of the starred quantities, (5.56).

5.4 Kinetic theory

Relativistic kinetic theory (Lindquist, 1966, Ehlers, 1971, Stewart, 1971, de Groot, van Leeuwen and van Weert, 1980, Bernstein, 1988) provides a self-consistent microscopically based treatment of matter and radiation. This is a natural unifying framework to deal with a gas of particles ranging from hydrodynamic (collision-dominated) to free-streaming (collision-free) behaviour. The photon gas undergoes a transition from hydrodynamic tight coupling with matter, through the non-equilibrium process of decoupling from matter, to non-hydrodynamic free streaming. The transition is characterized by the evolution of the photon mean free path from effectively zero to effectively infinity. This whole range of behaviour can be described by kinetic theory with Compton scattering (the Thomson approximation is adequate for cosmology). Free-streaming neutrinos are also described by kinetic theory. The baryonic matter that interacts with radiation can reasonably be described as a fluid.

We follow the 1+3 covariant kinetic theory formalism of Ellis, Matravers and Treciokas (1983b), Ellis, Treciokas and Matravers (1983), which builds on work by Ehlers, Geren and Sachs (EGS) (1968), Treciokas and Ellis (1971), Treciokas (1972) and Thorne (1981).

5.4.1 Distribution functions and the Boltzmann equation

In a gas of identical neutral particles, each 4-momentum p^a can be written $p^a = \hat{E} \hat{u}^a$ where $\hat{u}^a$ is the particle 4-velocity and $\hat{E}$ is the rest-frame energy. If u^a is the chosen fundamental 4-velocity, then $\hat{u}^a = \gamma(v)(u^a + v^a)$, where v^a is the particle's relative velocity, $u_a v^a = 0$. Then,

$$p^a = E(u^a + v^a) = E u^a + \lambda e^a, \quad \lambda = vE = \sqrt{E^2 - m^2}, \quad e_a e^a = 1, \tag{5.76}$$

where $E = -u_a p^a \,(= \gamma \hat{E})$ is the energy in the u^a-frame, λ is the magnitude of the 3-momentum, and e^a is the direction of relative motion. For massless particles, like photons, $\lambda = E$ and $v = 1$. For massive particles, like neutrinos or CDM particles, $\lambda = \gamma m v$.

If we can neglect polarization (or helicity), the gas can be described by a scalar-valued one-particle distribution function $f(x^a, p^b)$, which is the number of particles per unit phase space volume at the phase space point (x^a, p^b) (see Chapter 11 for the inclusion of polarization). For a given set of particles, their phase space volume is both Lorentz invariant (i.e. the same for all observers) and, in the absence of collisions, constant along their path. The collisionless evolution of the gas is thus described by the Liouville equation,

$$
\frac{\mathrm{d}f}{\mathrm{d}\tau} = p^a \frac{\partial f}{\partial x^a} + \frac{\mathrm{d}p^a}{\mathrm{d}\tau} \frac{\partial f}{\partial p^a} = 0. \tag{5.77}
$$

The particle world lines are geodesics with affine parameter τ, where $p^a = \mathrm{d}x^a/\mathrm{d}\tau$ and $\mathrm{d}p^a/\mathrm{d}\tau = -\Gamma^a{}_{bc} p^b p^c$. In the presence of collisions, we have the Boltzmann equation,

$$
\frac{\mathrm{d}f}{\mathrm{d}\tau} = C[f], \tag{5.78}
$$

where $C[f]$ is Lorentz invariant and determines the rate of change of f due to emission, absorption and scattering processes. For a collisionless gas, or a collisional gas in equilibrium due to detailed balancing, $C[f] = 0$.

The particle momentum and direction propagate as

$$
\frac{\mathrm{d}\lambda}{\mathrm{d}\tau} = -E^2 \dot{u}_a e^a - E\lambda \left(\sigma_{ab} e^a e^b + \tfrac{1}{3}\Theta \right), \tag{5.79}
$$

$$
\frac{\mathrm{d}e^{\langle a \rangle}}{\mathrm{d}\tau} = -\frac{E^2}{\lambda} s^a{}_b \dot{u}^b - E \left(\eta^a{}_{bc} \omega^c e^b + s^{ab}\sigma_{bc} e^c \right), \tag{5.80}
$$

where $s_{ab} := h_{ab} - e_a e_b$ is the screen-projection tensor, which projects into the two-dimensional screen perpendicular to the propagation direction e^a in the local rest-space of u^a. Note that $e_a \mathrm{d}e^{\langle a \rangle}/\mathrm{d}\tau = 0$, so that $e^a e_a = 1$ is preserved. In the FLRW limit, $\mathrm{d}\lambda/\mathrm{d}\tau = -E\lambda H$, so the momentum redshifts as $1/a$. Also, $\mathrm{d}e^{\langle a \rangle}/\mathrm{d}\tau = 0$, so that e^a is constant. In the real universe this is no longer so and (5.80) then describes the effect of gravitational lensing.

The distribution function can be expanded in covariant spherical multipoles (Thorne, 1981, Ellis, Matravers and Treciokas, 1983b),

$$
f(x, p) = \sum_{\ell=0}^{\infty} F_{A_\ell}(x, E) e^{A_\ell} = F(x, E) + F_a(x, E)e^a + F_{ab}(x, E)e^a e^b + \cdots, \tag{5.81}
$$

where the multipole tensors $F_{A_\ell} = F_{\langle a_1 \ldots a_\ell \rangle}(E)$ are projected (orthogonal to u^a), symmetric and tracefree (PSTF), and thus are irreducible under three-dimensional rotations. Equation (5.81) is equivalent to an expansion in spherical harmonics, but has the advantage of being fully covariant. The inversion is

$$
F_{A_\ell}(x, E) = \frac{1}{\Delta_\ell} \int f(x, E, e) e_{\langle A_\ell \rangle} \, \mathrm{d}\Omega \quad \text{where} \quad \Delta_\ell := \frac{4\pi 2^\ell (\ell!)^2}{(2\ell+1)!}, \tag{5.82}
$$

which follows from the identity

$$
\int e_{\langle A_\ell \rangle} e^{\langle B_{\ell'} \rangle} \, \mathrm{d}\Omega = \Delta_\ell h^{\langle B_\ell \rangle}_{\langle A_\ell \rangle} \delta_{\ell\ell'} = \Delta_\ell h^{\langle b_1}_{\langle a_1} \ldots h^{b_\ell \rangle}_{a_\ell \rangle} \delta_{\ell\ell'}. \tag{5.83}
$$

Propagation equations for the multipoles follow from substituting (5.81) into the Boltzmann equation, using (5.79) and (5.80), and taking the PSTF part (Ellis, Matravers and

Treciokas, 1983b),

$$
\begin{aligned}
E\dot{F}_{\langle A_\ell\rangle} &- \frac{\lambda^2}{3}\frac{\partial F_{A_\ell}}{\partial E}\Theta + \frac{\ell+1}{2\ell+3}\lambda\overline{\nabla}^a F_{aA_\ell} + \lambda\overline{\nabla}_{\langle a_\ell}F_{A_{\ell-1}\rangle} \\
&+ \ell E F_{a\langle A_{\ell-1}}\eta_{a_\ell\rangle}{}^{ab}\omega_b - \left[\lambda E\frac{\partial F_{\langle A_{\ell-1}}}{\partial E} - (\ell-1)\frac{E^2}{\lambda}F_{\langle A_{\ell-1}}\right]\dot{u}_{a_\ell\rangle} \\
&- \frac{\ell+1}{2\ell+3}\left[(\ell+2)\frac{E^2}{\lambda}F_{aA_\ell} + \lambda E\frac{\partial F_{aA_\ell}}{\partial E}\right]\dot{u}^a \\
&- \frac{\ell}{2\ell+3}\left[3E F_{a\langle A_{\ell-1}} + 2\lambda^2\frac{\partial F_{a\langle A_{\ell-1}}}{\partial E}\right]\sigma_{a_\ell\rangle}{}^a \\
&- \frac{(\ell+1)(\ell+2)}{(2\ell+3)(2\ell+5)}\left[(\ell+3)E F_{abA_\ell} + \lambda^2\frac{\partial F_{abA_\ell}}{\partial E}\right]\sigma^{ab} \\
&- \left[\lambda^2\frac{\partial F_{\langle A_{\ell-2}}}{\partial E} - (\ell-2)E F_{\langle A_{\ell-2}}\right]\sigma_{a_{\ell-1}a_\ell\rangle} = C_{A_\ell}[f].
\end{aligned}
\tag{5.84}
$$

Here $C_{A_\ell}[f]$ are the collision multipoles. The terms containing derivatives with respect to E arise from the redshifting of the particle's energy, as governed by (5.79). The isotropic expansion sources anisotropy at multipole ℓ from multipole ℓ, acceleration sources anisotropy at ℓ from $\ell\pm 1$, shear sources anisotropy at ℓ from both $\ell\pm 2$ and ℓ and vorticity sources anisotropy at ℓ from ℓ.

We shall use the multipole propagation equations in Chapter 11 to analyse the evolution of anisotropies in the CMB. Here we note that in an FLRW spacetime, the homogeneity and isotropy of the spatial hypersurfaces enforce the vanishing of all multipoles $\ell > 0$, and the isotropy and homogeneity of the monopoles: $f(x, p) = F(t, E)$. In addition, all terms on the left-hand side of (5.84) vanish except for the first two. The hierarchy of equations collapses to

$$
\frac{\partial F}{\partial t} - H\lambda\frac{\partial F}{\partial\lambda} = C[F].
\tag{5.85}
$$

For a collisionless gas, or a collisional gas in equilibrium due to detailed balancing, the solution is $F(t,\lambda) = F(a(t)\lambda)$, which reflects the fact that $a\lambda$ is conserved along the particle world lines.

5.4.2 Bulk properties of the gas

Macroscopic averages over the microscopic distribution function define the bulk properties of the gas, such as energy density, number density, entropy, etc.

The particle 4-current density (5.5) is

$$
N^a(x) = \int p^a f(x, p)\mathrm{d}P, \quad \mathrm{d}P = \frac{\mathrm{d}^3 p}{E} = \frac{\lambda^2}{E}\mathrm{d}\lambda\,\mathrm{d}\Omega.
\tag{5.86}
$$

Here dP is the Lorentz-invariant volume element on the positive-energy mass shell $p_a p^a = -m^2$. We decompose N^a as in (5.5); on using the identity,

$$\frac{(\ell+1)}{4\pi} \int e^{A_\ell} d\Omega = \begin{cases} 0 & \ell \text{ odd}, \\ h^{(a_1 a_2} h^{a_3 a_4} \dots h^{a_{\ell-1} a_\ell)} & \ell \text{ even}, \end{cases} \tag{5.87}$$

the number density and particle drift vector are given by

$$n = \Delta_0 \int_0^\infty \lambda^2 F \, d\lambda, \tag{5.88}$$

$$n^a = \Delta_1 \int_0^\infty \lambda^2 v F^a \, d\lambda, \quad v = \frac{\lambda}{E}, \tag{5.89}$$

where Δ_ℓ is defined in (5.82). The propagation equation for n follows from integrating the $\ell = 0$ moment of (5.84) over $\lambda^2 d\lambda$:

$$\dot{n} + \Theta n + \overline{\nabla}^a n_a + \dot{u}^a n_a = \Delta_0 \int_0^\infty \lambda v \, \bar{C}[f] d\lambda, \tag{5.90}$$

where $\bar{C}[f]$ is the collision monopole. This is a generalization of (5.6). The right-hand side represents non-conservation of particle number due to interactions, the left-hand side is the divergence of N^a, and we can rewrite (5.90) as

$$\nabla_a N^a = \int C[f] dP. \tag{5.91}$$

The entropy 4-current density is defined (for classical statistics) by (Ehlers, 1971)

$$S^a(x) = N^a(x) - \int p^a f(x, p) \ln f(x, p) dP, \tag{5.92}$$

and its divergence is

$$\nabla_a S^a = -\int \ln f \, C[f] dP. \tag{5.93}$$

Entropy is generated if the right-hand side is non-zero, since $\nabla_a S^a$ is the entropy production density. We can define an equilibrium distribution as one with $\nabla_a S^a = 0$. In particular, a collisionless gas is in equilibrium since $C[f]$ is identically zero. A gas in collision-dominated equilibrium has $C[f] = 0$ due to detailed balancing.

The energy–momentum tensor is

$$T^{ab}(x) = \int p^a p^b f(x, p) dP. \tag{5.94}$$

Decomposing T^{ab} in the usual way, we find that

$$\rho = \Delta_0 \int_0^\infty \lambda^2 E F \, d\lambda, \quad p = \frac{\Delta_0}{3} \int_0^\infty \lambda^2 E v^2 F \, d\lambda, \tag{5.95}$$

$$q^a = \Delta_1 \int_0^\infty \lambda^2 E v F^a \, d\lambda, \tag{5.96}$$

$$\pi^{ab} = \Delta_2 \int_0^\infty \lambda^2 E v^2 F^{ab} \, d\lambda. \tag{5.97}$$

The propagation equations for these quantities follow from integrating (5.84) with $\lambda^2 d\lambda$ and appropriate powers of the velocity-weight $v := \lambda/E$. For the energy and momentum densities,

$$\dot{\rho} + \Theta(\rho + p) + \overline{\nabla}^a q_a + 2\dot{u}^a q_a + \sigma^{ab}\pi_{ab} = \Delta_0 \int_0^\infty \lambda^2 \bar{C}[f] d\lambda, \qquad (5.98)$$

$$\dot{q}_{\langle a \rangle} + \tfrac{4}{3}\Theta q_a + (\rho + p)\dot{u}_a + \overline{\nabla}_a p + \overline{\nabla}^b \pi_{ab}$$

$$+ (\eta_{abc}\omega^c + \sigma_{ab})q^b + \dot{u}^b \pi_{ab} = \Delta_1 \int_0^\infty \frac{\lambda^3}{E} C_a[f] d\lambda. \qquad (5.99)$$

The terms on the right-hand sides represent energy and momentum exchange through interactions, modifying (5.11), (5.12). The left-hand sides are $\nabla^b T_{bc}$ projected along u^c and $h^c{}_a$ respectively. It follows that

$$\nabla_b T^{ab} = \int p^a C[f] dP. \qquad (5.100)$$

The propagation equations for the energy and momentum densities do not form a closed system even when there are no interactions – because of the isotropic pressure (which is in general related to ρ via a dynamical equation of state), and the anisotropic stress, which needs a propagation equation. The required information is contained in the Boltzmann equation, in terms of the given collision term. This equation can be recast as a two-dimensional, infinite hierarchy for the moments of f integrated over energy with positive (integer) velocity-weights (Ellis, Matravers and Treciokas, 1983b, Lewis and Challinor, 2002). These integrated moments contain (5.95)–(5.97) as a subset.

The two-dimensional hierarchy simplifies in a number of important special cases. For photons ($m = 0, \lambda = E$) and for relativistic neutrinos (at temperatures $T \gg m$), $\lambda \sim E$, the hierarchy becomes one-dimensional. In this case, we can define the bolometric multipoles,

$$I_{A_\ell}(x) = \Delta_\ell \int_0^\infty E^3 F_{A_\ell}(x, E) dE. \qquad (5.101)$$

Then the lowest three multipoles determine the energy–momentum tensor: $I = \rho$, $I_a = q_a$, $I_{ab} = \pi_{ab}$. For non-relativistic matter the hierarchy is genuinely two-dimensional, but it can be truncated at low velocity weight (provided that the typical free-streaming distance per Hubble time is small compared to the size of the inhomogeneity). This truncation scheme can be used to study the effect of velocity dispersion on linear structure formation (Maartens, Triginer and Matravers, 1999, Lewis and Challinor, 2002). For tightly coupled collisional matter, such as the CMB in the pre-recombination era when Thomson scattering is efficient, truncation can also be performed (but at much higher multipoles). This is because anisotropies at multipole ℓ are suppressed by $(v_* k t_{\text{coll}}/a)^\ell$, where a/k is the scale of inhomogeneity, t_{coll} is the collision time and v_* is a typical particle speed.

5.4.3 Collision term

In Boltzmann's approximation, the gas is not too far from equilibrium, and not too dense or too cold, so that particles which are about to collide are not correlated. Implicit here is a mean-field type approximation that avoids directly dealing with long-range gravitational interactions, i.e. the gas particles move in between collisions in the gravitational field generated collectively by themselves (and possibly other sources).

Collisions (assumed here to be binary) conserve energy–momentum, so that $p^a + p'^a = p''^a + p'''^a$. The probability of collision (and hence the cross-section) is determined by a Lorentz-invariant function $W(pp' \to p''p''')$, and then the Boltzmann collision integral is (Ehlers, 1971)

$$C[f] = \tfrac{1}{2} \int dP' \int dP'' \int dP''' (f''f''' - ff')W, \tag{5.102}$$

where f' denotes $f(x, p')$ and similarly for f'', f'''. In general one expects that collisions will drive the gas towards equilibrium, provided that the rate of interaction is high enough (e.g. in cosmology, higher than the expansion rate).

Collision-dominated equilibrium is achieved if there is detailed balancing, i.e. $f''f''' = ff'$, or equivalently, $\ln f(x, p)$ is an additive collision-invariant. For any scalar $\alpha(x)$ and vector $\beta^a(x)$, we have that $\alpha(x) + \beta_a(x)p^a$ is an additive collision invariant for elastic binary collisions. Thus an equilibrium distribution function is given by

$$f(x, p) = \left\{ \exp[-\alpha(x) - \beta_a(x)p^a] + \epsilon \right\}^{-1}, \tag{5.103}$$

where ϵ takes the values 0 (classical particles), $+1$ (fermions) and -1 (bosons). In order to ensure that $f \to 0$ as $E \to \infty$, β_a must be a non-spacelike future-directed vector, and we can use it to define the preferred (equilibrium) 4-velocity, i.e. $\beta_a = \beta u_a$ ($\beta > 0$). Then $\beta_a p^a = -\beta E$. It follows that (Exercise 5.4.2)

$$\alpha_{,a} = 0, \quad \beta_{(a;b)} = \chi g_{ab}, \tag{5.104}$$

where $\chi = 0$ if $m > 0$. Thus we have the important theorem:

Theorem 5.1 Collision-dominated equilibrium
is only possible in relativistic spacetime

- *for massive particles, if spacetime is stationary (i.e. admits a timelike Killing vector);*
- *for massless particles if spacetime admits a timelike conformal Killing vector (as does RW spacetime).*

In particular, a massive gas in an expanding RW spacetime cannot be in collisional equilibrium, and must therefore have non-zero bulk viscosity.
Note that any collision-free gas in any spacetime is in equilibrium, since the entropy production vanishes by (5.93).

For the CMB in cosmology, the collision integral is not of the simple Boltzmann form since photons interact with electrons (and much more weakly with protons) via Compton scattering in the Thomson regime. The collision term for Thomson scattering (neglecting polarization) is,

$$C[f] = \sigma_T n_e E_b \left[\bar{f}(x, p) - f(x, p) \right], \tag{5.105}$$

where $E_b = -p_a u_b^a$ is the photon energy relative to the baryonic (i.e. baryon-electron) frame u_b^a, and $\bar{f}(x, p)$ determines the number of photons scattered into the phase space volume element at (x, p). The differential Thomson cross-section is proportional to $1 + \cos^2 \alpha$, where α is the angle between initial and final photon directions in the baryonic frame. Thus $\cos \alpha = e_b^a e'_{ba}$, where e'_{ba} is the initial and e_b^a is the final direction, so that

$$p'^a = E_b \left(u_b^a + e'^a_b \right), \quad p^a = E_b \left(u_b^a + e_b^a \right), \tag{5.106}$$

where we have used $E'_b = E_b$, which follows since the scattering is elastic. Then $\bar{f}$ is given by

$$\bar{f}(x, p) = \frac{3}{16\pi} \int f(x, p') \left[1 + \left(e_b^a e'_{ba} \right)^2 \right] d\Omega'_b. \tag{5.107}$$

The exact forms of the photon energy and direction in the baryonic frame are

$$E_b = E \gamma_b \left(1 - v_b^a e_a \right), \tag{5.108}$$

$$e_b^a = \frac{E}{E_b} \left[e^a + \gamma_b^2 \left(v_b^b e_b - v_b^2 \right) u^a + \gamma_b^2 \left(v_b^b e_b - 1 \right) v_b^a \right]. \tag{5.109}$$

Exercise 5.4.1 Derive (5.90) and (5.91).

Exercise 5.4.2 For an equilibrium distribution (5.103),
(a) show that N^a, S^a, T^{ab} have perfect-fluid form with $u^a = \beta^a / \beta$;
(b) use $df / d\tau = 0$ to derive (5.104).

Exercise 5.4.3 Prove the identity

$$V_{\langle b} S_{A_\ell \rangle} = V_{\langle b} S_{A_\ell \rangle} - \left(\frac{\ell}{2\ell + 1} \right) V^c S_{c(A_{\ell-1}} h_{a_\ell b)}, \quad S_{A_\ell} = S_{\langle A_\ell \rangle}. \tag{5.110}$$

Using (5.87), show that for any projected vector v^a:

$$v^a e_a f = \tfrac{1}{3} F_a v^a + \left[F v_a + \tfrac{2}{5} F_{ab} v^b \right] e^a + \left[F_{\langle a} v_{b \rangle} + \tfrac{3}{7} F_{abc} v^c \right] e^{\langle a} e^{b \rangle} + \cdots$$

$$= \sum_{\ell \geq 0} \left[F_{\langle A_{\ell-1}} v_{a_\ell \rangle} + \left(\frac{\ell+1}{2\ell+3} \right) F_{A_\ell a} v^a \right] e^{\langle A_\ell \rangle}. \tag{5.111}$$

(We use the convention that $F_{A_\ell} = 0$ for $\ell < 0$.)

Exercise 5.4.4 Derive (5.108) and (5.109).

5.5 Electromagnetic fields

Electromagnetism is central to cosmology because (a) we observe by electromagnetic radiation, which is the geometric optics limit of the electromagnetic field; (b) magnetic fields are present in galaxies and clusters and play a key role in star and galaxy formation and evolution – and they could even be significant on cosmological scales in the early universe;

and (c) electromagnetism provides a model for many important features of the gravitational field.

5.5.1 Electromagnetic field tensor

The electromagnetic field tensor (or Faraday tensor) is $F_{ab} = F_{[ab]}$. For an observer moving with 4-velocity u^a, it is measured as an electric field,

$$E_a = F_{ab}u^b = E_{\langle a \rangle},\qquad(5.112)$$

and a magnetic field,

$$B_a = \tfrac{1}{2}\eta_{acd}F^{cd} = F^*_{ab}u^b = B_{\langle a \rangle},\qquad(5.113)$$

where F^*_{ab} is the dual. These completely represent the field,

$$F_{ab} = 2u_{[a}E_{b]} + \eta_{abc}B^c.\qquad(5.114)$$

These equations, giving the 1+3 splitting of the field relative to the velocity u^a, contain the usual transformation properties of the electromagnetic field when we change to a different 4-velocity; see Exercise 5.5.2. We write the magnitudes of these 3-vectors as $E^2 = E^a E_a$, $B^2 = B^a B_a$.

The Lorentz force experienced by a particle with electric charge e and 4-velocity V^a is $F_a = eF_{ab}V^b$. The particle equation of motion is $V^b\nabla_b V^a = (e/m)F_{ab}V^b$, where m is its mass. Substituting from (5.114) gives the usual expression for this acceleration in terms of electric and magnetic fields. For a charged fluid moving with 4-velocity u^a, the momentum conservation equation will have a term corresponding to the Lorentz force (see below).

5.5.2 Maxwell equations

The Maxwell equations are

$$\nabla_b F^{ab} = J^a,\quad \nabla_{[a}F_{bc]} = 0,\qquad(5.115)$$

where the 4-current is

$$J^a = \mu u^a + j^a,\quad j_a u^a = 0.\qquad(5.116)$$

Here μ is the charge density and j_a the 3-current measured by u^a. (We use Heaviside–Lorentz units, in which $\mu_0 = 1 = \epsilon_0$.)

Because of the Riemann tensor symmetries, these equations imply the conservation of current,

$$\nabla_a J^a = 0.\qquad(5.117)$$

Making a 1+3 split of these equations, using the definitions (5.112), (5.113) of the electric and magnetic fields, we find that Maxwell's equations gain many kinematic terms due to the motion of the observers who measure E^a and B^a:

$$\overline{\nabla}_a E^a = \mu - 2\omega_a B^a \,, \tag{5.118}$$

$$\overline{\nabla}_a B^a = 2\omega_a E^a \,, \tag{5.119}$$

$$\dot{E}_{\langle a \rangle} = \left(\sigma_{ab} + \eta_{abc}\omega^c - \tfrac{2}{3}\Theta h_{ab}\right)E^b + \eta_{abc}\dot{u}^b B^c + \text{curl } B_a - j_a, \tag{5.120}$$

$$\dot{B}_{\langle a \rangle} = \left(\sigma_{ab} + \eta_{abc}\omega^c - \tfrac{2}{3}\Theta h_{ab}\right)B^b - \eta_{abc}\dot{u}^b E^c - \text{curl } E_a \,, \tag{5.121}$$

and the current conservation equation becomes

$$\dot{\mu} + \Theta\mu + \overline{\nabla}_a j^a + \dot{u}_a j^a = 0 \,. \tag{5.122}$$

These equations reduce to the usual form of Maxwell's equations for a set of Minkowski observers ($\dot{u}^a = \omega^a = \sigma_{ab} = \Theta = 0$). The wave equations for E^a and B^a that follow from these equations will also contain many kinematic terms that will vanish in the Minkowski case.

5.5.3 Maxwell energy–momentum tensor

The energy–momentum tensor of an electromagnetic field is

$$T_{\text{em}}^{ab} = -F^{ac}F_c{}^b - \tfrac{1}{4}g_{ab}F_{cd}F^{cd}, \tag{5.123}$$

with $g_{ab}T_{\text{em}}^{ab} = 0$. Substituting from (5.114), we see that the field has the stress tensor of a radiative imperfect fluid with

$$\rho_{\text{em}} = \tfrac{1}{2}(E^2 + B^2) = 3p_{\text{em}} \,, \tag{5.124}$$

$$q_{\text{em}}^a = \eta^{abc}E_b B_c \,, \quad \pi_{\text{em}}^{ab} = -E^{\langle a}E^{b\rangle} - B^{\langle a}B^{b\rangle} \,. \tag{5.125}$$

The momentum density q_{em}^a is the Poynting vector. The energy conditions (5.16) and (5.17) are satisfied.

It follows from Maxwell's equations (5.115) that

$$\nabla_b T_{\text{em}}^{ab} = -F^{ab}J_b \,, \tag{5.126}$$

so if there is no 4-current (i.e. no interaction between the electromagnetic field and other matter), then the energy and momentum of the field by itself are conserved.

The total energy–momentum tensor, $T_{\text{b}}^{ab} + T_{\text{em}}^{ab}$, for a combination of a charged baryonic perfect fluid and an electromagnetic field is conserved, so that $\nabla_b T_{\text{b}}^{ab} = F^{ab}J_b$. This leads to

$$\dot{\rho} + \Theta(\rho + p) = E^a j_a \,, \tag{5.127}$$

$$(\rho + p)\dot{u}_a + \overline{\nabla}_a p = \mu E_a + \eta_{abc}j^b B^c \,. \tag{5.128}$$

The terms on the right of the momentum equation are the Lorentz force density.

5.5.4 Relativistic magnetohydrodynamics

Magnetic fields are a crucial ingredient in the universe, whose presence is ubiquitous, but whose origin remains to be fully explained. Here we focus on the covariant approach to cosmic magnetism (see Barrow, Maartens and Tsagas (2007) for a review).

After inflation, the universe is a good conductor – even when the number density of free electrons drops dramatically during recombination, its residual value is enough to maintain high conductivity in baryonic matter. As a result, B-fields of cosmological origin have remained frozen into the expanding baryonic fluid during most of their evolution. Magnetic effects on structure formation can thus be analysed within the ideal magnetohydrodynamics approximation. Ohm's law in the rest frame of the fluid has the covariant form

$$j_a = \varsigma E_a, \tag{5.129}$$

where the 3-current is defined in (5.116) and ς is the conductivity. In the ideal MHD limit, non-zero spatial currents arise for $E_a \rightarrow 0$ and $\varsigma \rightarrow \infty$. Then the energy–momentum tensor of the magnetic field simplifies to

$$T_B^{ab} = \tfrac{1}{2} B^2 u^a u^b + \tfrac{1}{6} B^2 h^{ab} + \pi_B^{ab}, \quad \pi_B^{ab} = -B^{\langle a} B^{b \rangle}. \tag{5.130}$$

The B-field corresponds to an imperfect fluid with energy density $\rho_B = B^2/2$, isotropic pressure $p_B = B^2/6$ and anisotropic stress π_B^{ab}. The anisotropic stress has unit eigenvectors parallel and orthogonal to B^a, with eigenvalues $-2B^2/3$ and $+B^2/3$ respectively. This means that the field exerts a negative pressure along its own field lines. We can think of this as following from the 'tension' in the field lines – the magnetic field lines tend to straighten. The field exerts an enhanced positive pressure orthogonal to its field lines.

Maxwell's equations reduce to one propagation equation (the magnetic induction equation) and three constraints:

$$\dot{B}_{\langle a \rangle} = \left(\sigma_{ab} + \eta_{abc} \omega^c - \tfrac{2}{3} \Theta h_{ab} \right) B^b, \tag{5.131}$$

$$\text{curl } B_a + \eta_{abc} \dot{u}^b B^c = j_a, \tag{5.132}$$

$$2\omega^a B_a = \mu, \quad \overline{\nabla}^a B_a = 0. \tag{5.133}$$

The right-hand side of (5.131) is due to the relative motion of the neighbouring observers and guarantees that the magnetic force lines always connect the same matter particles, so that the field remains frozen-in with the highly conducting fluid. (This is similar to the way vorticity gets frozen into the fluid, see Section 6.2.) Equation (5.132) shows how the spatial currents are responsible for keeping the field lines frozen-in with the matter. (In the MHD limit, the magnetic field is not sourced by currents, as confirmed by (5.131).)

The magnetic induction equation (5.131) leads to an evolution equation for the energy density of the field (Exercise 5.5.6),

$$(B^2)^{\cdot} = -\tfrac{4}{3} \Theta B^2 - 2\sigma_{ab} \pi_B^{ab}. \tag{5.134}$$

This shows that in a highly conducting medium, $B^2 \propto \ell^{-4}$, unless there is substantial anisotropy, in which case the B-field behaves as a dissipative radiative fluid. In a spatially homogeneous, radiation-dominated universe with weak overall anisotropy, the shear term on the right-hand side means that the ratio B^2/ρ_γ is no longer constant but displays a slow 'quasi-static' logarithmic decay (Zel'dovich, 1970, Barrow, 1997).

Total energy conservation for a perfect fluid plus magnetic field is given by $\dot{\rho} + \Theta(\rho + p) = 0$; it has no magnetic terms since the magnetic energy density is separately conserved, which is guaranteed by the magnetic induction equation (5.131). The total momentum conservation gives

$$\left(\rho + p + \tfrac{2}{3}B^2\right)\dot{u}^a + \overline{\nabla}^a p = -\eta^{abc}B_b \text{ curl } B_c - \pi_B^{ab}\dot{u}_b, \tag{5.135}$$

where ρ is the fluid energy density. The B curl B term in (5.135) is the magnetic Lorentz force, which is always normal to the B-field lines and may be decomposed as (Exercise 5.5.6)

$$\eta_{abc}B^b \text{ curl } B^c = \tfrac{1}{2}\overline{\nabla}_a B^2 - B^b \overline{\nabla}_b B_a. \tag{5.136}$$

The last term is the result of the magnetic tension. Insofar as this tension stress is not balanced by the pressure gradients, the field lines are out of equilibrium and there is a non-zero Lorentz force acting on the particles of the fluid.

Exercise 5.5.1 Complex notation. Show that if we define complex quantities $\mathcal{G}_{abcd} = g_{ac}g_{bd} - g_{ad}g_{bc} + i\eta_{abcd}$, $\mathcal{F}_{ab} = F_{ab} + \tfrac{i}{2}\eta_{ab}{}^{cd}F_{cd} = F_{ab} + iF_{ab}^*$, and $\mathcal{E}^a = E^a + iB^a$, then (5.114) is equivalent to $\mathcal{F}_{ab} = \mathcal{G}_{abcd}u^c\mathcal{E}^d$.

Exercise 5.5.2 We can split the electromagnetic field relative to a different 4-velocity $\tilde{u}_a$, as in (5.18), i.e. $F_{ab} = 2\tilde{u}_{[a}\tilde{E}_{b]} + \tilde{\eta}_{abc}\tilde{B}^c$. Show that

$$\tilde{\eta}_{abc} = \gamma\eta_{abc} + \gamma\left(2u_{[a}\eta_{b]cd} + u_c\eta_{abd}\right)v^d, \tag{5.137}$$

and that

$$\tilde{E}_a = \gamma\left(E_a + \varepsilon_{abc}v^b B^c + v^b E_b u_a\right), \tag{5.138}$$

$$\tilde{B}_a = \gamma\left(B_a - \varepsilon_{abc}v^b E^c + v^b B_b u_a\right), \tag{5.139}$$

which generalize the special relativity transformation laws.

Exercise 5.5.3 Show that in the case of an FLRW universe, Maxwell's equations reduce to the standard form for flat-spacetime in terms of the rescaled electric, magnetic and current vectors,

$$\hat{E}^a = a^2 E^a, \quad \hat{B}^a = a^2 B^a, \quad \hat{J}^a = a^2 J^a, \tag{5.140}$$

where a denotes the scale length ℓ in FLRW. Thus a source-free solution in Minkowski spacetime gives a solution in a flat FLRW spacetime by this rescaling (this is essentially due to the conformal invariance of Maxwell's equations and the conformal flatness of FLRW universes). Determine the wave equation for $\hat{E}^a$ that follows from the equations.

Exercise 5.5.4 The second Maxwell equation guarantees the local existence of a 4-potential A_a for the electromagnetic field: $F_{ab} = 2\nabla_{[a}A_{b]}$. Show

(a) that the potential is fixed up to a gauge transformation $A_a \to A_a + f_{,a}$, where f is an arbitrary function of position, which can be used to impose the Lorentz gauge: $\nabla_a A^a = 0$;

(b) that when this gauge is imposed, there is still such a gauge freedom provided f satisfies the harmonic condition $f_{;ab}g^{ab} = 0$, and
(c) that then Maxwell's equations reduce to

$$\nabla^b \nabla_b A^a + R^{ab} A_b = J^a . \tag{5.141}$$

Exercise 5.5.5 Derive the 1+3 Maxwell and current conservation equations (5.118)–(5.122).

Exercise 5.5.6 In relativistic MHD:

(a) Show that there can be a non-zero magnetic field even if there is a non-zero charge density, provided that the fluid rotates, but that if the charge density is zero, either the vorticity must vanish or the magnetic field must be orthogonal to the vorticity vector.
(b) Derive the evolution equation (5.134), the MHD equation (5.135), and the decomposition (5.136).

5.6 Scalar fields

Quantum scalar fields play a central role in particle physics and string theory (see Section 20.3), and in inflationary cosmology. The quantum equations for the inflaton field are relevant for an analysis of its fluctuations, which we discuss in Section 12.2. For the background dynamics of inflation, as well as the classical evolution of its fluctuations, it is sufficient to consider a classical scalar field, which we briefly discuss here.

A minimally coupled scalar field φ has Lagrangian density

$$L_\varphi = -\sqrt{g}\left[\tfrac{1}{2}\nabla_a\varphi\nabla^a\varphi + V(\varphi)\right], \tag{5.142}$$

where $V(\varphi)$ is the potential that describes self-interaction of the scalar field. The energy–momentum tensor then has the form

$$T_\varphi^{ab} = \nabla^a\varphi\nabla^b\varphi - \left[\tfrac{1}{2}\nabla_c\varphi\nabla^c\varphi + V(\varphi)\right]g^{ab}, \tag{5.143}$$

and its conservation leads to the Klein–Gordon equation,

$$\nabla^a\nabla_a\varphi - V'(\varphi) = 0. \tag{5.144}$$

(When $\nabla_a\varphi = 0$, (5.143) reduces to $T_\varphi^{ab} = -V(\varphi)g^{ab}$, and $\nabla_b T_\varphi^{ab} = 0$ implies that $\nabla_a V(\varphi) = 0$; thus $V(\varphi)$ is an effective cosmological constant and φ is not a dynamical scalar field.)

In a covariant 1+3 description of scalar fields, one first needs to assign a 4-velocity to the φ-field. Provided that $\nabla_a\varphi$ is timelike, $\nabla_a\varphi$ is normal to the spacelike surfaces $\varphi(x^a) = \text{const}$, and the canonical 4-velocity is

$$u_a = -\frac{1}{\dot\varphi}\nabla_a\varphi, \tag{5.145}$$

where $\dot\varphi = u^a\nabla_a\varphi \neq 0$. This means that $\dot\varphi^2 = -\nabla_a\varphi\nabla^a\varphi > 0$ and $u_a u^a = -1$, as required.

It follows immediately from (5.145) that the flow is irrotational,

$$\omega_a = 0,\tag{5.146}$$

and

$$\overline{\nabla}_a\varphi = 0,\tag{5.147}$$

which is a key feature of the covariant analysis. Also (see Exercise 5.6.2),

$$\dot{u}_a = -\frac{1}{\dot{\varphi}}\overline{\nabla}_a\dot{\varphi}.\tag{5.148}$$

The energy–momentum tensor (5.143) has perfect-fluid form, with

$$\rho_\varphi = \tfrac{1}{2}\dot{\varphi}^2 + V(\varphi),\ \ p_\varphi = \tfrac{1}{2}\dot{\varphi}^2 - V(\varphi).\tag{5.149}$$

Scalar fields do not generally behave like barotropic fluids. This is underlined by the fact that the adiabatic sound speed is not the true or effective sound speed, namely the maximal speed of propagation of field fluctuations, as shown in Section 10.2.5. The effective sound speed is in fact the speed of light,

$$c_s^2 := \frac{\dot{p}_\varphi}{\dot{\rho}_\varphi} \neq c_{s\,\mathrm{eff}}^2 = 1.\tag{5.150}$$

The last equality follows from the fact that spatial fluctuations in pressure and density in the rest frame are equal (see Section 10.2.5), i.e.

$$\overline{\nabla}_a p_\varphi = \overline{\nabla}_a \rho_\varphi,\tag{5.151}$$

which is a consequence of (5.149).

The Klein–Gordon equation becomes

$$\ddot{\varphi} + \Theta\dot{\varphi} + V'(\varphi) = 0,\tag{5.152}$$

which is the energy conservation equation. The conservation of momentum is identically satisfied by virtue of (5.148).

Note that if $\dot{\varphi} = 0$, we have the exceptional equation of state $p_\varphi + \rho_\varphi = 0$; then the scalar field is constant, and acts as a cosmological constant or vacuum energy (5.52). Equation (5.152) then shows that the potential V must be flat, when evaluated at φ: $(\partial V/\partial\varphi)(\varphi) = 0$. This will be a good description of the behaviour when the scalar field is *potential dominated*: that is, when $\dot{\varphi}^2 \ll V(\varphi)$. In this case the energy inequality $\rho_\varphi + 3p_\varphi \geq 0$ will be violated leading to the possibility of an accelerating expansion of the universe. This is crucial to inflationary theory in the early universe (Section 9.4) and may possibly also play a role in the dynamics of dark energy in the late universe (Section 14.2). If $V = 0$, we have the exceptional equation of state $p = \rho$; then $\dot{\varphi} = \mathrm{const}/\ell^3$. This will be a good description of the behaviour when the scalar field is *kinetic dominated*: that is, when $\dot{\varphi}^2 \gg V(\varphi)$.

For a non-negative potential energy, $V \geq 0$, it follows from (5.149) that

$$-1 \leq w_\varphi := \frac{p_\varphi}{\rho_\varphi} \leq 1\ \ (V \geq 0).\tag{5.153}$$

If we allow $V < 0$, as in the 'ekpyrotic' model (Khoury *et al.*, 2001), then these bounds can be violated, and, in particular, one can achieve extreme kinetic domination, $w_\varphi \gg 1$. However, it is not clear that this is physically realistic.

Exercise 5.6.1 Show that the conservation equations (5.11) and (5.12) are identically satisfied as a result of (5.143) and (5.144); and conversely, that (5.144) is satisfied if the conservation equations are, provided that $\dot{\varphi} \neq 0$.

Exercise 5.6.2 Determine the kinematic quantities for the 4-velocity vector defined above.

5.7 Quantum field theory

In the very early universe, but after the time that quantum gravity dominated (see Section 20.2.1), there will probably be an era when the universe is dominated by quantum fields but gravity can be treated as a classical theory.

Quantum field theory (QFT) in curved spacetime, in which one continues to treat spacetime classically, had spectacular success in the derivation of Hawking radiation and the consequent understanding of the thermodynamics of black holes (Hawking, 1975), but also has difficulties (Wald, 1994). The usual discussions of QFT assume a well-defined initial vacuum state (distinct vacua give unitarily inequivalent theories), which curved spaces generally do not have. Those discussions describe states in terms of particles, but in curved space the particles observed depend on the observer's motion. Hence a reformulation of the usual theory is required, especially for the interacting field case, which is relevant for example to reheating after inflation (one such reformulation uses an algebraic approach attempting to formulate predictions in terms of probabilities: see e.g. Hollands and Wald (2005) and references therein). However, many discussions of QFT in cosmology are carried out using, in effect, flat space QFT. Another issue is the nature of the correspondence between quantum and classical theories; the usual inflationary scenarios make assumptions about this transition. Discussions of this issue can be found in, e.g., Brandenberger (1985), Sakagami (1988), Padmanabhan, Seshadri and Singh (1989) and Brandenberger, Laflamme and Mijic (1991). Padmanabhan (1993) even phrases the usual treatment as *defining* the classical fluctuations to be the same as the quantum ones.

The two most important applications of QFT in cosmology are to phase transitions in the early universe, and quantum fluctuations of the inflaton field. Since our theme is the role of relativity in cosmology, we give only a brief introduction to the ideas of phase transitions, and refer the reader to the books and papers cited and references therein. Quantum fluctuations during inflation are discussed in Section 12.2.

A phase transition occurs when a more ordered but less symmetric state becomes energetically favourable, for instance when a liquid with rotational symmetry cools and forms a solid with a (necessarily reduced) crystallographic symmetry. Typically the solution does not share the symmetry of the governing equations, and is only one of a set of possible solutions of the same energy which can be mapped into one another by the governing equations' symmetry. Different solutions may occur at neighbouring points, as for example when a ferromagnet cools below its Curie temperature and has different magnetization directions in neighbouring domains. A clear discussion of phase transitions in cosmology can be found in Coles and Lucchin (2003).

Consider first the simple case of a one-dimensional state space of a field Ψ with two energy minima which can be taken to be at $\pm\Psi_0$. There will be domains where $\Psi = \Psi_0$ and domains where $\Psi = -\Psi_0$. At the boundary between two different domains, in order for there to be a continuous value for the field concerned there must be small regions where $-\Psi_0 < \Psi < \Psi_0$, i.e. the field is not in a minimum energy state. These are called *domain walls*. Similarly for a field with a two-dimensional state space with a ring of minimum energy solutions surrounding the origin, a circle in physical space may take those same state space values but, for continuity, points inside the circle must then have states not of minimal energy: this gives *strings* (not the same as the superstrings of Section 20.2.1). Going up in dimension gives *monopoles* and *textures*. These are generically referred to as *topological defects*. A priori, any of them might occur in the early universe. The density of defects is usually estimated from a correlation length ξ, which depends on the particular phenomenon under discussion but which Kibble (1987) argued had to be less than the particle horizon. Use of these ideas may require some caution, since particle horizons (Section 7.9.1) are not disjoint (MacCallum, 1982), and there may be correlations on larger scales (Wald, 1993). However, the inconsistency between the high estimated number density of monopoles, which goes like ξ^{-3}, and observation was a motivation for the introduction of inflation, which reduces this density.

Symmetry breaking may be induced by an external magnetic field, or spontaneously, i.e. brought about by a gradual change of internal parameters of the matter concerned. In cosmology, changes due to reduction of the ambient temperature arise and are considered spontaneous phase transitions. When a phase transition happens, there may be an era of 'supercooling' in which the field stays in the no longer energetically favourable maximally symmetric state. This is referred to as a *false vacuum*. When the field finally moves towards minimal energy, the energy released produces reheating (which can also arise in other ways).

At high temperatures, $T > 10^{15}$ GeV, there is believed to be a Grand Unified Theory (GUT) with a symmetry group sufficiently large to include the known symmetries $SU(3) \times SU(2) \times U(1)$ of the strong, weak and electromagnetic forces. Then when $T \approx 10^{15}$ GeV, there is a phase transition in which the symmetry of the quantum field reduces to $SU(3) \times SU(2) \times U(1)$, the symmetry of the Standard Model of particle physics (see Section 9.6.3). The symmetry breaking can form (magnetic) monopoles.

The second expected transition, at 0.1–1 TeV, is the electroweak symmetry breaking which leads to lepton masses, followed at 200–300 MeV by the QCD phase transition at which quark confinement arises and which ushers in the hadron era. The extremely large gap between 10^{15} and 10^3 GeV is often referred to as the 'desert', as nothing is expected to happen there, in the normal model of elementary particles and fields.

6

Dynamics of cosmological models

In this chapter, we examine general dynamical relations that hold in any cosmological model, initially without restricting the equation of state.

We begin by making a 1+3 decomposition of the field equations (3.15), using the decomposition (5.9) of T_{ab}. We find they are equivalent to

$$R_{ab}u^a u^b = 4\pi G(\rho + 3p) - \Lambda, \tag{6.1}$$

$$R_{ab}u^a h^b{}_c = -8\pi G q_c, \tag{6.2}$$

$$R_{ab}h^a{}_c h^b{}_d = [4\pi G(\rho - p) + \Lambda]h_{cd} + 8\pi G \pi_{cd}. \tag{6.3}$$

We now link this dynamics with the kinematics of Chapter 4. Remember that we have to satisfy all 10 of these field equations, whereas in the Newtonian case we only have to satisfy one, the Poisson equation,

$$\Phi^{,i}{}_{,i} = 4\pi G \rho_N - \Lambda,$$

basically because equations that are dynamical field equations in the GR case reduce to geometrical identities in the Newtonian case (compare Section 3.4)

6.1 The Raychaudhuri–Ehlers equation

The most important field equation in terms of kinematic quantities, the *Raychaudhuri–Ehlers equation*, is obtained from substituting (4.46) into (6.1). It gives the evolution of Θ along the fluid flow lines (Raychaudhuri, 1955, Ehlers, 1961):

$$\dot{\Theta} + \tfrac{1}{3}\Theta^2 + 2(\sigma^2 - \omega^2) - \dot{u}^a{}_{;a} + 4\pi G(\rho + 3p) - \Lambda = 0. \tag{6.4}$$

This equation is the fundamental equation of gravitational attraction. To see its implications, we rewrite it in the form

$$3\ddot{\ell}/\ell = -2(\sigma^2 - \omega^2) + \overline{\nabla}_a \dot{u}^a + \dot{u}_a \dot{u}^a - 4\pi G(\rho + 3p) + \Lambda \tag{6.5}$$

which follows from the definition (4.35) of the scale factor ℓ. This equation for the curvature $\ddot{\ell}$ of the curve $\ell(\tau)$ directly shows that shear, energy density and pressure tend to make matter collapse, as they tend to make the $\ell(\tau)$ curve bend down, while vorticity and a positive cosmological constant tend to make matter expand, as they tend to make the $\ell(\tau)$ curve bend up; the acceleration terms are of indefinite sign. It also shows that $(\rho + 3p) = \rho_N(1 + \epsilon + 3p/\rho_N)$ is the active gravitational mass density of a fluid obeying the description in Section 5.2; hence any increase in the internal energy or the pressure increases its active gravitational mass density.

For example, in the case of a static star, $\Theta = \omega = \sigma = 0$ and we can neglect the cosmological constant on this scale, so the equation reduces to $\dot{u}^a{}_{;a} = 4\pi G(\rho + 3p)$, where the acceleration is determined from the pressure gradient by (5.39). We obtain $(\overline{\nabla}_a p)/(\rho + p))_{;a} = -4\pi G(\rho + 3p)$, the basic balance equation between gravitational attraction and hydrostatic pressure for a star. In the corresponding Newtonian equations, $\rho + 3p \to \rho_N$ and $\rho + p \to \rho_N$; due to these differences, gravitational collapse is much less severe in Newtonian theory than in GR.

6.1.1 Static FLRW universe models

In the case of a static FLRW universe model, $\Theta = \omega = \sigma = 0 = \dot{u}^a$, so (6.5) becomes

$$4\pi G(\rho + 3p) = \Lambda. \tag{6.6}$$

Thus *static universes with ordinary matter are only possible if* $\Lambda > 0$ (Einstein, 1917). Then, given the equation of state $p = p(\rho)$ and the cosmological constant, there is a unique radius a_s for this Einstein static solution, at which the gravitational attraction caused by the matter and the repulsion caused by the cosmological constant balance.

This universe is unstable (Eddington, 1930) because if we increase a, so $a > a_s$, ρ decreases but Λ stays constant, and hence $\ddot{a} > 0$ and the universe expands to infinity; while similarly $a < a_s \Rightarrow \ddot{a} < 0$ and the universe collapses. This instability[1] leads us to believe that *the universe should either be expanding or contracting, but not static*; indeed, failure to perceive this in the 1920s can be regarded as one of the major lost opportunities in the history of cosmology (all the major figures in cosmology at that time 'knew' the universe was static, see Ellis (1989)).

The corresponding Newtonian model satisfies $4\pi G\rho_N = \Lambda$; the same qualitative results hold as in GR (i.e. $\Lambda > 0$ and the model is unstable).

6.1.2 The first singularity theorem

The fundamental singularity theorem follows immediately from the Raychaudhuri equation (Tolman and Ward, 1932, Raychaudhuri, 1955).

Theorem 6.1 (Irrotational geodesic singularities) *If* $\Lambda \leq 0$, $\rho + 3p \geq 0$ *and* $\rho + p > 0$ *in a fluid flow for which* $\dot{u} = 0$, $\omega = 0$ *and* $H_0 > 0$ *at some time* s_0, *then a spacetime singularity, where either* $\ell(\tau) \to 0$ *or* $\sigma \to \infty$, *occurs at a finite proper time* $\tau_0 \leq 1/H_0$ *before* s_0.

Proof: The proof is simple: if $\ddot{\ell} = 0$, then $\ell \to 0$ a time $t_H = 1/H_0$ ago; however, with the given conditions, $\ddot{\ell} < 0$, so following the curve $\ell(\tau)$ back in the past, it must drop below the straight line $\ell = H_0\ell_0(t - t_H)$ and reach arbitrarily small positive values of ℓ at a time less

---

[1] As one cannot perturb the actual universe, 'instability' here refers to the difficulty of setting exactly the correct initial conditions for a static universe rather than its instability to perturbation from a previous state.

than $1/H_0$ ago (unless some other spacetime singularity intervenes before $\ell \to 0$, which can happen only if the shear diverges first).

In the exceptional case where the shear diverges first, a conformal spacetime singularity (see Section 6.7) will occur where $\ell \neq 0$: because of (6.5), that can occur only in very exceptional circumstances, and we know of no physically relevant example where it happens. In the general case where $\ell \to 0$, the matter world lines converge at a finite time in the past and a spacetime singularity develops if $\rho + p > 0$, for then as the universe contracts, the density and pressure increase indefinitely, implying the spacetime curvature also does. For ordinary matter this will additionally imply that $T \to \infty$, that is, the universe originates at a hot big bang. Furthermore, an age problem becomes possible; for if we observe structures in the universe, such as stars, globular clusters, or galaxies, that are older than $1/H_0$, there is a contradiction with the assumptions of the theorem (for the universe must be older than its contents!) $\qquad\qquad\qquad\qquad\qquad\qquad\qquad\qquad\qquad\qquad\qquad\qquad\qquad\qquad\qquad\qquad$ $\square$

Similarly the argument implies that the universe must experience a very rapid evolution through its hot early phase. The straight line estimate $\ddot{\ell} = 0 \Rightarrow H = \ell_0 H_0/\ell$; however, the high densities will cause a considerable steepening of the $\ell(\tau)$ curve at early times, leading to the inequality $H > \ell_0 H_0/\ell$. For example at the time of decoupling the scale function ℓ_d obeys $\ell_d/\ell_0 \approx 1/1000$, which implies a Hubble parameter $H_d > 1000 H_0$. Similarly at the time of nucleosynthesis $\ell/\ell_0 \approx 10^{-8}$, showing that then $H > 10^8 H_0$.

Application: This result applies in particular to an expanding FLRW universe, where (using (2.65)) $a = \ell \to 0$ and a hot big bang must occur (the shear is zero in this case, so a conformal singularity cannot occur). The proof makes clear that *an increase in pressure does not resist the occurrence of the singularity*, but rather decreases the age of the universe and so makes the age problem worse (the pressure increases the active gravitational mass and there are no pressure gradients to resist the collapse).

This is the basic singularity theorem, on which further elaborations are built. How can one avoid the singularity? It is clear that shear anisotropy makes the situation worse. On the face of it, there are five possible routes to avoid the conclusion: a positive cosmological constant; acceleration; vorticity; an energy condition violation; or alternative gravitational equations. We consider them in turn.

(1) *Cosmological constant* $\Lambda > 0$. In principle this could dominate the matter at small ℓ and turn the universe around. However, in practice this cannot happen because we have seen galaxies and quasars up to a redshift over 6, implying (see Chapter 13) that the universe has expanded by at least a ratio of 7 before now. If it had bounced, then at the turnaround the density would have been greater than the present density by a factor of at least $7^3 = 343$; so the cosmological constant would have to be equivalent to a large energy density in order to dominate the Raychaudhuri equation then. This is well outside the values consistent with observation (see Section 13.2). If we accept that the microwave background radiation indicates that the universe has expanded by at least a factor of 1000, the argument is even more overwhelming; the cosmological constant would have to be equivalent to more than 10^9 times the present matter density to dominate the

Raychaudhuri equation then! (Observationally acceptable alternatives to the cosmological constant as models for dark energy, such as 'quintessence' (Section 14.2), do not upset this general argument.)

(2) *Pressure inhomogeneity* (acceleration) and

(3) *Rotational anisotropy* (the effect of 'centrifugal force').

These both involve abandoning the FLRW geometry. On the face of it, they could succeed; however the powerful Hawking–Penrose singularity theorems strongly restrict the allowable cases where they might in fact succeed, because of the CMB observations which show, for universes that are approximately FLRW, that the conditions of those more general theorems hold (see Section 6.7 and Hawking and Ellis (1973)).

(4) *Violation of the energy condition* $\rho + 3p > 0$. The energy condition (5.17) is obeyed by all normal matter, but a false vacuum (5.52) can violate it and so in principle cause a turnaround of the universe, avoiding an initial singularity. However, we only expect (5.52) to become relevant above temperatures of at least 10^{12} K. Thus even if violating the energy condition could enable us to avoid the initial singularity, the turnaround would only take place under extraordinarily extreme conditions when quantum effects are expected to be dominant. Hence we can rephrase the conclusion: a viable non-singular universe model cannot obey the laws of classical physics at all times in the past.

(5) *Other gravitational field equations*. Finally, we have of course assumed Einstein's field equations here. An alternative theory of gravity might hold that avoids the singularity, as for example in the Steady State universe and its variants (Bondi, 1960, Hoyle, Burbidge and Narlikar, 1993, 1994) effectively by introducing negative energy terms into the Raychaudhuri equation. In particular at very early times quantum gravitational effects are expected to become important, possibly causing effective energy condition violations.

Thus the prediction of a singularity is a classical prediction; physically, we may assume that as we follow the evolution back into the past, the universe cannot avoid entering the quantum gravity domain. We do not yet have any reliable idea of what this implies (see Chapter 20).

In Newtonian theory, the discussion is as above except for one important point: in this case rotation *can* enable the universe to avoid the initial singularity (unlike the GR case). This is shown by the existence of spatially homogeneous rotating and expanding but shearfree Newtonian universe models, in which the rotation spins up to enable the universe to avoid the initial singularity (see e.g. Heckmann and Schucking (1955)), whereas such universes cannot exist in relativity theory (Gödel, 1952, Ellis, 1967); see Section 6.2.2 below.

6.1.3 Evaluation today

We obtain very useful information by evaluating the Raychaudhuri equation at the present time. To express this, we define some parameters as follows. The *deceleration parameter* is

$$q_0 = -\left(\frac{\ddot{\ell}}{\ell}\right)_0 \frac{1}{H_0{}^2} \tag{6.7}$$

(not to be confused with energy flux q^a), which is a dimensionless version of the second derivative $\ddot{\ell}$, with the sign chosen so that a positive value corresponds to deceleration. The energy density of matter, pressure and the cosmological constant are represented by the dimensionless parameters

$$\Omega_0 = \frac{8\pi G\rho_0}{3H_0{}^2}, \quad \Omega_{p0} = \frac{8\pi Gp_0}{3H_0{}^2}, \quad \Omega_{\Lambda 0} = \frac{2\Lambda_0}{3H_0{}^2}. \tag{6.8}$$

It then follows directly from the Raychaudhuri equation that

$$2q_0 = \frac{4}{3}\left(\frac{\sigma_0^2}{H_0^2} - \frac{\omega_0^2}{H_0^2}\right) - \frac{2(\overline{\nabla}_a \dot{u}^a + \dot{u}_a \dot{u}^a)}{3H_0^2} + \Omega_0 + 3\Omega_{p0} - \Omega_{\Lambda 0}. \tag{6.9}$$

One may define the total density parameter $\Omega_{\text{tot}} = \Omega + 3\Omega_p - \Omega_\Lambda$. If the rotation, shear, acceleration and pressure terms are small today compared with the others, as is highly plausible, then

$$2q_0 \simeq \Omega_0 - \Omega_{\Lambda 0}, \tag{6.10}$$

where the error is of the magnitude of the terms neglected in passing from the previous equation; and if we also assume $\Lambda = 0$, then $2q_0 \simeq \Omega_0$. These become exact equations in an FLRW universe with vanishing pressure.

These direct relations between the deceleration and density parameters are pivotal in observational cosmology. The relation (6.10) can be used to determine Λ from observations of q_0 and Ω_0. Only recently have such observations achieved sufficient accuracy to give strong limits. The data outlined in Section 13.2 provide estimates $\Omega_0 \approx 0.27$ and $\Omega_{\Lambda 0} \approx 0.73$, supporting the arguments for the presence of dark matter and dark energy.

The Newtonian discussion is the same, except that there is no pressure contribution to (6.9).

6.1.4 First integrals

In the FLRW case, the Raychaudhuri equation (6.5) for $\ell = a$ becomes

$$3\ddot{a}/a = -4\pi G(\rho + 3p) + \Lambda. \tag{6.11}$$

Now the conservation equation (5.38) implies $(a^2\rho)\dot{} = -a\dot{a}(\rho + 3p)$. Thus provided $\dot{a} \neq 0$ we can multiply (6.11) by $a\dot{a}$ and integrate to find

$$3\dot{a}^2 - 8\pi G\rho a^2 - \Lambda a^2 = \text{const}, \tag{6.12}$$

which is just the *Friedmann equation* which governs the time evolution of FLRW universe models (discussed further in Chapter 9). The constant is the curvature of the three-spaces $t = \text{constant}$ (see (6.23) and (6.55) below). If the shear or vorticity are non-zero and we know, from some geometric constraints, σ^2 or ω^2 as a function of ℓ, we could integrate (6.5) similarly to obtain a generalized Friedmann equation for these more general universe models. Equation (6.12) also holds for Newtonian RW cosmological models provided one replaces the total energy density ρ by its Newtonian limit, the mass density ρ_N.

6.2 Vorticity conservation

Before looking at the rest of the field equations, we examine the evolution of vorticity, since its presence or absence radically affects the treatment of the remaining equations.

6.2.1 Vorticity propagation

From (4.43) we can obtain the form

$$(\ell^2 \omega)^{\cdot \langle e \rangle} = \ell^2 \sigma^e{}_d \omega^d - \tfrac{1}{2}\ell^2 \operatorname{curl} \dot{u}^e \qquad (6.13)$$

for the *vorticity propagation equation* which is subject to the constraint (4.44); here the acceleration is determined by the momentum conservation equation (5.12). If the flow is that of a barotropic perfect fluid, there is an acceleration potential r (given by the second of (5.41)), and (6.13) becomes

$$(r\ell^2 \omega)^{\cdot \langle e \rangle} = r\ell^2 \sigma^e{}_d \omega^d, \qquad (6.14)$$

which is the basis of the usual (Kelvin–Helmholtz) vorticity conservation laws. Firstly, it implies the permanence of vorticity:

A: For a perfect fluid with barotropic equation of state, $\omega \neq 0$ at one point on a world line $\Rightarrow \omega \neq 0$ at every point on that world line.

Thus vorticity can be generated only with a non-barotropic or imperfect fluid, by irreversible processes such as viscosity (through their effect on the acceleration). To understand the implications further, we have to substitute back for $\sigma^f{}_d$ in terms of $u^f{}_{;d}$, finding

$$(r\ell^3 \omega)^{\cdot \langle e \rangle} = u^e{}_{;d}(r\ell^3 \omega^d), \qquad (6.15)$$

which shows that $X^e = (r\ell^3 \omega^e)$ is a relative position vector (it satisfies the equation $h^f{}_e X^e{}_{;d} u^d = u^e{}_{;d} X^d$). In the case of a linear equation of state (5.49), by Exercise 5.2.4, this becomes

$$(\ell^{-3(w-1)} \omega)^{\cdot \langle e \rangle} = u^e{}_{;d}(\omega^d \ell^{-3(w-1)}), \qquad (6.16)$$

showing that $X^e = \ell^{-3(w-1)} \omega^e$ is a relative position vector (always pointing to the same neighbouring particle). Thus,

B1: Vortex lines consist at all times of the same particles, that is, the vorticity is frozen into the fluid flow and

B2: The distance of neighbouring particles in the vorticity direction is proportional to $\omega \ell^{3(1-w)}$.

Now the volume of a section of a vortex tube of length δl and cross-sectional area δF is $\delta V = \delta l \, \delta F \propto \ell^3$, so we can rewrite this in the form

$$\omega \propto \delta l^w \delta F^{w-1}, \quad \omega^2 / \rho \propto \delta l^{3w+1} \delta F^{3w-1}, \qquad (6.17)$$

		General case		Almost isotropic	
Table 6.1 Scaling factors for vorticity, using a linear equation of state					
Matter	w	$\omega \propto \cdots$	$\omega^2/\rho \propto \cdots$	$\omega \propto \cdots$	$\omega^2/\rho \propto \cdots$
dust	0	δF^{-1}	$\delta l\, \delta F^{-1}$	ℓ^{-2}	ℓ^{-1}
radiation	1/3	$\delta l^{1/3}\delta F^{-2/3}$	δl^2	ℓ^{-1}	ℓ^2
stiff	1	δl	$\delta l^4\, \delta F^2$	ℓ	ℓ^8

which is valid for a general expansion. The term ω^2/ρ measures the importance of rotation in the Raychaudhuri equation. In the case of an almost isotropic expansion, $\delta F \propto \ell^2$, so in that case the previous equation reduces to

$$\omega \propto \ell^{3w-2}, \quad \omega^2/\rho \propto \ell^{9w-1}. \tag{6.18}$$

Thus we obtain Table 6.1. This shows that in an almost isotropic expansion, vorticity 'spins up' as we go back in the past for all ordinary matter: the dynamical importance of vorticity increases as a radiation fluid expands, but decreases for dust. However, we must beware of taking this as the general behaviour; if there is significant distortion, the behaviour is given by the middle columns in the table, not the last ones. The relation between shear and vorticity contained in the above equations is not simple; we must, for example, clearly distinguish between the rotation of the fluid and of the shear eigenvectors. The simplest behaviour will be if the vorticity vector is at all times a shear eigenvector.

6.2.2 Warning

The simplest conceivable rotating and expanding case is a shearfree expansion of dust: $\sigma = 0 = \dot{u}^a$, $\Theta \neq 0$, $\omega \neq 0$. In this case, $\omega = \Omega/\ell^2$, $\dot{\Omega} = 0$; then we can integrate (6.5) to get a generalized Friedmann equation,

$$3\dot{\ell}^2 - 8\pi G\rho\ell^2 - \Lambda\ell^2 + 2\Omega^2/\ell^2 = \text{const}, \tag{6.19}$$

which suggests we can have a solution of Einstein's equations in which a build-up of rotation does indeed stop the initial singularity of the universe, centrifugal force causing a bounce at early times.

However, there is no such relativistic expanding solution! The problem is that we have so far only used one of the ten Einstein equations. Until we have examined all 10 field equations, we cannot claim to have a solution. In this case, the other equations show no such solution can exist (Ellis, 1967, Senovilla, Sopuerta and Szekeres, 1998). In the Newtonian case, on the other hand, such solutions do exist (Heckmann and Schucking, 1955).

The Newtonian discussion of vorticity conservation is similar but simpler: the results are identical to the 'dust' case discussed above.

Exercise 6.2.1 (a) Show that vorticity *can* be generated in the case of matter with a 'perfect fluid' stress tensor, provided that $\eta^{abc}\overline{\nabla}_b p\,\overline{\nabla}_c\rho \neq 0$ (in the barotropic situation above, this quantity vanishes).
(b) Indicate how viscosity can generate vorticity.

Exercise 6.2.2 Examine vorticity conservation in the case of a fluid obeying the 'cosmological constant' equation of state $p = -\rho$. (N.B. the momentum conservation equation is degenerate in this case.)

Exercise 6.2.3 Magnetic field conservation and evolution. In the case of observers who measure a pure magnetic field, the Maxwell equations are (5.132)–(5.133) while (5.131) implies

$$(\ell^2 B)^{\cdot\langle c\rangle} = (\omega^c{}_b + \sigma^c{}_b)\ell^2 B^b,$$

in close analogy with the equations governing vorticity. Show from these equations that $(\ell^3 B^a)$ is a relative position vector, so that the magnetic field conservation laws (similar to the vorticity conservation above) are,

A: If $E^a = 0$, magnetic field lines cannot be created or destroyed.

B: The magnetic field lines are frozen into the fluid, i.e. the integral curves of B^a consist at all times of the same particles.

C: As the fluid evolves, the field strength B changes inversely with the cross-sectional area δF of the magnetic field tubes: $B \propto 1/\delta F$.

(Note that μ can be non-zero, even when there is no electric field, if $\omega^a \neq 0$.)

6.3 The other Einstein field equations

As shown by the result quoted in Section 6.2.2, we must consider all 10 Einstein field equations, or an equivalent system of equations. We have so far considered only one: the '(00)' equation. For brevity we describe the remaining equations using the numbering that would follow from use of a tetrad with $\boldsymbol{u} = \boldsymbol{e}_0$.

The $(0i)$ equations

Substituting from (6.2) into (4.48) we get

$$0 = 8\pi G q_{\langle a\rangle} - \tfrac{2}{3}\overline{\nabla}_a \Theta + \overline{\nabla}^b \sigma_{ab} - \text{curl}\, \omega_{\langle a\rangle} + 2\eta_{abc}\omega^b \dot{u}^c . \tag{6.20}$$

These are a further three of the field equations.

In the Newtonian case, these equations are the identities

$$\omega^{ij}{}_{,j} - \sigma^{ij}{}_{,j} + \tfrac{2}{3}\Theta^{,i} = 0. \tag{6.21}$$

In the FLRW case these equations are identically satisfied because the symmetry implies $q^a = 0$, $\Theta = \Theta(t)$.

The (ij) equations

For the remaining six equations the situation is radically different for $\omega = 0$ and $\omega \neq 0$. We shall deal with $\omega = 0$ first.

6.3.1 The vorticity-free case

The Gauss equation

When $\omega = 0$, we can examine the geometry of the uniquely defined family of surfaces orthogonal to u^a, embedded in the four-dimensional spacetime. Their metric (the first fundamental form, compare Section 2.1) is h_{ab}, while their second fundamental form is $\Theta_{ab} = \overline{\nabla}_{(a} u_{b)}$. $\overline{\nabla}$ is now the covariant derivative in these surfaces. This follows because it (i) is linear, (ii) obeys the Liebniz rule, (iii) commutes with contraction (by h^{ac}), (iv) has zero torsion (being the projection of a torsion-free connection) and (v) preserves the three-dimensional metric: $\overline{\nabla}_c h_{ab} = h_c{}^g h_a{}^e h_b{}^f h_{ef;g} = 0$, each of these properties following because the corresponding property holds for the four-dimensional covariant derivative.

Thus, contracting the Gauss equation (2.114) and using the propagation equation (4.42) and the Raychaudhuri equation (6.5), we find

$$^3R_{ab} = \overline{\nabla}_{(a} \dot{u}_{b)} - \ell^{-3} (\ell^3 \sigma)^{\cdot}{}_{\langle ab \rangle} + \dot{u}_{\langle a} \dot{u}_{b\rangle} + 8\pi G \pi_{ab} \tag{6.22}$$
$$+ \tfrac{2}{3} (\sigma^2 - \tfrac{1}{3}\Theta^2 + \Lambda + 8\pi G \rho) h_{ab} \,,$$

showing how the matter tensor (ρ, π_{ab}) directly affects the Ricci curvature of the three-dimensional space, with correction terms due to the embedding. This gives us the last six field equations when $\omega = 0$ (if we know $^3R_{ab}$, this is an equation for the rate of change of shear along the flow lines).

If we contract again (or contract (2.114) twice) we obtain

$$^3R = 2(\sigma^2 - \tfrac{1}{3}\Theta^2 + \Lambda + 8\pi G \rho) \,, \tag{6.23}$$

which is a generalized Friedmann equation (or the 'Hamiltonian constraint' of Section 3.3.3). This gives the 3-space Ricci scalar in terms of the matter energy density and cosmological constant, corrected by embedding terms; it is a generalization of the first integral equation (6.12) we obtained previously. We can give an equation for the time derivative of ^3R that follows from (6.23), (6.5) and (5.11):

$$(^3R - 2\sigma^2)^{\cdot} = \tfrac{2}{3}\Theta(6\sigma^2 - {}^3R - 2\dot{u}^a{}_{;a}) - 16\pi G(\pi^{ab}\sigma_{ab} + q^a{}_{;a} + 2q^a \dot{u}_a), \tag{6.24}$$

where the second bracket on the right vanishes for a 'perfect fluid'.

In the case of a three-dimensional space, the full Riemann tensor is determined by the Ricci tensor and the Ricci scalar as in (2.82). Thus, with the previous equations, $^3R_{abcd}$ is completely characterized in terms of the expansion, shear, acceleration, energy density and anisotropic pressure. Thus (Ehlers, 1961) this fully expresses Einstein's intention of having the curvature of *space* determined by the matter content of spacetime.

In the case of irrotational flows, we now have the full set of 10 field equations: the Raychaudhuri equation (6.5), the $(0j)$ equations (6.20) and the (ij) equations (6.22), which can conveniently be split into their trace (6.23) and their tracefree part

$$^3R_{\langle ab \rangle} = \overline{\nabla}_{\langle a} \dot{u}_{b\rangle} - \ell^{-3} (\ell^3 \sigma)^{\cdot}{}_{\langle ab \rangle} + \dot{u}_{\langle a} \dot{u}_{b\rangle} + 8\pi G \pi_{ab} \,. \tag{6.25}$$

However, it should be noted that we have not introduced any covariant quantities from which an expression for $^3R_{ab}$ can be calculated, so at present the $^3R_{ab}$ are extra variables lacking an independently derived evolution equation. We thus need to introduce extra variables for the spatial components of the connection that will give a complete set of equations. This can be accomplished in a convenient way by using the tetrad equations described in Section 2.8. Nevertheless, even without this completed set of variables and equations, we can attain many useful understandings and results from the covariant equations, as in Exercises 6.3.1 and 6.3.2.

Non-rotating fluids: comoving coordinates

In the case of non-rotating fluid, there is an obvious choice of comoving coordinates, with the time t given by (4.36), leading to constant time surfaces orthogonal to the fluid flow. Then (with this choice $a_i = 0$ in Exercise 5.2.3) the metric has the form

$$ds^2 = h_{ij} \, dx^i dx^j - \frac{1}{r^2(x^\mu)} \, dt^2, \ u_\nu = -\frac{1}{r(x^\mu)} \, \delta_\nu^0. \tag{6.26}$$

The Gauss relation determines the geometry of the surfaces $t =$ const. In terms of the approach above, we obtain this form by noting that $\omega = 0 \Rightarrow a_{[i,j]} = 0$, so there is a function $f(x^j)$ such that $a_i = f_{,i}$. Making a coordinate transformation $x^{0'} = x^0 + f(x^j)$ with this choice of f we obtain (6.26) from (5.43).

In general, the time term is *proportional* to an exact differential, with proportionality factor $r(x^\mu)$ (and Exercise 4.6.2 characterizes the spatial variation of $dt/d\tau$); the time term *is* an exact differential if $\dot{u}^a = 0 \Leftrightarrow r = r(t)$. In the latter case (5.41) no longer determines r, and we can renormalize the time coordinate ($t \to t' = t'(t)$) to set $r = 1$; we then have normalized comoving coordinates (compare Section 4.1).

6.3.2 Rotating fluids

When $\omega \neq 0$ we have more difficulty in giving a covariant form for the remaining field equations, especially one for which we can pose a satisfactory initial value problem. There is no set of hypersurfaces orthogonal to the flow. If we attempt to use Cauchy surfaces (Section 3.3) not orthogonal to the flow we may find that such surfaces do not exist globally, and there may be closed timelike lines giving a causality problem (see e.g. the discussion of the Gödel solution in Hawking and Ellis (1973)).

There are two ways of proceeding. The first is to choose a family of surfaces $t =$ const and then to decompose the field equations along the normals to these surfaces as in the case of a non-rotating fluid. However, as these cannot be the fluid flow lines, a perfect fluid will appear to be an imperfect fluid in this frame, and, for example, the conservation equations will now be more complicated. The other approach is to use the decomposition orthogonal to the fluid flow, discussed in the rest of this chapter. In both cases, completion of the equations will require introduction of extra variables such as the Weyl tensor and/or a tetrad and associated rotation coefficients.

Correspondingly, two choices of coordinate systems are available. One is to use coordinates that, instead of being comoving with the fluid, are based on a chosen set of hypersurfaces $t = $ const and are comoving with the normals to those hypersurfaces. The other possibility is to choose comoving coordinates; the time surfaces then cannot be chosen orthogonal to the fluid flow lines, $g_{0j} \neq 0$, and the Gauss formalism above is not directly applicable.

It is here that a big difference between Newtonian theory and Einstein's theory is apparent: in Newtonian theory there are always unique spatial sections and their metric is flat. No causality problems arise because of the unique time coordinate that orders events along all timelike curves.

Non-comoving description

As described above, a hypersurface-orthogonal description of rotating universes is possible if we abandon coordinates that are comoving with the fluid; this description can also be used when $\omega = 0$ if there is some special reason to do so (e.g. if there is a surface of symmetry that is not orthogonal to the fluid flow lines). It may also be best in a multi-fluid situation, where we appropriately choose some frame (in general, not comoving with any of the fluids) and express the energy–momentum tensors and field equations relative to this choice. Associated with this, one may use an orthogonal tetrad formalism with (e.g.) the normal n^a to some chosen hypersurfaces, rather than the fluid flow vector u^a, as the timelike tetrad vector. Then the tilt angle (or relative velocity) between the u^a and n^a is an important variable. This approach is used, for example, for the tilted Bianchi universes in Chapter 18.

To set this up, first choose the time surfaces. The 1+3 decomposition used previously applies in this case, but based on n^a rather than u^a, so n^a replaces u^a in the equations above. In the following, we use a tilde to denote that the decomposition is relative to n^a rather than u^a and a prime to denote differentiation along the normal lines, replacing the dot denoting covariant differentiation along the fluid flow lines. For example, (4.10) becomes $\tilde{h}_{ab} = g_{ab} + n_a n_b$ and (4.38) becomes $n_{a;b} = \tilde{\sigma}_{ab} + \frac{1}{3}\tilde{\Theta}\tilde{h}_{ab} - \tilde{n}'_a \tilde{n}_b$. We can then use the Gauss formalism as above, relative to the 4-velocity vector n^a. However, the fluid flow lines will then not be orthogonal to these surfaces, i.e. the fluid will move relative to them $(u^a \neq n^a \Leftrightarrow \tilde{u}_j \neq 0)$. Consequently a perfect fluid will appear to be imperfect in this frame; if we decompose relative to n^a rather than u^a, its effective equations of state will be given by (5.20)–(5.22). The fluid conservation equation will not appear simple in this frame; we require the general form (5.11)–(5.12) rather than the simple form (5.38)–(5.39), and cannot use the simple integrations (5.41)–(5.42) (as the expansion $\tilde{\Theta}$ of the normals is not related in a simple way to the expansion Θ of the fluid).

With this choice, the coordinates take the form (6.26) but r is not now the fluid acceleration potential (for the fluid is not at rest in these coordinates). The equations above hold along the normals (which necessarily have zero vorticity: $\tilde{\omega} = 0$) with $\tilde{\rho}, \tilde{q}_a, \tilde{p}$ and $\tilde{\pi}_{ab}$ respectively the total energy density, energy flux, isotropic pressure and shearfree anisotropic pressure from all constituents of the matter. There are various choices of time t to simplify the equations further:

(1) *Use of scalar quantities*. A time can be determined by some well-defined physical or geometrical scalar, e.g. density $\Leftrightarrow \tilde{\rho} = \tilde{\rho}(t)$; alternatively, we can choose time so that the pressure $\tilde{p}$, temperature $\tilde{T}$, or scalar field Φ are constant on these surfaces. A special case is when we choose the surfaces to be the surfaces of symmetry in Bianchi models, which are necessarily surfaces of constant energy density and pressure.

(2) *Constant normal expansion*. The time function is chosen so that $\tilde{\Theta} = \tilde{\Theta}(t)$. This is 'York time' (Smarr and York, 1978, Eardley and Smarr, 1979).

(3) *Zero shear surfaces*. The time function is chosen so that $\tilde{\sigma}_{ab} = 0$ where possible (but this implies a restricted set of spacetimes, since zero-shear surfaces are only possible for very restricted Weyl tensors (van Elst and Ellis, 1998)) or by using minimal shear surfaces (compare Bardeen (1980)).

(4) *Geodesically parallel surfaces*. The time function is chosen so that $n^{|a} = 0$ (acceleration terms vanish). (Globally problems will then occur, by the Raychaudhuri singularity theorem applied to the normals. Thus such coordinates will often be singular even if spacetime is not.) We still have freedom to specify the initial surface in this case.

(5) *Proper time*. The time coordinate is chosen to measure proper time along the fluid flow lines, from some arbitrarily chosen initial surface.

In general one cannot satisfy more than one of these conditions; in special cases, a choice will imply one or more of the others. Which choice is best will depend on the geometrical situation we are investigating.

Comoving description

In a comoving approach, we cannot use as variables the curvature ${}^3R_{ab}$ of 3-spaces orthogonal to the fluid flow (as there are no such 3-spaces). The basic strategy is to introduce extra variables in addition to the variables discussed previously, to close the equations. Apart from simply introducing coordinates in the traditional way and developing equations for the metric tensor components, there are various possibilities (not mutually exclusive):

(a) to introduce a covariant quantity ${}^3\tilde{R}_{ab}$ analogous to ${}^3R_{ab}$ and which reduces to ${}^3R_{ab}$ when $\omega = 0$, but does not have such a natural interpretation (compare Section 6.5);

(b) to use covariant Weyl tensor components and the Bianchi identities as developed in Section 6.4, possibly with extra spatial vectors defined covariantly from the problem: an example of this approach is given by Sopuerta (1998);

(c) to introduce a complete set of variables for the connection components, by introducing an orthonormal tetrad (see Section 6.5), with the timelike vector chosen to be the fluid flow vector, and its rotation coefficients (or equivalently commutator coefficients) then regarded as primary variables.

To complete the description one will also usually introduce specific coordinates adapted to the fluid flow (see below). In the end this approach will generally be required to complete both approaches (a) and (b) above, for they usually give many but not all of the equations required for completion of a solution. An example of this approach is Ellis (1967).

When $\omega \neq 0$ and comoving fluid coordinates are used as above, we cannot set $a_i = 0$ in (5.43) by a gauge transformation. Instead we can choose the $t = x^0$ origin, $f(x^i)$, so that at each point in some hypersurface $t = \text{const}$, the vector ω^a lies in that surface, i.e. $\omega^0 = \frac{1}{2}\eta^{0ijk}a_i a_{j,k} = 0$; because a_i is independent of t, this will then be true at all times. With this choice, $\epsilon^{ijk}a_i a_{j,k} = 0$, showing that there exist functions $y(x^i)$, $z(x^i)$ such that $a_i = yz_{,i}$; these functions must be independent, as otherwise $\omega_{ij} = 0$. Using the freedom of initial labelling to choose $x^3 = z$ and $x^2 = x^2(y,z)$ where $\partial x^2/\partial y \neq 0$, we have

$$a_i(x^j) = y(x^2,x^3)\delta_i^3, \quad \partial y/\partial x^2 \neq 0. \tag{6.27}$$

Thus we can always obtain coordinates (5.43)–(5.48) with a_i given by (6.27). With this choice the vorticity relation (4.44) and propagation equations (6.14) are both identically satisfied. It is possible to additionally choose $x^2 = y \Rightarrow a_i dx^i = x^2 dx^3$, but this may be too restrictive for some applications.

Two additional points about the approaches just discussed are worth making. When the fluid rotates, so there is no family of hypersurfaces orthogonal to the fluid flow, a further approach may be better than any of the above, namely:

(d) use non-orthonormal basis vectors, with the fluid flow as the timelike vector and the spacelike vectors lying in some uniquely chosen spacelike hypersurfaces. Then the associated coordinates can be chosen to be comoving, but the angles between the basis vectors will not be constant; thus the dynamical variables will include some metric tensor components and their derivatives (which are all constant when an orthonormal tetrad is used) as well as commutator components.

Only experience can tell which is best in any particular case; suitable coordinate choices can be introduced as appropriate to a specific problem and the tetrad choice can be tied in to these coordinates.

Finally in each of the cases (b)–(d), one has, by choice of variables, cast the geometric and dynamical equations into the form of a (generically infinite-dimensional) *dynamical system* (see e.g. Bogoyavlenskii (1985) and Wainwright and Ellis (1997)), possibly subject to constraints. It will be finite-dimensional if we are examining families of high-symmetry spacetimes, which define *invariant sets* or *involutive subsets* within the full infinite-dimensional space of cosmological models (i.e. are such that given initial data in the subset, the evolution remains within that subset). We shall discuss these aspects further in Chapters 18 and 19.

We shall look at specific examples of the use of the equations for either rotating or non-rotating cases in the sequel. Now we turn to general properties of the equations that are true for both rotating and non-rotating universes.

Exercise 6.3.1 Consider 'dust', (5.50), with $\omega = 0$, $^3R_{\langle ab\rangle} = 0$.

(a) Deduce from (6.25) that $\sigma_{ab} = \Sigma_{ab}/\ell^3$, $\dot{\Sigma}_{ab} = 0$. Now find a first integral of the Raychaudhuri equation that generalizes (6.12). What is the relation of this equation to (6.23)?

(b) Show that (6.24) again leads to the relation $^3R = {}^3R_0/\ell^2$.

Exercise 6.3.2 Suppose $\pi_{ab} = q^a = 0 = \sigma_{ab} = \omega_{ab} = \dot{u}^a$ (an FLRW universe).

(a) Show from (6.25) that $^3R_{\langle ab \rangle} = 0$. Prove that in this case,
$^3R_{ab} = \frac{1}{3}{}^3Rh_{ab}$ and $R_{abcd} = \kappa(g_{ac}g_{bd} - g_{ad}g_{bc})$ where $\kappa = {}^3R/6$.

(b) Show from (6.24) that $^3R = 6K/a^2$, $\dot{K} = 0$, and hence derive the Friedmann equation from (6.23).

Exercise 6.3.3 What are the components of the vorticity vector and tensor with the choice (6.27)? Characterize the remaining coordinate freedom preserving this form, and determine the optimally related tetrad basis. (See King and Ellis (1973).)

Exercise 6.3.4 Use the comoving approach for the rotating case and deduce properties of $^3\tilde{R}_{ab}$. (Using $\overline{\nabla}$, work out the corresponding 'curvature tensor', Ricci tensor and contracted Bianchi identities, being particularly careful not to incorrectly assume symmetries). A small research problem is that to close the equations we need the time-development of this tensor; can this be found?[2]

Problem 6.1 Determine the spatial sections in an FLRW universe that most closely have the properties of Newtonian space sections. Demonstrate how the properties of Newtonian time follow from the FLRW geometry in an appropriate limit. (Compare Ellis and Matravers (1985) and references therein).

6.4 The Weyl tensor and the Bianchi identities

Using (6.3), (4.51) gives

$$\ell^{-2}(\ell^2\sigma)\dot{}_{\langle ab \rangle} = \overline{\nabla}_{\langle a}\dot{u}_{b \rangle} + \dot{u}_{\langle a}\dot{u}_{b \rangle} - \omega_{\langle a}\omega_{b \rangle} - \sigma_{\langle a}{}^c\sigma_{b \rangle c} - E_{ab} + 4\pi G\pi_{ab}, \qquad (6.28)$$

which shows how the electric part of the Weyl tensor and the anisotropic fluid pressures cause distortion in a fluid (i.e. act as tidal forces). Another way to see this is to write the geodesic deviation equation for the case when u^a is geodesic, this being

$$\ddot{\xi}^a = -(E^a{}_b - 4\pi G\pi^a{}_b)\xi^b - \left[\frac{4}{3}\pi G(\rho + 3p) - \frac{1}{3}\Lambda\right]\xi^a. \qquad (6.29)$$

Equation (4.52) shows how the magnetic part H_{ab} is determined from the curls of the shear and vorticity tensors. From these equations, we can in principle obtain present-day limits on E_{ab} and H_{ab}.

The Newtonian analogues of (6.28) and (4.52) are

$$\ell^{-2}(\ell^2\sigma_{ij})\dot{} + \omega_j\omega_i + \sigma_j{}^k\sigma_{ki} + \frac{1}{3}h_{ij}(2\sigma^2 - \omega^2 - a^k{}_{,k}) + E_{ij} = a_{(i,j)}, \qquad (6.30)$$

$$(\omega_{(i}{}^{j;k} + \sigma_{(i}{}^{j;k})\eta_{m)jk} = 0. \qquad (6.31)$$

The second of these is the equation showing the vanishing of the Newtonian analogue of the magnetic part of the Weyl tensor.

---

[2] It would be preferable to do this covariantly; it certainly should be possible in a tetrad formalism, compare Ellis (1967) or MacCallum (1973) and Section 6.5.

In general we can introduce E_{ab} and H_{ab} as extra variables, characterizing the nature of spacetime curvature; in the irrotational case, we can regard E_{ab} as an alternative variable to ${}^3R_{\langle ab\rangle}$.

The Bianchi identities

The form (2.86) of the Bianchi identities is very analogous to the Maxwell equations (5.115) and current conservation equation (5.122).

As in the case of Maxwell's equations we can separate these equations out to obtain a set of Maxwell-like equations for **E** and **H**. These equations are most easily obtained using the complex form analogous to Exercise 5.5.1 for the electromagnetic case and equivalent to Exercise 4.7.1:

$$\mathcal{C}_{abcd} := C_{abcd} + \tfrac{i}{2}\eta_{ab}{}^{ef}C_{efcd} = \mathcal{G}_{abpq}\mathcal{G}_{cdrs}u^p u^r \mathcal{E}^{qs}, \text{ where} \tag{6.32}$$

$$\mathcal{E}_{ab} := E_{ab} + i\,H_{ab}, \ \mathcal{G}_{abcd} := 2g_{a[c}g_{d]b} + i\,\eta_{abcd}.$$

We obtain, taking $8\pi G = 1$ for brevity, and using the four-dimensional covariant notation combined with the angle brackets of Section 4.4,

$$\overline{\nabla}^b E_{ba} = \eta_{abc}\sigma^b{}_d H^{dc} - 3H_{ab}\omega^b \tag{6.33}$$

$$+\tfrac{1}{3}\overline{\nabla}_a\rho - \tfrac{1}{2}\overline{\nabla}^b\pi_{ba} - \tfrac{1}{3}\Theta q_a + \tfrac{1}{2}\sigma_{ab}q^b + \tfrac{3}{2}\eta_{abc}\omega^b q^c,$$

$$\dot{E}_{\langle ab\rangle} = \operatorname{curl} H_{ab} + 2\dot{u}^c\eta_{cd(a}H_{b)}{}^d - \Theta E_{ab} + 3\sigma_{c\langle a}E^c{}_{b\rangle} - \omega_{c\langle a}E^c{}_{b\rangle}$$

$$-\tfrac{1}{2}\overline{\nabla}_{\langle a}q_{b\rangle} - \dot{u}_{\langle a}q_{b\rangle} - \tfrac{1}{2}\dot{\pi}_{\langle ab\rangle} - \tfrac{1}{6}\Theta\pi_{ab} - \tfrac{1}{2}\sigma^c{}_{\langle a}\pi_{b\rangle c} \tag{6.34}$$

$$-\tfrac{1}{2}\omega^c{}_{\langle a}\pi_{b\rangle c} - \tfrac{1}{2}(\rho + p)\sigma_{ab},$$

$$\overline{\nabla}^b H_{ba} = -\eta_{abc}\sigma^b{}_d E^{dc} + 3E_{ab}\omega^b \tag{6.35}$$

$$+(\rho + p)\omega_a - \tfrac{1}{2}\eta_{abc}\sigma^b{}_d\pi^{dc} - \tfrac{1}{2}\pi_{ab}\omega^b - \tfrac{1}{2}\operatorname{curl} q_a$$

$$\dot{H}_{\langle ab\rangle} = -\operatorname{curl} E_{ab} - 2\dot{u}^c\eta_{cd(a}E_{b)}{}^d - \Theta H_{ab} + 3\sigma_{c\langle a}H^c{}_{b\rangle} \tag{6.36}$$

$$-\omega_{c\langle a}H^c{}_{b\rangle} + \tfrac{1}{2}\operatorname{curl}\pi_{ab} + \tfrac{1}{2}\sigma^c{}_{(a}\eta_{b)cd}q^d - \tfrac{3}{2}\omega_{\langle a}q_{b\rangle},$$

in close analogy with the Maxwell equations (5.118)–(5.121), the matter terms corresponding to source terms. These equations show how the propagation of the gravitational field (the Weyl tensor) is governed by the matter distribution, although not all derivatives of the Weyl tensor are thus determined (Maartens, Ellis and Siklos, 1997, Pareja and MacCallum, 2006). As in the Maxwell case, they can be used to obtain wave equations for E^{ab} and H^{ab}.

The Newtonian analogues of these equations are, $E^{ij}{}_{;j} = \tfrac{1}{3}\rho^{,i}$ (corresponding to (6.33)), $E^{(i}{}_{k,m}\eta^{j)km} = 0$, (corresponding to (6.36)) and $\sigma_{[j}{}^{[i}{}_{,m]}{}^{,k]} + \tfrac{2}{3}h^{[i}{}_{[j}\Theta^{,k]}{}_{,m]} = 0$ (corresponding to the combination of (6.34) and (6.35), see Ellis (1971a)).

Implied equations

Given the set of equations so far, propagation equations are implied for other quantities of interest. For instance, we already have propagation equations for all the kinematic quantities except the acceleration, but, for suitable equations of state, that equation can be derived by taking the time derivative of the momentum conservation equation. (When no equation of state has been given, $\dot{p}$ is a free function.) Considering the case of a perfect fluid for simplicity, (5.38) and (5.39) imply the acceleration propagation equation,

$$h_a{}^c(\dot{u}_c)^{\cdot} = \dot{u}_a \Theta\left(\frac{\mathrm{d}p}{\mathrm{d}\rho} - \frac{1}{3}\right) + h_a{}^b\left(\frac{\mathrm{d}p}{\mathrm{d}\rho}\Theta\right)_{,b} - \dot{u}_c(\omega^c{}_a + \sigma^c{}_a). \tag{6.37}$$

Two other quantities of considerable interest are the spatial gradients of the energy density and expansion; these characterize spatial inhomogeneity of a universe model. Propagation equations can be determined for these quantities in a straightforward way. In the perfect fluid case, the first is

$$\ell^{-4}h_c{}^a(\ell^4\overline{\nabla}_a\rho)^{\cdot} = -(\rho + p)\overline{\nabla}_c\Theta - (\omega^a{}_c + \sigma^a{}_c)\overline{\nabla}_a\rho , \tag{6.38}$$

obtained by taking the spatial gradient of the energy conservation equation. The second is

$$\ell^{-3}h_c{}^a(\ell^3\overline{\nabla}_a\Theta)^{\cdot} = {}^3\tilde{R}\dot{u}_c - (\sigma^b{}_c + \omega^b{}_c)\overline{\nabla}_b\Theta \tag{6.39}$$
$$+ \overline{\nabla}_c[-4\pi G\rho - 2\sigma^2 + 2\omega^2 + \overline{\nabla}_d\dot{u}^d + \dot{u}^d\dot{u}^d] ,$$

$${}^3\tilde{R} := -\tfrac{1}{3}\Theta^2 - 2\sigma^2 + 2\omega^2 + \overline{\nabla}_c\dot{u}^c + \dot{u}_c\dot{u}^c + 8\pi G\rho + \Lambda , \tag{6.40}$$

obtained by taking the spatial gradient of the Raychaudhuri equation. (Note that when $\omega = 0$, ${}^3\tilde{R}$ is the 3-space Ricci scalar.) In a certain sense there is no new information in these equations. However, it is useful to have explicitly the information they contain, even if it was implicitly given by the previous equations; for example the latter two play a central role later in our discussion of density perturbations of FLRW universe models.

Exercise 6.4.1 Show that when $\omega = 0$, the electric part of the Weyl tensor is related to the tracefree part of the 3-Ricci tensor by

$$E_{ab} = {}^3R_{\langle ab\rangle} - \sigma_{\langle a}{}^c\sigma_{b\rangle c} + \tfrac{1}{3}\Theta\sigma_{ab} - 4\pi G\pi_{ab} .$$

Exercise 6.4.2 Show that when $\omega = 0$, equation (6.33) is equivalent to the three-dimensional Ricci identities $2\overline{\nabla}^b{}^3R_{ab} = \overline{\nabla}_a{}^3R$.

Exercise 6.4.3 (a) Show that if a perfect-fluid-filled spacetime is conformally flat ($E_{ab} = H_{ab} = 0$) and $\rho + p > 0$, then $\sigma_{ab} = \omega^a = 0 = \overline{\nabla}_a\rho = 0$.
(b) Show that if additionally $p = p(\rho)$, then $\dot{u}^a = 0$; thus this must be an FLRW universe.

6.5 The orthonormal 1+3 tetrad equations

As mentioned in Section 6.3, tetrads provide one way of completing the set of variables and equations, using the methods described in Section 2.8. In a tetrad approach, one can calculate the rotation coefficients from (2.105) and then obtain the Ricci tensor and the EFE from (2.108), the formulae involving first derivatives of the rotation coefficients. If the rotation coefficients are regarded as the variables then, for consistency, the commutation coefficients must satisfy the Jacobi identities (2.103). One can call this a *minimal tetrad formalism*. It is often a good approach to use in examining the consistency of the field equations and obtaining solutions when particular assumptions have been made about spacetime geometry; for example Ellis (1967) determined the LRS dust models this way and Ellis and MacCallum (1969) the orthogonal perfect fluid Bianchi models.

Alternatively, one can consider a larger set of quantities as variables, e.g. the rotation coefficients and the Ricci and Weyl tensor components, and make use of the (second) Bianchi identities as equations for the curvature tensor components. This can be called an *extended tetrad formalism*. Applications of the Newman–Penrose formalism (see e.g. Stephani *et al.* (2003)) often take this approach, e.g. for studying gravitational radiation or algebraically special solutions (Section 2.7.6): it has been used to study fluids (e.g. Ozsváth (1965) and Allnutt (1981)).

Which is the most useful approach will depend on the problem being tackled; if one first imposes symmetries and then solves, the first may be better, while if one imposes restrictions on the Weyl tensor or its derivatives, the second may be more appropriate.

In both cases, if an orthonormal tetrad is used, as it will be below, many of the tetrad equations will be direct tetrad translations of covariant equations that have featured above (the primary covariant equations), because many of the rotation coefficients will be kinematic quantities (as defined in Chapter 4). However, some of the tetrad equations (particularly those related to spatial variations) may not have direct analogues in that set of equations. Thus the tetrad equations are more complete: they are sufficient to guarantee existence of a solution, whereas the primary covariant equations are necessary but not always sufficient.

There are situations, however, in which *all the tetrad equations are covariant equations*. This will happen whenever the tetrad vectors have been uniquely defined in a covariant manner; for example if the timelike vector is the fluid velocity u^a and the spacelike vectors are unique eigenvectors of the shear tensor. Then all the rotation coefficients are covariantly defined quantities and so all the equations are invariant relations. Using the approach to classification of spacetimes outlined, with its application to cosmology, in Section 17.2, one finds such a unique covariant choice is possible in general spacetimes but may not be possible in all cases.

We now give more specific details of these approaches (MacCallum, 1973, van Elst and Uggla, 1997).

6.5.1 A minimal tetrad formalism

In a minimal tetrad formalism based on an orthonormal tetrad with timelike vector chosen as the fluid flow vector, the *unknowns* are

(1) the tetrad components,
(2) the tetrad rotation coefficients (or equivalently, the commutator coefficients) and
(3) the matter fields.

The *equations* are

(1) the tetrad equations (giving their components relative to a coordinate system),
(2) the Jacobi identities for the commutator coefficients,
(3) the EFE written as differential equations for the commutator coefficients, based on the Ricci identities for all four tetrad vectors and
(4) the energy–momentum conservation equations for the matter, i.e. the contracted Bianchi identities, together with any matter field equations required.

The idea is to solve as far as possible for the rotation coefficients first, and only then to solve the tetrad equations for their explicit components; the Jacobi identities are integrability conditions guaranteeing the existence of tetrad components corresponding to the commutator coefficients.

Because we are taking $e_0 = \boldsymbol{u}$, the commutation coefficients with a 0 index are almost all specified by the kinematic quantities. The exceptions are the $\gamma^i{}_{j0}$; to parametrize these we introduce the components of the rotation of the tetrad with respect to a Fermi-propagated frame,

$$\Omega_{ij} = e_i \cdot \dot{e}_j \Leftrightarrow \Omega^k = \tfrac{1}{2}\eta^{kij} e_i \cdot \dot{e}_j \tag{6.41}$$

(beware that this standard notation risks confusion with the Ω of (6.8)). The purely spatial coefficients are decomposed using

$$\tfrac{1}{2}\gamma^i{}_{km}\eta^{jkm} = n^{ij} + a^{ij}, \tag{6.42}$$

where $n^{ij} = n^{(ij)}$ and $a^{ij} = a^{[ij]} = \eta^{ijk}a_k$, so that $a_j = \tfrac{1}{2}\gamma^k{}_{jk}$ and

$$\gamma^i{}_{jm} = \eta_{jmk}n^{ik} + 2a_{[j}\delta^i_{m]} \, .$$

In the irrotational case, these determine the three-dimensional Ricci tensor $^3R_{ij}$; in general, together with the other rotation coefficients they completely determine the Weyl tensor components E_{ab} and H_{ab}.

The corresponding commutator expressions are

$$[\mathbf{e}_0, \mathbf{e}_i] = \dot{u}_i \mathbf{e}_0 - [\Theta_i{}^k + \eta^k{}_{ij}(\omega^j - \Omega^j)]\mathbf{e}_k \tag{6.43}$$

$$[\mathbf{e}_i, \mathbf{e}_k] = -2\eta_{ikj}\omega^j \mathbf{e}_0 + (\eta_{ikj}n^{jm} + 2a_{[i}\delta^m_{k]})\mathbf{e}_m \, . \tag{6.44}$$

Note that spatial indices in the orthonormal tetrad can be raised and lowered arbitrarily, but raising or lowering the index 0 introduces a factor -1.

The Jacobi identities (the second set of equations required) can be labelled by indices $\{ab\}$ for the equation obtained from

$$R^a{}_{cdf}\eta^{bcdf} = 0.$$

The $\{00\}$ and $\{0i\}$ equations are the tetrad forms of (4.44) and (4.43), namely,

$$0 = (\partial_i - 2a_i - \dot{u}_i)\omega^i \, , \tag{6.45}$$

$$\partial_0\omega^i = -\tfrac{2}{3}\Theta\omega^i + \sigma^i{}_j\omega^j + \tfrac{1}{2}n^i{}_j\dot{u}^j - \eta^{ijk}[\tfrac{1}{2}(\partial_j - a_j)\dot{u}_k + \omega_j\Omega_k] \, . \tag{6.46}$$

The remaining 12 equations give as the $\{i0\}$ equation and as the skew and symmetric parts of $\{ij\}$ respectively:

$$0 = (\partial_j - 2a_j)n^{ij} - \tfrac{2}{3}\Theta\omega^i - 2\sigma^i{}_j\omega^j + \eta^{ijk}(\partial_j a_k + 2\omega_j \Omega_k), \qquad (6.47)$$

$$\partial_0 a^i = -(\sigma^{ij} + \tfrac{1}{3}\Theta\delta^{ij} - \omega^{ij} + \Omega^{ij})a_j \qquad (6.48)$$
$$+ \tfrac{1}{2}(\partial_j + \dot{u}_j)(\sigma^{ij} - \tfrac{2}{3}\Theta\delta^{ij} - \omega^{ij} + \Omega^{ij}),$$

$$\partial_0 n^{ij} = [2\sigma^{(i}{}_k - \tfrac{1}{3}\Theta\delta^{(i}{}_k + \omega^{(i}{}_k - \Omega^{(i}{}_k]n^{j)k} + \delta^{ij}(\partial_k + \dot{u}_k)(\omega^k - \Omega^k)$$
$$- [\partial_k + \dot{u}_k][\eta^{km(i}\sigma^{j)}{}_m + \delta^{k(i}(\omega^{j)} - \Omega^{j)})]. \qquad (6.49)$$

The G_{00} and G_{0i} field equations are the tetrad versions of (6.5) and (6.20) respectively which read,

$$\partial_0\Theta = -\tfrac{1}{3}\Theta^2 - 2\sigma^2 + 2\omega^2 + (\partial_i + \dot{u}_i - 2a_i)\dot{u}^i - 4\pi G(\rho + p) + \Lambda, \qquad (6.50)$$

$$0 = 8\pi Gq^i - \tfrac{2}{3}\delta^{ij}\partial_j\Theta + (\partial_j - 3a_j)\sigma^{ij} - \eta^{ijk}n_{jm}\sigma^m{}_k \qquad (6.51)$$
$$- \eta^{ijk}(\partial_j - a_j)\omega_k + n^i{}_j\omega^j + 2\eta^{ijk}\omega_j\dot{u}_k.$$

To express the remaining Einstein equations we follow the first method of Section 6.3 for the rotating case, i.e. introduce a quantity $^3\tilde{R}_{ij}$ which will be the three-dimensional curvature if ω^a is zero (and so gives the usual treatment in that case). The tracefree part and trace of this are

$$^3\tilde{R}_{\langle ij\rangle} = \partial_{\langle i}a_{j\rangle} + 2n_{k\langle i}n^k{}_{j\rangle} - n^k{}_k n_{\langle ij\rangle} - \eta^{km}{}_{\langle i}(e_{|k|} - 2a_{|k|})n_{j\rangle m}, \qquad (6.52)$$

$$^3\tilde{R} = 4\partial_i(a^i) - 6a_i a^i - n_{kj}n^{kj} + \tfrac{1}{2}(n^k{}_k)^2. \qquad (6.53)$$

In terms of these quantities the PSTF part of the G_{ij} equations is

$$\partial_0\sigma^{ij} = -\Theta\sigma^{ij} + [\delta^{k(i}\partial_k + \dot{u}^{(i} + a^{(i}]\dot{u}^{j)} + 2\eta^{km(i}[2\Omega_k\sigma^{j)}{}_m - \sigma^{j)}{}_k\dot{u}_m]$$
$$+ 2\omega^{(i}\Omega^{j)} - {}^3\tilde{R}^{\langle ij\rangle} + 8\pi G\pi^{ij}, \qquad (6.54)$$

and the trace can be combined with the G^{00} equation to give

$$0 = 8\pi G\rho - \tfrac{1}{3}\Theta^2 + \sigma^2 - \omega^2 - 2\omega_i\Omega^i - \tfrac{1}{2}{}^3\tilde{R} + \Lambda. \qquad (6.55)$$

Note that in this treatment (6.52) and (6.53) are regarded as defining the 3-space quantities, which are not to be treated as extra variables.

Finally, we have the contracted Bianchi identities,

$$\partial_0\rho = -(\rho + p)\Theta - \pi^{ij}\sigma_{ij} - (\partial_i - 2a_i + 2\dot{u}_i)q^i, \qquad (6.56)$$

$$\partial_0 q_i = -\tfrac{4}{3}\Theta q_i - \sigma_i{}^j q_j - (\rho + p)\dot{u}_i - \partial_i p - (\partial_j - 3a_j + \dot{u}_j)\pi_i{}^j$$
$$+ \eta_i{}^{jk}[(\omega_j + \Omega_j)q_k + n_{jm}\pi^m{}_k], \qquad (6.57)$$

together with whatever matter field equations or equations of state are needed to completely specify the matter terms. Note that if the matter terms are specified independently these become (like the remaining Bianchi identities in this treatment) true identities, and hence trivial. In practice the matter specification usually gives q_a independently of (6.57) and therefore that equation becomes an equation for $\dot{u}^i$.

To summarize, the variables to be solved for are

$$(\dot{u}^i, \omega_i, \Theta, \sigma_{ij}, \Omega_i, a_i, n_{ij}, \rho, q_i, p, \pi_{ij}).$$

We have evolution equations for $(\omega_i, \Theta, \sigma_{ij}, a_i, n_{ij}, \rho, q_i)$; equations of state are needed to give the remaining fluid variables, which then usually determine the propagation of $\dot{u}^i$, while Ω_i will in general come from a tetrad choice (the choice is so far fixed only up to a position-dependent spatial rotation). When the fluid equations do not result in a specification of $\dot{u}^i$, we encounter an indeterminacy corresponding to the freedom of choice of the lapse function in the Hamiltonian formalism; however, this may sometimes be determined by consistency conditions for the full set of equations. We discuss this below in connection with the Newtonian limit.

6.5.2 An extended tetrad formalism

In an extended tetrad formalism, there are extra variables and equations; the *unknowns* additional to those of Section 6.5.1 are,

(4) the Weyl tensor components, and possibly (as discussed more fully in Section 6.6.2)
(5) their covariant derivatives;

and the extra *equations* are

(5) the (second) Bianchi identities, which are differential equations for the Weyl tensor components, and possibly
(6) the 'super-Bianchi' identities, differential equations for the Weyl tensor derivatives.

Such a formalism also amounts to treating the Ricci tensor components, i.e. the energy–momentum variables, as independent quantities subject to the Bianchi identities (which are equations rather than strictly identities, for these choices of variables).

The idea now is to solve as far as possible for the Weyl tensor (and perhaps its derivatives) first and only then to solve for the rotation coefficients and tetrad components; the existence of the rotation coefficients is guaranteed by the Bianchi identities (see Section 6.6.2). In practice one often solves the equations in some mixed order.

The tetrad formulae for the electric and magnetic parts of the Weyl tensor are as follows:

$$E_{ij} = -\partial_0(\sigma_{ij}) - \tfrac{2}{3}\Theta\sigma_{ij} + (\partial_{\langle i} + \dot{u}_{\langle i} + a_{\langle i})\dot{u}_{j\rangle} - \sigma_{k\langle i}\sigma^k{}_{j\rangle}$$
$$+ \eta^{km}{}_{\langle i}\left[2\sigma_{j\rangle m}\Omega_k - n_{j\rangle k}\dot{u}_m\right] - \omega_{\langle i}\omega_{j\rangle} + 4\pi G\pi_{ij} , \qquad (6.58)$$

$$H_{ij} = (\partial_{\langle i} + 2\dot{u}_{\langle i} + a_{\langle i})\omega_{j\rangle} + \tfrac{1}{2}n^k{}_k\sigma_{ij} - 3n^k{}_{\langle i}\sigma_{j\rangle k}$$
$$+ \eta^{km}{}_{\langle i}\left[(\partial_{|k|} - a_{|k|})\sigma_{j\rangle m} - n_{j\rangle k}\omega_m\right] . \qquad (6.59)$$

Combining (6.58) with the field equation (6.54), one easily derives

$$(E_{ij} + 4\pi G\pi_{ij}) = \tfrac{1}{3}\Theta\sigma_{ij} - \sigma_{k\langle i}\sigma^k{}_{j\rangle} - \omega_{\langle i}\omega_{j\rangle} - 2\omega_{\langle i}\Omega_{j\rangle} + {}^3\tilde{R}_{\langle ij\rangle} . \qquad (6.60)$$

From this expression it can be seen that the 'electric' part of the Weyl tensor is closely related to ${}^3\tilde{R}_{\langle ij\rangle}$ (compare Exercise 6.4.1).

The Bianchi identities in orthonormal tetrad form can be found in MacCallum (1973) or van Elst and Uggla (1997). They follow directly from the covariant forms (6.33)–(6.36) given above.

6.6 Structure of the 1+3 system of equations

The variables of the extended system of 1+3 covariant equations derived above, relative to a chosen four-velocity u^a, are

$$(\dot{u}^a, \omega_a, \Theta, \sigma_{ab}, \rho, q_a, p, \pi_{ab}, E_{ab}, H_{ab}),$$

all of which are scalars, spatial vectors or PSTF tensors: they have 32 independent components altogether. The equations are the components of the Ricci identity (4.41) for u^a, into which where possible the Einstein field equations (3.15) have been substituted, and the components of the second Bianchi identities. We can readily divide these equations into 'evolution' equations (those specifying a time-derivative, in the sense of Section 4.3, of one of the basic variables) and (differential) constraint equations involving spatial derivatives along directions orthogonal to u^a.

The extended system thus obtained contains four first Bianchi (Jacobi) identities giving $\dot{\omega}^{\langle e\rangle}$ and $\overline{\nabla}^b \omega_b$ ((6.13) and (6.45)), five constraint relations for the curvature components H_{ab} (4.52) and four equations giving $\dot{\Theta}$ and q^a ((6.4) and (6.20)), all of these coming from the Ricci identity and the Einstein equations, five linear combinations (6.28) of the Ricci identities (4.51) and the field equations (6.3), giving $\dot{\sigma}_{\langle ab\rangle}$, and all 20 independent Bianchi identities ((5.11) and (6.33)–(6.36)), giving $\dot{\rho}, \dot{q}^a, \dot{E}_{\langle ab\rangle}, \dot{H}_{\langle ab\rangle}, \overline{\nabla}^a E_{ab}$, and $\overline{\nabla}^b H_{ab}$. There are therefore 12 Jacobi identities and six independent linear combinations of the Ricci identity (4.51) and the field equations (6.3) missing, compared with, for example, a general tetrad system. To solve the Einstein equations fully these missing equations, or extra equations which can serve as their equivalents, have to be satisfied. (The completeness of the minimal orthonormal tetrad equations was discussed in Section 6.5.)

There are evolution equations for 23 of the variables in the set. Thus there are nine quantities, to whit $(\dot{u}^a, p, \pi_{ab})$, for which we have no evolution equation (though (6.34) could be re-arranged to give $\dot{\pi}_{ab}$ rather than $\dot{E}_{ab}$) until equations of state for the matter are specified. When this is done, the momentum conservation equation (6.20) usually becomes an equation for $\dot{u}_a$ rather than $\dot{q}_a$: in particular, for a perfect fluid only p remains to be specified. In general one might expect that the process leading to the specification of u^a (see Section 5.1.1) would also lead to a specification of $\dot{u}^a$, since if a completely free choice of u^a is allowed, $\dot{u}^a$ could have any value.

Of the equations we do have, the $\dot{\Theta}$ equation has been studied in detail in Section 6.1, and the $\dot{\omega}^a$ equation, which we note requires knowledge only of σ_{ab} and $\dot{u}^a$, in Section 6.2. From the $\dot{\sigma}_{ab}$ equation we note that ω^a has a tidal effect and can create shear, while the E_{ab} terms are tidal forces. Since gravitational waves are possible, distant matter and boundary conditions can create a non-zero Weyl tensor contribution which then induces shear which in turn feeds into the Raychaudhuri equation and tends to cause convergence of the timelike world lines of fundamental observers.

6.6.1 The initial value problem

It is awkward to pose a well-formulated initial value problem using these equations if $\omega^a \neq 0$, since the spatial derivatives are not then derivatives in a hypersurface and so the 'constraint' equations are not really constraints on initial values.

The structure is seen most clearly in the case of *zero rotation*, for then we can use the Gauss equation relating the fluid flow properties to the geometry of the orthogonal 3-spaces. The time development of the expansion and shear are given by (6.5) and (6.25) respectively, corresponding to six field equations; the remaining field equations are (6.20) and (6.23), which are constraint equations relating the expansion and shear to the energy density, and momentum density, and must be preserved under the time development of the fluid flow.

As discussed in Section 3.3.1, the nature of this evolution is such as to preserve the constraint equations: solving these on some surface $t = t_0$ gives a solution of the initial value problem. They will remain true at later times if initially true, provided the energy–momentum conservation equations (i.e. contracted Bianchi identities), and hence the associated conservation of mass and momentum, are true everywhere (compare Section 5.1); for a perfect fluid these will have the simple form (5.38), (5.39), determining the fluid acceleration and the rate of change of the energy density along the flow lines, which are orthogonal to the naturally defined surfaces $t = \text{const}$ (see (4.36)). Thus we obtain a solution everywhere within the Cauchy development (see Section 3.3) of the initial surface. However, in general this domain will be limited, because singularities will eventually occur. This issue is discussed in Section 6.7.

Essentially the same structure will hold in the case of rotating solutions; this is most easily seen by using a non-comoving description based on a suitable choice of time surfaces.

6.6.2 Consistency of the equations

Perhaps the clearest way to formulate this issue is to consider the 1+3 equations as a part of a full orthonormal tetrad formalism (as developed in Section 6.5). If we start with the equations defining connection and curvature, the integrability conditions are just the first and second Bianchi identities.[3] However, we actually substitute from some of these equations into others. If we take as our variables the tetrad components, commutation coefficients and Riemann tensor components, we shall have equations giving each of these three sets in terms of the previous ones, plus versions of the Bianchi identities. Edgar (1980) showed how the equations usually then taken give integrability conditions which are in general combinations of the commutator equations and the Bianchi identities. In this discussion it is assumed (sometimes implicitly) that the connection is metric and that the commutation coefficients and Riemann tensor components have the usual symmetries.

---

[3] Integrability conditions are necessary but not usually sufficient to guarantee existence and uniqueness of solutions, which depend also on analytic properties of the equations and data.

There are two main variants of this general discussion. One is to take a different set of variables. Those used above contain redundant terms (being part of an extended tetrad formalism rather than a minimal one) so a smaller set, such as just tetrad components, commutators and matter variables, could be used. Equally one could add the components of any number of derivatives of the Riemann tensor as extra variables, together with the equations relating them to the previous variables and the equations (third and higher Bianchi and Ricci identities) they must satisfy.

So far we have assumed a completely general spacetime and choice of tetrad. Additional equations and integrability conditions may arise if one imposes additional conditions, such as specific equations of state. One could distinguish between two different types of choice of equality, or set of equalities, as follows.

First, one could constrain the tetrad choice in a way that is compatible with the general case under study. The derivatives of the constraint will then give additional conditions. However, the processes of imposing the constraint and solving the equations inevitably commute; that is to say one could in principle have solved the equations in a general basis and then chosen the tetrad by applying a Lorentz transformation at each point (i.e. algebraically). Therefore the extra equations in this case must just be specializations of the differential equations necessarily satisfied by a tetrad solving the general tetrad equations.

In the cosmological context an example of a condition of this type is given by the requirement of zero heat flux, $q_a = 0$, corresponding to a particular choice of u^a (assuming the energy–momentum obeys the usual energy conditions which guarantee the existence of a timelike eigenvector of the Ricci tensor). When using a tetrad other such choices are that the spatial tetrad vectors are eigenvectors of either σ_{ab} or E_{ab}, which gives them a helpful invariant significance.

The second type of equality or set of equalities that can be imposed is one which gives a consistent specialization of the general system of equations but which is not generally possible just by choice of tetrad: for example, the conditions that a spacetime is a vacuum, of Petrov type D, and has shearfree and geodesic principal null congruences. Such conditions define, within the space of tetrad variables, an *invariant set* of data in involution (compare Section 6.3.2). For this to be true, the time-derivatives must preserve the conditions imposed. In cosmology, the assumption that the matter content is irrotational 'dust' (with no acceleration) is such a set of conditions.

One has to check that such sets of conditions really are consistent. Edgar (1980) has shown, for example, that tetrad equations remain consistent when the vacuum condition is used in general relativity.

In the cosmological context, while we can impose any geometrical restrictions we wish on the initial data (provided they are compatible with the initial data constraints), we cannot assume the same restrictions will necessarily remain true as the fluid flow develops. If we suppose some particular restrictions hold at all times, this assumption may or may not be compatible with the field equations (compare the warning in Section 6.2.2).

A number of papers have discussed the consistency of the 1+3 equations in particular cases (often for the irrotational dust or perfect fluid case): see e.g. van Elst (1996), Maartens (1997), Maartens, Lesame and Ellis (1998), Velden (1997). The general (and successful) aim of these papers is to show that the equations are consistent in the sense that if the constraint equations are taken to be true at some instant, then the evolution equations guarantee that the constraints are true at all times. One can in fact show that this consistency of the 1+3 system follows from the commutators and first and second Bianchi identities without new calculations; see MacCallum (1998) for the dust case.

One often wants to work out such a set of consistent conditions starting from some incomplete set of conditions. An example is given by the investigations, in cosmology, aimed at deciding what solutions, if any, satisfy conditions such as the vanishing of the electric or magnetic parts of the Weyl tensor (see Section 19.8).

6.6.3 First-order symmetric hyperbolic form

The field equations have both a wave nature (gravitational waves are possible, allowing transfer of energy and momentum at the speed of light) and a conservative nature (constraint equations, satisfied in an initial surface, remain satisfied at later times because of the conservation equations and so energy and momentum can be bound to localized objects). We have discussed the second aspect (i.e. conservation of the constraints); now we turn to the first.

Because of the wavelike nature of the equations (their characteristics are null surfaces), a solution is determined only within the Cauchy development of initial data (as defined in Chapter 3). This hyperbolic nature of the above set of equations is not obvious. To show it, we need to transform them to a symmetric hyperbolic normal form. This involves taking suitable linear combinations of the tetrad variables defined above, obtaining a collection of dependent field variables u^A which are functions of a set of local spacetime coordinates $\{x^\mu\}$ and are such that the resulting equations are of the form

$$M^{AB\mu} \partial_\mu u_B = N^A , \tag{6.61}$$

where the objects $M^{AB\mu} = M^{AB\mu}(x^\nu, u^C)$ and $N^A = N^A(x^\mu, u^B)$ denote four *symmetric* matrices and a vector, respectively, each acting in a space of dimension equal to the number of dependent fields. The set of equations (6.61) is *hyperbolic* if the contraction $M^{AB\mu} n_\mu$ with the coordinate components of some past-directed timelike 1-form n_μ yields a positive-definite matrix; it is *causal* if this contraction is positive-definite for *all* past-directed timelike 1-forms n_μ (Geroch and Lindblom, 1990). If it satisfies all these conditions, it is said to be a First Order Symmetric Hyperbolic (FOSH) evolution system. Standard theorems then guarantee the existence of solutions to the time evolution equations.

The set of characteristic 3-surfaces $\phi = \text{const}$ underlying a FOSH evolution system can be interpreted as a collection of wavefronts with phase function ϕ across which certain physical quantities may be discontinuous. The associated characteristic eigenfields propagate along so-called bicharacteristic rays *within* these 3-surfaces at velocities v, which represent their slopes relative to the direction of **u** (Courant and Hilbert, 1962). The characteristic condition the associated vector fields **ξ** must satisfy is

$$0 = \det [M^{AB\mu} \xi_\mu]. \tag{6.62}$$

This determines the characteristic velocities and eigenfields.

In van Elst and Ellis (1999) (see also Ellis (2005)) it is shown from the tetrad equations given above that cosmological models ($\mathcal{M}$, **g**, **u**) with barotropic perfect fluid matter source fields form a FOSH evolution system. This shows their hyperbolic nature, as desired. The nature of the resulting characteristics and associated variables is of interest.

6.6.4 Solving the equations

When solving the EFE there are two especially important points.

Firstly (compare Section 5.2.3), one should clearly state what matter content one is assuming for spacetime (vacuum, perfect fluid, electromagnetic field, scalar field, etc.) when obtaining solutions, and also specify any needed equations of state for the chosen form of matter (for example in the perfect fluid case, one should state what relations hold between the energy density ρ and the pressure p). Until this has been done, the EFE do not describe a well-defined physical situation (compare Section 5.2.3).

Secondly, it is important to note that when obtaining solutions of the EFE, *one must always make certain that all ten equations are satisfied*. Unless this has been verified, one is not in a position to claim one has a correct solution of the field equations. (This may sound a trivial remark but it is surprisingly often overlooked.)

We examine particular solutions under assumptions about the spacetime symmetries, which determine their geometry and hence the nature of the solutions, as well as their matter content. Looking at this in the cosmological context is the burden of the following chapters.

6.7 Global structure and singularities

This chapter is concerned with generic relations holding in all realistic cosmological models. Apart from the generally valid equations developed up to this point, there is a further set of remarkable generic[4] results that have been established as a result of the pioneering work of Roger Penrose, Stephen Hawking, Bob Geroch, and Brandon Carter. These relate to global properties of spacetime and the existence of singularities.

There is only space here to give the briefest summary of these results, which are presented in depth in Hawking and Ellis (1973), see also Wald (1984), Tipler, Clarke and Ellis (1980) and Joshi (1996).

6.7.1 Existence theorems for singularities

According to GR, singularities occur in cosmology not only in the context of high-symmetry models such as the FLRW models, but also for realistic anisotropic and inhomogeneous models of the universe in which the energy condition (5.17) is satisfied. The latter will be

---

[4] That is, they are not based on any specific exact solutions.

true for classical fields; it is precisely because this condition is an inequality rather than an exact equation that these theorems have their power.

The case of irrotational dust was discussed in Section 6.1.2; the same singularity theorem holds as in the FLRW special case, irrespective of the degree of anisotropy or inhomogeneity in the spacetime. However, acceleration (due to pressure gradients) or vorticity could in principle upset this prediction. Examination of specific classes of models failed to find specific realistic cosmological models where the singularity was avoided in this way,[5] but analytic examination of the full set of covariant or tetrad equations failed to provide a proof that all realistic (anisotropic and inhomogeneous) cosmologies are singular.

This situation was dramatically changed by Roger Penrose's pioneering work on black hole singularities (1965), giving a theorem predicting the existence of singularities in realistic gravitational collapse cases. This was extended to the case of cosmology by Stephen Hawking and others, proving a series of theorems culminating in the major singularity theorem of Hawking and Penrose (1970).

Global concepts

The key to these developments is the study of global properties of spacetimes, rather than local properties (Hawking and Ellis, 1973, Tipler, Clarke and Ellis, 1980).

First, one has to work with atlases rather than just local coordinate systems, allowing global coordinate coverage of spacetimes with complicated topologies.

Second, one has to delineate the nature of causality in spacetime: determining domains of influence, domains of dependence (the Cauchy development of data on a surface, see Section 3.3.1), and the boundaries of these regions (null cones and Cauchy horizons for example). Causal boundaries are generated by null geodesics, which generically develop self-intersections and caustics; but when caustics occur, this signifies that the boundary has necessarily come to an end (the caustic points are beyond places where self-intersections occur, and so lie inside the domain of dependence rather than on the causal boundary).

Thirdly, one can characterize limits of geodesic connectivity from points and surfaces; these are related to the Cauchy development of a surface.

And fourthly, one defines a spacetime as being singular if there exist inextendible geodesics in the spacetime: that is, the spacetime runs into some kind of edge that prevents continuation of geodesics. This is of particular physical significance when the geodesics are timelike or null: then the possible histories of particles come to an end (to the future) or start at a beginning (in the past), in contradiction to the situation in non-singular spacetimes, where particle histories can continue for ever.

The specific key elements in the proof of existence of such singularities were (a) use of the timelike and null versions of the Raychaudhuri equation for families of irrotational geodesics with suitable energy conditions, implying intersection of these geodesics after a finite distance or time, and (b) very careful analysis of the causal properties of spacetime and the domains of dependence of initial data on spacelike surfaces (these domains being bounded by null horizons generated by null geodesics). Under very general circumstances

---

[5] Newtonian dust models that are shearfree, expanding and rotating apparently give such universes; but they have no GR counterparts, see Sections 6.2.2 and 6.8.2.

characterizing both a black hole geometry and the situation arising after refocusing of our past light-cone in a realistic cosmology, and providing causal violations are avoided, these theorems showed the existence of an edge to spacetime, implied by the existence of incomplete geodesics, i.e. incomplete particle histories. The proof did not, however, determine the nature of the singularity, i.e. did not necessarily imply an infinite matter density would arise.

Furthermore, the existence of the CMB alone was adequate evidence of the refocusing of our past light-cone, which is the central geometrical feature implying existence of closed trapped surfaces, as in the black hole case, so leading to prediction of existence of a singularity in the cosmological context (provided the energy conditions are satisfied). Thus GR implies existence of an edge to spacetime associated with a singularity at the start of the universe. As long as the energy condition is satisfied, singularity avoidance is only possible – given suitable causality and energy conditions – if matter is concentrated in such small isolated regions that reconvergence of our past light-cone is avoided; and that would imply either major CMB anisotropies that are not observed, or lack of enough matter to cause the observed blackbody spectrum of that radiation.

The ways of avoiding singularities in general are the same as those discussed in the simpler cases of Section 6.1.2. The conclusion is again that to avoid singularities would require some effect of quantum gravity, or an alternative classical field theory of gravity.

6.7.2 Classification of singularities

Because of their nature (proof by contradiction), these theorems prove geodesic incompleteness, rather than showing that the energy density diverges. They do not determine the nature of the singular behaviour near the origin of the universe. We shall define a singularity as a boundary of spacetime where either the curvature diverges (see discussion below) or geodesic incompleteness occurs. The relation between these two kinds or aspects of singularities is still not fully clear; but often they will occur together. For our purposes, existence of either indicates there is a problem with spacetime at that boundary; hence our definition.

The nature of singularities in classical cosmological solutions is very varied. Specific examples are discussed later as we discuss specific exact solutions with anisotropy and inhomogeneity. We consider here the broader classification of cosmological curvature singularities into *scalar curvature singularities* and *non-scalar singularities*. In both cases the curvature tensor diverges, but in the first case, this is associated with divergence of a scalar quantity associated with the curvature tensor; in the second case, it is only associated with divergence of the components of the curvature tensor in a parallelly propagated orthonormal frame. A further type of spacetime singularity is *quasi-regular singularities*, where the curvature is perfectly regular but conical singularities occur. We do not consider them further here, as they are not significant in cosmology, except as idealized representations of cosmic strings; but this idealization is so rigid as to not correspond to the way cosmic strings are usually believed to behave.

Scalar curvature singularities

These occur when at least one scalar polynomial expression formed from the curvature tensor, the metric tensor, and other uniquely defined covariant tensors (such as the matter 4-velocity) diverges. These singularities can be of various kinds; in particular either the Ricci tensor or the Weyl tensor (or both) may diverge. It is not known in general whether singularities, in the incomplete causal geodesics sense, occur where scalars formed from derivatives of these various tensors diverge. It also could be that neither the Ricci tensor nor the Weyl tensor gives a divergent scalar polynomial and the only such are mixed ones. We do not know of such an example, however.

Ricci scalar divergence implies that matter or field energies and densities are unbounded, and hence so are curvature invariants; thus the spacetime itself is singular. The classic example is the standard Big Bang at the start of FLRW universe models for ordinary matter, where $R_{ab}u^a u^b$ and $T_{ab}u^a u^b$ diverge for a uniquely defined unit timelike vector field u^a. Note that the Ricci scalar R need not diverge there, because it is proportional to $T = 3p - \rho$ and so will take the value zero in a radiation-dominated era, even if ρ and p diverge. However, the scalar $T^{ab}T_{ab} = (\rho + p)^2 + 3p^2$ will diverge in this case. The Weyl tensor is regular at any such singularity in FLRW models, because it is exactly zero in all these models; so they present pure Ricci singularities. This case is of course very special; in general the Weyl tensor will also diverge. Many examples show the variety of scalar singularities that might exist at the beginning of the universe in more general spacetimes. There is also a possibility of 'sudden singularities' and 'rip singularities' at late times (Barrow and Tsagas, 2005, Cattoën and Visser, 2005, Barrow and Lip, 2009), but these seem to occur only for hypothetical matter with implausible physical properties, so we shall not discuss them further.

Weyl scalar divergence implies that the gravitational field diverges. This can happen even if the matter density does not diverge; the classic example is the singularity at $r = 0$ in the spatially homogeneous, anisotropic domain in the maximally extended Schwarzschild solution (with non-zero mass M). The Ricci tensor is regular at this singularity, because it is zero everywhere in a Schwarzschild solution. However, the conformal curvature scalar invariant $C^{abcd}C_{abcd} = M/r^6$ diverges there, so this is a pure Weyl singularity. Another example is given by the singularities in the Kasner spatially homogeneous but anisotropic vacuum solutions (see Section 18.2), which are also pure Weyl singularities. Generically a singularity will be a Weyl singularity if a vacuum exists there, which is not the case in the real universe, but may be asymptotically true at the start or end of the universe. Indeed in many models matter is negligible at both these times.

General scalar curvature singularities will have both the Ricci and the Weyl tensors singular, characterized by both $R_{ab}R^{ab}$ and $C_{abcd}C^{abcd}$ (and/or possibly some mixed scalars) diverging. The associated singularities can be *spacelike* or *timelike* in character, as in the FLRW and Reissner–Nordström cases respectively (the latter case is empty except for a Maxwell field, but similar cases can occur in the LTB models (see Section 15.1)), and in special cases could be null in nature. In the early universe they may be *matter dominated*, as in the standard FLRW case, often called *velocity dominated* in anisotropic situations such as the Bianchi I case (Eardley, Liang and Sachs, 1971), or *curvature dominated*, as for example in the Bianchi IX case.[6] And they may have a variety of geometric structures: they may be isotropic, as in the FLRW case, cigar-like or pancake-like, as in the generic and special Kasner (Bianchi I) cases respectively (Thorne, 1967), or chaotically oscillating with numerous Kasner-like epochs occurring in different directions, as in the Bianchi IX 'Mixmaster' solutions (Misner, 1969a, Hobill, Burd and Coley, 1994).

Belinski, Lifshitz and Khalatnikov (1971) argued that generic singularities will be oscillatory and locally like the Bianchi IX spatially homogeneous solutions. This may be true for spacelike singularities (see Section 19.10.1), although a rigorous proof has yet to be given. However, it seems unlikely that this will be so for solutions with timelike singularities, whose behaviour and indeed physical significance is quite unlike that of spacelike singularities (Tomita, 1978, Liang, 1979).

A complete categorization of the full range of singular possibilities, and when they will be likely to occur, has not yet been given. That is a worthwhile endeavour. One should note of course that these analyses are based on GR, the classical theory of gravity, and could be modified when quantum gravity effects are taken into account. Nevertheless these understandings of the classical possibilities set the stage for understanding the quantum gravity options.

Non-scalar singularities

These are quite different in character. In specific kinds of spatially homogeneous models (tilted Class B Bianchi models), it is possible for there to be a dramatic change in the nature of the solution where the surfaces of homogeneity change across a null surface from being spacelike to being timelike (at early times). Associated with this is a singularity where all scalar quantities are finite but components of the matter energy–momentum tensor diverge when measured in a parallelly propagated frame. This is discussed further in Section 18.6.1.

Exercise 6.7.1 Describe and discuss the structure of the set of equations in the Newtonian case corresponding to those considered above.

Exercise 6.7.2 Examine the validity of the singularity theorems in the case of scalar–tensor theories.

Exercise 6.7.3 Examine the relation between Weyl scalar divergence singularities and conformal invariance of the spacetime.

6.8 Newtonian models and Newtonian limits

It is of course important that the equations discussed here reduce to Newtonian equations in an appropriate limit, and so relate to the Newtonian approach used in most astrophysical studies. In this chapter we have shown the Newtonian analogues of many of the equations

---

[6] These and other examples of singularities in spatially homogeneous solutions will be discussed in Chapter 18.

given; despite major differences in the nature of the theories, these equations are in the main very similar in structure to the GR relations, allowing a direct comparison of various features of the two theories (Ellis, 1971b). However,

- the relativistic equations include extra terms involving the pressure and acceleration;
- the Newtonian equations are not of symmetric hyperbolic form: they allow instantaneous communication of information;
- there are particular cases where there are no simple Newtonian analogues to the relativistic variables, in particular the 3-space Ricci curvature $^3R_{ab}$ and the magnetic part of the Weyl tensor H_{ab}.

Thus when these terms are significant, we expect, on taking appropriate limits, either to get the Newtonian equations plus correction terms due to relativistic effects (e.g. modifications due to spatial curvature), or to find there is no Newtonian correspondence (as in the case of gravitational radiation). In the case of the Newtonian analogues of the FLRW universe models, however, there are no such problems, because of the very high symmetry of these models; the Newtonian analogues to these models have very similar equations to the GR ones, apart from extra pressure terms mentioned above.

6.8.1 Newtonian cosmology

Newton failed to develop a viable cosmological model because of the problem of ill-defined or infinite forces occurring if there is an infinite distribution of matter (see Norton (1998)). The Newtonian version of the FLRW models was derived by Milne and McCrea (Milne, 1934, McCrea and Milne, 1934) *after* the GR models had been discovered.

The key to developing the Newtonian cosmological models is the use of a potential (rather than force) formulation, with generalized boundary conditions and a generalized understanding of the nature of acceleration (hence they are not strictly Newtonian models, but rather are based in a generalized form of Newtonian theory). One uses a convective derivative (represented by a 'dot') to correspond to the GR covariant derivative along fluid world lines, and one represents the combined effects of gravitation and inertia through an 'acceleration' vector a_j, defined as in Exercise 4.3.2, which vanishes for the case of 'free fall', i.e. motion under gravity and inertia alone. Then the momentum conservation equation for a perfect fluid with density ρ_N and pressure p takes the form

$$\rho_N a_j + p_{,j} = 0 \tag{6.63}$$

(from (5.13)). The matter and energy conservation are separate equations, the former taking the form

$$\dot{\rho}_N + \rho_N \Theta = 0, \tag{6.64}$$

where $\Theta = 3\dot{\ell}/\ell$ is the fluid expansion, and the gravitational field equations are

$$\Phi^{,i}_{\ ,i} + \Lambda = 4\pi G \rho_N, \tag{6.65}$$

where G is the gravitational constant and Λ the Newtonian equivalent of a cosmological constant.

In the case of a fluid for which the shear, vorticity and acceleration a_j all vanish, we obtain the Newtonian versions of the FLRW models, which are discussed in Section 9.9. The Friedmann and Raychaudhuri equations for these follow immediately, with two differences from the GR case. First, the constant K in the Friedmann equation is simply a constant of integration, with no relation to spatial curvature. Second, the source term in the Raychaudhuri equation is ρ_N rather than $\rho + 3p/c^2$. Finally, one should note that the Newtonian potential for these models diverges at infinity, rather than obeying the usual Newtonian limits (which are tailored to isolated systems). This emphasizes the fact that 'Newtonian' cosmological models are not compatible with Newtonian gravitational theory as usually stated in textbooks (see the discussion in Section 3.4).

The Newtonian versions of anisotropic spatially homogeneous cosmologies can be developed along similar lines (Heckmann and Schucking (1955, 1956), Hibler (1976)), with no analogue of the magnetic part of the Weyl tensor H_{ab}, but a good analogue of its electric part E_{ab}, see (6.30).

Occurrence of singularities

Comparison of the Newtonian and Relativistic equations shows why the singularity issue is much worse in GR than in Newtonian Gravitational Theory (NGT).

- Firstly, the momentum conservation equation in GR has an inertial mass density $\rho + p$ instead of ρ_N as in NGT; hence the same pressure gradient generates less response in GR than in NGT. Thus pressure is less effective in counteracting inhomogeneities in GR than in NGT.
- Secondly, the Raychaudhuri equation in GR has an inertial mass density $\rho + 3p$ instead of ρ_N as in NGT; hence the same mass of matter generates a greater gravitational attraction in GR than NGT.
- Thirdly, the energy conservation equation in GR has a prefactor $\rho + p$ to the volume expansion instead of ρ_N as in NGT, hence the same fluid contraction generates a larger density increase in GR than in NGT.
- Fourthly, fluid rotation cannot spin up to resist the gravitational attraction in the same way in GR as in NGT: hence there are NGT non-singular rotating cosmologies with no GR counterparts (see Section 6.2.2).

Taken together, these effects make it much more difficult to resist singularity occurrence in GR than in NGT. Finally, in GR the result is a full spacetime singularity involving diverging spacetime curvature and indeed an end to space and time, instead of just a density divergence, as in NGT. Thus the consequences are far more catastrophic in GR.

Structure formation

Local structure formation in the expanding universe is modelled initially by linearly perturbed FLRW models, as discussed in Chapter 10. The Newtonian version of the perturbation equations can be developed exactly in parallel with the 1+3 covariant perturbation equations, see Ellis (1990). But the situation then goes nonlinear as density inhomogeneities build up.

The linearized equations are no longer applicable, and the usual procedure is then to turn to numerical solution of Newtonian equations for astrophysical systems, except in a few cases where strong gravitational fields lead to black hole formation, when relativistic methods are needed. Thus much of astrophysical cosmology is based on Newtonian rather than GR equations, tacked on to models of an earlier phase of structure formation handled via perturbed FLRW models.

However, the viewpoint of this book is that NGT exists as a valid gravitational theory only as an approximation to GR, because the latter is the correct classical theory of gravity. Thus rather than assuming they are valid a priori, one should derive the appropriate Newtonian equations from the GR equations, as an approximation that works under suitable circumstances, but not always, and in general only locally. Hence in approaching astrophysics, it is important to understand the Newtonian limit of GR models.

6.8.2 The Newtonian limit

The weak field limit of Newtonian gravitational attraction in the expanding universe, as used in structure formation calculations in a cosmological context, will come from looking at the dynamics of almost-FLRW universe models (Peebles, 1980, Bertschinger, 1992). However, as remarked before, we are concerned also with nonlinear Newtonian theory and the way this is a limit of the relativistic theory, because the physical effects are nonlinear in many astrophysical contexts. It remains an unsolved problem to show satisfactorily how the nonlinear Newtonian versions of the equations can be derived in a suitable limit from the relativistic theory.

Issues arise both because the number of gravitational field equations is quite different in the two cases (see Section 3.4), and because the GR equations are hyperbolic and have to be reduced to elliptic equations in the Newtonian limit. Consequently the nature of the initial value problem and associated boundary conditions is quite different in the two cases.

Newtonian-like solutions

To obtain a Newtonian-like form of the equations, one can assume existence of a quasi-Newtonian (Eulerian) non-comoving reference 4-velocity $n^a (n_a n^a = -1)$, such that its shear and vorticity vanish:

$$\sigma_{ab}(n) = 0, \ \omega_{ab}(n) = 0 \tag{6.66}$$

(see van Elst and Ellis (1998)). This implies that in this frame there is no magnetic part of the Weyl tensor relative to n^a:

$$H_{ab}(n) = 0. \tag{6.67}$$

Therefore the covariant gravitational equations become ODEs rather than hyperbolic equations, and no gravitational waves can occur, corresponding to the situation in Newtonian theory. Furthermore there is an acceleration potential Φ such that

$$\dot{n}^a := n^a{}_{;b} n^b = \overline{\nabla}_a \Phi, \ E_{ab} = \overline{\nabla}_{\langle a} \overline{\nabla}_{b \rangle} \Phi, \tag{6.68}$$

where $\overline{\nabla}_a$ is the covariant derivative projected orthogonal to n^a.

The conditions (6.66) imply strong restrictions on the spacetimes; the associated integrability conditions are not fully solved, and this remains an interesting problem for investigation. The Newtonian-like dynamical equations for this situation are given by van Elst and Ellis (1998), who relate this approach to Bertschinger's study of linear energy density inhomogeneities (Bertschinger, 1992), which uses a similar quasi-Newtonian frame. The key point is that the Newtonian equation (6.65) does not come from the Raychaudhuri equation, as one might initially have expected, but rather from the 'div E' equation (6.33) for the electric part of the Weyl tensor, when there is a potential for the Weyl tensor (compare (6.68)). This equation has the elliptic kind of form that is associated with Newtonian gravitational theory and so leads to the Newtonian correspondence.

Local velocities

Peculiar velocities of matter associated with large-scale motions can be studied by using such a quasi-Newtonian frame associated with a quasi-Newtonian observer with 4-velocity u^a. Matter with 4-velocity $\tilde{u}^a$ will have a velocity v^a relative to this reference frame:

$$\tilde{u}^a \simeq u^a + v^a, \ \ v^a u_a = 0, \ \ v_a v^a \ll 1. \tag{6.69}$$

The way this leads to the equations used for 'Great Attractor' studies of local velocities and to the Zel'dovich approximation for gravitational collapse is shown respectively in Ellis, van Elst and Maartens (2001) and Ellis and Tsagas (2002). An in-depth development of how this works overall, and is related to CMB anisotropies, is given in Tsagas, Challinor and Maartens (2008) (and see also Zibin and Scott (2008)). Broadly speaking one does indeed get satisfactory Newtonian-like derivations of the peculiar velocity equations and the nature of gravitational collapse as indicated by the Zel'dovich analysis. However, as discussed in Chapter 16, one should be wary of assuming that these quasi-Newtonian coordinates can be used globally in realistic cosmological models that represent large-scale structures and voids imbedded in an expanding almost-FLRW background. Their validity may be local rather than global.

Shear-free dust solutions

A warning against the assumption that a Newtonian limit of this kind is without problems in the cosmological context has already been mentioned in Section 6.2.2. It is an exact theorem that shearfree dust solutions of Einstein's field equations cannot both expand and rotate, i.e.

$$\sigma = 0 \, , \quad p = 0 \quad \Rightarrow \quad \theta \omega = 0 \, ; \tag{6.70}$$

see Ellis (1967), and, for generalizations, Stephani *et al.* (2003), Section 6.2.1. However, shearfree solutions of the corresponding Newtonian equations do exist which can both expand and rotate; compare Narlikar (1963). Consequently, the Newtonian limit is singular. Consider a sequence $GR\ (i)_{\sigma=0}$ of relativistic shearfree dust solutions with a limiting solution $GR\,(0)_{\sigma=0}$ that constitutes the Newtonian limit of this sequence. The latter solution will necessarily satisfy (6.70) because every solution $GR\,(i)_{\sigma=0}$ in the sequence does so. The corresponding exact Newtonian solution $NGT(0)_{\sigma=0}$ will therefore also necessarily satisfy (6.70). The Newtonian solutions $NGT(j)_{\sigma=0}$ that do not satisfy (6.70) are clearly not obtainable as limits of any sequence of relativistic solutions $GR\,(j)_{\sigma=0}$. Assuming Einstein's field equations represent the genuine theory of gravitational interactions in the physical universe, this result tells us that not all Newtonian solutions are acceptable approximations.

An important application of this result is as follows: Narlikar (1963) has shown that shearfree and expanding Newtonian cosmological solutions can have vorticity that spins up as the universe decreases in size and hence causes a 'bounce' (the associated centrifugal forces avoid a singularity). This would be a counter-example to the cosmological singularity theorems quoted above, if there were GR analogues of these singularity-free cosmological solutions; but the shearfree theorem in Ellis (1967) shows that there are no such GR solutions. Thus this is a case where the Newtonian models are very misleading. The Newtonian limit is singular in such cases; so we need to be cautious about that limit in other situations of astrophysical and cosmological interest.

Exercise 6.8.1 Show that the equation in Newtonian theory corresponding to (6.5) is the same except that $(\rho+3p)\to\rho_N$ (the active gravitational mass density is just the mass density).

Observations in cosmological models

The test of a cosmological model is how well it reproduces and predicts astronomical observations of objects at cosmological distances. Thus it is important to determine what features we are able to measure by such observations, and how they are related to the cosmological model.

7.1 Geometrical optics and null geodesics

The basis of astronomical observations is the geometric optics limit of Maxwell's equations, supplemented by the quantum mechanical concept of a photon. The photon viewpoint enters into detector design (in the case of very distant galaxies, individual photons are detected), but the geometric optics approximation is used to describe the propagation of radiation through a curved spacetime. Information is conveyed to us along light-rays which are null geodesics on the future light-cone of the emitter and the past light-cone of the observer.

7.1.1 Geometric optics approximation

Maxwell's equations (5.115) in a source-free region can be written as (Exercise 5.5.4)

$$F_{ab} = 2\nabla_{[a}A_{b]}, \ \nabla_b\nabla^b A_a + R_{ab}A^b = 0, \ \nabla_a A^a = 0. \tag{7.1}$$

The last equation defines the Lorentz gauge, imposed by using the gauge freedom $A_a \rightarrow A_a + f_{,a}$. The remaining freedom in A_a is then a similar gauge transformation but with $\nabla^a\nabla_a f = 0$.

Now we assume that there are solutions of these equations of the form

$$A_a = g(\psi)\alpha_a + \text{small tail terms}, \tag{7.2}$$

where (a) $g(\psi)$ is an arbitrary function of the phase ψ, and (b) g varies rapidly compared with the amplitude α_a, in the sense that

$$\left| g' k_{[a}\alpha_{b]} \right| \gg \left| g \nabla_{[a}\alpha_{b]} \right|, \tag{7.3}$$

where $g' := \partial g / \partial \psi$ and we have defined the propagation vector k_a as

$$k_a := \psi_{,a} \ \Rightarrow \ k_{[a;b]} = 0. \tag{7.4}$$

Condition (a) represents the essential feature that arbitrary information can be propagated by the signal,[1] while (b) is the condition that the solution represents a high-frequency wave with a relatively slowly varying amplitude. One can show that at a large distance from an isolated system in an almost-FLRW expanding universe, the radiation field does in fact have this form (Hogan and Ellis, 1989).

Substituting (7.2) into (7.1), ignoring the tail terms, and equating to zero separately the coefficients of g, g', and g'' (since $g(\psi)$ is arbitrary), we find

$$k^a k_a = 0 \quad \Leftrightarrow \quad k^a \nabla_a \psi = 0, \tag{7.5}$$

$$2k^b \nabla_b \alpha_a = -\alpha_a \nabla_b k^b, \tag{7.6}$$

$$\nabla_b \nabla^b \alpha_a + R_{ab}\alpha^b = 0, \tag{7.7}$$

$$\alpha_a k^a = 0, \quad \nabla_a \alpha^a = 0. \tag{7.8}$$

We now consider the implications of these equations.

Null geodesics

Firstly, (7.5) shows k_a is null; then the first of (7.8) shows α_a is spacelike (if it were parallel to k_a, then by (7.11) below it would generate no electromagnetic field). Secondly, (7.5) implies $k^a \nabla_b k_a = 0$, so, by (7.4),

$$k^b \nabla_b k^a = 0, \tag{7.9}$$

i.e. the light-rays $x^\mu(v)$ tangent to $k^\mu = dx^\mu/dv$ are null geodesics, as expected by the foundational principles of GR.

It follows that light-rays are differentially bent by an inhomogeneous gravitational field; thus a curved spacetime will in general distort optical images (Jordan, Ehlers and Sachs, 1961, Penrose, 1968). By (7.5), $d\psi/dv = 0$, so $\psi(x^\mu)$ is constant along these light-rays, and the surfaces $\psi = $ const are the future light-cones of the emitter's world line. It follows further that $dg(\psi)/dv = 0$, that is, the signal function g is a constant on each light-ray, showing that the arbitrary information expressed in the function g is propagated unchanged along these rays.

As we look down the past light-cone, observations of distant objects will necessarily see them as they were in the past. This is an essential limitation and a difficulty in interpreting observations. It is hard to draw a meaningful comparison with Newtonian theory, since Maxwell's equations are incompatible with the symmetries of Newtonian geometry, and there is no experimentally viable purely Newtonian theory of light propagation.

Polarization

The direction of the amplitude α_a is parallely propagated along the light-rays by (7.6). Thus the state of polarization of the light is completely unaffected by the curvature of

---

[1] The fact that g is a function of ψ alone implies that this function is isotropic at the emitter. This is not a serious limitation because α_a can vary with direction at the emitter, so the case of a source emitting radiation anisotropically is included.

spacetime. More precisely, there are two independent solutions of (7.6) along the light-rays (because the spacelike vectors orthogonal to k^a span a two-dimensional surface). Any numerical parameters describing the polarization represent the relative magnitude of these components of the solution, and will be constant along these null geodesics because each satisfies the same equation, (7.6). Furthermore any spatial directions associated with the polarization are determined from the vectors α_a, which are parallelly propagated along the light-rays, so such spatial directions will also be parallelly propagated along the light-rays.

Amplitudes

Equation (7.6) also shows that the square magnitude $\alpha^2 = \alpha^a \alpha_a$ propagates along the light-rays via

$$\frac{\mathrm{d}}{\mathrm{d}v}\alpha^2 = -\alpha^2 \nabla_a k^a. \tag{7.10}$$

This determines the intensity properties of the radiation, as we shall see later.

By (7.7) and (7.8), the amplitude α_a satisfies the same relations as the full potential A_a. These equations will play no further part in the present discussion; their essential effect is to show that we cannot in general omit tail terms (propagating off the light-cone) if (7.2) is to be an exact solution of (7.1).[2] However, these terms are small if spacetime curvature is small, and they do not affect photon propagation.

Electromagnetic field

From (7.1) and (7.2),

$$F_{ab} \approx g'(k_a \alpha_b - \alpha_a k_b). \tag{7.11}$$

This determines, by (5.112) and (5.113), the electric and magnetic fields. If we use the remaining gauge freedom to set α_a orthogonal to u^a at the point of observation, then

$$E_a \approx g'\alpha_a(-k_b u^b), \ \ B_a \approx g'\eta_{abc}k^b\alpha^c \quad \text{at the observer.} \tag{7.12}$$

This shows the standard radiation pattern: E_a and B_a are at each instant equal in magnitude, and are orthogonal to each other and to the radiation propagation direction $k_{\langle a \rangle}$.

Splitting the ray 4-vector

We can split the ray 4-vector into parts parallel and orthogonal to the observer's 4-velocity:

$$k^a = (-u_b k^b)(u^a + e^a), \ \ e^a e_a = 1, \ \ e^a u_a = 0, \tag{7.13}$$

here $e^a = k^{\langle a \rangle}/(-u_b k^b)$ is the propagation direction of the light-ray, as measured by u^a. The factor $-u_b k^b$ is proportional to the photon frequency v as measured by u^a-observers.

---

[2] Spacetimes in which electromagnetic waves can propagate without tails are either conformally flat or conformally plane-wave spacetimes (Friedlander, 1975, McLenaghan and Sasse, 1996); FLRW models are conformally flat, but perturbed FLRW are not.

The proportionality constant can be fixed by the freedom to rescale the affine parameter. We will choose $-u_b k^b = v$. A small increment dv in the affine parameter v will correspond to a displacement $k^a dv$, measured by the observer to be a time difference δt and a spatial distance δl, where, compare (4.13),

$$|\delta t| = |\delta l| = (-k_a u^a) dv. \tag{7.14}$$

FLRW case

The implications of the geodesic equation (7.9) for cosmology are fundamental. For a RW metric (2.65), one can integrate these equations in three ways. Firstly, directly, by calculating $\Gamma^\mu_{\nu\sigma}$ and then integrating (7.9) to find $x^\mu(v)$. Secondly, by finding k^a from the fact that there is a first integral $\xi_A := \xi^a_A k_a$ for each Killing vector field ξ^a_A in FLRW. Thirdly, and most economically, by using the spacetime symmetry to determine the null geodesics directly from (2.65). We follow the last approach here.

By spatial homogeneity and isotropy, any null geodesic is equivalent to a radial null geodesic through the origin of coordinates (we can choose the origin to lie on the null geodesic, because all spatial locations are identical; and all null geodesics through one point are equivalent to all the others, because of isotropy). From (2.65), radial null geodesics are characterized by $dt^2 = a^2(t) dr^2$. Then

$$u(t_e, t_0) := r_e - r_0 = \int_{t_e}^{t_0} \frac{dt}{a(t)} = \int_{a_e}^{a_0} \frac{da}{a^2 H}, \tag{7.15}$$

determines the past light-cone of an arbitrary point in an FLRW universe with given matter content and curvature.

Exercise 7.1.1 Derive in detail the polarization results sketched above.

Exercise 7.1.2 Integrate (7.9) in an FLRW geometry (for radial geodesics only) to find $k^\mu(v)$ and $k^\mu(x^\nu)$. Show from this that $k^\mu u_\mu = -1/a(t)$ and $dv = a(t) dt = a^2(t) dr$ on a radial null geodesic, and calculate $\nabla_\mu k^\mu$.

Problem 7.1 Find the equations for non-radial null geodesics in a RW spacetime (needed to calculate observational relations for a source not at the origin). Hint: try changing coordinates, or using the Killing vector fields.

7.2 Redshifts

The *redshift z* of a source as measured by an observer is defined in terms of the wavelength λ of light by

$$z := \frac{\lambda_o - \lambda_e}{\lambda_e} = \frac{\Delta\lambda}{\lambda_e}, \tag{7.16}$$

where o refers to the observer and e to the emitter. The measurement of redshifts is done by identifying absorption or emission lines for particular elements in the spectra of distant objects, measuring their observed wavelength, and comparing this with the known (laboratory) wavelength of these lines for a source at rest. The interpretation depends on assuming these spectra were the same in the past, i.e. that atomic physics is unchanged over cosmological time scales.

The rate of change of any signal $g(\psi)$ measured by an observer moving with 4-velocity $u^\mu = dx^\mu/d\tau$ is $dg/d\tau = g'k_\mu u^\mu$. If observers u_1^a, u_2^a measure the rate of change of the same signal $g(\psi)$, these will be in the ratio $(k_a u_1^a)/(k_b u_2^b)$. By (7.16),

$$1 + z = \frac{\lambda_o}{\lambda_e} = \frac{\nu_e}{\nu_o} = \frac{(u_a k^a)_e}{(u_b k^b)_o}. \tag{7.17}$$

This determines the redshift from the 4-velocity vectors $u^a|_o$, $u^a|_e$, and the tangent vector k^a to the null geodesic. The relation is true no matter what the separation of the emitter and observer, and holds independent of any interpretation of the redshift as 'Doppler' or 'gravitational'.

The major characteristic of the redshift effect is that *the fractional change in wavelength is the same for all wavelengths*; if this is not true in some spectra, then the explanation cannot be a simple redshift phenomenon. It must be emphasized that the effect is essentially a *time dilation* effect, as is clear from the above derivation: *all* observed phenomena in the source will appear to be slowed down in the same ratio (e.g. if a quasar has redshift 3, then any observed variations in its luminosity will be seen to occur at a rate four times slower than they are happening at the source). We usually refer to the effect in terms of (spectral) redshift because this happens to be the most reliable way of measuring effective time dilation.

7.2.1 Linear redshift relation in cosmology

Suppose the emitter and observer are fundamental observers. Then the change in $u^a k_a$ occurring in a parameter displacement dv along a null geodesic is $d(u^a k_a) = k^b \nabla_b (u^a k_a) dv = (u_{a;b} k^a k^b) dv + u_a (k^b \nabla_b k^a) dv$. The second term vanishes by (7.9). Substituting from (4.38) and (7.13),

$$d(u^a k_a) = (\Theta_{ab} e^a e^b + \dot{u}_a e^a)(u^c k_c)^2 dv, \tag{7.18}$$

where e^a is the ray propagation direction. The change $d\lambda$ in wavelength in the parameter distance dv is given by $d\lambda/\lambda = -d(u_a k^a)/(u_b k^b)$, so the change of redshift along the null geodesic segment is (Ehlers, 1961)

$$\frac{d\lambda}{\lambda} = \left(\frac{1}{3}\Theta + \sigma_{ab} e^a e^b + \dot{u}_a e^a\right) dl = (dl)\dot{} + \dot{u}_a e^a \, dl, \tag{7.19}$$

where the second equality follows from (4.32). The redshift has been split relative to the fundamental 4-velocity u^a into a 'Doppler' part (the first term, determined by the expansion tensor) and a 'gravitational' part (the second term, determined by the acceleration vector). Furthermore, we see how this redshift–distance relation varies with direction in the sky: the angular dependence of the terms due to Θ (isotropic monopole), $\dot{u}_a$ (dipole) and σ_{ab}

(quadrupole) are different, so we can in principle observe these quantities directly from this relation, by measuring redshifts at different points in the sky and estimating the corresponding distances from the observed brightness of the sources (see below), assuming there are sufficient sources within the range where the linearized approximation holds.

For the preferred static observers in a static spacetime, we shall have only the acceleration term, giving the usual prediction of gravitational redshifts. An interesting question is whether there could be significant gravitational redshifts in a cosmological context.[3]

FLRW case

In an FLRW model, there will only be an isotropic Doppler contribution, giving the standard prediction of cosmological redshift z in an expanding universe (see Chapter 13):

$$1 + z = \frac{a(t_0)}{a(t_e)}, \tag{7.20}$$

where t_0 is the time of observation and t_e the time of emission. This result may be obtained in various ways:

(1) integrating (7.19) with $\sigma_{ab} = 0 = \dot{u}_a$;
(2) showing directly that $k^a u_a = -a^{-1}$ from Exercise 7.1.2 and (7.17);
(3) considering light pulses emitted at t_e and $t_e + \delta t_e$, received at t_0 and $t_0 + \delta t_0$, and using (7.15) for each light-ray to calculate $\delta t_0 / \delta t_e$ (note: r_e and r_0 are constant since source and observer are both comoving).

7.2.2 Different contributions to redshift

Real sources move relative to the fundamental 4-velocity, and may also have gravitational fields sufficient to alter the observed redshifts. Because redshifts are due to time dilation, these effects are multiplicative; that is, if z_{eD} is the Doppler shift due to the relative motion of the source and z_{eG} the redshift due to its gravitational field, and similarly for the observer, and if z is the cosmological redshift determined by (7.19), then the total redshift z_{tot} is given by

$$(1 + z_{\mathrm{tot}}) = (1 + z)(1 + z_{eD})(1 + z_{eG})(1 + z_{oD})(1 + z_{oG}). \tag{7.21}$$

Unfortunately it is only the total redshift z_{tot} that can be measured from spectra; there is no direct way that these different contributions to the redshift of a particular source can be separated out. What we have to rely on is that (by definition of the average velocity) the source Doppler shifts will cancel out if we observe sufficient sources in some spacetime region, and the observer redshift can be measured from the CMB temperature anisotropy, while we believe we can estimate the gravitational redshifts on physical grounds. The problem lies in identifying which sources lie 'in the same spacetime region'; it is clear that we can only obtain correct results if we can identify which distant objects (seen together in

---

[3] Ellis, Maartens and Nel (1978) developed a non-standard model based on this idea, but it is not a realistic universe model.

one image of the sky) are in fact part of the same cluster, i.e. lie close together in space. It is difficult not to end up with a circular argument (using the redshifts to claim they do lie in a cluster).

It is possible to argue that these steps can all be done in a reasonable way, separating out the different redshift contributions, but nevertheless this ambiguity remains a problem and underlies arguments about anomalous redshift effects (Flesch and Arp, 1999, Arp and Carosati, 2007). It also leads to the 'finger of god' distortion of galaxy clustering in redshift (see Section 12.3.5).

Exercise 7.2.1 Deduce from the above the standard radial 'Doppler' result $1 + z = \exp(-\beta) = \sqrt{(1+V)/(1-V)}$, where $V = \tanh\beta$ is the radial speed of the emitter relative to the observer, assuming they are so close to each other that we can use the flat-space approximation. (Hint: $u_e^a = \cosh\beta\, u_o^a + \sinh\beta\, e^a$, where $e^a e_a = 1$, $e_a u^a = 0$, and $k^a \propto u^a - e^a$.) Carry out an analogous calculation to determine the transverse Doppler effect (emitter motion is transverse to the line of sight).

Exercise 7.2.2 Integrate (7.19) and evaluate (7.17) in an FLRW universe.

7.3 Geometry of null geodesics and images

The observer's *screen space* is the two-dimensional space in the rest frame of u^a orthogonal to k^a. It is spanned by orthonormal vectors e_1^a and e_2^a, orthogonal to both u^a and e^a (see (7.13)). This represents the surface of a screen on which images conveyed by the light-rays are displayed. The metric tensor of screen space is

$$s^{ab} = h^{ab} - e^a e^b = e_1^a e_1^b + e_2^a e_2^b, \quad e_{Ia} e_J^a = \delta_{IJ}, \quad e_I^a e_a = 0 = e_I^a u_a, \tag{7.22}$$

(where $I, J = 1, 2$) which is a projection tensor into screen space: $s^a{}_b s^b{}_c = s^a{}_c$, $s^a{}_a = 2$, $s_{ab} u^b = 0$, $s_{ab} k^b = 0$.

For a given light-ray and k^a, the screen space depends on the observer 4-velocity. Consider a vector X^a representing a displacement in the image that links the light-rays at the extremities of the image. As in the timelike case (Section 4.5), this vector will be a connecting vector linking the light-rays, and so satisfying the differential equation

$$k^b \nabla_b X^a = X^b \nabla_b k^a. \tag{7.23}$$

Placing a screen or equivalent detector orthogonal to the rays projects the image into the screen space, in effect by projecting each connecting vector X^a into a relative position vector $\widehat{X}^a := s^a{}_b X^b$ of points in the actual image. This vector still connects the same light-rays but is in the instantaneous rest-space of the observer; the observed displacement on the screen will connect the same pair of points.

Varying the 4-velocity, $u^a \to \tilde{u}^a$, will only change the relative position vector by a multiple of the null vector k^a: i.e. $\widehat{X}^a \to (\widehat{X})^{\tilde{}a} = \widehat{X}^a + \alpha k^a$ for some scalar α. Then the scalar product of any two such relative position vectors representing displacements in the image is unchanged: $(\widehat{X})^{\tilde{}a}(\widehat{Y})^{\tilde{}}_a = \widehat{X}^a \widehat{Y}_a$, because k^a is null and orthogonal to both. Thus (Jordan, Ehlers and Sachs, 1961) *the shape and size of any image is independent of the motion of the observer.* (It is easiest to see this by thinking of the shadow cast by an opaque object, but the result will be true for any image formation process where the information is conveyed along null geodesics.)

7.3.1 Optical kinematic quantities

The propagation of images along the light-rays is characterized by optical kinematic quantities defined in the screen space – which are closely analogous to the kinematic quantities for timelike curves (Section 4.6). Since $k_a = \psi_{,a}$, the light-rays will have no vorticity (Section 6.3.1). Furthermore, the light-rays are geodesic, so there is no acceleration. Thus only the null expansion and shear remain:

$$\widehat{\Theta}_{ab} = s_a{}^c s_b{}^d \nabla_c k_d = \widehat{\sigma}_{ab} + \tfrac{1}{2}\widehat{\Theta} s_{ab}, \ \ \widehat{\Theta} = \nabla_a k^a, \ \widehat{\sigma}_{ab} s^{ab} = 0. \tag{7.24}$$

The null expansion $\widehat{\Theta}$ measures the area rate of expansion of images:

$$\frac{\mathrm{d}}{\mathrm{d}v} \delta S = \tfrac{1}{2}\widehat{\Theta}\, \delta S. \tag{7.25}$$

The null shear measures the rate of distortion of images.

The rate of change of the null expansion along the light-rays in turn is determined by the null version of the Raychaudhuri equation:

$$\frac{\mathrm{d}}{\mathrm{d}v}\widehat{\Theta} = -\tfrac{1}{2}\widehat{\Theta}^2 - \widehat{\sigma}_{ab}\widehat{\sigma}^{ab} - R_{ab}k^a k^b, \tag{7.26}$$

analogous to (4.45). This shows how matter (which directly determines the Ricci tensor through the field equations) will tend to converge the light-rays. For example, for a perfect fluid, from (5.37),

$$R_{ab}k^a k^b = 8\pi G(\rho + p)(u^a k_a)^2, \tag{7.27}$$

which will be positive if (5.16) is satisfied. It will be zero for a Λ-dominated universe.

The rate of change of the null distortion is determined by

$$\frac{\mathrm{d}}{\mathrm{d}v}\widehat{\sigma}_{ab} = -\widehat{\Theta}\widehat{\sigma}_{ab} - C_{acbd}k^c k^d, \tag{7.28}$$

analogous to (4.51). As in the timelike case, anisotropy (which must occur if there is inhomogeneity) sources Weyl curvature, which causes distortion through (7.28) and hence 'light bending' – which in turn causes convergence through (7.26). This distortion is referred to as *gravitational lensing*. In Sections 12.4 and 12.5 we shall discuss the various cases of lensing: weak lensing, strong lensing and microlensing.

The similarity of the timelike and null cases is due to the fact that both are essentially determined from the geodesic deviation equation, or its generalization to non-geodesic curves: see Section 2.6. The null versions differ from the timelike ones as follows: the factors $\tfrac{1}{3}$ change to $\tfrac{1}{2}$ (screen space is two-dimensional, rest space is three-dimensional); the null version presented here only covers geodesic and rotation-free curves (the case applicable to geometric optics), while the timelike case includes completely general motions; matter cannot directly cause distortion of light-rays, while an imperfect fluid can do so in the case of timelike curves.

The clearest way of representing lensing effects is through the geodesic deviation equation for null geodesics (Lewis and Challinor, 2006, de Swardt, Dunsby and Clarkson, 2010a). By (7.13), $k^a = v(u^a + e^a)$, and then we find (Example 7.3.1)

$$\frac{\delta^2 X^a}{\delta v^2} = -\frac{1}{2} R_{bc} k^b k^c \widehat{X}^a - v^2 \left(2\widehat{E}^{ab} - \widehat{E}^c{}_c s^{ab} + 2\widehat{H}^{c\langle a} \eta^{b\rangle}{}_c \right) X_b$$
$$- \left[\left(E_{bc} + \eta^d{}_b H_{cd} \right) k^c - \frac{1}{2} v \left(\pi_{bc} e^c - q_b \right) \right] \widehat{X}^b k^a, \tag{7.29}$$

where a hat denotes a projection into screen space (via s_{ab}), and $\eta_{ab} = \eta_{abc} e^c$ is the alternating tensor in screen space. This is the basic equation for gravitational lensing, a key tool in present-day cosmology, discussed in Chapter 12.

Exercise 7.3.1 Derive (7.29). (de Swardt, Dunsby and Clarkson, 2010a)

Exercise 7.3.2 Show that for FRLW with dust and Λ,

$$-k^a u_a dv = -dt$$
$$= \frac{dz}{(1+z)[H_0^2(1+2q_0 z) + (\Lambda/3)\{2z - 1 + (1+z)^{-2}\}]^{1/2}}. \tag{7.30}$$

Exercise 7.3.3 Using the result of Exercise 7.1.2, integrate (7.25)–(7.28) in an FLRW universe for a congruence:
(a) with vanishing shear;
(b) with non-zero shear (relevant to gravitational lensing).

7.4 Radiation energy and flux

In the geometric optics case given by (7.11), the electromagnetic stress tensor (5.123) takes the form

$$T_{ab} \approx \alpha^2 (g')^2 k_a k_b. \tag{7.31}$$

By (7.5), this is the energy–momentum tensor of particles moving at the speed of light, sometimes called 'null dust'. We can regard this as the (classical) stress tensor of photons conveying energy from the source to the observer. The conservation law $\nabla_b T^{ab} = 0$ is equivalent to (7.9) and (7.10), as we might expect, because the source-free Maxwell equations imply energy–momentum conservation for the electromagnetic field.

From (7.31), an observer with 4-velocity u^a finds the instantaneous flux across a surface perpendicular to k^a to be the same as the instantaneous energy density of the radiation and as the pressure exerted by the radiation in the ray propagation direction e^a [(7.13)]. All are equal to

$$\mathcal{F} = T_{ab} u^a u^b \approx \alpha^2 (g')^2 (k_a u^a)^2. \tag{7.32}$$

(The pressure orthogonal to k^a is zero.) However, what is measured in practice by an observation is not $\mathcal{F}$ but some convolution with the response function of a detector leading to a weighted time-average of $\mathcal{F}$ over a large number of high-frequency oscillations. More precisely, the observed flux F (the rate at which radiation crosses a unit area of surface per unit time) is the convolution of $\mathcal{F}$ over the time of observation with the detector frequency response function; the result can be written as $G(\psi)\alpha^2(k_a u^a)^2$ where $G(\psi)$ is a slowly varying function of ψ. The measured flux can therefore be written

$$F = G(\psi)\alpha^2(k_a u^a)^2, \tag{7.33}$$

where $G(\psi)$ is constant along the null geodesics and α^2 obeys equation (7.10).[4] F may also be given as the source's *apparent magnitude* m, defined by

$$m := -2.5\log_{10} F + \text{const.} \tag{7.34}$$

7.4.1 Image intensity and apparent size

Consider a bundle of null geodesics diverging from a source at some instant (and so lying in a light-cone $\psi = \text{const}$). We wish to determine how the flux of radiation (and so image intensity) varies along these light-rays. If we combine (7.10) and (7.25), we find that $d(\alpha^2 \delta S) = 0 \Leftrightarrow \alpha^2 \delta S = \text{const}$, i.e. the magnitude α^2 varies inversely as the cross-sectional area δS of the bundle. Furthermore the flux F, and hence the factors α^2, $(k_a u^a)^2$ in (7.33) are measured at the observer. Thus by (7.17), $F = G(\psi)(\alpha_e^2 \delta S_e / \delta S_o)(k_a u^a)_e^2(1+z)^{-2}$, i.e.

$$F\delta S = C \frac{d\Omega}{(1+z)^2}, \tag{7.35}$$

where C is constant along the bundle of null geodesics (depending only on the source properties at the time of emission and the detector response function) and $d\Omega$ is the solid angle subtended by the null geodesics at the source (which is clearly a constant along the geodesics).

In physical terms, this result may be understood in the following way: each photon from the source has an energy $h\nu$. The total energy conveyed by these photons per unit time is proportional to (1) the number of photons arriving per unit time, leading to one factor $(1+z)^{-1}$ (because the rate at which they are measured to arrive is in that ratio to their rate of emission), and (2) the energy per photon, which depends on the frequency of the photons at the observer, leading to the second factor $(1+z)^{-1}$ (which is the ratio of the frequency at the observer to the frequency at the emitter). In addition to these factors, the flux F of energy observed (the energy arriving per unit area per unit time) is proportional to (3) $1/\delta S$, because photons are conserved along the bundle of null geodesics (so the same number of photons are spread out over a larger area as the null rays diverge).

When the energy condition $\rho + p > 0$ holds, the spacetime curvature tends to cause the bundle of null geodesics to converge, by (7.27) and (7.26). If there is sufficient matter present to start reconverging the null geodesics, so that the cross-sectional area δS is the same at two points far apart (say P and Q), then the factor α^2 will be the same at these two points. Thus the source will seem anomalously bright to an observer at Q; if the observers

---

[4] From now on we ignore corrections due to tail terms, and so replace $\approx$ by $=$.

at Q and P both adjust their velocities so as to see the same redshift, they will both measure the same flux of radiation from the source, although one is much further from it than the other. Near a point where the null geodesics are refocused, this gravitational lens effect can in principle produce very high fluxes. It is now observed to occur in many gravitationally lensed galaxies, and enables us to see lensed galaxies at much greater distances than those not lensed.

7.4.2 Source luminosity

The constants in (7.35) still have to be related to the source characteristics. The *luminosity* L of the source is defined as the total rate of emission of radiant energy. In principle, L at some instant t_e would have to be measured by enclosing the source in a 2-sphere S and measuring the rate at which radiation emitted at time t_e crosses each surface element dS of the sphere. Then we form the integral

$$L = \int_S (1+z)^2 F \, dS. \tag{7.36}$$

By (7.35), this is a constant, independent of the choice of the 2-sphere and of its motion; it is just the source luminosity.

In practice, we observe the flux from the source along some bundle of geodesics which subtends a small solid angle $d\Omega$ at the source. Consider a sphere lying in the locally Minkowski spacetime near the source, surrounding it and centred on it (this implies that the sphere's 4-velocity is the same as that of the source, so $z = 0$ on this sphere), and on which the bundle of geodesics has cross-sectional area dS. Then on this sphere, $F \, dS = C \, d\Omega$ (measurable by local observations of flux across the area dS). From (7.35) and (7.36), $L = \int_S C \, d\Omega$. Assuming the source radiates isotropically, the value of C on this sphere (the flux emitted per unit solid angle) is the same everywhere, so that $C = L/4\pi$ (the fraction of the total luminosity emitted into the solid angle $d\Omega$). Replacing C in (7.35) by this relation, we find

$$F \, dS = \frac{L}{4\pi} \frac{d\Omega}{(1+z)^2}. \tag{7.37}$$

When the source radiates anisotropically (e.g. if some local mechanism causes substantial beaming), then the radiation emitted in a particular solid angle $d\Omega$ is not simply proportional to $L \, d\Omega$ (as more radiation is emitted in some directions than others), and the relation between F and L has to be modified accordingly to take this anisotropy into account.

7.4.3 Area distances and luminosity distance

We define the *galaxy area distance* r_G by the relation

$$dS_G = r_G^2 d\Omega_G, \tag{7.38}$$

where the subscript G denotes a bundle of light-rays diverging from the source, $d\Omega_G$ is its solid angle at the source and dS_G is its cross-sectional area at the observer – see Figure 7.1.

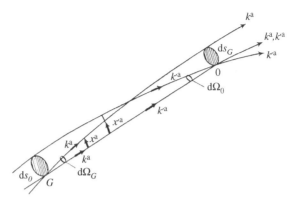

Fig. 7.1 Areas and angles at observer O and galaxy G.

Then we can rewrite (7.36) as

$$F = \frac{L}{4\pi} \frac{1}{(1+z)^2 r_G^2}. \tag{7.39}$$

These are convenient ways of expressing the observed flux in a curved spacetime as an 'inverse square' law. One should not, however, be misled by this apparent simplicity: the crucial point is how the area distance defined by (7.38) relates to other measures of distance along the light-rays (coordinate distance, affine parameter distance or redshift). We shall look at specific examples of this later, e.g. in discussing the FLRW universe models.

The basic problem with the galaxy area distance r_G is that, from its definition (7.38), it is not directly observable: an observer O can measure dS_G (the area of O's detector surface determines the bundle of rays), O cannot – without knowing r_G – determine the solid angle $d\Omega_G$ into which this radiation was emitted. We can define a closely related quantity, the *observer area distance* r_O, which is in principle directly measurable (for objects of known intrinsic size). This is obtained by considering a bundle of rays diverging from the observer to the emitter, subtending a solid angle $d\Omega_O$ at the observer and with cross-sectional area dS_O (the subscript O denotes the ray bundle diverging from the observer) – see Figure 7.1. Then r_O is defined by

$$dS_O = r_O^2 d\Omega_O. \tag{7.40}$$

This is in principle measurable for an object of known type, because the observer can measure $d\Omega_O$, the solid angle in the sky subtended by the object, and estimate dS_O, its cross-sectional area.

The quantity r_O is often simply called the *area distance*, or the *angular diameter distance*, sometimes denoted D_A:

$$D_A = r_O. \tag{7.41}$$

The quantity r_G is related to the *luminosity distance* D_L, defined by

$$D_L := (1+z)r_G, \tag{7.42}$$

so that (7.39) becomes

$$F = \frac{F_G}{D_L^2}, \quad F_G := \frac{L}{4\pi}. \tag{7.43}$$

This distance has the advantage that it is in principle directly observable by flux measurements. Using (7.34) we can rewrite (7.39) as

$$\mu := m - M = 5\log_{10} D_L + \text{const}, \quad M := -2.5\log_{10} L, \tag{7.44}$$

where m is the observed (apparent) magnitude and M is the intrinsic (absolute) magnitude, and μ is known as the distance modulus.

7.4.4 Reciprocity or distance-duality relation

We now have two apparently independent area distances, r_G and r_O (or equivalently D_L and D_A), between a given galaxy and an observer, the first defined along a future-directed ray bundle from source to observer, the second along a past-directed ray bundle from observer to source. These share common rays that link points in the object to points in the detector. In fact, provided that photons are conserved between source and observer, there is a conserved quantity along such a common geodesic Γ, for connecting vectors of the past-going and future-going families of geodesics, and consequently there is a simple relation between the area distances.

To see this, let the two families of geodesics have tangent vectors k^a, k'^a respectively, coinciding on the common geodesic Γ (see Figure 7.1). Let v, v' be affine parameters and X^a, X'^a be connecting vectors for k^a, k'^a respectively. Each of X^a, X'^a satisfies the geodesic deviation equation along Γ. It follows that

$$X'^a k^b \nabla_b X_a - X^a k^b \nabla_b X'_a = \text{const along } \Gamma. \tag{7.45}$$

Evaluating this constant at O (where $X'^a = 0$) and at G (where $X^a = 0$), we find

$$(X'^a k^b \nabla_b X_a)_G = -(X^a k^b \nabla_b X'_a)_O. \tag{7.46}$$

Now this is true for all pairs of such connecting vectors X^a, X'^a. Choosing two pairs of such vectors that are orthogonal at both O and G, we obtain the relation,

$$\mathrm{d}S_G \, \mathrm{d}\Omega_O (k^a u_a)^2|_O = \mathrm{d}S_O \, \mathrm{d}\Omega_G (k^a u_a)^2|_G. \tag{7.47}$$

Hence we obtain (Etherington, 1933)

Theorem 7.1 Reciprocity or distance-duality relation.
If photons are conserved, area and luminosity distances obey

$$r_G = (1+z)r_O \quad \Leftrightarrow \quad D_L = (1+z)^2 D_A. \tag{7.48}$$

Observational tests of the distance-duality between D_L and D_A are an important probe of a fundamental law in cosmology (Bassett and Kunz, 2004). Currently, these tests do not reveal any statistically significant deviation from (7.48).

Reciprocity shows that the two distances are the same apart from redshift factors (which essentially result from the special-relativistic transformation law for solid angles). If there is a gravitational lens effect leading to an anomalously large source apparent size, this is accompanied by an anomalously large radiation flux. If we have the situation that light is refocused, so that the angular diameter of an object of a given size decreases to a minimum and then starts increasing again as it is moved further down the past light-cone of the observer, then the flux received in a given solid angle at the observer from the near and far sources can be the same (up to redshift factors).

We can rewrite (7.39) in the form

$$F = \frac{F_G}{D_L^2} = \frac{F_G}{(1+z)^2 D_A^2}. \tag{7.49}$$

Different powers of $(1+z)$ will occur in luminosity relations depending on the distance definition used – which has been a source of considerable confusion in the past. The essential point is that (unlike the case of a flat spacetime) there are different possible definitions of distance in a curved spacetime, depending on what observation we have in mind to use to determine the separation of emitter and observer.

7.4.5 FLRW area and luminosity distances

In FLRW models we can determine any one of these distances as a function of redshift and cosmological parameters, so that these observational relations can be written in explicit form in terms of the Hubble rate and density parameters.

Consider a bundle of past-directed rays diverging from the event $P := \{r = 0, t = t_0\}$, bounded by the four rays (θ, ϕ), $(\theta + d\theta, \phi)$, $(\theta, \phi + d\phi)$, $(\theta + d\theta, \phi + d\phi)$ (note that the coordinates θ and ϕ are constant along radial null geodesics). They subtend a solid angle $d\Omega_O = \sin\theta d\theta d\phi$. From (2.65) with $dt = 0$, $dr = 0$, they will span an area $dS_O = a^2(t_e) f^2(u) d\Omega_O$ at time t_e on the past light-cone of P, where u is the comoving radial distance defined in (7.15). Comparing with (7.40) we see that the area distance is given by

$$D_A = a(t_0)(1+z)^{-1} f\left(\int_0^z \frac{dz'}{H(z')}\right), \tag{7.50}$$

where we have used $a(t_0) = a(t_e)(1+z)$. Similarly, for geodesics diverging from the source at time t_e the luminosity distance is given by

$$D_L = a(t_0)(1+z) f\left(\int_0^z \frac{dz'}{H(z')}\right). \tag{7.51}$$

It follows that these formulae obey distance-duality (7.48).

Observational tests based on D_L, D_A are discussed in Section 13.2. In some cases, one can write these distances as simple functions of z. For a dust universe with $\Lambda = 0$, we can use (7.15) to determine u, and we find the Mattig relations,

$$D_A(z) = \frac{2}{H_0 \Omega_{m0}^2 (1+z)^2} \left[\Omega_{m0}z - (2 - \Omega_{m0})\left(\sqrt{1 + \Omega_{m0}z} - 1\right)\right]. \tag{7.52}$$

An important consequence of (7.52) is *refocusing of the past light-cone*: for any specific value of Ω_{m0} there will be some redshift z_* for which the area distance is a maximum because the cross-sectional area of the past light-cone is a maximum there. The universe as a whole acts a giant gravitational lens making everything at larger distances appear anomalously large. All objects at higher redshifts will subtend the same angular size as a similar object at some lower redshift, and so will be assigned the same area as that closer object. Hence they will also be anomalously luminous. For the Einstein–de Sitter universe ($K = 0$), we find $z_* = 1.25$; for lower density universes it has larger values.

Exercise 7.4.1 Reciprocity/distance-duality: fill in the details of the above derivation (Ellis, 1971a).

Exercise 7.4.2 Derive (7.52) directly from the geodesic deviation equation for null geodesics (Ellis and van Elst, 1999a).[5] Show that for the critical density (Einstein–de Sitter) case,

$$D_A(z) = 2H_0^{-1}(1+z)^{-3/2}\left[(1+z)^{1/2} - 1\right]. \tag{7.53}$$

Exercise 7.4.3 Show that when $p = 0 = \Lambda$, the luminosity distance is related to the redshift by $2(1+z) = \Omega_{m0}(1 + H_0 D_L) + (2 - \Omega_{m0})\sqrt{1 + 2H_0 D_L}$.

Exercise 7.4.4 Work out the redshift value at which objects in an Einstein–de Sitter universe will appear to have the same angular size as if they were on the last scattering surface.

Exercise 7.4.5 Using proper spatial distance and proper time as (non-comoving) coordinates, give a two-dimensional plot of the actual shape of the past light-cone in an Einstein–de Sitter universe, showing how it reaches a maximum spatial size as one goes to the past and then refocuses to a point at the initial singularity (Ellis and Rothman, 1993).

Problem 7.2 Consider what happens to the reciprocity relation when the null geodesics go through a caustic.

7.5 Specific intensity and apparent brightness

The above analysis does not in fact correspond to what is actually measured. Firstly, it corresponds to measuring bolometric flux (the energy emitted at all wavelengths) – but real detectors measure radiation in a restricted wavelength band. Secondly, imaging detectors respond to radiant energy received per unit solid angle, i.e. the intensity of the radiation. We consider these in turn.

---

[5] It can also be derived from the null Raychaudhuri equation (7.26), but it is simpler via the linear geodesic deviation equation.

7.5.1 Specific flux

Most detectors measure radiation in a very narrow wavelength band (e.g. U, B, V bands). Bolometric detectors attempt to capture all visible and infrared radiation, but it is not possible for a single detector to measure radiation in all wavebands, including both radio and X-rays as well as visible. Indeed astronomy is split into sub-disciplines by the wavebands measured.

To allow for this, we represent the source spectrum by a function $\mathcal{I}(\nu)$, where $L\mathcal{I}(\nu)\mathrm{d}\nu$ is the rate at which radiation is emitted by the source at frequencies between ν and $\nu + \mathrm{d}\nu$. The function $\mathcal{I}(\nu)$ is normalized via $\int_0^\infty \mathcal{I}(\nu)\mathrm{d}\nu = 1$. Then we rewrite (7.49) as

$$\frac{\overset{\cdot}{F}D_L^2}{F_G} = \int_0^\infty \mathcal{I}(\nu_G)\mathrm{d}\nu_G = (1+z)\int_0^\infty \mathcal{I}(\nu_O(1+z))\,\mathrm{d}\nu_O. \tag{7.54}$$

Then the flux measured in the frequency range $(\nu, \nu + \mathrm{d}\nu)$ by the observer is

$$F_\nu \mathrm{d}\nu = \frac{F_G(1+z)}{D_L^2}\mathcal{I}(\nu(1+z))\mathrm{d}\nu. \tag{7.55}$$

We call F_ν the *specific flux* of the radiation. It is often assumed that over some wavelength range, $\mathcal{I} \propto \nu^{-\alpha}$ where α is a constant spectral index. For many optical sources at wavelengths $\gtrsim 5000\,\text{Å}$, $\alpha \approx 2$ and for many radio sources, $0.7 \le \alpha \le 0.9$. (An alternative way of allowing for the effect of the source spectrum is to introduce a *K-correction*, representing the difference between the flux and the specific flux.)

7.5.2 Specific intensity

So far we have implicitly assumed the sources observed are point sources. In practice we usually observe extended sources, as for example in the case of imaging cameras, and the instrument responds to the flux per unit solid angle, i.e. the *intensity* of radiation from the source. Even photometer and spectrograph measurements involve an aperture which determines an effective solid angle of measurement, so they also correspond to intensity measurements.

Thus, what is actually measured pointwise in an image is the *specific intensity* I_ν: the intensity in a specific frequency range. Considering a source of area $\mathrm{d}S_O$, we find from (7.55) that I_ν is given by

$$I_\nu \mathrm{d}\nu := \frac{F_\nu \mathrm{d}\nu}{\mathrm{d}\Omega_O} = I_G \frac{\mathcal{I}(\nu(1+z))\mathrm{d}\nu}{(1+z)^3}, \quad I_G := \frac{F_G}{\mathrm{d}S_O}, \tag{7.56}$$

where I_G is the source surface brightness (an intrinsic property of the source), and we have used (7.40)–(7.42). This important result, a direct consequence of the reciprocity/distance-duality, shows that *the measured specific intensity is independent of the area distance of the source* – it depends only on the redshift. This generalizes to arbitrary curved spacetimes the standard laboratory result that apparent surface brightness is independent of distance from the object observed (since the inverse square law for the intensity cancels with the change in its observed solid angle). To measure the specific flux from an extended source, we have to integrate this expression over the observed image, whose apparent size is determined by (7.40) together with detection limits.

7.5.3 Absorption and emission

There may be absorption or emission of radiation by intervening matter along the line of sight between the source G and the observer O. Consider the change in the specific intensity $\mathcal{I}_\nu$ as radiation at the event A with affine parameter value v propagates to the nearby event with affine parameter $v + dv$ on the bundle of rays from G to O. By (7.56) we can represent the change in $\mathcal{I}_\nu$ due to geometrical and redshift effects alone by the differential equation $d\mathcal{I}_{\nu'}/dv = 3(1+z)^{-1}\mathcal{I}_{\nu'}dz/dv$, where $\nu' := \nu(1+z)$ is the frequency of radiation at A. When redshifted to O, this is observed at frequency ν. Let $S(v,\nu)d\nu > 0$ be the rate of emission of radiation by each source at A per unit solid angle in the frequency range ν to $\nu + d\nu$, $n_s(v)$ be the number density of sources at A, $n_a(v)$ be the number density of particles scattering or absorbing radiation at A, and $\sigma(v,\nu)$ be the interaction cross-section of these particles at frequency ν. Allowing for these processes in the volume $dl\,dS_0 = (-u_a k^a)dv\,dS_0$ at A, the change in $\mathcal{I}_{\nu'}$ along the geodesic can be represented by the differential equation,

$$\frac{d\mathcal{I}_{\nu'}}{dv} - \frac{3\mathcal{I}_{\nu'}}{1+z}\frac{dz}{dv} + n_a(v)\sigma(v,\nu')\mathcal{I}_{\nu'} - u_a k^a = n_s(v)S(v,\nu')(-u_a k^a), \qquad (7.57)$$

where z and $u_a k^a$ are regarded as known functions of v.

Integrating this equation along the geodesic from the source G ($v = 0$) to the observer ($v = v_*$), we find the specific intensity at O is

$$\mathcal{I}_\nu = \int_0^{v_*} \frac{n_s(v)S(v,\nu(1+z))}{(1+z)^3} \exp[-p(v,\nu)](-u_a k^a)(v)dv$$
$$+ \frac{\mathcal{I}_{\nu(1+z_*)}(0)}{(1+z_*)^3} \exp[-p(v_*,\nu)], \qquad (7.58)$$

where the optical depth $p(v,\nu)$ between A and O for radiation observed at O at frequency ν is

$$p(v,\nu) = \int_0^v n_a(v')\sigma(v',\nu(1+z'))(-u_a k^a)(v')dv'. \qquad (7.59)$$

These equations determine the specific intensity of radiation we observe in any direction in the sky. The second term in (7.58) represents radiation propagating to us from the event G at affine parameter value $v = 0$, attenuated by absorption, while the first term represents integrated emission and absorption from all sources and absorbers lying between G and the observer O.

FLRW case

For simplicity we take $p = 0 = \Lambda$ and assume that the radiation sources and absorbing particles are conserved ($n(z) = n(0)(1+z)^3$). If we can ignore absorption, on using

Example 7.3.2 the contribution to $\mathcal{I}_\nu$ from sources up to a redshift z_* is

$$\mathcal{I}_\nu = \frac{n_s(0)}{H_0} \int_0^{z_*} \frac{S(z, \nu(1+z))}{(1+z)^3 \sqrt{1+\Omega_{m0}z}} dz. \tag{7.60}$$

One can work out from this the effect of line emission, or of simple spectra (regarded as built up as an integral of line emissions). Using these equations, detailed analysis of integrated radiation from sources gives vital information on their number density and evolution.

For absorption, the optical depth up to a redshift z in a dust FLRW universe is

$$p(z, \nu) = \frac{n_a(0)}{H_0} \int_0^z \frac{(1+z)\sigma(z, \nu(1+z))}{\sqrt{1+\Omega_{m0}z}} dz. \tag{7.61}$$

One can work out from this the effect of line absorption, or simple absorption processes regarded as built up as an integral of line absorbers. In the particular case of Thomson scattering, which is wavelength and redshift independent, so that $\sigma(z, \nu) = \sigma_T = \text{const}$, one can integrate (7.57) directly to get

$$p(z, \nu) = \frac{2\sigma_T n_a(0)}{3H_0 \Omega_{m0}^2} [(3\Omega_{m0} + \Omega_{m0}z - 2)(1+\Omega_{m0}z)^{1/2} - (3\Omega_{m0} - 2)]. \tag{7.62}$$

This will represent the case of an ionized intergalactic medium. Detailed analysis of absorption effects, including line spectra such as the Lyman-α forest, gives vital information on the temperature history and spatial distribution of the intergalactic medium.

7.6 Number counts

Number counts relate to the number dN of sources detected in a bundle of rays, for a small affine parameter displacement v to $v + dv$ at an event A. This corresponds to a distance $dl = (k^a u_a)dv$ in the rest frame of a comoving galaxy at A if k^a is the tangent vector to the past directed null geodesics (so $k^a u_a > 0$). The cross-sectional area of the bundle is $dS_0 = D_A^2(v)d\Omega$ if the geodesics subtend a solid angle $d\Omega$ at the observer, so the corresponding volume at A is $dV = dl\, dS_0 = (k^a u_a)dv D_A^2(v)d\Omega$. Hence if the number density of sources at A is n per unit proper volume, and $f_d(v)$ is the fraction of sources at distance v that are detected by means of the observational technique used, the number detected is $dN = f_d(v)n(v)D_A^2(v)d\Omega(k^a u_a)dv$. The total number $N(v_*)$ of sources observed up to some affine parameter distance v_* is then

$$N(v_*) = d\Omega \int_0^{v_*} f_d(v)n(v)D_A^2(v)(k^a u_a)dv. \tag{7.63}$$

To turn this into an observational relation we need to relate dv to a redshift increment for comoving observers, or a magnitude increment for a class of standard candles. If we can estimate the mass per galaxy (or other object observed), then number counts enable us to estimate the density of matter contributed by these objects to the overall matter density of the universe.

FLRW case

In an FLRW universe, if the number of sources is conserved then $n = n_0 a_0^3 / a^3$ and (7.63) gives $dN = f_d(v) n_0 a_0^3 f^2(u) d\Omega du$, on using (7.50) and Exercise 7.1.2. If $f_d(v)$ can be regarded as a constant (e.g. the sources are so close that we detect them all, $f_d = 1$), then

$$N(u) = 4\pi f_d n_0 a_0^3 \int_0^u f^2(u') du' \qquad (7.64)$$

is the number of sources seen in all directions at distances up to u. Using (9.10), the integral respectively takes the forms $\{(2\sqrt{K}u - \sin 2\sqrt{K}u)/2\sqrt{K}, u^3/3, (\sinh 2\sqrt{-K}u - 2\sqrt{-K}u)/2\sqrt{-K}\}$ for $\{K > 0, K = 0, K < 0\}$.

Verifying this relation confirms the spatial homogeneity of the universe, and in principle enables us to determine the sign of the spatial curvature K. In practice the statistical uncertainties, together with source evolution (which affects the detection probabilities), prevent this from being a useful test of K.

7.7 Selection and detection issues

The quantity $f_d(v)$ (the fraction of sources at distance v that are detected) is crucial to number counts, and indeed to all cosmological statistics, e.g. in the observed magnitude–redshift relations. But sources have not only to be detected, they have also to be identified as belonging to the relevant class of sources, hence a selection process is also important. Some key issues are as follows.

- Detection and selection take place on the basis of properties of images, rather than directly on the basis of source properties. Hence to examine these effects, one needs to map object properties (e.g. source luminosity and scale size) to image properties (e.g. apparent magnitude and image size).
- As shown in an illuminating manner by Disney (1976) in a discussion of visibility of low surface brightness galaxies, the primary detection criterion is specific intensity of incident light at the detector. This is determined by the surface brightness distribution at the source, which in turn is related to both the source luminosity and size. Hence one cannot adequately discuss selection criterion on the basis of one source characteristic (e.g. magnitude) alone. This also implies that as far as detection limits are concerned, an evolution of source size is twice as important as luminosity evolution.
- In terms of source classification, image size (apparent angle) is also important. Hence it is very useful to set up an observational map between source characteristics for sources at a given redshift in a given cosmology (the source plane), and corresponding image characteristics for a specific detection system (the image plane), allowing one to map back selection and detection effects from the image plane to the source plane (Ellis, Perry and Sievers, 1984).

These effects are routinely handled by observers carrying out major surveys such as the SDSS and 2dFGRS. However, they are not often clearly explicated in theoretical discussions of cosmological observations. A key issue is that the way these corrections are handled is likely to be very model dependent, implicitly introducing assumptions about source properties and perhaps even about the cosmological models. Thus they should be made explicit rather than handled behind the scenes as merely an incidental feature of astronomical data reduction, for they crucially influence the outcomes of source surveys, and indeed the statistics of any cosmological observations. It would be to the benefit of the field if clear up-to-date discussion of these effects and how they are handled were to be available.

Exercise 7.7.1 Re-express (7.64) in terms of H_0, Ω_{m0}, $\Omega_{\Lambda 0}$ and D_A. Using the expression (7.52) for $D_A(z)$, this gives $N(z)$.

7.8 Background radiation

We receive radiation from all directions in the sky from both discrete sources and background radiation (due to unresolved sources, intergalactic gas, and the primordial universe itself) which arrives at Earth at all wavelengths. We have looked at observations of discrete sources in some detail above. Much detailed information on the matter–radiation interaction in the early universe is encoded in the background radiation, for example about the density of hot intergalactic gas (that emits X-rays) and of neutral hydrogen.

Here we shall only look at two key issues: the way blackbody radiation propagates in a general curved spacetime, and the issue of the total amount of integrated radiation to be expected.

7.8.1 Blackbody radiation

It follows immediately from (7.56) that radiation emitted as blackbody radiation remains blackbody. Defining $g(\nu) = I_G \mathcal{I}(\nu)/\nu^3$, we can rewrite this equation as

$$I_\nu d\nu = g(\nu(1+z))\nu^3 d\nu, \tag{7.65}$$

which is the specific intensity of radiation at each frequency ν for any observer who measures the source redshift as z. If blackbody radiation is emitted by a source at temperature T_e, then at the source,

$$I_\nu d\nu = f(\nu/T_e)\nu^3 d\nu, \tag{7.66}$$

where $f(\nu/T_e)$ is the Planck function for blackbody radiation at a temperature T_e. Comparing these expressions at the source (where $z = 0$) shows $g(\nu) = f(\nu/T_e)$, so we can rewrite (7.65) as

$$I_\nu d\nu = f(\nu/T)\nu^3 d\nu, \quad T := T_e(1+z)^{-1}. \tag{7.67}$$

Thus,

Theorem 7.2 Blackbody spectra
Every observer measures a blackbody spectrum for blackbody radiation, but with the temperature decreased by a factor $(1+z)$.

   This remarkable result (which in effect follows from the reciprocity relation) is not as celebrated as it should be. Like Hawking's black hole radiation result, it combines GR, quantum mechanics and statistical mechanics to give a simple but important result. It is a key result for cosmology, as it underlies our understanding of the CMB observations.

7.8.2 Integrated radiation

The basic equation determining the total radiation received is (7.58). Omitting the absorption terms, this takes the form

$$\mathcal{I}_\nu = \int_0^{v_*} \frac{n_s(v)S(v,v(1+z))}{(1+z)^3}(-u_a k^a)\mathrm{d}v + \frac{\mathcal{I}_{\nu(1+z_*)}(0)}{(1+z_*)^3}. \tag{7.68}$$

Because of the finite age of the universe, the integral is only taken to the LSS and not to infinity. Because of the expansion of the universe, the redshift factors reduce the radiation from each distant source to much less than the emitted intensity. And the amount of radiation emitted by each source is limited: $S(v,v(1+z))$ is not too large so that in fact the second (initial surface) term is the dominant term in the received radiation; and that term corresponds to 3000 K, diluted to 3 K by the expansion of the universe since that radiation was emitted at a redshift of about 1100.

   Detailed examination of (7.68) is used by astronomers to interpret the background radiation received on Earth at all frequencies, and not just the 3 K blackbody radiation. This provides valuable information on the radiation history of unresolved sources in the sky. This is discussed further in Section 11.7. It also gives a resolution of Olbers' paradox (Section 21.1.1).

7.9 Causal and visual horizons

A fundamental feature affecting our observational situation is the limits arising because causal influences cannot propagate at speeds greater than the speed of light. Thus the region that can causally influence us is bounded by our past light-cone. Combined with the finite age of the universe, this leads to the existence of a *particle horizon* limiting the part of the universe with which we can have had causal connection, and a *visual horizon* (lying inside the particle horizon) limiting the domain about which we can have any observational evidence.

7.9.1 Particle horizons

A particle horizon comprises the limiting world lines of the furthest matter that ever intersects our past light-cone (Rindler, 1956, 2001, Penrose, 1968, Tipler, Clarke and Ellis, 1980). This is the limit of matter that we can have had any kind of causal contact with since the start of the universe. This depends on the time at which we want the answer to that question: at later and later times in our history, we can see more and more of the universe. Geometrically, the world lines comprising the particle horizon are those world lines that intersect our past light-cone in the limit as we go back to the start of the universe.

In an FLRW universe, from (7.15) it is characterized by the comoving radial distance,

$$u_{ph} = \int_0^{t_0} \frac{dt}{a(t)}. \tag{7.69}$$

The present physical distance to the matter comprising the horizon is

$$d_{ph} = a(t_0)u_{ph}. \tag{7.70}$$

The key question is whether the integral (7.69) converges or diverges as we go to the limit of the initial singularity where $a \to 0$. This integral diverges in the case of the Milne universe with $a(t) = t$; hence there is no particle horizon in that model. But that is not a realistic universe model, because it is empty.

Particle horizons will exist in FLRW cosmologies with ordinary matter and radiation, for u_{ph} will be finite in those cases. For example in the Einstein–de Sitter universe, $u_{ph} = 3t_0^{1/3}$, $d_{ph} = 3t_0 = 2/H_0$. We shall then have had causal contact with only a fraction of what exists, and hence shall only have seen part of what is out there, with one exception: this is not the case if we live in a 'small universe', with spatially compact sections so small that light has had time to traverse right around the whole universe since its start. This case is discussed further in Section 9.1.6. Here we assume we are not in a small universe.

Penrose's powerful use of conformal methods (Hawking and Ellis, 1973, Tipler, Clarke and Ellis, 1980) gives a very clear geometrical picture of the nature of horizons. These methods are based on the use of conformally flat coordinates, so that light-cones are the same as in flat spacetime (but spatial distances and proper times are distorted). In the case of RW universes, one can derive these diagrams by using the conformal time coordinate $\tau = \int dt/a(t)$ so that the metric (2.65) becomes[6]

$$ds^2 = a^2(\tau)\left[-d\tau^2 + dr^2 + f^2(r)(d\theta^2 + \sin^2\theta d\phi^2)\right]. \tag{7.71}$$

The equation for radial null geodesics is then given by

$$ds^2 = 0 = d\theta = d\phi \Rightarrow dr = \pm d\tau. \tag{7.72}$$

When coordinates (τ, r) are used for radial sections ($\theta = \text{const}, \phi = \text{const}$) of the spacetime, the null geodesics (and hence the light-cones) are at $\pm 45°$, as in the case of Minkowski spacetime in canonical coordinates (see Figure 7.2).

---

[6] These are conformally flat coordinates for the metric in the case $K = 0 \Rightarrow f(r) = r$. A conformal factor dependent on the spatial coordinates as well is needed to get the conformally flat form when $K \neq 0$.

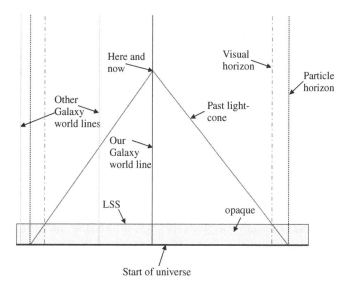

Fig. 7.2 Particle horizon and visual horizon of an event 'here and now', for an FLRW universe in conformally flat coordinates. The galaxy to the far left cannot have been seen by us (it is outside the visual horizon), nor can we have had any causal contact with it (it is outside the particle horizon). World lines of matter comprising the visual horizon are shown as dot-dash lines; WMAP images this mattter at the LSS.

Matter world lines are vertical lines (as we are using comoving spatial coordinates) and the surfaces of constant time $t = $ const (the surfaces of homogeneity in spacetime) are horizontal lines; but τ is not proper time along the world lines, and spatial distances are completely distorted. For a radiation equation of state at early times, the boundary $t = 0$ of the spacetime is a spacelike surface, and the particle horizon for the observer at $r = 0$ at time t_0 is the set of matter world lines through the points where the observer's past light-cone intersects the initial singularity at $t = 0$. We can in principle have received radiation from all matter this side of the particle horizon, but none from matter beyond it; indeed we cannot have interacted in any way with such matter, and have no information whatever about it (although we feel its Coulomb fields). This is an absolute limit on communication and causal effects in the expanding universe.

Equation (7.70) gives the size of the particle horizon at the present time. We are at each moment surrounded by a 2-sphere of this radius, representing the limits of any possible causal interaction at the present time; we shall call this the *causal limit sphere*. We have already (in principle) received information from all matter this side of this 2-sphere, and can have received none from any matter the other side. The 2-sphere is the intersection of the world lines comprising the particle horizon with the past light-cone. But in an FLRW model it can also be considered as the intersection of our creation light-cone with the surface of constant time $t = t_0$. Note here that we cannot talk as if space were partitioned into disjoint horizons: there is nothing special about the horizon for any specific fundamental observer, rather there is a particle horizon for every fundamental observer (MacCallum, 1982).

The horizon always grows, because (7.69) shows that u_{ph} is a monotonically increasing function of t_0. Despite contrary statements in the literature, *it is not possible for matter to leave the horizon once it has entered.* In a (perturbed) FLRW model, once causal contact has taken place, it remains until the end of the universe. Particle horizons may not exist in non-FLRW universes, for example Bianchi (anisotropic) models (Misner, 1969a). In universes with closed spatial sections, a supplementary question arises: is the scale of closure smaller than the horizon scale? There may be a finite time when causal connectivity is attained, and particle horizons cease to exist. In standard $K > 0$ FLRW models, this occurs just as the universe reaches the final singularity; if, however, there is a positive cosmological constant or other effective positive energy density field, it will occur earlier. If the scale of closure is smaller than the visual horizon, one has the case of a 'small universe', mentioned above.

The importance of these horizons is two-fold: they represent absolute limits on what is testable in the universe (Ellis, 1975, 1980), and they underlie causal limitations relevant in the origin of structure and uniformity, and so affect the formation of structure in the early universe (see Chapter 12).

7.9.2 Visual horizons

Clearly we cannot obtain any observational data on what is happening beyond the particle horizon. However, we cannot even see that far, because the universe was opaque to all wavelengths before decoupling. Our view of the universe is limited by the *visual horizon,* comprising the world lines of the furthest matter we can observe – namely, the matter that emitted the CMB at the LSS (Ellis and Stoeger, 1988, Ellis and Rothman, 1993). This occurred at the time of decoupling $t = t_{\mathrm{dec}}$ ($z_{\mathrm{dec}} \approx 1100$), and so the visual horizon is characterized by $r = u_{vh}$, where from (7.15),

$$u_{vh} = \int_{t_{\mathrm{dec}}}^{t_0} \frac{\mathrm{d}t}{a(t)} < u_{ph}. \tag{7.73}$$

Indeed the LSS delineates our visual horizon in two ways, made clear in Figure 7.2: we are unable to see to *earlier times* than its occurrence (because the early universe was opaque for $t < t_{\mathrm{dec}}$), and we are unable to detect matter at *larger distances* than that we see on the LSS (we cannot receive radiation from matter at comoving coordinate values $r > u_{vh}$). Analogous to the causal limit sphere, the visual horizon at the present time is represented by a *visual limit sphere*: a 2-sphere of matter around us lying inside the particle horizon's causal limit sphere, such that we have already (in principle) seen all matter this side of this sphere, and can have seen none of the matter the other side. The picture we obtain of the LSS by measuring the CMB from satellites such as COBE and WMAP is just a view of the matter comprising the visual horizon, viewed by us at the time in the far distant past when it decoupled from radiation.

The position of the visual horizon is determined by the geometry since decoupling. Visual horizons do indeed exist, unless we live in a small universe, spatially closed with the closure scale so small that we can have seen right around the universe since decoupling, as already mentioned. There is no change in these visual horizons if there was an early inflationary period, for inflation does not affect the expansion or null geodesics during this later period.

The major consequence of the existence of visual horizons is that many present-day speculations about the super-horizon structure of the universe – e.g. the chaotic inflationary theory – are not observationally testable, because one can obtain no definite information whatever about what lies beyond the visual horizon (Ellis, 1975, 1980). This is one of the major limits to be taken into account in our attempts to test cosmological models.

Unless we live in a small universe, *the universe itself is much bigger than the observable universe*. There may be many galaxies – perhaps an infinite number – at a greater distance than the horizon that we cannot observe by any electromagnetic radiation. Furthermore, no causal influence can reach us from matter more distant than our particle horizon – the distance light can have travelled since the creation of the universe – so this is the furthest matter with which we can have had any causal connection (Rindler, 1956, Hawking and Ellis, 1973, Tipler, Clarke and Ellis, 1980). We can hope to obtain information on matter lying between the visual horizon and the particle horizon by neutrino or gravitational radiation observatories; but we can obtain no reliable information whatever about what lies beyond the particle horizon.

We can in principle feel the gravitational Coulomb effect of matter beyond the horizon because of the force it exerts (for example, matter beyond the horizon may influence velocities of matter within the horizon, even though we cannot see it). This is possible because of the constraint equations of GR, which are in effect instantaneous equations valid on spacelike surfaces.[7] However, we cannot uniquely decode that signal to determine what matter distribution outside the horizon caused it: a particular velocity field might be caused by a relatively small mass near the horizon, or a much larger mass much further away (Ellis and Sciama, 1972). Claims about what conditions are like on very large scales – i.e. much bigger than the Hubble scale – are unverifiable (Ellis, 1975), for we have no observational evidence as to what conditions are like beyond the visual horizon. The situation is like that of an ant surveying the world from the top of a sand dune in the Sahara desert. Her world model will be a world composed only of sand dunes – despite the existence of cities, oceans, forests, tundra, mountains, and so on beyond her horizon.

7.9.3 Event horizons

There are also *event horizons* in some cosmological models (Rindler, 1956, Tipler, Clarke and Ellis, 1980, Rindler, 2001). They are the limiting past light-cones of all events on the observer's world line in the far future, separating the spacetime events that will ever be observable by a particular fundamental observer, from those that will not. By (7.15) the radial coordinate size of these limiting past light-cones at time t_0 in an RW universe that lasts forever is given by

$$u_{eh} = \int_{t_0}^{\infty} \frac{dt}{a(t)},\qquad(7.74)$$

so the question is whether this integral diverges or not. Roughly speaking, it diverges if ordinary matter dominates the late universe, so no event horizons exist in that case (which

---

[7] Section 3.3.1 explains why these are valid at any late time in a solution of the EFE.

corresponds to future infinity being null). The observer will eventually see all spacetime events. It converges if a cosmological constant dominates the late universe, so event horizons exist in that case (which corresponds to future infinity being spacelike). There are then many events the observer will never be able to see, no matter how long he or she lives. If the universe ends in a second singularity in the future (a big crunch) at a time t_{bc}, then the future limit of the integral (7.74) must be taken as t_{bc}. The integral will then be finite for all ordinary matter, so there will be event horizons in these cases.

The definition of event horizon in the cosmological case agrees with the black hole one, which can be thought of in terms of geodesics outgoing from the black hole, rather than the past cones of external observers. While event horizons are central to the study of black holes, they are of little significance in cosmology as they refer to the far future of the universe, which is never attained in any finite time (unless the universe re-collapses, in which case no observers can exist at late times). They have no relevance to present-day causal limits or observational possibilities.

Their existence in de Sitter universes is sometimes used as the basis of calculating Hawking–Gibbons blackbody radiation in an inflationary era in the early universe (which then has quantum fluctuations that provide the seeds for galaxy formation at much later times). However, this must be done with caution, in that if the de Sitter phase ever comes to an end (as is required for the present-day epoch of the universe to come into existence) then event horizons may not in fact exist; and whether they do exist or not is independent of the properties of the early inflationary phase of the universe.

7.9.4 Hubble sphere

In the literature on the inflationary universe it is common to refer to the 'horizon', defined as the characteristic radius $R_H(t) = H^{-1}(t)$. It is often stated to be the limit from which causal influences can propagate, due to the special relativity limit that no cause can propagate faster than the speed of light. However, in fact this scale has nothing to do with the speed of propagation of physical effects (van Oirschot, Kwan and Lewis, 2010); rather it is a characteristic scale for the relative importance of expansion of the universe in relation to other physical effects. As such it plays an important role in the generation and evolution of perturbations in the early universe (see Section 12.2), but calling it a 'horizon' is misleading. It is preferable to call it the Hubble sphere, or the Hubble horizon.

Exercise 7.9.1 Show that particle horizons exist for an FLRW universe with only matter and radiation (i.e. a HBB model), and determine u_{ph}, d_{ph} in this case. Show that particle horizons do not occur in a de Sitter universe.

Exercise 7.9.2 Explain how in an Einstein–de Sitter universe, the particle horizon size can be $3t_0$ when the age of the universe is only t_0 (Ellis and Rothman, 1993).

Exercise 7.9.3 Show that an event horizon occurs in a de Sitter universe. Deduce that one will occur in realistic universe models, if the present day acceleration of the universe is caused by a cosmological constant. Determine if a realistic universe model can have both a particle and an event horizon (van Oirschot, Kwan and Lewis, 2010).

Exercise 7.9.4 Show that at any instant t, the Hubble sphere is the radius where objects receding from the origin according to Hubble's law, $v_{rec} = H(t)a(t)$, are instantaneously receding at the speed of light. Those further out are receding faster than light, those closer in at a lesser speed. Explain why this does not violate special relativity (Harrison, 2000, Ellis and Rothman, 1993, Davis and Lineweaver, 2004, van Oirschot, Kwan and Lewis, 2010).

8

Light-cone approach to relativistic cosmology

The standard approach to cosmology is a model-based approach: find the simplest possible model of spacetime that can accommodate the observational data. An alternative is a *direct observational approach*. The first method determines observational relations and parameters from a model; the second attempts to determine a model from observational relations. We introduce the latter method in this chapter, and it will also feature in Chapter 15; the former is essentially used in the rest of this book.

As mentioned before, a fundamental feature of cosmology is that there is only one universe, which we cannot experiment on: we can only observe it, and moreover, on a cosmological scale, only from one specific spacetime event. Observations therefore give direct access only to our past light-cone, at one cosmological time. How can we then devise and test suitable cosmological models?

8.1 Model-based approach

In the standard approach, one chooses a family of models first, characterized by as few free parameters and free functions as possible. Then one fixes these parameters and functions in order to reproduce astronomical observations as accurately as possible. Therefore this is in fact a form of light-cone best-fitting procedure: one is obtaining a best-fit of the chosen family of models to the real universe via comparison of observational relations predicted by the model with actual observations.

Traditionally, this is applied almost exclusively to the FLRW models. The merit of the approach is that it has good explanatory power, which serves as a vindication of the chosen models. In particular, it provides a framework that explains the origin of the elements, of the CMB, and the basics of structure formation, as explained elsewhere in this book. It gives an exciting link from cosmology to quantum theory, nuclear physics and elementary particle physics.

There is necessarily a non-uniqueness to the procedure. We could have chosen other models, for example Bianchi I models in which the shear dies away rapidly enough to not affect the observations. These are less special than FLRW models, which are a priori infinitely improbable within the mathematical family of models, because of their exact symmetries. One can advocate FLRW models on an Occam's razor basis (they are simpler than any other), which is necessarily a philosophical rather than observational criterion. One could try to use a fuller range of options to test if any of them fit better. If the process is broadened to include a wide range of alternatives, this provides a setting in which we can evaluate the FLRW models relative to these other models. Specific alternatives we might wish to examine include:

- *Small universes*, that appear homogeneous but are globally different from standard FLRW models (Section 9.1).
- *Almost isotropic Bianchi models*, that may be either almost isotropic at all late times (e.g. Bianchi I) or have a temporary but long-lived almost isotropic epoch near a saddle point in their phase plane (e.g. Bianchi VII) (Section 18.5).
- *Lemaître–Tolman–Bondi* (LTB) spherically symmetric models, where we are located near the centre (Chapter 15).

However, there will always be other such models we have not examined: might not one of them give a better fit? In an era of precision cosmology, as one pushes the limits of the observational tests, there will always be anomalies that need to be resolved, and more generic models may provide an answer.

This raises the question: can we do away with choosing a model a priori, and attempt to construct the model directly from observations? This is the purpose of the direct observational approach.

8.2 Direct observational cosmology

In the direct observational approach, one starts with a generic metric, and then tries to progressively restrict its geometry directly by use of observational data. Thus one tries to determine the spacetime geometry by seeing what is actually there, rather than by starting off with a chosen restricted family of models.

This approach is in fact a venerable one in astronomy and cosmology, for it is the way that large-scale motions and large-scale structures such as voids and walls were discovered (often in the face of resistance from the astronomical establishment). However, in these cases it is done in essence using a Newtonian model of the local region of the universe: it is not done relativistically. A more relativistic version is embodied in what is undertaken through large-scale surveys such as 2dFGRS and SDSS, and through gravitational lensing observations. However, insofar as these are relativistic, they are usually done in the context of perturbed FLRW models.

A general relativistic version of the direct approach was pioneered by McCrea (1935, 1939) (see also Florides and McCrea (1959)), and then developed systematically in a major paper by Kristian and Sachs (1966), using a 1+3 covariant decomposition. These papers, however, use a power-series description, and so are of restricted applicability. The method was extended to generic spacetimes by Ellis *et al.* (1985) (based on Maartens (1980), Nel (1980), summarized in Ellis (1984); for other versions, see Dautcourt (1983a,b)). In these papers it was shown that, for suitable matter content, *in principle the spacetime metric can be determined directly from astronomical observations,* without assuming a specific model model first, as in the standard approach.

The direct approach was however formulated for a baryonic universe. The strong evidence for dark matter, and the growing evidence for late-time acceleration, which is typically interpreted as dark energy, completely change the picture. In effect, *the direct approach can only work, in a complete sense, for a baryonic universe.* The fundamental reason is that *the dominant cosmological components – dark matter and dark energy – cannot be directly observed.* Unlike luminous baryonic matter, the dark components are only manifest via their gravitational effect. This unavoidably means that we must impose a model for these dark components – not merely their physical properties, but how they relate spatially to observed matter – in order to determine their distribution via observations.

We can formulate two important corollaries to these points: *the direct approach is in principle feasible if*

- the late-time acceleration can be explained in terms of inhomogeneities or a cosmological constant (rather than a spatially varying field), and dark matter can be directly detected at cosmological distances from its relation to baryons; or,
- a modified theory of gravity is constructed that does not require dark matter or dark energy.

The distribution of dark matter is mapped by weak lensing surveys (Massey *et al.*, 2007). But to relate the measured projected potential on the sky in each redshift bin to the dark matter, we require a specific model, such as a perturbed FLRW model. The dark matter 4-velocity is usually assumed to be aligned with that of baryonic matter – but this is also based on a perturbed FLRW model. From now on we shall assume that the CDM velocity is the same as the baryonic velocity, and that we know the primordial ratio of CDM density to baryonic density, as well as the bias factor that relates the concentrations of CDM and baryons in clustered matter; thus we assume (Clarkson and Maartens, 2010) that

$$\rho_{\rm c} \text{ is known from } \rho_{\rm b} \text{ and } u_{\rm b}^a = u_{\rm c}^a =: u^a. \tag{8.1}$$

A possible dark energy component, either Λ or a dynamical field, cannot be measured via direct observations (see Section 8.4.1 below). In order to pursue the direct observational approach, we shall need to assume a knowledge of dark energy – or to follow an alternative approach that there is in fact no dark energy (see Chapter 15). Here we shall assume for simplicity that there is dark energy in the form of Λ, and that its value is known from non-cosmological physics:

$$\Lambda \text{ known independently of cosmological observations.} \tag{8.2}$$

8.2.1 Observational coordinates: metric and kinematics

Observational coordinates are fully adapted to the actual process of observation – principally, the fact that cosmological observations are made via electromagnetic signals that propagate along the past light-cone of the observer (i.e. of our galaxy); see Figure 8.1. Cosmological data are determined on the past light-cone of the observer, and not on

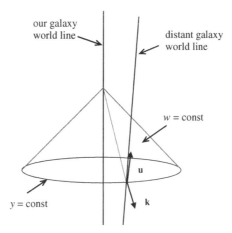

Fig. 8.1 Observational coordinates.

spatial surfaces of constant proper time from the big bang. This is the reason that spatial homogeneity is not directly observable, unlike isotropy about our world line.

We define $x^0 = w$ such that $w = $ const are the past light-cones $C^-(w)$ of events on the observer world line C (note that w is not differentiable at C). We normalize w on C to be proper time: $\mathrm{d}s^2|_C = -\mathrm{d}w^2$. Then w is determined up to translation, $w \to w+$ const; this freedom is fixed by choosing a value w_0 corresponding to here-and-now.

The past light-cones $C^-(w)$ are generated by the (past-directed) geodesic ray 4-vector (see Section 7.1):

$$k_\mu = \partial_\mu w, \ \ k^\mu = \frac{\mathrm{d}x^\mu}{\mathrm{d}v}, \ \ k^\mu k_\mu = 0 = k^\nu \nabla_\nu k^\mu, \tag{8.3}$$

where v is an affine parameter (note that k^μ is not defined on C). We fix the affine freedom in v by choosing $v = 0$ on C and

$$k_\mu u^\mu \to 1 \text{ as } v \to 0, \tag{8.4}$$

where u^μ is the 4-velocity of matter. The radial coordinate $x^1 = y$ measures distance down the null geodesics that rule the past light-cones – and this distance incorporates both a spatial distance from C and a time difference from the observer.

The coordinates $x^I = (\theta, \phi)$ are then chosen to label the null geodesics in each past light-cone, so that $k^\mu \partial_\mu x^I = 0$, i.e. $k^I = 0$. Then

$$k^\mu = B^{-1} \delta^\mu{}_1, \ B := \frac{\mathrm{d}v}{\mathrm{d}y}, \ k_\mu = \delta_\mu{}^0. \tag{8.5}$$

Various choices for y are possible, including $y = v$ ($B = 1$), and $y = z$, the redshift, which is given from (8.3)–(8.5) as

$$1 + z = k_\mu u^\mu = \frac{\mathrm{d}w}{\mathrm{d}\tau} = u^0, \tag{8.6}$$

where τ is proper time on the world line of the emitter. We choose $y = z$ on the observed past light-cone $C^-(w_0)$, and then define y in the rest of spacetime by dragging it off to the future and past, i.e. by requiring that y is comoving with the matter:

$$y = z \text{ on } C^-(w_0) \text{ and } u^\mu \partial_\mu y = 0 \;\Rightarrow\; u^1 = 0. \tag{8.7}$$

The observational coordinates $x^\mu = (w, y, \theta, \phi)$ cover the part of spacetime that is observable from C. At each event in this region, w gives the time of observation, θ, ϕ give the direction of observation, and y is a representation of distance to the observer. We note also that the coordinates may give a many-to-one representation in parts of the observable region – if the light-cone develops caustics, due either to gravitational lensing or to compactness of spatial sections in a small universe, it will only be 1–1 near the origin.

The metric in observational coordinates then takes the form

$$ds^2 = -A^2 dw^2 + 2B \, dy \, dw + 2C_I dx^I dw + D^2 \left(d\Omega^2 + L_{IJ} dx^I dx^J \right), \tag{8.8}$$

and the (geodesic) matter 4-velocity is given by

$$u^0 = 1 + z, \; u^1 = 0, \; u^I = (1+z)V^I, \; V^I := \frac{dx^I}{dw}, \tag{8.9}$$

$$u_0 = -(1+z)^{-1} + u^I C_I, \; u_1 = (1+z)B, \; u_I = g_{IJ}u^J + (1+z)C_I. \tag{8.10}$$

The normalization $u_\mu u^\mu = -1$ leads to

$$A^2 = (1+z)^{-2} + 2C_I V^I + g_{IJ} V^I V^J. \tag{8.11}$$

The metric and 4-velocity components have direct physical meaning in observational coordinates. In particular, D is the area distance for the central observer (Section 7.4.3), L_{IJ} determines the lensing shear (image distortion) of individual objects as measured by the central observer, and V^I are the transverse velocities of sources (proper motions) measured by the central observer. Note that we have defined L_{IJ} and V^I in a covariant way and *not* as perturbations of some background quantities. The number of sources observed at $w = w_0$ in a solid angle $d\Omega_0$, from redshift z to $z + dz$, is [(7.63)]

$$dN = f_d n D^2 (1+z) B \, d\Omega_0 \, dz, \tag{8.12}$$

where we have used (8.5), (8.7). Here f_d is the selection function, giving the fraction of sources actually detected, and n is the source number density.

The expansion and shear of light-rays in the screen space [(7.24)] are

$$\widehat{\Theta} = \frac{2}{BD} \frac{\partial D}{\partial y}, \; \widehat{\sigma}_{\mu\nu} = \delta_\mu{}^I \delta_\nu{}^J \frac{D^2}{2B} \frac{\partial}{\partial y} L_{IJ}. \tag{8.13}$$

Note that L_{IJ} is not tracefree ($g^{IJ} L_{IJ} \neq 0$), but it has only two degrees of freedom. This follows since the definition of area distance requires that $\det g_{IJ} = D^4 \sin^2\theta$, which implies $(L_{23})^2 = (1 + L_{22})L_{33} + \sin^2\theta L_{22}$. This condition then ensures that the shear is tracefree: $g^{IJ}\widehat{\sigma}_{IJ} = 0$.

The expansion, shear and vorticity of the matter are complicated expressions (Maartens, 1980, Maartens and Matravers, 1994); their limiting behaviour is given below.

In order to ensure that the null surfaces $w = $ const in the metric (8.8) have regular vertices along C ($y = 0$), we need to impose regularity conditions along C. These are derived by constructing null-geodesic-based normal coordinates (Ellis *et al.*, 1985) (an extension of the spatial geodesic approach of Manasse and Misner (1963)). For the metric components, regularity requires the following limiting behaviour:

$$D = D_1 y + O(y^2), \quad A = 1 - D_{1,0} y + O(y^2), \quad B = D_1 + O(y), \quad (8.14)$$

$$C_I = D_{1,I} y + O(y), \quad L_{IJ} = L_{IJ2} y^2 + O(y^3), \quad (8.15)$$

where $D_1 = D_1(w, x^I)$, $L_{IJ2} = L_{IJ2}(w, x^K)$, and for the redshift and affine parameter:

$$z = H^{\mathrm{obs}} D_1 y + O(y^2), \quad v = (H^{\mathrm{obs}})^{-1} z + O(z^2). \quad (8.16)$$

Here $H^{\mathrm{obs}}(w, x^I)$ is the observed effective 'Hubble' parameter measured from distance–redshift observations along C – which in general is anisotropic. It reduces to the Hubble parameter in an FLRW spacetime. By (8.16), $H^{\mathrm{obs}} = (\mathrm{d}z/\mathrm{d}v)_{v=0}$, and then from (7.19) we obtain (MacCallum and Ellis, 1970, Humphreys, Maartens and Matravers, 1997, Clarkson, 2000)

$$H^{\mathrm{obs}} = \tfrac{1}{3}\Theta + \dot{u}_a e^a + \sigma_{ab} e^a e^b. \quad (8.17)$$

For the transverse velocities,

$$V^I = V_0^I + O(y), \quad V_0^I D_{1,I} = D_{1,0} - D_1 H^{\mathrm{obs}}, \quad (8.18)$$

where $V_0^I = V_0^I(w, x^J)$, and the second equality arises from the normalization condition (8.11). The transverse velocity components do *not* in general vanish along C – regularity on C is maintained since the vector fields $V^I \partial/\partial x^I$ vanish along C. The matter density obeys

$$\rho_{\mathrm{m}} = \rho_{\mathrm{m}0} + O(y), \quad (8.19)$$

where $\rho_{\mathrm{m}0,I} = 0$ since ρ_{m} is a physical scalar along C.

Choosing $y = z$ on $C^-(w_0)$, the light-ray and matter kinematic scalars behave near C as follows:

$$\widehat{\Theta} = 2H_0^{\mathrm{obs}} z^{-1} - 2(H_0^{\mathrm{obs}})^2 D_2 + O(z), \quad \widehat{\sigma}_{IJ} = \widehat{\sigma}_{IJ1} z + O(z^2), \quad (8.20)$$

$$\Theta = \left[3H_0^{\mathrm{obs}} + (\sin\theta)^{-1}\left(\sin\theta\, V_0^I\right)_{,I} \right] + O(z), \quad (8.21)$$

$$\sigma_{ij} = \sigma_{ij0} + O(z), \quad \omega_i = \omega_{i0} + O(z), \quad (8.22)$$

where $H_0^{\mathrm{obs}} = H^{\mathrm{obs}}(w_0, x^I)$ and $D_2, \widehat{\sigma}_{IJ1}, V_0^I, \sigma_{ij0}, \omega_{i0}$ are functions of w_0, x^K. Any covariantly defined finite scalar must be independent of x^I along C: in particular, this implies that the magnitudes of shear and vorticity are isotropic on C. Since $\Theta_{,I}|_C = 0$, the anisotropy of the Hubble parameter is determined by the transverse velocities (Maartens, 1980, Maartens and Matravers, 1994):

$$H_{0,I}^{\mathrm{obs}} = -\tfrac{1}{3}\left[(\sin\theta)^{-1}\left(\sin\theta\, V_0^J\right)_{,J} \right]_{,I}. \quad (8.23)$$

Exercise 8.2.1 Consider the case of spherical symmetry (isotropy of the spacetime about C). Then we must have isotropic observations, and will find

$$A_{,I} = B_{,I} = D_{,I} = C_I = L_{IJ} = 0, \quad z_{,I} = V^I = H^{\text{obs}}_{,I} = \rho_{m,I} = 0. \tag{8.24}$$

Show that (a) the matter is irrotational, $\omega_\mu = 0$; (b) the null shear vanishes, $\widehat{\sigma}_{IJ} = 0$; (c) the matter expansion and shear are

$$\Theta = \frac{1}{A}\left(\frac{B_{,0}}{B} + 2\frac{D_{,0}}{D}\right), \quad \sigma = \frac{2}{\sqrt{3}A}\left|\frac{B_{,0}}{B} - \frac{D_{,0}}{D}\right|, \tag{8.25}$$

and the shear vanishes along C: $\sigma|_C = 0$ (recall that $2\sigma^2 = \sigma_{\mu\nu}\sigma^{\mu\nu}$); (d) vanishing matter acceleration implies

$$A^2 A_{,1} = B A_{,0} - A B_{,0}. \tag{8.26}$$

Exercise 8.2.2 Show that the RW metric $ds^2 = -dt^2 + a^2(t)[dr^2 + f^2(r)d\Omega^2]$ and 4-velocity are transformed to observational form,

$$ds^2 = A^2(w-r)\left[-dw^2 + 2drdw + f^2(r)d\Omega^2\right], \quad u^\mu = A^{-1}\delta^\mu_0, \tag{8.27}$$

via the coordinate change $w = r + \int dt/a(t)$, where $A(w-r) = a(t)$.

8.3 Ideal cosmography

In cosmography, we try to determine as much as possible without using any theory of gravity.

8.3.1 Observational data on the past light-cone

The world lines of discrete cosmological sources (galaxies, clusters, SNIa) pierce the past light-cone $C^-(w_0)$ of the observer, and their signals reach the observer via null geodesics of $C^-(w_0)$. Cosmological observations are effectively made at one time instant w_0, and they directly determine the redshift z, which is a convenient choice for the radial distance y on $C^-(w_0)$. Then:

- Given the intrinsic properties and evolution of sources, observations in principle also determine
  (a) the area distance $D_A(w_0, z, x^I)$ or equivalently, $D_L(w_0, z, x^I)$;
  (b) the lensing distortion of images, $L_{IJ}(w_0, z, x^K)$.
  (In practice, we use standard candles and statistical analysis, supplemented by astrophysical modelling and simulations, in the absence of knowledge of intrinsic properties and source evolution.)
- The number counts $N(w_0, z, x^I)$ of galaxies (including clusters of galaxies) are also in principle directly observable, and are related to the total baryonic matter, given the selection function and the source masses (including the 'missing' baryons in gas).

Then using (8.1) to include CDM, the observed number counts (8.12) determine $B(w_0, z, x^I) \rho_{\mathrm{m}}(w_0, z, x^I)$.

- In principle observations over extended time scales determine the instantaneous transverse velocities $V^I(w_0, z, x^J)$ of discrete sources. (In practice, this requires observation over significant time scales in order to trace the path of the source on the celestial sphere.)

8.3.2 Limits to ideal cosmography

It follows that, in principle and for idealized observations, we can directly determine the following quantities on $C^-(w_0)$ down to some maximum observed redshift $z_*(x^I)$:

$$\text{Idealized data} \Rightarrow \{u^\mu, B\rho_{\mathrm{m}}, g_{IJ}\} \text{ on } C^-(w_0),\ 0 \leq z \leq z_*(x^I). \tag{8.28}$$

But this is insufficient to determine the spacetime geometry of the past light-cone, because we need C_I and we cannot separate out B and ρ_{m}. What we need in order to fully determine $g_{\mu\nu}$ on $C^-(w_0)$, are $B = \mathrm{d}v/\mathrm{d}z$ and C_I: knowledge of C_I, together with (8.28), determines A, by (8.11). This means that, without gravitational field equations, we are unable to fully determine the spacetime on the past light-cone (down to z_*), even assuming perfect information from discrete-source observations (Ellis *et al.*, 1985). As a consequence, it is also impossible to test gravity theories directly.

Theorem 8.1 Limits of cosmography
Even with perfect observations, cosmography (no gravitational field equations) cannot determine the spacetime geometry on our past light-cone.

Cosmological testing of gravity theories
Observations cannot directly test GR on cosmological scales, or test any alternative theories of large-scale gravity. Any such tests are based on model-dependent assumptions about spacetime and its contents – and they test those models as much as they test theories of gravity.

(See Chapters 13 and 14.)

8.4 Field equations: determining the geometry

Analysis of the EFE on the past light-cone and in its neighbourhood (Ellis *et al.*, 1985) – together with assumptions (8.1) and (8.2) – then shows the following remarkable set of results.

8.4.1 Ideal cosmological data: the past light-cone

It turns out that the idealized data set (8.28) is precisely what is needed for the EFE to determine B and C_I on the past light-cone (Ellis *et al.*, 1985). There is not too much data (i.e. the system is not over-determined) and not too little (i.e. the system is not under-determined).

Theorem 8.2 EFE determine the past light-cone
Given the data set (8.28) – based on idealized observations of luminous sources and the assumptions (8.1)–(8.2) on dark matter and dark energy – Einstein's field equations on the past light-cone (i.e. those equations without derivatives transverse to $C^-(w_0)$), uniquely determine the matter distribution (ρ_m, u^μ) and geometry $(g_{\mu\nu})$ of the observable part of $C^-(w_0)$.

Note that the reconstruction of the geometry of the past light-cone from the cosmological data depends critically on the assumptions made about CDM and Λ. For different assumptions we get different answers. In particular:

Cosmological constant undetermined
Astronomical observations cannot determine Λ in a model-independent way.

That is why the standard determination of dark energy (see Chapter 13) is completely model dependent, and one can match the SNIa (and possibly other) observations by choosing a suitable family of cosmological models without dark energy (see Chapter 15).

8.4.2 Prediction to the past of the light-cone

The next step is to integrate the EFE into the causal past of the observable region of the past light-cone $C^-(w_0)$ (i.e. the region of spacetime which can be reached from there by past directed timelike or null curves). As shown by Ellis *et al.* (1985):

Theorem 8.3 EFE determine interior of past light-cone
Given the matter distribution and metric on $C^-(w_0)$, the Einstein equations transverse to $C^-(w_0)$ propagate the metric to the past of $C^-(w_0)$, and thus determine the spacetime within a region of the interior of $C^-(w_0)$.

This is a rather intriguing theoretical result: it is not clear why this should be the case. It implies the possibility of a long-term direct observational programme to determine the geometry of the observable part of the universe – the past light-cone from the present back to the LSS – directly, without making any assumptions about its geometry. However, it does need assumptions about the CDM and dark energy, i.e. (8.1) and (8.2).

The CMB anisotropies determine features of the LSS itself. At prior times, data such as element abundances serve to restrict the spacetime geometry, even though it is hidden from us. Determining the geometry of the LSS can be done in an inverse way; the CMB fluctuations are indeed direct indications of conditions on the LSS. Determining the geometry at the time of nucleosynthesis cannot be done this way; but then it properly belongs to physical cosmology rather than observational cosmology, in the sense we are using the terms here.

8.4.3 Prediction to the future of the light-cone

Things are different when we try to determine the spacetime to the future of $C^-(w_0)$, from data on $C^-(w_0)$. We want to propagate the data off the light-cone in both directions of time.

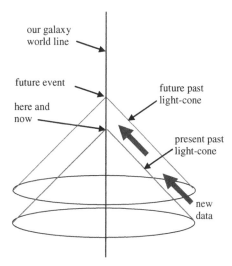

Fig. 8.2   Predicting to the future from data on the past light-cone: new data can nullify predictions.

We can do so to the past, as stated above: the available data on $C^-(w_0)$ are sufficient to determine the geometry off it to the past, as we can integrate the Einstein equations uniquely off $C^-(w_0)$ with those data as the starting point. However, integration to the future is quite different, because new information can come in from points to the future of $C^-(w_0)$ and change any prediction we can make on the basis of data on $C^-(w_0)$ alone. The situation is inherently unpredictable: we simply do not have enough data available to predict to the future (unless we live in a small universe, where we have already seen around the whole universe). This is illustrated in Figure 8.2.

Theorem 8.4 EFE cannot determine the future of $C^-(w_0)$
If we do not inhabit a small universe, it is not possible to uniquely predict conditions to the future of $C^-(w_0)$, since more data are required for that purpose than are available to us – we are unable to capture any information that propagates towards us along past light-cones to the future of $C^-(w_0)$.

For example, as time progresses the particle horizon expands, and may come to encompass a vast wall where different domains in a chaotic inflationary universe meet. As the physics in these domains may in principle be quite different – involving different matter content, and perhaps even different values of the fundamental constants of nature – one might expect violent electromagnetic and gravitational radiation to be emitted by such a clash of expanding universe domains with completely different physics. Gamma rays and gravitational radiation could pour down on us from the sky without any warning, because these events would not have been seen by us before this happened. Thus in chaotic inflationary universes, where a boundary between expanding universe domains with different physics might appear across the visual horizon and dramatically interfere with local physics, we cannot even guarantee that the Moon will rise tomorrow: unpredicted gravitational waves could tear it away from the Earth.

Predictability condition

Unless we live in a small universe, we can only predict the future evolution of the universe from the data observationally available to us if we make a 'non-interference assumption': that no unexpected influences will appear across our visual horizon as time progresses, and alter the predictions we make on the basis of the available data.

Why do we not normally notice this restriction? Because in the usual approach, we assume an almost-FLRW geometry, implicitly assuming statistical spatial homogeneity, which means the data beyond what we can see are very similar to what we can see. But this is an unverifiable assumption, compatible with old forms of the cosmological principle, but excluding, for example, chaotic inflation. This result also emphasizes the very special nature of small universes, discussed in Section 9.1.6 below: these are the only universes where we can strictly predict to the future from what we can see now.

Problem 8.1 Make precise the nature of the non-interference conditions needed in order to predict the future, as explained above, in a realistic cosmological context.

8.5 Isotropic and partially isotropic observations

The analysis in the general case is complicated, and the details may be found in Ellis *et al.* (1985). Some exact results can be derived when the observations of matter on the past light-cone are isotropic or partially isotropic.

8.5.1 Isotropic matter distribution on the past light-cone

A fundamental result applies to the case when we assume that observations of discrete sources on the past light-cone are isotropic. Intuitively one expects that the spacetime should be isotropic in this event. However, the result is not at all obvious – and furthermore, it is not clear how many of the observables need to be isotropic for the result to hold.

The original result (Maartens, 1980, Ellis *et al.*, 1985, Maartens and Matravers, 1994) neglected Λ and CDM, and we can incorporate both in the way explained above (Clarkson and Maartens, 2010). Without adopting the Copernican Principle, we have the following result:

Theorem 8.5 Matter isotropy on light-cone → spatial isotropy
If an observer comoving with the matter measures isotropic area distances, number counts, bulk velocities, and lensing, in a dust universe with Λ, then the spacetime is isotropic about the observer's world line (i.e. LTB).

Isotropy of bulk velocities seen by the observer is equivalent to vanishing proper motions (tranverse velocities) on the observer's sky. Isotropy of lensing means that there is no distortion of images. Thus by (8.28), isotropic observations imply that

$$D_{,I} = 0 = (B\rho_{\mathrm{m}})_{,I}\,, \quad V^I = 0 = L_{IJ}\,. \tag{8.29}$$

Momentum conservation ($\dot{u}^a = 0$) then gives

$$B_{,0} + [\ln(1+z)]_{,0}B = (1+z)^{-3}z_{,1}, \tag{8.30}$$

$$C_{I,0} + [\ln(1+z)]_{,0}C_I = (1+z)^{-3}z_{,I}. \tag{8.31}$$

The integrability condition on $w = w_0$ (where $y = z$) is

$$C'_I + (1+z)^{-1}C_I = B_{,I}, \tag{8.32}$$

where a prime denotes $\partial/\partial z$. The radial field equation in $C^-(w_0)$ gives

$$(B^{-1})' + (\ln D)'B^{-1} = (1+z)^2 \frac{DB\rho_{\mathrm{m}}}{2D'}. \tag{8.33}$$

Since the right-hand side is isotropic, it follows that $(B^{-1})_{,I} = 0$ on $w = w_0$, using the central condition (8.14). Then (8.32) shows that $C_I = 0$ on $w = w_0$, using the central condition (8.15). Finally, $A_{,I} = 0$ from (8.11).

Thus the metric on $C^-(w_0)$ is isotropic – and the interior of $C^-(w_0)$ must also be isotropic, in order to evolve to an isotropic state at $w = w_0$. It is clear from the proof that there is no redundancy in the assumptions – we need isotropy of all four observables.

If we adopt the Copernican Principle, it follows that all observers see isotropy, so that spacetime is isotropic about all galactic world lines – and hence spacetime is FLRW. This result then becomes a 'matter' alternative to the EGS theorem (Section 11.1), as a basis for FLRW (Maartens and Matravers, 1994):

Theorem 8.6 Matter isotropy on light-cones → FLRW
In a dust region of a universe with Λ, if all fundamental observers measure isotropic area distances, number counts, bulk velocities and lensing, then the spacetime is FLRW in that region.

In essence, this is the Cosmological Principle, but derived from observed isotropy and not from assumed spatial isotropy.

8.5.2 Isotropy of lensing and velocities

If we assume only isotropy of transverse velocities and lensing for one observer, then $V^I = 0 = L_{IJ}$, but $D_{,I}$ and $(B\rho_{\mathrm{m}})_{,I}$ may be non-zero. Analysis of equations (8.30)–(8.33), together with the light-cone field equations with ∂_I derivatives, shows that the spacetime is not isotropic about the observer, although the anisotropy is restricted (Maartens, 1980, Maartens and Matravers, 1994):

Theorem 8.7 Isotropic lensing and velocities
If an observer comoving with the matter measures isotropic bulk velocities and lensing, in a dust universe with Λ, then Einstein's equations enforce isotropy of the past light-cone only to $O(z)$ and isotropy of the area distance only to $O(z^2)$.

By (8.23), the observed Hubble rate is isotropic along C, i.e. $H_{0,I}^{\mathrm{obs}} = 0$. A series solution of the field equations on $C^-(w_0)$, obeying the central conditions (8.14)–(8.19), shows in detail where anisotropy is allowed (Maartens, 1980, Maartens and Matravers, 1994):

$$D = (H_0^{\mathrm{obs}})^{-1}z + \alpha_1 z^2 + D_3(\theta,\phi)z^3 + O(z^4), \tag{8.34}$$

$$D_3 := \beta_1 + \beta_2 \cos\theta + \beta_3 \sin\theta \sin\phi + \beta_4 \sin\theta \cos\phi, \tag{8.35}$$

$$B = (H_0^{\mathrm{obs}})^{-1} + 2\alpha_1 z + \left[3D_3 + \tfrac{1}{4}\rho_{\mathrm{m}0}(H_0^{\mathrm{obs}})^{-3}\right]z^2 + O(z^3), \tag{8.36}$$

$$C_I = D_{3,I}z^3 + O(z^4), \tag{8.37}$$

$$\rho_{\mathrm{m}} = \rho_{\mathrm{m}0} + H_0^{\mathrm{obs}}\left(\alpha_2 - 2\alpha_1\rho_{\mathrm{m}0} + 18\alpha_1^{-2}D_3\right)z + O(z^2), \tag{8.38}$$

where the αs and βs are constants. Note that the matter density may be anisotropic at $O(z)$.

8.5.3 Partial isotropy of area distances

If we have full isotropy of area distances, number counts, transverse motions and lensing for all observers, then spacetime is FLRW. There is a stronger result, based only on distances, and in fact only requiring isotropy up to third order in redshift (Hasse and Perlick, 1999):

Theorem 8.8 Isotropic distances to $O(z^3) \rightarrow FLRW$
In a dust region of a universe with Λ, if all fundamental observers measure isotropic area distances up to third order in a redshift series expansion, then the spacetime is FLRW in that region.

The proof relies on series expansions in a general spacetime, using the method of Kristian and Sachs (1966) (see also MacCallum and Ellis (1970), Humphreys, Maartens and Matravers (1997), Clarkson (2000)):

$$D = (K^a K^b \nabla_a u_b)_o^{-1} z + O(z^2), \tag{8.39}$$

where $K^a = -(u^a + e^a) = -k^a/(-u_b k^b)$ is a past-directed ray vector at the observer (see (7.13)). The higher-order terms involve $K^a K^b K^c \nabla_a \nabla_b u_c$, $K^a K^b K^c K^d \nabla_a \nabla_b \nabla_c u_d$ and $R_{ab}K^a K^b$ (Clarkson, 2000, Clarkson and Maartens, 2010). Since $(K^a K^b \nabla_a u_b)_o = [k^a \nabla_a (u_b k^b)]_o = (\mathrm{d}z/\mathrm{d}v)_o$, it follows from (7.19) that

$$\left(K^a K^b \nabla_a u_b\right)_o = \left[\tfrac{1}{3}\Theta + \dot{u}_a e^a + \sigma_{ab}e^a e^b\right]_o = H_o^{\mathrm{obs}}, \tag{8.40}$$

where the second equality is (8.17). Therefore isotropy at $O(z)$ for all observers enforces $\dot{u}_a = 0 = \sigma_{ab}$. With these conditions, the $O(z^2)$ term reduces to

$$\left(K^a K^b K^c \nabla_a \nabla_b u_c\right)_o = \tfrac{1}{3}\left[\Theta^2 + 4\pi G\rho_{\mathrm{m}} - \Lambda - 2\omega_a \omega^a\right]_o$$
$$+ \left[\left(\mathrm{curl}\,\omega_a - \tfrac{1}{3}\overline{\nabla}_a \Theta\right)e^a + \left(E_{ab} - \tfrac{1}{2}\pi_{ab} + \omega_{\langle a}\omega_{b\rangle}\right)e^a e^b\right]_o, \tag{8.41}$$

where a q_a term has been eliminated using (6.20). Isotropy at $O(z^2)$ then imposes 3 curl $\omega_a = \overline{\nabla}_a \Theta$, and $2E_{ab} = \pi_{ab} - 2\omega_{\langle a}\omega_{b\rangle}$. The complicated $O(z^3)$ term then leads to $\omega_a = \overline{\nabla}_a \Theta = 0$, and $\pi_{ab} = E_{ab} = 0$, and thus the spacetime is FLRW.

It is an open but important question how the Copernican results above translate to the realistic case of almost-isotropy – i.e. is the spacetime almost-FLRW? (This could be compared to the almost-EGS results, Section 11.1, based on almost-isotropy of the CMB.)

8.5.4 Determining a spherically symmetric geometry

The way to specifically carry out the direct observational approach indicated above in the spherically symmetric case has been pursued by Maartens *et al.* (1996), Araujo (1999), Lu and Hellaby (2007), McClure and Hellaby (2008), Araujo *et al.* (2008), Hellaby and Alfedeel (2009) and van der Walt and Bishop (2010). Lu and Hellaby (2007) show how to set up a numerical programme for determining the metric of the universe from observational data, particularly addressing the numerical problems at the vertex and those caused by the maximum in the area distance. Hellaby and Alfedeel (2009) give a presentation of the mathematical solution in terms of four arbitrary functions. McClure and Hellaby (2008) simulate observational uncertainties and improve the previous numerical scheme to ensure that it will be usable with real data as soon as observational surveys are sufficiently deep and complete.

8.5.5 Verifying an FLRW geometry

Using such a scheme, one should in principle be able to test spatial homogeneity of the universe from astronomical observations. Given isotropy, what are the observational data that can prove the universe has FLRW geometry? This is answered in principle by a result of Ellis *et al.* (1985):

Theorem 8.9 FLRW from area distances and number counts
A dust isotropic universe with cosmological constant is FLRW if, and only if, the area distances and number counts as functions of redshift take exactly the standard FLRW forms (7.50) and (7.64).

(The original result neglected CDM and Λ.) This gives a precise theoretical answer to the question posed: but it is very difficult to use in practice, because of evolutionary effects (sources may change with time) and observational problems (such as selection effects). Note that it is also model dependent: it is valid only for GR and a specific matter model. One can, however, find model-independent direct tests of spatial homogeneity in cases where the universe appears spherically symmetric about us (Clarkson, Bassett and Lu, 2008), as discussed in Chapter 15.

Exercise 8.5.1 Fill in the details of the proof that matter isotropy on the light-cone implies isotropic (LTB) geometry. Then prove that FLRW forms for $N(z)$ and $D(z)$ imply homogeneity (FLRW geometry).

8.6 Implications and opportunities

The direct observational approach attempted to answer the questions: what is the real information in the cosmological data? What can we deduce from those data alone, without a-priori assumptions about the geometry or matter distribution? These are intrinsically important questions to pose. It was a similar approach that led to detection of the great walls and voids that characterize the large-scale structure of the universe. But the approach is seriously compromised by the presence of dark matter and dark energy. And even if we make assumptions – (8.1) and (8.2) – to incorporate these dark components, the direct approach suffers from having no explanatory power. At most it just records what is there, in contrast to the standard model, which gives causal explanations (at a statistical level) for why that kind of structure is there.

However this approach is useful as a complement to the usual approach, because it throws light on various theoretical features of cosmology that may otherwise be obscured – some of these have been highlighted above. It allows examination of issues that one cannot easily look at with the direct approach, because the assumed RW geometry imposes spatial homogeneity, which then changes the nature of causality in these models. The symmetry changes a hyperbolic set of equations into a system of ordinary differential equations, with completely different causal properties.

8.6.1 Direct observational cosmology with galaxy surveys

There is a major difficulty in determining the density of matter at each event, because the mass-to-light ratio can be very variable: what you see is not all that is there. Indeed we run directly into the problem of dark matter: much of the matter present does not emit any radiation at all, and so is not observable. Gravitational lensing appears to be the answer for determining the matter distribution indirectly – but this relies on a perturbed RW model. It may be possible to derive partial constraints on CDM beyond the RW framework, and this deserves further investigation. The issue of dark energy is equally problematic, as emphasized above. We have to deal with the problem of the dark components via the assumptions (8.1) and (8.2).

Massive surveys of the galaxy distribution, such as 2dFGRS, SDSS and the upcoming DES, directly map the (visible) matter on our past light-cone, and therefore provide a new opportunity (not possible at the time that the direct approach was formulated) to pursue the approach (within the context of the necessary assumptions about the dark components). The survey data can be treated as if the galaxies lie on the spatial surface defined by the time of observation. The higher the redshift of the survey, the greater the errors introduced by this approximation. Clustering statistics in a statistically homogeneous universe are defined on spatial surfaces of constant time – whereas in fact the observed clustering statistics are on the null surface of the past light-cone. Correlation functions have been defined on the past light-cone (Yamamoto and Suto, 1999), and other GR corrections to the galaxy power spectrum have also been computed (Yoo, Fitzpatrick and Zaldarriaga, 2009, Yoo,

2010, Bonvin and Durrer, 2011, Challinor and Lewis, 2011). An important challenge can be presented for theoretical cosmology:

Galaxy survey challenge
Develop the analysis of light-cone statistics and selection effects as far as possible without making assumptions on the light-cone geometry, and see how far we can go towards determining that geometry from massive galaxy redshift surveys, given the necessary assumptions on CDM and Λ.

8.6.2 Transverse velocities

We have seen above what data are needed to determine the geometry of the universe on the past light-cone. Apart from the problem of the dark components, there is another major difficulty in obtaining these data: the velocities of the matter. While we can measure the radial component of velocities, determined by redshift, it is almost impossible to measure the transverse components V^I. There is no problem in principle: all one has to do is measure the apparent motions of distant objects across the celestial sphere, relative to a local non-rotating rest frame. In practice the motions involved are so small that this is not feasible at present, or in the foreseeable future, for objects at cosmological distances.

Missing velocity data
The transverse velocities that we need to fix the motion of distant matter are not measured in practice.

These are key data for anisotropic models. They also relate to other aspects of the spacetime geometry (Maartens, 1980, Maartens and Matravers, 1994):

Theorem 8.10 Transverse velocities and anisotropy
Anisotropy in the observed Hubble parameter implies that the transverse velocities are non-zero.
Non-zero shear or vorticity along C imply non-zero transverse velocities.

The first result follows from (8.23), and the second from an expansion of the kinematic quantities about the centre (Maartens, 1980, Maartens and Matravers, 1994). Hence transverse velocities encode data about anisotropy in the Hubble parameter and the shear and vorticity tensors. They of course vanish in FLRW models, so lack of these data is no problem in that case.

In effect the default position is to assume the transverse velocities vanish on average in all viable universe models, with local radial peculiar velocities being of the same magnitude as local transverse peculiar velocities; but this may not be true. The mean pairwise velocities of galaxies are determined by radial and transverse components. In the FLRW framework, the pairwise velocity distribution is deduced from the radial peculiar velocities (Ferreira *et al.*, 1999), but the procedure does not apply to more general models of the universe.

The practical implication is that it is worthwhile to try the best that we can to determine these transverse velocities through very precise measurements of proper motions of objects at cosmological distances, for these are major missing data in present day cosmology. This may be possible with VLBI (Titov, 2009) or through future surveys. Just as we had great surprises from some radial large-scale motions measured, we could perhaps also be in for a surprise as regards the transverse components.

8.6.3 Levels of uncertainty

What is the real degree of uncertainty resulting from the limits on cosmological data? Two points are worth making here (Ellis, 1975, 1980).

(1) Firstly, an obvious point but which needs emphasizing: *uncertainty increases with redshift,* i.e. with distance down the past light-cone. This is because the images are both fainter and smaller, and more intervening matter may obscure and distort what is happening at greater distances. Thus there are contours of increasing uncertainty in spacetime, with uncertainty growing down the light-cone and with proper-time off the light-cone.

The implication is that *our models become more and more theory dependent, and less and less observationally based, as we look to higher and higher redshifts.* The direct observational approach becomes more difficult the further back into the past one goes. In a sense an exception to this rule is the LSS itself, because of the detailed maps we are obtaining of the CMB anisotropies, which directly reflect conditions on the LSS. However, they too are interfered with by intervening matter (notably through the Rees–Sciama effect, the Sunyaev–Zel'dovich effect, and by gravitational lensing); hence one cannot interpret them uniquely without understanding all the intervening material.

(2) Secondly, there is generically an *ambiguity of determination of the spacetime from the observations:* two or more quite different models may be able to explain the same observations, depending on the matter content of the spacetime. This ambiguity in the spherically symmetric case is made explicit by the theorems of Mustapha, Hellaby and Ellis (1999), where the key focus is on the possibility of a time evolution of the sources observed. Unless one can limit this time evolution by astrophysical data, one cannot obtain unique cosmological information. Indeed what usually happens in this context is the other way round: it is assumed that the universe is spatially homogeneous, and the unknown time evolution of radio sources is then determined on the basis of this cosmological assumption (Ellis, 1975). Cosmology is not probed by these observations, rather they are used to determine astrophysical data on the basis of cosmological assumptions.

This ambiguity also has important applications in terms of interpreting the SNIa data, that are usually taken to indicate that the recent universe is accelerating, where the key issue is whether Λ is zero or not. This issue is discussed further in Chapters 14 and 15.

8.6.4 Averaging the light-cone

As in all cosmological models, one should carefully state what scale is being represented by the proposed model: will it only attempt a large-scale, smoothed out model, or will it try to provide a more detailed one, characterizing inhomogeneities in detail? This issue of averaging will be discussed in more detail in Chapter 16, and here we just make one comment: in the real universe, strong gravitational lensing leads to existence of a huge number of caustics in the past light-cone $C^-(w_0)$ at small scales, as every star, galactic core and dense cluster of galaxies will cause multiple images and associated caustics to occur in $C^-(w_0)$. Each caustic results in light-rays, that were the boundary of the past of here-and-now, plunging into the interior of the past (see the diagrams in Ellis, Bassett and Dunsby (1998)). Thereafter they still represent the paths of light-rays, but are no longer part of the causal boundary of the past, $J^-(w_0)$. This caustic structure is only visible when examined on small scales; when viewed on a large scale, these details are not visible, but the result is that $C^-(w_0)$ has an effective thickening of its surface. Evolving the field equations off $C^-(w_0)$ to the past is very tricky once this occurs.

Problem 8.2 Assess the best cosmologically relevant measures of proper motions that will be possible with future technology.

Problem 8.3 Estimate the levels of uncertainty that are encountered as one pursues the direct observational approach to earlier and earlier time.

PART 3

THE STANDARD MODEL AND EXTENSIONS

9 Homogeneous FLRW universes

FLRW cosmological models are those universes which are everywhere isotropic about the fundamental velocity (technically: there is a G_3 group of isotropies acting about every spacetime point which leaves the fundamental velocity invariant).[1] This will be the case if and only if the observations of every fundamental observer are isotropic at all times. This implies further symmetries of these universes: as well as being isotropic about each event, they are *spatially homogeneous*: all physical properties are the same everywhere on spacelike surfaces orthogonal to the fluid flow (technically: there is a G_3 group of isometries acting simply transitively on these surfaces). This will be proved in the sequel, but geometrically the result is clear: spheres of constant density centred on one point P are only consistent with spheres of constant density centred on other points Q and R if the density is constant.[2]

Because of these exact symmetries, these spacetimes cannot themselves be realistic models of the observed universe: they do not represent any of the inhomogeneities associated with the astronomical structures we see all around us. Realistic models of the observed universe are provided by perturbed FLRW universes, which are almost isotropic about every point, and hence are almost spatially homogeneous (they are inhomogeneous on small scales but homogeneous on large scales). *The 'almost FLRW' models are the standard models of cosmology at the present time* (considered in the following chapter). The FLRW models discussed in this chapter are the background models used to construct those more realistic cosmological models.

It is remarkable that the FLRW models provide a very good approximation to the observed universe despite their very high symmetry (as implied by the above, they are invariant under a group G_6 of isometries). The justification for assuming this symmetry for the background models is that we observe the universe to be isotropic about us to a high degree of approximation, once we (1) average over large enough scales (i.e. on scales significantly larger than clusters of galaxies) and (2) allow for our peculiar velocity relative to the average motion of matter in the universe (in practice, relative to the microwave background radiation). Thus on cosmological scales, there is no particular direction we can point to and say, 'The centre of the universe is over there'. There are then two possibilities: either (a) the universe is spatially inhomogeneous, and we are near a distinguished place about which it looks spherically symmetric, or (b) we are at a typical place in the universe, which is isotropic for every observer, and consequently is spatially homogeneous.

---

[1] Here and in the sequel, 'isotropy' means *spatial* isotropy, not spacetime isotropy.

[2] We need three points because an inhomogeneous curved 3-space can have *two* different (antipodal) centres of spherical symmetry, but no more.

The usual choice is to prefer the latter possibility, on the grounds of some form of *Copernican principle*: the assumption that we are not at a privileged position in the universe (Bondi, 1960, Weinberg, 1972). We shall later reconsider this issue (and we discuss option (a) in Chapter 15). However, in this chapter we accept the argument, and so examine in depth the FLRW universes on the understanding that they do indeed gives us good models of the observed universe domain, when suitably averaged over inhomogeneities. Particular FLRW models of importance in cosmology are the Einstein static universe, de Sitter universe, Milne universe and Einstein–de Sitter universe; we shall describe them in this chapter. These are based on a fluid description of the matter in the universe. One can also include scalar fields in FLRW models, or use a kinetic theory description of the matter. We shall also deal with these possibilities here.

9.1 FLRW geometries

The Robertson–Walker (RW) geometries are everywhere isotropic about the fundamental world lines. These geometries are employed in the Friedmann–Lemaître (FL) world models, which originally were considered only with pressure-free matter plus a cosmological constant; but they have since been used with much more general matter content. We shall refer to all universe models with a RW geometry and some suitably specified matter content determining the dynamical evolution via the EFE as *FLRW models*.[3]

9.1.1 Consequences of isotropy

Geometric isotropy about all the fundamental world lines clearly implies zero shear, vorticity and acceleration everywhere:

$$\sigma_{ab} = 0, \ \omega^a = 0, \ \dot{u}^a = 0, \tag{9.1}$$

for otherwise these quantities would pick out preferred directions in the 3-space orthogonal to u^a. Therefore, from (4.37), there is a normalized proper time t which is a potential for u^a, i.e. $u_a = -t_{,a}$, and is unique up to $t \to t + \text{const}$. The surfaces of spatial homogeneity in these universes are $t = \text{const}$, which are orthogonal to the fluid flow lines. This much follows purely from geometry.

The total matter content of necessity has to have perfect fluid form,

$$\pi_{ab} = 0, \ q_a = 0, \tag{9.2}$$

as follows from the assumption of isotropy: if q_a or π_{ab} were non-zero, the stress tensor and hence the Ricci tensor would be anisotropic. Then $p = p(t)$, or there would be an anisotropy via a pressure gradient; hence $\overline{\nabla}_a p = 0$, which implies $\dot{u}^a = 0$ from the momentum conservation equation (confirming the vanishing of the acceleration). From the $(0, i)$

---

[3] In contrast to *kinematic world models*, where no field equations are employed.

field equations, $q_a = 0 \Rightarrow \overline{\nabla}_a \Theta = 0$. Putting this together, all the non-zero kinematic and stress–energy scalars are functions only of the time t:

$$\rho = \rho(t),\ p = p(t),\ \Theta = \Theta(t)\ \Leftrightarrow\ \overline{\nabla}_a \rho = \overline{\nabla}_a p = \overline{\nabla}_a \Theta = 0, \tag{9.3}$$

expressing the spatial homogeneity of these universes on the surfaces $t = $ const (all physical quantities are constant on these surfaces). Then:

FLRW definition:
A universe is FLRW if and only if (9.1)–(9.3) *hold everywhere.*

Alternatively, we can characterize these models directly from isotropy of observations. On the one hand, if a cosmological model is isotropic about each point, then astronomical observations will be isotropic everywhere also. Conversely, suppose all cosmological observations are isotropic about each observer. Then the restrictions (9.1) follow directly from measured isotropy of the magnitude–redshift relation (implying vanishing σ_{ab} and $\dot{u}^a$), number counts (implying $\overline{\nabla}_a \rho = 0$, which can only happen if $\omega_a = 0$), and vanishing of proper motions (implying vanishing ω_a and σ_{ab}). The first and third of (9.3) follow because otherwise anisotropies would be observed in the magnitude–redshift relation. Equations (9.2) then follow from the Gauss equation and the $(0,i)$ field equations, respectively. Finally $\dot{u}^a = 0$ implies $\overline{\nabla}_a p = 0$ from the energy–momentum conservation equation (as the matter is a perfect fluid with $\rho + p \neq 0$). Thus:

Theorem 9.1 Isotropic observations
A universe model is FLRW if and only if all astronomical observations by all fundamental observers are isotropic at all times.

The 4-velocity u^a about which the universe is spatially isotropic is unique, provided the universe is not also *spacetime* isotropic (i.e. additionally invariant under boosts). That characterizes spacetimes of constant curvature, invariant under a group G_{10} of isometries (see Sections 2.7.3 and 9.3.1 and Chapter 17). From (5.37), this exceptional case can happen only when $\rho + p = 0$, for otherwise (when $\rho + p \neq 0$), u^a is uniquely defined as the timelike eigenvector of the Ricci tensor.

9.1.2 Spacetime geometry

We have seen that the spatially homogeneous surfaces $t = $ const are orthogonal to the fluid flow lines. We define the scale factor a from a chosen constant value a_1 on some initial surface $t = t_1$ by the relation $\dot{a}/a = \Theta/3$; then from (4.35), $\Theta = \Theta(t) \Rightarrow a = a(t)$. Thus,

$$u_{a;b} = \frac{1}{3}\Theta(t)h_{ab} = \frac{\dot{a}}{a}h_{ab},\ \ a(t) = \alpha \exp\left[\int_{t_1}^{t} \frac{1}{3}\Theta(t')dt'\right], \tag{9.4}$$

where $\alpha = $ const. Now using comoving coordinates (t, x^i) as in Section 6.3.1, and writing $h_{ij} = a^2(t)f_{ij}(x^\mu)$ (see (6.26)), from (5.48) the condition $\sigma_{ij} = 0$ implies $f_{ij,0} = 0$. Thus $ds^2 = -dt^2 + a^2(t)f_{ij}(x^k)dx^i dx^j$, $u^\mu = \delta_0^\mu$. This shows the splitting of the spacetime metric into parts parallel and orthogonal to u^μ (compare (4.13)); the orthogonal part $a^2(t)f_{ij}(x^k)$ is the metric of the spatially homogeneous 3-spaces.

The scale factor $a(t)$ describes how all spatial distances change as the universe evolves. To make this explicit, consider a curve γ_1 joining the world lines C_1, C_2 of two fundamental observers in a surface $t = t_1$, and given in terms of the comoving coordinates by $x^\mu(v) = (t_1, \lambda^i(v))$. The distance d_1 measured between C_1, C_2 along this curve will be

$$d_1 = \int_{C_1}^{C_2} a(t_1) \left[f_{ij}(x^k) \frac{d\lambda^i}{dv} \frac{d\lambda^j}{dv} \right]^{1/2} dv. \tag{9.5}$$

At any later time t_2 the distance d_2 between the same world lines along the corresponding curve γ_2 given (in comoving coordinates) by the same functions $\lambda^i(v)$, i.e. $x^\mu(v) = (t_2, \lambda^i(v))$, will be given by the corresponding expression with t_1 replaced by t_2, and so will be related to d_1 by

$$d_2 = [a(t_2)/a(t_1)]d_1. \tag{9.6}$$

Thus as a increases, all comoving lengths scale proportionately to $a(t)$ and we have an isotropic expansion about every point, but with no centre because the universe is spatially homogeneous (and the expansion is not an expansion *into* anything: there is no spatial edge to the universe that can expand into any space 'outside' the universe, for the universe is all that there is!). The mapping of 3-spaces $t = $ const into each other defined by the fluid flow is a *conformal mapping*, i.e. preserves angles as well as ratios of lengths.

From the above, a distance d_1 between fundamental particles in the initial surface $t = t_1$ scales with $a(t)$ in the sense that at later times this corresponds to the distance $d(t) = a(t)d_1$. Clearly the speed of motion in the surface $t = $ const is defined by

$$v(t) := \dot{d}(t) = \dot{a}(t)d_1 = H(t)d(t), \quad H(t) := \dot{a}(t)/a(t). \tag{9.7}$$

This shows how the Hubble expansion law may be interpreted as an exact law of recession in the surfaces $t = $ const. Note that this is a notional speed that does not correspond to the transfer of information, and so can exceed the speed of light (Ellis and Rothman, 1993), and it cannot be directly measured by any astronomical observation we can carry out.

Conformal structure

It is clear from (4.51)–(4.52) that in a (perfect fluid) FLRW universe, $E_{ab} = 0 = H_{ab} \Rightarrow C_{abcd} = 0$; thus these universes are conformally flat. Conversely, if $C_{abcd} = 0$ and $\rho + p \neq 0$, then by (6.33) $\overline{\nabla}_a \rho = 0$, by (6.34) $\sigma_{ab} = 0$, and by (6.35) $\omega^a = 0$. Hence if the matter is a perfect fluid with barotropic equation of state, $\overline{\nabla}_a p = 0$ also, and the conservation equations show $\dot{u}^a = 0$. Then we have an FLRW universe.[4]

Theorem 9.2 Conformal flatness
A barotropic perfect fluid universe is FLRW if and only if it is conformally flat (i.e. $C_{abcd} = 0$).

---

[4] We do not need the barotropic condition if the matter is in geodesic motion.

This result is helpful in examining the propagation of electromagnetic waves in FLRW universes, as well as the global properties of these spacetimes (in particular, the nature of their horizons, discussed in Section 7.8).

9.1.3 3-space geometries

From the Gauss equation in the form (6.25):

$$\pi_{ab} = \sigma_{ab} = 0 \Rightarrow {}^3R_{\langle ab \rangle} = 0 \Leftrightarrow {}^3R_{ab} = \tfrac{1}{3}{}^3Rh_{ab}, \tag{9.8}$$

so the three-dimensional Ricci tensor is isotropic: this is also clear directly from the assumption of isotropy, as otherwise there would be preferred spatial directions, the eigendirections of the 3-Ricci tensor. Exercise 6.3.2 then shows that the homogeneous hypersurfaces orthogonal to the fluid flow are spaces of constant curvature $K/a^2(t)$ where K is a constant and, as in Section 2.7.7, we can set K to be 1, 0 or -1 and obtain from (2.90) the metric form and 4-velocity:

$$ds^2 = -dt^2 + a^2(t)\left[dr^2 + f^2(r)\left(d\theta^2 + \sin^2\theta d\phi^2\right)\right], \quad u^\mu = \delta^\mu{}_0,$$

$$f(r) := (\sinh r, r, \sin r) \text{ for } K = (-1, 0, +1). \tag{9.9}$$

(Other choices of radial coordinate are also widely used, e.g. Weinberg (1972), Peebles (1971) and (9.10)).

Conversely, if the metric and 4-velocity take the form (9.9) in a set of coordinates $x^\mu = (t, r, \theta, \phi)$, then (9.1) and (9.4) hold. The field equations then show that (9.2) and (9.3) follow, and the universe is an FLRW model, so:

Theorem 9.3 FLRW coordinates
A universe model is FLRW if and only if coordinates can be found so that the metric and 4-velocity are given by (9.9).

The nature of these 3-spaces of constant curvature follows immediately from this derivation. From (2.90), the 2-sphere S_d at distance $d = a(t_1)r$ from the origin of coordinates in $t = t_1$ has surface area $A_1 = 4\pi a^2(t_1)f^2(r)$. Thus we can imagine testing the geometry of the space sections by comparing the radii and surface areas of spheres centred on some point p (which we choose as the origin of coordinates, but in fact is an arbitrary point in the space: there is nothing special about this point).

In the flat-space case ($K = 0 \Rightarrow {}^3R_{abcd} = 0$), the familiar Euclidean relation holds: $A \propto r^2$, and the 3-spaces continue to infinity (one can attain arbitrarily large distances from the original point, and the volume of these 3-surfaces is unbounded). In the hyperbolic case ($K < 0$), the area increases faster with distance than in the Euclidean case, as $A \propto \sinh^2 r$. Again these 3-spaces are unbounded; this is the three-dimensional case analogous to the two-dimensional Lobachevski plane of constant negative curvature (which can be mapped into the interior of the unit circle).

In the elliptic case ($K > 0$), $A \propto \sin^2 r$: the area increases slower than in the Euclidean case, reaches a maximum when $r = d/a(t_1) = \pi$, and thereafter decreases to zero as $r \to 2\pi$

at a point q 'antipodal' to the centre p. To see what is happening, consider geodesics γ_1, γ_2 leaving p in opposite directions. They intersect each sphere S_d in points r_1, r_2 respectively that are antipodal to each other in S_d; therefore as $r \to 2\pi$, these geodesics approach q from precisely opposite directions. Hence if one moves from p along the geodesic γ_1, after approaching q and passing through it one continues along the path of the geodesic γ_2 and then arrives back at p; and this happens whatever direction is chosen for γ_1. Thus *when $K > 0$ the space sections are necessarily closed and of finite volume.* The situation is exactly modelled by the two-dimensional surface of an ordinary sphere, which is the two-dimensional analogue of the three-dimensional space of constant positive curvature. This is why Einstein preferred this case to the other possibilities. indeed it was his motivation for investigating his static universe with closed spatial sections: it solves the problem of boundary conditions for local physical systems (what are the boundary conditions on physical fields at infinity?) This problem vanishes when there is no infinity, and periodic boundary conditions are imposed by the topology.

When we choose $K = 0, \pm 1$ as in (9.9), this implies that K and r are dimensionless and hence a has dimension length. It is common practice in cosmology to take a as dimensionless. Then we are free to normalize its current value, e.g. to unity: $a_0 = 1$. With dimensionless a, it follows that r has dimension length and K has dimension $(\text{length})^{-2}$, and is given by the curvature scale ${}^3R_0$ – see (A.6). Then we have that the metric function $f(r) = r$ for $K = 0$, while for $K \neq 0$,

$$f(r) = \begin{cases} K^{-1/2} \sin\left(\sqrt{K}\, r\right) & \text{for } K > 0 \\ (-K)^{-1/2} \sinh\left(\sqrt{-K}\, r\right) & \text{for } K < 0 \end{cases} \tag{9.10}$$

$$|K| = \left({}^3R_0\right)^{-2} = \left(a_0 H_0 \sqrt{|\Omega_{K0}|}\right)^{-2},$$

where $\Omega_{K0} := -K/(a_0 H_0)^2$ [see (9.16)].

9.1.4 Symmetry properties

The point p was an arbitrary point in the surface $t = t_1$; we could equally have chosen any other point p' as the origin of coordinates, and (because K is constant) would have obtained the identical metric components and geodesic behaviour centred on that point. Thus the spatial sections, with metric (2.90), (2.93), are completely homogeneous: all points are equivalent to each other. From (9.3) and (9.4), the scale factor and expansion are also constant on $t = t_1$ (which is any one of the surfaces orthogonal to the 4-velocity u^a), so the spacetime itself (with metric (9.9)) is spatially homogeneous (we already know that all physical scalars, e.g. the density and pressure, depend only on the time coordinate t which labels the surfaces). As all physical and geometrical quantities are identical at all points of each surface $t = \text{const}$, these are surfaces of homogeneity of the cosmological model. We have therefore shown that 'isotropy everywhere' implies spatial homogeneity of the spacetime.

The property of homogeneity can be formalized in various ways; most commonly this is done in terms of continuous symmetry groups and associated Killing vectors (Section 2.7):

Theorem 9.4 FLRW symmetries
FLRW universe models are uniquely characterized as invariant under a symmetry group G_6 acting on spacelike 3-spaces, with a G_3 simply transitive subgroup of isometries and a G_3 isotropy group around every point.

Note that FLRW models have Bianchi symmetries for particular Bianchi types (depending on the curvature K). (See Chapters 17, 18.)

A FLRW universe is *not* homogeneous on other spacelike sections than the geometrically preferred surfaces $t =$ const; however, these homogeneous surfaces do not relate in a simple way either to astronomical observations, or to the Newtonian limit. As to the first, spatial sections of instantaneity determined by radar do not coincide with these spatial sections if the universe is expanding; and cosmological observations (down the past null cone) cut across these surfaces, so (as we discuss later) observationally verifying spatial homogeneity is not easy. As to the second, one can claim (Ehlers, 1973) that the 'almost-Newtonian' spacetime sections experienced by an observer O are those space sections generated by geodesics orthogonal to O's world lines. In an evolving FLRW universe model, these surfaces are not the surfaces $t =$ const, and the density and pressure are not constant on these surfaces.

If the metric is as in (9.9) but the matter 4-velocity is different (i.e. not orthogonal to the surfaces of homogeneity), we can claim to have a spacetime with FLRW geometry but (provided $\rho + p \neq 0$) the matter content is not a perfect fluid (Coley and Tupper, 1983). This illustrates the fact that the energy–momentum tensor by itself does not force a particular physical interpretation of its nature. However, the combinations of sources required look somewhat contrived, even though each constituent may be of a familiar type, and the universe will not appear isotropic to observers moving with this 4-velocity. There is no physical reason to choose this model.

9.1.5 Topology

The discussion so far has mainly related only to local properties, but it should be realized that different global connectivities are possible in each case. If $K = 0$, the spatial sections are locally flat and we can change to Cartesian coordinates (x, y, z) in the standard way; the metric will then be $ds^2 = -dt^2 + a^2(t)dx^2$. It is usual to assume these coordinates have the standard infinite range: $-\infty < x, y, z < \infty$; then the space sections $t =$ const are without boundary and of infinite volume, and there is an infinite amount of matter in the universe. However, there are many other possibilities. The simplest is the *torus topology*, where there are scales L_i such that if the point p has coordinates x^i it is identified with every point q with coordinates $(x + nL_x, y + mL_y, z + pL_z)$ where (m, n, p) are arbitrary integers. In this case each space section $t =$ const is without boundary but is of finite volume, and there is a finite amount of matter in the universe, which has closed ('compact') spatial sections. The universe is no longer simply connected, as is the case for its 'natural' topology, when the 3-spaces are isomorphic to Euclidean space E^3. There are many other possible topologies for flat spatial sections (see e.g. Wolf (1972), Ellis (1971b)), including generalizations of the Möbius strip. Thus as well as giving the spacetime metric, we need to specify its global connectivity (its 'topology') in order to fully specify its geometry.

The case $K < 0$ is similar: the 'natural' topology of the space sections is that of Euclidean 3-space, but there are many other (in fact, infinitely many) possibilities allowing finite, closed spatial sections (Thurston, 1997).

When $K > 0$, things are fundamentally different. Considering the geodesics and coordinates described above in this case, where now $f(r) = \sin r$, if no identifications are made we obtain the compact geometry of a 3-sphere, as described above (which is simply connected). There are still various other topologies possible: for example, each antipodal point q could be the *same* as the original point p; but all of them are necessarily compact (Ellis, 1971b).

Thus spatial sections may have an 'unnatural' topology (i.e. not be simply connected), but alternative topologies are more probable (on any ordinary measure) than 'simple' topologies and are suggested by string theory approaches to fundamental physics. These models have a length scale that is indeterminate on the basis of present-day physics, and so just has to be set as initial data with no known deeper cause; but that is part of the larger problem that we have no idea what kind of mechanism – if any – determines the topology of the spatial sections of the universe. The standard assumption that they are simply connected is a theoretical prejudice that may or may not be true.

Thus the usual assumption that $K < 0$ and $K = 0$ models are 'open', with infinite spatial sections, is not necessarily true. However, $K > 0$ models are necessarily closed.

For any compact FLRW universe model, at any time t_* there are two important length scales:

- the *minimal closed comoving length* L_1: a 3-sphere of radius greater than $a(t_*)L_1$ in the surface $t = t_*$ intersects itself at least once, while one of radius less than $a(t_*)L_1$ does not intersect itself;
- the *complete closed comoving length* L_2: a 3-sphere of radius greater than $a(t_*)L_2$ in the surface $t = t_*$ intersects itself in every direction, while one of radius less than $a(t_*)L_2$ does not intersect itself in at least one direction.

A key question then is how this relates to horizons in the universe.

Causally closed universes

When the comoving particle horizon $u_{ph}(t_*)$ at time t_* is less than L_1, then causal horizons are broken in at least one direction from that time on, while if $u_{ph}(t_*)$ at time t_* is less than L_2, then causal horizons are broken in all directions from that time on. Given a compact universe, depending on the dynamics, these possibilities may never occur (for example, in a dust-filled FLRW model with $K > 0$ and S^3 spatial topology), or they may occur at some finite time – the horizon breaking time (for example, in an FLRW model with $\Lambda > 0$ and $K > 0$ with S^3 spatial topology). In an inflationary universe model with compact spatial sections, they can occur very early in the history of the universe; after that time, all particles are in causal contact with each other. This then solves the boundary problem for local physics raised by Einstein and Wheeler, and completely solves the horizon problem: no new information enters the past light-cone of any particle at later times.[5]

It also eliminates many divergences in physics because it implies a long-wavelength cutoff to all physical effects. An intriguing question is whether this cutoff might have observable consequences. N-body simulations of structure formation in cosmology effectively assume such a cutoff via periodic boundary conditions. This is done to solve computational problems rather than being taken as a model of how the universe really is; but it has to affect the largest wavelength structures predicted by the theory.

9.1.6 'Small universes'

The further question for a closed universe in observational terms is whether or not we can see right round the universe. If the scale L_1 is less than the visual horizon u_{vh} at some time t_*, the universe is so small that we can see round the universe in at least one direction from then on, while if the scale L_2 is less than u_{vh} the universe is so small that we can see round the universe in all directions from then on. We call the latter case a *small universe*: by definition, that is a universe which closes up on itself spatially for topological reasons, and does so on such a small scale that we have seen right round the universe since the time of decoupling.

These universes are of course a subclass of causally closed universes. In this case we can see all the matter that exists (there are no visual horizons or matter beyond the horizon), with multiple images of many objects occurring (Ellis and Schreiber, 1986); indeed a universe that appears to consist of a vary large number of galaxies can actually consist of a relatively small number of galaxies that are imaged many times over. Then the universe gives the appearance of an unbounded homogeneous universe, even if it is really a 'small' (something like 300 to 800 Mpc) inhomogeneous block; thus this provides an explanation for the apparent homogeneity of the universe (it looks homogeneous because we are seeing the same thing over and over again!).

Checking if the universe is a small universe or not is an important task; there is a quite different relation of humanity to universe in this case, because the entire universe is observable, which is otherwise false (Ellis and Schreiber, 1986). These are thus the only cosmologies where we have all the data needed to predict to the future (Section 8.5). This possibility is observationally testable in various ways, discussed in Section 13.4.3. One should note here that small universes are compatible with inflation: nothing in the inflationary scenario determines the topology of the spatial sections, so it is compatible with the creation of structure via inflation (Section 12.2).

Exercise 9.1.1 Show the tangent vector β^a to each dragged along curve γ in an FLRW universe is a relative position vector, and obeys the equation $\dot{\beta}^a = H\beta^a$. Integrate to show $\beta^a = a(t)K^a$, $\dot{K}^a = 0$, $K_a u^a = 0$. Confirm from these equations that $\dot{e}^a = 0$, $\delta l^{\cdot} = H(t)\delta l$ as required by specialization of (4.32)–(4.33).

---

[5] This is not the case for inflation in a universe without compact spatial sections, for then new information is always entering the particle horizon as time evolves.

Exercise 9.1.2 Find the equations for the surfaces of constant time in FLRW that a fundamental observer determines by radar. Show they coincide with the surfaces of constant density only if the universe is static. If $a(t) = t^{2/3}$, what are these surfaces? (Ellis and Matravers, 1985).

Exercise 9.1.3 Explain why it is no accident that the 2-sphere example of a curved surface serves as an exact model of the 3-sphere case. [Consider the surface $\phi = \text{const}$ in the 3-space.]

Exercise 9.1.4 Determine the other possible topologies for spatially closed universes with flat spatial sections. (See Ellis (1971b). Note that we do not suggest trying the $K < 0$ case: it is very difficult!)

9.2 FLRW dynamics

9.2.1 Dynamical equations

The basic equations governing the dynamics of FLRW universe models have already been derived. Collecting them together, the scale factor $a(t)$, energy density $\rho(t)$ and pressure $p(t)$ are related by the Raychaudhuri equation (6.11), energy conservation (5.38) and the Friedmann equation:

$$3\frac{\ddot{a}}{a} = -4\pi G(\rho + 3p) + \Lambda, \tag{9.11}$$

$$\dot{\rho} + 3H(\rho + p) = 0, \tag{9.12}$$

$$H^2 = \frac{8\pi G}{3}\rho + \frac{\Lambda}{3} - \frac{K}{a^2}. \tag{9.13}$$

The Friedmann equation is the 3-curvature equation (6.23) and is also a first integral of the other two equations when $\dot{a} \neq 0$ (see (6.12)). The momentum conservation equation will be identically satisfied, as will the other eight field equations provided the metric has the form discussed in Section 9.1.

Theorem 9.5 FLRW dynamics
Given an FLRW universe model described by (9.9), when $\dot{a} \neq 0$, only the conservation equation (9.12) and Friedmann equation (9.13) need be satisfied; then all ten Einstein field equations will be satisfied.

This follows because the Raychaudhuri equation will then be a consequence of (9.12), (9.13), and the other eight field equations are trivial. This will be true irrespective of the equation of state; but to have a determinate set of equations, we must give suitable equations of state for the matter (as discussed in Chapter 5). Note that although $p = p(t)$ and $\rho = \rho(t)$ the equation of state need not be barotropic (i.e. of the form $p = p(\rho)$) – it could be of the form $p = p(\rho, s)$ where $s = s(t)$ is the entropy of the matter, determined by further equations of state. It is this extra freedom that allows the expansion and collapse phases of a realistic model to have different behaviours.

The matter content of an FLRW universe necessarily has a perfect fluid form. However, this does not mean the fluid has to have a 'perfect fluid' equation of state. Indeed we can suppose for example that the fluid has equations of state (5.33)–(5.35) with non-zero coefficients of viscosity λ, heat conductivity κ and bulk viscosity ζ. Since $\sigma_{ab} = 0, \overline{\nabla}_a T = 0$ and $\dot{u}_a = 0$ in an FLRW universe, these equations of state imply the stress–energy tensor must take the perfect fluid form (5.37) in an FLRW universe; non-zero bulk viscosity means that the fluid is not barotropic. Thus even in this case we will arrive at the same dynamical equations for the universe as above. The same applies if we have a kinetic theory description of matter (Section 9.5), or if scalar fields dominate (Section 9.7.3). In each case (when we have a RW geometry) we will necessarily have a perfect fluid form for the energy–stress tensor, with effective energy density and pressure related to the matter properties via suitable equations of state, which embody the physics of the situation. The same gravitational dynamical equations will apply in all cases. The key point is that ρ should represent the sum total of all matter contributions to the energy density, whatever they are; then the above equations will be universally valid.

There is an ambiguity with Λ. One can regard it as an extra term in the EFE, as indicated above, giving an extra degree of freedom in the relation between the matter and the geometry; or (following W. H. McCrea) one can regard it as a contribution to the matter term: as mentioned above, it is a fluid with equation of state $p = -\rho$. We will adopt whichever of these equivalent viewpoints is convenient for specific analyses.

9.2.2 Density parameters and dynamical properties

We utilize the standard definitions:

$$\Omega := \frac{8\pi G \rho}{3H^2}, \quad \Omega_\Lambda := \frac{\Lambda}{3H^2}, \quad q := -\frac{1}{H^2}\frac{\ddot{a}}{a}, \tag{9.14}$$

which are the dimensionless density parameters and deceleration parameter. The Friedmann equation then can be written

$$\frac{K}{a^2 H^2} = \Omega_{\text{total}} - 1 \equiv \Omega + \Omega_\Lambda - 1, \tag{9.15}$$

showing that $K > 0$ ($= 0, < 0$) if $\Omega_{\text{total}} > 1$ ($= 1, < 1$) respectively. Defining a positive density parameter for curvature, (9.15) becomes

$$\Omega + \Omega_\Lambda + \Omega_K = 1, \quad \Omega_K = -\frac{K}{a^2 H^2}. \tag{9.16}$$

This leads to a representation of the matter, curvature and cosmological constant in a 'cosmic triangle' (Bahcall *et al.*, 1999).

The density parameter Ω as presented here represents the contribution to the energy density of all matter and fields present: baryons, CDM, photons, neutrinos, but not the cosmological constant. It is often useful to separate out the matter and radiation contributions, to give $\Omega = \Omega_{\text{m}} + \Omega_{\text{r}}$; they will have different variations with time. We can further split the matter into CDM and baryons, $\Omega_{\text{m}} = \Omega_{\text{c}} + \Omega_{\text{b}}$ and the radiation into photons and neutrinos, $\Omega_{\text{r}} = \Omega_\gamma + \Omega_\nu$.

Whatever the physics, we can define an effective (generically time-dependent) equation of state parameter $w(t) = p/\rho$ (note that this can be defined for a mixture of fluids, such as matter plus radiation; one can also choose to include the cosmological constant). The Raychaudhuri equation is then

$$q = \tfrac{3}{2}\Omega\left(w + \tfrac{1}{3}\right) - \Omega_\Lambda. \qquad (9.17)$$

Present-day values

The present-day values of these quantities (denoted as usual by a subscript 0) are related by

$$2q_0 = \Omega_0(1 + 3w_0) - 2\Omega_{\Lambda 0}, \qquad (9.18)$$

from (6.9), where $\Omega_0 + \Omega_{\Lambda 0} + \Omega_{K 0} = 1$. Observed redshifts plus distance estimates show $H_0 \sim 10^{-10}\mathrm{yr}^{-1} > 0$. As both q_0 and Ω_0 are in principle observable, we can use these equations to determine $\Omega_{\Lambda 0}$ (and so Λ) and $\Omega_{K 0}$.

Accelerating or decelerating?

Suppose we represent Λ as a component of the cosmic fluid. Then (9.17) shows that $w = -1/3$ is a critical value separating decelerating periods ($q > 0$, $w > -1/3$) from accelerating periods ($q < 0$, $w < -1/3$). Following Barrow (1993), we shall say the universe is *inflationary* when $q < 0$, for this is the essential feature which (if continued for long enough) enables the horizon size to grow large relative to the visible region of the universe. It is certainly satisfied during a period of exponential expansion. If at some time the dominant dynamical feature of the universe is a cosmological constant, then it will be in an inflationary phase.

Exercise 9.2.1 Derive equations (9.11)–(9.13) directly from (9.4). (Directly calculate the restricted form of the Raychaudhuri equation, conservation equations, and Gauss equation when the fluid only expands.)

Exercise 9.2.2 Suppose we examine an FLRW universe model from a 4-velocity that is tilted relative to the surfaces of homogeneity. What are the effective equation of state of the matter and field equations relative to this 4-velocity? (Coley and Tupper, 1983).

9.3 FLRW dynamics with barotropic fluids

If the total energy density is composed of matter and radiation, we can write $\rho = \rho_\mathrm{m} + \rho_\mathrm{r}$, and correspondingly $\Omega = \Omega_\mathrm{m} + \Omega_\mathrm{r}$. For present-day values $\Omega_{\mathrm{r}0} \sim 10^{-4}$, $\Omega_{\mathrm{m}0} \sim 0.3$, $\Omega_{\Lambda 0} \sim 0.7$, we obtain

$$2q_0 \simeq \Omega_{\mathrm{m}0} - 2\Omega_{\Lambda 0}, \quad \Omega_{K 0} \simeq 1 - \Omega_{\mathrm{m}0} - \Omega_{\Lambda 0}. \qquad (9.19)$$

For best-fit current observationally determined values of the parameters, see Chapter 13.

Dimensionless form

It is convenient to rewrite the Friedmann equation in terms of these dimensionless quantities and the normalized scale factor $y := a(t)/a_0$. The general result for a universe with non-interacting matter and radiation is

$$\dot{y}^2 = H_0^2 \left[\Omega_{m0} y^{-1} + \Omega_{r0} y^{-2} + \Omega_{\Lambda 0} y^2 + \Omega_{K0} \right], \tag{9.20}$$

which is essentially quartic in y. In the pure matter case ($\Omega_r = 0 = \Omega_\Lambda$) this reduces to $\dot{y}^2 = H_0^2 [2q_0/y - (2q_0 - 1)]$, while in the pure radiation case ($\Omega_m = 0 = \Omega_\Lambda$) it reduces to $\dot{y}^2 = H_0^2 [q_0/y^2 - (q_0 - 1)]$. One application is to the age t_0 of the universe, given by

$$t_0 = \int_0^1 \frac{dy}{\dot{y}}. \tag{9.21}$$

Unique solutions to these equations, and so for $a(t)$, are obtained from the initial data set $\{a_1, H_1^2, \Omega_{m1}, \Omega_{r1}, \Omega_{\Lambda 1}\}$ at an arbitrary time t_1. Note, however, that one-parameter families of these parameters will represent the *same* cosmological model as seen at different times (these parameters will vary with the chosen time t_1). It is not obvious a priori which sets of parameter values represent the same model.[6] This is made clearest by the phase diagram representations of their evolution (Section 9.4 below).

Dynamics: $\Lambda = 0$

Consider first the case when $\Lambda = 0$ and the energy conditions are satisfied. Following the universe back into the past, there is a singular origin before the HBB era. If $K < 0$ or $K = 0$ the universe expands forever in the future because by (9.13) $\dot{a}$ is never zero; the scale function $a(t)$ is then unbounded in the future (Figure 9.1). If $K > 0$ it will reach a maximum value of a where $\dot{a} = 0$ and then recollapse to a future singularity, where the density and temperature again increase without limit and spacetime again (at least on a classical view) comes to an end. The physics of the collapse phase will be somewhat different from the expansion phase (Rees, 1999) but essentially all complex objects will eventually be broken down again to their constituent elementary particles in the Hot Big Crunch in the future, as the photons successively become more energetic than each binding energy.

It is noteworthy that, when $\Lambda = 0$, the question of whether the universe recollapses in the future or not is the same as whether it has positively curved space sections or not (see (9.13)), and whether it is a high-density universe ($\Omega \geq 1$) or not. Table 9.1 summarizes the situation: here the first three columns are valid for any FLRW model; the last column is the situation in the case that the universe is dominated by pressure-free matter at recent times.

It is *not* true that a 'closed' universe (with $\Lambda = 0$) will necessarily collapse in the future; for example, we can have a $K = 0$ universe, with a torus topology as discussed above, that is closed but expands forever. However, it is true that, on the one hand, an 'open' (infinite) universe must have $K = 0$ or $K < 0$, and so cannot be a high-density universe: it must have

---

[6] This is a special case of the equivalence problem for cosmological models; see Section 17.2.

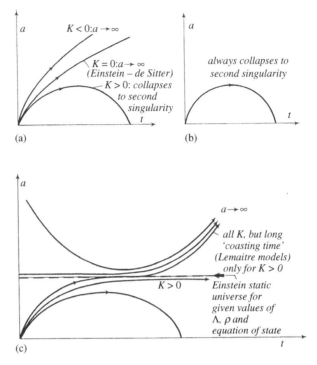

Fig. 9.1   Scale factor $a(t)$ for: (a) $\Lambda = 0$, (b) $\Lambda < 0$, (c) $\Lambda > 0$.

Table 9.1 The three kinds of behaviour when $\Lambda = 0$			
$K > 0$ (spherical geometry)	$\Omega > 1$	recollapses	$q_0 > 1/2$
$K = 0$ (flat geometry)	$\Omega = 1$	just escapes	$q_0 = 1/2$
$K < 0$ (hyperbolic geometry)	$\Omega < 1$	expands forever	$q_0 < 1/2$

$\Omega \leq 1$ and will expand forever, while on the other hand, every very high-density universe ($\Omega > 1$) is necessarily closed (has $K > 0$), and will collapse in the future.

Dynamics: $\Lambda < 0$

When $\Lambda < 0$, the universe necessarily collapses again in the future, irrespective of the value of K.

Dynamics: $\Lambda > 0$

The greatest variety of behaviour can occur when $\Lambda > 0$. If $K < 0$ or $K = 0$, the universe expands forever (for it would have done so if Λ had been zero; now Λ assists the expansion). If $K > 0$, the universe can (after starting from a big bang) turn around and recollapse, or expand forever, depending on the value of Λ and the density of matter. Then there are the unstable static solutions, and solutions asymptotic to them in the late future, either starting from a big bang (and separating those that expand forever from those that recollapse), or collapsing to a finite value from infinity. Finally there are models that collapse from infinity to a finite radius, and then re-expand to infinity; and that start asymptotically in the past at the Einstein static universe, and then expand forever.[7] Note particularly that there are some universes (the Eddington–Lemaître models) which expand through a HBB epoch and eventually escape to infinity, but before doing so are almost static, being very close to the Einstein static universe for a very long time.

9.3.1 Exact solutions

We have already obtained the Einstein static universe (Section 6.1.1). We now look at the expanding solutions, first for vanishing Λ and then for non-vanishing Λ. To obtain exact solutions, we have to specify the matter precisely. In this section, we consider non-interacting pressure-free dust ('baryons') and radiation, possibly with a cosmological constant. In later sections we consider other possibilities (scalar fields, a kinetic theory description, and irreversible processes). For most choices only a qualitative or numerical description is available.

Matter plus radiation ($\Lambda = 0$)

We consider first a non-interacting mixture of pressure-free matter and radiation, with $\Lambda = 0$. It is convenient to rewrite the Friedmann equation (9.20) for this case in the form

$$\dot{y}^2 = (a_0 y)^{-2}[\alpha_r^2 + 2\alpha_m y - K y^2], \tag{9.22}$$

where $\alpha_m := a_0^2 H_0^2 \Omega_{m0}/2$, $\alpha_r := \left(a_0^2 H_0^2 \Omega_{r0}\right)^{1/2}$. If we choose the form (9.9) for the metric, i.e. with $K = 0, \pm 1$ and where a has dimension length, then the general solution can be written in parametric form in terms of the dimensionless conformal time $\tau := \int \mathrm{d}t/a(t)$. We set $t = \tau = 0$ when $a = 0$, and obtain:

$$K > 0 : a = a_0\left[\alpha_m(1 - \cos\tau) + \alpha_r \sin\tau\right],$$
$$t = a_0\left[\alpha_m(\tau - \sin\tau) + \alpha_r(1 - \cos\tau)\right], \tag{9.23}$$
$$K = 0 : a = a_0\left[\tfrac{1}{2}\alpha_m \tau^2 + \alpha_r \tau\right], \quad t = \tfrac{1}{6}a_0\left[\alpha_m \tau^3 + 3\alpha_r \tau^2\right], \tag{9.24}$$
$$K < 0 : a = a_0\left[\alpha_m(\cosh\tau - 1) + \alpha_r \sinh\tau\right],$$
$$t = a_0\left[\alpha_m(\sinh\tau - \tau) + \alpha_r(\cosh\tau - 1)\right]. \tag{9.25}$$

It is interesting how in this parametrization the dust and radiation terms decouple; this solution includes generic pure dust solutions, $\alpha_r = 0$, and generic pure radiation solutions,

---

[7] As well as the time-symmetric versions of each solution, which we take for granted.

$\alpha_m = 0$. The general case represents a smooth transition from a radiation-dominated early era to a matter-dominated later era, and (if $K \neq 0$) on to a curvature-dominated era.

Einstein–de Sitter universe

Particular cases allowing simpler representation are of interest. The simplest expanding pure matter solution ($p = 0 \Rightarrow \rho_r = \Omega_r = 0$, $\Lambda = 0$) is the Einstein–de Sitter universe, spatially flat ($K = 0$) and self-similar:

$$a(t) = A_m(t - t_m)^{2/3}, \quad \dot{A}_m = 0 = \dot{t}_m. \tag{9.26}$$

It is a high-density universe, with age close to the Hubble time:

$$\Omega = 1, \, q_0 = 1/2, \tag{9.27}$$

$$t_0 = \tfrac{2}{3} H_0^{-1}. \tag{9.28}$$

Milne universe

The simplest expanding empty universe model is the Milne universe, characterized by $\rho_m = \rho_r = 0 \Rightarrow \Omega = 0$, $\Lambda = 0$, $q_0 = 0$, $K < 0$. The Friedmann equation is dominated by the 3-space curvature, and

$$a(t) = t - t_*, \tag{9.29}$$

$$t_0 = H_0^{-1}. \tag{9.30}$$

(The normalization factor in a is necessarily 1 because $K < 0$.) This is in fact the flat spacetime of Special Relativity with a cloud of test particles expanding uniformly in it; that is, gravity does not curve the spacetime at all (this is possible because, dynamically speaking, the universe is taken to be empty; and we know the FLRW family of models are conformally flat). This model is the asymptotic future state of low-density universes with vanishing Λ.

Models with $\Lambda \neq 0$

In general these are more complex than those discussed above. However, in the case of pure radiation plus a cosmological constant, the equation (9.20) only involves even powers of y; so, on defining $\xi = y^2$, simple analytic solutions exist again for all values of K.

de Sitter solution

The simplest solution with a cosmological constant is the empty de Sitter solution, characterized by $p = \rho = 0$, $\Lambda > 0$. The Raychaudhuri equation becomes $\ddot{a} = \Lambda a/3$, with solution

$$a = A \exp\left[\sqrt{\Lambda/3}(t - t_*)\right] + B \exp\left[-\sqrt{\Lambda/3}(t - t_*)\right], \tag{9.31}$$

where the constants A, B can be rescaled by choice of the constant t_*. The Friedmann equation then imposes $4AB\Lambda = 3K$. Thus,

$$K > 0 \Rightarrow a = \sqrt{3K/\Lambda}\cosh\sqrt{\Lambda/3}\,t, \tag{9.32}$$

$$K = 0 \Rightarrow a = A\exp\sqrt{\Lambda/3}\,t, \tag{9.33}$$

$$K < 0 \Rightarrow a = \sqrt{-3K/\Lambda}\sinh\sqrt{\Lambda/3}\,t. \tag{9.34}$$

These are all forms of a four-dimensional spacetime of positive constant curvature: $C_{abcd} = 0$, $R_{ab} - \frac{1}{4}Rg_{ab} = 0$, so the only non-zero curvature tensor component is the Ricci scalar $R > 0$. The model has the same maximal amount of symmetry as Minkowski spacetime, invariant under a 10-dimensional group of symmetries. The different FLRW forms with space sections of positive, zero or negative curvature take advantage of various subgroups of this full symmetry group. One can represent the de Sitter universe as a four-dimensional hyperboloid imbedded in a five-dimensional flat spacetime; only the first set of coordinates (9.32) covers the whole hyperboloid. Only a part of the hyperboloid is covered by the other coordinates, so they represent geodesically incomplete parts of this spacetime (Schrödinger, 1956, Hawking and Ellis, 1973).

The non-uniqueness of the 4-velocity that allows these different FLRW forms for the same spacetime arises because all timelike vectors are eigenvectors of the Ricci tensor (if we write the stress tensor in the perfect fluid form, it satisfies the exceptional equation of state $\rho + p = 0$) and it has many different spatially homogeneous spatial sections. Thus we have different universe models for the same spacetime (given by different choices of the fundamental velocity field u^a in that spacetime). As well as these FLRW forms, we can also choose coordinates representing part of the same spacetime in a static, inhomogeneous (and so non-FLRW) form.[8]

The exponentially expanding form of the model (9.33) is a self-similar solution, and is in a steady state: although the universe is expanding, the expansion does not vary with time: $H = \sqrt{\Lambda/3} = $ const, and no other physical invariant changes. This is only possible because it is an empty universe; the density also stays constant – at zero. Of course the cosmological constant term may be regarded as an energy density, e.g. of the vacuum.

Because the flat de Sitter universe is in a steady state, there is no origin and no end to the expansion of the universe. It is this form that became the *Steady State* universe in 1948, proposed by Bondi and Gold purely as a kinematic model satisfying the *Perfect Cosmological Principle* that it is unchanging in time as well as space (Bondi, 1960). Hoyle (1948) suggested a modification of Einstein's field equations, by addition of a creation field that would allow a non-zero density of matter to expand while the energy density remained constant because of the continuous creation everywhere of new matter. It predicts $q_0 = -1$. It was contradicted by the evidence of evolution of the density of radio sources in the past, and was finally dropped as a serious contender when the CMB was discovered.

Because it only covers half the full de Sitter hyperboloid, the flat-sliced model is geodesically incomplete in the past (Penrose, 1999). It is a singular universe, as it has a boundary at a finite distance from any spacetime point (Ellis and King, 1974). Inflationary models of

---

[8] This is analogous to the Rindler form of Minkowski spacetime.

the early universe (Section 9.4) are close to the exponentially expanding de Sitter model. Current data are consistent with a model dominated by positive Λ, and in this case our observed universe domain would be asymptotically de Sitter.

Anti-de Sitter solution

The anti-de Sitter universe is the companion spacetime of constant spacetime curvature, but with opposite curvature: $R < 0$. It can be represented in FLRW form by the metric

$$ds^2 = -dt^2 + \cos^2 t \left[d\chi^2 + \sinh^2 \chi (d\theta^2 + \sin\theta^2 d\phi^2) \right]. \tag{9.35}$$

This coordinate system, however, only covers part of the spacetime; unlike the de Sitter space, the whole of this spacetime is covered by a static coordinate system (Hawking and Ellis, 1973). This metric is not a good representation of the real universe because it requires $\Lambda < 0$, in contradiction to observations, but it seems to play a fundamental role in string theory, in particular because of the AdS/CFT correspondence (Section 20.3).

9.3.2 Early and late solutions

In this section, we consider only ordinary matter, i.e. we assume that $0 \leq w \leq 1$, but we include the possibility of non-zero Λ.

Solutions at large a: $\Lambda = 0$

If the universe expands to arbitrarily large values of a at late times, and $\Lambda = 0$, then $K = 0$ or $K < 0$. The asymptotic solution depends on the value of K.

  If $K < 0$, the curvature term will dominate the Friedmann equation (9.13) at late enough times. Thus the asymptotic form of equation (9.13) is just the asymptotic form of (9.20): $\dot{y} = 1/a_0$, leading to the Milne solution (9.29) as the asymptotic solution for the regime when the effect of the matter can be neglected relative to the curvature term. As the universe's expansion slows down at late times, most of the expansion, in terms of elapsed proper time, will take place when this asymptotic form is valid; thus for most of its history, the formula (9.30) would give a good estimate of the age of the universe.

  If $K = 0$, the matter term in (9.20) will dominate at late times, leading to the Einstein–de Sitter solution (9.26) as the asymptotic solution. Again the universe's expansion slows down at late times; thus for most of its history, the formula (9.28) would give a good estimate of the age of the universe.

Solutions at large a: $\Lambda > 0$

If $\Lambda \neq 0$, it must be positive for a long-term expansion of the universe, and the late-time asymptotic solution will be the exponential form (9.33) of the de Sitter universe (whatever the value of K), valid when Λ dominates the matter and curvature terms. In this case the age of the universe and the Hubble parameter are unrelated at most times.

Solutions in the HBB era

At early enough times near the initial singularity, the energy density term will dominate the Friedmann equation (9.13) (independent of the values of Λ or K). Thus the effective equation at early times will be $3\dot{a}^2 = 8\pi G\rho a^2$. For the HBB case of a radiation-dominated early universe ($w = 1/3$), we obtain the Tolman model,

$$a = A_{\rm r}(t - t_{\rm r})^{1/2}, \quad \rho = (3/32\pi G)(t - t_{\rm r})^{-2}. \tag{9.36}$$

The unique relation $T = (3/32\pi Ga_R)^{1/4}(t - t_{\rm r})^{-1/2}$ between temperature and time in the early universe leads to the standard nucleosynthesis predictions, in agreement with observations (Section 9.6.6) – with essentially no free parameters.

9.3.3 Combined solutions

For many purposes the dynamics of the universe are adequately described by (9.26) after decoupling, when it is matter dominated (until the cosmological constant takes over at fairly recent times), and by (9.36) at early times when it is radiation dominated (after an inflationary epoch and before equality). If we wish to describe a single model by (9.36) at early times and (9.26) at late times, then we need to use the freedom in $A_{\rm m}, t_{\rm m}, A_{\rm r}$ and $t_{\rm r}$, to ensure that at the changeover time $t_{\rm eq}$ both $a(t)$ and $\dot{a}(t)$ must be continuous (see Exercise 9.3.3). The same is true for a changeover from an inflationary era to radiation domination in the early universe, and from matter domination to an accelerated era in the late universe (Ellis, 1988).

Viscous eras

Irreversible events certainly occur in the early universe, for example during baryosynthesis, nucleosynthesis, the decoupling of matter and radiation and star formation. To illustrate the effect of irreversibility on dynamics, one can obtain exact solutions of the FLRW equations for simple cases of fluids with bulk viscosity (Treciokas and Ellis, 1971) (because of the RW symmetry, heat conduction and shear viscosity will have no dynamical effect). A more realistic treatment is based on kinetic theory (see Section 9.6.4).

Ages

There are simple inequalities between the age of the universe $t - 0$ and the Hubble constant H_0 when $\Lambda = 0$. It is reasonable to assume that at late times but before Λ dominates, the age depends essentially only on the time spent in the late 'dust' (pressure-free) epoch. Then in a low-density universe ($0 \leq \Omega < 1$), we have $2/3H_0 < t_0 \leq 1/H_0$; while in a high-density universe, $1 \leq \Omega$, $1/2H_0 \leq t_0 \leq 2/3H_0$. When $\Lambda > 0$, one can get ages much higher than $1/H_0$; and this is always a recourse when it seems that we are finding objects in the universe older than $1/H_0$. We can in each case find explicit formulae for the age from (9.21).

Exercise 9.3.1 Given suitable equations of state, determine the unique Einstein static value of a in terms of constants describing the amount of matter and radiation present.

Exercise 9.3.2 Obtain the condition separating the universe models that expand forever from those that recollapse, when $\Lambda > 0$, $K > 0$.

Exercise 9.3.3 Determine the conditions on A_m, A_r and t_m, t_r leading to continuity of $a(t)$ and $\dot{a}(t)$ at the transition from radiation domination to matter domination.

Exercise 9.3.4 Obtain exact formulae for the age of the universe t_0 in terms of the Hubble constant H_0 and deceleration parameter q_0 when $\Lambda = 0 = p$. [Separate formulae apply when $K = 0$, > 0, and < 0.] Prove the inequalities cited above.

9.4 Phase planes

It is useful to represent the dynamics of the universe in terms of various 'phase planes', showing the set of FLRW models in terms of pairs of suitable parameters and how these parameters evolve with time. Phase planes for the case $\Lambda = 0$ are given by Gott *et al.* (1976), characterizing the universe by the Hubble constant H_0 and density parameter Ω_0. Another version is given in Rindler (1977). Because we now have evidence that $\Lambda > 0$, while these help to understand the behaviour of families of FLRW models, they will only represent the real universe reasonably well for a limited part of its evolution, and will not relate these models well to present-day observational parameters.

A version allowing non-zero Λ was given by Stabell and Refsdal (1966), with plots of q_0 versus $\Omega_0/2$. Flow lines of the dynamical system ($p = 0$ solutions) and 'equal age' lines are plotted there. Lines of constant K and constant Λ are invariant curves in this diagram, which has the Einstein–de Sitter model ($q_0 = 1/2 = \Omega_0/2$) as a 'source', the Milne universe ($\Omega_0 = 0$, $q_0 = 0$) as an unstable limit point and the de Sitter universe ($\Omega_0 = 0$, $q_0 = -1$) as a 'sink'.

9.4.1 Matter, radiation and Λ

The extension of the Rindler diagram to the case of a non-interacting mixture of matter and radiation is given by Ehlers and Rindler (1989). This is a three-dimensional phase space because the matter and radiation vary differently. It shows how a skeleton of higher symmetry (self-similar) solutions act as sources, attractors and saddle points guiding the evolution of the more general solutions (see also Section 18.4).

This three-dimensional phase space, shown in Figure 9.2, contains various invariant planes, corresponding to first integrals of the dynamics. These are shown in the lower panels: (ii) is the plane of zero radiation (Stabell and Refsdal, 1966), (iii) is the plane of zero matter, (iv) is the plane of $\Lambda = 0$, and (v) is the plane of $K = 0$. The background dynamics of the real universe will differ only at times of significant matter–radiation interaction.

The phase planes show the Tolman (P_r) and Einstein–de Sitter (P_d) models as sources, and the Milne universe (M) as an attractor in the plane $\Lambda = 0$ but a saddle point in the three-dimensional phase space. The de Sitter universe (S) is an attractor in the three-dimensional

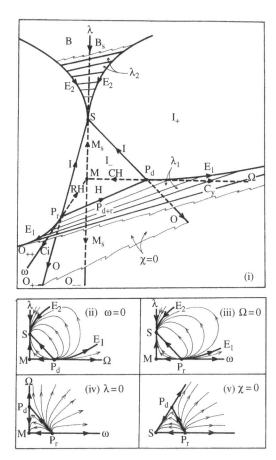

Fig. 9.2   FLRW evolution relating density parameters of matter, radiation and Λ. Here $\lambda, \Omega, \chi, \omega$ correspond to our $\Omega_\Lambda, \Omega_m, K, \Omega_r$. (From Ehlers and Rindler (1989). © RAS.)

phase space; that is what underlies the way the inflationary universe works (Section 9.4). Each of those exact solutions is a fixed point in the phase plane, hence each is a self-similar solution. The separatrix E_1 divides models that expand forever from those that recollapse, and E_2 divides universes with a singular start from those without an initial singularity (the collapse from infinity and bounce, due to the cosmological constant).

But there is a problem with this phase diagram: it cannot represent the whole trajectory if the universe recollapses because the variables used are singular there ($H \to 0 \Rightarrow \Omega \to \infty$), and it does not show what happens at infinity. One needs an extended and compactified phase plane to get complete histories (Section 18.4). An example is given in the next section.

9.4.2 Density parameter Ω versus scale factor a

We can obtain (Ω, a) phase planes for universes where the total pressure p and total energy density ρ (here, including the cosmological constant) are related by $p = w\rho$ provided $w = w(\Omega, a)$ (Madsen and Ellis, 1988).

The Raychaudhuri equation gives

$$\frac{d\Omega}{da} = H\Omega(\Omega - 1)(1 + 3w), \qquad (9.37)$$

valid for any w – but the equation gives a (Ω, a) phase plane flow if $w = w(\Omega, a)$ (and in particular if $w = w(a)$ or $w = $const). It immediately follows that both $\Omega = 0$ and $\Omega = 1$ are solutions of (9.37), no matter what form $w(a)$ takes; on the other hand if $w(a) = -1/3$ for all a (the critical equation of state such that $q = 0$), then $\Omega = \Omega_0$ is a solution for all values of Ω_0. Furthermore, combining these equations gives $d\Omega/da = -2q(1 - \Omega)/a$, showing that the signs of $d\Omega/da$ and q are the same when $\Omega > 1$, and $d\Omega/da = 0$ when $q = 0$.

During an epoch of constant w, when the universe is dominated by a simple one-component fluid, we find that $d^2\Omega/da^2 = (1 + 3w)(1 - \Omega)[(1 + 3w)(1 - 2\Omega) + 1]\Omega/a^2$. Apart from the special cases $w = -1/3$, $\Omega = 0$, and $\Omega = 1$, this vanishes when $\Omega = [2(3w + 1)]^{-1} + 1/2$. We can find the explicit solutions, either directly or by integrating the conservation equations to get $\rho = \rho_0 y^{-3(1+w)}$. We obtain

$$\Omega = \Omega_0 \Big[\Omega_0 - (\Omega_0 - 1)y^{1+3w}\Big]^{-1}. \qquad (9.38)$$

The behaviour is quite different if $w > -1/3$ or $w < -1/3$, see Figure 9.3 and 9.4. In the critical case $w = -1/3$, as mentioned above, $\Omega = \Omega_0$ is a solution for all Ω_0, so the phase curves in the (Ω, a) plane are simply horizontal lines.

In the case $w > -1/3$, it is awkward to see what happens at late times, so it is convenient to transform the variables to bring the infinities of both a and Ω to a finite value (e.g. change to (s, ω) where $s = \arctan(\ln a)$, $\omega = \arctan(\ln \Omega)$). To obtain the complete picture, we then have to adjoin to the axis where Ω runs from zero to infinity a further axis segment where it decreases back from infinity to zero (see Figure 9.3, bottom panel). The bottom half then represents the expansion phase of the universe and the top half the contraction phase (if there is one). Non-static solutions can be followed through turnaround points where $\dot{a} = 0$ (and so Ω is infinite) because there $H \to 0$, and $\Omega \to \infty$ like $1/H^2$; also, $\dot{\Omega} \to \infty$.

The complete plane is time symmetric, representing all solutions (expanding and contracting). We see that all expanding solutions start asymptotically (when $a \to 0$) at $\Omega = 1$. If $K < 0$, then the universe expands forever and $\Omega \to 0$. If $K = 0$, then the universe expands forever and $\Omega = 1$ at all times. If $K > 0$, the universe expands to a maximum radius where it turns around (at the point on the flow line where $\Omega = \infty$) and collapses to a second singularity where $\Omega \to 1$ again. This form of solution will hold in particular for a pure dust or pure radiation solution. The dust phase planes are illustrated in Figure 9.2.

In the case $w < -1/3$, the opposite behaviour happens (see Figure 9.4). Now the energy conditions are violated so the family of solutions with $K > 0$ collapse from infinity ($\Omega = 1$) to a finite minimum radius and then re-expand to infinity. The $K = 0$ solutions expand forever with $\Omega = 1$ always. The $K < 0$ solutions expand from $\Omega = 0$ to $\Omega = 1$; in fact, in all cases the future form has $\Omega = 1$ (the asymptotic de Sitter solution). It is this driving of

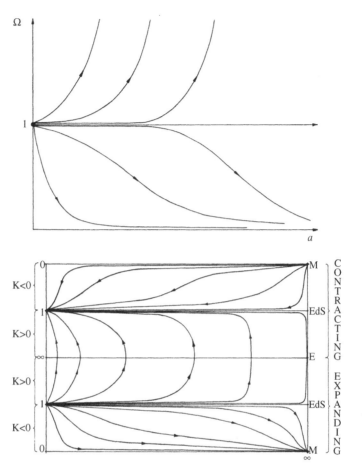

Fig. 9.3 Evolution of Ω with time for $w > -1/3$. Bottom panel: compactified version, showing the entire universe evolution. (From Madsen and Ellis (1988). © RAS.)

Ω to 1 that underlies the usual inflationary universe picture. There is a strong similarity between these behaviours, shown by inverting the plots right to left. This similarity is rooted in an exact symmetry: the flow equations are invariant under the transformation $y \rightarrow 1/y$, $1 + 3w \rightarrow -(1 + 3w)$.

Inflationary universe models correspond to a combination of these diagrams (see Figure 9.5). Suppose the universe starts at time $t = 0$ (which is not inevitable: it could have existed forever, see the discussion of the emergent universe below) and is initially in a radiation-dominated phase, then inflation starts at time t_i and ends at t_f. From $t = 0$ until $t = t_i$ the universe is a $w > -1/3$ model (as in Figure 9.3), from $t = t_i$ to $t = t_f$ it is a $w < -1/3$ model (as in Figure 9.4), and from $t = t_f$ either forever (in a universe without a cosmological constant) or until some late time t_Λ when a cosmological constant dominates again, the universe is a $w > -1/3$ model. Some interesting new features emerge, in particular there is now an unstable Einstein static universe that is a saddle point for the

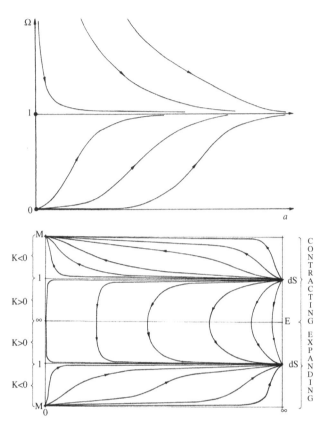

Evolution of Ω with time for $w < -1/3$. Bottom panel: compactified version, showing the entire universe evolution. (From Madsen and Ellis (1988). © RAS.)

solutions. There are also non-singular inflationary models that never encounter an initial singularity.

The present time t_0 will be some time greater than t_f; it is clear from the phase plane that in principle one can find inflationary universes with any value whatever for the density parameter Ω_0 at the present time t_0 (Ellis, 1988). The same result will hold in universes with a non-zero cosmological constant causing accelerated expansion at late times, when a further segment with $w < -1/3$ needs to be added to the model for $t > t_\Lambda$. This might be a realistic phase plane for the dynamics of the real universe. All of this is shown in Figure 9.5.

Exercise 9.4.1 Determine the (Ω, a) phase planes for the case of ordinary matter plus a positive Λ. Show that the Einstein static universe is a saddle point at the centre of this phase plane. It consists in effect of back to back copies of Figures 9.3 and 9.4, joined at the value of $a(t)$ where matter domination gives way to a cosmological constant-dominated epoch (Madsen and Ellis, 1988). This is the phase plane like Figure 9.5 for the case where there is additionally a late-time acceleration period driven by Λ.

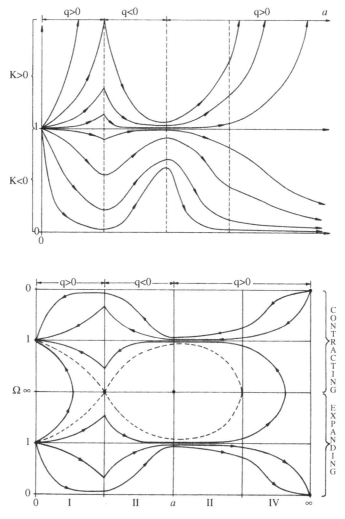

Fig. 9.5 Evolution of Ω for a universe with a radiation era, followed by inflation, followed by matter and radiation eras. Bottom panel: compactified version, showing the entire universe evolution. (From Madsen and Ellis (1988). © RAS.)

9.5 Kinetic solutions

Exact kinetic theory solutions may also be found. Here we give a particular example of interest. Recall (Section 5.4) that a distribution function $f(x, p)$ for collision-free particles at x with 4-momenta p evolves according to the Liouville equation $df/d\tau = 0$, and the matter tensor is determined by $T^{ab} = \int p^a p^b dP$ where dP is the volume element on the tangent space at x. For particles of a single rest mass $m = (-p^a p_a)^{1/2}$, we have a solution of the Liouville equation in a $K = 0$ FLRW model if

$$f = F(\xi), \quad \xi := (E^2 - m^2)^{1/2} a(t), \tag{9.39}$$

for arbitrary $F(\xi)$, where $E = -p^a u_a$ is the particle energy relative to u^a. Then the energy density and pressure are

$$\rho = \frac{4\pi}{a^4(t)} \int_0^\infty \xi^2 [\xi^2 + m^2 a^2(t)]^{1/2} F(\xi) d\xi, \tag{9.40}$$

$$p = \frac{4\pi}{3a^3(t)} \int_0^\infty \xi^4 [\xi^2 + m^2 a^2(t)]^{-1/2} F(\xi) d\xi. \tag{9.41}$$

From the Friedmann equation, we have a solution of all 10 EFE for an arbitrary choice of $F(\xi)$ if

$$t = \sqrt{\frac{3}{32\pi G}} \int_0^{a^2} \left[4\pi \int_0^\infty \xi^2 (\xi^2 + m^2 u)^{1/2} F(\xi) d\xi \right]^{-1/2} du. \tag{9.42}$$

(Ehlers, Geren and Sachs (EGS), 1968, Ellis, Matravers and Treciokas, 1983a).

Exercise 9.5.1 Determine the effective equation of state parameter $w = w(t)$ for the kinetic theory solution given above, (a) for $m \neq 0$; (b) for $m = 0$. What is the form of the relation $w = w(\rho)$?

Exercise 9.5.2 Determine the kinetic theory FLRW solutions for $K \neq 0$.

9.6 Thermal history and contents of the universe

In this section, we provide a brief and mainly qualitative overview of the current understanding of the contents and thermal history of the universe. For a more detailed discussion, see Mukhanov (2005), Durrer (2008) or Peter and Uzan (2009). Milestones in the history of the universe are illustrated in Table 9.2. For times $t > 10^{-10}$ s, the physics is known, based on GR and the Standard Model of particle physics, with its minimal extensions, e.g. to incorporate massive neutrinos. (However, note that the electroweak and quark–hadron transitions are still not well understood.) For $t < 10^{-10}$ s, the physics is uncertain, more so as we go further back. The Large Hadron Collider, which is in the initial stages of operation at the time of writing, is probing energies $\gtrsim$ TeV, at the interface of known and unknown physics.

9.6.1 The universe at $t < 10^{-10}$ s: uncertain physics

Around and before the Planck time, $t \lesssim t_P \approx 10^{-43}$ s, we expect that GR breaks down and gravity should become a quantum interaction. The nature of this quantum gravity era remains speculative in the continued absence of a satisfactory quantum gravity theory. We shall discuss some of the issues arising from this, and discuss the candidate quantum gravity theories such as string theory, in Chapter 20.

At energies below the Planck scale but above the electroweak unification scale, we expect that the electromagnetic, weak and strong interactions will be unified. There are candidate Grand Unified Theories, mainly based on supersymmetry – which relates bosons to fermions, so that each fermion has a boson superpartner, and vice versa. Supersymmetric

Table 9.2 History of the universe. (Numerical values are approximate. Adapted from Baumann (2009).)			
	Time	Energy	Redshift
Quantum Gravity era?	$< 10^{-43}$ s	10^{19} GeV	
Grand Unification?	$\sim 10^{-36}$ s	$\sim 10^{16}$ GeV	
Inflation & reheating?	$\gtrsim 10^{-34}$ s	$\lesssim 10^{15}$ GeV	
CDM decoupling?	$< 10^{-10}$ s	> 1 TeV	
Baryogenesis?	$< 10^{-10}$ s	> 1 TeV	
Electroweak unification	$\sim 10^{-10}$ s	$0.1-1$ TeV	
Quark–hadron transition	$\sim 10^{-4}$ s	$0.1-0.4$ GeV	
Neutrino decoupling	1 s	1 MeV	
Electron–positron annihilation	4 s	0.5 MeV	
Nucleosynthesis	200 s	0.1 MeV	10^8
Matter–radiation equality	10^4 yrs	1 eV	10^4
Photon decoupling	4×10^4 yrs	0.1 eV	1,100
Dark Ages	$10^5 - 10^8$ yrs		> 25
Reionization	10^8 yrs		$25-6$
Galaxy formation	$\sim 6 \times 10^8$ yrs		~ 10
Dark energy era	$\sim 10^9$ yrs		~ 2
Solar system	8×10^9 yrs		0.5
Today	14×10^9 yrs	1 meV	0

string theory provides a unification of the three interactions, but with a wide range of possible mechanisms and energy scales. The typical energy scale in GUTs is $M_{GUT} \sim 10^{16}$ GeV.

Currently the most successful phenomenology we have for understanding the very early universe is inflation, which is discussed in the following section. This is typically expected to take place at an energy scale $\lesssim 10^{15}$ GeV. Inflation provides a framework for understanding how the apparently causally disconnected regions of the observable universe happen to have the same CMB temperature, and it also predicts the generation of fluctuations that seed the growth of large-scale structure. However, it does not address the problems of grand unification or quantum gravity.

At the end of inflation, the observable universe is cold and essentially empty of matter: the universe is reheated and populated with particles via the decay of the inflaton field. Between reheating and the electroweak transition, a number of crucial processes are expected to occur, all of them beyond the reach of the Standard Model of particle physics, and all remaining uncertain at the time of writing. They include the problem of identifying the dark matter particle and the problem of baryogenesis.

One of the major problems in cosmology is to account for the matter/anti-matter asymmetry, i.e. the fact that we only observe matter in stars and galaxies (apart from high-energy collisions that can produce anti-particles, which rapidly annihilate). In the very early Universe, we expect that some mechanism generated a baryon asymmetry which led to the baryonic structures that we observe. This is known as the problem of baryogenesis. A baryogenesis mechanism can be based in the Standard Model, but it produces far too little asymmetry. Baryogenesis requires a process that is strongly non-equilibrium, that violates baryon number conservation, and that violates CP (charge conjugation and parity). Various models have been proposed, typically based on supersymmetry and GUTs.

Most extensions of the Standard Model are based on supersymmetry. The Minimal Supersymmetric Standard Model (MSSM) adds only those particles required by supersymmetry, i.e. the superpartners, and an enlarged Higgs sector. This has more than 100 undetermined parameters, since colliders have not yet probed the higher energy scales. This number of parameters can be strongly reduced by using ideas from GUTs and supergravity theories, leading to the Constrained MSSM (CMSSM). See Olive (2010) for a discussion.

9.6.2 Candidate particles for cold dark matter

The cosmological and astrophysical evidence for dark non-baryonic matter is strong (see Section 12.3.1 and 12.3.2). Although massive neutrinos are non-baryonic and dark, they are not cold, and they erase perturbations on large scales in contradiction to observations. The Standard Model of particle physics does not provide a suitable CDM candidate, i.e. a non-baryonic cold, stable and neutral particle. There are a number of candidate CDM particles based on supersymmetric and other extensions of the Standard Model (see Feng (2010) for a review).

The leading candidates are probably WIMPs – weakly interacting massive particles – that are stable supersymmetric partners. Supersymmetric extensions of the Standard Model include a number of light supersymmetric particles, such as the neutralino, sneutrino and gravitino, which interact with the $W^{\pm}, Z^0$ bosons but not the photon or gluons. Since these particles have not been detected at the time of writing, their masses and cross-sections are unknown, although limits may be imposed from collider and cosmological observations. The WIMPs can be either thermal relics – i.e. in equilibrium with the cosmic plasma before decoupling, or non-thermal relics, which are produced by a non-thermal mechanism.

Thermal relics are non-relativistic at the time of decoupling, i.e. when they fall out of equilibrium with the primordial plasma because the interaction rates that keep them in equilibrium fall below the Hubble expansion rate. They have mass $m \gg 1\,\text{keV}$, and have a relic density depending on their mass and cross-section. An example is shown in Figure 9.6. A candidate WIMP is the neutralino, whose mass is constrained by cosmology and collider experiments to be in the range $100\,\text{GeV} \lesssim m_\chi \lesssim 400\,\text{GeV}$.

Axions are an example of a non-thermal relic. They form a weakly interacting scalar field condensate, which enforces zero momentum, so that axions can behave as CDM even though their mass is very small, $m_a < 0.01\,\text{eV}$. Axions arise from a simple extension of the Standard Model, and in the presence of magnetic fields they can oscillate into photons.

Note that there are also candidate particles for 'warm dark matter', which are relativistic at decoupling but non-relativistic before matter–radiation equality, with mass of order $0.1 - 1\,\text{keV}$. Examples are sterile neutrinos and gravitinos.

9.6.3 The Standard Model: electroweak and quark–hadron transitions

The Standard Model applies for energies below the electroweak transition (though the electroweak and quark–hadron transitions are only partly understood, given the complexities

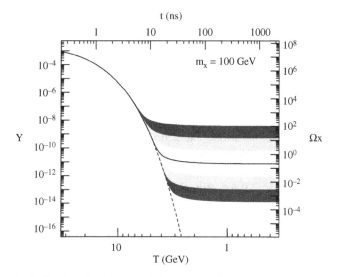

Fig. 9.6 Comoving number density Y and resulting thermal relic density $\Omega_{\chi 0}$ of a WIMP with $m = 100\,$GeV, that freezes out at $T \sim 5\,$GeV. The solid curve is for an annihilation cross-section that yields the correct relic density; shaded regions have cross-sections that differ by 10, 10^2 and 10^3 from this value. (Reprinted from Feng (2010), with permission from the Annual Review of Astronomy and Astrophysics. © 2010 by Annual Reviews, http://www.annualreviews.org .)

Table 9.3 Standard model of particle physics		
Family	Spin	Particles
baryons (qqq)	$n + \frac{1}{2}$, $n = 0, 1, 2, \ldots$	$p^+, n, \Delta, \ldots$
mesons $(q\bar{q})$	n, $n = 0, 1, 2, \ldots$	$\pi^{0,\pm}, K^{0,\pm}, \ldots$
leptons	$\frac{1}{2}$	e^-, μ^-, τ^-; massless: ν_e, ν_μ, ν_τ
gauge fields	1	$Z^0, W^\pm$; massless: γ, g^a

of strong coupling and non-equilibrium processes). It incorporates the strong, weak and electromagnetic interactions, with symmetry group

$$\mathrm{SU}(3)_c \times \mathrm{SU}(2)_L \times \mathrm{U}(1)_Y . \qquad (9.43)$$

The strong interaction is mediated by eight massless neutral gauge bosons g^a (gluons), the weak interaction by three massive charged bosons $Z^0, W^\pm$ (vector bosons), and the electromagnetic interaction by the massless neutral photon γ. Quarks and anti-quarks make up the baryons (fermionic) and mesons (bosonic) – collectively known as hadrons, which are the particles that feel the strong interaction (and also the weak interaction). Hundreds of hadrons have been observed so far. The leptons (fermionic) are the charged electron, muon, taon and their associated neutral massless neutrinos. Leptons feel the weak interaction. The overall structure is summarized in Table 9.3. It is now known that at least two of the neutrinos must have mass, so that an extension of the Standard Model is needed to incorporate this feature.

This model accounts for all particles observed in colliders and particle detectors, provided we introduce the Higgs mechanism to break the electroweak symmetry, $SU(2)_L \times U(1)_Y \to U(1)_{em}$. It is estimated that this happens at $\sim 100\,GeV$. At higher energies, the $W^\pm, Z^0$ bosons are massless, and interaction rates are rapid enough (i.e. $\Gamma \gg H$) to keep quarks and leptons in equilibrium. At lower energies $W^\pm, Z^0$ acquire mass, and the cross-section of the weak interaction decreases. This leads to neutrino decoupling at $\sim 1\,MeV$, as discussed below.

For $T \gtrsim 200\,MeV$, the quarks and gluons interact only weakly with each other. Below this temperature, the strong interaction increases in strength sufficiently to confine the quarks and gluons within hadrons.

9.6.4 Kinetics and thermodynamics of the hot Big Bang

The basic idea in understanding many of the key developments in the evolution of the universe is this: interaction rates which keep particles in equilibrium are determined by the temperature and by the number density of particles, since particles must be able to find each other to interact. As the universe expands, the number densities fall and therefore the interaction rates fall. Thus there is a tendency for species of particles to fall out of equilibrium and to decouple from the thermal plasma. This is characterized by the behaviour of the interaction rate Γ_I of species I relative to the Hubble rate H at temperature T, the plasma temperature:

$$\text{equilibrium: } \Gamma_I \gg H,\, T_I = T\,; \quad \text{decoupling: } \Gamma_I \lesssim H,\, T_I \neq T. \tag{9.44}$$

Note that, after electron–positron annihilation at $t \sim 1\,s$, the huge number of photons per baryon, $\sim 10^9$, means that $T = T_\gamma$, the temperature of the photons (with blackbody spectrum).

Particle species in equilibrium with the thermal plasma have Fermi–Dirac ($+$) or Bose–Einstein ($-$) distribution functions (5.103):

$$F_I(E,T) = \frac{1}{(2\pi)^3} \frac{g_I}{\exp\left[(E - \mu_I)/T(t)\right] \pm 1}, \tag{9.45}$$

where g_I is the degeneracy factor (determined by quantum statistics), μ_I is the chemical potential, and each particle energy is given by its 3-momentum and mass as $E = (p^2 + m_I^2)^{1/2}$. Then F_I determines the number density n_I, energy density ρ_I and pressure p_I as integrals over momentum space (see Section 5.4). The important limiting cases are

$$T \gg m, \mu: \quad n = c_1 g T^3,\quad \rho = c_2 g T^4,\quad p = \tfrac{1}{3}\rho, \tag{9.46}$$

$$T \ll m: \quad n = g\left(\frac{m}{2\pi}\right)^{3/2} T^{3/2} e^{(\mu - m)/T}, $$

$$\rho = \left(m + \tfrac{3}{2}T\right)n,\quad p = nT, \tag{9.47}$$

where

$$\text{bosons: } c_{1B} = \frac{\zeta(3)}{\pi^2},\quad c_{2B} = \frac{\pi^2}{30}, \tag{9.48}$$

$$\text{fermions: } c_{1F} = \tfrac{3}{4}c_{1B},\quad c_{2F} = \tfrac{7}{8}c_{2B}. \tag{9.49}$$

It follows that in the radiation era, the total density is

$$\rho_r(T) = \frac{\pi^2}{30} g_* T^4, \quad g_* = \sum_I \alpha_I g_I \left(\frac{T_I}{T}\right)^4, \tag{9.50}$$

where $\alpha_I = 1$ for bosons and $\alpha_I = \frac{7}{8}$ for fermions. g_* is the effective number of ultra-relativistic degrees of freedom. It changes with T, as various species decouple, from $g_* \sim 100$ after the electroweak transition, to $g_* = 2$ after electron–positron annihilation.

The temperature satisfies

$$T \propto \frac{1}{a} = 1 + z, \tag{9.51}$$

and the Hubble rate and cosmic time are

$$H(T) = \left(\frac{4\pi^3 G}{45}\right)^{1/2} g_*^{1/2} T^2 \approx 1.7 g_*^{1/2} \frac{T^2}{M_P}, \tag{9.52}$$

$$t(T) \approx 0.3 g_*^{-1/2} \frac{M_P}{T^2} \approx 2.4 g_*^{-1/2} \left(\frac{\text{MeV}}{T}\right)^2 \text{s}. \tag{9.53}$$

Photons are not conserved, since they can be created or annihilated in inelastic scatterings such as Brehmsstrahlung, $e + p \leftrightarrow e + p + \gamma$. This requires that $\mu_\gamma = 0$. If a particle is kept in equilibrium with its anti-particle by reactions of the form $I + \bar{I} \leftrightarrow \gamma + \gamma$, e.g. electrons and positrons, then it follows that $\mu_I = -\mu_{\bar{I}}$. At high temperatures, $T \gg m_I$, (9.46) implies that

$$n_I - n_{\bar{I}} \approx \frac{g_I}{6\pi^2} T^3 \left[\pi^2 \left(\frac{\mu_I}{T}\right) + \left(\frac{\mu_I}{T}\right)^3\right]. \tag{9.54}$$

Thus there is an asymmetry between particles and anti-particles. At lower temperatures, $T < m_I$, the particles annihilate to produce photons. Only a small excess of particles over anti-particles survives, given from (9.47) by

$$n_I - n_{\bar{I}} \approx 2 \left(\frac{m_I}{2\pi}\right)^{3/2} T^{3/2} e^{-m_I/T} \sinh \frac{\mu_I}{T}. \tag{9.55}$$

In the case of electrons, this small excess of surviving electrons corresponds to about 1 electron to 10^9 photons. Note that electrical neutrality of the universe implies that $n_p = n_e - n_{\bar{e}}$.

As discussed in Section 5.2, the entropy density s satisfies

$$(sa^3)^{\cdot} = -\frac{\mu}{T}(na^3)^{\cdot}, \quad s := \frac{\rho + p - \mu n}{T}. \tag{9.56}$$

In the cosmological case, either na^3 is constant (particle number conservation), or $\mu \ll T$, so that we have conservation of the entropy $S = sa^3$, and

$$s = \frac{2\pi^2}{45} q_* T^3, \quad q_* = \sum_I \alpha_I g_I \left(\frac{T_I}{T}\right)^3. \tag{9.57}$$

The photon number density is proportional to the entropy and therefore is a good measure of entropy. This follows from (9.46) and (9.57):

$$n_\gamma = \frac{45\zeta(3)}{\pi^4 q_*} s \approx \frac{1}{1.8 q_*} s .$$ (9.58)

Note that when all particles are in equilibrium, then $q_* = g_*$.

The process of decoupling of a species I is a non-equilibrium process. However, the final decoupled state, when the I particles are free-streaming, is another equilibrium state. In addition, the particles maintain the form of their distribution function, since only their 3-momentum redshifts, $p(t) = p(t_{\text{dec}}) a(t_{\text{dec}})/a(t)$. Thus for $t > t_{\text{dec}}$, the distribution satisfies $F_I(p,t) = F_I[a(t)p/a_{\text{dec}}, t_{\text{dec}}]$. If decoupling takes place when the species is relativistic, i.e. $T \gg m, \mu$, then

$$F_I(p, t > t_{\text{dec}}) = \frac{1}{(2\pi)^3} \frac{g_I}{\exp[E/T_I(t)] \pm 1}, \quad T_I(t) = T_{\text{dec}} \frac{a_{\text{dec}}}{a(t)} .$$ (9.59)

The decoupled temperature redshifts like the photon temperature, $T_I \propto a^{-1}$, and the entropy S_I is separately conserved. If the species becomes non-relativistic at a time $t_{\text{nr}} \gg t_{\text{dec}}$, then the distribution maintains the form above, with $E \approx m_I$. This is the case, for example, with massive neutrinos.

The total entropy S is constant, and the decoupled species has constant entropy S_I. Thus the entropy of the remaining species in equilibrium with the photons,

$$S - S_I = \frac{2\pi^2}{45} q_\gamma(T) T^3 a^3 , \quad q_\gamma(t) \equiv \sum_{J \neq I} \alpha_J g_J \left(\frac{T_J}{T} \right)^3 ,$$ (9.60)

is also constant. It follows that after I-decoupling, the temperature of the plasma is given by

$$T \propto q_\gamma^{-1/3} a^{-1} .$$ (9.61)

If the number of relativistic species does not change, then the temperature redshifts as a^{-1}. When a species becomes non-relativistic, its entropy is transferred to the relativistic species in thermal equilibrium, and the plasma undergoes a consequent heating in a short time. This increase in temperature is given by

$$T(t_{\text{dec}} + \epsilon) = \left[\frac{q_\gamma(t_{\text{dec}} - \epsilon)}{q_\gamma(t_{\text{dec}} + \epsilon)} \right]^{1/3} T(t_{\text{dec}} - \epsilon),$$ (9.62)

while the temperature of the decoupled species is given for $t > t_{\text{dec}}$ by

$$T_I = \left[\frac{q_\gamma(T)}{q_\gamma(T_{\text{dec}})} \right]^{1/3} T .$$ (9.63)

(We have assumed that q_I remains constant.)

9.6.5 Neutrinos

At high T, electron neutrinos are in equilibrium via the weak interactions

$$\nu_e + \bar{\nu}_e \leftrightarrow e + \bar{e}, \quad \nu_e + e \leftrightarrow \nu_e + e .$$ (9.64)

Similar interactions affect the ν_μ, ν_τ neutrinos, but since the number densities of μ and τ are negligible at $T = O(\text{TeV})$ compared to the density of electrons, ν_μ, ν_τ are coupled more weakly, and therefore decouple earlier than the electron neutrinos. At early times it is a good approximation to treat the neutrinos as massless.

The weak cross-section is $\sigma_w \sim G_F^2 T^2$, where G_F is the Fermi coupling constant, for neutrino energies $E \gg m_e$, and so the interaction rate is $\Gamma = n\langle \sigma_w v \rangle \sim G_F^2 T^5$. Using (9.52),

$$\frac{\Gamma}{H} \sim \left(\frac{T}{1\,\text{MeV}} \right)^3. \tag{9.65}$$

Thus neutrinos decouple at $T_{\text{dec}} \sim 1\,\text{MeV}$. The neutrino temperature remains equal to the photon temperature, $T_\nu = T \propto a^{-1}$, until the temperature drops below the electron mass. For $T_{\text{dec}} > T > m_e$, there are four fermion states ($g_e = g_{\bar{e}} = 2$) and two boson states ($g_\gamma = 2$) in equilibrium. When $T < m_e$, after electron–positron annihilation, only the photons contribute to q_γ. Thus,

$$q_\gamma(T > m_e) = \tfrac{11}{2}, \quad q_\gamma(T < m_e) = 2, \tag{9.66}$$

and conservation of entropy gives the heating of the plasma encoded in (9.63):

$$T_\gamma = \left(\tfrac{11}{4} \right)^{1/3} T_\nu \approx 1.4 T_\nu. \tag{9.67}$$

The cosmic neutrino background therefore has a current temperature of $1.95\,\text{K}$.

9.6.6 Nucleosynthesis (light elements)

Hydrogen constitutes about 75% of all observed baryonic matter in the universe, with helium accounting for most of the rest, and only trace contributions from other elements. The isotope deuterium and elements helium, lithium and beryllium are produced in the early universe, and heavier elements are synthesized much later in stars. Primordial nucleosynthesis may be analysed via the weak interaction and nuclear reactions, and we can predict the abundances of these light elements, and then compare with current observations. This is a crucial test of the HBB model.

Primordial nucleosynthesis, also known as BBN, is sensitive to: (1) g_*, the number of relativistic degrees of freedom, and hence to the number of neutrino species N_ν (with canonical value $N_\nu = 3$), and (2) the baryon–photon ratio η [(9.69)] and thereby the baryon density parameter $\Omega_{b0}h^2$.

The strong interaction binds neutrons and protons in atomic nuclei, but at high temperatures neutrons and protons are kept in equilibrium via the weak interactions

$$\nu_e + n \leftrightarrow p + e, \ \bar{e} + n \leftrightarrow p + \bar{\nu}_e, \ n \leftrightarrow p + e + \bar{\nu}_e. \tag{9.68}$$

In addition, the high entropy, reflected in the high number of photons relative to baryons, suppresses the formation of nuclei since free neutrons and protons are entropically favoured. Thus the baryon to photon ratio,

$$\eta \equiv \frac{n_b}{n_\gamma} \approx 5 \times 10^{-10} \left(\frac{\Omega_{b0}h^2}{0.02} \right), \tag{9.69}$$

is a critical parameter in the process of nucleosynthesis.

As the temperature drops the weak interactions (9.68) eventually fail to keep neutrons and protons in equilibrium. This happens at the 'freeze-out' temperature $T_f \sim 0.8\,\text{MeV}$, and the fraction of neutrons to protons is temporarily frozen at the current equilibrium value,

$$\frac{n_n}{n_p} \approx \exp{-\frac{(m_n - m_p)}{T_f}} \sim \frac{1}{5}. \tag{9.70}$$

This fraction of surviving neutrons determines the abundances of the light nuclei that can now form. The process is therafter affected by neutron decay $n \to p + e + \bar{\nu}_e$, with lifetime $\tau_n \sim 900\,\text{s}$. The freeze-out neutron fraction decreases exponentially for $t > t_f$ as

$$X_n = X_{n\,f}e^{-t/\tau_n}, \quad X_n \equiv \frac{n_n}{n_n + n_p}. \tag{9.71}$$

Free neutrons are nearly all captured in nuclei by $t \sim 250\,\text{s}$, so that neutron decay plays a substantial role.

Light nuclei begin to form when the temperature has dropped to $T \lesssim 0.1\,\text{MeV}$. Low number densities suppress reactions like $p + p + n + n \to {}^4\text{He}$, and so complex light nuclei must be produced through two-body reactions. The first step in the chain is deuterium production ($D = {}^2\text{H}$) via

$$p + n \to D + \gamma. \tag{9.72}$$

Until D has been formed in sufficient abundance, the production of helium and heavier elements like lithium is delayed. This is known as the deuterium bottleneck. Deuterium production is suppressed by photo-dissociation, i.e. effectively by the small value of η, and only becomes significant for

$$T \lesssim T_{\text{nuc}} \approx 0.09\,\text{MeV}. \tag{9.73}$$

The deuterium bottleneck opens up when the reactions

$$D + D \to {}^3\text{He} + n, \; D + D \to T(\equiv {}^3\text{H}) + p \tag{9.74}$$

become efficient. This happens when the D fraction reaches a critical value, after which this fraction drops as D is destroyed in the DD reactions. Helium-4 is produced when tritium or helium-3 combine with deuterium, and indirectly, when helium-3 captures a neutron to produce tritium. Most of the neutrons are fused into helium-4 by the reaction chains

$$np \to D \to T \to {}^4\text{He}, \; np \to D \to {}^3\text{He} \to T \to {}^4\text{He}. \tag{9.75}$$

The numerically computed evolution of the various species in the nucleosynthesis process is shown in Figure 9.7.

The final helium-4 abundance is thus determined by the available free neutrons at the time when the deuterium fraction reaches its critical (maximum) value. This neutron availability is itself determined by the number of relativistic species N_ν and the baryon density (equivalently, η).

- At a given temperature, the greater N_ν, the faster the universe expands – so that neutrons freeze out earlier and the freeze-out fraction $X_{n\,f}$ increases.
- Also, more relativistic species means the nucleosynthesis temperature is reached earlier – and so more neutrons avoid decay (see (9.71)).

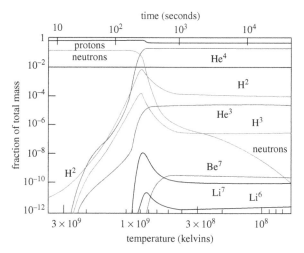

Fig. 9.7 Evolution of mass fractions of the key particles in BBN. (Reproduced from
http://aether.lbl.gov/www/tour/elements/early/ , courtesy of George Smoot.)

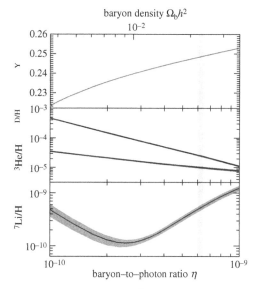

Fig. 9.8 Predicted BBN abundances versus baryon-to-photon ratio η. ^{4}He is given via Y_p, its density relative to the baryon
density. The lines give the mean values, and the bands mark 1σ uncertainties. The vertical band gives the WMAP 1σ
values for η. (From Cyburt, Fields and Olive (2008).)

- The greater the baryon density, i.e. the greater η, the earlier nucleonsynthesis begins, and
  so the greater the number of neutrons available.

Primordial nucleosynthesis is a complicated process because it involves a complex chain
of non-equilibrium particle and nuclear reactions, and requires numerical integration to
arrive at accurate results. The results are illustrated in Figure 9.8.

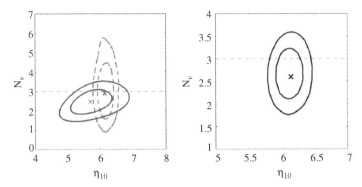

Fig. 9.9 *Left:* 68% and 95% contours in the $N_\nu, \eta_{10} (\equiv 10^{10}\eta)$ plane from nucleosynthesis (D & ^{4}He) (solid) and from the CMB and large-scale structure (dashed). Crosses give the best-fit values. *Right:* Joint constraints. (From Steigman (2010).)

The primordial abundances of D, ^{3}He and ^{4}He are in good agreement with the CMB data (WMAP) and large-scale structure data (SDSS), which constrain $\Omega_{b0}h^2$ and N_ν, as shown in Figure 9.9. There is also good agreement with spectroscopic observations of stars. However, there is a potentially serious discrepancy with the abundance of ^{7}Li. Part of the problem is the difficulty of measuring element abundances at low redshifts, and these measurements may also be sensitive to poorly understood astrophysics in stars.

9.6.7 Recombination and photon decoupling

After nucleosynthesis, the main ingredients of the cosmic plasma are $\gamma, e, p \equiv H^+$ and fully ionized helium, He^{2+} (other ionized light nuclei play a negligible role). Photons are strongly coupled to baryons via Thomson ($e-\gamma$) and Coulomb ($p-e$) interactions. As the temperature drops, the ionized nuclei begin to capture free electrons. Helium recombination takes place before hydrogen recombination, since its ionization potentials are greater. Helium recombination takes place in two stages:

$$He^{2+} + e \rightarrow He^+ \; (E_{I+} = 54.4\,\mathrm{eV}) \rightarrow He \; (E_I = 24.6\,\mathrm{eV}), \qquad (9.76)$$

where E_{I+}, E_I are the ionization energies. By $T \sim 5000\,\mathrm{K}$, helium is neutral and decouples from the radiation. At this temperature, the hydrogen is still fully ionized, and it plays the key role in the formation of the fossil CMB radiation.

For $T \lesssim 5000\,\mathrm{K}$, the reaction $p + e \leftrightarrow H + \gamma$ keeps the plasma in equilibrium. As the temperature drops further, this interaction becomes less effective, and the probability grows of electrons being captured by protons to form hydrogen. This is measured by the ionization fraction X_e, which satisfies the Saha equation, leading to

$$\frac{X_e^2}{1 - X_e} = \left(\frac{m_e T}{2\pi}\right)^{3/2} \frac{e^{-E_I/T}}{n_b}, \quad X_e \equiv \frac{n_e}{n_b}, \; n_b = n_p + n_H, \qquad (9.77)$$

where the hydrogen ionization energy and photon temperature are

$$E_I = m_e + m_p - m_H = 13.6 \,\text{eV}, \tag{9.78}$$

$$T = T_0(1+z) = 2.725(1+z)\,\text{K} \approx 2.3 \times 10^{-4}(1+z)\,\text{eV}, \tag{9.79}$$

and $n_b = \eta n_{\gamma 0}(1+z)^3$. Equation (9.77) shows that hydrogen recombination only occurs for $T \ll E_I$.

The Saha equation is based on equilibrium thermodynamics, and it predicts that the ionization fraction should continue to fall exponentially with temperature. However, like nucleosynthesis, recombination is a non-equilibrium process. In particular, hydrogen recombination produces a large number of nonthermal photons that distort the thermal radiation spectrum. A detailed analysis based on kinetic theory shows that the ionization fraction in fact freezes out: the residual electron fraction is

$$X_e(\infty) \approx 7 \times 10^{-3}. \tag{9.80}$$

During recombination, the electron density decreases rapidly, and the $e-\gamma$ interaction rate due to Thomson scattering, $\Gamma = n_e \sigma_T$, drops rapidly, so that the photons decouple soon afterwards. An estimate of the decoupling redshift is given by $\Gamma = H$. Using

$$\Gamma = 3 \times 10^{-26} X_e \frac{\Omega_{b0} h^2}{0.02}(1+z)^3 \,\text{eV}, \tag{9.81}$$

$$H^2 = \Omega_{m0} H_0^2 (1+z)^3 \left(1 + \frac{1+z}{1+z_{\text{eq}}}\right), \tag{9.82}$$

we find that the decoupling redshift is a solution of

$$(1 + z_{\text{dec}})^{3/2} = \frac{280}{X_e(\infty)} \left(\frac{\Omega_{b0} h^2}{0.02}\right)^{-1} \left(\frac{\Omega_{m0} h^2}{0.15}\right)^{1/2} \left(1 + \frac{1 + z_{\text{dec}}}{1 + z_{\text{eq}}}\right)^{1/2}. \tag{9.83}$$

We shall discuss decoupling and recombination again in Chapter 11.

9.6.8 The Dark Ages and the epoch of reionization

After recombination, the baryonic matter is effectively all in the form of neutral hydrogen and helium. From the decoupling redshift of $z = 1100$ down to a redshift $z \sim 200$, the gas temperature follows the CMB temperature since the residual ionization (9.80), although very small, is enough to maintain sufficient coupling via Compton scattering:

$$T_{\text{gas}} = T_\gamma = T_{\gamma 0}(1+z), \quad z \gtrsim 200. \tag{9.84}$$

Expansion and cooling eventually break this coupling and the gas temperature drops below the CMB temperature, evolving adiabatically as

$$T_{\text{gas}} \propto (1+z)^2, \quad 200 \gtrsim z \gtrsim 20. \tag{9.85}$$

For $z \lesssim 20$, the gas begins to be heated by emissions from the first stars, and eventually exceeds the CMB temperature,

$$\dot{T}_{\text{gas}} > 0, \quad z \lesssim 20. \tag{9.86}$$

After recombination, the baryonic pressure drops towards zero and gravity overcomes the counterbalancing effect of pressure. The baryonic gas falls into dark matter haloes, and over-densities grow as $\delta \sim a$. Because of the weakness of gravitational instability in an expanding background, it takes of the order of a few 100 Myr before the first stars form. Thus there is a period after recombination, the so-called Dark Ages, when baryonic matter is dark. The 'backlight' of the CMB radiation leads to emission and absorption features of the neutral hydrogen 21 cm hyperfine spin flip transition. The restframe frequency of 1420 MHz is red-shifted for the range $z \sim 10$–100 to ~ 140–14 MHz. This provides, in principle, a probe of the Dark Ages via massive radio telescopes, such as the planned Square Kilometre Array (SKA).

Current simulations indicate that the first stars condense from the gas in dark matter halos, at the late stage of the Dark Ages (estimated at around $z \sim 15 - 30$), eventually 'lighting up' the universe and reionizing it via ultraviolet radiation. The epoch of reionization stretches from the time of fully neutral gas to fully ionized gas. The observation of Lyman-α absorption by neutral gas (the Gunn–Peterson effect) of the light from distant quasars, and the WMAP constraints on Thomson scattering of CMB photons by reionized gas, lead to estimates that reionization stretches over the redshift range $11 \gtrsim z \gtrsim 6$.

Stars aggregate into galaxies and galaxies into clusters. This growth is suppressed (and may eventually end) when dark energy begins to dominate. In later chapters, we discuss in detail the topics of structure formation and dark energy.

9.7 Inflation

We have strong evidence that the universe was radiation dominated back to early times – at least back to nucleosynthesis ($t \sim 10^2$ s) and possibly back to the time of electroweak unification ($t \sim 10^{-10}$ s). In the HBB model of the universe, radiation domination persists all the way back to the inevitable singularity at $t = 0$. Variants of this model may have eras of differing equations of state at earlier times, but all share the property that the universe decelerates for $t > 0$.

There are certain puzzling features of a decelerating early universe that raise serious questions about initial conditions. We discuss these issues below and then we discuss how a short, accelerating era at very early times addresses these puzzles. The puzzles are often termed 'problems', and inflation is then presented as a solution to these problems. While this is a reasonable approach, we should note that there are various underlying assumptions that are worth making explicit. The key assumption is the reasonable expectation that models of the universe should not be sensitive to initial conditions. But there is no firm physical principle that underlies this expectation – it is an assumption (see Section 21.4.1 for further discussion). The HBB model rests on highly special initial conditions, which seem unnatural – but there is no physical principle yet known that rules out this possibility. It is conceivable that future developments in quantum gravity could explain what appear to be extreme fine-tunings. Furthermore, inflation itself is not completely free of initial conditions – for example, it requires a large enough patch where gradients are initially small enough.

Inflation does address the special initial conditions in an interesting and important way, even if it remains a phenomenological scenario that is yet to be rooted in a fundamental theory. In addition, inflation produces a mechanism for seeding structure formation (which we describe in more detail in Section 12.2) – and this mechanism was not constructed a priori to solve the problem of seeding structure formation. Instead, it was a prediction of the scenario. This is a real strength of the inflation model – and up to now, there is no real alternative to inflation for the origin of structure.

9.7.1 Some puzzles of a decelerating early universe

If we look in opposite directions on the sky and measure the CMB temperature, we find it is the same to 1 part in $\sim 10^5$. This suggests that a thermalization process operated before decoupling. However, in a decelerating radiation universe, thermalization could not have taken place across the CMB sky. We can see this as follows. The particle horizon at recombination is

$$L_{\rm rec} \equiv a_{\rm rec}\tau_{\rm rec} \equiv a_{\rm rec}\int_0^{t_{\rm rec}} \frac{d\tilde{t}}{a(\tilde{t})} \approx t_{\rm rec} \approx \frac{1}{H_{\rm rec}}, \tag{9.87}$$

where τ is the comoving particle horizon, and the approximations indicate that we neglect a multiplicative factor $\mathcal{O}(1)$ (in order to avoid complications of the transition from radiation to matter domination). This is the distance that light travels from the beginning of the universe at $t = 0 (= a = \tau)$, and represents the limit of causal interaction at the time of last scattering – i.e. particles that are separated by more than $L_{\rm rec}$ can never have been in causal communication. Points on the last scattering surface at opposite ends of the sky are separated today by a distance equal to the distance to the last scattering surface, $D_{\rm rec} \approx H_0^{-1}$ – which is much greater than the maximal causal separation, $D_{\rm rec} \gg L_{\rm rec}$. And yet, the particles at these locations at the time when the CMB distribution was frozen had never been in causal communication. This is illustrated in Figure 9.10.

It is often called the 'horizon problem'.

Another puzzle is often called the 'flatness problem' – which arises from the fact that if the universe is close to flat today, then the evolution of the curvature density parameter implies a severe fine-tuning of the curvature in the early universe. The curvature parameter in a HBB model evolves as the square of the comoving Hubble radius, $|\Omega_K| = |K|(aH)^{-2}$. For $w \equiv p/\rho = $ const,

$$(aH)^{-1} = H_0^{-1}a^{(1+3w)/2} \propto \tau \quad (w \neq -1). \tag{9.88}$$

Thus the comoving Hubble radius *grows* in a HBB model, since $w \geq 0$. In particular, this means that the curvature grows, $|\Omega_K| \propto |K|\tau^2$, and so it must be strongly suppressed in the past if it is small today – unless $K = 0$. If we take $w = 0$ (i.e. we approximate the universe as always matter dominated), then the curvature at nucleosynthesis for example is given by

$$|\Omega_K(a_{\rm nuc})| \approx \frac{a_{\rm nuc}}{a_0}|\Omega_K(a_0)| \approx 10^{-9}|\Omega_K(a_0)|. \tag{9.89}$$

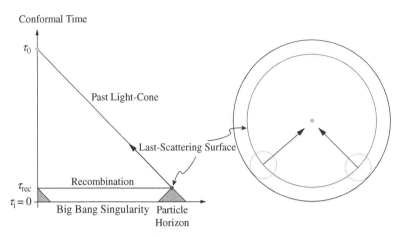

Fig. 9.10   Conformal diagram of the HBB model. (From Baumann (2009).)

Within the HBB framework, this problem can be resolved by assuming that there are extremely fine-tuned initial conditions on the curvature (including the possibility that $K = 0$).

Another puzzle is the 'monopole problem': phase transitions in the very early universe – which are associated with the breaking of symmetries as the temperature drops – can produce topological defects, such as monopoles. If a phase transition takes place at the Grand Unified Theory scale, $T \sim 10^{16}$ GeV, then the monopoles could dominate the energy density in the universe. Inflation can evade this problem if it takes place after the GUT phase transition, since the accelerated expansion disperses the monopoles and dramatically suppresses their energy density.

9.7.2 Inflation addresses horizon and flatness problems

Underlying both the horizon and flatness puzzles of the HBB model is the same key fact – *the comoving Hubble radius* (9.88) *grows for* $t > 0$. If instead there is a primordial era in which this scale *shrinks* with expansion, then neither of the features will require highly fine-tuned initial conditions. By (9.88), the condition for a shrinking comoving Hubble radius is $1 + 3w < 0$, which is precisely the condition for acceleration, $\ddot{a} > 0$.

Inflation is a period of 'slow-roll' acceleration (see Section 9.7.3) with H nearly constant (i.e. $w \approx -1$), so that

$$-\tau \approx (aH)^{-1}. \tag{9.90}$$

This means that the singularity $a = 0$ is pushed to $\tau = -\infty$, and the comoving Hubble radius $\approx -\tau$ decreases. Equation (9.90) breaks down by the end of inflation, and the brief reheating period, $\tau = 0$, leads into a HBB evolution in a radiation-dominated universe. As illustrated in Figure 9.11, the past light cones of all points on the last scattering surface intersect in the past, provided inflation lasts long enough ($\gtrsim 60$ e-folds) to shrink the comoving Hubble radius sufficiently.

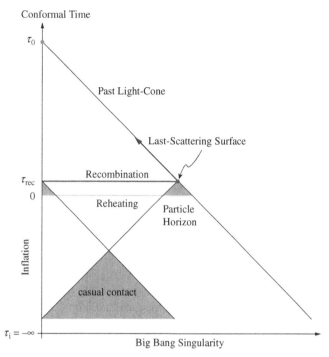

Fig. 9.11   Conformal diagram of the inflationary model. (From Baumann (2009).)

At the same time, the rapid decrease of the comoving Hubble radius reduces any non-zero primordial curvature. Provided inflation lasts long enough, this removes the fine-tuning that is illustrated in (9.89).

In addition to its ability to address the issues of the horizon and flatness, inflation has the further crucial feature that it incorporates a mechanism for generating the primordial inhomogeneities that seed structure formation, as discussed in Section 12.2. This may be thought of as a *prediction* of inflation, if we consider inflation as a construction for solving the horizon and flatness problems.

Finally, we note that while inflation alleviates the fine-tuning of the initial conditions of the HBB model, it is *not* a theory of initial conditions, and indeed it is not fully independent of initial conditions. For example, we need to assume that the initial inflaton velocity and initial inhomogeneities in the inflaton are small enough to allow inflation to begin. See also Sections 8.4.3 and 21.4.

9.7.3 Dynamics of inflation

Inflation may be defined equivalently as a period of accelerating expansion, or of a decreasing comoving Hubble scale (see Figure 12.1 in Section 12.2). Via the Friedmann equations, these conditions are in turn equivalent to a source of the gravitational field that violates the strong energy condition. In summary:

$$\ddot{a} > 0 \;\Leftrightarrow\; \frac{\mathrm{d}}{\mathrm{d}t}\left(\frac{H^{-1}}{a}\right) < 0 \;\Leftrightarrow\; w \equiv \frac{p}{\rho} < -\tfrac{1}{3}. \tag{9.91}$$

The violation of the strong energy condition is simple to achieve for a scalar field, whose dynamics were discussed in Section 5.6:

$$w = \frac{-V(\varphi) + \dot{\varphi}^2/2}{V(\varphi) + \dot{\varphi}^2/2} < -\frac{1}{3} \Rightarrow V(\varphi) > \dot{\varphi}^2. \tag{9.92}$$

(We assume that the potential energy is positive.) The dynamics of the field are governed by the Klein–Gordon equation,

$$\ddot{\varphi} + 3H\dot{\varphi} + V'(\varphi) = 0. \tag{9.93}$$

The special case of $\dot{\varphi} = 0$, which implies $V = $ const, is the extreme way of satisfying the inflation condition (9.92). This corresponds to a cosmological constant $\Lambda_{\text{inf}} = 8\pi G V$ that drives de Sitter expansion, $H = $ const. (Λ_{inf} should not be confused with the low-energy cosmological constant, $\Lambda \ll \Lambda_{\text{inf}}$, that acts as dark energy in the late universe.) In this case, $w = -1$, which is the limiting value for a scalar field ($-1 \leq w \leq 1$). If w is close to, but above, -1, then $\dot{\varphi}$ is small, but non-zero, and the expansion rate is close to de Sitter: the Hubble rate is nearly constant, but slowly decreasing. This is known as a 'slow-rolling' inflaton field. Inflationary models are typically of the slow-roll kind.

Qualitatively, slow-roll requires small inflaton velocity $\dot{\varphi}$ and acceleration $\ddot{\varphi}$, so that the Friedmann and Klein–Gordon equations become

$$H^2 \approx \frac{8\pi G}{3} V, \quad \dot{\varphi} \approx -\frac{V'}{3H}. \tag{9.94}$$

We can quantify the slow-roll property via two slow-roll parameters:

$$\epsilon = 4\pi G \frac{\dot{\varphi}^2}{H^2} = -\frac{\dot{H}}{H^2} = \frac{1}{H} \frac{dH}{dN} = \frac{3}{2}(1 + w), \tag{9.95}$$

$$\eta = -\frac{1}{H} \frac{\ddot{\varphi}}{\dot{\varphi}} = -\frac{1}{\dot{\varphi}} \frac{d\dot{\varphi}}{dN}, \tag{9.96}$$

where N is the number of e-folds before the end of inflation:

$$N \equiv \ln \frac{a_{\text{end}}}{a} = \int_{\varphi}^{\varphi_{\text{end}}} \frac{H}{\dot{\varphi}} d\varphi = \sqrt{4\pi G} \int_{\varphi}^{\varphi_{\text{end}}} \frac{d\varphi}{\sqrt{\epsilon}} \approx 8\pi G \int_{\varphi_{\text{end}}}^{\varphi} \frac{V}{V'} d\varphi. \tag{9.97}$$

Slow-roll is then characterized by $\epsilon \ll 1, |\eta| \ll 1$, and the end of inflation is defined by $\epsilon_{\text{end}} = 1$.

An alternative pair of slow-roll parameters is based on the potential, ensuring that the slope and curvature of V are small:

$$\epsilon_V = \frac{1}{16\pi G} \left(\frac{V'}{V}\right)^2 \approx \epsilon, \quad \eta_V = \frac{1}{8\pi G} \frac{V''}{V} \approx \eta + \epsilon. \tag{9.98}$$

Slow-roll inflation should last for $\gtrsim 60$ e-folds, with the large-scale CMB anisotropies (the Sachs–Wolfe effect, as discussed in Section 11.5) being seeded by fluctuation modes that exceed the Hubble scale near the beginning of this period, i.e. $N(\varphi_{\text{cmb}}) \sim 60$. As the inflaton rolls down the flat potential, it eventually picks up speed as the potential steepens, until the kinetic energy is sufficient to break the accelerating condition, i.e. $V(\varphi_{\text{end}}) = \dot{\varphi}_{\text{end}}^2$. This is illustrated in Figure 9.12.

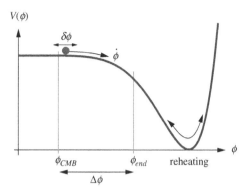

Fig. 9.12 Schematic of inflationary dynamics. (From Baumann (2009).)

At the end of inflation, the universe is a 'cold desert'. There must therefore be a mechanism for 'reheating' the universe and populating it with the matter and radiation that sources the HBB era. This is achieved within the simple inflation scenario by oscillations of the inflaton about the minimum of its potential. During these oscillations, the inflaton decays into the fields and particles of the radiation era. The coupling of the inflaton to these fields and particles is of course not known, since there is as yet no fundamental theory for the inflaton itself. But simple phenomenological models have been constructed that can achieve a rapid and efficient conversion of inflaton energy into matter and radiation (see Bassett, Tsujikawa and Wands (2006) for a review). Reheating is clearly a non-equilibrium process. The resulting non-thermal distributions of created particles are, however, rapidly thermalized by interactions, initiating the thermal plasma era of the HBB.

9.7.4 Simple models of inflation

Inflation takes place at energies well beyond those accessible to terrestrial experiments, and also to the Standard Model of particle physics and its current minimal extensions. Attempts to imbed inflation in string theory are ongoing at the time of writing, with no generally accepted and testable model on the horizon. Inflation remains at the level of phenomenology, and it is remarkable that the simplest single-field models can be successfully grafted onto the HBB model to produce a 'standard' model of cosmology. In order to accommodate the late-time acceleration of the universe within this framework, it is necessary to add the (late-time) cosmological constant to the inflationary potential: $V \to V + 8\pi G \Lambda$.

Single-field models may be divided according to the behaviour of the slow-roll parameters, as follows.

- *Large-field models* $(0 < \eta_V \leq \epsilon_V)$
  Inflation begins when φ is $\gtrsim M_P = G^{-1}$ away from its stable $(V'' < 0)$ minimum. The key example is 'chaotic' inflation, driven by power-law potentials,

$$V = V_n \left(\frac{\varphi}{M_P} \right)^n, \quad n = 2, 3, \ldots, \tag{9.99}$$

where V_n is a constant. Slow-roll conditions impose $\varphi > n M_P$.

- *Small-field models* $(\eta_V < 0 < \epsilon_V)$

The inflaton rolls away from an unstable minimum, as in the 'new inflation' models:

$$V = V_n \left[1 - \left(\frac{\varphi}{\mu} \right)^n \right], \tag{9.100}$$

where μ is a mass scale.

- *Hybrid models* $(0 < \epsilon_V < \eta_V)$

A typical potential is

$$V = V_n \left[1 + \left(\frac{\varphi}{\mu} \right)^n \right]. \tag{9.101}$$

Strictly, the hybrid models are two-field models, since a second field is required to end the inflation driven by φ. However, during inflation the second field is trapped in a minimum and plays no role, so that the hybrid potential is effectively single-field.

Constraints on these classes of inflation from the CMB and large-scale structure data are shown in Figure 9.13.

The classical dynamics of the inflaton dictates that the inflaton always rolls down its potential. However, quantum fluctuations can also drive the inflaton uphill, which has the effect of prolonging inflation and enlarging the volume of the region. In some regions the inflaton will remain high enough up the potential hill to maintain acceleration. This stochastic scenario, known as 'eternal' inflation, is used to motivate the idea of a 'multi-verse'. Tunneling is supposed to take place via the Coleman–de Luccia process in a de Sitter $K > 0$ universe and leads to new bubbles of ordinary matter in a $K < 0$ FLRW phase. There is a competition between the rate of nucleation and the rate of expansion, so that depending

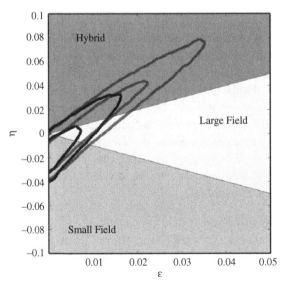

Fig. 9.13   Constraints on slow-roll parameters from the CMB (WMAP) (grey; red in colour version) and CMB + galaxy distribution (SDSS) (black). (From Peiris and Easther (2006).) A colour version of this figure is available online.

on the parameter values either (i) (high expansion rate) the bubbles never intersect and the resulting inflation pattern is eternal with a fractal structure, often called a multiverse; or (ii) (high nucleation rate) all bubbles intersect and eventually the entire universe (when its entire compact space section has nucleated) has left the de Sitter phase and inflation comes to an end; or (iii) (in between) all sorts of complex patterns of intersecting bubbles and inflating phases can occur (Sekino, Shenker and Susskind, 2010). However, the physics of the tunneling process is speculative; it is an extrapolation of known tunneling processes to situations where it may or may not occur. See, e.g. Freivogel *et al.* (2006), Vilenkin (2006), Ellis and Stoeger (2009a) for various views. There is an ongoing debate about probabilities in the multiverse (see Section 21.5).

The simple single-field models of inflation have Lagrangian density

$$\mathcal{L} = p(X, \varphi) = X - V(\varphi), \quad X \equiv -\tfrac{1}{2} \partial_\mu \varphi \partial^\mu \varphi, \tag{9.102}$$

where p is the scalar field pressure. More complicated models have been considered. Extensions of the Standard Model of particle physics, including string theory, typically include a number of scalar fields, which motivates the analysis of multi-field inflation. These models can generate isocurvature modes and non-Gaussianity, as discussed in Section 12.2. The Lagrangian is generalized to

$$p(X, \varphi_I) = X - V(\varphi_I), \quad X \equiv -\tfrac{1}{2} G^{IJ}(\varphi_K) \partial_\mu \varphi_I \partial^\mu \varphi_J, \tag{9.103}$$

where G_{IJ} is a metric in field space. Another generalization is to modify the canonical kinetic energy term, i.e. to consider functions $p(X, \varphi)$ more general than $X - V$. The simplest example is a 'phantom' scalar field, with $p = -X - V$, but this is quantum mechanically unstable. K-inflation models use more complicated non-standard kinetic terms to achieve inflation even when the potential is not flat. Dirac–Born–Infeld models have

$$p(X, \varphi) = \frac{1}{f(\varphi)} \left[\sqrt{1 - 2f(\varphi)X} - 1 \right] - V(\varphi), \tag{9.104}$$

in the simplest case. These models arise from certain string theory scenarios. Attempts to construct inflation within string theory are reviewed in Baumann (2009), Baumann and McAllister (2009).

Exercise 9.7.1 For the simple chaotic inflationary potential, $V = \tfrac{1}{2} m^2 \varphi^2$, show that:

$$\epsilon_V = \frac{1}{4\pi} \left(\frac{M_P}{\varphi} \right)^2 = \eta_V, \tag{9.105}$$

$$N(\varphi) \approx 2\pi \left[\left(\frac{\varphi}{M_P} \right)^2 - \left(\frac{\varphi_{\text{end}}}{M_P} \right)^2 \right], \quad \varphi_{\text{end}} \approx \frac{M_P}{\sqrt{4\pi}}, \tag{9.106}$$

$$\varphi_{\text{cmb}} \approx \sqrt{\frac{30}{\pi}} M_P \text{ where } N(\varphi_{\text{cmb}}) = 60. \tag{9.107}$$

Also show that the slow-roll dynamical equations have solution:

$$\varphi - \varphi_{\text{end}} = \frac{m M_P}{2\sqrt{3\pi}} (t_{\text{end}} - t), \quad a = a_{\text{end}} \exp \left[\frac{2\pi}{M_P^2} (\varphi_{\text{end}}^2 - \varphi^2) \right]. \tag{9.108}$$

9.8 Origin of FLRW geometry

The FLRW models are very exceptional within the family of all cosmological models, because of their very high symmetry (see Chapter 17). Why is the universe in which we live like this?

9.8.1 Origin of uniformity

Assuming we accept the arguments for uniformity and for validity of the FLRW metric, we are led to one of the major issues in cosmology: namely *why is the universe on average so smooth?* These models, because of their exact spatial homogeneity and isotropy, are, on almost any a-priori assignment of probabilities, infinitely improbable in the family of all cosmological models. Thus one of the major themes in cosmology is the attempt to explain this uniformity, which we discuss in Chapter 21.

9.8.2 Preservation of FLRW symmetry

Suppose a universe initially has the full FLRW symmetries, i.e. (9.1)–(9.3) hold on an initial surface $t = t_1$ (the matter moving orthogonally to this surface). Then provided the matter remains a perfect fluid the universe must remain an FLRW universe, basically because there is nothing in the initial data that can choose a preferred spatial direction, so the part of spacetime determined by these data will also have no preferred spatial directions; thus it will remain an FLRW universe.

A more formal proof, developing the idea that any symmetries in initial data will be preserved in development of those data, is discussed in Chapter 17; however it is also interesting to prove the result directly from the dynamic equations. Essentially, the subset of these equations governing the growth of anisotropies is a homogeneous set of equations: so if the anisotropies all vanish on an initial surface they will all vanish at later times. More precisely, if on $t = t_1$, we have

$$\omega_a = \sigma_{ab} = \dot{u}^a = 0 = \overline{\nabla}_a \rho = \overline{\nabla}_a p = \overline{\nabla}_a \Theta, \tag{9.109}$$

then the same will hold at all later and earlier times within the Cauchy development of those data; thus a perfect-fluid universe that initially has the FLRW symmetries will remain an FLRW universe.

If the matter for some reason does not remain a perfect fluid, then this result need not hold; for a specific example, based on a kinetic theory description of the matter where the spacetime has the FLRW symmetries while the particle distribution function does not, see Matravers and Ellis (1989). However, then there needs to be an anisotropy hidden in the particle distribution function; we do not regard this as physically plausible.

Exercise 9.8.1 Determine the time evolution equations for the quantities $\overline{\nabla}_a \rho, \overline{\nabla}_a p, \overline{\nabla}_a \Theta$. Hence prove the statement above: the quantities in (9.109) form an involutive set for FLRW initial data.

9.9 Newtonian case

The Newtonian version of the RW models was first given by Milne (1934), McCrea and Milne (1934), later explicated further by Heckmann and Schucking (1955, 1956) and Bondi (1960). The Newtonian analogues of the GR models are determined by the conditions $\omega_i = \sigma_{ij} = \dot{u}_i = 0$ which imply $\rho = \rho(t), p = p(t), \Theta = \Theta(t)$. The coordinates of any fluid particle are $x^i = \ell(t)c^i$ with $c^i = $ const. The basic dynamical equations are

$$\dot{\rho}_N + 3\rho_N \dot{\ell}/\ell = 0, \tag{9.110}$$

$$3\ddot{\ell}/\ell + 4\pi G \rho_N - \Lambda = 0, \tag{9.111}$$

$$E_{ij}(t) = 0, \tag{9.112}$$

$$3\dot{\ell}^2 - 8\pi G \rho_N \ell^2 - \Lambda \ell^2 = 10E, \quad E = \text{const}, \tag{9.113}$$

where the last one is a first integral. The gravitational potential satisfying the Poisson equation $\Phi^{\cdot i}{}_{,i} + \Lambda = 4\pi G \rho_N$ is (Bondi, 1960)

$$\Phi = [4\pi G \rho_N(t) - \Lambda] h_{ij} x^i x^j / 6, \tag{9.114}$$

where h_{ij} is the Newtonian spatial metric.

Several points are of interest, relative to the GR case.

Firstly, the gravitational potential (9.114) diverges at infinity, contrary to the usual boundary conditions in Newtonian gravitational theory. We have to drop the condition that $\Phi \to 0$ at infinity in order to attain these models. Indeed for more general Newtonian models, we also have to drop the condition that $E_{ij} \to 0$ at infinity.

Secondly, the time development of Newtonian cosmological models is not determined until some restriction is put on $E_{ij}(t)$, e.g. choosing some world line and determining $E_{ij}(t)$ as an arbitrary function along that world line (Heckmann and Schucking, 1956). In the particular case of FLRW analogues, the condition (9.112) is that $E_{ij}(t)$ be set to zero at all times. Note that this condition must be reset at each instant: the fact that it is true at one instant is no guarantee that it will be true at the next instant. This is quite unlike the GR case.

Thirdly, any non-zero pressure makes no difference whatever to the time development of these universe models. This is because these universes are spatially homogeneous, so there is no pressure gradient to influence the dynamics (see (9.110), (9.111)). By contrast in the GR case, because of the nature of the conservation equations, non-zero pressure influences the variation of the density as the scale factor changes. Furthermore, because of the nature of the Einstein equations, non-zero pressure influences the gravitational field causing the scale factor to change relative to the case with no pressure. Both effects are crucial in determining the thermal history of the universe.

Fourthly, the integration constant E in (9.113) has no relation to the spatial curvature, as is the case in GR.

Finally, this description is attained via a potential description of the gravitational dynamics. A force description runs into serious difficulties, as discovered by Newton: it is either ambiguous or divergent or both. This is why Newton never succeeded in creating a viable cosmological model. It was only in the 1930s (after the GR models had been discovered) that successful Newtonian cosmologies were derived by Milne and McCrea.

10

Perturbations of FLRW universes

The FLRW model provides a good description of the averaged dynamics of the universe on very large scales. As we move to smaller scales, the homogeneity of the FLRW model becomes an increasingly poor description of the universe, which contains inhomogeneities such as small over- and under-densities at early times, and stars and galaxies at later times.

Before the onset of structure formation, these inhomogeneities may be treated as small deviations from FLRW, i.e. we can use an almost-FLRW model – a linearly perturbed FLRW. Once structure formation is underway, we can continue to use the perturbed FLRW on scales above the comoving galaxy cluster scale. On smaller scales, nonlinear effects become increasingly important and we need to move beyond linear perturbation theory.

The growth of structure is based on gravitational instability, i.e. the tendency of over- and under-densities to be enhanced through the universally attractive nature of gravity. If $\delta\rho$ and $\delta\Theta$ denote the small deviations from the FLRW background density $\bar\rho$ and volume expansion $\bar\Theta = 3H$, then, denoting the normalized density perturbation $\delta\rho/\bar\rho$ by δ,

$$\rho = \bar\rho(1+\delta), \quad \Theta = 3H + \delta\Theta. \tag{10.1}$$

Energy conservation (5.11) and the Raychaudhuri equation (6.4) for dust then lead to the background equations at zero order, and at first order to

$$\delta_{,0} + \delta\Theta = 0 \text{ and } \delta\Theta_{,0} + 2H\delta\Theta = -4\pi G\bar\rho\delta, \tag{10.2}$$

where we have used $\dot{u}_a = 0$ from the momentum conservation equation. Eliminating $\delta\Theta$ leads to the evolution equation for small over-densities,

$$\delta_{,00} + 2H\delta_{,0} - 4\pi G\bar\rho\delta = 0. \tag{10.3}$$

If we neglect the cosmological constant, then the background variables are $H = 2t^{-1}/3$ and $\bar\rho = (6\pi Gt^2)^{-1}$. The solution is

$$\delta = A_+(\boldsymbol{x})t^{2/3} + A_-(\boldsymbol{x})t^{-1} = B_+(\boldsymbol{x})a + B_-(\boldsymbol{x})a^{-3/2}, \tag{10.4}$$

where $A_\pm, B_\pm$ are amplitudes of the growing $(+)$ and decaying $(-)$ modes.

We can also track the evolution of peculiar velocities during the growth of structure via $\delta\Theta$. The dust four-velocity is $u^a = \bar{u}^a + v^a$ where v^a is the small peculiar velocity relative to the background frame of $\bar{u}^a$, with $\bar{u}^a v_a = 0$. It follows that $\delta\Theta = \overline{\nabla}^a v_a = -\delta_{,0}$, so that

$$v^a = C_{1-}^a(\boldsymbol{x})t^{-1/3} + C_{2-}^a(\boldsymbol{x})t^{-2}. \tag{10.5}$$

This Newtonian approach to perturbations gives a flavour of what is involved in Newtonian growth of structure, but in order to track this process carefully from the primordial to the late universe, taking into account the fluctuations in all matter sectors as well as in the gravitational field, we need a systematic approach.

10.1 The gauge problem in cosmology

Any approach to the analysis of perturbations faces the so-called gauge problem, which reflects the fact that in perturbation theory we deal with two spacetime manifolds (Lifshitz, 1946, Lifshitz and Khalatnikov, 1963, Sachs and Wolfe, 1967, Bardeen, 1980, Kodama and Sasaki, 1984, Ellis and Bruni, 1989, Mukhanov, Feldman and Brandenberger, 1992, Malik and Wands, 2009), the physical spacetime $\mathcal{M}$, and $\overline{\mathcal{M}}$, a fictitious background FLRW spacetime. A gauge is a one-to-one correspondence $\overline{\mathcal{M}} \to \mathcal{M}$, between the two spacetimes. This point-identification map is generally arbitrary. When a coordinate system is introduced in $\overline{\mathcal{M}}$, the gauge carries it to $\mathcal{M}$. A change in the map $\overline{\mathcal{M}} \to \mathcal{M}$, keeping the background coordinates fixed, is known as a gauge transformation. This introduces a coordinate transformation in the physical spacetime, but also changes the event in $\mathcal{M}$ which is associated with a given event in the background $\overline{\mathcal{M}}$. Gauge transformations are therefore different from coordinate transformations which merely relabel events. The gauge freedom is usually expressed as a freedom of coordinate choice in $\mathcal{M}$, but it should be understood that it generally changes the point-identification between the two spacetimes.

Although we can always perturb away from a given background spacetime, recovering the smooth metric from a given perturbed one is not a uniquely defined process. This is a problem because it is always possible to choose an alternative background and therefore arrive at different perturbation values. Selecting an unperturbed spacetime from a given lumpy one corresponds to a gauge choice. Determining the best gauge is known as the fitting problem in cosmology and there is no unique answer to it (see Section 16.2).

By definition, the perturbation of any quantity is the difference between its value at some event in the real spacetime and its value at the corresponding event (associated via the gauge) in the background. Spacetime scalar quantities that have non-zero and position-dependent background values will lead to gauge-dependent perturbations. Following Stewart and Walker (1974) and Stewart (1990), we consider a one-parameter family of perturbed spacetimes $\mathcal{M}_\epsilon$ embedded as hypersurfaces in a 5-manifold $\mathcal{N}$. We define a point-identification map between $\overline{\mathcal{M}}$ and $\mathcal{M}_\epsilon$, by introducing in $\mathcal{N}$ a vector field X^A (with $A = 0, \ldots, 4$), which is everywhere transverse to the embeddings $\mathcal{M}_\epsilon$. Points lying along the same integral curves of X^A, which are parametrized by ϵ for convenience, will be regarded as the 'same'. Thus, selecting a specific vector field X^A corresponds to a choice of gauge. If Q_ϵ is some geometrical quantity defined on $\mathcal{M}_\epsilon$, with background value $\overline{Q}$, then the perturbation is

$$\delta Q = Q_\epsilon - \overline{Q} = \epsilon \overline{\mathcal{L}}_X Q_\epsilon + O(\epsilon^2). \tag{10.6}$$

Here Q_ϵ is the image in $\overline{\mathcal{M}}$ of the perturbed quantity (the pullback). This shows that even quantities that behave like scalars under coordinate changes will not remain invariant under gauge transformations. The value of δQ is entirely gauge dependent and therefore arbitrary.

For instance, one can select the gauge so that the surfaces of constant $\overline{Q}$ are the surfaces of constant Q_ϵ, thus setting $\delta Q = 0$ (Ellis and Bruni, 1989).

One way of addressing the gauge problem is by fixing the gauge. This can be problematic if a gauge choice turns out to contain residual gauge freedom. This is the case for the synchronous gauge introduced in the pioneering work of Lifshitz (1946). In order to avoid spurious gauge modes, one has to take care to compute only physically observable quantities. Alternatively, we can employ gauge-invariant variables (Bardeen, 1980, Kodama and Sasaki, 1984, Ellis and Bruni, 1989).

Gauge-independent quantities must remain invariant under gauge transformations between the background and the real spacetimes. According to (10.6), the only cases are scalars that are constant in the background or tensors that vanish (or are expressible as a linear combination with constant coefficients of products of Kronecker deltas) (Stewart and Walker, 1974). Given the symmetries of FLRW models, any tensor that describes spatial inhomogeneity or anisotropy must vanish in the background and therefore its linear perturbation will remain invariant under gauge transformations. This is the basis for the 1+3 covariant and gauge-invariant (CGI) approach to perturbations (Hawking, 1966, Lyth and Mukherjee, 1988, Ellis and Bruni, 1989).

An alternative approach starts from perturbations of the FLRW metric and energy–momentum tensors, and explicitly constructs combinations that are invariant under general gauge transformations (Bardeen, 1980, Kodama and Sasaki, 1984). We start by reviewing the metric-based approach to perturbations, and then we describe the CGI approach.

10.2 Metric-based perturbation theory

The standard perturbative formalism is a metric-based approach, which starts from an FLRW metric in suitable coordinates and defines perturbations away from that metric. This approach was introduced in general relativity by Lifshitz (1946), and a gauge-invariant version was developed by Bardeen (1980). For reviews, see Kodama and Sasaki (1984), Mukhanov, Feldman and Brandenberger (1992). We follow the notation and the more geometrical approach of Malik and Wands (2009).

10.2.1 Perturbations of the metric

We start with the FLRW metric in conformal time,

$$
\bar{g}_{\mu\nu} = a^2 \begin{pmatrix} -1 & 0 \\ 0 & \gamma_{ij} \end{pmatrix}, \tag{10.7}
$$

where γ_{ij} is the metric on the static hypersurface conformal to the homogeneous hypersurfaces with constant curvature K. We denote covariant derivatives with respect to γ_{ij} by a vertical bar. First-order perturbations of this metric, $g_{\mu\nu} = \bar{g}_{\mu\nu} + \delta g_{\mu\nu}$, can be split into scalar, vector and tensor parts, which are fields over the static hypersurface:

- *Scalar* perturbations are constructed from a scalar quantity or its derivatives, and any background quantities such as the 3-metric γ_{ij}. A generic first-order scalar metric perturbation is described by four scalars $\phi(\tau, x^i)$, $\psi(\tau, x^i)$, $B(\tau, x^i)$ and $E(\tau, x^i)$, where

$$\delta g_{00} = -2a^2\phi, \quad \delta g_{0i} = a^2 B_{|i}, \tag{10.8}$$

$$\delta g_{ij} = -2a^2 \left(\psi \gamma_{ij} - E_{|ij} \right). \tag{10.9}$$

Here, ϕ generalizes the Newtonian potential (since it determines particle acceleration in this metric), and ψ determines the perturbation of the 3-curvature of the static surfaces $\tau = \text{const}$.

- *Vector* perturbations are built from solenoidal (rotational or transverse) 3-vectors, $S_{[i|j]} \neq 0$, and have no scalar part. This rules out vector quantities that are constructed from scalars, which are irrotational or longitudinal, i.e. $B_{||ij|} = 0$. They are divergence-free, otherwise they would define a scalar field (non-locally, requiring a decay condition for $K \leq 0$, Stewart (1990)): so $\gamma^{ij} S_{i|j} = 0$. Symmetric 3-tensors which are constructed from vector perturbations must have no scalar part, so that they are trace-free. The vector metric perturbation is generically given in terms of solenoidal 3-vectors $S_i(\tau, x^j)$ and $F_i(\tau, x^j)$:

$$\delta g_{0i} = -a^2 S_i, \quad \delta g_{ij} = 2a^2 F_{(i|j)}. \tag{10.10}$$

- *Tensor* perturbations have no scalar or vector parts, so that they arise from symmetric, trace-free and divergence-free 3-tensors. The tensor metric perturbation $h_{ij}(\tau, x^k)$ is defined by[1]

$$\delta g_{ij} = a^2 h_{ij} \quad \text{where} \quad h_{[ij]} = 0 = \gamma^{ij} h_{ij} = \gamma^{jk} h_{ij|k} = 0. \tag{10.11}$$

Thus the most general linear metric perturbation is

$$ds^2 = a^2 \left\{ -(1+2\phi)d\tau^2 + 2(B_{|i} - S_i)d\tau dx^i \right.$$
$$\left. + \left[(1-2\psi)\gamma_{ij} + 2E_{|ij} + 2F_{i|j} + h_{ij} \right] dx^i dx^j \right\}. \tag{10.12}$$

There are 10 degrees of freedom in the perturbation variables, corresponding to the 10 metric components. The inverse metric tensor is

$$g^{\mu\nu} = a^{-2} \begin{pmatrix} -(1-2\phi) & B^{|i} - S^i \\ B^{|j} - S^j & (1+2\psi)\gamma^{ij} - 2E^{|ij} - 2F^{(i|j)} - f^{ij} \end{pmatrix}. \tag{10.13}$$

10.2.2 Gauge transformations

We can make a gauge transformation, based on a first-order change of coordinates, $x^\mu \to \tilde{x}^\mu$:

$$\tilde{\tau} = \tau + \xi^0(\tau, x^j), \quad \tilde{x}^i = x^i + \xi^{|i}(\tau, x^j) + \xi^i(\tau, x^j), \quad \xi^i_{\,|i} = 0. \tag{10.14}$$

The function ξ^0 determines the constant-τ hypersurfaces, i.e. the time-slicing, while $\xi^{|i}$ and ξ^i fix the spatial coordinates in these hypersurfaces. The choice of coordinates is arbitrary

---

[1] Not to be confused with the 1+3 projection tensor h_{ab}.

to first order and the definitions of the first-order metric and matter perturbations are thus gauge dependent.

Any four-dimensional scalar β is homogeneous in the background and can be written as $\beta(\tau, x^i) = \bar{\beta}(\tau) + \delta\beta(\tau, x^i)$. Under a gauge transformation (10.14),

$$\delta\tilde{\beta} = \delta\beta - \xi^0 \bar{\beta}'. \tag{10.15}$$

Physical scalars on the hypersurfaces, such as the curvature or $\delta\rho$, only depend on the choice of ξ^0, and are independent of the coordinates within the hypersurfaces, determined by ξ. The function ξ can only affect the components of 3-vectors or 3-tensors on the hypersurfaces and not 3-scalars.

Then to first order:

$$ds^2 = a^2(\tilde{\tau})\left\{ -\left[1 + 2\left(\phi - \mathcal{H}\xi^0 - \xi^{0\prime}\right)\right]d\tilde{\tau}^2 + 2\left(B + \xi^0 - \xi'\right)_{|i} d\tilde{\tau}d\tilde{x}^i \right.$$
$$- 2\left(S_i + \xi_i'\right)d\tilde{\tau}d\tilde{x}^i + \left[\left(1 - 2\{\psi + \mathcal{H}\xi^0\}\right)\gamma_{ij} + 2(E - \xi)_{|ij}\right.$$
$$\left.\left. + 2\left(F_{i|j} - \xi_{i|j}\right) + h_{ij}\right]d\tilde{x}^i d\tilde{x}^j \right\}, \tag{10.16}$$

where $\mathcal{H} = a'/a$. Thus the coordinate transformation (10.14) induces a change in the metric perturbation quantities.

$$\tilde{\phi} = \phi - \mathcal{H}\xi^0 - \xi^{0\prime}, \quad \tilde{\psi} = \psi + \mathcal{H}\xi^0, \tag{10.17}$$

$$\tilde{B} = B + \xi^0 - \xi', \quad \tilde{E} = E - \xi, \tag{10.18}$$

$$\tilde{F}_i = F_i - \xi_i, \quad \tilde{S}_i = S_i + \xi_i', \quad \tilde{h}_{ij} = h_{ij}. \tag{10.19}$$

10.2.3 Gauge-invariant quantities: metric

The two scalar gauge functions allow two of the metric scalar perturbations to be eliminated so that there should be two remaining gauge-invariant combinations. The transformations (10.17)–(10.18) show that

$$\Phi = \phi - \mathcal{H}\sigma - \sigma' \text{ where } \sigma = E' - B, \tag{10.20}$$

$$\Psi = \psi + \mathcal{H}\sigma, \tag{10.21}$$

are gauge-invariant forms of the Newtonian potential and the curvature perturbation. The quantity σ is the shear potential for constant-τ surfaces: see Exercise 10.2.3. Other gauge-invariant metric scalars can also be defined; for example (Exercise 10.2.2):

$$\mathcal{A} = \phi + \psi + \left(\frac{\psi}{\mathcal{H}}\right)', \quad \mathcal{B} = B - E' - \frac{\psi}{\mathcal{H}}, \tag{10.22}$$

$$\mathcal{Q} = \phi + \frac{1}{a}[a(v + B)]'. \tag{10.23}$$

Given the vector gauge freedom, there is one gauge-invariant metric vector perturbation, i.e. two degrees of freedom in a single transverse 3-vector. A convenient choice is

$$Q_i = S_i + F_i'. \tag{10.24}$$

Gauge transformations have no tensor mode, so that the tensor perturbation h_{ij} is automatically gauge invariant.

10.2.4 Matter perturbations and gauge-invariant quantities

The total energy–momentum tensor (5.9) in the energy frame is

$$T_{\mu\nu} = (\rho + p)\,u_\mu u_\nu + p g_{\mu\nu} + \pi_{\mu\nu}, \tag{10.25}$$

where the four-velocity,

$$u^\mu = \frac{1}{a}\frac{\mathrm{d}x^\mu}{\mathrm{d}\tau} = \bar{u}^\mu + \delta u^\mu, \tag{10.26}$$

is a linear perturbation of the background four-velocity $\bar{u}^\mu = a^{-1}\delta_0^\mu$. Using $g_{\mu\nu}u^\mu u^\nu = -1$, we find that

$$u^\mu = \frac{1}{a}\Big(1-\phi,\ v^{|i}+v^i\Big),\, u_\mu = a\Big(-1-\phi,\ v_{|i}+v_i+B_{|i}-S_i\Big). \tag{10.27}$$

Then,

$$T^0_{\ 0} = -(\rho+\delta\rho), \tag{10.28}$$

$$T^0_{\ i} = (\rho+p)\left(v_{|i}+v_i+B_{|i}-S_i\right), \tag{10.29}$$

$$T^i_{\ j} = (p+\delta p)\gamma^i_j + \pi^i_j, \tag{10.30}$$

$$a^{-2}\pi_{ij} = \Pi_{|i|j} - \tfrac{1}{3}\nabla^2\Pi\gamma_{ij} + \Pi_{(i|j)} + \Pi_{ij}, \tag{10.31}$$

where Π is the scalar potential for anisotropic stress, Π_i is the transverse vector potential for anisotropic stress, and Π_{ij} is the transverse traceless tensor mode of anisotropic stress. For convenience, we have dropped the overbars on background quantities.

The density, pressure and velocity perturbations are gauge dependent, while the scalar, vector and tensor parts of the anisotropic stress are all gauge invariant, since anisotropic stress vanishes in the background. Density and pressure are scalar quantities, which transform as in (10.15).

For $\delta\rho$, a useful gauge-invariant form is defined by

$$\Delta = \delta + \frac{\rho'}{\rho}(v+B), \tag{10.32}$$

where we have used (10.15). Other gauge-invariant density perturbations are

$$\delta\rho_\sigma = \delta\rho + \rho'(B-E'),\quad \delta\rho_\psi = \delta\rho + \frac{\rho'}{\mathcal{H}}\psi. \tag{10.33}$$

The velocity transforms as

$$\tilde{v} = v+\xi',\quad \tilde{v}^i = v^i+\xi^{i\prime}. \tag{10.34}$$

It follows that

$$\mathcal{V} = v+E', \tag{10.35}$$

is a gauge-invariant velocity potential for scalar perturbations.

10.2.5 Speed of sound and pressure perturbations

For a general medium, the effective, physical sound-speed $c_{s\,\text{eff}}$ is the propagation speed of acoustic scalar fluctuations in the rest frame, given by (see Kodama and Sasaki (1984) and Exercise 10.2.4):

$$c_{s\,\text{eff}}^2 = \left.\frac{\delta p}{\delta \rho}\right|_{\text{rf}} . \qquad (10.36)$$

In the rest frame, the medium has zero peculiar velocity and orthogonal world lines:

$$v|_{\text{rf}} = 0, \;\; B|_{\text{rf}} = 0 \;\Leftrightarrow\; u^i|_{\text{rf}} = 0 = u_i|_{\text{rf}} \;\Leftrightarrow\; T^i_0|_{\text{rf}} = 0 = T^0_i|_{\text{rf}} . \qquad (10.37)$$

This defines the comoving orthogonal gauge (or zero momentum gauge).

The pressure perturbation δp is in general composed of adiabatic and non-adiabatic parts:

$$\delta p = c_s^2 \delta \rho + \delta p_{\text{nad}}, \;\; c_s^2 := \frac{p'}{\rho'} = w + \frac{\rho}{\rho'}\, w' . \qquad (10.38)$$

where c_s is the adiabatic sound-speed (5.14). Since p_{nad} vanishes in the background, its perturbation δp_{nad} is gauge invariant.

For an adiabatic medium, such as a barotropic fluid, $\delta p_{\text{nad}} = 0$ and $c_s = c_{s\,\text{eff}}$. If $w = \text{const}$, then further we have $c_s^2 = w$. By contrast, for a non-adiabatic medium, $c_s \neq c_{s\,\text{eff}}$. An example is a scalar field φ (see Section 5.6). The rest frame is defined by the surfaces $\varphi = \text{const}$, since this is the frame where the scalar field energy–momentum tensor has perfect fluid form and zero momentum density. Thus the rest frame coincides with the uniform-field gauge, defined by $\delta\varphi = 0$. The constant field surfaces are orthogonal to the rest-frame four-velocity,

$$u^\mu|_{\text{rf}} = u^\mu|_{\delta\varphi=0} \propto \nabla^\mu \varphi , \qquad (10.39)$$

so that $\nabla_\mu \varphi$ reduces to a time derivative in this frame. Thus the kinetic energy density in the rest frame is $-\frac{1}{2}\nabla_\mu\varphi\nabla^\mu\varphi = \varphi'^2/(2a^2)$. Since $\delta\varphi = 0$ in the rest frame, we have $\delta V = 0$, where $V(\varphi)$ is the potential. The density and pressure perturbations are consequently equal in the rest frame (see (12.16) and (12.18)):

$$\delta\rho = -\dot{\varphi}^2 \phi = \delta p . \qquad (10.40)$$

Then by (10.36), the physical speed of sound is equal to the speed of light, independent of the form of $V(\varphi)$, whereas the adiabatic sound speed depends on $V(\varphi)$:

$$c_{s\,\text{eff}}^2 = 1 \text{ for any } V(\varphi), \;\; c_s^2 = 1 + \frac{2a^2 V_\varphi}{3\mathcal{H}\varphi'} . \qquad (10.41)$$

Fluid models for dark energy with constant w are at face value barotropic adiabatic models. But if we treat the dark energy strictly as an adiabatic fluid, then the sound speed c_s would be imaginary ($c_s^2 = w < 0$), leading to instabilities in the dark energy. In order to fix this problem, it is necessary to impose $c_{s\,\text{eff}}^2 > 0$ by hand, and it is natural to adopt the scalar field value (10.41). Then the dark energy fluid is non-adiabatic.

We can find a useful relation for the non-adiabatic pressure perturbation by making a gauge transformation, $x^{\mu} \to x^{\mu} + (\delta\tau, \delta x^i)$, from the rest frame gauge to a general gauge. This leads to

$$v + B = (v + B)\big|_{\text{rf}} + \delta\tau, \quad \delta p = \delta\big|_{\text{rf}} - p'\delta\tau, \quad \delta\rho = \delta\rho\big|_{\text{rf}} - \rho'\delta\tau. \tag{10.42}$$

Thus $\delta\tau = v + B$, and substituting into the pressure and density fluctuations, we obtain

$$\delta p = c_s^2 \delta\rho + \left(c_{s\,\text{eff}}^2 - c_s^2\right)\left[\delta\rho + \rho'\left(v + B\right)\right], \tag{10.43}$$

$$\delta p_{\text{nad}} = \left(c_{s\,\text{eff}}^2 - c_s^2\right)\rho\Delta, \tag{10.44}$$

where we have used (10.32).

10.2.6 Gauge-invariant quantities: curvature

The intrinsic spatial curvature on constant-τ surfaces is

$$^3R = \frac{6K}{a^2} + \frac{12K}{a^2}\psi + \frac{4}{a^2}\nabla^2\psi, \tag{10.45}$$

so that the perturbed 3-Ricci scalar is

$$\delta^3R = \frac{12K}{a^2}\psi + \frac{4}{a^2}\nabla^2\psi. \tag{10.46}$$

Thus the metric perturbation ψ determines the curvature of perturbed $\tau = \text{const}$ surfaces. δ^3R is gauge invariant for a flat FLRW background, since in that case ^3R vanishes in the background. However, Ψ is more useful, and is gauge invariant for flat and non-flat backgrounds.

Other gauge-invariant curvature perturbations are also useful – especially those which are conserved under certain broad conditions. Two such quantities are:

$$\mathcal{R} = \psi - \mathcal{H}(v + B), \tag{10.47}$$

$$\zeta = -\psi - \mathcal{H}\frac{\delta\rho}{\rho'}. \tag{10.48}$$

$\mathcal{R}$ coincides with the curvature perturbation in the comoving $v = 0$, orthogonal ($B = 0$) gauge (see (10.60) below), and $-\zeta$ coincides with the curvature perturbation on uniform-density hypersurfaces [(10.59)]. The explicitly gauge-invariant relation between these quantities follows on using (10.32):

$$\zeta = -\mathcal{R} - \frac{\mathcal{H}\rho}{\rho'}\Delta. \tag{10.49}$$

The generalized Poisson equation (10.74) shows that

$$\mathcal{R} = -\zeta \quad \text{on super-Hubble scales.} \tag{10.50}$$

The perturbed energy conservation equation (10.72) shows that

$$\zeta' = -\tfrac{1}{3}\nabla^2\mathcal{V} - \mathcal{H}\frac{\delta p_{\text{nad}}}{\rho + p}. \tag{10.51}$$

Combining these two results, we have the important result:

$$\zeta' = 0 = \mathcal{R}' \text{ on super-Hubble scales for adiabatic modes,} \qquad (10.52)$$

since $\delta p_{\mathrm{nad}} = 0$ and $\nabla^2 \mathcal{V}$ may be neglected.

The perturbed $(0i)$ field equation (10.69) in gauge-invariant form gives an explicitly gauge-invariant formula for $\mathcal{R}$:

$$\mathcal{R} = \Psi + \frac{2}{3} \frac{(\Psi' + \mathcal{H}\Phi)}{(1+w)\mathcal{H}}. \qquad (10.53)$$

This is very useful for relating the conserved curvature perturbation to the metric potentials, especially in the case of vanishing anisotropic stress, when $\Psi = \Phi$ (see below):

$$\mathcal{R} = \Phi + \frac{2}{3} \frac{\Phi' + \mathcal{H}\Phi}{(1+w)\mathcal{H}} \text{ when } \Pi = 0. \qquad (10.54)$$

This enables us to relate the amplitude of the Newtonian potential for perturbations re-entering the Hubble horizon during the radiation- or matter-dominated eras, to the amplitude of the curvature perturbation at horizon exit during inflation.

10.2.7 Specific gauges

Although we can work with gauge-invariant quantities, it may also often be convenient to choose a particular gauge. Confining attention to scalar perturbations, some of the gauges are as follows:

Newtonian or longitudinal gauge:

This is the gauge in which the metric is diagonal, so that

$$E = 0 = B \Rightarrow \phi = \Phi, \ \psi = \Psi, \ \delta\rho = \delta\rho_\sigma, \ v = \mathcal{V}, \qquad (10.55)$$

where the gauge-invariant $\delta\rho_\sigma$ and $\mathcal{V}$ are defined in (10.33) and (10.35). In addition, the shear of constant-τ surfaces, defined in (10.20), vanishes,

$$\sigma = -B + E' = 0. \qquad (10.56)$$

This gauge is closest to the Newtonian equations on small scales.

The extension of the gauge to include vector perturbations is called the Poisson gauge, which adds the condition $S_i = 0$.

Flat (or uniform curvature) gauge:

In this gauge the $\tau = \text{const}$ surfaces are unperturbed:

$$\psi = 0 = E \Rightarrow \phi = \mathcal{A}, \ B = \mathcal{B}, \ \delta\rho = \delta\rho_\psi, \ v = \mathcal{V}, \qquad (10.57)$$

where the gauge-invariant quantities $\mathcal{A}, \mathcal{B}$ and $\delta\rho_\psi$ are defined in (10.22) and (10.33). For a scalar field in the flat gauge, the field perturbation coincides with the gauge-invariant

Sasaki–Mukhanov variable Q:

$$\delta\varphi = Q := \delta\varphi + \frac{\varphi'}{\mathcal{H}}\psi. \tag{10.58}$$

Uniform density gauge:

The constant-τ hypersurfaces have unperturbed (total) density:

$$\delta\rho = 0 \;\Rightarrow\; \psi = -\zeta. \tag{10.59}$$

Comoving orthogonal gauge:

The fluid four-velocity (10.27) is comoving and normal to the constant-τ hypersurfaces, so that

$$B = 0 = v \;\Rightarrow\; \phi = Q, \;\; \psi = \mathcal{R}, \;\; \delta\rho = \rho\Delta, \tag{10.60}$$

where the gauge-invariant Q, $\mathcal{R}$ and Δ are defined respectively by (10.23), (10.47) and (10.32).

Synchronous gauge:

The metric has no perturbations in its time components:

$$\phi = 0 = B. \tag{10.61}$$

This simplifies the time evolution equations and hence is used in the CMB Boltzmann codes such as CMBFAST and CAMB (see Chapter 11). However, this does not determine the time-slicing unambiguously – there is a residual gauge freedom $\hat{\xi}^0 = C(x^i)/a$, and it is not possible to define gauge-invariant quantities in general using this gauge condition. In the Boltzmann codes, the residual ambiguity is removed by setting the CDM velocity to zero.

10.2.8 Perturbed Einstein and conservation equations: scalars

The first-order perturbed Einstein equations $\delta G^\mu_\nu = 8\pi G \delta T^\mu_\nu$ for scalar modes give two constraint and two evolution equations. In a general gauge, the (00) (energy) and (0i) (momentum) constraints are:

$$\left(\nabla^2 + 3K\right)\psi - 3\mathcal{H}(\psi' + \mathcal{H}\phi) + \mathcal{H}\nabla^2\sigma = 4\pi G a^2 \delta\rho, \tag{10.62}$$

$$\psi' + \mathcal{H}\phi + K\sigma = -4\pi G a^2(\rho + p)(v + B). \tag{10.63}$$

The (ij) evolution equations are:

$$\psi'' + 2\mathcal{H}\psi' - K\psi + \mathcal{H}\phi' + (2\mathcal{H}' + \mathcal{H}^2)\phi = 4\pi G a^2\left(\delta p + \tfrac{2}{3}\nabla^2\Pi\right), \tag{10.64}$$

$$\sigma' + 2\mathcal{H}\sigma - \phi + \psi = 8\pi G a^2\Pi. \tag{10.65}$$

(Recall that $\sigma = E' - B$.)

The perturbed conservation equations $\delta \nabla^\nu T_{\mu\nu} = 0$ yield evolution equations for the density perturbation and momentum,

$$\delta\rho' + 3\mathcal{H}(\delta\rho + \delta p) = (\rho + p)\left[3\psi' - \nabla^2(v + E')\right], \tag{10.66}$$

$$[(\rho + p)(v + B)]' + \delta p + \tfrac{2}{3}\left(\nabla^2 + 3K\right)\Pi = -(\rho + p)[\phi + 4\mathcal{H}(v + B)]. \tag{10.67}$$

We can re-express the perturbation equations (10.62)–(10.67) in terms of gauge-invariant variables. For a flat background ($K = 0$):

$$-\nabla^2\Psi + 3\mathcal{H}(\Psi' + \mathcal{H}\Phi) = -4\pi G a^2 \delta\rho_\sigma = 3(\mathcal{H}' - \mathcal{H}^2)(\Psi + \zeta), \tag{10.68}$$

$$\Psi' + \mathcal{H}\Phi = -4\pi G a^2(\rho + p)\mathcal{V} = \frac{\mathcal{H}' - \mathcal{H}^2}{\mathcal{H}}(\Psi - \mathcal{R}), \tag{10.69}$$

$$\Psi'' + \mathcal{H}(2 - 3c_s^2)\Psi' + \mathcal{H}\Phi' + [2\mathcal{H}' + (1 - 3c_s^2)\mathcal{H}^2]\Phi$$
$$= 4\pi G a^2\left(c_s^2\rho\Delta + \delta p_{\text{nad}} + \tfrac{2}{3}\nabla^2\Pi\right), \tag{10.70}$$

$$\Psi - \Phi = 8\pi G a^2\Pi, \tag{10.71}$$

$$\zeta' = -\tfrac{1}{3}\nabla^2\mathcal{V} - \mathcal{H}\frac{\delta p_{\text{nad}}}{\rho + p}, \tag{10.72}$$

$$\mathcal{V}' + \mathcal{H}\mathcal{V} + \Phi = \frac{-1}{\rho + p}\left(c_s^2\rho\Delta + \delta p_{\text{nad}} + \tfrac{2}{3}\nabla^2\Pi\right). \tag{10.73}$$

Note that (10.69) has been used to derive (10.70).

Combining (10.68) and (10.69) we arrive at a gauge-invariant generalization of the Newtonian Poisson equation:

$$\nabla^2\Psi = 4\pi G a^2\rho\Delta = 3(\mathcal{H}^2 - \mathcal{H}')(\zeta + \mathcal{R}). \tag{10.74}$$

We can also derive an evolution equation for Δ when $p = 0 = \Pi$:

$$\Delta'' + \mathcal{H}\Delta' - 4\pi G a^2\rho\Delta = 0. \tag{10.75}$$

10.2.9 Perfect fluid scalar modes: solutions

In the case of adiabatic perturbations ($\delta p_{\text{nad}} = 0$) with $K = 0$ and vanishing anisotropic stress ($\Pi = 0 \Leftrightarrow \Psi = \Phi$), we can derive a second-order evolution equation for the Newtonian potential,

$$\Phi'' + 3\mathcal{H}(1 + c_s^2)\Phi' + \left[2\mathcal{H}' + (1 + 3c_s^2)\mathcal{H}^2 - c_s^2\nabla^2\right]\Phi = 0. \tag{10.76}$$

Using (10.54) for the comoving curvature perturbation, this evolution equation may be rewritten in the form

$$\mathcal{R}' = \frac{2c_s^2}{3(1+w)\mathcal{H}}\nabla^2\Phi, \tag{10.77}$$

showing again the conservation of $\mathcal{R}$ on large scales.

For a perfect fluid with $w = \text{const}$, the scale factor evolves as

$$a \propto \tau^\nu, \quad \nu = \frac{2}{1+3w}, \quad c_s^2 = w = \text{const}. \tag{10.78}$$

Since the equations are linear, one can express each quantity Q in terms of functions harmonic on the spatial sections. For $K = 0$ this means a Fourier decomposition using wave vectors $\mathbf{k}$,

$$Q = \frac{1}{(2\pi)^3}\int Q_k \exp(i\mathbf{k}\cdot\mathbf{x})\mathrm{d}^3k, \tag{10.79}$$

where the inverse integral is taken over the spatial sections, presumed to be infinite, i.e. this formulation implies an assumption about the behaviour beyond the visual horizon of Section 7.9 (see MacCallum (1982)). Then (10.76) can be written in Fourier space, using $k := |\mathbf{k}|$, as,

$$\frac{\mathrm{d}^2 F}{\mathrm{d}x^2} + \frac{2}{x}\frac{\mathrm{d}F}{\mathrm{d}x} + \left[c_s^2 - \frac{\nu(\nu+1)}{x^2}\right]F = 0, \quad F := x^\nu\Phi_k, \ x := k\tau. \tag{10.80}$$

This is a spherical Bessel equation, leading to the general solution

$$\Phi_k = -\tfrac{3}{2}\nu^2 x^{-\nu} Z_\nu(c_s x), \quad Z_\nu := A j_\nu + B n_\nu, \tag{10.81}$$

where Z_ν denotes a linear combination of the spherical Bessel functions.

Now we can use the Poisson equation (10.74) to find the gauge-invariant comoving density perturbation, and (10.69) to find the gauge-invariant velocity perturbation:

$$\Delta_k = x^{2-\nu} Z_\nu(c_s x), \tag{10.82}$$

$$kV_k = -\tfrac{3}{4}x^{1-\nu}\left[Z_\nu(c_s x) - \frac{c_s x}{(\nu+1)}Z_{\nu-1}(c_s x)\right]. \tag{10.83}$$

Using the asymptotic behaviour of the spherical Bessel functions, these results lead to the large-scale ($c_s x \ll 1$, i.e. wavelength much greater than the acoustic horizon) and small-scale ($c_s x \gg 1$) solutions in Table 10.1.

Note that in the case of matter ($c_s = 0$), only the $c_s x \ll 1$ solutions in Table 10.1 are relevant.

It follows from Table 10.1 that the non-decaying mode of the potential is constant on super-acoustic scales (or on all scales for matter domination):

$$\Phi = \text{const when } c_s k\tau \ll 1. \tag{10.84}$$

In the radiation era, $\nu = 1$, while $\nu = 2$ for the matter era. Thus on large scales, the non-decaying mode of the density and velocity perturbations evolve during radiation domination as

$$\Delta_\mathrm{r} \propto a^2, \quad kV_\mathrm{r} \propto a, \tag{10.85}$$

Table 10.1 Solutions for the gauge-invariant potential, density and velocity perturbations (from Peter and Uzan (2009)).

	$c_s x \ll 1$ large scales	$c_s x \gg 1$ small scales
Φ_k	$\Phi_+ + \Phi_- x^{-1-2\nu}$	$\Phi_+ x^{-1-\nu} \cos[c_s x - \pi(\nu+1)/2]$
Δ_k	$\dfrac{-2(\Phi_+ x^2 - \Phi_- x^{1-2\nu})}{3\nu^2}$	$\dfrac{-2\Phi_+ x^{1-\nu} \cos[c_s x - \pi(\nu+1)/2]}{3\nu^2}$
$k\mathcal{V}_k$	$\dfrac{-2[\Phi_+ x - \Phi_-(1+\nu^{-1})x^{-2\nu}]}{3\nu(1+w)}$	$\dfrac{-2\Phi_+ x^{1-\nu} \cos[c_s x - \pi(\nu+1)/2]}{3\nu(1+w)}$

while during matter domination,

$$\Delta_{\mathrm{m}} \propto a, \ k\mathcal{V}_{\mathrm{m}} \propto \sqrt{a}, \tag{10.86}$$

on all scales.

10.2.10 Vector and tensor perturbations

The transverse momentum density for vector perturbations is

$$q_i = (\rho + p)(v_i - S_i), \tag{10.87}$$

and it satisfies the momentum conservation equation,

$$q_i' + 4\mathcal{H}q_i = -\left(\nabla^2 + 2K\right)\Pi_i, \tag{10.88}$$

where Π_i is the transverse vector potential for anisotropic stress. We see that we need non-zero Π_i to source q_i.

The gauge-invariant metric vector perturbation $Q_i = S_i + F_i'$ satisfies the $(0i)$ constraint and (ij) evolution equations:

$$\left(\nabla^2 + 2K\right)Q_i = -16\pi Ga^2 q_i, \tag{10.89}$$

$$Q_i' + 2\mathcal{H}Q_i = 8\pi Ga^2 \Pi_i. \tag{10.90}$$

The first equation can be thought of as the 'vector Poisson equation'. If $q_i = 0$ – as in the case of a scalar field – then $Q_i = 0$, i.e. there are *no* vector perturbations. Thus vector perturbations need to be actively sourced via anisotropic stress – otherwise they are zero or purely decaying. Examples of active sources are magnetic fields and topological defects.

Tensor perturbations satisfy the evolution equation (from the (ij) field equation)

$$h_{ij}'' + 2\mathcal{H}h_{ij}' + \left(2K - \nabla^2\right)h_{ij} = 8\pi Ga^2 \Pi_{ij}, \tag{10.91}$$

where Π_{ij} is the transverse traceless anisotropic stress. Tensor modes are therefore sourced or damped by anisotropic stress. In the absence of this stress, they evolve freely under gravity. Equation (10.91) is a wave equation, and it confirms that gravitational waves propagate at the speed of light.

For $K = 0$ and a perfect fluid with $w = \text{const}$ (see Exercise 10.2.9), the solutions are

$$h_{ij} = (k\tau)^{-\nu+1/2}\left[\alpha^+_{ij}J_{\nu-1/2}(k\tau) + \alpha^-_{ij}N_{\nu-1/2}(k\tau)\right]. \tag{10.92}$$

It follows that the non-decaying mode is constant on super-Hubble scales.

Exercise 10.2.1 Prove the gauge transformation formulae (10.16)–(10.18).

Exercise 10.2.2 Verify that (10.22), (10.23) and (10.33) are gauge invariant.

Exercise 10.2.3 Show that the unit time-like vector field orthogonal to constant-τ hypersurfaces is

$$N_\mu = a(-1 - \phi, 0),\ N^\mu = \frac{1}{a}(1 - \phi, -B^{|i} + S^i). \tag{10.93}$$

Define the shear in the usual way by decomposing $N_{\mu;\nu}$. Show that for scalar perturbations,

$$\sigma^N_{ij} = a\left(\sigma_{|ij} - \tfrac{1}{3}\gamma_{ij}\sigma^k_{|k}\right), \tag{10.94}$$

where the scalar shear potential is given in (10.20). Show that for vector and tensor perturbations, respectively,

$$\sigma^N_{ij} = a\left[S_{(i|j)} + F'_{(i|j)}\right],\ \sigma^N_{ij} = \tfrac{1}{2}ah'_{ij}. \tag{10.95}$$

Exercise 10.2.4 On small scales, where we can neglect the Hubble expansion, (10.76) reduces to $\Phi'' - c_s^2\nabla^2\Phi = 0$. This shows that c_s, defined in (5.14), is indeed the propagation speed of scalar fluctuations in an adiabatic fluid. Now generalize (10.76) to the non-adiabatic case, and thus verify that $c_{s\,\text{eff}}$, defined in (10.36) and satisfying (10.44), is indeed the propagation speed of scalar fluctuations, i.e. the effective sound-speed.

Exercise 10.2.5 Derive (10.45) for the 3-Ricci scalar of the $\tau = \text{const}$ hypersurfaces.

Exercise 10.2.6 For a scalar field, show that

$$\delta p_{\text{nad}} = -\frac{2a^2V_\varphi}{3\mathcal{H}\varphi'}\delta\rho. \tag{10.96}$$

Exercise 10.2.7 Verify (10.51) and (10.53).

Exercise 10.2.8 Use (10.62)–(10.67) to derive (10.68)–(10.75).

Exercise 10.2.9 Verify the results in Section 10.2.9 for scalar modes, and the solution (10.92) for tensor modes.

10.3 Covariant nonlinear perturbations

The 1+3 covariant approach developed in Chapters 4–6 is well suited to a gauge-invariant analysis of perturbations. This is based on early work by Hawking (1966), Lyth and

Mukherjee (1988) and Ellis and Bruni (1989), subsequently systematized by Ellis, Bruni, Dunsby and co-workers. Various generalizations and improvements, as well as further references, may be found in the review articles by van Elst and Ellis (1998) and Tsagas, Challinor and Maartens (2008), which we shall draw on. A key difference between the covariant and gauge-invariant (CGI) approach and the standard approach is that *CGI starts from the fully nonlinear equations, rather than from the background.* Thus we begin with the nonlinear case and then linearize, and this gives some advantages when considering nonlinear questions.

In the CGI approach, we first choose a fundamental 4-velocity, with comoving observers who will measure the physical quantities in the universe. There are various physically motivated choices of u^a, and a change in choice, $u^a \to \tilde{u}^a$ leads to a transformation in the frame-dependent physical quantities measured by observers (see Sections 5.1.1, 5.3).

For a given choice of u^a, we can perform a covariant 1+3 splitting of all physical and geometrical quantities, as in Chapters 4 and 5. All such quantities may be described by PSTF vectors and tensors:

$$V_a = V_{\langle a \rangle}, \quad S_{ab} = S_{\langle ab \rangle}. \tag{10.97}$$

Higher-rank PSTF tensors are needed in kinetic theory, as discussed in Section 5.4. Spatial inhomogeneities relative to u^a observers are not described by scalars (as in the standard approach), but by the spatially projected gradients of scalars. A key such variable is the comoving fractional gradient in the energy density,

$$\Delta_a = \frac{a}{\rho} \overline{\nabla}_a \rho, \quad a \equiv \ell, \tag{10.98}$$

where for later convenience we replace ℓ, defined in (4.35), by a. This gradient vanishes in spacetimes with homogeneous spatial sections, and thus satisfies the Stewart–Walker lemma for gauge–invariance (based on (10.6)). The second key quantity is the comoving volume-expansion gradient,

$$\mathcal{Z}_a = a \overline{\nabla}_a \Theta, \tag{10.99}$$

which gives a CGI description of velocity perturbations.

To deal with the evolution of projected gradients we use the identity (4.62):

$$(\overline{\nabla}_a f)^{\cdot} = \overline{\nabla}_a \dot{f} + (\dot{u}^b \overline{\nabla}_b f) u_a + \dot{u}_a \dot{f} - \tfrac{1}{3} \Theta \overline{\nabla}_a f - \sigma_{ab} \overline{\nabla}^b f + \eta_{abc} \omega^b \overline{\nabla}^c f. \tag{10.100}$$

10.3.1 Fluids

The comoving gradient of the energy conservation equation, using momentum conservation to eliminate the pressure gradient, and using the commutation identity (10.100) for time and spatial derivatives, leads to the evolution equation (Exercise 10.3.1)

$$\dot{\Delta}_{\langle a \rangle} = w\Theta \Delta_a - (1+w)\, \mathcal{Z}_a + \frac{a\Theta}{\rho}\left(\dot{q}_{\langle a \rangle} + \tfrac{4}{3}\Theta q_a\right) + \frac{a\Theta}{\rho}\left(\sigma_{ab} + \omega_{ab}\right)q^b$$

$$+ \frac{a\Theta}{\rho}\overline{\nabla}^b \pi_{ab} - (\sigma_{ba} + \omega_{ba})\Delta^b - \frac{a}{\rho}\overline{\nabla}_a\left(2\dot{u}^b q_b + \sigma^{bc}\pi_{bc}\right) - \frac{a}{\rho}\overline{\nabla}_a\overline{\nabla}^b q_b$$

$$+ \frac{a\Theta}{\rho}\pi_{ab}\dot{u}^b + \frac{1}{\rho}\left(\overline{\nabla}^b q_b + \dot{u}^b q_b + \sigma^{bc}\pi_{bc}\right)(\Delta_a - a\dot{u}_a). \tag{10.101}$$

Even if Δ_a is initially zero, it will become non-zero due to the various sources in this equation. One important source is $\mathcal{Z}_a$, whose evolution follows from the comoving gradient of the Raychaudhuri equation (Exercise 10.3.1):

$$\dot{\mathcal{Z}}_{\langle a \rangle} = -\tfrac{2}{3}\Theta \mathcal{Z}_a - 4\pi G\left[(1+3c_s^2)\rho\Delta_a + 3a\Gamma_a\right] - \tfrac{1}{2}a\left(\Theta^2 - \dot{\Theta}\right)\dot{u}_a$$

$$+ a\overline{\nabla}_a\overline{\nabla}^b\dot{u}_b - (\sigma_{ba} + \omega_{ba})\mathcal{Z}^b - 2a\overline{\nabla}_a\left(\sigma^2 - \omega^2\right) + 2a\dot{u}^b\overline{\nabla}_a\dot{u}_b$$

$$- a\left[2\left(\sigma^2 - \omega^2\right) - \overline{\nabla}^b\dot{u}_b - \dot{u}^b\dot{u}_b\right]\dot{u}_a. \tag{10.102}$$

Here $w = p/\rho$ and we have written the comoving pressure gradient in terms of its adiabatic and non-adiabatic parts:

$$a\overline{\nabla}_a p = c_s^2\rho\Delta_a + a\Gamma_a,\ \ \Gamma_a = \overline{\nabla}_a p_{\mathrm{nad}} = pa^{-1}\mathcal{E}_a, \tag{10.103}$$

where c_s is the adiabatic sound speed, and the dimensionless entropy gradient, $\mathcal{E}_a$ may be used in place of the gradient of the non-adiabatic pressure Γ_a. This is the covariant analogue of (10.38).

If we choose u^a as the energy frame 4-velocity, then we can set $q_a = 0$ in (10.101). For a perfect fluid or scalar field, we can also set $\pi_{ab} = 0$, so that (10.101) simplifies to

$$\dot{\Delta}_{\langle a \rangle} = w\Theta \Delta_a - (1+w)\,\mathcal{Z}_a - (\sigma_{ba} + \omega_{ba})\Delta^b. \tag{10.104}$$

In (10.102) we can set $\Gamma_a = 0$ if the fluid is adiabatic.

In the metric-based perturbative formalism, the curvature perturbation is conserved for the adiabatic growing mode on super-Hubble scales, as shown by (10.51). The geometric interpretation of this is via the perturbation δN of the expansion e-folds, $N = \ln a$, using the so-called 'separate universe' picture (Starobinsky, 1985, Wands *et al.*, 2000), which can be applied at second and higher orders of perturbation. A covariant version of this result is based on defining an appropriate spatial-gradient quantity, and leads to a simple geometric nonlinear conserved quantity for a perfect fluid (Langlois and Vernizzi, 2005).

Along each fluid particle world line, we can define a covariant e-fold function,

$$\alpha = \tfrac{1}{3}\int \Theta\, dt, \tag{10.105}$$

where t is proper time. Applying the identity (4.62) for commuting time and space derivatives,

$$\tfrac{1}{3}\overline{\nabla}_a\Theta = \mathcal{L}_u\overline{\nabla}_a\alpha - \dot{\alpha}\dot{u}_a, \tag{10.106}$$

where $\mathcal{L}_u$ is the Lie derivative along u_a, i.e. $\mathcal{L}_u(\overline{\nabla}_a f) = \overline{\nabla}_a \dot{f} - \dot{f} \dot{u}_a$. The projected gradient of the energy conservation law gives

$$\mathcal{L}_u(\overline{\nabla}_a \rho) + 3(\rho + p)\mathcal{L}_u(\overline{\nabla}_a \alpha) + \Theta \overline{\nabla}_a(\rho + p) = 0. \tag{10.107}$$

Using (10.103) and defining

$$\zeta_a := \overline{\nabla}_a \alpha - \frac{\dot{\alpha}}{\dot{\rho}}\overline{\nabla}_a \rho, \tag{10.108}$$

this becomes

$$\mathcal{L}_u \zeta_a = -\frac{\Theta}{(\rho + p)}\Gamma_a. \tag{10.109}$$

For adiabatic perturbations, $\Gamma_a = 0$ and

$$\mathcal{L}_u \zeta_a = 0, \tag{10.110}$$

so that ζ_a is a conserved quantity, in the adiabatic case, on all scales and at all perturbative orders. This is the CGI analogue of the metric-based curvature perturbation ζ.

10.3.2 Multiple perfect fluids

For a mixture of interacting perfect fluids (see (5.70)),

$$\nabla_b T_I^{ab} = Q_I^a, \quad \sum_I Q_I^a = 0, \tag{10.111}$$

where Q_I^a are the rates of energy–momentum density exchange. Defining the energy exchange $\mathcal{Q}_I = -Q_I^a u_a$ and momentum exchange $\mathcal{Q}_I^a = h^a{}_b Q_I^b$ in the fundamental (u^a) frame, (10.111) leads to

$$\dot{\rho}_I = -\Theta(\rho_I + p_I) - \overline{\nabla}_a q_I^a - 2\dot{u}_a q_I^a + \mathcal{Q}_I, \tag{10.112}$$

$$(\rho_I + p_I)\dot{u}^a = -\frac{c_{sI}^2 \rho_I}{a}\Delta_I^a - \frac{p_I}{a}\mathcal{E}_I^a$$
$$-\dot{q}_I^{\langle a \rangle} - \tfrac{4}{3}\Theta q_I^a + (\sigma^a{}_b + \omega^a{}_b)q_I^b + \mathcal{Q}_I^a, \tag{10.113}$$

where $c_{sI}^2 = \dot{p}_I / \dot{\rho}_I$, $q_I^a = (\rho_I + p_I)v_I^a$ (see Section 5.3) and $\mathcal{E}_I^a$ is the entropy gradient of the I-fluid. Note that

$$\sum_I \mathcal{Q}_I = 0 = \sum_I \mathcal{Q}_I^a. \tag{10.114}$$

Density inhomogeneity in the I-fluid, relative to the u_a-frame, is described by

$$\Delta_I^a = \frac{a}{\rho_I}\overline{\nabla}^a \rho_I. \tag{10.115}$$

Taking the time derivative of Δ_I^a and using the conservation laws (10.112) and (10.113), we find

$$
\dot{\Delta}_I^{\langle a \rangle} = w_I \Theta \Delta_I^a - (1 + w_I) \, \mathcal{Z}^a + \frac{a}{\rho_I} \Theta \left(\dot{q}_I^{\langle a \rangle} + \tfrac{4}{3} \Theta q_I^a \right) - \frac{a}{\rho_I} \overline{\nabla}^a \left(\overline{\nabla}_b q_I^b - \mathcal{Q}_I \right)
$$

$$
- \left(\sigma_b{}^a + \omega_b{}^a \right) \Delta_I^b - \frac{2a}{\rho} \overline{\nabla}^a \left(\dot{u}_b q_I^b \right) + \frac{a\Theta}{\rho_I} \left(\sigma^a{}_b + \omega^a{}_b \right) q_I^b
$$

$$
+ \frac{1}{\rho} \left(\overline{\nabla}_b q_I^b + 2 \dot{u}_b q_I^b - \mathcal{Q}_I \right) \left(\Delta_I^a - a \dot{u}^a \right) - \frac{a\Theta}{\rho_I} \mathcal{Q}_I^a . \tag{10.116}
$$

When the interaction term is specified, this describes the propagation of spatial inhomogeneities in the density distribution of the I-species. The nonlinear evolution of $\mathcal{Z}_a$ is governed by (10.102).

10.3.3 Magnetized fluids

General features of relativistic magnetohydrodynamics were discussed in Section 5.5.4.

Inhomogeneity associated with a magnetic field may be described via the comoving fractional gradient of magnetic energy density,

$$
\mathcal{B}_a = \frac{a}{B^2} \overline{\nabla}_a B^2 . \tag{10.117}
$$

In the presence of magnetic fields, the nonlinear evolution of spatial inhomogeneities in the density distribution of a single, highly conducting perfect fluid is described by (Exercise 10.3.2)

$$
\dot{\Delta}_{\langle a \rangle} = w \Theta \Delta_a - (1 + w) \, \mathcal{Z}_a + \frac{a\Theta}{\rho} \eta_{abc} B^b \text{curl } B^c
$$

$$
+ \tfrac{2}{3} c_A^2 a \Theta \dot{u}_a - (\sigma_{ba} + \omega_{ba}) \Delta^b + \frac{a\Theta}{\rho} \pi_{ab}^B \dot{u}^b , \tag{10.118}
$$

where $\pi_B^{ab} = -B^{\langle a} B^{b \rangle}$ is the magnetic anisotropic stress, and the Alfvén speed c_A is defined by

$$
c_A^2 := \frac{B^2}{\rho} . \tag{10.119}
$$

The nonlinear evolution equation for the expansion gradients is

$$
\dot{\mathcal{Z}}_{\langle a \rangle} = -\tfrac{2}{3} \Theta \mathcal{Z}_a - 4\pi G \left[\rho \Delta_a + B^2 \mathcal{B}_a \right] + 12\pi G a \eta_{abc} B^b \text{curl } B^c
$$

$$
+ a \overline{\nabla}_a \overline{\nabla}^b \dot{u}_b + 2a \dot{u}^b \overline{\nabla}_a \dot{u}_b + \left[\tfrac{1}{2} {}^3 R - 3 \left(\sigma^2 - \omega^2 \right) + \overline{\nabla}^b \dot{u}_b + \dot{u}_b \dot{u}^b \right] a \dot{u}_a
$$

$$
- (\sigma_{ba} + \omega_{ba}) \mathcal{Z}^b + 12\pi G a \pi_{ab}^B \dot{u}^b - 2a \overline{\nabla}_a \left(\sigma^2 - \omega^2 \right) . \tag{10.120}
$$

Finally, the nonlinear propagation of inhomogeneities in the magnetic energy density is given by (Exercise 10.3.2)

$$
\dot{\mathcal{B}}_{\langle a \rangle} = \frac{4}{3(1+w)} \dot{\Delta}_{\langle a \rangle} - \frac{4w\Theta}{3(1+w)} \Delta_a - \frac{4a\Theta}{3\rho(1+w)} \eta_{abc} B^b \operatorname{curl} B^c
$$

$$
- \frac{4}{3} a\Theta \left[1 + \frac{2B^2}{3\rho(1+w)} \right] \dot{u}_a - (\sigma_{ba} + \omega_{ba}) \mathcal{B}^b
$$

$$
+ \frac{4}{3(1+w)} (\sigma_{ba} + \omega_{ba}) \Delta^b - \frac{4a\Theta}{3\rho(1+w)} \pi_{ab}^B \dot{u}^b - \frac{2a}{B^2} \pi_B^{bc} \overline{\nabla}_a \sigma_{bc}
$$

$$
- \frac{2a}{B^2} \sigma^{bc} \overline{\nabla}_a \pi_{bc}^B + \frac{2}{B^2} \sigma_{bc} \pi_B^{bc} \mathcal{B}_a - \frac{2a}{B^2} \sigma_{bc} \pi_B^{bc} \dot{u}_a . \tag{10.121}
$$

This follows on using (5.134) and (10.120).

10.3.4 Scalar fields

The basic general properties of scalar fields are given in Section 5.6. The canonical 4-velocity – in which the energy–momentum tensor takes perfect fluid form – is orthogonal to $\varphi = \text{const}$ surfaces, so that

$$
\overline{\nabla}_a \varphi = 0 \;\Rightarrow\; \Delta_a \equiv \frac{a}{\rho_\varphi} \overline{\nabla}_a \rho_\varphi = \frac{a\dot{\varphi}}{\rho_\varphi} \overline{\nabla}_a \dot{\varphi} . \tag{10.122}
$$

This is like a (nonlinear) CGI version of the standard uniform-field gauge.

Using (5.151), we may adapt the equations (10.101) and (10.102) to a scalar field:

$$
\dot{\Delta}_{\langle a \rangle} = w\Theta \Delta_a - (1+w) \mathcal{Z}_a - (\sigma_{ba} + \omega_{ba}) \Delta^b , \tag{10.123}
$$

$$
\dot{\mathcal{Z}}_{\langle a \rangle} = -\tfrac{2}{3}\Theta \mathcal{Z}_a - 16\pi G \rho_\varphi \Delta_a - \frac{a}{3} \left(\Theta^2 - \dot{\Theta} \right) \dot{u}_a + a \overline{\nabla}_a \overline{\nabla}^b \dot{u}_b
$$

$$
- (\sigma_{ba} + \omega_{ba}) \mathcal{Z}^b - 2a \overline{\nabla}_a \left(\sigma^2 - \omega^2 \right) + 2a \dot{u}^b \overline{\nabla}_a \dot{u}_b
$$

$$
- a \left[2 \left(\sigma^2 - \omega^2 \right) - \overline{\nabla}^b \dot{u}_b - \dot{u}^b \dot{u}_b \right] \dot{u}_a . \tag{10.124}
$$

Finally, combining (5.148) and (5.151),

$$
\dot{u}_a = - \frac{\rho_\varphi}{a(\rho_\varphi + p_\varphi)} \Delta_a . \tag{10.125}
$$

Exercise 10.3.1 Using the identities in Section 4.8, derive (10.101), (10.102) and (10.109).

Exercise 10.3.2 Derive (10.118) and (10.121). (See Barrow, Maartens and Tsagas (2007).)

10.4 Covariant linear perturbations

In order to linearize the nonlinear equations of the previous section and Chapter 6, we first characterize the limiting spacetime, i.e. the unperturbed (zero-order) FLRW background.

- The first requirement is that the fundamental 4-velocity u^a should smoothly tend to the unique FLRW 4-velocity in the limit. This requires that u^a is chosen in a unique and invariant way – which is typically achieved by a physical criterion (e.g. choosing u^a as the energy-frame 4-velocity).
- Then a covariant characterization of the FLRW background is:
  energy–momentum: $\overline{\nabla}_a \rho = 0 = \overline{\nabla}_a p$, $q_a = 0$, $\pi_{ab} = 0$;
  kinematics: $\overline{\nabla}_a \Theta = 0$, $\dot{u}_a = 0 = \omega_a$, $\sigma_{ab} = 0$;
  curvature: $E_{ab} = 0 = H_{ab}$, $^3R_{\langle ab \rangle} = 0$;
  dynamics: $\Theta = 3H$, $3H^2 = 8\pi G \rho - 3K/a^2$, $\dot{\rho} + 3(1+w)H\rho = 0$.
- In the perturbed equations, we neglect all terms of quadratic and higher order in $\overline{\nabla}_a f$ (any f), q_a, π_{ab}, $\dot{u}_a$, σ_{ab}, ω_a, E_{ab}, H_{ab}, $^3R_{\langle ab \rangle}$ and in their time ($u^a \nabla_a$) and spatial ($\overline{\nabla}_a$) derivatives.

Here ρ and $p = w\rho$ refer to the total quantities (including any Λ term). Strictly we should write $\bar{\rho}$ to distinguish the background energy density from the perturbed ρ, but whenever ρ multiplies a perturbative quantity, only the background part of ρ contributes to first order. As discussed in Section 10.1, scalars like ρ that do not vanish in the background cannot be gauge-invariantly split into a perturbation and a background value. In the covariant approach, we avoid this gauge problem by not directly using the perturbations of scalars (e.g. $\delta\rho$), but instead by using their spatial gradients (e.g. $\overline{\nabla}_a \rho$). Since these vanish in the background, they are automatically gauge invariant.

However, the CGI perturbation formalism is not frame independent, since it depends on a choice of fundamental 4-velocity u^a. A change of fundamental 4-velocity (a linearized Lorentz boost),[2]

$$u^a \to \tilde{u}^a = u^a + v^a , \quad u_a v^a = 0, \tag{10.126}$$

leads to the transformations (see Exercises 5.1.1 and 10.4.1):

$$\tilde{\Theta} = \Theta + \overline{\nabla}_a v^a , \quad \tilde{\dot{u}}_a = \dot{u}_a + \dot{v}_a + H v_a , \tag{10.127}$$

$$\tilde{\omega}_a = \omega_a - \tfrac{1}{2}\text{curl } v_a , \quad \tilde{\sigma}_{ab} = \sigma_{ab} + \overline{\nabla}_{\langle a} v_{b \rangle} , \tag{10.128}$$

$$\tilde{\rho} = \rho , \quad \tilde{p} = p , \quad \tilde{q}_a = q_a - (\rho + p)v_a , \quad \tilde{\pi}_{ab} = \pi_{ab} , \tag{10.129}$$

$$\tilde{E}_{ab} = E_{ab} , \quad \tilde{H}_{ab} = H_{ab} . \tag{10.130}$$

This is the CGI analogue of a gauge transformation in the standard formalism.

In the coordinate metric-based approach, first-order perturbations are decomposed from the start into scalar, vector and tensor modes, using appropriate harmonics (i.e. eigenfunctions of the 3-Laplacian). The covariant approach does not depend on any initial splitting into harmonic modes and it is independent of any Fourier-type decomposition – although harmonic modes are necessary for quantitative calculations.

In the covariant perturbation formalism, all the perturbative quantities are PSTF rank-1 and rank-2 tensors (as in (10.97)). Higher-rank PSTF tensors are needed for CMB perturbations – this is discussed in Chapter 11. These PSTF rank-1 and rank-2 tensors contain

---

[2] From now on, equality is understood as holding to first order.

three-dimensional scalar, vector and tensor modes:

$$V_a = \overline{\nabla}_a V + \mathcal{V}_a, \quad S_{ab} = \overline{\nabla}_{\langle a} \overline{\nabla}_{b\rangle} S + \overline{\nabla}_{\langle a} \mathcal{S}_{b\rangle} + \mathcal{S}_{ab}, \tag{10.131}$$

$$\text{where } \overline{\nabla}^a \mathcal{V}_a = 0, \ \overline{\nabla}^a \mathcal{S}_a = 0 = \overline{\nabla}^b \mathcal{S}_{ab}, \ \mathcal{S}_{ab} = \mathcal{S}_{\langle ab\rangle}.$$

Note that all the quantities $V, \mathcal{V}_a, \ldots, \mathcal{S}_{ab}$ are gauge invariant, since they vanish in the background.

In particular, the scalar potential V vanishes in the background, and hence, by the identity (4.60),

$$\text{curl } \overline{\nabla}_a V = -2\dot{V}\omega_a = 0 \text{ to first order}, \tag{10.132}$$

since $V\omega_a$ is second order. By contrast, for a scalar that does not vanish in the background, such as ρ, we see that

$$\text{curl } \overline{\nabla}_a \rho = -2\dot{\rho}\omega_a = -2\dot{\rho}|_{\text{background}}\,\omega_a, \tag{10.133}$$

which is non-zero if $\omega_a \neq 0$, because of the non-zero background value of $\dot{\rho}$.

The modes of perturbations are characterized as follows:

- The scalar modes are characterized by the fact that all PSTF vectors and tensors are generated by scalar potentials:

$$\mathcal{V}_a = 0, \ \mathcal{S}_a = 0, \ \mathcal{S}_{ab} = 0 \ \text{ and curl } V_a = 0 = \text{curl } S_{ab}. \tag{10.134}$$

- For vector modes, all PSTF vectors are transverse (solenoidal) – including spatial gradients of scalars f such that $\dot{f}$ does not vanish in the background – and all PSTF tensors are generated by transverse vector potentials:

$$V_a = \mathcal{V}_a, \ S_{ab} = \overline{\nabla}_{\langle a} \mathcal{S}_{b\rangle}, \ \text{curl } \overline{\nabla}_a f = -2\dot{f}\omega_a. \tag{10.135}$$

Note that the vorticity supports only vector modes because of the constraint equation $\overline{\nabla}^a \omega_a =$ (see (10.143) below).
- Tensor modes are characterized by

$$\overline{\nabla}_a f = 0, \ V_a = 0, \ S_{ab} = \mathcal{S}_{ab}. \tag{10.136}$$

Note that to first order, $\dot{V}_a = \dot{V}_{\langle a\rangle}, \dot{S}_{ab} = \dot{S}_{\langle ab\rangle}$.

We can expand these modes in harmonic eigenfunctions $\mathcal{Q}_k$ of the scalar Laplace–Beltrami operator (Fourier modes in the case $K = 0$). For example, for scalar modes,

$$V = \sum_k V_{(k)} \mathcal{Q}_k, \ \overline{\nabla}_a V_{(k)} = 0, \ \overline{\nabla}^2 \mathcal{Q}_k = -\frac{k^2}{a^2}\mathcal{Q}_k, \ \dot{\mathcal{Q}}_k = 0. \tag{10.137}$$

Here $k = \nu$ for $K = 0$ and $k = \nu^2 + 1$ for $K < 0$, with $\nu \geq 0$ representing the comoving wavenumber. For $K > 0$, $k = \nu(\nu + 2)$ with $\nu = 1, 2, \ldots$.

10.4.1 Perturbed equations and identities

The 1+3 covariant evolution and constraint equations were given in Chapter 6 in general nonlinear form. Linearization about the FLRW limit gives the following equations, which include scalar, vector and tensor modes:

Evolution:

$$\dot{q}_a + \tfrac{4}{3}\Theta q_a + \rho(1+w)\dot{u}_a + \overline{\nabla}_a p + \overline{\nabla}^b \pi_{ab} = 0, \tag{10.138}$$

$$\dot{\omega}_a + \tfrac{2}{3}\Theta \omega_a + \tfrac{1}{2}\mathrm{curl}\,\dot{u}_a = 0, \tag{10.139}$$

$$\dot{\sigma}_{ab} + 2H\sigma_{ab} + E_{ab} - 4\pi G\pi_{ab} - \overline{\nabla}_{\langle a}\dot{u}_{b\rangle} = 0, \tag{10.140}$$

$$\dot{E}_{ab} + 3H E_{ab} - \mathrm{curl}\,H_{ab} + 4\pi G\rho(1+w)\sigma_{ab}$$
$$+ 4\pi G\,(\dot{\pi}_{ab} + H\pi_{ab}) + 4\pi G\overline{\nabla}_{\langle a}q_{b\rangle} = 0, \tag{10.141}$$

$$\dot{H}_{ab} + 3H H_{ab} + \mathrm{curl}\,E_{ab} - 4\pi G\mathrm{curl}\,\pi_{ab} = 0. \tag{10.142}$$

Constraint:

$$\overline{\nabla}^a \omega_a = 0, \tag{10.143}$$

$$\overline{\nabla}^b \sigma_{ab} - \mathrm{curl}\,\omega_a - \tfrac{2}{3}\overline{\nabla}_a\Theta + 8\pi G q_a = 0, \tag{10.144}$$

$$\mathrm{curl}\,\sigma_{ab} + \overline{\nabla}_{\langle a}\omega_{b\rangle} - H_{ab} = 0, \tag{10.145}$$

$$\overline{\nabla}^b E_{ab} + 4\pi G\overline{\nabla}^b \pi_{ab} - \frac{8\pi G}{3}\overline{\nabla}_a\rho + 8\pi GH q_a = 0, \tag{10.146}$$

$$\overline{\nabla}^b H_{ab} + 4\pi G\mathrm{curl}\,q_a - 8\pi G\rho(1+w)\omega_a = 0, \tag{10.147}$$

$${}^3R_{\langle ab\rangle} - E_{ab} - 4\pi G\pi_{ab} - H(\sigma_{ab} + \omega_{ab}) = 0. \tag{10.148}$$

The Gauss–Codazzi trace-free constraint (10.148) has a partner scalar constraint that gives ${}^3R$. But this is not gauge invariant, so we take its spatial gradient to get the gauge-invariant constraint,

$$\overline{\nabla}_a {}^3R - 16\pi G\overline{\nabla}_a\rho + 4H\overline{\nabla}_a\Theta = 0. \tag{10.149}$$

This defines a CGI curvature perturbation in the case $\omega_a = 0$ (when it is meaningful to talk of the curvature of the hypersurfaces orthogonal to u^a).

In these equations, the energy–momentum terms $\rho, p = w\rho, q_a$ and π_{ab} refer to the total source of the gravitational field. We perform the following replacements (defined in (10.98), (10.99) and (10.103)):

$$a\overline{\nabla}_a\rho = \rho\Delta_a, \; a\overline{\nabla}_a p = c_s^2\rho\Delta_a + a\Gamma_a, \; a\overline{\nabla}_a\Theta = \mathcal{Z}_a. \tag{10.150}$$

These comoving gradients contain both scalar and vector modes. For example, for the density inhomogeneity:

$$\text{scalar: } \Delta := a\overline{\nabla}^a \Delta_a = \frac{a^2}{\rho}\overline{\nabla}^2\rho, \; \text{vector: } \mathrm{curl}\,\Delta_a = 6a(1+w)H\omega_a. \tag{10.151}$$

The energy conservation and Raychaudhuri equations are evolution equations for scalars, so we need to take their spatial gradients to arrive at gauge-invariant equations. The linearization of (10.101) and (10.102) leads to:

$$\dot{\Delta}_a = 3wH\,\Delta_a - (1+w)\,\mathcal{Z}_a + \frac{3aH}{\rho}\,(\dot{q}_a + 4Hq_a)$$

$$- \frac{a}{\rho}\,\overline{\nabla}_a\overline{\nabla}^b\,q_b + \frac{3aH}{\rho}\,\overline{\nabla}^b\,\pi_{ab}, \tag{10.152}$$

$$\dot{\mathcal{Z}}_a = -2H\,\mathcal{Z}_a - 4\pi G\left[(1+3c_s^2)\rho\,\Delta_a + 3a\Gamma_a\right]$$

$$- a\left(3H^2 - \dot{H}\right)\dot{u}_a + a\overline{\nabla}_a\overline{\nabla}^b\,\dot{u}_b. \tag{10.153}$$

In deriving and manipulating CGI perturbative equations, we often need identities for commuting the time and spatial derivatives. Linearization of (10.100) and the other nonlinear identities (see Section 4.8 for details and references) leads to:

$$(a\overline{\nabla}_a f)\dot{} = a\overline{\nabla}_a \dot{f}, \quad (a\overline{\nabla}_a V_b)\dot{} = a\overline{\nabla}_a \dot{V}_b, \quad (a\overline{\nabla}_a S_{bc})\dot{} = a\overline{\nabla}_a \dot{S}_{bc}, \tag{10.154}$$

$$\overline{\nabla}_{[a}\overline{\nabla}_{b]}f = -\dot{f}\omega_{ab} \quad \text{or} \quad \mathrm{curl}\,\overline{\nabla}_a f = -2\dot{f}\omega_a, \tag{10.155}$$

$$\overline{\nabla}_{[a}\overline{\nabla}_{b]}V_c = -\frac{K}{a^2}V_{[a}h_{b]c}, \quad \overline{\nabla}_{[a}\overline{\nabla}_{b]}S^{cd} = -\frac{2K}{a^2}S_{[a}{}^{(c}h_{b]}{}^{d)}, \tag{10.156}$$

$$\overline{\nabla}^a\mathrm{curl}\,V_a = 0, \quad \overline{\nabla}^b\mathrm{curl}\,S_{ab} = \tfrac{1}{2}\mathrm{curl}\,\overline{\nabla}^b S_{ab}, \tag{10.157}$$

$$\mathrm{curl}\,\mathrm{curl}\,V_a = -\overline{\nabla}^2 V_a + \overline{\nabla}_a(\overline{\nabla}^b V_b) + \frac{2K}{a^2}V_a, \tag{10.158}$$

$$\mathrm{curl}\,\mathrm{curl}\,S_{ab} = -\overline{\nabla}^2 S_{ab} + \tfrac{3}{2}\overline{\nabla}_{\langle a}(\overline{\nabla}^c S_{b\rangle c}) + \frac{3K}{a^2}S_{ab}, \tag{10.159}$$

$$\overline{\nabla}^2(\overline{\nabla}_a f) = \overline{\nabla}_a(\overline{\nabla}^2 f) + \frac{2K}{a^2}\overline{\nabla}_a f + 2\dot{f}\,\mathrm{curl}\,\omega_a. \tag{10.160}$$

The FLRW background curvature term Ka^{-2} arises from the commutation of spatial derivatives via the spatial Ricci identity, using the fact that the background 3-Riemann tensor has constant curvature: ${}^3R_{abcd} = 6Ka^{-2}(h_{ac}h_{bd} - h_{ad}h_{bc})$.

We now consider various applications and solutions of these equations.

10.4.2 Barotropic fluid

For a single barotropic perfect fluid, with $p = p(\rho)$, (10.152) and (10.153) lead to

$$\dot{\Delta}_a = 3wH\Delta_a - (1+w)\mathcal{Z}_a, \tag{10.161}$$

$$\dot{\mathcal{Z}}_a = -2H\mathcal{Z}_a - 4\pi G\rho\Delta_a - \frac{c_s^2}{1+w}\left(\overline{\nabla}^2 + \frac{K}{a^2}\right)\Delta_a$$

$$- 6ac_s^2 H\mathrm{curl}\,\omega_a, \tag{10.162}$$

where we have used momentum conservation (10.138) and the commutation identities above. Equation (10.138) also gives the vorticity and shear propagation equations (10.139) and (10.140) as

$$\dot{\omega}_a = -\left(2 - 3c_s^2\right) H \omega_a,\tag{10.163}$$

$$\dot{\sigma}_{ab} = -2H\sigma_{ab} - E_{ab} - \frac{c_s^2}{a(1+w)}\overline{\nabla}_{\langle a}\Delta_{b\rangle}.\tag{10.164}$$

Thus vorticity decays with the expansion unless the barotropic medium has a sound speed $c_s > \sqrt{2/3}$.

The CGI curvature perturbation (10.149) evolves as

$$\left(a^3\overline{\nabla}_a{}^3R\right)^{\cdot} = \frac{4c_s^2}{(1+w)}Ha^2\overline{\nabla}^2\Delta_a,\tag{10.165}$$

so that it is conserved on large scales.

We take the comoving divergence of (10.161) and (10.162), and then eliminate $\overline{\nabla}^a\mathcal{Z}_a$, to find that

$$\ddot{\Delta} + \left(2 - 6w + 3c_s^2\right)H\dot{\Delta} - \left[4\pi G\left(1 + 8w - 6c_s^2 - 3w^2\right)\rho\right.$$
$$\left. - 12(w - c_s^2)\frac{K}{a^2} + c_s^2\overline{\nabla}^2\right]\Delta = 0.\tag{10.166}$$

This is the covariant analogue of the standard equation for $\delta = \delta\rho/\rho$. The last term on the right demonstrates the competing effects of gravitational attraction and pressure support, with collapse occurring when the quantity within the braces is positive. The physical wavelength of the mode is $\lambda = a/k$, so that gravitational contraction will take place only on scales larger than the critical (Jeans) length

$$\lambda_J \approx \frac{c_s}{\sqrt{4\pi G\left(1 + 8w - 6c_s^2 - 3w^2\right)\rho + 12(w - c_s^2)K/a^2}}.\tag{10.167}$$

Since curvature and dark energy are typically negligible after inflation, we set $K = 0 = \Lambda$ in the radiation era. With $w = 1/3 = c_s^2$ and $H = 1/(2t)$, we can solve (10.166) on super-Hubble scales, $k/aH \ll 1$, where the pressure support is negligible:

$$\Delta = \Delta_+ \left(\frac{t}{t_{\rm eq}}\right) + \Delta_- \left(\frac{t}{t_{\rm eq}}\right)^{-1/2},\tag{10.168}$$

with $\dot{\Delta}_{\pm} = 0$. During the radiation era, large-scale radiation density perturbations grow as $\Delta \propto a^2$. On sub-Hubble scales, $k/aH \gg 1$, pressure gradients can support against gravitational collapse and the solution oscillates:

$$\Delta_{(k)} = C(k)\exp\left[i\sqrt{3}\frac{k}{a_{\rm eq}H_{\rm eq}}\left(\frac{t}{t_{\rm eq}}\right)^{1/2}\right],\tag{10.169}$$

where the real part is understood.

After matter–radiation equality, for CDM we have $w = 0 = c_s^2$ and $H = 2/(3t + t_{\rm eq})$. We can neglect c_s for baryons after recombination. Then (10.166) leads to the scale-independent solution,

$$\Delta = \Delta_+ \left(\frac{t}{t_{eq}}\right)^{2/3} + \Delta_- \left(\frac{t}{t_{eq}}\right)^{-1}. \tag{10.170}$$

Matter density perturbations in the matter era grow as $\Delta \propto a$ on all scales.

Dissipative processes can modify the ideal-fluid evolution of perturbations. For example, radiation begins to deviate from perfect-fluid behaviour as the photon interaction rate with electrons drops below the expansion rate, and photon free-streaming effects become significant. This has an increasingly important effect on baryonic matter: as photons diffuse from high-density to low-density regions, they tend to drag baryons, erasing small-scale fluctuations. This is known as the Silk damping effect (Silk, 1967).

In the early universe, dark matter decoupled from thermal equilibrium with the plasma and entered a regime of collisionless motion. A collisionless gas is not a perfect fluid, and strictly is not a fluid at all (hydrodynamic behaviour requires interactions). But for massive particles, the dust approximation (i.e. a perfect 'fluid' with vanishing pressure) works for suitably large scales where free-streaming effects may be neglected. (Free-streaming damping with massive particles is known as Landau damping.) Below the free-streaming scale, small-scale structure does not grow as in the dust case, but is erased. The higher the velocity dispersion, the greater is the free-streaming scale, and therefore the greater is the minimum mass of fluctuations that can grow. For cold dark matter, the free-streaming masses are very low, and perturbations grow unimpeded by damping processes on all scales of cosmological interest.

10.4.3 Multiple perfect fluids

In the FLRW background all components are perfect fluids sharing the same 4-velocity u^a. Thus the peculiar velocities v_I^a are gauge invariant. Momentum conservation gives

$$a\rho_I (1+w_I) \dot{u}^a = -c_{sI}^2 \rho_I \Delta_I^a - p_I \mathcal{E}_I^a$$
$$-a\left(\dot{q}_I^a - 4Hq_I^a\right) + \mathcal{Q}_I^a, \tag{10.171}$$

where $c_{sI}^2 = \dot{p}_I/\dot{\rho}_I$, $w_I = p_I/\rho_I$ and $q_I^a = \rho_I(1+w_I)v_I^a$. Equation (10.116) linearizes to

$$\dot{\Delta}_I^a = 3\left(w_I - c_{sI}^2\right) H\Delta_I^a - 3w_I H\mathcal{E}_I^a - (1+w_I)\mathcal{Z}^a$$
$$-\frac{a}{\rho_I}\overline{\nabla}^a\left(\overline{\nabla}_b q_I^b - \mathcal{Q}_I\right) - \frac{1}{\rho}\mathcal{Q}_I\Delta_I^a$$
$$+\frac{1}{\rho(1+w)}\left[3(1+w_I)H - \frac{\mathcal{Q}_I}{\rho_I}\right]\left(c_s^2\rho\Delta^a + p\mathcal{E}^a + a\dot{q}^a + 4Haq^a\right). \tag{10.172}$$

The total and partial equations of state and speeds of sound are related by

$$w = \frac{1}{\rho}\sum_I \rho_I w_I, \quad c_s^2 = \frac{1}{\rho(1+w)}\sum_I c_{sI}^2\rho_I(1+w_I). \tag{10.173}$$

Using Equations (10.102) and (10.103), we obtain

$$\dot{\mathcal{Z}}_a = -2H\mathcal{Z}_a - 4\pi G\rho\Delta_a - \frac{c_s^2}{1+w}\left(\overline{\nabla}^2\Delta_a + \frac{K}{a^2}\Delta_a\right) \tag{10.174}$$

$$-\frac{w}{1+w}\left(\overline{\nabla}^2\mathcal{E}_a + \frac{K}{a^2}\mathcal{E}_a\right) - \frac{a}{\rho(1+w)}\overline{\nabla}_a\overline{\nabla}^b(\dot{q}_b + 4Hq_b)$$

$$+\frac{3a}{\rho(1+w)}\left[\frac{1}{2}\rho(1+w) - \frac{K}{a^2}\right](\dot{q}_a + 4Hq_a) - 6ac_s^2 H\,\mathrm{curl}\,\omega_a\,.$$

Equations (10.172) and (10.174) govern the evolution of scalar and vector modes in an almost-FLRW universe filled with several interacting and non-comoving perfect fluids. The vector mode can be removed by taking the divergence. Inhomogeneities in the total energy density ρ are related to those in the individual fluids by

$$\Delta^a = \frac{1}{\rho}\sum_I \rho_I\Delta_I^a\,. \tag{10.175}$$

Then (10.171) and (10.103) allow us to relate the individual and total non-adiabatic pressure gradients:

$$p\mathcal{E}^a = \sum_I p_I\mathcal{E}_I^a + \sum_I c_{sI}^2\rho_I\Delta_I^a - c_s^2\sum_I \rho_I\Delta_I^a, \tag{10.176}$$

where the effective total sound speed is given by (10.173) to zero order. Using this we can recast (10.176):

$$p\mathcal{E}_a = \sum_I p_I\mathcal{E}_I^a + \frac{1}{2(\rho+p)}\sum_{I,J}(\rho_I + p_I)(\rho_J + p_J)\left(c_{sI}^2 - c_{sJ}^2\right)\mathcal{E}_{IJ}^a, \tag{10.177}$$

$$\mathcal{E}_{IJ}^a = \frac{\Delta_I^a}{1+w_I} - \frac{\Delta_J^a}{1+w_J} = -\mathcal{E}_{JI}^a. \tag{10.178}$$

Thus the total non-adiabatic (entropy) perturbation is made up of intrinsic ($\mathcal{E}_I^a$) and relative ($\mathcal{E}_{IJ}^a$) contributions. The intrinsic contribution vanishes for a barotropic fluid, but not for a scalar field. The relative contribution for two fluids vanishes if their sound speeds are equal, or if their density perturbations are tuned in the ratio:

$$\frac{\Delta_I^a}{1+w_I} = \frac{\Delta_J^a}{1+w_J}, \tag{10.179}$$

which is the adiabatic condition.

If there are no interactions, $Q_I^a = 0$, then the comoving divergence of (10.172) gives the evolution for the I density perturbation:

$$\dot{\Delta}_I = 3\left(w_I - c_{sI}^2\right)H\Delta_I - 3w_I H\mathcal{E}_I - (1+w_I)\mathcal{Z} - \frac{a^2}{\rho_I}\overline{\nabla}^2\left(\overline{\nabla}_a q_I^a\right)$$

$$+\frac{3(1+w_I)H}{\rho(1+w)}\left(c_s^2\rho\Delta + p\mathcal{E}\right) + \frac{3a^2(1+w_I)H}{\rho(1+w)}\overline{\nabla}^a(\dot{q}_a + 4Hq_a), \tag{10.180}$$

where $\mathcal{E}_I = a\overline{\nabla}^a \mathcal{E}_I^a$ and $\mathcal{E} = a\overline{\nabla}^a \mathcal{E}_a$. Similarly, (10.174) leads to

$$\dot{\mathcal{Z}} = -2H\mathcal{Z} - 4\pi G\rho\Delta - \frac{c_s^2}{1+w}\left(\overline{\nabla}^2\Delta + \frac{3K}{a^2}\Delta\right)$$

$$-\frac{w}{1+w}\left(\overline{\nabla}^2\mathcal{E} + \frac{3K}{a^2}\mathcal{E}\right) + \frac{3}{2}a^2\overline{\nabla}^a(\dot{q}_a + 4Hq_a)$$

$$-\frac{a^2}{\rho(1+w)}\left[\overline{\nabla}^2\overline{\nabla}^a(\dot{q}_a + 4Hq_a) + \frac{3K}{a^2}\overline{\nabla}^a(\dot{q}_a + 4Hq_a)\right]. \quad (10.181)$$

We used the first-order identity (10.160), which gives $a\overline{\nabla}^a\overline{\nabla}^2\Delta_a = \overline{\nabla}^2\Delta + (2K/a^2)\Delta$, and analogous relations for $\mathcal{E}_a$ and $\mathcal{E}$.

In order to proceed, the total flux vector $q_a = \sum_I q_I^a = \sum_I \rho_I(1+w_I)v_I^a$ must be specified. Choosing the total energy frame $q_a = 0$, the system reduces to

$$\dot{\Delta}_I = 3\left(w_I - c_{sI}^2\right)H\Delta_I - 3w_I H\mathcal{E}_I - (1+w_I)\mathcal{Z}$$

$$-a(1+w_I)\overline{\nabla}^2 v_I + \frac{3(1+w_I)H}{1+w}\left(c_s^2\Delta + w\mathcal{E}\right), \quad (10.182)$$

$$\dot{\mathcal{Z}} = -2H\mathcal{Z} - 4\pi G\rho\Delta - \frac{c_s^2}{1+w}\left(\overline{\nabla}^2\Delta + \frac{3K}{a^2}\Delta\right)$$

$$-\frac{w}{1+w}\left(\overline{\nabla}^2\mathcal{E} + \frac{3K}{a^2}\mathcal{E}\right). \quad (10.183)$$

Here $v_I = a\overline{\nabla}^a v_I^a$ is the velocity perturbation, and its evolution follows from (10.171) (with $\mathcal{Q}_I^a = 0$) and (10.103):

$$\dot{v}_I = -\left(1 - 3c_{sI}^2\right)Hv_I - \frac{1}{a(1+w_I)}\left(c_{sI}^2\Delta_I + w_I\mathcal{E}_I\right)$$

$$-\frac{1}{a(1+w)}\left(c_s^2\Delta + w\mathcal{E}\right). \quad (10.184)$$

Radiation and CDM

A spatially flat almost-FLRW spacetime dominated by radiation and CDM has total energy density $\rho = \rho_r + \rho_c$, and total pressure $p = \rho_r/3$. The effective total equation of state parameter and sound-speed squared are

$$w = \frac{1}{3(1+y)}, \quad c_s^2 = \frac{4}{3(4+3y)}, \quad y := \frac{a}{a_{eq}}. \quad (10.185)$$

The radiation field is effectively homogeneous inside the sound horizon after averaging over acoustic oscillations, or on scales that are damped by diffusion. In this case, we can consider perturbations in the CDM only, i.e. $\rho\Delta \approx \rho_c\Delta_c$. By (10.182)–(10.184),

$$\dot{\Delta}_c = -\mathcal{Z} - a\overline{\nabla}^2 v_c, \quad \dot{\mathcal{Z}} = -2H\mathcal{Z} - 4\pi G\rho\Delta, \quad \dot{v}_c = -Hv_c, \quad (10.186)$$

where the last equation relies on $c_s^2\Delta + w\mathcal{E} = 0$. This follows from (10.177), which shows that $\mathcal{E} = -4\rho_c\Delta_c/(4\rho_r + 3\rho_c)$.

We can derive an equation for Δ_c, using the linear commutation law $(\overline{\nabla}^2 v_c)^{\cdot} = \overline{\nabla}^2 \dot{v}_c - 2H\overline{\nabla}^2 v_c$,

$$\Delta_c'' = -\frac{2+3y}{2y(1+y)}\Delta_c' + \frac{3}{2y(1+y)}\Delta_c, \tag{10.187}$$

where a prime denotes d/dy. By inspection, this equation admits a solution that is linear in y. The general solution can then be found as

$$\Delta_c = \mathcal{C}_+\left(1 + \frac{3}{2}y\right)$$

$$+ \mathcal{C}_-\left[\left(1 + \frac{3}{2}y\right)\ln\left(\frac{\sqrt{1+y}+1}{\sqrt{1+y}-1}\right) - 3\sqrt{1+y}\right]. \tag{10.188}$$

Thus $\Delta_c \propto a$ at late times, in agreement with a single-fluid Einstein–de Sitter model. Deep in the radiation era, $y \ll 1$, Δ_c is effectively constant. This stagnation, or freezing-in, of matter perturbations prior to equality is generic to models with a period of expansion that is dominated by relativistic particles, and is called the Meszaros effect (1974).

CDM and baryons

A purely baryonic matter content cannot explain the structure observed in the universe – baryonic density perturbations cannot grow fast enough from their amplitude at decoupling. The main reason is the tight coupling between photons and baryons in the pre-recombination era, which washes out baryonic perturbations. CDM is immune from photon drag, and CDM perturbations grow between equality and decoupling by a factor of $\sim a_{dec}/a_{eq}$. After decoupling the universe becomes effectively transparent to radiation and baryonic perturbations can start growing, driven by the CDM gravitational potential.

By (10.182)–(10.184), the baryon perturbations are governed by

$$\dot{\Delta}_b = -\mathcal{Z} - a\overline{\nabla}^2 v_b, \quad \dot{\mathcal{Z}} = -2H\mathcal{Z} - 4\pi G\rho\Delta, \quad \dot{v}_b = -Hv_b, \tag{10.189}$$

where $\rho = \rho_c + \rho_b \approx \rho_c$ and $\rho\Delta \approx \rho_c\Delta_c$. This system implies

$$\ddot{\Delta}_b + 2H\dot{\Delta}_b = 4\pi G\rho_c\Delta_c. \tag{10.190}$$

Since $\Delta_c \propto a$ after decoupling and $\rho_c \propto a^{-3}$, we find that

$$\Delta_b = \Delta_c\left(1 - \frac{a_{dec}}{a}\right), \quad a > a_{dec}. \tag{10.191}$$

This shows that $\Delta_b \to \Delta_c$ for $a \gg a_{rec}$: after decoupling, CDM accelerates the gravitational collapse of baryonic matter and therefore the onset of structure formation.

10.4.4 Magnetized fluids

Magnetic fields can imprint significant effects on the CMB and on early structure formation. In order to compute these effects, we need to incorporate magnetic fields into the perturbative formalism. For the standard metric-based approach and its applications to the CMB and structure formation, see for example Durrer (2007), Giovannini and Kunze (2008),

Sethi, Nath and Subramanian (2008), Paoletti, Finelli and Paci (2009), Subramanian (2010), Yamazaki *et al.* (2010). The covariant approach is reviewed in Barrow, Maartens and Tsagas (2007) (see also Betschart, Dunsby and Marklund (2004), Kobayashi *et al.* (2007), Kandus and Tsagas (2008)). Here we briefly describe the covariant approach.

Consider a spatially flat FLRW spacetime containing a sufficiently weak, statistically isotropic magnetic field: $\langle B_a \rangle = 0$, while $\langle B^2 \rangle \neq 0$, and $\langle B^2 \rangle / \rho \ll 1$ on all scales of interest (the angled brackets denote averaging over a suitable scale). The quantities B^2 and $B_a \overline{\nabla}_b B_c$ are first order. We can effectively think of B^a as 'half order', but B^a only arises in the perturbative equations in quadratic form.

The nonlinear inhomogeneity variable (10.117), i.e. $\mathcal{B}_a = a \overline{\nabla}_a B^2 / B^2$, is no longer suitable for the linear perturbative regime since B^2 vanishes in the background, and we define new quantities,

$$\mathcal{A}_a = \frac{a}{\rho} \overline{\nabla}_a B^2 \quad \text{with} \quad \mathcal{A} = a \overline{\nabla}^a \mathcal{A}_a , \tag{10.192}$$

which are first order. By linearizing the magnetic induction equation (5.131), we see that the magnetic energy density decays adiabatically as

$$B^2 \propto a^{-4} . \tag{10.193}$$

It follows that

$$\dot{\mathcal{A}} = (1 + 3w) H \mathcal{A} . \tag{10.194}$$

For a barotropic fluid with $w = \text{const}$, magnetized density perturbations evolve as

$$\dot{\Delta} = 3w H \Delta - (1 + w) \mathcal{Z} + \tfrac{3}{2} H \mathcal{A} . \tag{10.195}$$

This follows from (10.118) and (5.135). The direct magnetic effect on Δ arises via the magnetic pressure. The evolution of $\mathcal{Z}$ follows from (10.120),

$$\dot{\mathcal{Z}} = -2H \mathcal{Z} - 4\pi G \rho \Delta + 2\pi G \rho \mathcal{A} - \frac{c_s^2}{1 + w} \overline{\nabla}^2 \Delta - \frac{1}{2(1 + w)} \overline{\nabla}^2 \mathcal{A} . \tag{10.196}$$

Equations (10.194)–(10.196) lead to an evolution equation for Δ, with source terms in $\mathcal{A}$:

$$\ddot{\Delta} + (2 - 3w) H \dot{\Delta} - \left[4\pi G \left(1 - 2w - 3w^2 \right) \rho + w \overline{\nabla}^2 \right] \Delta$$
$$= \left[4\pi G (1 + w) \rho + \tfrac{1}{2} \overline{\nabla}^2 \right] \mathcal{A} , \tag{10.197}$$

where we have assumed $w = \text{const}$, and hence $c_s^2 = w$. This is the magnetized generalization of (10.166). The magnetic field acts as a source term that can seed density perturbations. On large scales, we can neglect $\overline{\nabla}^2 \mathcal{A}$ and the source term is decaying: $\rho \mathcal{A} \propto a^{-2}$, by (10.194).

The vector modes are also affected by the magnetic field. This is clearly seen via the magnetized vorticity evolution equation,

$$\dot{\omega}_a = -(2 - 3c_s^2) H \omega_a - \frac{1}{2(1 + w) \rho} B^b \overline{\nabla}_b \text{curl} \, B_a . \tag{10.198}$$

This shows how magnetic fields can generate vorticity, provided that curl B_a varies spatially along the magnetic field lines.

Magnetized tensor perturbations are considered below.

10.4.5 Scalar fields

Density perturbations of a minimally coupled scalar field are described by the comoving divergence (10.122): since $\overline{\nabla}_a \varphi = 0$, this is strictly only a measure of the inhomogeneity in the kinetic energy density,

$$\Delta = \frac{a^2 \dot{\varphi}}{\rho_\varphi} \overline{\nabla}^2 \dot{\varphi}. \tag{10.199}$$

Equations (10.123), (10.124) and (5.148) give

$$\ddot{\Delta} = -\left(2 - 6w + 3c_s^2\right) H \dot{\Delta} + \tfrac{1}{2}\left(1 + 8w - 3w^2 - 6c_s^2\right) H^2 \Delta$$
$$+ \frac{1}{a^2}\left[9(1 - w^2)K - 2k^2\right]\Delta, \tag{10.200}$$

where c_s is the adiabatic sound speed.

Standard slow-roll inflation corresponds to approximately exponential de Sitter expansion, with H and ρ_φ nearly constant. This is achieved when $\dot{\varphi}^2 \ll V(\varphi)$ and $|\ddot{\varphi}| \ll H|\dot{\varphi}|$. As we approach the de Sitter regime, (10.200) no longer depends on the background spatial curvature and

$$\ddot{\Delta} = -5H\dot{\Delta} - 6H\left[1 + \frac{1}{6}\left(\frac{k}{aH}\right)^2\right]\Delta, \tag{10.201}$$

where $H \approx \text{const}$. After the mode has crossed the Hubble radius, $k \ll aH$, the solution is

$$\Delta = C_1 e^{-2Ht} + C_2 e^{-3Ht}, \tag{10.202}$$

so that $\Delta \propto a^{-2}$ during inflation. Kinetic energy density fluctuations of the inflaton field will decay exponentially irrespective of their scale and the background curvature. But the large-scale curvature perturbation remains constant, as shown in Section 12.2.1.

In order to compute the large-scale curvature perturbation, we need to quantize the scalar field fluctuations and evaluate their amplitude at Hubble-crossing. In the metric-based approach, this is usually implemented via the Sasaki–Mukhanov variable Q (see Section 12.2.1). The covariant variable corresponding to this is given by (Pitrou and Uzan, 2007)

$$v_a = \frac{a\dot{\varphi}}{3H}\left(\int \overline{\nabla}_a \Theta \, dt - \overline{\nabla}_a \int \Theta \, dt\right). \tag{10.203}$$

This gradient variable corresponds to the variable $v = aQ$ in the metric-based approach.

10.4.6 Tensor perturbations

Gravitational waves are propagating fluctuations in the geometry of the spacetime fabric, usually described as weak perturbations of the background metric. The CGI approach is based instead on the propagating curvature, in the form of the electric and magnetic components of the Weyl tensor, which describe the free gravitational field (Hawking, 1966, Dunsby, Bassett and Ellis, 1997).

Pure tensor modes are transverse and tracefree, so that all physical PSTF rank-2 tensors are divergence-free:

$$\overline{\nabla}^b E_{ab} = \overline{\nabla}^b H_{ab} = \overline{\nabla}^b \sigma_{ab} = \overline{\nabla}^b \pi_{ab} = 0. \tag{10.204}$$

Using (10.138)–(10.153), we can show the following.

The condition (10.204) requires that

$$\overline{\nabla}_a \rho = \overline{\nabla}_a p = \overline{\nabla}_a \Theta = \omega_a = 0, \tag{10.205}$$

to linear order and at all times. These constraints are self-consistent (i.e. they are preserved in time) at the linear perturbative level, and they also guarantee that

$$\dot{u}_a = q_a = \overline{\nabla}_a \,{}^3R = 0. \tag{10.206}$$

The above constraints express the fact that all scalar modes (spatial gradients of physical scalars) and vector modes (transverse vectors) must vanish.

Then the only remaining nontrivial constraints are

$$H_{ab} = \text{curl}\, \sigma_{ab}, \quad {}^3R_{\langle ab \rangle} = H\sigma_{ab} - E_{ab}, \tag{10.207}$$

which show that the magnetic Weyl tensor is fully determined by the shear, and that the tracefree 3-Ricci tensor is also divergence-free.

For a magnetized fluid, we require in addition (Maartens, Tsagas and Ungarelli, 2001)

$$\overline{\nabla}_b \pi_B^{ab} = \tfrac{1}{3}\overline{\nabla}^a B^2 - B^b \overline{\nabla}_b B^a = 0, \tag{10.208}$$

at all times.

The energy density of gravitational radiation is determined by the pure tensor part h_{ij} of the metric perturbation,

$$\rho_{\text{gw}} = \frac{(h_{ij})'(h^{ij})'}{2a^2}, \tag{10.209}$$

where the prime denotes a conformal time derivative. In a comoving frame, with $u^a = \delta_0^a u^0$, we have (Goode, 1989)

$$\sigma_{ij} = a(h_{ij})', \quad \sigma^{ij} = a^{-3}(h^{ij})', \tag{10.210}$$

so that the CGI formula is

$$\rho_{gw} = \sigma^2. \tag{10.211}$$

The propagation equations (10.141) and (10.142) for a perfect fluid ($\pi_{ab} = 0$) are

$$\dot{E}_{ab} = -3HE_{ab} - 4\pi G\rho(1+w)\sigma_{ab} + \text{curl } H_{ab}, \tag{10.212}$$

$$\dot{H}_{ab} = -3HH_{ab} - \text{curl } E_{ab}. \tag{10.213}$$

The evolution of H_{ab} is determined by that of the shear; by (10.140),

$$\dot{\sigma}_{ab} = -2H\sigma_{ab} - E_{ab}. \tag{10.214}$$

The commutation law (10.156) for gradients of PSTF tensors and the zero-order expression $^3R_{abcd} = 6Ka^{-2}(h_{ac}h_{bd} - h_{ad}h_{bc})$, lead to the auxiliary relation

$$\text{curl } H_{ab} = \frac{3K}{a^2}\sigma_{ab} - \overline{\nabla}^2\sigma_{ab}. \tag{10.215}$$

Then (10.212) gives

$$\dot{E}_{ab} = -3HE_{ab} - 4\pi G\rho(1+w)\sigma_{ab} + \frac{3K}{a^2}\sigma_{ab} - \overline{\nabla}^2\sigma_{ab}. \tag{10.216}$$

The wave equation for the shear is

$$\ddot{\sigma}_{ab} = -5H\dot{\sigma}_{ab} - 4\pi G\rho(1-3w)\sigma_{ab} + \frac{K}{a^2}\sigma_{ab} + \overline{\nabla}^2\sigma_{ab}. \tag{10.217}$$

Introducing the tensor harmonics $Q_{ab}^{(k)}$, with

$$Q_{ab}^{(k)} = Q_{\langle ab\rangle}^{(k)}, \ \dot{Q}_{ab}^{(k)} = 0 = \overline{\nabla}^b Q_{ab}^{(k)}, \ \overline{\nabla}^2 Q_{ab}^{(k)} = -\frac{k^2}{a^2}Q_{ab}^{(k)}, \tag{10.218}$$

the shear modes satisfy

$$\ddot{\sigma}_{(k)} = -5H\dot{\sigma}_{(k)} - \left[4\pi G\rho(1-3w) - \frac{1}{a^2}(K-k^2)\right]\sigma_{(k)}. \tag{10.219}$$

Note that, in order to account for the different polarization states of gravitational radiation, one expands the tensor perturbations in terms of electric and magnetic parity harmonics (Challinor, 2000a). Nevertheless, the coupling between the two states means that (10.219) still holds.

For a spatially flat background and a radiation-dominated universe, on super-Hubble scales, we have $\ddot{\sigma}_{(k)} + 5H\dot{\sigma}_{(k)} = 0$, so that

$$\sigma_{(k)} = C_0 + C_1 t^{-5/2}. \tag{10.220}$$

On small scales the shear oscillates and decays. After equality, on super-Hubble scales,

$$\sigma_{(k)} = C_1 t^{-1/3} + C_2 t^{-2}, \tag{10.221}$$

so that after equality large-scale gravitational wave perturbations decay as $a^{-1/2}$.

Exercise 10.4.1 Derive the transformations (10.127) and (10.128) for the kinematic quantities.

Exercise 10.4.2 Show that in the decomposition (10.131), $\overline{\nabla}^a \overline{\nabla}^b S_{ab}$ is purely scalar, and curl $\overline{\nabla}^b S_{ab}$ is purely vector.

Exercise 10.4.3 Derive the Δ evolution equation (10.166).

Exercise 10.4.4 Derive the shear wave equation (10.217). Verify the solution (10.221).

11

The cosmic background radiation

A central pillar of modern cosmology is the near-isotropy of the CMB, compatible with a perturbed FLRW model of the universe. The small deviations from isotropy in the CMB temperature contain a wealth of information. Temperature anisotropies due to inhomogeneities were predicted by Sachs and Wolfe (1967) soon after the discovery of the CMB in 1965 by Penzias and Wilson. Shortly afterwards, polarization was predicted in models with anisotropy in the expansion rate around the time of recombination (Rees, 1968). The detailed physics of CMB fluctuations in almost-FLRW models was essentially understood for models with only baryonic matter by 1970 (Silk, 1968, Peebles, 1968, Zel'dovich, Kurt and Sunyaev, 1968, Peebles and Yu, 1970, Sunyaev and Zel'dovich, 1970). By the early 1980s, CDM was included (Peebles, 1982, Bond and Efstathiou, 1984). Further milestones included the effect of spatial curvature (Wilson, 1983), polarization (Kaiser, 1983, Bond and Efstathiou, 1984) and gravitational waves (Dautcourt, 1969, Polnarev, 1985). All of this work used the standard metric-based approach to cosmological perturbation theory, but CMB physics has also been studied extensively in the 1+3-covariant approach (Ellis, Matravers and Treciokas, 1983b, Ellis, Treciokas and Matravers, 1983, Stoeger, Maartens and Ellis, 1995, Maartens, Ellis and Stoeger, 1995a,b, Dunsby, 1997, Uzan, 1998, Challinor and Lasenby, 1998, 1999, Maartens, Gebbie and Ellis, 1999, Challinor, 2000a,b, Lewis, Challinor and Lasenby, 2000, Gebbie and Ellis, 2000, Gebbie, Dunsby and Ellis, 2000, Lewis, 2004a,b, Pitrou, 2009). This brings to the CMB the benefits of: (i) clarity in the physical meaning of the variables employed; (ii) covariant and gauge-invariant perturbation theory around a variety of background models; (iii) a good basis for studying nonlinear effects; and (iv) freedom to employ any coordinate system or tetrad.

In this chapter we review the physics of CMB temperature anisotropies, following mainly the 1+3-covariant approach, and using extensively the review by Tsagas, Challinor and Maartens (2008).

11.1 The CMB and spatial homogeneity: nonlinear analysis

It is a fundamental aspect of the standard model of cosmology that the FLRW background is spatially isotropic and homogeneous. This is explored in more detail in Section 9.8. The key evidence in support of this assumption is the high degree of isotropy of the CMB, which provides a probe of the universe's evolution back to the time of decoupling of photons. However, the CMB can only be observed from one world line, that of our Galaxy, and isotropy in itself would not allow us to distinguish an FLRW from an LTB model (observed from the centre). The point is that we need a supplementary assumption in order to fix the geometry of the background. The simplest assumption is a principle of 'democracy', i.e. that we do not occupy a special place in the universe. This Copernican Principle means that the CMB is isotropic for all observers if it is isotropic for us. Intuitively, isotropy for all observers should imply that the spacetime is FLRW, but the proof is not at all straightforward. It follows from a seminal (though under-recognized) theorem by Ehlers, Geren and Sachs (EGS) (1968). Note that the standard CMB anisotropy studies do *not* prove the result – they assume from the outset that the universe is almost FLRW. Indeed, it is not possible to tackle this problem via a perturbative approach – one needs to start from the nonlinear field and Liouville equations for a general spacetime.

The general nonlinear field equations in covariant form are covered in Chapter 6. The nonlinear Liouville equation for photons in an arbitrary spacetime corresponds to an infinite hierarchy of coupled nonlinear evolution equations for the covariant multipoles F_{A_ℓ} of the distribution function. The hierarchy is given by (5.84), in the massless and collisionless case, $\lambda = E, C_{A_\ell}[f] = 0$. These are evolution equations in phase space. In order to obtain evolution equations in spacetime, we multiply by E^3 and integrate over all energies. This leads to nonlinear evolution equations for the covariant intensity multipoles (Ellis, Treciokas and Matravers, 1983, Maartens, Gebbie and Ellis, 1999):

$$0 = \dot{I}_{\langle A_\ell \rangle} + \frac{4}{3}\Theta I_{A_\ell} + \frac{\ell}{(2\ell+1)}\overline{\nabla}_{\langle a_\ell} I_{A_{\ell-1}\rangle} + \overline{\nabla}^b I_{bA_\ell} + \frac{\ell(\ell+3)}{(2\ell+1)}\dot{u}_{\langle a_\ell} I_{A_{\ell-1}\rangle}$$
$$- (\ell-2)\dot{u}^b I_{bA_\ell} - \ell\omega^b \eta_{bc\langle a_\ell} I_{A_{\ell-1}\rangle}{}^c - (\ell-1)\sigma^{bc} I_{bcA_\ell}$$
$$+ \frac{5\ell}{(2\ell+3)}\sigma^b{}_{\langle a_\ell} I_{A_{\ell-1}\rangle b} - \frac{(\ell-1)\ell(\ell+2)}{(2\ell-1)(2\ell+1)}\sigma_{\langle a_\ell a_{\ell-1}} I_{A_{\ell-2}\rangle}, \qquad (11.1)$$

where I_{A_ℓ} are defined by (5.101). The monopole ($I = \rho_\gamma$) evolution equation is the energy conservation equation and the dipole ($I^a = q_\gamma^a$) evolution equation is the momentum conservation equation. The quadrupole ($I^{ab} = \pi_\gamma^{ab}$) gives (Stoeger, Maartens and Ellis, 1995, Maartens, Gebbie and Ellis, 1999):

$$\dot{\pi}_\gamma^{\langle ab\rangle} + \frac{4}{3}\Theta\pi_\gamma^{ab} + \frac{8}{15}\rho_\gamma\sigma^{ab} + \frac{2}{5}\overline{\nabla}^{\langle a}q_\gamma^{b\rangle} + 2\dot{u}^{\langle a}q_\gamma^{b\rangle} - 2\omega^c\eta_{cd}{}^{\langle a}\pi_\gamma^{b\rangle d}$$
$$+ \frac{10}{7}\sigma_c{}^{\langle a}\pi_\gamma^{b\rangle c} + \overline{\nabla}_c I^{abc} - \sigma_{cd}I^{abcd} = 0. \qquad (11.2)$$

The original EGS paper used a complicated combination of covariant and coordinate-based nonlinear analysis. Following Stoeger, Maartens and Ellis (1995), we present a shorter, more direct and transparent covariant analysis. EGS assumed that the only source of the gravitational field was the radiation, i.e. they neglected matter and assumed $\Lambda = 0$. Their result was generalized to include self-gravitating matter and dark energy by Clarkson and Maartens (2010) (extending previous results by Treciokas and Ellis (1971), Ferrando, Morales and Portilla (1992), Stoeger, Maartens and Ellis (1995), Clarkson and Barrett (1999), Räsänen (2009)):

Theorem 11.1 CMB isotropy + Copernican Principle $\rightarrow$ FLRW

In a region, if

- *collisionless radiation is exactly isotropic,*
- *the radiation four-velocity is geodesic and expanding,*
- *there are pressure-free baryons and CDM, and dark energy in the form of Λ, quintessence or a perfect fluid,*

then the metric is FLRW in that region.

Proof: For the fundamental 4-velocity we choose the radiation 4-velocity, i.e. $u^a = u_\gamma^a$, which has zero 4-acceleration and positive expansion:[1]

$$\dot{u}_a = 0, \ \Theta > 0. \tag{11.3}$$

Isotropy of the radiation distribution about u^a means that for fundamental observers, the photon distribution in momentum space depends only on components of the 4-momentum p^a along u^a, i.e. on the photon energy $E = -u_a p^a$:

$$f(x, p) = F(x, E), \ F_{a_1 \cdots a_\ell} = 0 \text{ for } \ell \geq 1. \tag{11.4}$$

All covariant multipoles of the distribution function beyond the monopole must vanish. In particular, it follows from (5.96) and (5.97), that the momentum density (from the dipole) and anisotropic stress (from the quadrupole) must vanish:

$$q_\gamma^a = 0 = \pi_\gamma^{ab}. \tag{11.5}$$

The radiation intensity octupole I_{abc} and hexadecapole I_{abcd} are also zero. Then the anisotropic stress evolution equation (11.2) enforces a shear-free expansion of the fundamental congruence:

$$\sigma_{ab} = 0. \tag{11.6}$$

We can also show that u^a is irrotational. Using (11.3), momentum conservation for radiation reduces to

$$\overline{\nabla}_a \rho_\gamma = 0. \tag{11.7}$$

Thus the radiation density is homogeneous relative to fundamental observers. Using energy conservation for radiation and the exact nonlinear identity (4.60) for the covariant curl of the gradient, we find

$$\text{curl } \overline{\nabla}_a \rho_\gamma = -2\dot{\rho}_\gamma \omega_a \ \Rightarrow \ \Theta \rho_\gamma \omega_a = 0. \tag{11.8}$$

By assumption $\Theta > 0$, and hence the vorticity must vanish:

$$\omega_a = 0. \tag{11.9}$$

---

[1] $\Theta \neq 0$ is essential: static spherically symmetric models are inhomogeneous but have isotropic CMB for all static observers (Ellis, Maartens and Nel, 1978).

Then we see from the curl shear constraint equation (4.52) that the magnetic Weyl tensor must vanish:

$$H_{ab} = 0. \tag{11.10}$$

Furthermore, (11.7) shows that the expansion must also be homogeneous. From the radiation energy conservation equation and (11.5), $\Theta = -3\dot{\rho}_\gamma/4\rho_\gamma$. Taking a covariant spatial gradient and using the time–space derivative commutation identity (4.62), we find

$$\overline{\nabla}_a \Theta = 0. \tag{11.11}$$

Then the shear divergence constraint (6.20) enforces the vanishing of the *total* momentum density in the fundamental frame,

$$q^a \equiv \sum_I q_I^a = 0 \Rightarrow \sum_I \gamma_I^2 (\rho_I^* + p_I^*) v_I^a = 0. \tag{11.12}$$

The second equality follows from (5.62), using the fact that the baryons, CDM and dark energy (in the form of quintessence or a perfect fluid) have no momentum density and anisotropic stress in their own frames,

$$q_I^{*a} = 0 = \pi_I^{*ab}, \tag{11.13}$$

where the asterisk denotes the intrinsic quantity (see Section 5.3). If we include other species, such as neutrinos, then the same assumption (11.13) applies to them. Except in unphysical special cases, it follows from (11.12) that

$$v_I^a = 0, \tag{11.14}$$

i.e. the bulk peculiar velocities of matter and dark energy (and any other self-gravitating species satisfying (11.13)) are forced to vanish – all species must be comoving with the radiation.

The comoving condition (11.14) then imposes the vanishing of the total anisotropic stress in the fundamental frame:

$$\pi^{ab} \equiv \sum_I \pi_I^{ab} = \sum_I \gamma_I^2 (\rho_I^* + p_I^*) v_I^{\langle a} v_I^{b \rangle} = 0, \tag{11.15}$$

where we have used (5.63), (11.13) and (11.14). Excluding unphysical special cases, the shear evolution equation (6.28) then leads to a vanishing electric Weyl tensor,

$$E_{ab} = 0. \tag{11.16}$$

Equations (11.12) and (11.15) now lead via the total momentum conservation equation (5.12) and the E-divergence constraint (6.33), to homogeneous total density and pressure:

$$\overline{\nabla}_a \rho = 0 = \overline{\nabla}_a p. \tag{11.17}$$

Equations (11.3), (11.6), (11.10), (11.11), (11.12), (11.15) and (11.17) constitute a covariant characterization of an FLRW spacetime. This establishes the generalized EGS theorem, extended from the original to include self-gravitating matter and dark energy. (We have also provided an alternative, 1+3 covariant, analysis.) It is straightforward to include other species such as neutrinos. The critical assumption needed for all species is the vanishing of the intrinsic momentum density and anisotropic stress, (11.13). The isotropy of the radiation and the geodesic nature of its 4-velocity then enforce the vanishing of (bulk) peculiar velocities v_I^a. We emphasize that one does *not* need to assume that the matter or other species are comoving with the radiation – it follows from the assumptions on the radiation. □

The original EGS result (i.e. without matter or dark energy) was generalized by Ellis, Treciokas and Matravers (ETM) (1983) to a much weaker assumption on the photon distribution: only the dipole, quadrupole and octupole need vanish. The key step is to show that the shear vanishes, without having zero hexadecapole. The quadrupole evolution equation (11.2) no longer automatically gives $\sigma_{ab} = 0$, and we need to find another way to show this. The elegant ETM trick is to return to the Liouville multipole equation, i.e. (5.84) in the massless collisionless case. The $\ell = 2$ equation, with $F_a = F_{ab} = F_{abc} = 0$, gives

$$\frac{12}{63} \frac{\partial}{\partial E} \left(E^5 \sigma^{ab} F_{abcd} \right) + E^5 \frac{\partial F}{\partial E} \sigma_{cd} = 0. \tag{11.18}$$

We integrate over E from 0 to ∞, and use the convergence property $E^5 F_{abcd} \to 0$ as $E \to \infty$. This gives

$$\sigma_{cd} \int_0^\infty E^5 \frac{\partial F}{\partial E} \, \mathrm{d}E = 0. \tag{11.19}$$

Integrating by parts, the integral reduces to $-5 \int_0^\infty E^4 F \, \mathrm{d}E$. Since $F > 0$, the integral is strictly negative, and thus we arrive at vanishing shear, $\sigma_{ab} = 0$. Then our proof above proceeds as before. Thus we have a generalization of the EGS–ETM theorem:

Theorem 11.2 CMB partial isotropy + CP → FLRW
In a region, if

- *collisionless radiation has vanishing dipole, quadrupole and octupole, $F_a = F_{ab} = F_{abc} = 0$,*
- *the radiation 4-velocity is geodesic and expanding,*
- *there are pressure-free baryons and CDM, and dark energy in the form of Λ, quintessence or a perfect fluid,*

then the metric is FLRW in that region.

This is the most powerful basis that we have – within the framework of the Copernican Principle – for background spatial homogeneity and thus an FLRW background model (see Section 9.8).

Although this theorem applies only to the 'background universe', its proof nevertheless requires a fully nonperturbative analysis.

In practice we can only observe approximate isotropy. Is the EGS result stable – i.e. does almost-isotropy of the CMB lead to an almost-FLRW universe? This would be the *realistic* basis for a perturbed FLRW model of the universe (assuming the Copernican Principle). Currently the result has only been established with further assumptions on the derivatives of the multipoles, by Stoeger, Maartens and Ellis (1995), Maartens, Ellis and Stoeger (1995b):

Theorem 11.3 CMB almost-isotropy + CP → almost-FLRW

In a region of an expanding universe with cosmological constant, if all observers comoving with the matter measure an almost isotropic distribution of collisionless radiation, and if some of the time and spatial derivatives of the covariant multipoles are also small, then the region is almost FLRW.

We emphasize that the perturbative assumptions are purely on the photon distribution, not on the matter or the metric – and one has to prove that the matter and metric are then perturbatively close to FLRW. Once again, a nonperturbative analysis is essential, since we are trying to prove an almost-FLRW spacetime, and we cannot assume an FLRW background a priori.

Almost-isotropy of the photon distribution means that $F_{a_1 \cdots a_\ell}(x, E) = \mathcal{O}(\epsilon)$ $(\ell \geq 1)$, where ϵ is a (dimensionless) smallness parameter. The intensity multipoles I_{A_ℓ} have dimensions of energy density and we therefore normalize them to the monopole $I = \rho_\gamma$: $I_{A_\ell}/I = \mathcal{O}(\epsilon)$. The task is to show that the relevant kinematical, dynamical and curvature quantities, suitably non-dimensionalized, are $\mathcal{O}(\epsilon)$. For example, the dimensionful kinematical quantities may be normalized by the expansion, $\sigma_{ab}/\Theta, \omega_a/\Theta$. The proof then follows the same pattern as our proof above of the exact EGS result – except that at each stage, we need to show that quantities are $\mathcal{O}(\epsilon)$ rather than equal zero. (Note that the almost-EGS result has not been proven for quintessence or perfect fluid dark energy, and this needs further investigation.)

However, in order to show this, we need smallness not just of the multipoles, but also of some of their derivatives. Smallness of the multipoles does not directly imply smallness of their derivatives, and we have to assume this (Nilsson *et al.*, 1999, Clarkson *et al.*, 2003, Räsänen, 2009). If all observers measure small multipoles, then it may be possible to show – perhaps using observations of galaxies in addition to the CMB – that the time and space derivatives on cosmologically significant scales must also be small. This remains an open question.

It may be possible to strengthen the above almost-EGS result by proving that it is sufficient for only the first three multipoles and their derivatives to be small. This would be an almost-EGS–ETM result, and would represent a more realistic foundation for almost-homogeneity than the almost-EGS result.

11.2 Linearized analysis of distribution multipoles

The multipoles $F_{A_\ell}(\ell \geq 1)$ and the projected gradient of the monopole, $\overline{\nabla}_a F$, are gauge-invariant measures of perturbations in the distribution function about an FLRW model. (As for most covariant and gauge-invariant perturbations, the variables do, however, depend on the choice of frame u^a.) For small departures from FLRW, the covariant and gauge-invariant variables will themselves be small and we can safely ignore products between small quantities.

In the FLRW background, the Liouville equation (5.77) for collisionless matter has general solution $f = \mathcal{F}(a\lambda)$ where $\mathcal{F}$ is an arbitrary function. We define the comoving momentum and energy,

$$q := a\lambda, \;\; \epsilon := aE, \;\; \epsilon^2 = q^2 + a^2 m^2, \tag{11.20}$$

where q is conserved in the background. We can then write the distribution function as $f(x^a, q, e^b)$, and the angular multipoles as $F_{A_\ell}(x^a, q)$. The spatial gradient of the scale factor obeys the linearized propagation equation

$$\dot{h}_a = \tfrac{1}{3} a \overline{\nabla}_a \Theta + a H \dot{u}_a, \;\; h_a := \overline{\nabla}_a a. \tag{11.21}$$

The multipoles of the Boltzmann equation (5.84) involve spacetime derivatives taken at fixed E (or λ). If we take the derivative at fixed q, then

$$\nabla_a F_{A_\ell}\Big|_\lambda = \nabla_a F_{A_\ell}\Big|_q + \left(\frac{1}{a} h_a - \frac{1}{3}\Theta a u_a \right) q \frac{\partial F_{A_\ell}}{\partial q}. \tag{11.22}$$

Then we can obtain the linearized multipole equations (Lewis and Challinor, 2002) (see Exercise 11.2.1),

$$\dot{F}_{A_\ell} + \frac{q}{\epsilon}\left(\frac{\ell+1}{2\ell+3} \right) \overline{\nabla}^b F_{bA_\ell} + \frac{q}{\epsilon} \overline{\nabla}_{\langle a_\ell} F_{A_{\ell-1} \rangle}$$

$$+ \delta_{\ell 1}\left(\frac{1}{a}\frac{q}{\epsilon} h_{a_1} - \frac{\epsilon}{q}\dot{u}_{a_1} \right) q \frac{\partial F}{\partial q} - \delta_{\ell 2}\sigma_{a_1 a_2} q \frac{\partial F}{\partial q} = \frac{a}{\epsilon} C_{A_\ell}[f], \tag{11.23}$$

where all spacetime derivatives are at fixed q. The dipole ($\ell = 1$) equation contains the variable

$$\mathcal{V}_a(q) := a\overline{\nabla}_a F\Big|_q + h_a q \frac{\partial F}{\partial q} = a\overline{\nabla}_a F\Big|_\lambda, \tag{11.24}$$

which obeys the evolution equation (see Exercise 11.2.1)

$$\dot{\mathcal{V}}_a = -\frac{aq}{3\epsilon}\overline{\nabla}_a\overline{\nabla}^b F_b + \dot{h}_a q \frac{\partial F}{\partial q} + \frac{a^2}{\epsilon}\dot{u}_a \bar{C}[f] + \frac{a^2}{\epsilon}\overline{\nabla}_a \bar{C}[f]\Big|_\lambda. \tag{11.25}$$

All spacetime derivatives are at fixed q, except for the last one. For a collisionless gas, (11.25) and the $\ell > 0$ multipole equations (11.23) form a closed system, given the kinematic equations. The collisionless forms of these equations are given in the synchronous and Newtonian gauges in Ma and Bertschinger (1995) and are used for numerical massive neutrino perturbations in Ma and Bertschinger (1995), Seljak and Zaldarriaga (1996), Dodelson, Gates and Stebbins (1996).

To solve (11.23) we decompose the spatial dependence of F_{A_ℓ} into scalar, vector and tensor parts which evolve independently in linear theory. In general, for $\ell > 2$, a rank-ℓ PSTF tensor can have higher-rank tensor contributions. But in linear theory there are no gravitational source terms for the higher-rank contributions. Therefore if we initialize in an early epoch when interactions are efficient in maintaining isotropy, the higher-rank contributions will not be present.

The potentials for the scalar, vector and tensor contributions can be expanded in terms of complete sets of harmonic eigenfunctions of the comoving projected Laplacian $a^2\overline{\nabla}^2$ (see Bruni, Dunsby and Ellis (1992)).

11.2.1 Scalar perturbations

The scalar component of F_{A_ℓ} is obtained by taking the PSTF part of ℓ projected derivatives of some scalar potential,

$$F_{A_\ell} = \overline{\nabla}_{\langle a_1} \cdots \overline{\nabla}_{a_\ell \rangle} S. \tag{11.26}$$

The potential S is expanded in terms of scalar-valued eigenfunctions, that satisfy

$$a^2\overline{\nabla}^2 Q^{(0)} + k^2 Q^{(0)} = 0, \quad \dot{Q}^{(0)} = 0. \tag{11.27}$$

These equations hold only at zero order, i.e. the harmonic functions are defined on the FLRW background. The superscript (0) denotes scalar perturbations, and for convenience we suppress the index (k) in $Q^{(0)}_{(k)}$. The allowed eigenvalues k^2 depend on the spatial curvature of the background model. Defining $\nu = k$ for $K = 0$ and $\nu^2 = (k^2 + K)/|K|$ for $K \neq 0$, where $6K/a^2$ is the curvature scalar of the FLRW spatial sections, the regular, normalizable eigenfunctions have $\nu \geq 0$ for open and flat models ($K \leq 0$). In Euclidean space, this implies all $k^2 \geq 0$. The $k = 0$ solutions are homogeneous and, therefore, do not appear in the expansion of first-order tensors, for example, $\Delta_a \equiv a\overline{\nabla}_a \rho/\rho$.

In open models, the modes with $\nu \geq 0$ form a complete set for expanding square-integrable functions, but they necessarily have $k \geq \sqrt{|K|}$ and so cannot describe correlations longer than the curvature scale (Lyth and Woszczyna, 1995). Super-curvature solutions (with $-1 < \nu^2 < 0$) can be constructed by analytic continuation. A super-curvature mode is generated in some models of open inflation (Bucher, Goldhaber and Turok, 1995).

In closed models ν is an integer ≥ 1 (Tomita, 1982, Abbott and Schaefer, 1986), and there are ν^2 linearly independent modes for each ν. The mode with $\nu = 1$ cannot be used to construct perturbations (its projected gradient vanishes globally), while the modes with $\nu = 2$ can only describe perturbations where all perturbed tensors with rank > 1 vanish (Bardeen, 1980).

For the scalar contribution to a rank-ℓ tensor like F_{A_ℓ}, we expand in rank-ℓ PSTF tensors $Q^{(0)}_{A_\ell}$ derived from the $Q^{(0)}$ via (Challinor and Lasenby, 1998, Gebbie and Ellis, 2000)

$$Q^{(0)}_{A_\ell} = \left(\frac{-a}{k}\right)^\ell \overline{\nabla}_{\langle a_1} \cdots \overline{\nabla}_{a_\ell \rangle} Q^{(0)}, \quad \dot{Q}^{(0)}_{A_\ell} = 0, \tag{11.28}$$

where the factor a^ℓ is necessary for the second equality. It follows that

$$Q^{(0)}_{A_\ell} = -\frac{a}{k}\overline{\nabla}_{\langle a_\ell} Q^{(0)}_{A_{\ell-1} \rangle}. \tag{11.29}$$

The multipole equation (11.23) also involves the divergence of $Q_{A_\ell}^{(0)}$, which satisfies (Challinor and Lasenby, 1999, Gebbie and Ellis, 2000)

$$\overline{\nabla}^{a_\ell} Q_{A_\ell}^{(0)} = \frac{k}{a} \frac{\ell}{(2\ell - 1)} \left[1 - (\ell^2 - 1) \frac{K}{k^2} \right] Q_{A_{\ell-1}}^{(0)}. \tag{11.30}$$

The curl satisfies (Challinor, 2000a)

$$\mathrm{curl}\, Q_{A_\ell}^{(0)} = 0, \tag{11.31}$$

where the curl of a general rank-ℓ PSTF tensor is defined by

$$\mathrm{curl}\, S_{A_\ell} = \eta_{bc\langle a_\ell} \overline{\nabla}^b S_{A_{\ell-1}\rangle}{}^c. \tag{11.32}$$

In closed models, the $Q_{A_\ell}^{(0)}$ vanish for $\ell \geq v$, so only modes with $v > \ell$ contribute to rank-ℓ tensors.

The decomposition of the distribution function into angular multipoles F_{A_ℓ}, and the subsequent expansion in the $Q_{A_\ell}^{(0)}$, combine to give a normal mode expansion which involves the objects $\overline{\nabla}_{\langle A_\ell\rangle} Q^{(0)} e^{A_\ell}$. For $K = 0$, with the $Q^{(0)}$ taken to be Fourier modes, this is equivalent to the usual Legendre expansion $P_\ell(\hat{\mathbf{k}} \cdot \mathbf{e})$ where $\hat{\mathbf{k}}$ is the Fourier wave vector (e.g. Ma and Bertschinger (1995)). In non-flat models, the expansion is equivalent to the Legendre tensor approach (Wilson, 1983). The advantage of separating the angular and scalar harmonic decompositions is that the former can be applied quite generally for an arbitrary cosmological model. Furthermore, extending the normal-mode expansions to cover polarization and vector and tensor modes in non-flat models is then rather trivial.

11.2.2 Vector perturbations

The vector component of F_{A_ℓ} is the PSTF part of $\ell - 1$ projected derivatives of a (projected) divergence-free vector potential: $F_{A_\ell} = \overline{\nabla}_{\langle A_{\ell-1}} V_{a_\ell\rangle}$. Such a potential may be expanded in PSTF rank-1 eigenfunctions of the Laplacian,

$$a^2 \overline{\nabla}^2 Q_a^{(\pm 1)} + k^2 Q_a^{(\pm 1)} = 0, \quad \overline{\nabla}^a Q_a^{(\pm 1)} = 0 = \dot{Q}_a^{(\pm 1)}. \tag{11.33}$$

The superscript (± 1) labels the two possible parities ('electric' and 'magnetic') of the vector harmonics (Tomita, 1982). These can be chosen so that

$$\mathrm{curl}\, Q_a^{(\pm 1)} = \frac{k}{a} \sqrt{1 + \frac{2K}{k^2}} Q_a^{(\mp 1)}, \tag{11.34}$$

which ensures that the parities have the same normalization. For vector modes we define $v = k$ for $K = 0$ and $v^2 = (k^2 + 2K)/|K|$ for $K \neq 0$. The regular, normalizable eigenmodes have $v \geq 0$ for flat and open models, while for closed models v is an integer ≥ 2.

We can differentiate the $Q_a^{(\pm 1)}$ to form PSTF tensor eigenfunctions:

$$Q_{A_\ell}^{(\pm 1)} = \left(\frac{-a}{k} \right)^{\ell - 1} \overline{\nabla}_{\langle A_{\ell-1}} Q_{a_\ell\rangle}^{(\pm 1)}, \quad \dot{Q}_{A_\ell}^{(\pm 1)} = 0. \tag{11.35}$$

They satisfy the same recursion relation (11.29) as the scalar harmonics. The projected divergence obeys (Lewis, 2004b)

$$\overline{\nabla}^{a_\ell} Q_{A_\ell}^{(\pm 1)} = \frac{k}{a} \frac{(\ell^2 - 1)}{\ell(2\ell - 1)} \left[1 - (\ell^2 - 2)\frac{K}{k^2} \right] Q_{A_{\ell-1}}^{(\pm 1)}, \tag{11.36}$$

and the curl gives

$$\text{curl } Q_{A_\ell}^{(\pm 1)} = \frac{1}{\ell} \frac{k}{a} \sqrt{1 + \frac{2K}{k^2}} \, Q_{A_\ell}^{(\mp 1)}. \tag{11.37}$$

(See Exercise 11.2.2.) As in the case of scalar perturbations, $Q_{A_\ell}^{(\pm 1)} = 0$ when $\ell \geq v$ in closed models.

11.2.3 Tensor perturbations

The tensor component of F_{A_ℓ} is the PSTF part of the $(\ell\text{-}2)$-th derivatives of a PSTF, divergence-free rank-2 tensor potential: $F_{A_\ell} = \overline{\nabla}_{\langle A_{\ell-2}} S_{a_{\ell-1} a_\ell \rangle}$. This potential may be expanded in the PSTF rank-2 eigenfunctions of the Laplacian,

$$a^2 \overline{\nabla}^2 Q_{ab}^{(\pm 2)} + k^2 Q_{ab}^{(\pm 2)} = 0, \quad \overline{\nabla}^b Q_{ab}^{(\pm 2)} = 0 = \dot{Q}_{ab}^{(\pm 2)}. \tag{11.38}$$

The superscript (± 2) labels the two possible parity states for the tensor harmonics (Thorne, 1980, Tomita, 1982, Challinor, 2000a). The states can be conveniently chosen so that

$$\text{curl } Q_{ab}^{(\pm 2)} = \frac{k}{a} \sqrt{1 + \frac{3K}{k^2}} \, Q_{ab}^{(\mp 2)}. \tag{11.39}$$

For tensor modes we define $v = k$ when $K = 0$ and $v^2 = (k^2 + 3K)/|K|$ if $K \neq 0$. The regular, normalizable eigenmodes have $v \geq 0$ for flat and open models, while for closed models v is an integer ≥ 3.

As for scalar and vector perturbations, we can form rank-ℓ PSTF tensors $Q_{A_\ell}^{(\pm 2)}$ by differentiation:

$$Q_{A_\ell}^{(\pm 2)} := \left(\frac{-a}{k} \right)^{\ell-2} \overline{\nabla}_{\langle A_{\ell-2}} Q_{a_{\ell-1} a_\ell \rangle}^{(\pm 2)}, \tag{11.40}$$

and they satisfy the same recursion relation (11.29) as the scalar harmonics. The projected divergence and curl are (Challinor, 2000a)

$$\overline{\nabla}^{a_\ell} Q_{A_\ell}^{(\pm 2)} = \frac{k}{a} \frac{(\ell^2 - 4)}{\ell(2\ell - 1)} \left[1 - (\ell^2 - 3)\frac{K}{k^2} \right] Q_{A_{\ell-1}}^{(\pm 2)}, \tag{11.41}$$

$$\text{curl } Q_{A_\ell}^{(\pm 2)} = \frac{2}{\ell} \frac{k}{a} \sqrt{1 + \frac{3K}{k^2}} \, Q_{A_\ell}^{(\mp 2)}. \tag{11.42}$$

As before, in closed models, $Q_{A_\ell}^{(\pm 2)} = 0$ for $\ell \geq v$.

Combining the angular and spatial expansions gives a set of normal-mode functions $\overline{\nabla}_{\langle A_{\ell-2}} Q_{a_{\ell-1} a_\ell \rangle}^{(\pm 2)} e_\ell^A$. This generalizes Wilson's approach (Wilson, 1983) for scalar perturbations to tensor modes.

Exercise 11.2.1
(a) Use (11.22) in (5.84), to prove (11.23).
(b) Use the gradient of the monopole of (11.23) to derive (11.25).

Exercise 11.2.2 Derive (11.36) and (11.37).

11.3 Temperature anisotropies in the CMB

The photon distribution function in the FLRW limit is the Planck distribution:

$$\bar{F}(q) = \left[\exp\left(\frac{q}{k_B T_0 a_0}\right) - 1\right]^{-1} = \left[\exp\left(\frac{E}{k_B T}\right) - 1\right]^{-1}, \qquad (11.43)$$

where $q = \epsilon = aE$ for massless particles, and the temperature is $T = T_0(a_0/a)$, for any fixed epoch a_0. The redshifting of energy ($E = q/a$) and temperature with expansion combine to preserve the Planck form of the background distribution, whether it is in collision-dominated or collisionless equilibrium.

In the perturbed universe, the distribution is $f = \sum F_{A_\ell} e^{A_\ell}$, where the monopole at zero order is the Planck distribution: $F = \bar{F}$. For scalar perturbations,

$$F_{A_\ell}(t,\mathbf{x},q) = -\frac{\pi}{\Delta_\ell} \frac{dF(q)}{d\ln q} \sum_k F_\ell^{(0)}(t,k) Q_{A_\ell}^{(0)}(t,k), \qquad \ell \geq 1, \qquad (11.44)$$

where the momentum-dependent prefactor is chosen so that the $F_\ell^{(0)}$ are independent of q. (Note that this is not the case for massive particles (Tsagas, Challinor and Maartens, 2008).) In the massless case, $\sum_k F_\ell^{(0)} Q_{A_\ell}^{(0)}$ are proportional to the multipoles of the temperature anisotropy.

For the gradient of the monopole, we define the harmonic coefficient $F_0^{(0)}$ via

$$a\bar{\nabla}_a F = -F \sum_k k F_0^{(0)} Q_a^{(0)}, \qquad (11.45)$$

so that

$$\Delta_a^\gamma = \frac{a}{\rho_\gamma} \bar{\nabla}_a \rho_\gamma = -\sum_k k F_0^{(0)} Q_a^{(0)}. \qquad (11.46)$$

The radiation momentum density and anisotropic stress are given by (see Exercise 11.3.1)

$$q_a^\gamma = \rho_\gamma \sum_k F_1^{(0)} Q_a^{(0)}, \qquad \pi_{ab}^\gamma = \rho_\gamma \sum_k F_2^{(0)} Q_{ab}^{(0)}, \qquad (11.47)$$

where ρ_γ is the radiation density.

The kinematic quantities are expanded as

$$h_a = -\sum_k kh\, Q_a^{(0)}, \qquad \dot{u}_a = \sum_k \frac{k}{a} A\, Q_a^{(0)}, \qquad \sigma_{ab} = \sum_k \frac{k}{a} \sigma\, Q_{ab}^{(0)}. \qquad (11.48)$$

After decoupling, the photon multipoles satisfy (11.23) and (11.25) with vanishing collision terms. Expanding in harmonics, we find (Lewis and Challinor, 2002)

$$\dot{F}_\ell^{(0)} + \frac{k}{a}\left\{\frac{\ell+1}{2\ell+1}\left[1-\left((\ell+1)^2-1\right)\frac{K}{k^2}\right]F_{\ell+1}^{(0)} - \frac{\ell}{2\ell+1}F_{\ell-1}^{(0)}\right\}$$
$$+ 4\delta_{\ell 0}\dot{h} + \delta_{\ell 1}\frac{4}{3}\frac{k}{a}(h+A) + \delta_{\ell 2}\frac{8}{15}\frac{k}{a}\sigma = 0. \tag{11.49}$$

Here we do not consider spectral distortions in the CMB, which are an important probe of the energetics of the universe (see e.g. Burigana and Salvaterra (2003)). If we neglect spectral distortions, the linearized CMB temperature anisotropy (and polarization brightness) are independent of energy, since linear perturbations in f inherit the spectral dependence $q\partial\bar{F}/\partial q$ of the Planck distribution (11.43). Then we can integrate over energy without loss of information, to define bolometric multipoles (5.101):

$$I_{A_\ell} = \Delta_\ell \int_0^\infty dE\, E^3 F_{A_\ell}, \quad \ell \geq 0. \tag{11.50}$$

The normalization ensures that the three lowest multipoles give the radiation dynamical quantities:

$$I = \rho_\gamma, \qquad I^a = q_\gamma^a, \qquad I^{ab} = \pi_\gamma^{ab}. \tag{11.51}$$

The fractional anisotropy in the CMB temperature is $\delta_T(e^a) = [T(e^a)-T]/T$. Then to first order,

$$\delta_T(e^a) = \frac{\pi}{I}\int_0^\infty dE\, E^3[f(E,e^a)-F(E,e^a)] = \frac{\pi}{I}\sum_{l\geq 1}\Delta_\ell^{-1}I_{A_\ell}e^{A_\ell}. \tag{11.52}$$

If the primordial perturbations are close to being Gaussian distributed, as in the case of simple inflationary models, then linearized CMB fluctuations are also close to Gaussian. If we further assume that the statistical properties of the fluctuations are invariant under the background symmetries, then the CMB power spectra fully characterize the statistics of the CMB anisotropies and polarization. The temperature power spectrum C_ℓ^T is defined in terms of the I_{A_ℓ} by

$$\left(\frac{\pi}{I}\right)^2\left\langle I_{A_\ell}I^{B_{l'}}\right\rangle = \Delta_\ell C_\ell^T \delta_\ell^{\ell'} h_{\langle A_\ell\rangle}^{\langle B_\ell\rangle}, \quad h_{\langle A_\ell\rangle}^{\langle B_\ell\rangle} := h_{\langle a_1}^{\langle b_1}\ldots h_{a_\ell\rangle}^{b_\ell\rangle}. \tag{11.53}$$

The angle brackets on the left-hand side denote a statistical average over the ensemble of fluctuations. Equation (11.53) is equivalent to the definition in terms of the variance of $a_{\ell m}$, where $\delta_T(e) = \sum_{\ell>0}a_{\ell m}Y_{\ell m}(e)$.

The temperature correlation function is (Exercise 11.3.2)

$$\left\langle\delta_T(e^c)\delta_T(e'^c)\right\rangle = \sum_{\ell\geq 1}\frac{(2\ell+1)}{4\pi}C_\ell^T P_\ell(\cos\theta), \tag{11.54}$$

where P_ℓ is a Legendre polynomial.

The splitting of the photon 4-momentum into energy and momentum depends on the choice of 4-velocity u^a. For a new velocity field $\tilde{u}^a = \gamma(u^a + v^a)$, where v^a is the projected relative velocity in the u^a frame and γ is the associated Lorentz factor, we have

$p^a = E(u^a + e^a) = \tilde{E}(\tilde{u}^a + \tilde{e}^a)$, where $u^a e_a = 0 = \tilde{u}^a \tilde{e}_a$. The energy and propagation directions in the $\tilde{u}^a$ frame are given by the Doppler and aberration formulae:

$$\tilde{E} = \gamma E(1 - e^a v_a) = E(1 - e^a v_a), \tag{11.55}$$

$$\tilde{e}^a = [\gamma(1 - e^b v_b)]^{-1}(u^a + e^a) - \gamma(u^a + v^a)$$
$$= e^a - v^a + (u^a + e^a)e^b v_b, \tag{11.56}$$

where the second equality in each case gives the linearized result. Using the invariance of $f(E, e^a)$, the bolometric multipoles transform as

$$\tilde{I}_{A_\ell} = \sum_{\ell'} \Delta_{\ell'}^{-1} I_{B_{\ell'}} \int d\Omega \, [\gamma(1 - e^b v_b)]^2 e^{B_{\ell'}} \tilde{e}_{\langle A_\ell \rangle} = I_{A_\ell}, \tag{11.57}$$

where the second equality holds to first order: therefore the multipoles are frame invariant in the linearized case.

Exercise 11.3.1 Prove (11.47).

Exercise 11.3.2 Use the result

$$e^{\langle A_\ell \rangle} e'_{\langle A_\ell \rangle} = \frac{(2\ell + 1)}{4\pi} \Delta_\ell P_\ell(\cos\theta), \quad e^a e'_a = \cos\theta, \tag{11.58}$$

to derive (11.54).

11.4 Thomson scattering

The dominant collisional process relevant to CMB anisotropies and polarization during recombination and reionization is Compton scattering. To an excellent approximation we can ignore electron recoil in the rest frame of the scattering electron, Pauli blocking and induced scattering. Then Compton scattering is very accurately approximated by classical Thomson scattering in the electron rest frame, with no change in the photon energy. Furthermore, we can neglect the small velocity dispersion of the electrons arising from their small finite temperature, and treat the problem as one of scattering off a cold gas of electrons. (Note that Compton scattering must be used for a hot gas, such as the intra-cluster gas that generates the Sunyaev-Zel'dovich effect; see Birkinshaw (1999) for a review.)

We denote the electron rest frame by $\tilde{u}^a$, with proper electron number density $\tilde{n}_e$. Then the projected collision tensor in the Thomson limit in that frame is (Challinor, 2000a)

$$\tilde{C}[f](\tilde{E}, \tilde{e}^a) = \tilde{n}_e \sigma_T \tilde{E} \left[-f(\tilde{E}, \tilde{e}^a) + \tilde{F}(\tilde{E}, \tilde{e}^a) + \tfrac{1}{10} \tilde{F}_{bc}(\tilde{E}, \tilde{e}^a) \tilde{e}^b \tilde{e}^c \right], \tag{11.59}$$

where σ_T is the Thomson cross-section and we have neglected polarization (see Section 11.6 for the case with polarization). This expression for the scattering term follows from inserting the multipole decomposition of the distribution into the kernel for Thomson in-scattering, and integrating over scattering directions. Scattering out of the phase-space element is described by $-\tilde{n}_e \sigma_T \tilde{E}^3 f$. In-scattering couples to the monopole and quadrupole in total intensity, and to the E-mode quadrupole. There is no change in energy density in the electron rest frame due to Thomson scattering, but there is momentum exchange if the radiation has a dipole moment.

Transforming to a general frame u^a, and keeping only first-order terms, (11.59) becomes

$$
C[f](E,e^a) = n_e \sigma_T E \Big[-f(E,e^a) + F(E,e^a)
$$

$$
- e^b v_b E \frac{\partial}{\partial E} F(E,e^a) + \tfrac{1}{10} F_{bc}(E,e^a)e^b e^c \Big], \qquad (11.60)
$$

where n_e is the electron density relative to u^a. Now the multipole expansion of the Boltzmann equation leads to the total intensity multipole equations,

$$
\dot{I}_{A_\ell} + 4H I_{A_\ell} + \overline{\nabla}^b I_{bA_\ell} + \frac{l}{(2\ell+1)} \overline{\nabla}_{\langle a_\ell} I_{A_{\ell-1}\rangle} + \frac{4}{3} I \dot{u}_{a_1}\delta_{l1} + \frac{8}{15} I \sigma_{a_1 a_2}\delta_{l2}
$$

$$
= -n_e \sigma_T \Big[I_{A_\ell} - I\delta_{l0} - \tfrac{4}{3} I v_{a_1}\delta_{l1} - \tfrac{1}{10} I_{a_1 a_2}\delta_{l2} \Big]. \qquad (11.61)
$$

The monopole moment of (11.61) does not vanish in the background and we use its projected gradient to characterize the perturbation in the radiation energy density:

$$
\dot{\Delta}^\gamma_a + \frac{a}{I} \overline{\nabla}_a \overline{\nabla}^b I_b + 4\dot{h}_a = 0, \qquad (11.62)
$$

where, to linear order, $3\dot{h}_a = a(3H\dot{u}_a + \overline{\nabla}_a\Theta)$ from (11.21). The above equation also follows from integrating (11.25) with $\lambda^3 \, d\lambda$ and noting that the linear Thomson collision term has no monopole.

Equations (11.61) and (11.62) provide a complete description of the linear evolution of the CMB anisotropies in the absence of polarization, in general almost-FLRW models. In particular, they are valid for all types of perturbation since no harmonic expansion has been made. We see that the highest rank of the source terms is $\ell = 2$, so that only scalar, vector and tensor modes can be excited.

11.5 Scalar perturbations

We expand the PSTF multipoles in the harmonic tensors $Q^{(0)}_{A_\ell}$, defined in (11.28):

$$
I_{A_\ell} = I \sum_k \left(\prod_{n=0}^{\ell} \kappa_n^{(0)} \right)^{-1} I_\ell^{(0)} Q^{(0)}_{A_\ell}, \qquad \ell \geq 1, \qquad (11.63)
$$

$$
\Delta^\gamma_a = \frac{a\overline{\nabla}_a I}{I} = -\sum_k k I_0^{(0)} Q^{(0)}_a, \qquad (11.64)
$$

where

$$
\kappa_\ell^{(m)} := [1 - K(\ell^2 - 1 - m)k^{-2}]^{1/2}, \qquad \ell \geq m, \qquad \kappa_0^{(0)} = 1. \qquad (11.65)
$$

In scalar harmonic form, the linearized multipole equations (11.61) become, on using (11.28), (11.30) and (11.48),

$$\dot{I}_\ell^{(0)} + \frac{k}{a}\left[\frac{(\ell+1)}{(2\ell+1)}\kappa_{\ell+1}^{(0)}I_{\ell+1}^{(0)} - \frac{\ell}{(2\ell+1)}\kappa_\ell^{(0)}I_{\ell-1}^{(0)}\right] + 4\dot{h}\delta_{\ell 0} + \frac{4}{3}\frac{k}{a}A\delta_{\ell 1}$$

$$+ \frac{8}{15}\frac{k}{a}\kappa_2^{(0)}\sigma\delta_{\ell 2} = -n_e\sigma_T\left[I_\ell^{(0)} - I_0^{(0)}\delta_{\ell 0} - \frac{4}{3}v\delta_{\ell 1} - \frac{1}{10}I_2^{(0)}\delta_{\ell 2}\right], \qquad (11.66)$$

for $\ell \geq 0$, where v is the harmonic coefficient of the baryon–electron velocity relative to u^a,

$$v_a = \sum v Q_a^{(0)}. \qquad (11.67)$$

These multipole equations hold for a general FLRW model and are fully equivalent to those obtained in Hu et al. (1998) using the total angular momentum method.

In closed models, $Q_{A\ell}^{(m)}$ vanishes for $\ell \geq \nu$, and therefore the same is true of $I_{A\ell}$. Power moves up the hierarchy as far as the $\ell = \nu - 1$ multipole, but is then reflected back down. This is enforced in (11.66) by $\kappa_\nu^{(m)} = 0$. The maximum multipole, and corresponding minimum angular scale, arise because of the focusing of geodesics in closed FLRW models.

Early computer codes to compute the CMB anisotropy integrated a carefully truncated version of the multipole equations directly. A major advance was made in Seljak and Zaldarriaga (1996), where the Boltzmann hierarchy was formally integrated, thus allowing a very efficient solution for the CMB anisotropy. This procedure was implemented in the CMB-FAST code,[2] and later, in parallelized derivative codes such as CAMB (Lewis, Challinor and Lasenby, 2000).[3]

The integral solution for the total intensity for general spatial curvature is (Zaldarriaga, Seljak and Bertschinger, 1998, Hu et al., 1998, Challinor, 2000a)

$$I_\ell^{(0)} = 4\int^{t_R} dt\, e^{-\tau}\left\{\left[-\frac{k}{a}\sigma + \frac{3n_e\sigma_T}{16\kappa_2^{(0)}}I_2^{(0)}\right]\left[\frac{1}{3}\Phi_\ell^\nu(x) + \frac{1}{(\nu^2+1)}\frac{d^2}{dx^2}\Phi_\ell^\nu(x)\right]\right.$$

$$\left. - \left(\frac{k}{a}A - n_e\sigma_T v\right)\frac{1}{\sqrt{\nu^2+1}}\frac{d}{dx}\Phi_\ell^\nu(x) - \left[\dot{h} - \frac{1}{4}n_e\sigma_T I_\ell^{(0)}\right]\Phi_\ell^\nu(x)\right\}, \qquad (11.68)$$

where, $\tau := \int n_e\sigma_T dt$ is the optical depth back along the line of sight, $x = \sqrt{|K|}\chi$ with χ the comoving radial distance (or, equivalently, conformal look-back time) along the line of sight, and $\Phi_\ell^\nu(x)$ are the ultra-spherical Bessel functions with $\nu^2 = (k^2 + K)/|K|$ for scalar perturbations.

In the linearized case, $I_\ell^{(0)}$ will depend linearly on the primordial perturbation ϕ_k via the transfer function:

$$I_\ell^{(0)} = T_\ell^T(k)\phi_k. \qquad (11.69)$$

The symmetry of the background ensures that the transfer functions depend only on the magnitude of the wavenumber k. The choice of ϕ_k is one of convention. For the adiabatic, growing-mode initial conditions that follow from single-field inflation, the convenient

---

[2] http://www.cmbfast.org
[3] http://camb.info/

choice is the (constant) curvature perturbation $\mathcal{R}_k$ on comoving hypersurfaces. For models with isocurvature fluctuations, the relative entropy gradient is appropriate. More generally, in models with mixed initial conditions having N degrees of freedom per harmonic mode, the transfer functions generalize to N functions per ℓ and k.

The power spectrum of ϕ may be defined via (Tsagas, Challinor and Maartens, 2008)

$$\left\langle \phi^2 \right\rangle = \int \frac{v\, dv}{(v^2 + 1)} \mathcal{P}_\phi(k), \tag{11.70}$$

and then the temperature power spectrum is

$$C_\ell^T = \frac{\pi}{4} \int \frac{v\, dv}{(v^2 + 1)} T_\ell^T(k) T_\ell^T(k) \mathcal{P}_\phi(k). \tag{11.71}$$

In flat and open models, $v\, dv/(v^2 + 1) = d\ln k$; in closed models, one replaces $v^2 + 1$ by $v^2 - 1$ and the integral becomes a discrete sum over integer v.

At large enough angular scales we can neglect anisotropic Thomson scattering, reionization and the finite width of the last scattering surface. The simple physics of scalar temperature anisotropies is apparent if we use the conformal Newtonian gauge, for which the shear of u^a vanishes. In this frame, the shear propagation equation becomes a constraint that determines the acceleration:

$$\overline{\nabla}_{\langle a} \dot{u}_{b \rangle} = E_{ab} - \tfrac{1}{2}\pi_{ab}. \tag{11.72}$$

The scalar modes of the electric Weyl tensor and anisotropic stress are

$$E_{ab} = \sum_k \frac{k^2}{a^2} \Phi_E Q_{ab}^{(0)}, \quad \pi_{ab} = \rho \sum_k \frac{k^2}{a^2} \Pi Q_{ab}^{(0)}. \tag{11.73}$$

Then the harmonic form of (11.72) is

$$A = -\Phi_E + \tfrac{1}{2}\rho a^2 \Pi, \tag{11.74}$$

and in the Newtonian gauge (10.55),

$$A = -\Phi, \quad \Phi_E = \tfrac{1}{2}(\Phi + \Psi). \tag{11.75}$$

Then (11.68) reduces to

$$\left[\frac{1}{4} I_\ell^{(0)} - \Phi \delta_{\ell 0} \right]_{\mathrm{dec}} = \left(\frac{I_0^{(0)}}{4} - \Phi \right) \Phi_\ell^v \bigg|_{\mathrm{dec}} + v_N \frac{1}{\sqrt{v^2 + 1}} \frac{d\Phi_\ell^v}{dx} \bigg|_{\mathrm{dec}}$$
$$+ \int_{t_{\mathrm{dec}}}^{t_0} (\dot{\Phi} + \dot{\Psi}) \Phi_\ell^v \, dt. \tag{11.76}$$

The temperature anisotropy is determined by three terms at decoupling: (1) intrinsic temperature variations $I_0^{(0)}/4$; (2) the Newtonian potential Φ which describes gravitational redshifting; and (3) Doppler shifts, where v_N^a is the baryon velocity relative to the zero-shear u^a. The integrated Sachs–Wolfe term in (11.76) arises because of the net blueshift as a photon crosses a decaying potential well. It contributes when the Weyl potential evolves in time, such as when dark energy starts to dominate the expansion dynamics at low redshift.

11.5.1 Tight coupling and the acoustic peaks

On comoving scales $\sim 30\,\mathrm{Mpc}$ or greater, photon diffusion due to the finite mean-free path to Thomson scattering can be ignored. In this limit, the dynamics of the source terms in (11.76) are those of a driven oscillator (Hu and Sugiyama, 1995). First we note from (11.61) that in the limit of tight-coupling,

$$I_a = \tfrac{4}{3} I v_a, \qquad I_{A_\ell} = 0 \text{ for } \ell \geq 2. \tag{11.77}$$

The CMB is therefore isotropic in the baryon rest frame and the linearized momentum evolution for the combined photon–baryon fluid gives

$$\dot{v}_a + \frac{HR}{(1+R)} v_a + \frac{1}{4(1+R)a} \Delta_a^\gamma + \dot{u}_a = 0, \; R := \frac{3\rho_b}{4\rho_\gamma}, \tag{11.78}$$

ignoring baryon pressure. The evolution of Δ_a^γ follows from (11.62):

$$\dot{\Delta}_a^\gamma + 4\dot{h}_a + \frac{4}{3} a \overline{\nabla}_a \overline{\nabla}^b v_b = 0. \tag{11.79}$$

Then (Exercise 11.5.1),

$$\Delta_a^{\gamma \prime\prime} + \frac{\mathcal{H}R}{(1+R)} \Delta_a^{\gamma \prime} - \frac{1}{3(1+R)} a^2 \overline{\nabla}_a \overline{\nabla}^b \Delta_b^\gamma = -4h_a'' - \frac{4\mathcal{H}R}{(1+R)} h_a' + \frac{4}{3} a^3 \overline{\nabla}_a \overline{\nabla}^b \dot{u}_b, \tag{11.80}$$

where $\mathcal{H} = aH$ is the conformal Hubble parameter. This equation is valid in any frame and describes a driven oscillator. The free oscillations are at frequency kc_s, where the sound speed is $c_s^2 = 1/[3(1+R)]$, and are damped by the expansion of the universe. In the Newtonian frame, we can express the driving terms on the right in terms of Φ and Ψ. Using (Exercise 11.5.1),

$$a\overline{\nabla}_{\langle a} \dot{h}_{b\rangle} = (a^2 \overline{\nabla}_{\langle a} \overline{\nabla}_{b\rangle} \Psi)^{\cdot} \;\; \Rightarrow \;\; \dot{h} = \dot{\Psi}, \tag{11.81}$$

we can recover the standard harmonic form of the oscillator equation in the Newtonian gauge (Exercise 11.5.1):

$$\Delta'' + \frac{\mathcal{H}R}{(1+R)} \Delta' + \frac{k^2}{3(1+R)} \Delta = -4\Psi'' - \frac{4\mathcal{H}R}{(1+R)} \Psi' - \frac{4}{3} k^2 \Phi. \tag{11.82}$$

For adiabatic initial conditions, the cosine solution of (11.82) is excited and all modes with $k \int^{\tau_{\mathrm{dec}}} c_s \, d\tau = n\pi$ are at extrema of their oscillation at last scattering. This gives a series of acoustic oscillations in the temperature power spectrum (Zel'dovich and Sunyaev, 1969). The first three have been observed by a combination of terrestrial experiments and the WMAP satellite (Dunkley et al., 2009). Figure 11.1 shows the temperature anisotropy data points from these observations, and the best-fit ΛCDM curve.

Examples of the CMB power spectra in a ΛCDM model are shown in Figure 11.2. The acoustic peaks are a rich source of cosmological information. Their relative heights depend on the baryon density (i.e. R) and matter density, since these affect the midpoint of the acoustic oscillation and the efficacy of the gravitational driving in (11.82) (Hu and

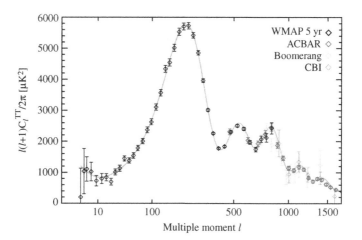

Fig. 11.1 Temperature power spectrum: data points from WMAP (5-year), Boomerang, CBI and ACBAR, and the best-fit ΛCDM curve. (From Tsagas, Challinor and Maartens (2008).)

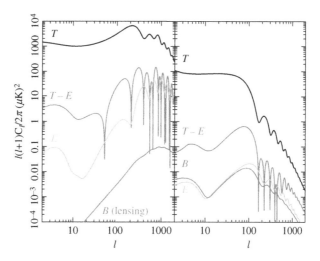

Fig. 11.2 Temperature and polarization (E and B modes) power spectra produced by adiabatic scalar perturbations (left) and tensor perturbations (right), for a tensor-to-scalar ratio $r = 0.28$ and optical depth to reionization of 0.08. The B-modes produced by gravitational lensing of the scalar E-mode polarization are also shown on the left. (From Tsagas, Challinor and Maartens (2008). A colour version of this figure is available online.

Sugiyama, 1995). The angular position of the peaks depends on the type of initial condition and on the angular diameter distance to last scattering. Moreover, the general shape of the spectra is related to the distribution of primordial power with scale, i.e. the power spectrum $\mathcal{P}_\phi(k)$.

On smaller scales photon diffusion becomes important. The breakdown of tight coupling has two important effects on the CMB. First, the acoustic oscillations are exponentially damped, as apparent in Figure 11.2. Second, anisotropies can start to grow in the CMB intensity and this produces linear polarization via Thomson scattering.

Exercise 11.5.1

(a) Use (11.78) and (11.79) to prove (11.80).
(b) Use (11.21) and (11.72), and the zero-shear $\dot{E}_{ab}$ equation, to derive (11.81).
(c) Finally, derive (11.82).

11.6 CMB polarization

As well as the total intensity, the polarization properties of the CMB are of great importance. Here we give a brief summary of the key features; further details may be found in Tsagas, Challinor and Maartens (2008).

   To include polarization, we describe the CMB photons by a one-particle distribution function that is tensor-valued: $f_{bc}(x^a, p^a)$ (Berestetskii, Lifshitz and Pitaevskii, 1982). This is a Hermitian tensor defined so that the expected number of photons contained in a proper phase-space element $d^3x d^3p$, and with polarization state ϵ^a is $\epsilon^{a*} f_{ab} \epsilon^b d^3x d^3p$. The complex polarization 4-vector ϵ^a is orthogonal to the photon momentum, $\epsilon^a p_a = 0$ (adopting the Lorentz gauge), and is normalized as $\epsilon_a^* \epsilon^a = 1$. The distribution function is also defined to be orthogonal to p^a so $f_{ab} p^a = 0$. For a photon in a pure polarization state ϵ^a, the direction of the electric field measured by u^a is $s^a{}_b \epsilon^b$ where $s_{ab} := h_{ab} - e_a e_b$ is the screen projection tensor (7.22).

   The (Lorentz-gauge) polarization 4-vector is only unique up to constant multiples of p^a, reflecting the remaining gauge freedom, but the *observed* polarization vector $s^a{}_b \epsilon^b$ is unique. To remove the residual gauge freedom from the distribution function f_{ab}, we can work directly with the screen-projected polarization tensor,

$$P_{ab} \propto E^3 s_a{}^c s_b{}^d f_{cd} . \tag{11.83}$$

The factor E^3 is included for convenience. We decompose P_{ab} into its irreducible components,

$$P_{ab}(E, e^d) = \tfrac{1}{2} I(E, e^d) s_{ab} + \mathcal{P}_{ab}(E, e^d) + \tfrac{1}{2} i V(E, e^d) \eta_{abc} e^c . \tag{11.84}$$

This defines the total intensity brightness, I, the circular polarization, V, and the linear polarization tensor $\mathcal{P}_{ab}$, which is PSTF and transverse to e^a. For quasi-monochromatic radiation with electric field $\mathrm{Re}[E^a(t) \exp(-i\omega t)]$, where ω is the angular frequency and the complex amplitude E^a varies little over a wave period, we have

$$P^{ab} \propto \langle E^a E^{b*} \rangle , \tag{11.85}$$

where the angle brackets denote time-averaging.

   The linear polarization is often described in terms of Stokes brightness parameters Q and U (Chandrasekhar, 1960) which measure the difference in intensity between radiation transmitted by a pair of orthogonal polarizers (for Q), and the same but after a right-handed rotation of the polarizers by 45 degrees about the propagation direction e^a (for U). If we introduce a pair of orthogonal polarization vectors e_1^a and e_2^a, orthogonal to u^a and e^a, and oriented so that $\{u^a, e_i^a, e^a\}$ form a right-handed orthonormal tetrad, we have

$$\mathcal{P}_{ab} e_i^a e_j^b = \frac{1}{2} \begin{pmatrix} I+Q & U+iV \\ U-iV & I-Q \end{pmatrix}. \tag{11.86}$$

The invariant $2\mathcal{P}^{ab}\mathcal{P}_{ab} = Q^2 + U^2$ is the squared magnitude of the linear polarization.

Since $I(E, e^c)$ and $V(E, e^c)$ are scalar functions on the sphere $e^a e_a = 1$ at a point in spacetime, their local angular dependence can be handled by an expansion in PSTF tensor-valued multipoles, as in (5.81):

$$I(E, e^c) = \sum_{\ell \geq 0} I_{A_\ell}(E) e^{A_\ell}, \quad V(E, e^c) = \sum_{\ell \geq 0} V_{A_\ell}(E) e^{A_\ell}. \tag{11.87}$$

For $\mathcal{P}_{ab}$, we use the fact that any STF tensor on the sphere can be written in terms of angular derivatives of two scalar potentials, P_E and P_B, as (Kamionkowski, Kosowsky and Stebbins, 1997)

$$\mathcal{P}_{ab} = \nabla_{\langle a}^{(2)} \nabla_{b \rangle}^{(2)} P_E + \epsilon^c{}_{\langle a} \nabla_{b \rangle}^{(2)} \nabla_c^{(2)} P_B, \tag{11.88}$$

where $\nabla_a^{(2)}$ and $\epsilon_{ab} = \eta_{abc} e^c$ are the covariant derivative and alternating tensor on the two-sphere. The scalar fields P_E and P_B are even and odd under parity respectively, and define the electric and magnetic parts of the linear polarization. Expanding P_E and P_B in PSTF multipoles, and evaluating the angular derivatives, leads to (Challinor, 2000a, Thorne, 1980)

$$\mathcal{P}_{ab}(E, e^c) = \sum_{\ell \geq 2} [\mathcal{E}_{abC_{\ell-2}}(E) e^{C_{\ell-2}}]^{\mathrm{tt}}$$
$$- \sum_{\ell \geq 2} [e_{d_1} \eta^{d_1 d_2}{}_{(a} \mathcal{B}_{b) d_2 C_{\ell-2}}(E) e^{C_{\ell-2}}]^{\mathrm{tt}}. \tag{11.89}$$

Here tt denotes the transverse (to e^a), trace-free part, so that in general $[J_{ab}]^{\mathrm{tt}} = s_a^c s_b^d J_{cd} - \frac{1}{2} s_{ab} s^{cd} J_{cd}$. The PSTF tensors $\mathcal{E}_{A_\ell}$ and $\mathcal{B}_{A_\ell}$ can be found by inverting (11.89):

$$\mathcal{E}_{A_\ell}(E) = M_\ell^2 \Delta_\ell^{-1} \int d\Omega \, e_{\langle A_{\ell-2}} \mathcal{P}_{a_{\ell-1} a_\ell \rangle}(E, e^c), \tag{11.90}$$

$$\mathcal{B}_{A_\ell}(E) = M_\ell^2 \Delta_\ell^{-1} \int d\Omega \, e_b \epsilon^{bd}{}_{\langle a_\ell} e_{A_{\ell-2}} \mathcal{P}_{a_{\ell-1} \rangle d}(E, e^c), \tag{11.91}$$

where $M_\ell := \sqrt{2\ell(\ell-1)/[(\ell+1)(\ell+2)]}$. The multipole expansion in (11.89) is the coordinate-free version of the tensor spherical harmonic expansion for CMB polarization in Kamionkowski, Kosowsky and Stebbins (1997). An alternative expansion, whereby $Q \pm iU$ is expanded in spin-weighted spherical harmonics, is also used (Seljak and Zaldarriaga, 1997).

The projected collision tensor in the Thomson limit in the electron rest frame, generalizing (11.59), is (Challinor, 2000a) (correcting two sign errors in the right-hand side of (3.7) of that reference)

$$\tilde{E}^2 \tilde{K}_{ab}(\tilde{E},\tilde{e}^c) = \tilde{n}_e \sigma_{\rm T} \left\{ \tfrac{1}{2} \tilde{s}_{ab} \left[-\tilde{I}(\tilde{E},\tilde{e}^c) + \tilde{I}(\tilde{E}) + \tfrac{1}{10} \tilde{I}_{d_1 d_2}(\tilde{E}) \tilde{e}^{d_1} \tilde{e}^{d_2} \right. \right.$$

$$\left. - \tfrac{3}{5} \tilde{\mathcal{E}}_{d_1 d_2}(\tilde{E}) \tilde{e}^{d_1} \tilde{e}^{d_2} \right] + \left[-\tilde{\mathcal{P}}_{ab}(\tilde{E},\tilde{e}^c) - \tfrac{1}{10} [\tilde{I}_{ab}(\tilde{E})]^{\rm tt} + \tfrac{3}{5} [\tilde{\mathcal{E}}_{ab}(\tilde{E})]^{\rm tt} \right]$$

$$\left. + \tfrac{1}{2} i \tilde{\eta}_{abd_1} \tilde{e}^{d_1} \left[-\tilde{V}(\tilde{E},\tilde{e}^c) + \tfrac{1}{2} \tilde{V}_{d_2}(\tilde{E}) \tilde{e}^{d_2} \right] \right\}. \tag{11.92}$$

In-scattering couples to the monopole and quadrupole in total intensity, and to the E-mode quadrupole. Linear polarization is generated by in-scattering of the quadrupoles in total intensity and E-mode polarization. Comparison with (11.89) shows that in the electron rest frame, the polarization is *generated* purely as an E-mode quadrupole. Circular polarization is decoupled from total intensity and linear polarization, so that in any frame the circular polarization will remain exactly zero if it is initially.

The equations for the energy-integrated multipoles of linear polarization, circular polarization and total intensity, are (Tsagas, Challinor and Maartens, 2008):

$$\dot{\mathcal{E}}_{A_\ell} + \tfrac{4}{3} \Theta \mathcal{E}_{A_\ell} + \frac{(\ell+3)(\ell-1)}{(\ell+1)^2} \overline{\nabla}^b \mathcal{E}_{bA_\ell} + \frac{\ell}{(2\ell+1)} \overline{\nabla}_{\langle a_\ell} \mathcal{E}_{A_{\ell-1}\rangle} - \frac{2}{(\ell+1)} {\rm curl}\, \mathcal{B}_{A_\ell}$$

$$= -n_e \sigma_{\rm T} \left[\mathcal{E}_{A_\ell} + \left(\tfrac{1}{10} I_{a_1 a_2} - \tfrac{3}{5} \mathcal{E}_{a_1 a_2} \right) \delta_{\ell 2} \right], \tag{11.93}$$

$$\dot{\mathcal{B}}_{A_\ell} + \frac{4}{3} \Theta \mathcal{B}_{A_\ell} + \frac{(\ell+3)(\ell-1)}{(\ell+1)^2} \overline{\nabla}^b \mathcal{B}_{bA_\ell} + \frac{\ell}{(2\ell+1)} \overline{\nabla}_{\langle a_\ell} \mathcal{B}_{A_{\ell-1}\rangle}$$

$$+ \frac{2}{(\ell+1)} {\rm curl}\, \mathcal{E}_{A_\ell} = -n_e \sigma_{\rm T} \mathcal{B}_{A_\ell}, \tag{11.94}$$

$$\dot{V}_{A_\ell} + \frac{4}{3} \Theta V_{A_\ell} + \overline{\nabla}^b V_{bA_\ell} + \frac{l}{(2\ell+1)} \overline{\nabla}_{\langle a_\ell} V_{A_{\ell-1}\rangle} = -n_e \sigma_{\rm T} \left(V_{A_\ell} - \tfrac{1}{2} V_{a_1} \delta_{\ell 1} \right), \tag{11.95}$$

$$\dot{I}_{A_\ell} + \frac{4}{3} \Theta I_{A_\ell} + \overline{\nabla}^b I_{bA_\ell} + \frac{l}{(2\ell+1)} \overline{\nabla}_{\langle a_\ell} I_{A_{\ell-1}\rangle} + \frac{4}{3} I A_{a_1} \delta_{\ell 1} + \frac{8}{15} I \sigma_{a_1 a_2} \delta_{\ell 2}$$

$$= -n_e \sigma_{\rm T} \left[I_{A_\ell} - I \delta_{\ell 0} - \tfrac{4}{3} I v_{a_1} \delta_{\ell 1} - \left(\tfrac{1}{10} I_{a_1 a_2} - \tfrac{3}{5} \mathcal{E}_{a_1 a_2} \right) \delta_{\ell 2} \right]. \tag{11.96}$$

The E- and B-mode multipoles are coupled by curl terms. In a general almost-FLRW cosmology, B-mode polarization is generated only by advection of the E-mode. This does not happen if the perturbations about FLRW are curl-free, as is the case for scalar perturbations. We thus have the important result that linear scalar perturbations do not generate B-mode polarization (Kamionkowski, Kosowsky and Stebbins, 1997, Seljak and Zaldarriaga, 1997).

To first order in the ratio of the mean-free time to the expansion time or the wavelength of the perturbation, the polarization is an E-mode quadrupole (Tsagas, Challinor and Maartens, 2008):

$$\mathcal{E}_{ab} \approx \frac{8}{45} \frac{I}{n_e \sigma_{\mathrm{T}}} \left(\sigma_{ab} + D_{\langle a} v_{b \rangle} \right). \tag{11.97}$$

For scalar perturbations, the polarization thus traces the projected derivative of the baryon velocity relative to the Newtonian frame. The peaks in the C_ℓ^E spectrum thus occur at the minima of C_ℓ^T as the baryon velocity oscillates $\pi/2$ out of phase with Δ. This behaviour can be seen in Figure 11.2. The large-angle polarization from recombination is necessarily small by causality, but a large-angle signal is generated by re-scattering at reionization (Page *et al.*, 2007).

11.7 Vector and tensor perturbations

Vector modes describe vortical motions of the cosmic fluids. They are not excited during inflation. Furthermore, due to conservation of angular momentum, the vorticity of radiation decays as $1/a$ and matter as $1/a^2$ so that vector modes are generally singular to the past (see Table 6.1). Vector modes are important in models with active sources such as magnetic fields (see Barrow, Maartens and Tsagas (2007) for a recent review) or topological defects (Turok, Pen and Seljak, 1998).

The CMB anisotropies from vector modes were first studied comprehensively in Abbott and Schaefer (1986). The full kinetic theory treatment was developed in the total-angular-momentum method in Hu and White (1997) and Hu *et al.* (1998). The 1+3-covariant treatment for the spatially flat case was given in Lewis (2004b). A systematic treatment, for general spatial curvature, may be found in Tsagas, Challinor and Maartens (2008).

The imprint of tensor perturbations, or gravitational waves, is implicit in the original work of Sachs and Wolfe (1967). The first detailed calculations for temperature were reported in Dautcourt (1969) and for polarization in Polnarev (1985). The E–B decomposition, which was already implicit in the early work of Dautcourt and Rose (1978), and the realization that B-mode polarization is a particularly sensitive probe of tensor modes, was developed in Kamionkowski, Kosowsky and Stebbins (1997) and Seljak and Zaldarriaga (1997). The effect of tensor modes on the CMB from the 1+3-covariant perspective, is discussed in Challinor (2000a) and Tsagas, Challinor and Maartens (2008).

11.8 Other background radiation

It is important to realize that while the blackbody CMB ($\simeq 1$ eV cm^{-3}) is the thermally dominant part of the cosmic background radiation, it is far from being the only component of that radiation. In fact we expect some kind of background radiation at all wavelengths;

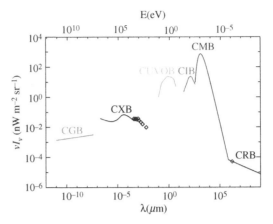

Fig. 11.3 Spectrum of the cosmic background radiations: radio (CRB), microwave (CMB), UV-optical (CUVOB), infrared (CIB), X-ray (CXB) and γ-ray (CGB). (Reprinted from Hauser and Dwek (2001), with permission from the Annual Review of Astronomy and Astrophysics. © 2001 by Annual Reviews, http://www.annualreviews.org .)

the CMB is just the microwave component (there will also be neutrino and gravitational wave backgrounds: these are discussed below).

11.8.1 Other electromagnetic background radiation

The overall observed background radiation spectrum is shown in Figure 11.3. Note that it is difficult to measure some wavelength bands because of galactic obscuration (the previous major problems in this regard due to our own atmosphere have been overcome by the advent of balloon and rocket-borne instruments), and complications arise in determining the spectrum as a whole because very different instruments have to be used to measure the radiation at different wavelengths. Furthermore, the ongoing problem is separating out 'background radiation' from that due to discrete sources. As resolution increases, what was 'background' may get resolved into discrete sources. Nevertheless there may be background radiation from intergalactic gas, as well as the radiation we are forced to classify as background radiation at any particular time because even though it is due to discrete sources, we are unable to resolve them.

We may conveniently classify the backgrounds as radio, microwave, optical, infrared, X-ray, and gamma ray. These radiation backgrounds at different wavelengths are related to each other through their interactions with each other and with matter, the latter depending crucially on the thermal history of intergalactic gas. In particular, if that gas is re-ionized at about the time of galaxy formation to about 10^5 K, it will emit X-rays, contributing to the X-ray background. But this is not the only origin of the X-ray background; as well as hot intergalactic gas, discrete sources contribute. The optical and infrared backgrounds are dominated by star formation processes in galaxies. High-energy electrons create radio waves in magnetic fields (which are ubiquitous), giving rise to a radio background. Dark matter particles may decay to create high-energy photons, which may assist with the origin of intergalactic gamma rays (still unknown).

Furthermore background radiation may interact with cosmic ray protons, giving a limit to the highest energy such protons we should detect, and with cosmic γ-rays, through pair production with the CMB photons; hence the cosmic ray spectrum is affected by the background radiation. Cosmic ray electrons interact with the CMB through inverse Compton scattering (the photon gains energy and the electron loses energy), relating the galactic radio background to X-rays. (Note that 'cosmic rays' are mostly from galactic rather than intergalactic sources.)

11.8.2 Neutrino background

As discussed in Section 9.6.5, the neutrinos free-stream after decoupling at $T \sim 1\,\text{MeV}$, and maintain their blackbody spectrum, up to small anisotropies, in qualitatively the same way as the photons. There is a neutrino background, much like the CMB – but it is unlikely to be detected. However, the neutrinos have important effects on the CMB and on large-scale structure formation. (See Lesgourgues and Pastor (2006) for a review.)

Treating the neutrinos as massless in the early universe is an excellent approximation, and on that basis we showed that

$$T_\gamma = \left(\tfrac{11}{4}\right)^{1/3} T_\nu \approx 1.4 T_\nu . \tag{11.98}$$

Thus the cosmic neutrino background would have a current temperature of $1.95\,\text{K}$.

In fact neutrinos are now understood to be massive – or more precisely, at least two of the three neutrino flavours are massive, and it appears that there can be oscillation from one type to another. Constraints on the squared mass differences from solar and atmospheric neutrino oscillation experiments give

$$(m_2 - m_1)^2 \sim 8 \times 10^{-5}\,\text{eV}, \ (m_3 - m_1)^2 \sim 2 \times 10^{-3}\,\text{eV}, \tag{11.99}$$

$$\sum m_I \gtrsim .05\,\text{eV}. \tag{11.100}$$

Constraints from the CMB and matter power spectra give

$$\sum m_I \lesssim 0.6\,\text{eV}. \tag{11.101}$$

Thus terrestrial and cosmological experiments provide a powerful pincer-like constraint on the neutrino masses, and in particular show that each of the masses is very light. The most massive neutrino becomes non-relativistic well after radiation–matter equality. We can estimate the non-relativistic redshift by setting the mean energy per neutrino equal to the mass. This gives

$$1 + z_{\text{nr}} \sim 945 \left(\frac{m_{I\,\text{max}}}{0.5\,\text{eV}}\right). \tag{11.102}$$

After the transition to non-relativistic energies, the neutrinos begin to have an effect on structure formation, since they can begin to cluster, above a minimum free-streaming scale, estimated as

$$k_{\text{fs}}(z) \sim \frac{0.3}{(1+z)^2} \left(\frac{m_I}{0.5\,\text{eV}}\right) \left[\Omega_{m0}(1+z)^3 + \Omega_\Lambda\right]^{1/2} h\,\text{Mpc}. \tag{11.103}$$

11.8.3 Gravitational wave background

As discussed in Section 12.2, inflation generates a primordial background of gravitational waves, with nearly scale-invariant spectrum, so that the background is present across all frequencies, although with very weak amplitude. In simple single-field models of inflation, the amplitude of this background is given relative to the scalar perturbations by the tensor-to-scalar ratio r [(12.55)]. In small-field and hybrid models the background would typically be negligibly small, but it could be substantial in large-field models (such as chaotic inflation).

In principle this background is a discriminator amongst inflation models, at least for simple single-field models. For more general multi-field inflationary models, constraints from the tensor background are more complicated. Nevertheless, some simple models could in principle be ruled out if their predicted strong signal is not seen.

The gravitational wave background cannot be accessed by ground-based gravity wave detectors because of the extreme weakness of the signal. But it is indirectly detectable via its generation of B-modes in the CMB polarization spectra – although this signal has to be disentangled from the weak-lensing B-mode signal. (A further possible complication is that cosmic strings also generate a B-mode.) WMAP placed only weak constraints, $r \lesssim 0.3$, on the amplitude. Planck is expected to significantly improve on this constraint. The CMB probes the background at wavelengths of roughly the present Hubble horizon, i.e. at frequencies $\sim 10^{-18}$ Hz. Space-borne laser interferometers, such as the proposed BBO and DECIGO, would probe the background at wavelengths of roughly the detector size, i.e. at frequencies ~ 0.1–1 Hz. Direct detection combined with indirect B-mode detection would significantly improve the constraints on the gravity wave background.

12 Structure formation and gravitational lensing

The primordial seeds of inhomogeneity, whose imprint is seen at last scattering in the CMB anisotropies, may be generated by quantum fluctuations during inflation in the very early universe. These seeds subsequently evolve from linear to nonlinear fluctuations via gravitational instability, and produce the large-scale matter distribution that is observed at lower redshifts. The previous chapter dealt with the CMB anisotropies. In this chapter we provide brief overviews of the primordial fluctuations from inflation, and then of the evolution of large-scale structure, as described via the power spectrum of matter. A key probe of the total matter (dark and baryonic) and its distribution is weak gravitational lensing by the large-scale structure of light from distant sources. We develop the theoretical framework for gravitational lensing and briefly describe how this is applied in cosmology. The following chapter will draw on this chapter and its predecessors to show how current observations constrain and describe the standard model of cosmology. We start with a summary of the statistical description of perturbations.

12.1 Correlation functions and power spectra

Perturbations on an FLRW background are treated as random variables in space at each time instant, and observations determine the statistical properties of these random distributions. (See Durrer (2008) for a more complete discussion.) A perturbative variable $A(x)$ at some fixed time is associated with an ensemble of random functions, each with a probability assigned to it. We define the 2-point correlation function $\langle A(x)A(x')\rangle$ as the average over the ensemble (incorporating the probability distribution). The random field is usually assumed to be statistically homogeneous and isotropic – i.e. invariant under translations and rotations (parity invariance is also usually applicable). Homogeneity and isotropy mean that the 2-point correlation function can be written as

$$\xi_A(x) = \langle A(x_0)A(x_0 + xn)\rangle, \tag{12.1}$$

so that ξ_A does not depend on the position x_0 or the (unit) direction n.

A fundamental limitation arises in cosmology – because there is only one universe to observe, i.e. there is only one realization of the stochastic process that generates the fluctuations whose consequences we observe. Therefore we cannot measure ensemble averages or expectation values, as we would in a repeatable laboratory experiment. What we can do when observing a fluctuation on a given scale λ is to average over many distinct regions of size $\sim \lambda$. An ergodic-type hypothesis allows us to replace the ensemble average by a spatial average over these regions. This is reasonable when the scale is much less than the observable part of the universe, i.e. for $\lambda \ll H_0^{-1}$. On larger scales, $\lambda = O(H_0^{-1})$, we are unable to average over many volumes – and thus the measured value could be quite far from the ensemble average. This is called the 'cosmic variance' problem.

In Fourier space (assuming a spatially flat background for simplicity),

$$A_{\boldsymbol{k}} = \int A(\boldsymbol{x}) e^{i\boldsymbol{k}\cdot\boldsymbol{x}}\, \mathrm{d}^3 x, \quad A(\boldsymbol{x}) = \frac{1}{(2\pi)^3} \int A_{\boldsymbol{k}} e^{i\boldsymbol{k}\cdot\boldsymbol{x}}\, \mathrm{d}^3 k. \tag{12.2}$$

The 2-point moment in Fourier space defines the power spectrum $P_A(k)$:

$$\langle A_{\boldsymbol{k}} A_{\boldsymbol{k}'} \rangle = (2\pi)^3 \delta^3(\boldsymbol{k} + \boldsymbol{k}') P_A(k), \tag{12.3}$$

where statistical isotropy means that P_A depends only on $|\boldsymbol{k}|$, and statistical homogeneity is reflected in the translation invariance encoded in the $\delta^3(\boldsymbol{k} + \boldsymbol{k}')$ term.

A Gaussian random field is characterized by the fact that all the odd-number moments vanish (e.g. $\langle A_{\boldsymbol{k}} A_{\boldsymbol{k}'} A_{\boldsymbol{k}''} \rangle = 0$), while all the even-number moments are determined by the 2-point moment (or equivalently, the power spectrum).

A convenient alternative definition of the power spectrum (with a different normalization) is

$$\mathcal{P}_A = \frac{k^3}{2\pi^2} P_A. \tag{12.4}$$

The power spectrum $P_A(\boldsymbol{k})$ in Fourier space and the real-space 2-point correlation function, $\xi_A(\boldsymbol{x})$, are a Fourier pair,

$$\xi_A(x) = \frac{1}{(2\pi)^3} \int P_A(k) e^{i\boldsymbol{k}\cdot\boldsymbol{x}}\, \mathrm{d}^3 k. \tag{12.5}$$

With statistical isotropy, this leads to

$$\xi_A(r) = \frac{1}{2\pi^2} \int_0^\infty k^2 j_0(kr) P_A(k)\, \mathrm{d}k = \int_0^\infty \frac{1}{k} j_0(kr) \mathcal{P}_A(k)\, \mathrm{d}k, \tag{12.6}$$

where $j_0(z) = \sin z / z$ is a spherical Bessel function.

For a Gaussian perturbation field, $\langle A(\boldsymbol{x}) \rangle = 0$. The variance is

$$\sigma_A^2 = \langle A^2(\boldsymbol{x}) \rangle = \xi_A(0) = \frac{1}{2\pi^2} \int_0^\infty k^2 P_A(k)\, \mathrm{d}k = \int_0^\infty \mathcal{P}_A(k) \frac{\mathrm{d}k}{k}. \tag{12.7}$$

Thus $\mathcal{P}_A$ is the contribution to $\langle A^2 \rangle$ per unit logarithmic interval in k. These integrals may diverge in either or both of the short-wavelength (ultraviolet) and long-wavelength (infrared) regimes. For example, for the matter density perturbation, $\mathcal{P}_\delta$ grows with k, and an ultraviolet blow-up arises because the underlying model – pressure-free matter without peculiar velocities – breaks down on small enough scales, where multi-streaming must arise to avoid unphysical shell-crossings. Therefore an ultraviolet cutoff is needed to make the integral converge: $k \leq k_{\mathrm{max}}$. Here k_{max}^{-1} is much smaller than cosmological scales, and also smaller than the smoothing scale imposed by the finite resolution of astronomical observations. In the case of the curvature perturbation, $\mathcal{P}_\zeta$ is almost scale-invariant, leading to logarithmic divergences at both ends of the integral. Then we require both ultraviolet and infrared cutoffs.

If $\bar{P}_A$ is the observed average, then the cosmic variance is defined as $(\Delta P_A)^2 = \langle \bar{P}_A^2 \rangle - \langle P_A \rangle^2$. The cosmic variance can be estimated as (Lyth and Liddle, 2009)

$$(\Delta P_A(k))^2 \sim \frac{1}{(\delta k/k)(kL)^3} P_A(k)^2, \tag{12.8}$$

where δk is the resolution in k and L is the scale of the observed region. For scales much smaller than L, the cosmic variance is small, but for scales approaching L, cosmic variance becomes large and degrades the statistical significance of the observations.

The three-dimensional spatial correlation function can be projected on the sky to produce a two-dimensional angular correlation function. Consider the case of the matter distribution, with $A = \delta$. The angular correlation function $w(\alpha)$ determines the probability of finding galaxies at angular separation α. It is related to the matter power spectrum by the Limber formula,

$$w(\alpha) = \int_0^\infty k P(k, z) \mathcal{G}(k\alpha) \, dk, \tag{12.9}$$

where the kernel $\mathcal{G}$ is given in the small-angle approximation by

$$\mathcal{G}(k\alpha) = \frac{1}{2\pi} \int J_0(k\alpha\chi) \left(\frac{1}{n} \frac{dn}{dz} \right)^2 F \, d\chi, \tag{12.10}$$

$$F^2 = \left(1 + H_0^2 \chi^2 \Omega_K \right) \frac{H(z)}{H_0}. \tag{12.11}$$

Here χ is the comoving angular diameter distance in the FLRW background, and dn/dz is the redshift distribution of galaxies. Galaxy surveys measure $w(\alpha)$, but the key desired information is the power spectrum. This poses the problem of inverting (12.9) to extract $P(k, z)$.

Non-Gaussianity in A is signalled by the fact that there are non-zero higher-order correlation functions that are not determined by the 2-point correlation function. It is typically described via the bispectrum B_A, which is the Fourier transform of the real-space 3-point correlation function:

$$\langle A_{k_1} A_{k_2} A_{k_3} \rangle = (2\pi)^3 \delta^3(\boldsymbol{k}_1 + \boldsymbol{k}_2 + \boldsymbol{k}_3) B_A(k_1, k_2, k_3). \tag{12.12}$$

The amplitude and shape of B_A are determined by the mechanisms producing the non-Gaussianity in a particular model. An initially Gaussian field that is coupled to gravity will develop non-Gaussianity via the nonlinearity of the gravitational interaction. Non-Gaussianity in the density perturbations and metric potentials leave an imprint on the CMB anisotropies and the distribution of large-scale structure, and observations may be used to place limits on non-Gaussianity and thereby to constrain various models.

12.2 Primordial perturbations from inflation

In Section 9.7 we discussed inflation in the background FLRW model, and showed how it addresses key issues in the kinematics and dynamics of the standard model of cosmology.

There is an equally important facet of inflation, i.e. the fact that inflation naturally generates primordial inhomogeneities. These are imprinted on the hot plasma of the Big Bang, and leave a fossil imprint on the CMB at the time that radiation decouples from matter. After decoupling, as the universe expands and cools, the inhomogeneities grow by gravitational instability, to eventually produce the stars and larger-scale structure. In addition to density fluctuations, inflation also naturally generates primordial tensor perturbations.

The origin of the primordial perturbations is the quantum fluctuations of the inflaton field. A systematic treatment of quantized fluctuations may be found in Bartolo *et al.* (2004), Mukhanov (2005), Lyth and Liddle (2009) and Baumann (2009). (Note that there remain various open questions, including the problem of defining the quantum to classical transition and the 'trans-Planckian' problem. We shall not discuss these problems; see, e.g. Martin and Brandenberger (2001), Vaudrevange and Kofman (2007), Kiefer and Polarski (2009).) Here we give a classical description of inflaton perturbations, with qualitative indications of the quantum treatment.

The basic idea is that quantum fluctuations of the inflaton field behave like one-dimensional quantum harmonic oscillators (with time-varying mass). Zero-point fluctuations of a quantum harmonic oscillator induce a non-zero variance of the oscillator amplitude, $\langle \hat{x}^2 \rangle = \hbar/2\omega$. Similarly, the inflaton zero-point fluctuations generate a non-zero variance $\langle \delta\varphi^2 \rangle$. The fluctuation modes (with comoving wavenumber k) are stretched from their original small scale (assumed to be above the Planck scale) by the rapid accelerating expansion of the universe, until their wavelength ak^{-1} exceeds the Hubble scale (when they are assumed to become classical fluctuations).

Quantum inflaton fluctuations are generated with significant amplitude only for modes with wavelength near the Planck scale. While the mode is sub-Hubble, its amplitude decays, $|\delta\varphi_k| \sim a^{-1}$ for $k > aH$. However, when the mode wavelength is stretched beyond the Hubble scale, $k < aH$, the amplitude is effectively frozen and preserved. Without inflation, the wavelength would not cross the Hubble radius, and thus the amplitude would decay away and there would be no seeds for structure formation.

While a mode's wavelength is sufficiently smaller than the Hubble radius, it evolves like a plane wave in Minkowski spacetime, since the spacetime curvature is effectively negligible. Once the wavelength is super-Hubble, the evolution is frozen. After inflation, the Hubble scale begins to grow more rapidly than the wavelength, and so eventually the mode's wavelength falls below the Hubble scale, i.e. the mode 're-enters' the Hubble 'horizon'. Inflaton perturbations are coupled to curvature perturbations, which subsequently, after inflation, couple to density perturbations. When a mode re-enters the Hubble horizon during the radiation or matter era, it is unfrozen and density perturbations in the matter begin to grow. The process that links the observed large-scale structure at late times to the microphysics of primordial inflation is illustrated schematically in Figure 12.1. Modes that cross the Hubble radius earlier in inflation re-enter the Hubble radius later, and with larger wavelength. The cosmologically relevant modes are those with wavelength $\gtrsim 1\,\mathrm{Mpc}$ that re-enter in the matter era. These modes leave the Hubble horizon about 60 e-folds before the end of inflation.

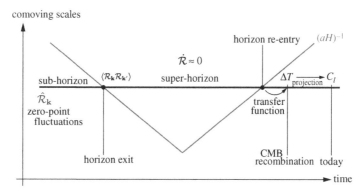

Fig. 12.1 Creation and evolution of perturbations during inflation. (From Baumann (2009).)

12.2.1 Evolution and amplitude of scalar perturbations

Inflaton fluctuations are coupled to metric scalar perturbations. The perturbed metric in Newtonian gauge (10.55) and for a flat background is

$$ds^2 = -(1 + 2\Phi)dt^2 + a^2(1 - 2\Psi)d\boldsymbol{x}^2. \tag{12.13}$$

In the Newtonian limit Φ is the Newtonian potential. Ψ is the curvature perturbation of the $t = $ const surfaces. The difference in the metric perturbations is sourced by anisotropic stress via (10.65), $\Phi - \Psi = -16\pi G a^2 \Pi$.

During inflation, the gravitational field is sourced by the inflaton field, with energy–momentum tensor given by

$$T^\mu{}_\nu = \varphi^{,\mu}\varphi_{,\nu} - \left[V(\varphi) + \tfrac{1}{2}\varphi^{,\gamma}\varphi_{,\gamma}\right]\delta^\mu{}_\nu. \tag{12.14}$$

The inflaton propagates according to the Klein–Gordon equation,

$$(-g)^{-1/2}\left[(-g)^{1/2}\varphi^{,\mu}\right]_{,\mu} - V_\varphi = 0. \tag{12.15}$$

Writing φ as $\varphi(t) + \delta\varphi(t,\boldsymbol{x})$, we find (see Exercise 12.2.1):

$$\delta T^0{}_0 = -\delta\rho_\varphi = -\left(\dot{\varphi}\delta\dot{\varphi} + V_\varphi\delta\varphi - \dot{\varphi}^2\Phi\right), \tag{12.16}$$

$$\delta T^0{}_i = (\rho_\varphi + p_\varphi)\partial_i v_\varphi = -\dot{\varphi}\partial_i\delta\varphi, \tag{12.17}$$

$$\delta T^i{}_j = \delta p_\varphi \delta^i{}_j = (\dot{\varphi}\delta\dot{\varphi} - V_\varphi\delta\varphi - \dot{\varphi}^2\Phi)\delta^i{}_j, \tag{12.18}$$

and hence that the anisotropic stress vanishes,

$$\Pi_\varphi = 0 \;\;\Rightarrow\;\; \Psi = \Phi. \tag{12.19}$$

The perturbed Klein–Gordon equation is (see Exercise 12.2.1)

$$\delta\ddot{\varphi} + 3H\delta\dot{\varphi} - \nabla^2\delta\varphi + V_{\varphi\varphi}\delta\varphi + 2V_\varphi\Phi - 4\dot{\varphi}\dot{\Phi} = 0. \tag{12.20}$$

The $(0i)$ perturbed Einstein equation becomes

$$\dot{\Phi} + H\Phi = 4\pi G\dot{\varphi}\delta\varphi, \tag{12.21}$$

on using (12.17). The coupled equations (12.20) and (12.21) determine $\delta\varphi$ and Φ.

In terms of the variable

$$\tilde{Q} = aQ, \quad Q = \delta\varphi + \frac{\dot{\varphi}}{H}\Phi, \tag{12.22}$$

the perturbed Klein–Gordon equation (12.20) becomes (see Exercise 12.2.1)

$$\tilde{Q}'' + \left(k^2 - \frac{a''}{a} + M^2 a^2\right)\tilde{Q} = 0, \quad M^2 := V_{\varphi\varphi} - \frac{8\pi G}{a^3}\left(\frac{a^3}{H}\dot{\varphi}^2\right)^{\cdot}, \tag{12.23}$$

where a prime denotes a conformal time derivative. Equation (12.23) has the form of an oscillator equation in Minkowski spacetime, with variable effective mass. In the slow-roll regime, $M^2 \approx 3(\eta - 2\epsilon)H^2$ and $a \approx -[H\tau(1-\epsilon)]^{-1}$. Since $\dot{\epsilon}, \dot{\eta}$ are second order in slow-roll, we can treat M^2/H^2 as constant. Then the equation reduces to

$$\tilde{Q}'' + \left(k^2 - \frac{\nu^2 - \frac{1}{4}}{\tau^2}\right)\tilde{Q} = 0, \quad \nu = \frac{3}{2} + \epsilon - \eta, \tag{12.24}$$

with general solution in terms of Hankel functions:

$$\tilde{Q} = \sqrt{-\tau}\left[C_1(k)H_\nu^{(1)}(-k\tau) + C_2(k)H_\nu^{(2)}(-k\tau)\right]. \tag{12.25}$$

At this point, we need to invoke quantum theory to determine the correct normalization. In the ultraviolet regime, $k \gg aH$ (or $-k\tau \gg 1$), we should recover the Minkowski vacuum state (characterized as the minimum energy state), $\exp(-ik\tau)/\sqrt{2k}$. With the large-argument limit of the Hankel functions, this leads to

$$aQ = \frac{\sqrt{\pi}}{2}e^{i(2\nu+1)\pi/4}\sqrt{-\tau}H_\nu^{(1)}(-k\tau). \tag{12.26}$$

A more convenient variable is the comoving curvature perturbation, which is proportional to Q, and is conserved on large scales:

$$\mathcal{R} := \Phi + \frac{H}{\dot{\varphi}}\delta\varphi = \frac{H}{\dot{\varphi}}Q, \quad \dot{\mathcal{R}} = O\left(\frac{k^2}{a^2 H^2}\right). \tag{12.27}$$

In the super-Hubble regime, $-k\tau \ll 1$, we find that

$$|Q| \approx \frac{H}{\sqrt{2k^3}}\left(\frac{k}{aH}\right)^{3\epsilon - \eta + 3/2}. \tag{12.28}$$

It follows that

$$\mathcal{P}_{\mathcal{R}} \approx \left(\frac{H^2}{2\pi\dot{\varphi}}\right)^2\left(\frac{k}{aH}\right)^{3\epsilon - \eta + 3/2} \approx \left(\frac{H^2}{2\pi\dot{\varphi}}\right)^2\Bigg|_{\mathrm{hc}}, \tag{12.29}$$

where the last expression is evaluated at Hubble crossing, $k = aH$.

Note that on super-Hubble scales $\mathcal{R} = -\zeta$, so that

$$\mathcal{P}_\zeta = \mathcal{P}_{\mathcal{R}}. \tag{12.30}$$

Furthermore, (12.27) shows that in slow-roll, $\mathcal{R} \approx H\delta\varphi/\dot{\varphi}$, so that, by (12.29),

$$\mathcal{P}_{\delta\varphi} \approx \left(\frac{H}{2\pi}\right)^2_{\text{hc}}. \tag{12.31}$$

We can also use the slow-roll approximations to find (see Exercise 12.2.2)

$$\mathcal{P}_{\mathcal{R}} \approx 8\pi G \left.\frac{H^2}{8\pi\epsilon}\right|_{\text{hc}} \approx (8\pi G)^2 \left.\frac{V}{24\pi^2\epsilon}\right|_{\text{hc}}. \tag{12.32}$$

12.2.2 Connecting inflation to the CMB

Perturbations in the inflaton field generated during inflation eventually seed the anisotropies that are imprinted in the CMB at last scattering, as well as the density perturbations that grow later into stars and galaxies. From the time of generation inside the Hubble radius at $t \sim 10^{-34}$ s to the decoupling of matter and radiation at $t \sim 400,000\,\text{yr}$, the perturbations evolve through three eras and two transitions: from inflation to radiation domination, across the reheating transition, then from radiation domination to matter domination across the matter–radiation equality. The reheating transition is typically a violently non-equilibrium process, whereas matter–radiation equality is an instantaneous moment in a smooth transition. Nevertheless, we are able to track the evolution of perturbations *without regard to the details of these transitions*.

This remarkable feature arises from the conservation of the growing mode of the curvature perturbation on super-Hubble scales, as discussed in Section 10.2.6. On these scales, the curvature perturbation does not 'feel' any of the details of reheating, and does not notice the transition to matter domination. This constancy allows us easily to track the evolution of the super-Hubble Newtonian potential and density perturbation, which are not conserved.

Since we can neglect anisotropic stresses, we have from (10.54) that

$$\mathcal{R} = \Phi + \frac{2}{3}\frac{\Phi' + \mathcal{H}\Phi}{(1+w)\mathcal{H}}. \tag{12.33}$$

For adiabatic perturbations, and with vanishing anisotropic stress, Φ is governed by (10.76):

$$\Phi'' + 3\mathcal{H}(1+c_s^2)\Phi' + \left[2\mathcal{H}' + \mathcal{H}(1+3c_s^2) - c_s^2\nabla^2\right]\Phi = 0. \tag{12.34}$$

On super-Hubble scales we can neglect the Laplacian term (and it is exactly zero for dust on all scales). For an adiabatic fluid, such as radiation or dust matter, (10.38) shows that $c_s^2 = w - w'/\mathcal{H}(1+w)$. This relation will also apply on large scales to an inflaton field. Then we can rewrite the evolution equation for super-Hubble scales as

$$\Phi'' + \left[(1+w)\mathcal{H} - \frac{w'}{3(1+w)}\right]\Phi' - \frac{w'}{1+w}\mathcal{H}\Phi = 0. \tag{12.35}$$

If we can neglect w', then $c_s^2 = w = \text{const}$. This is a good approximation during slow-roll inflation, when $w = \epsilon - 1$, in the radiation era, and in the matter era. During the transitions between these eras, w' is non-zero – but these transitions happen on a time scale that is very small compared to the characteristic (light-crossing) time of the wavelength of the super-Hubble modes. With these conditions, the evolution equation (12.35) has solution

$$\Phi = A_+ + A_- \int a^{-(1+w)} d\tau, \ \ A'_{\pm} = 0. \tag{12.36}$$

The A_- mode is decaying and we can neglect it. Thus the growing (i.e. non-decaying) mode, $\Phi = A_+$, is constant in time – and we can neglect the Φ' term in (12.33) to get:

$$\Phi = \frac{3(1+w)}{(5+3w)} \mathcal{R}. \tag{12.37}$$

For modes which remain super-Hubble across a transition from era 1 to era 2, we have

$$\Phi_2 = \frac{(5+3w_1)(1+w_2)}{(5+3w_2)(1+w_1)} \Phi_1, \tag{12.38}$$

since $\mathcal{R}_2 = \mathcal{R}_1$. The jump in Φ is positive when $w_2 > w_1$ and negative in the opposite case.

For a mode that is super-Hubble from the time of Hubble exit in inflation, when $\mathcal{R} = \mathcal{R}_{\text{prim}}$ (the primordial value), to the time of Hubble re-entry in the matter era, we have:

$$\Phi_{\text{inf}} = \tfrac{3}{2}\epsilon \mathcal{R}_{\text{prim}}, \ \ \Phi_{\text{rad}} = \tfrac{2}{3}\mathcal{R}_{\text{prim}}, \ \ \Phi_{\text{matt}} = \tfrac{3}{5}\mathcal{R}_{\text{prim}}. \tag{12.39}$$

12.2.3 Non-Gaussianity

The microscopic quantum generation of fluctuations is inherently Gaussian – the representative quantum harmonic oscillators have random phases. However, the modes do interact gravitationally, and so non-Gaussianity does emerge, reflected in non-zero higher-order correlations. For single-field inflation this effect is very small, and effectively negligible in most cases. Inflation thus provides a natural mechanism for laying down a Gaussian distribution of microscopic fluctuations, which are automatically stretched to macroscopic scales and hence serve as seeds for the growth of near-Gaussian density perturbations.

The non-Gaussianity in the Newtonian potential due to gravitational nonlinear coupling may be described in the weak coupling case by a simplification of (12.12) (Bartolo *et al.*, 2004),

$$\langle \Phi(\boldsymbol{k}_1)\Phi(\boldsymbol{k}_2)\Phi_{NL}(\boldsymbol{k}_3)\rangle = (2\pi)^3 \delta^3(\boldsymbol{k}_1 + \boldsymbol{k}_2 + \boldsymbol{k}_3) 2 f_{NL} P_\Phi(\boldsymbol{k}_1) P_\Phi(\boldsymbol{k}_2) \tag{12.40}$$

where Φ is the primordial Gaussian potential, the nonlinearity parameter f_{NL} is treated as constant, and the evolved nonlinear Φ is

$$\Phi_{NL}(\boldsymbol{x}) = f_{NL}\left[\Phi(\boldsymbol{x})^2 - \langle\Phi(\boldsymbol{x})^2\rangle\right]. \tag{12.41}$$

(Note that this expression guarantees that $\langle\Phi_{NL}(\boldsymbol{x})\rangle = 0$.)

For simple single-field models of inflation, $f_{NL} = O(10^{-2})$, while for multi-field and curvaton models, $f_{NL} = O(1-10)$ (Bartolo *et al.*, 2004, Malik and Wands, 2009). Constraints from the CMB and galaxy surveys give $-1 \lesssim f_{NL} \lesssim 70$. Since $|\Phi| \sim 10^{-5}$, these limits mean that the deviation from Gaussianity is extremely small, $\lesssim 0.1\%$. Any detection of $|f_{NL}| > 1$ would rule out the simplest single-field inflation models.

12.2.4 Isocurvature (or entropy) modes

Simple inflation models also generate adiabatic fluctuations on scales. A scalar field, unlike a perfect fluid, does support intrinsic entropy (non-adiabatic) perturbations (see Exercise 12.2.3):

$$\delta p_{\mathrm{nad}} := \delta p_\varphi - \frac{\dot{p}_\varphi}{\dot{\rho}_\varphi}\delta\rho_\varphi = 2V_\varphi\left(\delta\rho_\varphi + 3H\dot{\varphi}\delta\varphi\right) = 2V_\varphi\rho_\varphi\Delta_\varphi \propto \nabla^2\Phi. \tag{12.42}$$

Clearly $\nabla^2\Phi \to 0$ on large scales, so that we can ignore the entropy perturbations.

The inflaton fluctuations on large scales, where gradients can be neglected, correspond to a local shift (forward or backward) along the background trajectory in phase space, and thus are associated with perturbations in the time. The inflaton decays into radiation and matter, and so its fluctuations affect the total density in different locations after the end of inflation, but they cannot produce variations in the relative density between components:

$$\delta\rho_I \approx \dot{\rho}_I\delta t \quad \to \quad \frac{\delta\rho_I}{\dot{\rho}_I} = \frac{\delta\rho_J}{\dot{\rho}_J}. \tag{12.43}$$

We can make this more precise by defining the relative entropy perturbations

$$S_{IJ} := 3H\left(\frac{\delta\rho_I}{\dot{\rho}_I} - \frac{\delta\rho_J}{\dot{\rho}_J}\right) = 3(\zeta_J - \zeta_I). \tag{12.44}$$

The last equality follows from the definition of the curvature perturbation on uniform I-density slices,

$$-\zeta_I = \Psi + H\frac{\delta\rho_I}{\dot{\rho}_I}. \tag{12.45}$$

Pure adiabatic modes, as in the case of single-field inflation, are then characterized by $S_{IJ} = 0$ for all I, J. In this case, the total curvature perturbation, $-\zeta = \Psi + H\delta\rho/\dot{\rho}$, is equally shared by all components,

$$\zeta_I = \zeta_J = \cdots = \zeta. \tag{12.46}$$

The total curvature perturbation on large scales evolves as

$$\dot{\zeta} = -\frac{H}{\rho + p}\delta p_{\mathrm{nad}}, \quad \delta p_{\mathrm{nad}} = \delta p - \frac{\dot{p}}{\dot{\rho}}\delta\rho. \tag{12.47}$$

Thus it is conserved on large scales in the pure adiabatic case.

Relative perturbation modes between components correspond to isocurvature perturbations, with $S_{IJ} \neq 0$. A pure isocurvature mode on large scales corresponds to the case where the density perturbations of the components compensate each other so as to produce a zero initial total curvature perturbation,

$$\zeta|_{\mathrm{init}} = 0. \tag{12.48}$$

In multi-field models, intrinsic entropy in each field vanishes on large scales by (12.42), but relative isocurvature modes may be naturally generated,

$$S_{IJ} \propto \frac{\delta\varphi_I}{\dot{\varphi}_I} - \frac{\delta\varphi_J}{\dot{\varphi}_J}. \tag{12.49}$$

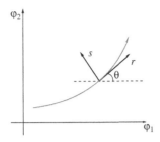

Fig. 12.2   Adiabatic (r) and entropy (s) fields in (φ_1, φ_2) space. (From Malik and Wands (2009). © Elsevier (2009).)

The imprint of these primordial modes on the CMB anisotropies is model dependent, but for typical simple models of isocurvature, CMB observations place stringent upper limits on the allowable isocurvature contribution (Komatsu *et al.*, 2011).

For simplicity, consider the case of two fields. A suitable local rotation in φ_I field space (see Figure 12.2) leads to a decomposition into instantaneous adiabatic (r) and entropy (s) fields. Then the curvature perturbation is related to the adiabatic perturbations δr along the trajectories, via $\mathcal{R} = H(\delta r/\dot{r}) + \Phi$, and it evolves as (Gordon *et al.*, 2001)

$$\dot{\mathcal{R}} = 2\frac{H}{\dot{r}}\dot{\theta}\,\delta s + \frac{H}{\dot{H}}\frac{k^2}{a^2}\,\Phi. \qquad (12.50)$$

Thus even on large scales, the curvature perturbation is not conserved in the presence of relative entropy modes, $\delta s := S_{12}$, if the trajectories are curved ($\dot{\theta} \neq 0$).

12.2.5 Tensor perturbations

If the mass term in (12.23) is not small, i.e. $M^2/H^2 > 1$, then the vacuum fluctuations are suppressed on cosmological scales. Significant quantum fluctuations are generated in all light fields during inflation, including the metric field. These latter fluctuations include tensor perturbations: for a flat background,

$$\mathrm{d}s^2 = -\mathrm{d}t^2 + a^2\left(\delta_{ij} + h_{ij}\right)\mathrm{d}x^i\mathrm{d}x^j,\ h_i{}^i = 0 = \partial^j h_{ij}. \qquad (12.51)$$

We decompose into Fourier modes, $h_{ij}(t,\boldsymbol{k}) = h(t)e_{ij}^{\pm}(\boldsymbol{k})$, where $\pm$ refers to the two polarizations. The classical evolution equation for the amplitude h is (10.91). Using (12.19), we obtain

$$\ddot{h} + 3H\dot{h} + \frac{k^2}{a^2}h = 0. \qquad (12.52)$$

This is the same as the wave equation for a massless scalar, and it follows from (12.31) that $\mathcal{P}_h \propto (H_{\mathrm{hc}}/2\pi)^2$.

The normalization follows from a quantization of the canonical variable. The result is

$$\mathcal{P}_h = 64\pi G\left(\frac{H_{\mathrm{hc}}}{2\pi}\right)^2, \qquad (12.53)$$

where we have incorporated a factor 2 for the two polarizations. In slow roll (Exercise 12.2.2),

$$\mathcal{P}_h = 2(8\pi G)^2 \frac{V_{hc}}{3\pi^2}. \tag{12.54}$$

The tensor-to-scalar ratio is given by (12.53) and (12.32):

$$r := \frac{\mathcal{P}_h}{\mathcal{P}_{\mathcal{R}}} \approx 16\epsilon. \tag{12.55}$$

Exercise 12.2.1 Derive (12.16)–(12.20) and (12.23). Verify that (12.25) is a solution of the last of these equations.

Exercise 12.2.2 Derive the slow-roll power spectra relations (12.32) and (12.54).

Exercise 12.2.3 Derive (12.42).

12.3 Growth of density perturbations

Density perturbations in matter are the progenitors of stars, galaxies and clusters. If the matter is pressure-free, i.e. dust, and if it dominates the background, then perturbations grow like the scale factor, $\delta \propto a$. Nonlinear structure corresponds to $\delta \gtrsim 1$, after which the collapsing region decouples from the cosmic expansion.

12.3.1 Evidence for cold dark matter: cosmological

Baryonic matter, despite being non-relativistic after matter–radiation equality, is tightly coupled to the radiation via Thomson and Coulomb scattering, up until the brief period of recombination. This means that δ_b cannot grow like a until last scattering. If $\delta_b \sim 10^{-5}$ at last scattering, then how does it grow to a nonlinear value by today:

$$\delta_{b0} = \frac{a_0}{a_{dec}} \delta_{b\,dec} \sim 10^{-2} \,? \tag{12.56}$$

This puzzle provides a strong cosmological motivation for the existence of non-relativistic matter (hence 'cold') that does not couple to radiation (hence 'dark') – and that can cluster well before last scattering. The cold dark matter (CDM) would need to dominate over luminous matter in order to grow structures quickly enough. Thus cosmology indicates that $\Omega_c > \Omega_b$. In fact, the joint constraints from the CMB anisotropies, galaxy surveys (in the form of baryon acoustic oscillations and weak lensing) and SNIa magnitudes (see Figure 13.1), show that

$$\Omega_{c0} \sim 0.25\,, \ \Omega_{b0} \sim 0.05 \ \Rightarrow \ \Omega_{c0} \sim 5\Omega_{b0}\,. \tag{12.57}$$

Note: we are assuming that GR is correct, and that the observable universe is a perturbed FLRW model.

Cold dark matter would then start to form potential wells before equality, so that by the time the baryonic matter is released from the grip of radiation, it experiences a speeded-up clustering via infall into the CDM halos.

Primordial nucleosynthesis places strong bounds on the total baryonic content of the universe, which are consistent with CMB and large-scale structure data:

$$\eta = \frac{n_b}{n_\gamma} \sim 10^{-10}, \quad \Omega_{b0}h^2 \sim 0.02.$$ (12.58)

Cosmology therefore further suggests that the CDM needs to be non-baryonic – or at least predominantly non-baryonic, since there could be a small fraction of dark baryonic matter, such as primordial black holes or brown dwarfs.

The cosmological motivation for non-baryonic CDM is fairly strong, if we assume that GR is the correct theory of gravity. This is further backed up by astrophysical evidence, which we shall briefly discuss.

12.3.2　Evidence for cold dark matter: astrophysical

Strong indirect evidence for CDM comes from the observation of circular orbital velocities of stars in the flat disks of spiral galaxies. The Newtonian limit of GR gives a Keplerian velocity that is determined by the total mass enclosed within the sphere containing the orbit:

$$v^2(r) = \frac{GM_{<r}}{r}, \quad M_{<r} = 4\pi \int_0^r \rho(\tilde{r})\tilde{r}^2 \mathrm{d}\tilde{r}.$$ (12.59)

The surface brightness of spiral galaxies is observed to fall off exponentially. Thus the stellar contribution to $M_{<r}$ should lead to $M_{<r\,\mathrm{stars}} \to$ const, and hence we expect the velocity curve to fall off as $v \propto 1/\sqrt{r}$. The galactic gas also makes a contribution to $M_{<r}$, which can be estimated by observations of the HI 21 cm and molecular lines. This baryonic gas contribution also produces a fall-off of velocity with distance.

In contrast to the predicted rotation curves, spiral galaxies are observed to have rotation curves that approach a plateau with increasing radius, $v \to$ const. The rotation curves differ in the interior regions, where different characteristics of the galactic bulges complicate the dynamics, but all share this feature of asymptotically constant v. This indicates the presence of a non-baryonic CDM with mass profile that falls off much slower than that of baryonic matter:

$$M_{<r\,c} \to r, \quad \rho_c \propto r^{-2} \text{ for large } r.$$ (12.60)

The asymptotic (and maximal) constant orbital velocity, v_∞, is related to the luminosity via the empirical Tully–Fisher relation for spiral galaxies:

$$L \propto v_\infty^4.$$ (12.61)

In summary: the rotation of spiral galaxies is incompatible with Newtonian gravity if only the stars, gas and dust in the galaxy are taken into account. The contradiction may be resolved if the galaxies are embedded in huge CDM halos.

Clusters of galaxies give independent evidence of CDM. The hot, X-ray-emitting gas at the centres of clusters would diffuse out of the clusters in less than a Hubble time if it were only bound by its self-gravity and that of the galaxies. The CDM creates a potential well deep enough to trap the intracluster gas. The CDM contribution to the total cluster mass can be estimated from its trapping effect on the gas. Other estimates of the CDM fraction come from lensing and from the virial theorem. These estimates show that

$$\frac{M_{\text{gas}}}{M_{\text{c}}} \sim 15 \pm 5\%, \quad \frac{M_{\text{stars}}}{M_{\text{c}}} \sim 3 \pm 2\%. \tag{12.62}$$

In addition, weak lensing by clusters shows that the gravitational potentials of clusters are deeper than can be explained only via their baryonic content.

Merging clusters have also given evidence of the predominance of CDM. In particular, the so-called 'bullet cluster' (see Section 12.5.5) has moved through a larger cluster, and the location of the gas (traced by X-ray measurements) is segregated from the location of the galaxies. The gravitational potential measured by lensing observations traces the galaxy distribution rather than the dominant gas distribution. A dominant CDM, whose potential well has trapped the galaxies, can account for this mismatch.

Finally, particle physics is able to provide a range of possible candidate CDM particles, in particular weakly interacting massive thermal relics (see Section 9.6.2). Currently there are various experiments underway attempting to detect CDM particles (see Feng (2010) for a review):

- Via *production* of CDM particles in terrestrial particle accelerators (such as the Large Hadron Collider, LHC).
- Via *direct detection* in terrestrial underground and underwater searches for elastic scattering events with nucleons (such as the Cryogenic Dark Matter Search experiment, CDMS).
- Via *indirect detection* in astrophysical signatures of CDM annihilation in very high-energy environments, such as around the super-massive black hole of the galaxy. The signal could be carried, for example, in cosmic rays that are produced and monitored by experiments such as PAMELA, ATIC, FERMI and HESS.

12.3.3 Effective CDM from modified gravity?

All of the above arguments and conclusions implicitly assume that GR holds on all scales, from galaxies to the Hubble radius. The CDM paradigm is compelling, but remains unproven – and it is useful, even necessary, to develop competing paradigms.

It is in principle possible that there is no CDM at all, and that CDM arises simply as an indicator of the failure of GR. However, in practice it has proved impossible, so far, to produce a covariant modified gravity theory that can avoid the need for CDM while at the same time being consistent with observations across the vast range of scales, from the Solar System, through galaxies and clusters, to the Hubble radius.

A modified Newtonian dynamics (MOND) was formulated by Milgrom to recover the Tully–Fisher relation (12.61). The modification kicks in at low accelerations, $\alpha \lesssim \alpha_0 \sim 10^{-8} \, \text{cm s}^{-2}$. (Surprisingly, this small characteristic acceleration α_0 corresponds to a length scale of order the current Hubble radius.) Despite the ad hoc nature of this modification, it is remarkably successful at accounting for rotation curves with a single universal parameter $\alpha_0 \approx 1.2 \times 10^{-8}\,\mathrm{cm\,s^{-2}}$.

However for galaxy clusters, the typical accelerations are higher than the galaxy α_0, and are thus in the regime where MOND recovers the Newtonian limit of GR. In order to account for the effects that CDM accounts for, the MOND parameter α_0 has to be at least twice as big in clusters as in galaxies, and even larger for elliptical galaxies.

MOND therefore fails to replace CDM on scales of clusters. Furthermore, as an ad hoc modification of Newtonian theory, it is unable to deal with cosmological scales. Indeed, the existence of an absolute acceleration scale also creates difficulties for consistency with Solar System constraints.

Proposals for relativistic modifications of GR that reproduce the success of MOND for galaxies have been made, in particular, Bekenstein's tensor–vector–scalar (TeVeS) theory and the simpler sub-class of Einstein–Aether vector–tensor theories (for reviews and recent work, see Ferreira and Starkman (2009), Bekenstein (2010), Zuntz *et al.* (2010)). These theories must introduce additional degrees of freedom in the gravitational interaction – scalar and vector graviton modes in addition to the tensor mode of GR. In some sense, this is simply replacing one kind of dark matter with another kind. However, if a theory could be found that avoids the need for dark matter and dark energy via the same mechanism, then that may be counted as a successful alternative. Currently, the indications are that the Einstein–Aether theories can easily replace dark energy, but are unable simultaneously to reproduce all of the key features of dark matter's contribution to background expansion and structure formation (Zuntz *et al.*, 2010). (See also Section 14.3.)

Apart from the huge theoretical difficulties facing any candidate modified gravity theory that attempts to replace CDM, there is also the fact that developments in particle physics *independently* predict various CDM candidates. If CDM candidates are detected, or ruled out, by experiments, then this will have major implications for attempts to modify GR.

12.3.4 Matter power spectrum

The perturbed metric in Newtonian gauge is given by (12.13). The gauge-invariant comoving matter density perturbation $\Delta = \delta - 3aHv$ obeys the Poisson and evolution equations, (10.74) and (10.75):

$$k^2 \Phi = 4\pi G a^2 \rho \Delta, \tag{12.63}$$

$$\ddot{\Delta} + 2H\dot{\Delta} - 4\pi G \rho \Delta = 0. \tag{12.64}$$

These equations are exact on all scales, if perturbations are linear and purely due to matter, and if there are no anisotropic stresses.

The solution of (12.63) in Fourier space is

$$\Delta_k(z) = -\frac{2}{3} \frac{k^2}{H_0^2} \frac{1}{\Omega_{m0}} \frac{g(z)}{1+z} \Phi_k. \tag{12.65}$$

Here $g = D_+(z)/a$, where $D_+ = \Delta_+/\Delta_{\mathrm{in}}$ is the growing mode amplitude of the density perturbation, given by the solution of (12.64). The growth suppression factor g is due to dark energy: in Einstein–de Sitter models, $g = 1$, so that with dark energy, $g \to 1$ for $z \gg 1$ and $g(0) \approx 0.75$. An exact expression is given in Exercise 12.3.1. Note that massive neutrinos, which become non-relativistic after matter–radiation equality, introduce a k-dependence to g.

The Newtonian potential is determined by the primordial curvature perturbation. The curvature perturbation is constant on super-Hubble scales. For a simple power-law spectrum,

$$\mathcal{P}_\zeta(k) = \frac{k^3}{2\pi^2} P_\zeta(k) = \mathcal{P}_\zeta(k_0) \left(\frac{k}{k_0}\right)^{n_s-1}, \tag{12.66}$$

where k_0 is a (super-Hubble) pivot scale. The Newtonian potential is related to the curvature perturbation with a scale-dependence that takes account of the matter–radiation transition and the re-entry of the Hubble horizon at different times by different scales. Those scales that re-enter the Hubble horizon before matter–radiation equality grow more slowly than the scales that enter after equality, as discussed in Chapter 10. This differential processing is encoded in the transfer function $T(k)$ (Lyth and Liddle, 2009), and

$$\Phi_k = -\tfrac{3}{5}T(k)\zeta_k. \tag{12.67}$$

The transfer function is complicated by the role of baryons, but the asymptotic (CDM-dominated) behaviour is given by

$$T(k) \sim \begin{cases} 1 & k \ll k_{\mathrm{eq}}, \\ (k_{\mathrm{eq}}/k)^2 \ln(k/k_{\mathrm{eq}}) & k \gg k_{\mathrm{eq}}. \end{cases} \tag{12.68}$$

Using (12.66), we arrive at

$$\mathcal{P}_\delta(k,z) = \mathcal{P}_\zeta(k_0) \left(\frac{2k^2}{5H_0^2 \Omega_{\mathrm{m0}}}\right)^2 \left(\frac{k}{k_0}\right)^{n_s-1} \frac{g^2(k,z)}{(1+z)^2} T^2(k), \tag{12.69}$$

where $\mathcal{P}_\zeta(k_0) \sim 10^{-9}$ is determined by CMB large-angle anisotropies. From (12.69), we have

$$\mathcal{P}_\delta(k,z) = 2.4 \times 10^{-9} \left(\frac{2k^2}{5H_0^2 \Omega_{\mathrm{m0}}}\right)^2 \left(\frac{k}{k_0}\right)^{n_s-1} \frac{g^2(k,z)}{(1+z)^2} T^2(k), \tag{12.70}$$

neglecting running of the spectral index. See Figure 12.3 for a typical transfer function and power spectrum, showing the characteristic scale $k_{\mathrm{eq}} \sim 1/(100\,\mathrm{Mpc})$ defined by the comoving size of the Hubble horizon at matter–radiation equality.

There are two further complications that need to be catered for.

- First, galaxy surveys detect only baryonic matter, and thus measure the galaxy power spectrum. This is related to the total matter power by a *bias* function,

$$\mathcal{P}_g(k,z) = b^2(k,z)\mathcal{P}_\delta(k,z). \tag{12.71}$$

In principle, b is determined by the hydrodynamics of baryons falling into CDM potential wells, but in practice this is still well beyond current understanding and we have to use simulations and empirical relations to approximate b on different scales.

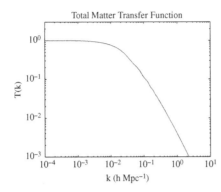
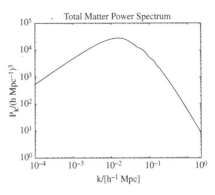

Fig. 12.3   Matter transfer function (left) and power spectrum (right) in a typical ΛCDM model.

- Second, in practice the measurements of the density distribution have a minimum resolution, and so we must introduce a *smoothing scale R*, below which structure is not detected. This is done via a window function $W(kR)$ (in Fourier space) that suppresses fluctuations on scales $k^{-1} < R$, i.e.

$$\mathcal{P}_g(k, z; R) = W^2(kR)\mathcal{P}_g(k, z). \tag{12.72}$$

For a real-space top-hat window function, in Fourier space $W(q) = 3[q^{-3}\sin q - q^{-2}\cos q]$. The power is usually normalized via the variance of the mass fluctuation within spheres of radius $R = 8h^{-1}$ Mpc:

$$\sigma_8^2 = \left\langle \left(\frac{\delta\rho}{\rho}\right)\right\rangle^2_{R=8/h} = \int \frac{dk}{k} |W(kr)|^2 \mathcal{P}_\delta(k, 0). \tag{12.73}$$

This expression ignores the bias factor. The CMB and weak lensing give $\sigma_8 \approx 0.8$.

An example of the observed power spectrum using different probes is shown in Figure 12.4.

There is an important point about the relativistic nature of galaxy clustering. In a statistically homogeneous universe, correlation functions are defined on spatial surfaces of constant time – whereas in fact the observed clustering statistics are on the null surface of the past light-cone. Correlation functions have been defined on the past light-cone (Yamamoto and Suto, 1999), and other GR corrections to the galaxy power spectrum have also been computed (Yoo, Fitzpatrick and Zaldarriaga, 2009, Yoo, 2010, Challinor and Lewis, 2011, Bonvin and Durrer, 2011). These corrections are small for low-redshift surveys, but become increasingly important at higher redshifts.

Finally, we note that non-Gaussianity in the primordial power spectrum can have a significant influence on large-scale structure. For non-Gaussianity of the local form, (12.41), the halo power spectrum is boosted on large scales where GR corrections need to be taken into account (Bruni *et al.*, 2011). Clustering of halos is enhanced for $f_{NL} > 0$ and decreased for $f_{NL} < 0$.

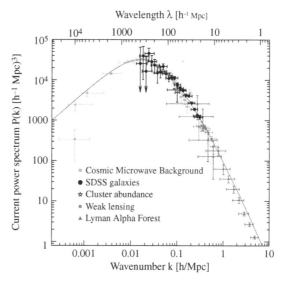

Fig. 12.4 Matter power spectrum from various probes; solid curve is the ΛCDM best fit. (From Tegmark *et al.* (2004). Reproduced courtesy Max Tegmark/SDSS collaboration and by permission of the AAS.) A colour version of this figure is available online.

12.3.5 Peculiar velocities and redshift-space distortions

The evolution of density fluctuations during structure formation sources coherent motions in the matter. In the linear regime, this is governed by

$$\boldsymbol{\nabla} \cdot \boldsymbol{v} = -a\dot{\delta}, \tag{12.74}$$

where we have neglected the metric potential term $\dot{\Phi}$ in the Newtonian limit. These peculiar velocities introduce a radial anisotropic distortion in redshift-space via a Doppler effect. The redshift-space distortions thus provide a handle on the peculiar velocity. In the linear regime (i.e. on sufficiently large scales), the distortion is a 'squashing' in the radial (line of sight) direction, while in the nonlinear regime there is a stretching ('finger of god') effect.

On large scales, the peculiar velocity of an infalling shell is small compared to its radius, and the shell appears squashed. On smaller scales, not only is the radius of a shell smaller, but also its peculiar infall velocity tends to be larger. For the shell that is just at turnaround, its peculiar velocity cancels the Hubble expansion, and it appears collapsed to a single velocity in redshift space. On even smaller scales, shells that are collapsing in proper coordinates appear inside out in redshift space. The combination of collapsing shells with previously collapsed, virialized shells, gives rise to the 'finger-of-god' shape. This is illustrated in Figure 12.5.

We consider the linear case, and assume a distant-observer (plane-parallel) limit and a scale-independent bias, $\delta_g(k,z) = b(z)\delta(k,z)$. Then the distortions are encoded in the parameter

$$\beta(z) = \frac{f(z)}{b(z)}, \quad f := \frac{\mathrm{d}\ln\delta}{\mathrm{d}\ln a}, \tag{12.75}$$

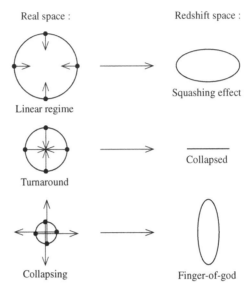

Real space :

Redshift space :

Linear regime

Squashing effect

Turnaround

Collapsed

Collapsing

Finger-of-god

Fig. 12.5   Peculiar velocities lead to redshift space distortions. (From Fig. 2 of Hamilton (1998), reproduced with kind permission from Springer Science+Business Media B.V.)

which governs the relation between the real-space and redshift-space fluctuations (Hamilton, 1998),

$$\delta_g^{\rm rs}(k_{\parallel}, k_{\perp}, z) = \left[1 + \beta(z) \left(\frac{k_{\parallel}}{k} \right)^2 \right] \delta_g(k, z), \qquad (12.76)$$

where $\parallel, \perp$ denote radial and transverse. (Note the assumption that there is no bias in the velocities, i.e. $v_g = v$.)

The linear growth factor f obeys the evolution equation (see Exercise 12.3.2)

$$\frac{\mathrm{d}f}{\mathrm{d}\ln a} + \frac{1}{2} \left(1 - \frac{\mathrm{d}}{\mathrm{d}\ln a} \ln \Omega_{\rm m} \right) f + f^2 = \frac{3}{2} \Omega_{\rm m}, \qquad (12.77)$$

and a good approximation to the solution of this equation is (Linder, 2005)

$$f(z) \approx \Omega_{\rm m}(z)^{\gamma}, \quad \gamma = 0.55 + 0.05 [1 + w(z = 1)]. \qquad (12.78)$$

A measurement of $\beta(z)$ via the redshift-space distortion then gives an estimate of $\Omega_{\rm m}(z)$ if we know the bias $b(z)$, or a measure of the bias if we know $\Omega_{\rm m}(z)$.

12.3.6 Baryon acoustic oscillations

Before last scattering, the tightly coupled photon–baryon plasma oscillates under the competing effects of gravitational collapse and radiation pressure. The resultant acoustic waves in the plasma travel at the sound speed $c_s = (1 + R)^{-1/2}$, where $R = 3\rho_{\rm b}/\rho_{\gamma}$. At baryon decoupling, the pressure on the baryons disappears and the baryon acoustic wave is frozen in, while photons stream freely. This leads to an enhanced baryon over-density at the distance travelled by a sound wave up to decoupling – and hence to a baryon acoustic oscillation (BAO) peak in the galaxy correlation function. The gravitational effect of the baryon over-density is also imprinted on the CDM perturbation. (Note that the feature occurs as a peak in the real-space correlation function, but as an oscillation in the power spectrum in Fourier space.) Unlike the acoustic oscillations in radiation, which are imprinted in the CMB TT power spectrum at the single redshift of decoupling (Section 11.5), the BAO feature in the galaxy distribution evolves with redshift. This makes it a highly effective probe of the background geometry and evolution of the universe (Eisenstein *et al.*, 2005, Cole *et al.*, 2005, Gaztañaga, Cabré and Hui, 2009). (See Bassett and Hlozek (2010) for a review.)

The BAO scale is the sound horizon at decoupling (Lyth and Liddle, 2009):

$$r_s(z_{\mathrm{dec}}) = \int_{z_{\mathrm{dec}}}^{\infty} \frac{c_s}{H} \, \mathrm{d}z$$

$$= \frac{1}{\sqrt{\Omega_{\mathrm{m}0} H_0^2}} \frac{2}{\sqrt{3 z_{\mathrm{dec}} R_{\mathrm{dec}}}} \ln \left[\frac{\sqrt{1 + R_{\mathrm{dec}}} + \sqrt{R_{\mathrm{eq}} + R_{\mathrm{dec}}}}{1 + \sqrt{R_{\mathrm{eq}}}} \right]. \tag{12.79}$$

Note that baryon decoupling actually occurs after photon decoupling because of a rela-tively low baryon density. The WMAP5 results (Komatsu *et al.*, 2009) show that $z_{\mathrm{b\,dec}} \approx 1020$, $z_{\gamma\,\mathrm{dec}} \approx 1090$ and $r_s(z_{\mathrm{b\,dec}}) \approx 153 \,\mathrm{Mpc}$, $r_s(z_{\gamma\,\mathrm{dec}}) \approx 147 \,\mathrm{Mpc}$.

The emergence of the BAO peak is illustrated in Figure 12.6 via the evolution of the radial mass profile of a pointlike over-density. At high redshift before decoupling ($z = 6824$ in Figure 12.6), the photons and baryons travel outwards as a pulse. Close to decoupling ($z = 1440$), a wake is induced in the CDM from the pulse of baryons and relativistic species. The neutrinos are streaming out of the perturbation. After recombination ($z \le 848$), the photons stream out from the baryonic perturbation, while CDM and baryons attract each other at the near-centre and BAO peaks.

The three-dimensional nature of the BAO feature means that on average there is an enhancement of clustering at spheres of comoving radius $r_s(z_{\mathrm{dec}})$ centred on any galaxy. Observations along and transverse to the line of sight to a galaxy at redshift z sample the clustering along those directions. This allows for independent extraction of the area distance (transverse) and the Hubble rate (line of sight) at z:

$$r_s(z_{\mathrm{dec}}) = (1 + z) D_A(z) \Delta\theta_s, \quad r_s(z_{\mathrm{dec}}) = \frac{\Delta z}{H(z)}, \tag{12.80}$$

where $\Delta\theta_s$ is the angle subtended at the observer by the physical sound horizon $r_s/(1 + z)$ and Δz is its redshift extent along the radial direction. As discussed above, redshift-space distortions introduced by peculiar velocities need to be corrected for in the radial relationship. A cut through a simulated galaxy correlation function is shown in Figure 12.7.

Although the baryon feature is imprinted in the linear regime, and its key properties are accessible via a linear analysis, the ongoing evolution of structure formation means that nonlinear distortions will arise at lower redshifts. In other words, the baryon acoustic peak is not a 'pure' geometric observable, but is processed by structure formation. However, the degree of processing is very small at high redshifts, and small even at lower redshifts.

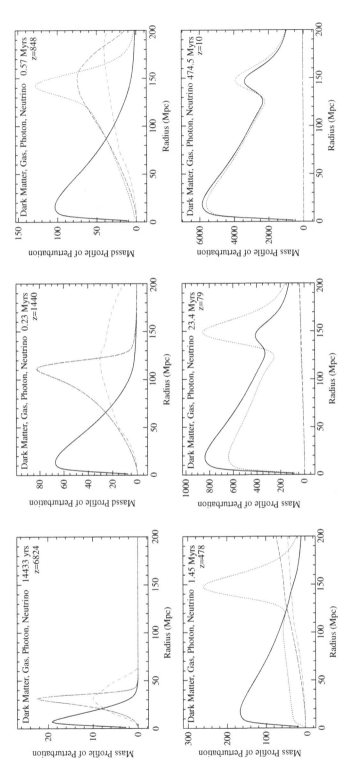

Fig. 12.6 Evolution of radial mass profile versus comoving radius, leading to BAO peak, for dark matter (solid line), gas (dotted line, blue in colour version), photons (dashed line, red) and neutrinos (dot-dash line, green). (From Eisenstein, Seo and White (2007). Reproduced by permission of the AAS.) A colour version of this figure is available online.

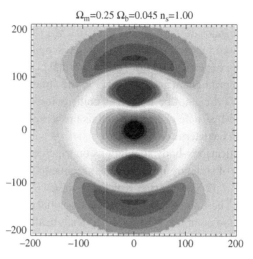

Fig. 12.7 Clustering in redshift space for a $z = 0.15 - 0.30$ slice, as predicted for a typical ΛCDM model. (From Gaztañaga, Cabré and Hui (2009). © RAS.) A colour version of this figure is available online.

Nonlinear effects suppress the amplitude and broaden the width of the baryon acoustic peak, and shift the peak location to slightly smaller scales. Both of these effects may be qualitatively understood as the consequence of nonlinear gravitational attraction due to surrounding galaxies. Detailed analytical and numerical investigations show that the effects can be computed, at least for standard cosmological models (Eisenstein, Seo and White, 2007, Shoji, Jeong and Komatsu, 2009).

12.3.7 Nonlinear structure formation

A crucial aspect of structure formation is the development of nonlinearity, as matter over-densities collapse beyond $\delta \sim 1$ on small scales. The nonlinear enhancement of the matter power spectrum on small scales is evident in Figure 13.4. In order to track the weakly nonlinear regime, which is important in accurate modelling of redshift-space distortions, weak lensing and baryon acoustic oscillations, nonlinear perturbation theory is sufficient. To go further and track the features of the galaxy distribution, one needs to move beyond perturbation theory.

Cold dark matter begins to cluster long before baryonic matter, and it also dominates over baryonic matter. It thus plays the role of 'scaffolding' around which baryonic matter aggregates to form galaxies: CDM 'halos' form gravitational potential wells into which baryons fall and form galaxies and clusters. The process is hierarchical, proceeding from smaller to larger stuctures through mergers. Although baryons are outweighed by CDM, they can affect CDM via gravitational interaction. For example, at the cores of clusters, the baryon concentration itself draws in more CDM. Some general properties of halos can be predicted via analytical or semi-analytical models, such as the Press–Schechter spherical or Sheth–Tormen elliptical models. The elliptical models are more realistic – weak lensing reveals that halos are not spherical. In the Sheth–Tormen model, about 50% of the CDM mass is in halos of mass greater than $10^{10} M_\odot$ (Massey, Kitching and Richard, 2010).

At a more fundamental level, the microscopic physical properties of the CDM particle (the nature of decoupling from the primordial plasma, the particle mass, cross-section, etc.) should imprint characteristic scales in the structure formation process. On small enough scales, CDM cannot be treated as collisionless, and this leads to a minimum halo mass. If the CDM particle is detected and its properties are determined, then these characteristic scales can be determined.

Sophisticated N-body simulations on supercomputers have been developed to track the evolution of structure from the linear regime to the present day (see Springel *et al.* (2005), Dolag *et al.* (2008), Boylan-Kolchin *et al.* (2009), Diemand and Moore (2011) for recent reviews). These simulations provide an essential guide to understanding structure formation, which is too complex to be analysed analytically. Given the huge range of mass scales – from a CDM particle, with typical mass $O(100)\,\mathrm{GeV} \sim 10^{-55} M_\odot$, to a cluster halo, with typical mass $\sim 10^{14} M_\odot$ – it is not possible to track hierarchical structure formation from the clustering of primordial CDM particles all the way up to galaxies and clusters. Instead the effective 'particle' mass in simulations is of the order of a galactic mass.

These simulations reveal the complex, evolving topology of the galaxy distribution, including the key features of walls, filaments, quasi-spherical regions with galaxy clusters that are connected by filaments, and voids (underdense regions with sparse galaxy population). Some features of the cosmic web can be qualitatively understood via the Zel'dovich approximation (Mukhanov, 2005). This approximation to the fully nonlinear Newtonian equations shows that collapse of structure takes place in different forms, according to the eigenvalues of the expansion tensor, α, β, γ. When the eigenvalues have similar magnitudes, quasi-spherical collapse results: if all eigenvalues are positive, these are over-dense regions (clusters and super-clusters), while if all eigenvalues are negative, they are underdense (voids). If $\gamma \ll \alpha \sim \beta$, then collapse is two-dimensional, forming cigar-like collapse regions (filaments). If $\beta, \gamma \ll \alpha$, collapse is one-dimensional and pancake-like regions form (walls).

The N-body simulations provide invaluable tools for understanding the galaxy distribution and testing the cosmological model. Empirical relations based on simulations are essential for probing the nonlinear regime in weak lensing and other observations. The N-body codes evolve CDM 'particles' (with mass of order a galactic mass) via the Newtonian equations. Their clustering creates halos, and then these halos are populated by baryonic galaxies, using phenomenological prescriptions that attempt to mimic the infall of baryons into the halos.

A serious limitation resides in the inherently Newtonian nature of the N-body simulations of large-scale structure. As these simulations extend over great fractions of the Hubble volume, general relativistic effects should become more important, and it is not clear whether the Newtonian analysis can be 'fixed' to take account of these effects. At a fundamental level, the N-body simulations are *not* self-gravitating – since they fix the background spacetime upon which large-scale structure performs its (Newtonian) evolution. However, the development of self-gravitating simulations is an enormous computational challenge. Progress in numerical GR computations of black holes and neutron stars provides a potentially valuable starting point for a future programme to tackle GR computations of structure formation.

At the time of writing there remain a number of potential inconsistencies between the N-body output and the observed features on smaller scales. Furthermore, the understanding of galaxy evolution and the process of forming larger and larger structures also presents a number of currently unresolved puzzles. The condensation of gas into stars is not only highly nonlinear, but also involves the complex astrophysics of the magnetohydrodynamic and radiative effects of baryon infall – heating, cooling, dissipation, turbulence, feedback from AGNs and supernovae, role of magnetic fields. These complexities can often only be dealt with by empirical prescriptions that are incorporated into hydrodynamic codes, whose output is then compared with observations.

A key limitation of N-body and hydrodynamic simulations is the 'black box' problem, i.e. that we do not always understand the underlying physical mechanism well enough, so that we do not always have physical guidance to interpret, check and correct the output. This is an inevitable consequence of attempting to model such a complex nonlinear process while the requisite astrophysics is not yet well enough developed.

12.3.8 Origin of vorticity

Vorticity in structures is most likely to originate from hydrodynamic effects during gravitational collapse. The alternative is a cosmic 'seed' for vorticity, which can counteract the adiabatic decay of cosmic vorticity governed by (6.14) and the Kelvin–Helmholtz conservation law. The only possible seeds are topological defects, if they exist. But the vector perturbations generated by defects do not transfer to the cosmic medium except at nonlinear order in the presence of dissipation and turbulence – and hence the effect is negligible. The basic reason is angular momentum conservation in the cosmic medium (Hollenstein *et al.*, 2008).

During gravitational collapse, dissipative effects arise, and these can act as sources of vorticity. In particular, shock fronts form – either triggered by outgoing blasts meeting infalling gas, or through differentially collapsing shells – and the non-adiabatic pressure has a gradient that is not aligned with the density gradient. From (6.13), the only possible source terms for ω_a are in the curl $\dot{u}_a$ term. This is determined by the curl of the momentum conservation equation (5.12) when $q_a = 0 = \pi_{ab}$. Using identity (4.60), we find a source term that is non-zero on shock fronts (Exercise 12.3.3):

$$\dot{\omega}_{\langle a \rangle} + \left(\frac{2}{3} - c_s^2\right)\Theta\omega_a - \sigma_{ab}\omega^b = \frac{1}{2(\rho + p)^2}\eta_{abc}\overline{\nabla}^b p\,\overline{\nabla}^c\rho. \tag{12.81}$$

The sourcing of vorticity via $\eta_{abc}\overline{\nabla}^b p\,\overline{\nabla}^c\rho$ is known as the Biermann mechanism – the same mechanism generates magnetic fields. When q_a and π_{ab} are non-zero via shear viscosity and heat flux, this also provides source terms in the vorticity evolution equation.

Exercise 12.3.1 Show that (12.64) may be rewritten as

$$D'' + \left(\frac{H'}{H} + \frac{3}{a}\right)D' - \frac{3}{2}\frac{\Omega_{m0}}{a^5}D = 0, \tag{12.82}$$

where prime denotes d/da and $D(a) = \Delta(a)/\Delta(a_{in})$. Verify that $D_- = H$ is a solution in ΛCDM, and show that the growing mode is

$$D_+(z) = \frac{5}{2}\frac{H(z)}{H_0}\Omega_{m0}\int_z^\infty \frac{1+\tilde{z}}{[H(z)/H_0]^3}\,d\tilde{z}. \tag{12.83}$$

Verify that the growth suppression factor is given by

$$g(z) \propto {}_2F_1\left[1,\frac{1}{3};\frac{11}{6};-\frac{\Omega_{\Lambda 0}}{\Omega_{m0}}\frac{1}{(1+z)^3}\right]. \tag{12.84}$$

Exercise 12.3.2 Show that the linear growth factor f obeys the evolution equation (12.77), and verify that (12.78) is an approximate solution.

Exercise 12.3.3 Derive (12.81).

12.4 Gravitational lensing

Light bending, one of the classical tests of GR discussed in all introductory texts, had an extra importance because it was the only one predicted by the theory before it was observed. However, the first detection of lensing at cosmological scales (Walsh, Carswell and Weymann, 1979) seems to have surprised many, despite the possibility having been discussed by several authors (see Schneider, Ehlers and Falco (1992) for an interesting review of the early development of the subject).

Gravitational lensing is now recognised as a major tool for exploring the distribution of mass in the universe and other cosmological parameters, in particular since the first weak lensing surveys in 2000. It is also used in studying, for example, the physics of quasars, the internal structure of galaxies and the detection of planets. The many observations include multiple images of a single source, giant luminous arcs, 'Einstein rings' (where the image fills an annulus round the lens), and arclets, as well as microlensing events. Here we can only summarize, so we refer the reader to books or substantial review articles, e.g. Wambsganss (2001), Petters, Levine and Wambsganss (2001), Mollerach and Roulet (2002), Perlick (2004), Schneider, Kochanek and Wambsganss (2006), Jetzer, Mellier and Perlick (2010) for more details and fuller bibliography.

The name 'light bending' implies that there is a straight line with which the bent one can be compared. Since in GR the null geodesics themselves are the (generalization of) 'straight lines', and there is thus no intrinsic comparator, such a comparison must use some fictitious background spacetime: compare Chapter 10.

For the Sun, gravitational light bending is calculated from the spherically symmetric Schwarzschild solution, and the comparator is flat space. Assuming the observer and source are effectively at infinity, the bending for a ray well outside the Schwarzschild radius of a spherical mass M is approximately

$$\widehat{\boldsymbol{\alpha}} = 4GM\frac{(\boldsymbol{\xi}-\boldsymbol{\xi}')}{|\boldsymbol{\xi}-\boldsymbol{\xi}'|^2}, \tag{12.85}$$

where $\boldsymbol{\xi}$ is a two-dimensional position vector perpendicular to the line of sight in a plane at the distance of the lensing mass, which is taken to be centred at $\boldsymbol{\xi}'$. The plane of such vectors is called the lens plane.

Small values for light bending imply that the impact parameters $|\boldsymbol{\xi} - \boldsymbol{\xi}'|$ (and perhaps the lensing objects themselves) are much bigger than the radii of black holes of the same mass would be, so (12.85) is usually derived from a weak-field approximation. Lensing by compact objects at small impact parameters is more complicated: for example, there may be closed null orbits to which incoming rays can approximate, leading to the possibility that rays will wind round the lensing object multiple times (see Bozza (2010) for a review).

Since an exact formula for geodesics in Schwarzschild can be given in terms of an integral, (12.85) can be derived from it, higher-order approximations are straightforward, and exact results for small impact parameters can be found in this case.

The light bending by the Sun for a given source as seen from the Earth is time dependent, because the Sun moves in the sky, so one can make a comparison observation with our past light-cone at a different time. At grazing incidence the bending for a distant source is about 1.75 arcsec. Since the Sun subtends an angle of about 1° at the Earth, this implies that rays from us touching the Sun and being bent by it cross again at about six light days behind the Sun, not far at all in astronomical terms. Such a crossing point of distinct geodesics from a point p is called a *conjugate point* of p.

This example shows that conjugate points are far from unusual and can arise quite close by. The light-cone develops caustics, cusps and folds at conjugate points, with the effect that it may then intersect a source world line several times, giving multiple images of the same object, at least one of them being significantly magnified (see e.g. Stewart (1994), Schneider, Ehlers and Falco (1992), Perlick (2004)). These in general come from different source world line points, i.e. different emission times. Such behaviour does not require strong fields or large bending of the rays, but is usually called strong lensing; 'weak lensing' means that a single image is seen and, usually, that its distortion is mild. Data can also be derived from an intermediate regime, described as flexion, where the distortion varies substantially across a single image (see e.g. Massey, Kitching and Richard (2010)): such an image typically forms an arc. The different cases are illustrated in Figure 12.8.

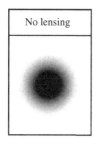

No lensing	Weak lensing	Flexion	Strong lensing
	Large-scale structure	Substructure, outskirts of halos	Cluster and galaxy cores

Fig. 12.8 The various regimes of gravitational lensing image distortion. (From Massey, Kitching and Richard (2010).)

The geometry of lensing is the same in all metric theories of gravity in which light travels along null geodesics, but the relation to the sources of the metric depends on the particular theory. Viable alternative gravity theories must reproduce the observed bending by the Sun. Almost all treatments of lensing use ray optics, not a full wave optics, compare Section 7.1.

Many of the applications of gravitational lensing concern light from discrete sources bent in the gravitational fields of other discrete bodies like the Sun (see e.g. Wambsganss (2001)). The lensing objects are not truly lenses, in that they do not produce a focused image. A single galaxy of about $10^{12} M_\odot$ can produce multiple images with separations about 3 arcsec, and clusters of size $10^{14} M_\odot$ separations of an arcminute. Some interesting and advanced mathematics is needed to properly determine information on the numbers of images and their magnifications (Petters and Werner, 2010). For brevity, we do not develop the theory of strong lensing here.

Images formed by galaxies or clusters will be of more distant galaxies or QSOs, and both sources and lenses will be at cosmological distances. In these cases the lenses themselves are small compared with those distances, and are thus often approximated as 'thin', i.e. lying in a plane.[1]

These cosmological scales also imply that the comparison spacetime should be curved, FLRW models being the first to study. As well as the focusing by the lens, beams are cosmologically focused (see e.g. (7.52)): indeed, this effect is important in inferring the spatial curvature from CMB observations (see Section 13.3.1). More recently lensing due to diffuse over- or under-densities, rather than discrete lenses, has been used to study the dark matter distribution (see Section 12.5.3).

Neither weak nor strong lensing usually involves large deflections, i.e. lensing by a strong gravitational field, and most applications to astronomical objects also do not require strong curvature in the source (although source galaxies may contain black holes). Cases involving strong curvatures may require the rigorous approach using the 'exact lens equation' (Frittelli and Newman, 1999), which writes spacetime position on the geodesic as a function of the observer's proper time, angles on the observer's celestial sphere and a radial distance (or affine parameter distance), although this equation can only be given explicitly in some special examples (see Perlick (2004) for a review of this method).

Finding detailed numerical or analytic approximations to the metric, and then tracing rays in those solutions (see e.g. Holz and Linder (2005)), can very accurately approximate the true behaviour, but to do this in fully detailed models is a formidable task. For many cases less accurate approximation treatments are adequate: we shall refer to some and give details of one, relevant to weak lensing in a perturbed FLRW model, which uses the $1 + 3$ and covariant ideas of this book to derive the quasi-Newtonian 'lensing equation', (12.86). This derivation is not entirely rigorous, a problem linked to the more general issue of finding a satisfactory way of obtaining Newtonian limits in cosmology (see Section 3.4).

---

[1] Kling, Newman and Perez (2000) show that although this works well for a single lens, the iterative method they introduced works better for two neighbouring lenses: the difference is below the observational errors in simple cases, so we do not give its details.

12.4.1 The lensing equation

The lensing equation gives the deflection caused by a thin lens, i.e. the angle between the ray arriving at the observer and the direction in which such a ray would have come in the absence of the lens. It assumes that the lens is (quasi-)stationary and that the weak field correction to the metric is well approximated by integrating along the unperturbed ray (this, by analogy with quantum physics, is called the Born approximation). For a flat-space comparator, one can simply 'add up' (i.e. integrate over) the elements of a lens, treating each part as a Schwarzschild deflector and considering the different lens planes to be identified. Then the input and output values of the ray's spatial direction e^a are related by

$$\widehat{\alpha} := \mathbf{e}_{\text{in}} - \mathbf{e}_{\text{out}} = 4G \int_{\mathbb{R}^2} \frac{(\boldsymbol{\xi} - \boldsymbol{\xi}')\Sigma(\boldsymbol{\xi}')}{|\boldsymbol{\xi} - \boldsymbol{\xi}'|^2} \mathrm{d}^2\xi', \tag{12.86}$$

where Σ is the surface mass density in the lens plane (i.e. the integral of the mass density with respect to distance perpendicular to the lens plane). We note that one can write (12.86) as

$$\widehat{\alpha} = \nabla\widehat{V}, \qquad \widehat{V} \equiv 4G \int_{\mathbb{R}^2} \Sigma(\boldsymbol{\xi}') \ln|\boldsymbol{\xi} - \boldsymbol{\xi}'| \mathrm{d}^2\xi'. \tag{12.87}$$

Note that for large angular scale analyses (e.g. to study the dark matter distribution) one must replace the flat lens plane by (part of) the celestial sphere, though this is not needed in lensing of discrete sources.

Writing D_s for the distance of the source, D_d for the distance of the lens (the deflector) and $D_{ds} = D_s - D_d$ for the distance between the deflector and the source, these distances being well defined in the comparator flat space, we can write (12.86), in terms of angles $\boldsymbol{\beta}$ and $\boldsymbol{\theta}$ for the unlensed and lensed positions described by position vectors on the celestial sphere (see Figure 12.9), as

$$\boldsymbol{\beta} = \boldsymbol{\theta} - \frac{D_{ds}}{D_s}\widehat{\alpha}(\boldsymbol{\xi}), \tag{12.88}$$

or in terms of the distance $\boldsymbol{\eta} = D_s\boldsymbol{\beta}$ from the source to the optical axis and the closest approach $\boldsymbol{\xi} = D_d\boldsymbol{\theta}$ to the deflector, assuming the angles and distances are measured from

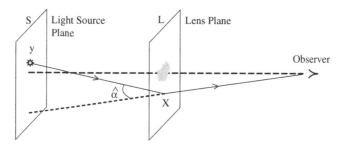

Fig. 12.9 A schematic of single-plane gravitational lensing. A pointlike light source is at y on the light source plane S. A light ray from the source is deflected through an angle $\widehat{\alpha}$ by the gravitational influence of the lens on the lens plane L. (From Fig. 1 of Petters and Werner (2010), reproduced with kind permission from Springer Science+Business Media B.V.)

this axis,

$$\eta = \frac{D_s}{D_d}\boldsymbol{\xi} - D_{ds}\widehat{\boldsymbol{\alpha}}(\boldsymbol{\xi}). \tag{12.89}$$

This can be rewritten in terms of some fiducial scale ξ_0 as

$$\boldsymbol{y} = \boldsymbol{x} - \boldsymbol{\alpha}(\boldsymbol{x}), \tag{12.90}$$

where $\boldsymbol{x} = \boldsymbol{\xi}/\xi_0$, $\boldsymbol{y} = D_d\boldsymbol{\eta}/D_s\xi_0$ and $\boldsymbol{\alpha} = D_d D_{ds}\widehat{\boldsymbol{\alpha}}(\xi_0\boldsymbol{x})/D_s\xi_0$. In this notation one can relate $\frac{1}{2}(\boldsymbol{y}-\boldsymbol{x})^2 - V(\boldsymbol{x})$, where $\boldsymbol{\alpha} = \nabla V$, to the arrival times of rays, which can be used in Fermat's principle to (re)derive the lensing equation.

In this form we have a *lensing map* from the lens plane to a source plane, the screen space at the distance D_s along the ray: this map need not be continuous or differentiable (Perlick, 2004). Note that when strong lensing occurs the emission is not from a single instant of time and one should replace the source plane by a timelike surface of successive planes.

The equation (12.88) can be extrapolated to a thin lens in a Robertson–Walker background by replacing the flat space distances by observer area distances. Note that the conformal flatness of FLRW universes ensures that the behaviour of rays away from the lens is simple. Alternatively one can calculate lensing using the Swiss cheese models of Section 16.4.1 (Kantowski, 1969b).

12.4.2 The null geodesic deviation equation

The null geodesic deviation equation (NGDE) treats lenses and images sufficiently small for the rays to be treated as neighbouring rays of a congruence, as introduced in Section 7.3: this is the only approximation involved. The Weyl tensor terms in (7.28) are important even when the Ricci tensor contribution in (7.26), i.e. the matter within the beam, is negligible. Calculating the effect of those terms in an inhomogeneous universe is in general too hard to be done exactly.

For the lensing application we use the null geodesic deviation equation in the form (7.29). We note that Λ does not appear there. Recently a derivation similar to that of (12.86), but using the Kottler (often called Schwarzschild–de Sitter) solution, showed that Λ nevertheless affects lensing, although the cosmological effect remains unclear (see Ishak, Rindler and Dossett (2010), Kantowski, Chen and Dai (2010) and references therein).

The NGDE (7.29) can be projected into the screen space (Lewis and Challinor, 2006, de Swardt, Dunsby and Clarkson, 2010a), giving

$$s^a{}_b \frac{\delta^2 X^b}{\delta v^2} = -4\pi G v^2 \left(\rho + p + \pi_{bc}e^b e^c - 2q_b e^b\right)\widehat{X}^a$$
$$-2v^2(\widehat{E}^{\langle ab\rangle} + \widehat{H}^{c\langle a}\eta^{b\rangle}{}_c)\widehat{X}_b, \tag{12.91}$$

where by analogy with PSTF tensors in three dimensions we define $\widehat{E}^{\langle ab\rangle}$ by

$$\widehat{E}^{\langle ab\rangle} \equiv \widehat{E}^{ab} - \frac{1}{2}\widehat{E}^c{}_c s^{ab}. \tag{12.92}$$

The projected equation (12.91) is equivalent to (7.26) and (7.28). The matrix of coefficients of $\widehat{X}^c$ on the right of this equation has been studied by Seitz, Schneider and Ehlers (1994).

12.4.3 The lensing equation from the null geodesic deviation equation

We now relate the lensing equation to the NGDE in a perturbed FLRW model. Following a number of authors, e.g. Sasaki (1993), Pyne and Birkinshaw (1996), Holz and Wald (1998) and de Swardt, Dunsby and Clarkson (2010a), we use a longitudinal gauge, which is a quasi-Newtonian gauge (see Sections 6.8 and 10.2). Thus surfaces of constant η are chosen so that their normals are shear and rotation free (but not necessarily geodesic). This implies that $H_{ab} = 0$, from (4.52). We ignore vector and gravitational wave (tensor) perturbations: since vector perturbations decay as the universe expands, while waves will only affect an image transiently and linearized effects should anyway be superposable, this seems reasonable. We assume that the matter content is dust, so from (10.55) and (10.71) $\Phi = \phi = \psi$ (more generally, for any metric theory in the same approximation scheme, the same formulae hold with $2\Phi \rightarrow \Phi + \Psi$).

As we are dealing with weakly (linearly) perturbed FLRW universes we can write $\tilde{u}^a = u^a + v^a$, as in Chapter 10, where u^a gives the surface normals, and $\tilde{u}^a$ is the fluid velocity. We use the tilde to denote quantities evaluated in the $\tilde{u}^a$ frame, and we retain only terms of order v^a. Given that the $\tilde{u}^a$ frame is that of comoving dust, the transformations (10.129) give

$$q_a = \rho v_a, \quad \pi_{ab} = 0, \tag{12.93}$$

to linear order. Now from the Maxwell–Weyl div-H constraint (10.147), curl $q_a = 0$, which implies curl $v_a = 0 = \omega^a$. The vorticity propagation equation (10.139) then implies that

$$\dot{u}_a = \overline{\nabla}_a \Phi, \tag{12.94}$$

where $\Phi = \ln r$ using (10.143) and the acceleration potential (see also Exercise 4.6.2). The quantity Φ will be identified with the Newtonian gravitational potential in an appropriate limit, as the notation suggests. Equation (10.144) then gives

$$8\pi G q_a = 8\pi G \rho v_a = \tfrac{2}{3}\overline{\nabla}_a \Theta \tag{12.95}$$

Linearized shear propagation (10.140) gives the constraint

$$E_{ab} = \overline{\nabla}_{\langle a} \overline{\nabla}_{b\rangle} \Phi. \tag{12.96}$$

This agrees with the Newtonian form for E_{ij}, (6.30), after substitution from Section 4.3.

To show this identification is correct we need to show Φ obeys the relativistic Poisson equation (10.62), on using $\sigma = 0$ (Newtonian gauge) and $\phi = \psi = \Phi$ (no anisotropic stress), i.e.

$$\left(\overline{\nabla}^2 + \frac{3K}{a^2}\right)\Phi - 3H(\dot{\Phi} + H\Phi) = 4\pi G \delta \rho. \tag{12.97}$$

The constraint (12.96) must hold identically under time and spatial derivatives. In order to take the spatial divergence, we need the linearized identity (Maartens, 1998, equation (46))

$$\overline{\nabla}^b \overline{\nabla}_{\langle a} V_{b\rangle} = \frac{1}{2}\overline{\nabla}^2 V_a + \frac{1}{6}\overline{\nabla}_a \overline{\nabla}^b V_b + \frac{K}{a^2} V_a, \tag{12.98}$$

which follows from (10.156). Together with the linearized identity (10.160) for commuting the Laplacian and the gradient, the divergence of (12.96) leads to

$$\overline{\nabla}^b \overline{\nabla}_{\langle a} \overline{\nabla}_{b \rangle} \Phi = \frac{2}{3} \overline{\nabla}_a \overline{\nabla}^2 \Phi + \frac{2K}{a^2} \overline{\nabla}_a \Phi. \tag{12.99}$$

Using the divergence constraint (10.146), we find

$$\overline{\nabla}_a \left[\left(\overline{\nabla}^2 + \frac{3K}{a^2} \right) \Phi - 4\pi G \rho + \frac{1}{3} \Theta^2 \right] = 0. \tag{12.100}$$

Now we differentiate (12.96) in the u^a direction, using the linearized commutators of Section 10.4.1. We obtain

$$\overline{\nabla}_{\langle a} \overline{\nabla}_{b \rangle} \left(\dot{\Phi} + \tfrac{1}{3} \Theta \right) + H \overline{\nabla}_{\langle a} \overline{\nabla}_{b \rangle} \Phi = 0. \tag{12.101}$$

A solution of this equation, taken to be the relevant one, is[2]

$$\overline{\nabla}_a \left(\dot{\Phi} + \tfrac{1}{3} \Theta + H \Phi \right) = 0. \tag{12.102}$$

Then (12.100) becomes

$$\overline{\nabla}_a \left[\left(\overline{\nabla}^2 + \frac{3K}{a^2} \right) \Phi - 3H(\dot{\Phi} + H\Phi) - 4\pi G \rho \right] = 0, \tag{12.103}$$

which implies (12.97).

To make the link with the lensing equation, we investigate the deflection as the ray passes the lens. We approximate v and $\widehat{X}^a$ as constant during the deflection. The proper distance corresponding to dv is $d\ell = v\, dv$, so we have,

$$s^a{}_b \frac{\delta^2 X^b}{\delta \ell^2} = -4\pi G \rho \widehat{X}^a - 2 \widehat{E}^{\langle ab \rangle} \widehat{X}_b. \tag{12.104}$$

Since the NGDE is linear in $\widehat{X}^c$ we can treat separately the contributions from the background ρ, which replaces flat space distances in (12.88) by observer area distances, and the perturbation terms. The coefficients on the right side of (12.104) from the perturbation are (assuming we can take a locally flat approximation so that derivatives commute)

$$-\nabla^2 \Phi s^a{}_b - 2 \widehat{\nabla}^a \widehat{\nabla}_b \Phi + \tfrac{2}{3} \nabla^2 \Phi s^a{}_b + (\nabla^2 \Phi - e^c e^d \nabla_c \nabla_d \Phi - \tfrac{2}{3} \nabla^2 \Phi) s^a{}_b$$

$$= -2 \widehat{\nabla}^a \widehat{\nabla}_b \Phi - s^a{}_b e^c e^d \nabla_c \nabla_d \Phi = -2 \widehat{\nabla}^a \widehat{\nabla}_b \Phi - s^a{}_b \frac{\partial^2}{\partial \ell^2} \Phi.$$

For a source and observer both far away from the lensing region, one can take infinite limits of the integration in ℓ, so integrating (12.104) now gives

$$\left[s^a{}_b \frac{\delta X^b}{\delta \ell} \right] = -\left(\int_{-\infty}^{\infty} 2 \widehat{\nabla}_a \widehat{\nabla}_b \Phi + s^a_b \frac{\partial^2}{\partial \ell^2} \Phi\, d\ell \right) \widehat{X}^b$$

$$= -\left(\int_{-\infty}^{\infty} 2 \widehat{\nabla}_a \widehat{\nabla}_b \Phi\, d\ell \right) \widehat{X}^b + \left[\frac{\partial \Phi}{\partial \ell} \right]_{-\infty}^{\infty} \widehat{X}^a$$

$$= -\left(\int_{-\infty}^{\infty} 2 \widehat{\nabla}_a \widehat{\nabla}_b \Phi\, d\ell \right) \widehat{X}^b.$$

---

[2] It may be the case, but has not been proved, that more general solutions differ only in gauge.

Since $\widehat{X}^a$ is an infinitesimal vector transverse to the rays, this should agree with the infinitesimal variation of $\widehat{\alpha}$ with position in the lens plane, i.e. we need to compare it with

$$\widehat{\alpha}_{a,b}\widehat{X}^b = 4G \int_{\mathbb{R}^2} \left(\frac{\delta_{ab}}{|\boldsymbol{\xi} - \boldsymbol{\xi}'|} - \frac{(\xi_a - \xi_a')(\xi_b - \xi_b')}{|\boldsymbol{\xi} - \boldsymbol{\xi}'|^3} \right) \Sigma(\boldsymbol{\xi}') \mathrm{d}^2\xi' \widehat{X}^a. \tag{12.105}$$

We have, in the local flat space,

$$\Phi = -G \int_{\mathbb{R}^3} \frac{\delta\rho' \, \mathrm{d}^3 r'}{|\boldsymbol{r} - \boldsymbol{r}'|}. \tag{12.106}$$

Taking the derivatives in the lens plane, we find

$$-2\widehat{\nabla}_a \widehat{\nabla}_b \Phi = 2G \int_{\mathbb{R}^3} \left(\frac{\delta_{ab}}{|\boldsymbol{r} - \boldsymbol{r}'|^2} - 2\frac{(\xi_a - \xi_a')(\xi_b - \xi_b')}{|\boldsymbol{r} - \boldsymbol{r}'|^4} \right) \delta\rho' \, \mathrm{d}^3 r' .$$

The integrals with respect to ℓ are, since $|\boldsymbol{r} - \boldsymbol{r}'|^2 = (\ell - \ell')^2 + |\boldsymbol{\xi} - \boldsymbol{\xi}'|^2$,

$$\int_{-\infty}^{\infty} \frac{\mathrm{d}\ell}{|\boldsymbol{r} - \boldsymbol{r}'|^2} = \frac{2}{|\boldsymbol{\xi} - \boldsymbol{\xi}'|}, \quad \int_{-\infty}^{\infty} \frac{\mathrm{d}\ell}{|\boldsymbol{r} - \boldsymbol{r}'|^4} = \frac{1}{|\boldsymbol{\xi} - \boldsymbol{\xi}'|^3}, \tag{12.107}$$

so, writing $\Sigma(\boldsymbol{\xi}') = \int \delta\rho(\boldsymbol{r}') \, \mathrm{d}\ell$,

$$-\int_{-\infty}^{\infty} 2\widehat{\nabla}_a \widehat{\nabla}_b \Phi = 4G \int_{\mathbb{R}^2} \left(\frac{\delta_{ab}}{|\boldsymbol{\xi} - \boldsymbol{\xi}'|} - \frac{(\xi_a - \xi_a')(\xi_b - \xi_b')}{|\boldsymbol{\xi} - \boldsymbol{\xi}'|^3} \right) \Sigma(\boldsymbol{\xi}') \mathrm{d}^2\xi', \tag{12.108}$$

which agrees with (12.105).

The actual value of $\widehat{\alpha}$ depends on the constant of integration of this derivative, corresponding to the choice of a ray taken to be undeflected. In applications relative deflections are the important quantities, so identifying such a ray is not necessary. But the method of Pyne and Birkinshaw (1996) instead considers a perturbed solution of the geodesic equations in the perturbed metric, thus in principle allowing computation of the deflection directly for each ray, or, alternatively, allowing direct calculation of the deflection of the central geodesic of the congruence. In the thin lens limit or in considering magnification and other properties of small images the results are the same.

We have thus linked the lensing equation with the perturbative lensing in a longitudinal gauge (remember this has to be added to the focusing by the averaged cosmological density ρ). In terms of scalar perturbations in a general gauge as described in Section 10.2, the perturbative contribution to (12.88) can be written

$$\widehat{\alpha} = -\int \widehat{\nabla}(\Phi + \Psi) \mathrm{d}v, \tag{12.109}$$

where $\mathrm{d}v = \mathrm{d}^2\xi'$ and Φ and Ψ are projected into the lens plane.

The treatment above is based on the 1+3 formalism. For spacetimes with preferred two-dimensional spaces, such as spherically symmetric spacetimes, a further expansion to a

1+1+2 formalism has been developed by de Swardt, Dunsby and Clarkson (2010b). Note that the preferred two-dimensional space is in general not the same as the screen space.

12.4.4 Lenses and images

To characterize the apparent size and shape of an image, the magnification matrix defined from (12.90) as $\mathcal{A}_{ab} = \partial y_a / \partial x^b = \delta_{ab} - \alpha_{a,b}$ is used. From the links made above between the optical scalars, the NGDE, and the lensing equation, one can see that the 2×2 matrix $\alpha_{a,b}$ can be written as

$$\left(\begin{array}{cc} \kappa + \gamma_1 & \gamma_2 - \omega \\ \gamma_2 + \omega & \kappa - \gamma_1 \end{array} \right), \tag{12.110}$$

where κ, $\gamma_1 + i \gamma_2$ and ω can be obtained by integrating the optical scalar equations (7.26) and (7.28) with respect to ℓ and are called the convergence, shear and (field) rotation. From (the linearized forms of) (7.26) and (7.28) it follows that the convergence depends on the surface density and the shear can be expressed in terms of a similarly defined surface Weyl tensor (Kling and Keith, 2005). In the derivation used above, $\omega = 0$ because $\boldsymbol{\alpha}$ is a gradient, but this need not be true when our approximations are inappropriate. The magnification of an image is defined as $1/\det \mathcal{A}$.

Lensing events by weak fields do not alter redshifts significantly and so do not change the specific intensity, (7.56). The peculiar velocities of source and observer are taken into account in specific intensity and observer area distance. The flux received from a given source thus varies only with the solid angle subtended at the observer by the image, and is given by the magnification, integrated across the source. It formally becomes infinite in strong lensing where the determinant passes through zero, and changes sign. The curves in the lens plane on which this happens are called *critical curves*. The real effect would be finite because close to such a critical curve the actual wave nature of light has to be considered, but in practice the fact that the sources are not point sources is more important in ensuring finite answers.

The shear gives the change of shape of an image. A method initially due to Kaiser and Squires (1993) showed that one can reconstruct the mass distribution in a thin lens from the shear. This method has been developed by several authors, in particular to deal with the finite resolution of real data; a variety of algorithms and statistical estimators have been used (see the reviews cited earlier). The method writes the surface density Σ (or convergence κ) as an integral of the shear (surface Weyl tensor) over the surface: Kling and Keith (2005) showed, using the formulation of Miralda-Escudé (1996), that this is simply an integral form of one of the (linearized) Bianchi identities.

Lensing also causes a time delay in the observations, partly due to the changed path length of the geodesics, and partly due to gravitational redshift. The understanding of (12.90) in terms of arrival time can be helpful here. The time delay is measurable if we at another time see the unlensed source, as in the Solar System where the gravitational contribution has been well tested, or when strong lensing results in multiple images and a true variation in the source gives rise to corresponding but time delayed variations in the different images.

12.5 Cosmological applications of lensing

12.5.1 Lensing and $m-z$ relations

We observe the visible matter in the universe to be significantly clumped, and expect its dark matter to also be clumped. We will tend to see images only when there is no visible matter contributing to the Ricci tensor within a beam, i.e. no corresponding Ricci tensor contribution in the NGDE. However, the clumps cause a non-zero Weyl tensor which has a net focusing effect, counteracting the absence of Ricci focusing. If the lens itself is not static, it can produce a non-zero net gravitational redshift. The combination of these effects can affect the measured $m-z$ relation. This important matter is discussed in Chapter 16.

12.5.2 Discrete sources and lenses

The first astronomical gravitational lens phenomena to be detected, other than the solar light bending, were double or multiple images of distant QSOs. Nowadays lensing objects, galaxies or clusters, can be detected both by such strong lensing and by weak lensing.

The lensing of distant galaxies or QSOs by closer ones gives information in three main ways. First, both strong and weak lensing can be used to infer the distribution of matter in the lens: this will include the dark matter in the lens as well as stars, dust and gas. Knowing the mass distribution, the virial theorem then allows one to infer the velocity dispersion. Early models used first spherical lenses and then ellipsoidal lenses. Now quite complicated forms for the lenses are considered, and matched in detail to observations. As well as the information this provides on the structure and evolution of galaxies or clusters, it constrains the distribution of dark matter.

There are several criteria for identifying strong lensing resulting in multiple images from the same source. One would expect to find similar spectra in all wave bands, the same redshift, and the same line emission, up to limits due to the different paths taken by the light and perhaps different material encountered, and ideally one would want to be able to see the lensing object and to be able to correlate time variations of the sources. Not all of these criteria are satisfied in all strong lensing candidates. Among the complications are that one or more of the images may show variability due to microlensing, for instance by stars in the lensing galaxy, and the spectra may appear to be different due to different absorption by intervening matter or because the lens differentially magnifies regions emitting in different ways, e.g. continuum- and line-emission regions. Strong lensing of galaxy–galaxy pairs gives constraints on the dark matter fraction and density profiles.

With weak lensing the difficulty is to know what the source was like in order to infer the lensing, if studying single images. This was recently addressed by a competition to compare different methods for inferring shear from images, using 3×10^7 simulated images (Bridle et al., 2010), whose results may lead to improved data analyses in future.

In practice a statistical approach is used, based on assumptions about the distant source population: for example, that spiral galaxies in the lensed region do not have aligned rotation axes. Provided many galaxies are imaged by the same lens, one can determine the shear

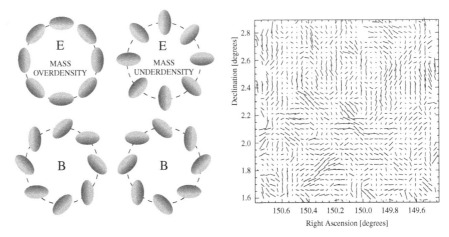

Fig. 12.10 The statistical signals sought by measurements of weak gravitational lensing are slight but coherent distortions in the shapes of distant galaxies.

(Left): A tangential, circular pattern of background galaxies is produced around a foreground mass overdensity. Physical gravitational lensing produces only these '*E*-mode' patterns. Measurements of the '*B*-mode' patterns illustrated below provide a free test for residual systematic defects.

(Right): The observed ellipticities of half a million distant galaxies within the 2 square degree Hubble Space Telescope COSMOS survey (Massey *et al.*, 2007). Each tick mark represents the mean ellipticity of several hundred galaxies. The *B*-mode signal is consistent with zero. (From Massey, Kitching and Richard (2010).)

field in the lens plane by averaging over many images. Shear has better signal to noise than magnification, but it needs about 100 images to get $S/N > 1$. Examples of the data obtained are given in Figures 1.9 and 12.10. For reviews of weak lensing see Heavens (2009), Massey, Kitching and Richard (2010) and Huterer (2010).

Lensing shows that the total mass of a galaxy is typically between 20 times the mass in stars, if the latter is $6 \times 10^{10} M_\odot$, and 50 times for $2-3 \times 10^{11} M_\odot$. This gives us information about the fraction of baryons in stars. One can also compare the matter distribution within objects with theory: most observations favour a flatter central region than expected from CDM predictions but more astrophysical modelling is needed to establish firm results.

Lensing can also reveal clusters behind foreground clusters, and the presence of multiple sets of emission lines at differing redshifts in a galaxy may show it is lensing several more distant sources.

An example of determination of the mass distribution in the lens is given by the recent measurement of the mass of the X-ray selected cluster XMMUJ2235.3-2557 from its lensing effect on more distant objects (Jee *et al.*, 2009): it has a redshift $z = 1.4$ and a mass, assuming a standard Navarro–Frenk–White mass profile and cosmological parameters from the WMAP data, of $(8.5 \pm 1.7) \times 10^{14} M_\odot$, which occurs with $\lesssim 1\%$ probability in a standard ΛCDM model.

The second type of information, mainly from strong lensing, comes from the ability to determine properties of sources at high redshifts at which, without the magnification provided by the lens, they would be unobservable. The lens is being used as a telescope.

One drawback of this method is that the sources need not be typical of the population of, say, galaxies at high redshift.

Finally, when multiple images of a variable source are observed and one can unambiguously identify the variations in two or more images, the time delay between the arrival of the same signal in the images gives a measure of the Hubble parameter. If we have a good model for the lens this can be used to obtain a value for the Hubble constant. The reason is if we made a change of length scale, and scaled masses the same way, the geometry of the light rays would agree (being conformally invariant) but the time delay would be scaled. Hence the measured time delay sets the scale, i.e. the Hubble constant.

A recent analysis using 18 time-delay lenses (Paraficz and Hjorth, 2010) gave a result 66^{6}_{-4}, consistent with other methods and even better agreement was found when using only five lenses where isothermal models could be assumed. Suyu et $al.$ (2010), using radio monitoring of the four images in the lens B1608+656 coupled with a detailed model of the lensing object using a high-resolution image from the Hubble Space Telescope and stellar velocity dispersion measurements, obtained a value $H_0 = 70.6 \pm 3.1$ (without a prior of $K = 0$) from this one lens. Combining their data with a $K = 0$ prior and WMAP results gave $H_0 = 69.7^{+4.9}_{-5}$ and $w = -0.94^{+0.17}_{-0.18}$. These uncertainties are comparable with those from baryon acoustic oscillation data (see Section 13.2), showing the potential of lensing results.

Weak lensing and flexion data are used to map galactic and cluster halos: since dark matter is believed to constitute a large part of the total mass of galaxies and clusters, this enables us to probe the relation of visible and dark matter without assumptions about 'bias'. The theory of virialized halos leads to the Navarro–Frenk–White profile and lensing observations are in good agreement with this.

12.5.3 Large-scale structure, dark matter and dark energy

We again refer for fuller information to the surveys of Heavens (2009), Massey, Kitching and Richard (2010) and Huterer (2010).

On the large scale, Weyl tensor terms encode the power spectrum of the distribution of the matter and so lensing can be used to determine this. The multipolar decomposition is in good agreement with other measures. In recent years the two-dimensional data of the shear field has been made three-dimensional (lensing 'tomography') by using the photometric redshifts from redshift surveys as a proxy for distance. By this means Massey et $al.$ (2007) were able to map the large-scale distribution of dark matter. These data are degenerate in the Ω_m–σ_8 plane but the degeneracy is resolved when CMB data are used. In principle observations of the mass at different redshifts could be used to test conservation of mass on cosmological time scales.

Weak lensing can also give constraints on the neutrino masses. (So far, direct observation of neutrinos gives good measures for mass differences but not for the absolute value.) The reason is that even if neutrinos are not massless, their low mass would imply a streaming out of mass concentrations leading to a reduction of small-scale power in the spectrum. Kristiansen, Elgarøy and Dahle (2007) give a limit $M_\nu < 1.43\text{eV}$.

The direct effects of dark energy on lensing are less important than the indirect effect via the observer area distance and growth rate of perturbations. For example, Kilbinger *et al.* (2009), who cite earlier analyses, give bounds $-1.18 < w < -0.88$ on the equation of state of dark energy, at 95% confidence, from lensing combined with the WMAP5 and SNIa data.

12.5.4 Lensing and the CMB

For fuller details of the results summarized here, and entry points into the literature, see Hanson, Challinor and Lewis (2010). As they say, from the point of view of CMB observations, lensing is a contaminant.

One might think that the effect of the lensing by stars, galaxies, clusters and so on, each causing caustics and folds in the past null cone, would be that the area we see on the last scattering surface would be significantly greater than that in a uniform FLRW universe, changing the interpretation of the CMB measurements. This indeed would happen if every line of sight passed near enough to a lensing object (a star, a gas cloud, a concentration of dark matter). But in fact only a small fraction of the celestial sphere is covered by stars, galaxies or clusters: astronomical images are misleading in this respect (the same point affects the Olbers' paradox arguments, Harrison (2000)).

Detailed estimates give an RMS deflection of CMB rays of 2.7 arcmin, with a coherence length of a few degrees (against a 10 arcmin scale of the $\ell = 1000$ modes in the CMB). Because there are more of them, smaller lenses contribute most, and the peak contribution comes from lenses at redshift $z \sim 2$. This results in a small but non-negligible effect on the CMB power spectrum (which completely characterizes the CMB assuming it is a Gaussian field, with no phase correlations), and the introduction of a vortex-like B-mode polarization which could confuse searches for the similar effect on polarizations due to gravitational waves. Note that the lensing does not introduce additional polarization: it just realigns existing polarizations. These effects can be considered as adding non-Gaussianity and anisotropy to the predicted observations.

The lensing will be correlated with large-scale temperature anisotropies caused by the (late-time) integrated Sachs–Wolfe effect, and this will lead to a contribution in the bispectrum indicating non-Gaussianity.

Conversely, the CMB measurements can be used to calculate a maximum likelihood estimator for the lensing distribution. In principle CMB measurements on small angular scales could also be used to reconstruct the mass distribution of individual clusters.

As yet the experimental data give at best about a 3σ detection of a lensing signal in the CMB, not yet a very convincing level, but future experiments should improve this substantially, and, conversely, will need to remove the lensing contribution when analysing data for other cosmological phenomena.

12.5.5 Testing cosmological gravity

Lensing can be used to test theories of gravity under the assumption of an almost RW metric (compare Section 8.3.2). For example, Smith (2009) used it to bound the PPN parameter γ and thence constrain scalar–tensor and $f(R)$ theories, but (Schwab, Bolton and Rappaport, 2010) the present data on this are not good enough for significant conclusions unless independent constraints on the lensing galaxies can be given.

The observations of the 'Bullet cluster' (Clowe, Gonzalez and Markevitch, 2004) show it is really two colliding clusters which have passed through one another. Lensing shows a mass distribution significantly different from the distribution of the hot interacting gas shown by X-ray emission: a number of similar but less spectacular examples have been found (Shan *et al.*, 2010). This is evidence against, or at least, requiring to be reconciled with, alternative gravity theories as an alternative to dark matter. Ferreras *et al.* (2009) concluded that strong lensing and galactic rotation curve data together rule out the alternative theory TeVeS (Section 12.3.3) as a way of removing the need for dark matter. Bean and Tangmatitham (2010) have examined constraints on variations of the governing equations arising from combining the SNIa, BAO, ISW, galaxy survey and weak lensing data: these are tight at last scattering but less so at later times.

Looking for aligned pairs of galaxies, Morganson *et al.* (2010) found no evidence for lensing by cosmic strings and thus put limits of $G\mu < 2.3 \times 10^{-6}$ and $\Omega_s < 2.1 \times 10^{-5}$ on the string tension and density, with stronger limits if some doubtful candidates were not strings.

Weak lensing surveys could be used to distinguish different theories by their effects on observer area distance and the growth rate of perturbations, for example to distinguish DGP theory (Section 14.3) and GRT. However, to do so one would need a good understanding of nonlinear clustering.

12.5.6 Microlensing and its uses

Microlensing occurs when the angular scale of the lens is smaller than that of the lensed object. Lensing is then not detectable as light bending or multiple images, but as variations in total brightness, possibly as intensity peaks in an image. It occurs in multiply imaged quasars, where the stars of the lensing galaxy cause variations in the quasar images, and in observations of stars in our own Galaxy. Since the lens typically moves across the image, one needs lensing formulae which take into account source and lens motion: Kopeikin and Schäfer (1999) provided these using a Lienard–Weichert formula for the light propagation.

One can also do pixel lensing, where several sources (stars) are grouped in a single pixel and one measures the aggregated microlensing. There is also 'millilensing' (Massey, Kitching and Richard, 2010) by, for example, substructures within a galaxy: observations so far suggest there is more substructure than predicted.

A number of searches have been conducted to find small objects in our Galaxy or neighbouring galaxies by their lensing effect on more distant stars (Mollerach and Roulet, 2002, Carr *et al.*, 2010), shown by a light curve characteristic of an object passing in front of the source. The inferred masses and densities of such objects constrain any contribution to dark matter in the form of such bodies. The conclusion, from a large number of observations, is that 'a substantial contribution of compact objects to a standard halo is now clearly excluded' (Moniez, 2010). This includes primordial black holes and masses up to $30 M_\odot$ (Carr *et al.*, 2010). Thus there is no large component of the dark matter in this form.

Microlensing can also be used to detect extrasolar planets (Dominik, 2010), which are of great interest for other reasons: over 400 such planets are now known, of which, so far, 10 have been found from microlensing.

In the case of quasars, one might obtain information about the size of the continuum emission and line emission regions, and the brightness profile, as well as the lensing objects. The best-studied example is Q2237+0305, 'Huchra's lens', where changes of brightness are obvious in the four distinct images.

Another application may be in searching for cosmic superstrings (Chernoff and Tye, 2007).

Overall, gravitational lensing started out as a theoretical prediction of GR, and now is a crucial tool in detecting different forms of dark matter as well as in other astrophysical studies.

13 Confronting the Standard Model with observations

The basic observational concepts and relations in cosmology were introduced in Chapter 7. Here we consider how this works out when applied to the standard models of cosmology: how do current observations support and refine the use of perturbed FLRW models, discussed in Chapters 9 to 12? This is crucial in determining how acceptable these models are as models of the real universe: observational testing is the core of scientific cosmology.

Extraordinary progress has been made in this regard in recent decades. Multi-wavelength observations (radio to γ-ray) of vast numbers of sources and various backgrounds have been made, gathering massive amounts of data. This has been based on developments in telescopes (ground-based, balloon-borne, space-based), detectors (photomultipliers, CCDs, fibre optics, adaptive optics, interferometric spectrometers, etc.), and high-performance computing power. We cannot go into those developments here, but simply refer to Lena, Lebrun and Mignard (2010) for a survey.

An important feature of the observational constraints is that they have two separate aspects. On the one hand, we can test the *background* model by observations that probe the expansion history, such as the magnitude–redshift diagram, and, on the other hand, we can test *perturbations* about the background by observations that probe the CMB anisotropies and the growth of structure.

For the background, we use galaxies, supernovae and the CMB as markers of the geometry of the background model; for this purpose we are only interested in their capacity to provide standard candles and standard rulers. For the perturbations, we need probes of the formation and evolution of CMB anisotropies and large-scale structure. The perturbed model is tested via its statistical predictions, and the best-fit parameters from tests of the background model should be used in testing the perturbations. If the perturbed model passes the observational tests, we will confirm the crucial consistency test:

Observational consistency *between the background model and the perturbed model.*

This incorporates a test of GR, because the link between background and perturbations is the theory of gravitation. For example, on super-Hubble scales, for adiabatic perturbations and neglecting anisotropic stress, the evolution of the perturbations is entirely determined by the background via (Bertschinger, 2006)

$$2\Phi'' - \frac{H''}{H'}\Phi' - \left(\frac{H'}{H} + \frac{H''}{H'}\right)\Phi = 0, \quad f' := \frac{df}{d\ln a}, \qquad (13.1)$$

where $\Phi (= \Psi)$ in Newtonian gauge is given by (12.13).

Before discussing these two kinds of observations and their relation to each other, we first consider the basis for supposing that perturbed FLRW models are good descriptions of the real universe.

13.1 Observational basis for FLRW models

The standard FLRW universe models are simple and have tremendous explanatory power. Furthermore, their major physical predictions – such as the existence of blackbody CMB and specific light element production in the early universe – seem confirmed. To what degree do observational data uniquely indicate these universe models for the expanding universe geometry?

As discussed in detail in Chapter 9, the background FLRW model is isotropic about every point, and hence is exactly spatially homogeneous. The perturbations discussed in Chapter 10 are assumed to be statistically compatible with these symmetries. Thus a key issue is,

Universe geometry: *is the universe (on a large enough scale) spatially homogeneous and isotropic?*

On small scales it is clearly neither. The FLRW description of the real universe implicitly assumes two fundamental components: (1) An *averaging scale* such that the universe is spatially homogeneous and isotropic above that scale (Ellis, 1984). At the time of writing the scale has been put at $O(70h^{-1})$ Mpc (Sarkar *et al.*, 2009), but note the competing claim of no homogeneity up to $100h^{-1}$ Mpc (Sylos Labini *et al.*, 2009). (2) A prescription for how the idealized metric form (9.9) is related to the real ('lumpy') universe that we see around us – which may be called the 'fitting problem' in cosmology (Ellis and Stoeger, 1987). These two issues, and the problems that arise through the implied averaging process, are discussed in Section 16.1.

A direct observational proof of large-scale homogeneity on this basis is very difficult (Ellis, 1980). As a consequence, the deduction of spatial homogeneity is usually made on the basis of the observed isotropy of matter and radiation about us (when averaged on a sufficiently large scale), together with the Copernican Principle. We now discuss the different approaches used to deduce homogeneity of the universe.

13.1.1 Theoretical approaches

Cosmological Principle

Spatial homogeneity is the simplest case and apparently we do not need anything more complex on the basis of current data. This can be encoded in a philosophical principle as the foundational basis of our cosmological models:

Cosmological Principle: *the universe is spatially homogeneous and isotropic.*

This is essentially an a-priori prescription for initial conditions for the universe, embodying the idea that the universe is necessarily simple. For decades following Milne's espousal of this principle (Milne, 1935, Gale, 2007), it was taken as a basic principle of cosmology (Bondi, 1960, Weinberg, 1972). It was even extended to the proposal of the 'Perfect Cosmological Principle' (Bondi, 1960): that the universe is space–time homogeneous instead of only spatially homogeneous. This was the basis of the Steady State universe models (Section 9.3.1), which were, however, ruled out by astronomical data.

The cosmological principle is sometimes justified on the basis of a closely related but weaker assumption:

Copernican Principle: *the Earth does not occupy a privileged position within the universe as a whole.*

This is the culmination of the Copernican revolution whereby the Earth was displaced from the being at the centre of the universe to being an average planet orbiting an average star situated in an average galaxy at a typical location in the universe. It is often taken as justifying the more technical Cosmological Principle as formulated above, which leads directly to the FLRW family of cosmological models.

However, because of their exact symmetries, on any reasonable measure the FLRW models are of zero probability within the family of all possible cosmological models. Thus this assumption implies that the universe is of an extremely special, hence highly fine-tuned, nature. From the late 1930s to the 1960s, this was taken as reasonable. However, in the late 1970s, the philosophical tide turned: it became common to assume the universe has a generic rather than special geometric nature. Indeed it was no longer assumed there were any specific cosmological principles at all. Instead, general physical principles, such as high probability and low entropy, would be applied to the universe itself (and not merely to the matter in it). Furthermore the abundance of data opened up the possibility of observational testing of spatial homogeneity. The isotropy of observations allows a more empirically based use of the Copernican Principle, as discussed below.

Physical arguments

One can claim that physical processes such as inflation (Section 9.7) make the existence of almost-FLRW regions highly likely, indeed much more probable than any alternative. This is a potentially viable argument, but it amounts to replacing an observational test by a theoretical argument based on a physical process that is yet to be grounded in an established fundamental theory. In addition, the result depends on an unknown measure as well as some fine-tuned initial conditions. However, it is strongly bolstered because predictions for the detailed pattern of CMB anisotropy (Hu and Sugiyama, 1995), based on the inflationary universe theory (Section 12.2), have been confirmed (Komatsu *et al.*, 2009). It is in principle conceivable that for example spherically symmetric inhomogeneous models (with or without inflation) can produce similar patterns of anisotropy. Such patterns could also emerge if suitable primordial initial conditions occur without a previous inflationary phase.

Uniform thermal histories

From an astrophysical viewpoint, a good reason for believing in spatial homogeneity is that we see the same kinds of objects everywhere we look in the sky. If conditions out there were different, surely we would see different kinds of objects as a result? Thus from uniformity in the nature of the objects we see in the sky (e.g. the same types of galaxy at large distances as nearby), it is reasonable to deduce they must have all undergone essentially the same thermal history. The aim is to prove a *Postulate of Uniform Thermal Histories:* observed homogeneity of structures implies spatial homogeneity of the universe (Bonnor and Ellis, 1986). This is a reasonable conjecture, because development of astrophysical structures depends on spacetime curvature, which (given the matter content) determines the thermal history of the universe.

However, turning this idea into a proper test of homogeneity has not succeeded so far: indeed it is not clear if this can be done, because some (rather special) counter-examples to this conjecture have been found (Bonnor and Ellis, 1986). Nevertheless the approach could be used to give evidence on spatial homogeneity. For example, observations showing that element abundances at high redshift in many directions are the same as locally (Pettini, 1999, Pettini, Lipman and Hunstead, 2005, Sigurdson and Furlanetto, 2006) are very useful in constraining inhomogeneity by showing that conditions in the very early universe at the time of nucleosynthesis must have been the same at distant locations in these directions. Similarly, if ages of distant objects were incompatible with local age estimates, this would be a possible indication of inhomogeneity (Jain and Dev, 2006).

13.1.2 Observational approaches

The FLRW models require both isotropy and spatial homogeneity. Isotropy of observations is well established: considered on a large enough angular scale, astronomical observations are very nearly isotropic about us, both for sources and background radiation. The CMB has a spectacularly high observed degree of isotropy, after allowance for the motion of the Earth, Sun and Galaxy through the universe (which combine to give a dipole variation): the temperature variations around the sky in the CMB are of order $|\delta T / T| < 10^{-5}$. The most detailed results now are those from the WMAP satellite (Jarosik *et al.*, 2011). They are consistent with the variations expected from the density perturbations which will later form the observed galaxies and clusters. Galaxy surveys do not currently have the sky coverage and depth for detailed limits on anisotropy, but tests for quadrupolar anisotropy in SDSS luminous red galaxies find no statistically significant deviation from isotropy (Pullen and Hirata, 2010).

Because isotropy applies to all observations, this establishes that in the observable region of the universe, to high accuracy both the spacetime structure and the matter distribution are isotropic about us. The generic models compatible with isotropy are spherically symmetric universe models, with vanishing pressure at late times (i.e. after decoupling) but possibly with a cosmological constant. The exact models of this kind are well known: they are the Lemaître–Tolman–Bondi models, discussed in Chapters 15 and 19. In general they will be spatially inhomogeneous, with our Galaxy located at or near the centre. This contradicts the Copernican Principle, but it is certainly possible in principle (see Chapter 15).

Is there convincing observational evidence for spatial homogeneity in addition to the spherical symmetry?

Number counts

The first observational evidence for spatial homogeneity was from galaxy number counts carried out by Hubble (1936) (compare Peebles (1971)). But the later radio source counts did not verify spatial homogeneity, indeed they are only compatible with FLRW models if we assume major source evolution occurs in either numbers or source flux. As yet we have no good astrophysical argument for what such evolution should be, and it is customary to run the argument backwards – assume that spatial homogeneity is known in some other way, and deduce the source evolution required to make the observations compatible with this geometric assumption (Ellis, 1975). It is always possible to find a source evolution that will achieve this (Mustapha, Hellaby and Ellis, 1999). Number counts by themselves do not show that the universe is spatially homogenous.

FLRW observational relations

If we could show that the source observational relations had the unique FLRW form, as in (7.52), and (7.64), as a function of redshift, this would establish spatial homogeneity in addition to the isotropy, and hence an FLRW geometry (Ellis *et al.*, 1985); see Section 8.6.2. However, the observational problems mentioned above – specifically, unknown source evolution – prevent us from carrying this through by observations of distant discrete sources: we cannot measure distances reliably enough because galaxies, quasars and radio sources are not good standard candles. Astrophysical cosmology could resolve this in principle, but is unable to do so in practice. Indeed (as just mentioned) the actual situation is the inverse: taking radio-source number-count data at face value, without allowing for source evolution, contradicts a RW geometry. Thus attempts to observationally prove spatial homogeneity in this way fail.

13.1.3 Best current argument for homogeneity

As mentioned above, observations confirm the key feature of isotropy about our spacetime location. This is a very special feature to observe. Surely we cannot be the only observers in the universe for whom this is true? That would make us very special in the class of all observers. But if it is true for arbitrary observers, then the universe of necessity has to be spatially homogeneous: this is an exact theorem (Walker, 1944, Ehlers, 1961, Ellis, 1971a).

Theorem 13.1 Homogeneity from isotropy
If all observers see an isotropic universe, then spatial homogeneity follows; in fact homogeneity follows if only three separated observers see isotropy.

This is the argument for spatial homogeneity that is most generally accepted. We cannot observe the universe from any other point, so we cannot observationally establish that far distant observers see an isotropic universe, and we need to assume isotropy of *all* observations.

A powerful enhancement, also assuming the Copernican Principle, follows from a theorem by Ehlers, Geren and Sachs (EGS) (1968), based only on radiation. The EGS theorem and its generalization by Ellis, Treciokas and Matravers (1983) (ETM) are discussed in Section 11.1. The original results assumed that the only source of the gravitational field was the radiation, i.e. matter and dark energy were neglected. The results are generalized to include self-gravitating matter and dark energy by Clarkson and Maartens (2010), leading to Theorem 11.2, repeated here:

Theorem 13.2 CMB partial isotropy + CP $\rightarrow$ FLRW
In a region, if

- *collisionless radiation has vanishing dipole, quadrupole and octupole,*
- *the radiation four-velocity is geodesic and expanding,*
- *there are pressure-free baryons and CDM, and dark energy in the form of Λ, quintessence or a perfect fluid,*

then the metric is FLRW in that region.

In practice we can only observe approximate isotropy. As discussed in Section 11.1, the more realistic basis for a spatially homogeneous universe (with the Copernican assumption) is Theorem 11.3 (Stoeger, Maartens and Ellis, 1995, Maartens, Ellis and Stoeger, 1995b), repeated here:

Theorem 13.3 CMB almost-isotropy + CP $\rightarrow$ almost-FLRW
In a region of an expanding universe with Λ, if all observers comoving with the matter measure an almost isotropic distribution of collisionless radiation, and if some of the time and spatial derivatives of the covariant multipoles are also small, then the region is almost FLRW.

These results are currently the most persuasive observationally based arguments we have for spatial homogeneity. This still relies on plausible philosophical assumptions. The deduction of spatial homogeneity follows not directly from astronomical data, but because we add to the observations a philosophical principle that is plausible but as yet untested. Experiments have been proposed to test the Copernican Principle by looking for violations of isotropy at events down our past light-cone. These include looking for spectral distortions of CMB photons scattered by ionized gas (Goodman, 1995, Caldwell and Stebbins, 2008): such distortions are induced by anisotropies in the CMB as seen by distant observers, and so provide in principle a neat way of confirming the Copernican assumption as used here. A similar test uses the kinematic Sunyaev–Zel'dovich effect in clusters to observe the dipole around distant observers (García-Bellido and Haugbølle, 2008). CMB polarization measurements may also be able to probe distant anisotropy (Kamionkowski and Loeb, 1997). The almost-EGS theorem then gives a framework for probing inhomogeneities via such observations.

Domains of validity

In which spacetime regions does this argument establish a RW-like geometry? The CMB probes the state of the universe from the time of decoupling to the present day, within the visual horizon. The argument from CMB isotropy can legitimately be applied for that epoch. However, it does not necessarily imply isotropy of the universe at much earlier or much later times. For example, as discussed in Chapter 18, there are Bianchi universes which admit intermediate isotropization (Wainwright and Ellis, 1997, Wainwright *et al.*, 1998): they can mimic a RW geometry arbitrarily closely for an arbitrarily long time. No matter how strong the bounds from CMB anisotropy measurements and data on element abundances, anisotropic modes can dominate at even earlier times as well as at late times (long after the present). If inflation took place, this conclusion could be reinforced, since inflation washes out any information about very early universe anisotropies and inhomogeneities in a very efficient way.

As well as this time limitation on when we can regard homogeneity as established, there are major spatial limitations. The above argument does not apply far outside the visual horizon, for we have no reason to believe the CMB is highly isotropic there, and no data from there. If chaotic inflation is correct, conditions there are not the same. Indeed that applies to all observational efforts to establish spatial homogeneity: they cannot succeed outside the visual horizon, unless we adopt the Copernican assumption.

Problem 13.1 An FLRW model is an approximation: it is only valid through a process of coarse-graining (see Chapter 16), and there are errors and statistical uncertainties in the data. Given this context, the 'proofs' that the universe is well described by an FLRW model are only approximately applicable. What observations would be sufficient to *disprove* this model as a good model of the observed region of the universe? (If one cannot give such criteria, then the model is not a scientifically testable hypothesis. But there are such tests: see Sections 13.4 and 15.6.5.)

Problem 13.2 Develop the Postulate of Uniform Thermal Histories idea further: does it give us a genuine test of homogeneity? If so, what limits does it give?

13.2 FLRW observations: probing the background evolution

Here we discuss the main observational tests that probe the background evolution, i.e. the expansion history $H(z)$:

$$\frac{H^2(z)}{H_0^2} = \Omega_{r0}(1+z)^4 + \Omega_{m0}(1+z)^3 + \Omega_{K0}(1+z)^2$$

$$+ \Omega_{de0} \exp 3 \int_0^z \left[\frac{1+w(z')}{1+z'} \right] dz' \tag{13.2}$$

where $w(z)$ is the dark energy equation of state function. Using six high-precision distance-determination methods, a recent estimate of H_0 is (Freedman and Madore, 2010)

$$H_0 = 100h \, \text{km/s/Mpc}, \quad h = 0.73 \pm 0.02 \, (\text{random}) \pm 0.04 (\text{systematic}). \tag{13.3}$$

13.2.1 Standard candles: supernovae

Early attempts to use galaxies as standard candles for relating luminosity to distance, or to use radio source number counts to calibrate the background expansion rate, failed – essentially because of evolution in these sources. Currently the best standard candles that we have are supernovae of type Ia (SNIa). Attempts have also been made to use Gamma-Ray Bursts (GRBs) as standard candles, based on various correlations in their properties (Wright, 2007). Although the GRBs have not yet been shown to be good standard candles, they lead to results that are consistent with those from SNIa.

Calibrations of SNIa light curves lead to an empirically determined intrinsic luminosity. The underlying physics is highly nonlinear and complex, and is not properly understood – and the empirical approach may be undermined by evolutionary and other systematic effects. Nevertheless, intensive investigation of a range of systematics has been made and is ongoing. Currently, SNIa are the most reliable standard candles we have amongst astrophysical objects. The luminosity distance as a function of redshift in FLRW spacetime is given by (7.51); using (9.10), this becomes

$$D_L = \frac{1+z}{H_0 \sqrt{-\Omega_{K0}}} \sin\left(\sqrt{-\Omega_{K0}} \int_0^z \frac{dz'}{H(z')/H_0}\right), \tag{13.4}$$

and the apparent magnitude is related to the absolute magnitude and luminosity distance by (7.44):

$$m = 5\log_{10}(D_L/1\,\text{Mpc}) + M + 42.38 - 5\log_{10} h. \tag{13.5}$$

The observed (m) and intrinsic (M) magnitudes constrain the expansion rate $H(z)$ via D_L for a given Ω_{K0} and $w(z)$ in (13.2). There is clearly a degeneracy between curvature and dynamics, and breaking this degeneracy requires independent observational constraints (see below). A recent compilation of magnitude–redshift data is shown in Figure 1.2, where the distance modulus is $\mu = m - M$. This shows that the universe is expanding more slowly at greater distances. If there are no systematic errors in the data, then in an FLRW model, the implication is that the expansion is accelerating, requiring dark energy. Figure 13.1 shows how such data constrain the density parameters in (7.44) for $w = -1$.

13.2.2 Standard rulers in matter: BAO

As discussed in Section 12.3.6, the BAO scale in the matter power spectrum provides a comoving imprint of the sound horizon at decoupling. This is in principle an excellent standard ruler, able to probe greater redshifts than SNIa. In addition, it may be understood on the basis of linear analysis, and is independent of the highly nonlinear astrophysics of individual objects like SNIa or galaxies.

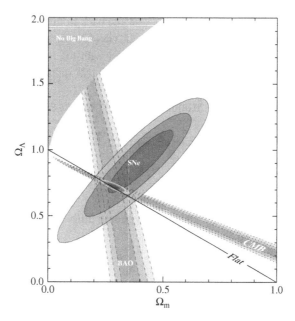

Fig. 13.1 Constraints (68%, 95% and 99% CL contours) in the (Ω_{m0}, $\Omega_{\Lambda 0}$) plane from SNIa, BAO and CMB. (From Kowalski *et al.* (2008). Reproduced by permission of the AAS.)

The baryon ruler must be determined statistically, thus requiring very large-volume surveys to achieve the precision needed. The first detection of the BAO scale used the 2dF (Cole *et al.*, 2005) and SDSS (Eisenstein *et al.*, 2005) galaxy surveys; see Figure 1.7 for the real-space feature and Figure 13.4 for the feature in Fourier space. This has provided a powerful new probe of the expansion history, independent of SNIa data (see Figure 13.1).

As explained in Section 12.3.6, the 2-point correlation function (12.1) for galaxies contains information on the radial (line of sight, ν) and transverse (σ) BAO features. Averaging over orientations θ, where $\nu = \sigma \cos \theta$, we get the monopole:

$$\xi_0(r) = \int_0^1 \xi(\sigma, \nu) \sin \theta d\theta, \quad r^2 = \sigma^2 + \nu^2. \tag{13.6}$$

The monopole leads to a constraint on an averaged distance measure D_V or an equivalent dimensionless parameter d (Eisenstein *et al.*, 2005, Cole *et al.*, 2005):

$$D_V(z) = \left[\frac{z(1+z)^2 D_A^2(z)}{H(z)} \right]^{1/3}, \quad d(z) = \frac{r_s(z_{\text{dec}})}{D_V(z)}, \tag{13.7}$$

where r_s is defined in (12.79). Recent results, based on the final SDSS data release 7 and the 2dF data, give (Percival *et al.*, 2010)

$$d(0.2) = .1905 \pm .00061, \quad d(0.35) = .1097 \pm .0036. \tag{13.8}$$

If the average is not taken, then additional information can in principle be extracted from observing the BAO peak in radial and transverse directions (Gaztañaga, Cabré and Hui, 2009, Kazin *et al.*, 2010). Firstly, a check for consistency between the two versions of the peak provides a test of isotropy of the background expansion. Secondly, $D_A(z)$ and $H(z)$ may be independently determined, via (12.80); in particular, a new direct handle on the evolution of $H(z)$ will be in contrast with the integrated form of $H(z)$ that is given by SNIa. Current data do not yet have sufficient statistical power to exploit these possibilities.

13.2.3 Standard rulers in matter: equality and damping scales

There are two other scales imprinted in the matter distribution: the comoving Hubble scale at matter–radiation equality,

$$k_{\text{eq}}^{-1} = \frac{1}{a_{\text{eq}} H_{\text{eq}}}, \tag{13.9}$$

and the Silk damping scale from photon diffusion (Lyth and Liddle, 2009),

$$k_S^{-1} \approx \frac{1}{a} \left(\frac{t}{n_e \sigma_T} \right)^{1/2}. \tag{13.10}$$

For the flat ΛCDM model, $k_{\text{eq}}^{-1} \approx 100\text{Mpc}$, and at decoupling, $k_S^{-1} \approx 10\,\text{Mpc}$. The equality scale marks the turn-over peak of the matter power spectrum.

13.2.4 Standard rulers from the CMB

The sound horizon at photon decoupling provides an even cleaner standard ruler than the baryon acoustic ruler, which is not processed by structure formation and therefore requires no nonlinear corrections. However, it is available only at one redshift. The CMB provides two distance ratios, $(1 + z_{\text{dec}})D_A(z_{\text{dec}})/r_s(z_{\text{dec}})$ and $D_A(z_{\text{dec}})/H^{-1}(z_{\text{dec}})$. Typically these are represented by the angular scale of the sound horizon, ℓ_A, and the 'shift' parameter, S,

$$\ell_A = \pi \frac{(1 + z_{\text{dec}})D_A(z_{\text{dec}})}{r_s(z_{\text{dec}})}, \quad S = \sqrt{\Omega_{\text{m}0} H_0^2} (1 + z_{\text{dec}}) D_A(z_{\text{dec}}). \tag{13.11}$$

Note that $S\sqrt{1 + z_{\text{dec}}}$ is only an approximation to $D_A(z_{\text{dec}})/H^{-1}(z_{\text{dec}})$.

These observables are sensitive to the parameters of the background, but they do not change much for simple models of dark energy. The WMAP 7-year results (Komatsu $et\ al.$, 2011) give

$$\ell_A = 302.09 \pm 0.76, \quad S = 1.725 \pm 0.018. \tag{13.12}$$

(This allows $\Omega_{K0} \neq 0$, but assumes negligible isocurvature and tensor perturbations, and negligible deviation from a power-law primordial spectrum of scalar perturbations.) The combined constraints from the CMB standard rulers, baryon acoustic oscillations and SNIa are shown in Figures 13.1 and 13.2. The BAO provides the most stringent constraint on Ω_{K0}.

13.2.5 Luminosity distance from gravitational waves

A new possible probe of cosmic distances arises from putative future space-based gravitational wave experiments such as the Big Bang Observer (BBO). Although BBO is conceived

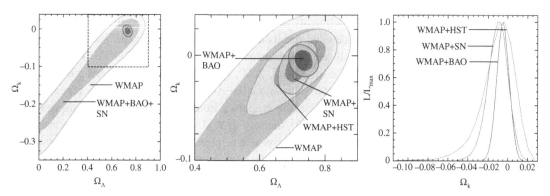

Fig. 13.2 Constraints (68% and 95% CL contours) on $\Omega_{\Lambda 0}$, Ω_{K0}. *Left:* WMAP5-only (light blue in the colour version) compared with WMAP+BAO+SN (purple). *Middle:* Blow-up of region in left panel, showing WMAP-only (light blue), WMAP+HST (grey), WMAP+SN (dark blue), WMAP+BAO (red). *Right:* Constraint on Ω_{K0} from WMAP+HST, WMAP+SN, and WMAP+BAO. (From Komatsu *et al.* (2009). Reproduced by permission of the AAS.) A colour version of this figure is available online.

primarily as a probe of the primordial gravitational wave background, a spin-off of this goal could give high-precision results on distances, in tandem with electromagnetic observations (Cutler and Holz, 2009). In order to measure the primordial background, BBO would need first to detect and remove the signals of $O(10^5)$ compact star binaries out to redshift $z \sim 5$. Since the binary signals have amplitude $\propto D_L^{-1}$, they will provide accurate measures of the luminosity distance. Electromagnetic detection of the host galaxy would then determine the redshift to the object. Given the high number of objects distributed out to high redshifts, this could in principle provide much better cosmological constraints than SNIa. In addition, the luminosity distance enters the signal amplitude in a way that is independent of the detailed astrophysics of the sources, so this is in principle also a cleaner measure than those based on SNIa. Given the high number of expected sources, the dispersion of the measured luminosities, which is dominated by lensing magnification, could also provide a probe of weak lensing comparable to that of dedicated weak lensing experiments (Cutler and Holz, 2009).

13.3 Almost FLRW observations: probing structure formation

The evolution of the radiation and matter perturbations on an FRLW background can be tracked via observations of the large-scale structure. This provides crucial constraints and tests that are complementary to those arising from observations of the background model evolution.

13.3.1 CMB anisotropies

The first observations of CMB temperature anisotropies, of order $|\delta T/T| \sim 10^{-5}$ (after removal of the dipole), came from the COBE satellite (Bennett *et al.*, 1996). A series of subsequent sensitive experiments have confirmed and improved the data, and detected polarization. The most detailed results currently are those from the WMAP experiment, supplemented by the small-scale experiments ACBAR, CBI, VSA, BOOMERANG, QUAD (Dunkley *et al.*, 2009, Komatsu *et al.*, 2009, Hinshaw *et al.*, 2009, Larson *et al.*, 2011). Table 13.1 gives the parameter values for a flat ΛCDM cosmology, and for curved models with constant-w dark energy. The table is adapted from Hinshaw *et al.* (2009), using WMAP 5-year data (Dunkley *et al.*, 2009), then using WMAP5+BAO+SN data (Komatsu *et al.*, 2009). The updates from the WMAP 7-year data release (Larson *et al.*, 2011, Komatsu *et al.*, 2011) include only small changes from the best-fit values, with significant improvement in errors on some parameters. The even more sensitive PLANCK experiment was launched in 2009 and its data are expected to further improve the constraints on various cosmological parameters and models.

The primary anisotropies are laid down at last scattering, $z \sim 1100$, and the smallest observable comoving scale is determined by Silk damping: $k^{-1} \gtrsim k_S^{-1} \sim 10\,\mathrm{Mpc}$. Secondary anisotropies are generated as the photons travel to us through the evolving large-scale matter distribution. These affect very large scales via the (linear) integrated Sachs–Wolfe effect, and very small scales via the Sunyaev–Zel'dovich effect, lensing, and other nonlinear effects. Detailed analyses of the CMB temperature anisotropies and polarization spectra were given in Chapter 11. Here we summarize the key features in the power spectra and how they constrain various cosmological parameters. A comprehensive discussion may be found in Durrer (2008).

The perturbed temperature can be described by the brightness function,

$$\Theta(\tau,\boldsymbol{x},\boldsymbol{e}) := \frac{\delta T(\tau,\boldsymbol{x},\boldsymbol{e})}{T(\tau)} = \sum_{\ell,m} \Theta_{\ell m}(\tau,\boldsymbol{x}) Y_{\ell m}(\boldsymbol{e}), \tag{13.13}$$

where $\boldsymbol{e}$ is the observed direction of the photon. The brightness multipoles are given by

$$\Theta_\ell(\tau,k) = \frac{1}{2(-i)^\ell} \int_{-1}^{1} \mathrm{d}\mu\, P_\ell^L(\mu)\Theta(\tau,k,\mu), \quad \mu := \frac{\boldsymbol{k}\cdot\boldsymbol{e}}{k}, \tag{13.14}$$

where P_ℓ^L are the Legendre polynomials.

The $\Theta_{\ell m}$ evolve according to the Boltzmann equation. The observed anisotropy today is given by $a_{\ell m} = \Theta_{\ell m}(\tau_0,\boldsymbol{x}_0)$, which are statistically homogeneous and isotropic random variables:

$$\langle a_{\ell m} a_{\ell' m'}^* \rangle = \delta_{\ell\ell'}\delta_{mm'} C_\ell, \quad C_\ell := \langle |a_{\ell m}|^2 \rangle. \tag{13.15}$$

The variance is related to the power spectrum of Θ_ℓ via

$$C_\ell = 4\pi \int_0^\infty \frac{\mathrm{d}k}{k} \mathcal{P}_{\Theta_\ell}(k) = 4\pi \int_0^\infty \frac{\mathrm{d}k}{k} T_\ell^2(k)\mathcal{P}_\zeta(k), \tag{13.16}$$

where T_ℓ is the transfer function determining the multipoles from the curvature perturbation: $\Theta_\ell = T_\ell(k)\zeta_k$. The 2-point correlation function is related to the variance C_ℓ via

$$C(\alpha) := \langle \Theta(\tau_0,\boldsymbol{x}_0,\boldsymbol{e}_1)\Theta(\tau_0,\boldsymbol{x}_0,\boldsymbol{e}_2) \rangle = \sum_\ell \frac{2\ell+1}{4\pi} P_\ell^L(\cos\alpha) C_\ell, \tag{13.17}$$

Table 13.1 Cosmological parameter summary			
Description	Symbol	WMAP-only	WMAP+BAO+SN
Parameters for standard ΛCDM model (flat, no tensors, no running)			
Age of universe	t_0	13.69 ± 0.13 Gyr	13.72 ± 0.12 Gyr
Hubble constant	H_0	$71.9^{+2.6}_{-2.7}$ km/s/Mpc	70.5 ± 1.3 km/s/Mpc
Baryon density	Ω_{b0}	0.0441 ± 0.0030	0.0456 ± 0.0015
CDM density	Ω_{c0}	0.214 ± 0.027	0.228 ± 0.013
Dark energy density	$\Omega_{\Lambda 0}$	0.742 ± 0.030	0.726 ± 0.015
Curvature perturbation ($k_0 = 0.002$/Mpc)	$\mathcal{P}_\zeta(k_0)$	$(2.41 \pm 0.11) \times 10^{-9}$	$(2.445 \pm 0.096) \times 10^{-9}$
Matter perturbation amplitude at $8h^{-1}$ Mpc	σ_8	0.796 ± 0.036	0.812 ± 0.026
Scalar spectral index	n_s	$0.963^{+0.014}_{-0.015}$	0.960 ± 0.013
Redshift at equality	z_{eq}	3176^{+151}_{-150}	3253^{+89}_{-87}
Angular diameter distance to equality	$(1 + z_{eq})D_A(z_{eq})$	14279^{+186}_{-189} Mpc	14200^{+137}_{-140} Mpc
Redshift of decoupling	z_{dec}	1090.51 ± 0.95	1090.88 ± 0.72
Age at decoupling	t_{dec}	380081^{+5843}_{-5841} yr	376971^{+3162}_{-3167} yr
Angular diameter distance to decoupling	$(1 + z_{dec})D_A(z_{dec})$	14115^{+188}_{-191} Mpc	14034^{+138}_{-142} Mpc
Comoving sound horizon at decoupling	$r_s(z_{dec})$	146.8 ± 1.8 Mpc	$145.9^{+1.1}_{-1.2}$ Mpc
Acoustic scale at decoupling	$\ell_A(z_{dec})$	$302.08^{+0.83}_{-0.84}$	302.13 ± 0.84
Reionization optical depth	τ_{reion}	0.087 ± 0.017	0.084 ± 0.016
Redshift of reionization	z_{reion}	11.0 ± 1.4	10.9 ± 1.4
Parameters for extended models			
Curvature density ($w = -1$)	Ω_{K0}	$-0.099^{+0.100}_{-0.085}$	$-0.0050^{+0.0060}_{-0.0061}$
Equation of state ($w = $const, $\Omega_{K0} = 0$)	w	$-1.06^{+0.41}_{-0.42}$	$-0.992^{+0.061}_{-0.062}$
Tensor to scalar ratio (no running)	$r(k_0)$	< 0.43 (95% CL)	< 0.22 (95% CL)
Running of spectral index (no tensors)	$dn_s(k_0)/d\ln k$	-0.037 ± 0.028	-0.028 ± 0.020
Neutrino density	$\Omega_\nu h^2$	< 0.014 (95% CL)	< 0.0071 (95% CL)
Neutrino mass	$\sum m_\nu$	< 1.3 eV (95% CL)	< 0.67 eV (95% CL)
Light neutrino families	N_{eff}	> 2.3 (95% CL)	4.4 ± 1.5

where $\cos\alpha = \boldsymbol{e}_1 \cdot \boldsymbol{e}_2$. The uncertainty in the C_ℓ due to cosmic variance is given by

$$\frac{\Delta C_\ell}{C_\ell} = \sqrt{\frac{2}{2\ell+1}}, \tag{13.18}$$

so that cosmic variance is a severe limitation on large scales (small ℓ), as expected. The temperature measurements made by WMAP lead to the C_ℓ curve in Figure 13.3, which shows how the error bars are dominated by cosmic variance at large angles.

On large scales, the scalar brightness function at the observer is

$$\Theta(\boldsymbol{e}) \approx \delta_{\gamma\,\mathrm{dec}} + \Phi_{\mathrm{dec}} + \int_{\mathrm{dec}}^{0} d\nu(\Phi' + \Psi'), \tag{13.19}$$

where ν is an affine parameter along the photon path and $\boldsymbol{x}_{\mathrm{dec}} \approx \tau_0 \boldsymbol{e}$ (with $\boldsymbol{x}_0 = 0$). The integral term is known as the integrated Sachs–Wolfe (ISW) effect, and the remaining terms constitute the 'ordinary' SW effect. On large scales and for adiabatic initial conditions, $\delta_\gamma = 4\delta_m/3 = -8\Phi/3$, so that the ordinary SW effect gives

$$\Theta^{SW}_{\mathrm{adi}} = \tfrac{1}{3}\Phi_{\mathrm{dec}} = -\tfrac{1}{2}\delta_{m\,\mathrm{dec}} = \tfrac{1}{5}\zeta_{\mathrm{dec}}. \tag{13.20}$$

The near scale-invariance of ζ leads to the SW 'plateau'. Note also that hot spots in the primordial CMB ($\Theta > 0$) correspond to *under-densities* in the matter. For isocurvature initial conditions, $\Theta^{SW}_{\mathrm{iso}} = 2\Phi_{\mathrm{dec}}$.

The ISW is made up of an early-time contribution (due to non-negligible radiation present around decoupling), and a more significant late-time contribution due to dark energy. This latter contribution – which would be absent in a matter-dominated universe (since in that

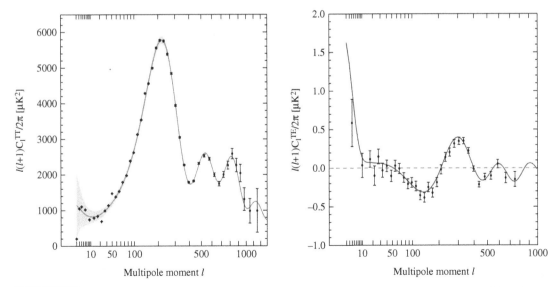

Fig. 13.3   Temperature (TT) and temperature-polarization (TE) power spectra for WMAP7 data. (From Larson *et al.* (2011). Courtesy WMAP Science Team. Reproduced by permission of the AAS.)

case $\Phi = \Psi = \text{const}$) – is responsible for the rise in the plateau on the largest scales, seen in the theoretical curve in Figure 13.3.

On intermediate scales, C_ℓ is determined by the acoustic peaks that arise from sound waves in the photon–baryon plasma before decoupling. The location of the peaks (for adiabatic modes) is given by $c_s k_n \tau_{\text{dec}} = n\pi$. The comoving linear scale π/k_n subtends the angle α_n, and thus

$$\ell_n \approx \frac{\pi}{\alpha_n} = D_A(\tau_{\text{dec}})k_n = n\pi \sqrt{3(1+R)} \frac{D_A(\tau_{\text{dec}})}{\tau_{\text{dec}}}. \tag{13.21}$$

On small scales, $\ell \gtrsim 1000$, the primordial C_ℓ show a (Silk) damping tail, with $\ell_S \approx \tau_0 k_S/\sqrt{2} \approx 1200$.

The acoustic peak locations, (13.21), and heights strongly depend on $\Omega_{b0}h^2$, $\Omega_{m0}h^2$ and D_A, and therefore are sensitive to the curvature. Open models $\Omega_{K0} > 0$ shift the peaks to smaller scales and vice versa for closed models. But this is subject to an important degeneracy. Changes in the curvature Ω_{K0} which keep the quantities $\Omega_{b0}h^2$, $\Omega_{m0}h^2$ and D_A fixed, keep the peak locations fixed and have negligible effects on the CMB anisotropies (except at the very largest scales, which are very weakly constrained because of large cosmic variance). This degeneracy means that the CMB does not strongly constrain the curvature on its own, as shown in Figures 13.1 and 13.2. Additional information is needed – e.g. a value of h from distance measurements or a value of $\Omega_{m0}h^2$ from galaxy surveys. Indeed, any observation that measures a distance to a single redshift z_1 cannot constrain Ω_{K0}, since the distance depends also on $H(z)$ all the way to z_1. Thus it is necessary to combine at least two distance measures to different redshifts. The results from combining WMAP data with BAO and SNIa are shown in Figures 13.1 and 13.2. With the WMAP 7-year data, together with BAO and H_0 data, the constraint on Ω_{K0} with Λ as dark energy is (Komatsu et al., 2011)

$$\Omega_{K0} = -.0023^{+0.0054}_{-0.0056}, \quad w = -1. \tag{13.22}$$

If w can vary from -1, but remains constant, then the constraints from WMAP7 and BAO and SN data are

$$\Omega_{K0} = -.0057^{+0.0067}_{-0.0068}, \quad w = -0.999^{+0.057}_{-0.056}. \tag{13.23}$$

CMB power on large scales can be boosted by tensor modes, or by a spectral index of primordial scalar perturbations with $n_s < 1$ (i.e. redder than the scale-invariant limit). The scalar power spectrum is given in terms of the curvature perturbation, for a nearly power-law behaviour, as

$$\mathcal{P}_\zeta(k) = \frac{k^3}{2\pi^2} P_\zeta(k) = \mathcal{P}_\zeta(k_0) \left(\frac{k}{k_0} \right)^{n_s - 1 + \alpha_s}, \quad \alpha_s = \frac{1}{2} \frac{dn_s}{d\ln k}, \tag{13.24}$$

where $k_0 = 0.002 \, \text{Mpc}^{-1}$ is a pivot scale and α_s is the 'running' of the spectral index. If there is no running, and if tensor modes are neglected, then Table 13.1 shows that for ΛCDM, the CMB on its own gives $n_s \approx 0.96$ and $\mathcal{P}_\zeta(k_0) \approx 2.4 \times 10^{-9}$. WMAP7 does not change these values at two significant figures (Komatsu et al., 2011).

When running is allowed (but tensors are neglected), WMAP7 together with BAO and H_0 data give $n_s = 1.008 \pm 0.042$ and $\alpha_s = -0.022 \pm 0.020$ (Komatsu et al., 2011). If tensors are included (without running), then $n_s \approx 0.97$, reflecting the degeneracy (i.e. tensors can compensate for large-scale power that is reduced as n_s is increased). When tensors and running are included, $n_s \approx 1.07$, $\alpha_s \approx -0.04$.

A future detection of B-mode polarization (which cannot be generated by scalar perturbations) would in principle lead to a detection of the tensor modes, and thus be able to break this degeneracy. The primordial gravitational waves are governed by (12.51)–(12.55). For a power-law spectrum with tensor spectral index n_t,

$$\mathcal{P}_h(k) = \mathcal{P}_h(k_0) \left(\frac{k}{k_0} \right)^{n_t}. \tag{13.25}$$

For single-field slow-roll inflation, the tensor-to-scalar ratio $\mathcal{P}_h(k_0)/\mathcal{P}_\zeta(k_0)$ is

$$r = 16\epsilon = -8n_t, \tag{13.26}$$

showing that the spectrum is nearly scale-invariant. Then the upper limit on r from WMAP7, with BAO and H_0 data, is $r < 0.24$ (95% confidence), assuming no running in n_s (running weakens this to $r < 0.49$) (Komatsu *et al.*, 2011).

There is also an important degeneracy between baryon content and spectral index: increasing $\Omega_{b0}h^2$ suppresses the second acoustic peak and enhances Silk damping, both of which can be compensated by increasing n_s (i.e. more power on smaller scales).

A higher optical depth to last scattering, $\tau_{\text{reion}} = \sigma_T \int_{\text{dec}}^{t_0} n_e dt$, suppresses power on small scales but leaves large scales unaffected, thus introducing further degeneracies. There is an independent constraint on τ_{reion} via the E-mode polarization, which is boosted on large scales by re-ionization, as shown in Figure 13.3. This is a clear signal of reionization, and in ΛCDM places constraints on the optical depth and reionization redshift,

$$\tau_{\text{reion}} = 0.87 \pm 0.014, \quad z_{\text{reion}} = 10.4 \pm 1.2, \tag{13.27}$$

from WMAP7 with BAO and H_0 data (Komatsu *et al.*, 2011) (see Table 13.1).

13.3.2 Matter distribution

The two main (and completed) spectroscopic galaxy surveys at the time of writing are the 2dF-GRS (2-degree Field Galaxy Redshift Survey) and SDSS (Sloan Digital Sky Survey). 2dF observed 230,000 galaxies and their redshifts, and SDSS observed 100 million luminosities, 1 million galaxy redshifts, and 100,000 quasar redshifts, and also linked this with a quasar survey.[1] The redshifts can be used as distance indicators to give three-dimensional pictures of the distribution. This idea was first carried out in the Centre for Astrophysics (CfA) survey (de Lapparent, Geller and Huchra, 1986). As well as giving a measure of the galaxy power spectrum, these surveys also reveal the topology of large-scale structure in the form of clusters, filaments, walls and voids. The matter power spectrum from SDSS Data Release 5 (main galaxy and Luminous Red Galaxy) is shown in Figure 13.4. Note the increase in power due to nonlinear effects for $k > 0.06\,h/\text{Mpc}$. The inset shows the BAO feature, discussed in detail in Section 13.2.2.

---

[1] See http://www.sdss.org/

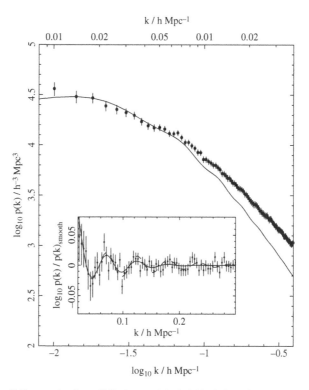

Fig. 13.4 Power spectrum from SDSS, assuming flat Λ CDM, $\Omega_{m0} = 0.24$ (solid line). *Inset:* Data points confirming BAO. Theoretical curves with (solid) and without (dashed) nonlinear correction on small scales. (From Percival *et al.* (2007). Reproduced by permission of the AAS.)

New galaxy surveys, such as DES (Dark Energy Survey), will extend the redshift depth and sky coverage, leading to greater precision in measurements of the matter power spectrum and its BAO feature.

Peculiar velocities

As discussed in Section 12.3.5, galaxy peculiar velocities cause distortions in the redshift space in which galaxy survey data resides. These distortions then become a valuable source of information about the peculiar velocities and thereby the matter distribution that sources them. The redshift-space over-density (12.76) contains the factor $\beta(z)$ (see (12.75)), which encodes the bias (assumed scale independent) and the background density. Thus measurements of the redshift-space power spectrum allow in principle a determination of the bias if $\Omega_m(z)$ is known, or of the latter if the bias is known (see, e.g. Guzzo *et al.* (2008)).

Neutral hydrogen

After last scattering, $z < 1100$, most electrons are trapped in hydrogen and other atoms, and neutral hydrogen clouds undergo gravitational collapse (assisted by CDM potential wells) to form the first stars. These stars in turn reionize the universe during the period

$6 \lesssim z \lesssim 30$. The period between recombination and reionization is the 'Dark Ages' described in Section 9.6.8. Although there are no light-emitting objects, there is the back-light of the CMB, leading to absorption and emission by hydrogen gas at the 21-cm rest wavelength of the hyperfine transition of the ground state. Thus the 21-cm line provides the best current probe of the dark ages – and will also give valuable information about the process of reionization (Furlanetto, Oh and Briggs, 2006, Barkana and Loeb, 2007). The wavelength of the 21-cm line is redshifted into the low-frequency radio range for $z \gtrsim 6$. Currently, radio arrays have not yet achieved the volume-coverage or precision necessary to map the matter power spectrum in the dark ages. This will require capacity of the order of the proposed Square Kilometre Array (SKA), which could usher in H21 tomography.

Galaxy cluster counts

Galaxy clusters are tracers of the evolving dark matter halos that drive large-scale structure formation. Their abundance provides, in principle, an independent estimate of Ω_{m0} and σ_8. Lensing by clusters, and X-ray and Sunyaev–Zel'dovich measurements help to place constraints on the cluster mass, but this is very sensitive to the nonlinear gas astrophysics of clusters (Hoekstra, 2007, Rykoff *et al.*, 2008, Mantz *et al.*, 2008). In addition, there are difficulties in determining the selection function.

Lyman-α forest

Neutral hydrogen absorption lines (corresponding to the Lyman-α resonance) in quasar spectra give a measure of the matter distribution on small scales, which in principle provides constraints on the primordial power spectrum (Kim *et al.*, 2007). In practice, this is complicated by nonlinear astrophysics.

13.3.3 Weak lensing

Gravitational lensing of light from distant sources by intervening matter distributions provides a powerful probe of the matter distribution, i.e. CDM and baryons together, as discussed in Sections 12.4 and 12.5. In principle, weak lensing is a more powerful probe of the matter distribution than galaxy redshift surveys, since it is independent of the bias between galaxies and CDM. In practice, weak lensing surveys are more difficult to implement because of the many systematic uncertainties involved in measuring shapes and hence magnification and shear.

Weak gravitational lensing probes an integrated gradient of the metric potential, via the deflection angle formula (12.109):

$$\hat{\alpha} = -\int \widehat{\nabla}(\Phi + \Psi)\mathrm{d}v, \tag{13.28}$$

where $\widehat{\nabla}$ is the gradient operator in the image plane normal to the line of sight. Weak lensing tomography, i.e. probing a range of redshifts to produce a 'stack' of slices, provides in principle a three-dimensional map of the total matter distribution (Massey *et al.*, 2007)

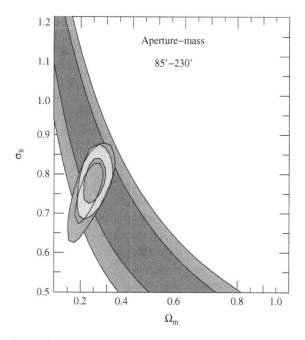

Fig. 13.5 Joint constraints (central region, lightest shading, orange in colour version) on measures of the total matter in a flat
Λ CDM universe, from weak lensing (linear scales) (CFHTLS survey) (dark shading with lighter higher confidence
band, purple and blue) and CMB (WMAP3) (mid-grey shading, green). (From Fu *et al.* (2008). Reproduced with
permission © ESO.) A colour version of this figure is available online.

(see Figure 1.9), in contrast with the three-dimensional map of luminous matter provided
by galaxy surveys (see Figure 1.6). However, space-based surveys will be necessary to
achieve a higher signal-to-noise ratio. A further difficulty is that lensing on smaller scales
is sensitive to nonlinear effects, requiring complicated and intensive analysis to extract the
lensing signal. Recent results from the Canada–France–Hawaii Telescope Legacy Survey
(CFHTLS) on linear scales are shown in Figure 13.5.

13.4 Constraints and consistency checks

Currently the principal observational probes of the standard cosmological model are provided by CMB anisotropies, galaxy redshift surveys and supernova luminosities. Other
observations provide complementary information and often act as useful consistency
checks; some are more than that, they are crucial requirements for viability of an almost-
FLRW model. For example, BAO measurements (which are derived from galaxy surveys)
provide a powerful confirmation of the model of the primordial plasma and the evolution of
structure. 21-cm radio surveys will extend this independent check to much higher redshifts.
Weak lensing provides a measure of dark matter, independent of the CMB, and thereby also
probes the consistency of the dark matter model. Here we briefly discuss some other probes
(see also Section 15.6.5).

CMB–matter correlations

The large-angle anisotropies in the CMB temperature encode a signature of the formation of structure via the integrated Sachs–Wolfe effect: from (13.19),

$$\left.\frac{\delta T}{T}\right|_{ISW} = \int_{dec}^{0} d\nu (\Phi' + \Psi'). \tag{13.29}$$

The ISW arises since photons are blue shifted when they fall into a gravitational potential and redshifted when they climb out of it. Hence if the potential varies during this time, photons acquire a net energy shift, reflected in the CMB temperature map. During matter domination, the potential is constant at large scales (in the Newtonian gauge). As matter domination is undermined by the growing strength of dark energy, the ISW term becomes significant. The correlation between CMB and matter power is therefore a sensitive indicator of the dark energy, and provides a consistency check of the presence of dark energy within the standard cosmological model (Boughn and Crittenden, 2004, Giannantonio *et al.*, 2008). Note that the ISW and weak lensing are determined by the same combination of metric potentials, $\Phi + \Psi$. In the standard model we can neglect anisotropic stress after decoupling, and this combination reduces to 2Φ. If we consider modifications of GR, then a gravitational effective anisotropic stress arises, so that $\Psi \neq \Phi$, and the ISW and weak lensing become powerful probes of this (see Section 14.3).

CMB and number count dipole alignment

In an almost-FLRW universe, there must be a $\sim 2\%$ number count dipole parallel to the CMB dipole for all cosmological sources, due to our motion relative to the cosmological rest frame (Ellis and Baldwin, 1984). If this is not true we cannot live in a RW geometry with the CMB coming from the surface of last scattering. This effect has been confirmed (Blake and Wall, 2002).

Big-bang nucleosynthesis

As described in Section 9.6.6, BBN at $T \sim 0.1\,\text{MeV}$ produces primarily helium-4, with small amounts of other stable nuclei, deuterium, helium-3 and lithium-7. The primordial abundances should then be reflected in observations at later epochs, and provide an important consistency test of the standard model. Furthermore, the abundances depend on the baryon density. The CMB provides an independent probe of the baryon density, and it is in very good agreement with the BBN value – see Figure 9.9. There remains at the time of writing some discrepancy in the case of lithium, whose detection in stars is at about 50% of the value inferred from WMAP. The discrepancy could be due to complicated astrophysics or systematics, or could signal some modification of high-energy physics.

Ages

The standard model predicts $t_0 \approx 13.7\,\text{Gyr}$ (see Table 13.1). A strong consistency test of this prediction is that the ages of objects in the universe must be $< t_0$. The oldest detected

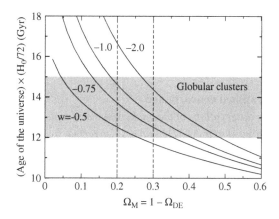

$\Omega_M = 1 - \Omega_{DE}$

Fig. 13.6 Age of the universe for different w (constant). Shaded region shows ages of the oldest globular clusters. Vertical dashed lines show Ω_{m0} values consistent with CMB and large-scale structure data. (Reprinted from Frieman, Turner and Huterer (2008), with permission from the Annual Review of Astronomy and Astrophysics. © 2008 by Annual Reviews, http://www.annualreviews.org .)

cosmological objects are globular clusters. In order to determine their ages, one needs accurate distance measures, and also accurate modelling of stellar populations. Current results indicate a consistency, as illustrated in Figure 13.6.

CMB temperature

In an expanding universe, the adiabatic cooling of the CMB leads to the relation $T(z) = T_0(1 + z)$ for the average CMB temperature, where $T_0 = 2.725$. Probes of the temperature at different redshifts, using molecular temperature determinations, quasar absorption spectra and the Sunyaev–Zel'dovich effect, confirm this law (Luzzi *et al.*, 2009).

GZK limit

Ultra-high cosmic rays ($E \gtrsim 10^{20}$ eV) travelling through the CMB will be above the threshold for photohadronic particle production (e.g. protons will induce pair production and pion production), and will consequently lose energy. This puts a limit on how far the highest energy extragalactic cosmic rays can propagate from their source, and should lead to a suppression of cosmic ray flux at the Earth at ultra-high energies (Greisen, 1966, Zatsepin and Kuz'min, 1966). It was claimed that this GZK cutoff has been observationally detected (Abbasi *et al.*, 2008, Abraham *et al.*, 2008), but the result is controversial at the time of writing.

Exercise 13.4.1 Which of the above tests are crucial tests of the background model, and which are rather tests of the matter distribution in the model (and hence potentially fixable by changing the matter model rather than the background)?

13.5 Concordance model and further issues

Putting this all together, we can state:

Observational concordance
Current observations of the background and of perturbations about it are indeed consistent, in the context of a ΛCDM universe governed by GR.

In particular, there is no significant evidence as yet for deviation from $w = -1$ (see e.g. Perivolaropoulos (2010), Larson *et al.* (2011), Komatsu *et al.* (2011)).

Modifications to GR induce an effective anisotropic stress, so that $\Psi - \Phi \neq 0$ (Section 14.3): there is currently no evidence for this. Other signatures of modified gravity (see Section 14.3) have also not been detected, so that currently there is no evidence for modifications to GR (Reyes *et al.*, 2010, Daniel *et al.*, 2010).

However, it should be pointed out that a number of assumptions have gone into these tests, even within the framework of a perturbed RW model. For example, a flat background is typically assumed. While the assumptions may be reasonable, it is worth testing them.

13.5.1 The need to consider curvature

The sign of the spatial curvature parameter K is a crucial property of the cosmological model. Despite the convenience of setting $K = 0$, data analyses should consider Ω_{K0} as a free parameter to be determined along with the other parameters. There are important degeneracies between K and other parameters, as discussed in Section 13.3.1.

Inflation predicts $\Omega_K \to 0$, and *not* $\Omega_K \equiv 0$. This distinction may seem academic, but it has far-reaching implications. For example, $K > 0$ ($\Omega_K < 0$) means that the background FLRW universe has closed spatial sections and so is necessarily finite, with a finite amount of matter and finite number of galaxies. This is conceptually completely different from the cases $K \leq 0$ which allow an infinite amount of matter and infinite number of galaxies (Ellis and Brundrit, 1979). And it is the only case with firstly a possibility of a bounce in the past (given suitable energy conditions), and secondly a possibility of a maximum radius and recollapse in the future.

Indeed, it is impossible for the observations discussed above to prove that $K = 0$, since the observable Ω_{K0} can only be interpreted statistically. By contrast, it would be possible to show that Ω_{K0} is statistically negative or positive.

13.5.2 Large-scale anomalies

Some anomalies in large-scale features of the CMB and the galaxy distribution have been identified (see Antoniou and Perivolaropoulos (2010) for a summary with references).

- *CMB:* The normals to the quadrupole and octupole planes are aligned approximately with the dipole, in apparent conflict with statistical isotropy. In addition, there is missing power on angular scales $\gtrsim 60°$.
- *Galaxy bulk flow:* The bulk flow (dipole moment) of peculiar velocities for a low redshift galaxy sample extends on scales $O(100)h^{-1}$ Mpc, with amplitude > 400 km/s – much larger than expected in ΛCDM.

At the time of writing, it is unclear whether these anomalies indicate real problems with the standard model, or whether they can be accommodated within the model via a better understanding of the highly complicated observational uncertainties and statistical subtleties that are involved in the analysis. In particular, more detailed analysis is needed of WMAP scanning and foreground removal and calibration. On the large-scale velocities, other samples at higher redshift do not find anomalously large flows.

13.5.3 Testing small universes

Small universes are discussed in Section 8.3.2. The simplest cases are spatially flat, with a toroidal topology, but these are highly exceptional (Cornish, Spergel and Starkman, 1998). Much more complex cases are possible; indeed there are an infinite number of possible topologies in the non-flat cases (Ellis, 1971b, Lachièze-Rey and Luminet, 1995). Can we observationally test for them?

In principle one can test for a small universe by direct observational identification of multiple images of the same object (including our own Galaxy) (Ellis and Schreiber, 1986). However, in practice this is very difficult: each image will be at a different redshift, seeing the object at a different stage in its history and effectively from a different direction. (Note the difference from ordinary gravitational lensing, where redshifts of different images are the same.) Only really distinctive large-scale structures might be identifiable via multiple images. The problem is to distinguish a statically homogeneous set of almost identical objects, from images of effective repetitions of the same objects.

More promising is to examine source statistics: a small universe will result in effectively periodic structures in the matter distribution, in principle identifiable via peaks in the spatial power spectrum (Uzan, Lehoucq and Luminet, 2000). However, such peaks may not occur in a small universe, depending on details of the structures and our location relative to them (Gomero *et al.*, 2002). We also expect a cutoff in large-scale power because a maximal scale exists in the universe – but it only occurs in models that can be characterized as 'well-proportioned' (Weeks *et al.*, 2004).

CMB anisotropies are more promising. Details of the CMB anisotropies are model dependent (Levin, 2002, Riazuelo *et al.*, 2004), but there is an important general prediction: whatever the topology, there will be identical circles of fluctuations in the CMB sky (Cornish, Spergel and Starkman, 1998) because one will see the same points in the intersection of our past light-cone with the surface of last scattering, in different directions in the sky. The precise configuration of such identical circles in the CMB sky is then uniquely related to the spatial topology, and would enable us to determine that topology. This effect has been searched for, with negative results so far (Cornish *et al.*, 2004, Shapiro Key *et al.*, 2007). But it is important to note that in general the circles will not be antipodal (Riazuelo *et al.*, 2004), so more general searches are needed than have so far been conducted.

An interesting model is the Poincáre dodecahedral ('soccer ball') universe (Luminet *et al.*, 2003). It can explain the low quadrupole observed in the CMB spectrum (Aurich, Lustig and Steiner, 2005) and is supported by some other data (Roukema *et al.*, 2008), but does not explain the quadrupole–octopole alignment (Weeks and Gundermann, 2007). It has been claimed that this possibility has been ruled out by the circles in the sky criterion (Shapiro Key *et al.*, 2007), but this is disputed (Caillerie *et al.*, 2007).

If it were ever proved that we live in a small universe, it would be a major discovery about the geometry of the universe, with major philosophical and observational implications, as well as ruling out many currently popular models (e.g chaotic inflation). This possibility should therefore be seriously tested. One particularly interesting point is that if we do indeed live in a small universe, this gives the one genuine possibility of detecting if the universe has flat spatial sections (Mota, Rebouças and Tavakol, 2010). Such a detection is not possible by the methods listed above in this chapter, since at most we can show that Ω_{K0} is either positive or negative, but not prove it is exactly zero.

13.5.4 Effective domain of dependence for our Galaxy

We discussed the nature of causal horizons in the standard models in Section 7.9. However, these horizons – based on the light-cone – do not in fact show what domains of spacetime are really important for the development of structure in our local cosmic neighbourhood. The effective domain of dependence for local conditions is much smaller than indicated by the past light-cone (Ellis and Stoeger, 2009a), as shown in Figure 13.7.

The limits from the event horizon are limits on what can be influenced by particles and forces acting at the maximal speed of light. However, freely propagating photons, massless neutrinos, and gravitons coming from cosmological distances have very little influence on our Galaxy (indeed we need very delicate experiments to detect them). Massive particles will travel much slower, and before decoupling, information travels by sound waves at $c_s < 1$ in the photon–baryon plasma. The characteristics for pressure-free scalar and vector perturbations are timelike curves, moving at zero velocity relative to the matter; while density perturbations associated with pressure can move at the speed of sound, only tensor perturbations can travel at the speed of light.

Thus the true domain that influences us significantly is the small region round our past world line characterized after decoupling by the comoving scale from which matter coalesced into our Galaxy: a present distance of about 1–2 Mpc, corresponding to an observed angle of about 0.6 arcmin on the LSS. This is the effective horizon size at that time. Events inside this horizon (which is generated by timelike curves) can have had a significant effect on our local neighbourhood; those outside did not. Before decoupling, it would have been

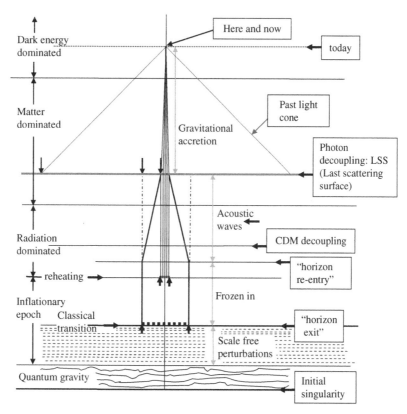

Fig. 13.7 Spacetime domains that significantly affect the Milky Way. The three downward pointing arrows on the LSS show (from left to right) the size of the visual horizon, and the comoving size of the realistic domain of influence between the start of the inflationary era and 'horizon entry' in the radiation dominated era, and the same at decoupling of matter and radiation (NB: not to scale!). The two inner upward-pointing arrows show the size of the matter horizon at the origin of the Milky Way matter when reheating takes place at the end of inflation. (From Ellis and Stoeger (2009), © RAS.)

limited by the sound horizon rather than the particle horizon. The *matter horizon* is indicated in the figure (growing into the past, from the present day). It lies inside the effective horizon, because local conditions have been affected by sound waves as well as by accretion of matter.

14 Acceleration from dark energy or modified gravity

14.1 Overview of the problem

If we describe the universe as a perturbed RW model (with or without GR), then the data provide compelling evidence that the expansion of the universe has been accelerating since a redshift $z \sim 1$. The mathematically simplest model that produces such acceleration is ΛCDM, but alternatives have been proposed. This chapter provides a brief overview of accelerating perturbed RW models, with and without GR, and the theoretical problems that they face (based on Durrer and Maartens (2008); see also Ellis *et al.* (2008), Frieman, Turner and Huterer (2008), Bean (2010), Linder (2010), Ruiz-Lapuente (2010), Amendola and Tsujikawa (2010) for recent reviews).

In the standard model, the data indicate that the present cosmic energy budget is given by

$$\Omega_{\Lambda 0} \approx 0.75, \quad \Omega_{m0} \approx 0.25, \quad \Omega_{K0} \approx 0, \quad \Omega_{r0} \approx 10^{-4}. \tag{14.1}$$

Early indications of a positive Λ arose in the 1980s. The inflationary model of the early universe predicted $\Omega_K \to 0$, but repeatedly the observations of matter were producing estimates of $\Omega_{m0} \lesssim 0.3$. In addition, based on the then current Einstein–de Sitter model, the age of the universe was less than the estimated ages of the oldest stars. Attempts to find enough missing CDM to bring Ω_{m0} up to the critical value of 1 were unsuccessful, and Λ was being invoked by some as a resolution to the problem. Observational evidence against the Einstein–de Sitter model was mounting, and the supernova surveys of the late 1990s provided strong direct evidence. The field equations give the dimensionless acceleration as

$$\frac{1}{H_0^2} \frac{\ddot{a}}{a} = \Omega_{\Lambda 0} - \frac{1}{2} \Omega_{m0}(1+z)^3 - \Omega_{r0}(1+z)^4. \tag{14.2}$$

Together with (14.1), this gives the dramatic conclusion that the universe is currently accelerating.

This conclusion is based on a perturbed FLRW model (governed by GR). In this case the distance to a given redshift z, and the time elapsed since that redshift, are tightly related via the only free function of this geometry, $a(t)$. If the universe instead is isotropic around us but not homogeneous, i.e. if it resembles a Lemaître–Tolman–Bondi solution with our galaxy cluster at the centre, then this tight relation between distance and time for a given redshift would be lost and present data would not necessarily imply acceleration or the need for dark energy. This possibility is discussed in Chapter 15. In a more general scenario, anisotropic inhomogeneity could be generated by nonlinear averaging and backreaction effects, and the link between observations and dark energy would be weakened and could be broken: this is discussed in Chapter 16.

Here we assume a perturbed RW model. The simplest way to explain acceleration is then a cosmological constant, i.e. the ΛCDM model. Even though Λ can be considered as just another gravitational constant (in addition to Newton's constant G), it enters the Einstein equations in exactly the same way as a contribution from the vacuum energy, i.e. via a Lorentz–invariant energy–momentum tensor $T^{\mathrm{vac}}_{\mu\nu} = -(\Lambda/8\pi\,G)g_{\mu\nu}$. The only observable signature of both a cosmological constant and vacuum energy is their effect on spacetime – and so a vacuum energy and a classical cosmological constant cannot be distinguished by observation. Therefore the 'classical' notion of Λ is effectively indistinguishable from quantum vacuum energy.

Even though the absolute value of vacuum energy cannot be calculated within quantum field theory, *changes* in the vacuum energy (e.g. during a phase transition) can be calculated, and they do have a physical effect – for example, on the energy levels of atoms (Lamb shift), which is well known and well measured. Furthermore, differences of vacuum energy in different locations, e.g. between or on one side of two large metallic plates, have been calculated and their effect, the Casimir force, is well measured. There is no doubt about the reality of vacuum energy. For a field theory with cutoff energy scale E_c, the vacuum energy density scales with the cutoff as $\rho_{\mathrm{vac}} \sim E_c^4$, corresponding to a cosmological constant $\Lambda_{\mathrm{vac}} = 8\pi\,G\rho_{\mathrm{vac}}$. If $E_c = M_{\mathrm{p}}$, this yields a naïve contribution to the 'cosmological constant' of $\Lambda_{\mathrm{vac}} \sim 10^{38}\,\mathrm{GeV}^2$, whereas the measured effective cosmological constant is the sum of the 'bare' cosmological constant and the contribution from the cutoff scale,

$$\Lambda_{\mathrm{eff}} = \Lambda_{\mathrm{vac}} + \Lambda \sim 10^{-83}\,\mathrm{GeV}^2 . \tag{14.3}$$

Hence a cancellation of about 120 orders of magnitude is required. This is called the *fine-tuning* or *size* problem of dark energy: a cancellation is needed to arrive at a result which is many orders of magnitude smaller than each of the terms. It is possible that the quantum vacuum energy is much smaller than the Planck scale. But even if we set it to the lowest possible supersymmetry scale, $E_{\mathrm{susy}} \sim 1\mathrm{TeV}$, arguing that at higher energies vacuum energy exactly cancels due to supersymmetry, the required cancellation is still about 60 orders of magnitude. The problem that simple quantum field theory estimates of the magnitude of the vacuum energy are between 60 and 120 orders of magnitude bigger than the observed value indicates a profound disjuncture between quantum field theory and general relativity.

A reasonable attitude towards this open problem is the hope that quantum gravity will explain this cancellation, by showing that the vacuum does not gravitate, $\Lambda_{\mathrm{vac}} \equiv 0$. In this event, one may be able to argue that Λ is a genuinely gravitational constant, not connected to the vacuum energy: even though it appears in the same form as a vacuum energy, quantum gravity will have shown that the vacuum energy does not gravitate. Alternatively, non-gravitating vacuum energy can be imposed at a phenomenological level (using ideas due to Weinberg (1989)), by postulating that the gravitational field equations are the *trace-free* Einstein equations (known as 'unimodular gravity'), which do not 'see' a vacuum energy–momentum tensor. The conservation equations must then be separately imposed – and as a consequence, the full Einstein equations are recovered, but with a Λ term that is not related to the vacuum (see e.g. Ellis *et al.* (2010)). However, this interpretation of Λ as a 'classical' constant like the gravitational coupling constant $8\pi G$ also leads to an 'unnaturalness' problem. The dimensionless number that can be formed from these two gravitational constants is $8\pi G \Lambda \sim 10^{-120}$ – which seems implausibly small.

The unexpected observational result for Λ leads to a second problem, *the coincidence problem*: given that $\rho_\Lambda = \Lambda_{\mathrm{eff}}/8\pi G = \mathrm{const}$, while $\rho_{\mathrm{m}} \propto (1+z)^3$, why is ρ_Λ of the order of the *present* matter density $\rho_{\mathrm{m}}(t_0)$? It was completely negligible in most of the past and will entirely dominate in the future. Without this special coincidence of values, the large-scale structure in the universe would not be observed in its current form – indeed, an early domination of Λ could prevent all structure formation. This argument may also be presented in an alternative way. The key scale in the matter power spectrum is the matter–radiation equality scale $\sim 100\,\mathrm{Mpc}$ (which defines the turn-around in the spectrum). The onset of nonlinearity occurs when the density perturbation obeys $\delta \gtrsim 0.3$. For a comoving scale of 100 Mpc, this occurs at redshift $z_{\mathrm{nl}} \sim 1$ – which is coincident with the redshift where deceleration ends and acceleration begins:

$$z_{\mathrm{nl}}(k = 1/100\,\mathrm{Mpc}) \sim z_{\mathrm{acc}} \sim 1 \,. \tag{14.4}$$

These problems with Λ have spurred the search for other explanations of the accelerated expansion. Instead of a cosmological constant, one may introduce a time-varying component, with an equation of state such that $w < -1/3$, $w \neq -1$. Such a dynamical 'dark energy' component has the potential to address the coincidence problem – but the fine-tuning problem requires an investigation that goes beyond cosmological dynamics. So far, no theoretically consistent model of dark energy has been proposed which can yield a convincing or natural explanation of either of these problems.

Alternatively, it is possible that there is no dark energy field, but instead the late-time acceleration is a signal of a *gravitational* effect. In GR, this requires that the impact of inhomogeneities somehow acts to produce acceleration, or the appearance of acceleration – and necessarily this means that the universe is not perturbed FLRW.

If we stay with perturbed RW, then the alternative to dark energy is that gravity itself is weakened on large scales, i.e. that there is an 'infrared' modification to GR which accounts for the late-time acceleration. The classes of modified gravity models which have been widely investigated are scalar–tensor models and brane-world models. Schematically, one is modifying the geometric side of the field equations, $G_{\mu\nu} \to G_{\mu\nu} + G_{\mu\nu}^{\mathrm{dark}}$, rather than the matter side, $T_{\mu\nu} \to T_{\mu\nu} + T_{\mu\nu}^{\mathrm{dark}}$, as in the GR approach. Modified gravity represents an intriguing possibility for resolving the theoretical crisis posed by late-time acceleration. However, it turns out to be extremely difficult to modify GR at low energies in cosmology, without violating observational constraints from cosmological and Solar System data, or without introducing ghosts and other instabilities into the theory. Up to now, there is no convincing alternative to the GR dark energy models – which themselves are not very convincing.

14.2 Dark energy in an FLRW background

Here we briefly consider the main forms of dark energy within GR. Within GR, various dynamical alternatives to ΛCDM have been investigated, in an attempt to address the coincidence problem – none of the alternatives addresses the vacuum energy problem. Dynamical dark energy models replace the constant $\Lambda/8\pi G$ by the evolving energy density of a dark energy component. In order to produce acceleration at late times, the equation of state is constrained by $w < -1/3$.

14.2.1 Dark energy as vacuum energy

If Λ is treated as vacuum energy, then we face the problems of accounting for the incredibly small and highly fine-tuned value of the vacuum energy, encapsulated in (14.3).

String theory provides a tantalizing possibility in the form of the 'landscape' of vacua. There appear to be a vast number of vacua admitted by string theory, with a broad range of vacuum energies above and below zero. The idea is that our observable region of the universe corresponds to a particular small positive vacuum energy, whereas other regions with greatly different vacuum energies will look entirely different. This multitude of regions forms in some sense a 'multiverse' – an interesting but highly speculative idea, which we discuss in Chapter 21.

14.2.2 Fluid models

Phenomenological fluid models of dark energy are difficult to motivate. Adiabatic fluid models are typically unstable to perturbations, since the adiabatic speed of sound is usually imaginary for negative w:

$$c_s^2 = \frac{\dot{p}}{\dot{\rho}} = w - \frac{\dot{w}}{3H(1+w)}.$$
(14.5)

In particular, for constant w models, $c_s^2 < 0$ and the model is physically unviable. Constant-w models of dark energy must be *non-adiabatic*, i.e. the effective speed of sound (which governs the growth of inhomogeneities in the fluid) is not equal to the adiabatic speed of sound: $c_{s\,\text{eff}}^2 \neq c_s^2$. The speed of sound must be positive to avoid unphysical growth of dark energy inhomogeneities, and usually one sets $c_{s\,\text{eff}} = 1$ (the quintessence value) by hand. (See Section 10.2.5.)

It is possible to evade this constraint in an adiabatic fluid ($c_{s\,\text{eff}} = c_s$) if $\dot{w}$ is sufficiently negative, as can be seen from (14.5). For example, the 'Chaplygin gas' fluid model has equation of state $p = -A/\rho^\alpha$, where A and α are constants, $0 < \alpha \leq 1$. This model has real c_s. (See Exercise 14.2.1.)

14.2.3 Quintessence

A scalar field φ, with (standard) Lagrangian (5.142) is a self-consistent model that naturally avoids the problems of fluid models: the effective sound speed is 1, for any potential $V(\varphi)$, and it is not equal to the adiabatic sound speed, which depends on V (see (10.41)). This shows that a scalar field is intrinsically non-adiabatic. In an FLRW spacetime, using (5.149) and (5.152),

$$\rho_\varphi = \tfrac{1}{2}\dot{\varphi}^2 + V(\varphi), \qquad p_\varphi = \tfrac{1}{2}\dot{\varphi}^2 - V(\varphi), \tag{14.6}$$

$$\ddot{\varphi} + 3H\dot{\varphi} + V'(\varphi) = 0, \tag{14.7}$$

$$H^2 + \frac{K}{a^2} = \frac{8\pi G}{3}\left(\rho_r + \rho_m + \rho_\varphi\right). \tag{14.8}$$

The field rolls down its potential and the dark energy density varies through the history of the universe. 'Tracker' potentials have been found for which the field energy density follows that of the dominant matter component, thus opening up the possibility of solving the coincidence problem. However, while these models are insensitive to initial conditions, they do require a strong fine-tuning of the parameters of the Lagrangian to secure recent dominance of the field, and hence do not evade the coincidence problem.

More generally, the quintessence potential, somewhat like the inflaton potential, remains arbitrary, until and unless fundamental physics selects a potential. There is currently no natural choice of potential. There is no compelling reason as yet to choose quintessence above the Λ model of dark energy. Quintessence models do not seem more natural, better motivated or less contrived than Λ. Nevertheless, they are a viable possibility and computations are straightforward. Therefore, they remain an interesting target for observations to shoot at. And it may turn out that developments in particle physics do select a candidate potential in the future.

14.2.4 Interacting quintessence

It is possible that quintessence and cold dark matter, as fields beyond the Standard Model of particle physics, interact with each other, but not with baryonic matter or photons (or only extremely weakly). This does not violate the tight current constraints from fifth force experiments, because these experiments only probe baryonic matter. It could lead to a new approach to the coincidence problem, since a coupling in the dark sector may provide a less unnatural way to explain why acceleration kicks in when $\rho_m \sim \rho_{de}$.

In the presence of coupling, the energy conservation equations in the background become

$$\dot{\varphi}\left[\ddot{\varphi} + 3H\dot{\varphi} + V'(\varphi)\right] = Q, \tag{14.9}$$

$$\dot{\rho}_{dm} + 3H\rho_{dm} = -Q, \tag{14.10}$$

where Q is the rate of energy exchange. For a given model of Q, it is usually possible to choose parameters so that the model remains consistent with the geometric data (from CMB, SNIa and BAO) that constrain the background expansion history. The perturbations of the full energy–momentum conservation equations, $\nabla_\nu T_c^{\mu\nu} = Q_c^\mu = -Q_{de}^\mu = -\nabla_\nu T_{de}^{\mu\nu}$, show that there is a momentum transfer as well as an energy transfer. Analysis of the perturbations typically leads to more stringent constraints, with some forms of coupling being ruled out by instabilities.

14.2.5 Non-standard scalar fields

Another possibility is a scalar field with a non-standard kinetic term in the Lagrangian, for example,

$$L_\varphi = F(\varphi, X) - V(\varphi) \text{ where } X := -\tfrac{1}{2} g^{\mu\nu} \partial_\mu \varphi \partial_\nu \varphi. \qquad (14.11)$$

The standard Lagrangian has $F(\varphi, X) = X$. Some of the non-standard F models may be ruled out on theoretical grounds. An example is provided by 'phantom' fields, with negative kinetic energy density (ghosts), $F(\varphi, X) = -X$. They have $w < -1$, so that their energy density *grows* with expansion. This bizarre behaviour is reflected in the instability of the quantum vacuum for phantom fields.

Another example is a 'k-essence' field, which has $F(\varphi, X) = \varphi^{-2} f(X)$. This theory has no ghosts, and it can produce late-time acceleration. The effective sound speed of the field fluctuations for the Lagrangian in (14.11) is

$$c_{s\,\text{eff}}^2 = \frac{F_{,X}}{F_{,X} + 2X F_{,XX}}. \qquad (14.12)$$

For a standard Lagrangian, $c_{s\,\text{eff}}^2 = 1$. But for the class of F that produce accelerating k-essence models, it turns out that there is always an epoch during which $c_{s\,\text{eff}}^2 > 1$, so that these models may be ruled out according to standard causality requirements.

For models not ruled out on theoretical grounds, there is the same general problem as with quintessence, i.e. that no model is better motivated than ΛCDM, none is selected by fundamental physics and any choice of model is more or less arbitrary. Quintessence then appears to at least have the advantage of simplicity – although ΛCDM has the same advantage over quintessence.

14.2.6 Dark sector degeneracy

Finally, it is important to note a fundamental limitation that operates for all dark energy models. When investigating generic dark energy models we always have to keep in mind that since both dark energy and dark matter are only detected gravitationally, we can only measure the total energy–momentum tensor of the dark component,

$$T_{\mu\nu}^{\text{dark}} = T_{\mu\nu}^{\text{de}} + T_{\mu\nu}^{\text{c}}. \qquad (14.13)$$

Hence, if we have no information on the equation of state of dark energy, there is a degeneracy between the dark energy equation of state $w(z)$ and $\Omega_{\text{m}}(z)$. Without additional assumptions, we cannot measure either of them by purely gravitational observations. This dark sector degeneracy becomes even worse if we allow for interactions between dark matter and dark energy.

Exercise 14.2.1 Compute the adiabatic speed of sound for the Chaplygin gas, $p = -A/\rho^\alpha$, with $0 < \alpha \le 1$, and determine whether $c_s^2 > 0$.

Exercise 14.2.2 Analyse the dynamics of quintessence with an exponential potential. Compare this with the case of a phantom field with exponential potential.

Exercise 14.2.3 Using (14.5) and (14.12), find the adiabatic and effective sound speeds for quintessence, phantom fields and k-essence fields.

14.3 Modified gravity in a RW background

Late-time acceleration from nonlinear effects of structure formation is an attempt, within GR, to solve the coincidence problem without a dark energy field. The modified gravity approach shares the assumption that there is no dark energy field – but it generates the acceleration via 'dark gravity', i.e. *a weakening of gravity on the largest scales*, due to a modification of GR itself. In the modified gravity approach, it is assumed (but not shown) that the vacuum energy does not gravitate.

Could the late-time acceleration of the universe be a gravitational effect? A historical precedent is provided by attempts to explain the anomalous precession of Mercury's perihelion by a 'dark planet', named Vulcan. In the end, it was discovered that a modification to Newtonian gravity was needed.

A consistent modification of GR requires a covariant formulation of the field equations in the general case, i.e. including inhomogeneities and anisotropies. It is not sufficient to propose ad hoc modifications of the Friedmann equation, of the form $f(H^2) = 8\pi G\rho/3$ or $H^2 = 8\pi G g(\rho)/3$, for some functions f or g. Such a relation only allows us to compute the background observations – but we *cannot* compute the density perturbations without knowing the covariant parent theory that leads to such a modified Friedmann equation. And we also cannot compute the Solar System predictions.

It is very difficult to produce infrared corrections to GR that meet all the minimum requirements:

• Theoretical consistency (in the sense discussed below in Section 14.4).
• Late-time acceleration consistent with SNIa luminosity distances, BAO, the CMB shift parameter and other data that constrain the expansion history.
• A matter-dominated era with an evolution of the scale factor $a(t)$ that is consistent with the requirements of structure formation.
• Density perturbations that are consistent with the observed growth factor, matter power spectrum, peculiar velocities, CMB anisotropies and weak lensing power spectrum.
• Stable static spherical solutions for stars, and consistency with terrestrial and Solar System observational constraints.
• Consistency with binary pulsar period data.

One of the major challenges is to compute the cosmological perturbations for structure formation in a modified gravity theory. In GR, the perturbations are well understood (see Section 10.2.8). The perturbed metric in Newtonian gauge is (12.13) in any modified gravity theory that is a metric theory. In GR, the difference $\Psi - \Phi$ is sourced by anisotropic stresses, $\Psi - \Phi = 8\pi G a^2 \Pi$, and vanishes if the gravitational field is entirely due to non-relativistic matter or a perfect fluid. In modified gravity this will no longer hold: even in the absence of matter anisotropic stress, there is an effective *gravitational* anisotropic stress:

$$\Psi - \Phi = 8\pi G a^2 \Pi_{\text{modg}}. \tag{14.14}$$

On super-Hubble scales (and for adiabatic perturbations, but including the possibility of anisotropic stresses), the evolution of the perturbations is entirely determined by the background (and the anisotropic stresses which relate the potentials Φ and Ψ):

$$\Psi'' + \Phi'' - \frac{H''}{H'}\Psi' + \left(\frac{H'}{H} - \frac{H''}{H'}\right)\Phi = 0, \tag{14.15}$$

where a prime denotes $d/d\ln a$. This generalizes (13.1), and holds in both GR and modified gravity.

The primordial anisotropies in the CMB, imprinted at last scattering, will not carry a signature of modified gravity since the deviations from GR only emerge much later, during structure formation. The large-angle anisotropies in the CMB temperature, however, encode a signature of the formation of structure. They are determined by the propagation of photons along the geodesics of the perturbed geometry. For adiabatic perturbations one obtains on large scales the expression (13.29) for the integrated Sachs–Wolfe (ISW) effect, where the integral is along the (unperturbed) trajectory of the light-ray from last scattering to today. The same combination $\Phi + \Psi$ also determines the weak lensing signal, with the deflection angle given by (12.109).

The ISW effect arises from the fact that the photons are blue shifted when they fall into a gravitational potential and redshifted when they climb out of it. Hence if the potential varies during this time, they acquire a net energy shift. In a modified gravity theory, the ISW and lensing relations (13.29) and (12.109) still apply.

In GR, the gauge-invariant comoving matter density perturbation (10.32), $\Delta = \delta - 3aHv$, obeys the Poisson and evolution equations (12.63) and (12.64). In modified gravity, these equations will be modified. The modifications can be parameterized in different ways. For simplicity, we can represent the changes in the GR equations (12.63) and (12.64) in terms of two modifications to the Newton constant:

$$k^2\Psi = -4\pi G(1 + \alpha_{\text{modg}})a^2\rho\Delta, \tag{14.16}$$

$$\ddot{\Delta} + 2H\dot{\Delta} - 4\pi G(1 + \beta_{\text{modg}})\rho\Delta = 0. \tag{14.17}$$

14.3.1 *f(R)* theory

GR has a unique status as a four-dimensional theory where gravity is mediated by a massless spin-2 particle, and the field equations are second order. Consider modifications to the Einstein–Hilbert action of the general form

$$\int d^4x \sqrt{-g}\, R \rightarrow \int d^4x \sqrt{-g}\, f(R, R_{\mu\nu}R^{\mu\nu}, C_{\mu\nu\alpha\beta}C^{\mu\nu\alpha\beta}), \tag{14.18}$$

where $R_{\mu\nu}$ is the Ricci tensor, $C_{\mu\nu\alpha\beta}$ is the Weyl tensor and $f(x_1, x_2, x_3)$ is an arbitrary (at least three times differentiable) function. Since the curvature tensors contain second derivatives of the metric, the resulting equations of motion will in general be fourth order, and gravity is then carried also by massless spin-0 and spin-1 fields in general.

However, Ostrogradski's theorem on dynamics with higher than second-order time derivatives applies: there is in general a ghost instability. There is actually only one way out, which is the case $\partial_2 f = \partial_3 f = 0$, i.e. f may only depend on the Ricci scalar. The reason is that in the Ricci scalar R, only a single component of the metric appears with second derivatives. In this case, the consequent new degree of freedom can be fixed completely by the g_{00} constraint, so that there is no ghost instability in $f(R)$ theories.

Therefore, the only acceptable low-energy generalizations of the Einstein–Hilbert action of the form (14.18) are $f(R)$ theories, with $f''(R) \neq 0$ (see Capozziello and Francaviglia (2008) and Sotiriou and Faraoni (2010) for reviews). The field equations are

$$f'(R)R_{\mu\nu} - \tfrac{1}{2}f(R)g_{\mu\nu} - \left[\nabla_\mu\nabla_\nu - g_{\mu\nu}\nabla^\alpha\nabla_\alpha\right]f'(R) = 8\pi G T_{\mu\nu}, \qquad (14.19)$$

and standard energy–momentum conservation holds, $\nabla_\nu T^{\mu\nu} = 0$. The trace of the field equations is a wave-like equation for f', with source term $T = T_\mu{}^\mu$:

$$3\nabla^\alpha\nabla_\alpha f'(R) + Rf'(R) - 2f(R) = 8\pi G T. \qquad (14.20)$$

This equation is important for investigating issues of stability in the theory, and it also implies that Birkhoff's theorem does not hold.

There has been a revival of interest in $f(R)$ theories due to their ability to produce late-time acceleration. (Starobinsky constructed an inflationary model with $f(R) = R + \alpha R^2$ in the 1980s.) However, it turns out to be extremely difficult for this simplified class of modified theories to pass the observational and theoretical tests. A simple example of an $f(R)$ model is $f(R) = R - \mu/R$. For $|\mu| \sim H_0^4$, this model successfully achieves late-time acceleration as the μ/R term starts to dominate. But the model strongly violates Solar System constraints, can have a strongly non-standard matter era before the late-time acceleration, and suffers from nonlinear matter instabilities.

In $f(R)$ theories, the gravitational interaction is mediated by a spin-0 scalar as well as the spin-2 tensor degree of freedom. Indeed, $f(R)$ theories are a special case of scalar–tensor theories (see Section 14.3.2). This spin-0 field is precisely the cause of the problem with Solar System constraints in most $f(R)$ models, since the requirement of late-time acceleration leads to a very light mass for the scalar. The modification to the growth of large-scale structure due to this light scalar may be kept within observational limits. But on Solar System scales, the coupling of the light scalar to the Sun and planets induces strong deviations from the weak-field Newtonian limit of GR, in obvious violation of observations.

In the Brans–Dicke form (14.30) of the general scalar–tensor action (14.24), the $f(R)$ scalar has an associated Brans–Dicke parameter that vanishes, $\omega_{BD} = 0$, whereas Solar System and binary pulsar data currently require $\omega_{BD} > 40,000$. The Brans–Dicke action has a kinetic term but no potential, whereas the scalar–tensor form of $f(R)$ has a potential but no kinetic term. The potential in the $f(R)$ action is what allows one to evade the Solar System/binary pulsar constraints – since it provides a mechanism for giving mass to the scalar.

The only way to evade the Solar System/binary pulsar problem is to increase the mass of the scalar near massive objects like the Sun, so that the Newtonian limit can be recovered, while preserving the ultralight mass on cosmological scales. This *chameleon* mechanism can be used to construct models that evade Solar System/binary pulsar constraints. However, the price to pay is that additional parameters must be introduced, and the chosen $f(R)$ tends to look unnatural and strongly fine-tuned. An example is

$$f(R) = R + \lambda R_0 \left[\left(1 + \frac{R^2}{R_0^2} \right)^{-n} - 1 \right], \qquad (14.21)$$

where λ, R_0, n are positive parameters. This model and others that successfully avoid local constraints via a chameleon mechanism, need to mimic the background evolution of a GR dark energy model very closely in order to produce late-time acceleration.

Cosmological perturbations in $f(R)$ theory are well understood. The modification to GR produces an effective anisotropic stress

$$\Psi - \Phi = 8\pi G a^2 \Pi_{f(R)} \propto \frac{f''(R)}{f'(R)}, \qquad (14.22)$$

and deviations from GR are conveniently characterized by the dimensionless parameter

$$B = \frac{dR/d\ln a}{d\ln H/d\ln a} \frac{f''(R)}{f'(R)}. \qquad (14.23)$$

Models like (14.21) with a chameleon mechanism to evade local constraints, can match the observations of expansion history, large-angle CMB anisotropies (see Figure 14.1) and linear matter power spectrum, for appropriate choices of parameters. However, there may also be problems with singularities in the strong gravity regime, which could be incompatible with the existence of neutron stars – another unintended, and unexpected, consequence of the scalar degree of freedom, this time at high energies.

Regardless of a possible high-energy singularity problem, $f(R)$ models that pass the Solar System and late-time acceleration tests are valuable working models for probing the

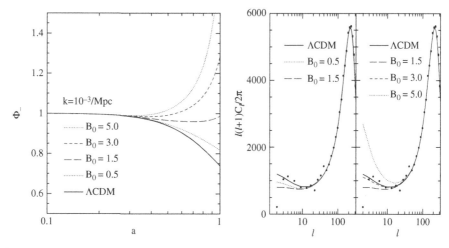

Fig. 14.1   *Left:* ISW potential $(\Phi + \Psi)/2$ for $f(R)$ models, where B_0 indicates the strength of deviation from GR [(14.23)]. *Right:* Large-angle CMB anisotropies for the same models. (Reprinted with permission from Song, Peiris and Hu (2007). Copyright by the American Physical Society.)

features of modified gravity theories and for developing tests of GR itself. In order to pursue this programme, one needs to compute not only the linear cosmological perturbations and their signature in the growth factor, the matter power spectrum and the CMB anisotropies – but also the weak lensing signal. For this, we need the additional step of understanding the transition from the linear to the nonlinear regime. Scalar–tensor behaviour on cosmological scales relevant to structure formation in the linear regime must evolve to Newtonian-like behaviour on small scales in the nonlinear regime – otherwise we cannot recover the GR limit in the Solar System. This means that the standard fitting functions in GR cannot be applied, and we require the development of N-body codes in $f(R)$ theories.

14.3.2 Scalar–tensor theories

The general scalar–tensor theory (see Peter and Uzan (2009)), which pre-dates but may also be motivated via low-energy string theory, has an action of the form

$$S = \frac{1}{16\pi G} \int d^4x \, \sqrt{-g} \left[F(\varphi)R - Z(\varphi)g^{\mu\nu}\partial_\mu\varphi\partial_\nu\varphi - 2U(\varphi) \right]$$
$$+ S_m[g_{\mu\nu}; \text{matter fields}], \tag{14.24}$$

where φ is the spin-0 field that supplements the spin-2 graviton (note that φ is dimensionless in the above expression). The matter fields are minimally coupled to the metric, as in GR, and there is no coupling to the scalar φ. This minimal coupling defines the Jordan frame, and it means that instruments in experiments (which are made of matter fields) are not affected by the local value of φ. Therefore in the Jordan frame, experimental measurements will have the same interpretation as in GR.

One of the dimensionless functions F and Z can always be eliminated by a re-definition of the scalar field; the simplest choice is $Z = 1$, which gives the standard form to the kinetic term of φ – after a re-definition, $\varphi \to \sqrt{8\pi G}\varphi$, which gives the scalar field the usual dimension. F determines the strength of gravitational interaction by effectively modifying G – and we require $F > 0$ to keep gravity attractive. Z should also be positive to avoid a ghost kinetic term. The scalar field has potential $U/8\pi G$.

The field equations arise from varying (14.24) with respect to $g_{\mu\nu}$, while variation with respect to φ and the matter fields give, respectively, the dynamical equation for φ and the matter conservation equations:

$$F(\varphi)G_{\mu\nu} = 8\pi\, GT_{\mu\nu} + Z(\varphi)\left[\partial_\mu\varphi\partial_\nu\varphi - \tfrac{1}{2}g_{\mu\nu}\partial_\alpha\varphi\partial^\alpha\varphi\right]$$
$$+ \nabla_\mu\nabla_\nu F(\varphi) - g_{\mu\nu}\nabla^\alpha\nabla_\alpha F(\varphi) - g_{\mu\nu}U(\varphi), \tag{14.25}$$

$$Z(\varphi)\nabla^\alpha\nabla_\alpha\varphi = U'(\varphi) - \tfrac{1}{2}F'(\varphi)R - \tfrac{1}{2}Z'(\varphi)\partial_\alpha\varphi\partial^\alpha\varphi, \tag{14.26}$$

$$\nabla_\nu T^{\mu\nu} = 0. \tag{14.27}$$

The Klein–Gordon type equation (14.26) can be rewritten without R – see (14.42).

In the context of late-time acceleration, these models are also known as 'extended quintessence'. Since there is one more function than in quintessence, it is always possible to reproduce the same background expansion history of any given quintessence model. The modified Friedmann equation is

$$3F\left(H^2 + \frac{K}{a^2}\right) - 3\dot{F}H = 8\pi G\rho + \frac{1}{2}Z\dot{\varphi}^2 + U\,. \tag{14.28}$$

The perturbations of a quintessence model cannot, however, be reproduced. For example, the scalar–tensor theory has an effective gravitational anisotropic stress given by

$$\Psi - \Phi = 8\pi Ga^2\,\Pi_{\mathrm{st}} = \frac{F'(\varphi)}{F(\varphi)}\,\delta\varphi\,. \tag{14.29}$$

The Brans–Dicke form of scalar–tensor theory has

$$F_{BD}(\varphi) = \varphi\,, \quad Z_{BD}(\varphi) = \frac{\omega_{BD}(\varphi)}{\varphi}\,, \tag{14.30}$$

and Brans–Dicke theory itself is the special case with

$$U_{BD} = 0\,. \tag{14.31}$$

$f(R)$ theory is also a special case of (14.24), which may be seen as follows. We set

$$F(\varphi) = f'(\varphi)\,, \quad Z(\varphi) = 0\,, \quad U(\varphi) = \tfrac{1}{2}\left[\varphi f'(\varphi) - f(\varphi)\right]\,. \tag{14.32}$$

Then this defines the scalar–tensor form of the $f(R)$ action, and reproduces the original $f(R)$ action if we make the identification,

$$\varphi = R \quad [\text{with } f''(\varphi) \neq 0]\,. \tag{14.33}$$

(Note that the case $f''(\varphi) = 0$ is just GR.)

Varying-G theories are another example of scalar–tensor theories, since the effective gravitational coupling is

$$G_{\mathrm{eff}}(\varphi) = \frac{G}{F(\varphi)}\,. \tag{14.34}$$

The Jordan frame is the one in which matter is minimally coupled, and the gravitational scalar is non-minimally coupled to curvature. The Einstein frame puts the gravitational action into Einstein–Hilbert form, so that it has an uncoupled Einstein–Hilbert term, $R/16\pi G$; as a consequence, matter is then generally non-minimally coupled. These frames are mathematically equivalent under a conformal transformation of the metric and a re-scaling of the scalar field:

$$\tilde{g}_{\mu\nu} = F(\varphi)g_{\mu\nu}\,, \quad \left(\frac{\mathrm{d}\tilde{\varphi}}{\mathrm{d}\varphi}\right)^2 = \frac{3}{2}\left[\frac{\mathrm{d}\ln F(\varphi)}{\mathrm{d}\varphi}\right]^2 + \frac{Z(\varphi)}{F(\varphi)}\,. \tag{14.35}$$

However, the frames are not physically equivalent – only one frame, the Jordan frame, respects the weak equivalence principle. The action (14.24) becomes

$$S = \frac{1}{16\pi G}\int \mathrm{d}^4x\,\sqrt{-g}\left[\tilde{R} - \tilde{g}^{\mu\nu}\partial_\mu\tilde{\varphi}\partial_\nu\tilde{\varphi} - 2\tilde{U}(\tilde{\varphi})\right]$$
$$+ S_{\mathrm{m}}\left[A^2(\tilde{\varphi})\,\tilde{g}_{\mu\nu}; \text{matter fields}\right]\,, \quad A(\tilde{\varphi}) := \frac{1}{\sqrt{F(\varphi)}}\,. \tag{14.36}$$

The gravitational action in (14.36) suggests that we have removed the spin-0 gravitational degree of freedom, since $\tilde{\varphi}$ appears in the gravitational action in exactly the same form as a standard minimally coupled scalar field in GR. However, the spin-0 scalar has *not* been downgraded to a standard scalar field. Its influence has simply been shifted, and is now reflected in its non-minimal coupling to the matter fields – the matter action is no longer defined in terms of the metric only, but of $A^2(\tilde{\varphi})$ times the metric. The convenience of the Einstein frame is that the field equations have the Einstein form, but this comes at the price of non-minimal coupling in the matter equations – and the consequent difficulties of correctly defining and interpreting measurements and observations in the Einstein frame. On the other hand, the Einstein frame is the appropriate frame for investigating the Cauchy problem and the regularity of the theory.

The coupling to matter appears in the Einstein frame as a violation of energy–momentum conservation:

$$\tilde{\nabla}^{\nu}\tilde{T}_{\mu\nu} = \frac{\mathrm{d}\ln A(\tilde{\varphi})}{\mathrm{d}\tilde{\varphi}}\tilde{T}^{\alpha}{}_{\alpha}\partial_{\mu}\tilde{\varphi}, \quad \tilde{T}_{\mu\nu} = A^2 T_{\mu\nu}. \tag{14.37}$$

In the Einstein frame, the gravitational coupling is the constant G, unlike the Jordan frame where $G \to G_{\mathrm{eff}}$, which varies in time and space. By contrast, particle masses are constant in the Jordan frame, but variable in the Einstein frame.

The $f(R)$ action in the Jordan frame can be transformed to the Einstein frame via

$$\tilde{g}_{\mu\nu} = f'(\varphi)g_{\mu\nu}, \quad \tilde{\varphi} = -\sqrt{12\pi G}\ln f'(\varphi). \tag{14.38}$$

In terms of $\tilde{g}_{\mu\nu}$ and $\tilde{\varphi}$ the gravitational Lagrangian then becomes a standard Einstein–Hilbert plus scalar field Lagrangian, with potential

$$V(\tilde{\varphi}) = e^{-\tilde{\varphi}/\sqrt{3\pi G}}\left[\varphi(\tilde{\varphi})\,e^{\tilde{\varphi}/\sqrt{12\pi G}} - f(\varphi(\tilde{\varphi}))\right]. \tag{14.39}$$

14.3.3 Vector–tensor theories

As discussed in Section 12.3.3, a tensor–vector–scalar theory TeVeS has been developed by Bekenstein as a covariant relativistic theory that contains a MOND-like limit. The additional degrees of freedom mean that in principle TeVeS can avoid the need for both dark matter and dark energy. However, it is not clear that all the observations can be satisfied in this case (see Ferreira and Starkman (2009), Skordis (2009), Bekenstein (2010) for reviews). In a sense, the full scalar and vector degrees of freedom, on top of the standard tensor degree of freedom, represent too much arbitrariness – perhaps not in principle different from the apparent arbitrariness of dark matter and dark energy contributions. Of course, if a simple version could be found, with reduced freedom of functions and parameters, that would be very interesting.

In fact a vector–tensor sub-class of theories has recently been investigated as a possible candidate. The action for these Einstein–Aether theories is

$$S = \frac{1}{16\pi G}\int \mathrm{d}^4x\,\sqrt{-g}\left[R + M^2 F(\mathcal{K}) + \lambda\left(\mathcal{A}_{\mu}\mathcal{A}^{\mu} + 1\right)\right]$$
$$+ S_{\mathrm{m}}[g_{\mu\nu}; \text{matter fields}], \tag{14.40}$$

where M is a mass scale, and λ is a Lagrange multiplier which enforces that $\mathcal{A}^\mu$ is a unit timelike vector field, which is necessary for compatibility with RW geometry in the background. The dynamics for the gravitational vector $\mathcal{A}^\mu$ are encoded in

$$\mathcal{K} = M^{-2} K^{\mu\nu}{}_{\alpha\beta} \nabla_\mu \mathcal{A}^\alpha \nabla_\nu \mathcal{A}^\beta , \tag{14.41}$$

where $K^{\mu\nu}{}_{\alpha\beta} = c_1 g^{\mu\nu} g_{\alpha\beta} + c_2 \delta^\mu_\alpha \delta^\nu_\beta + c_3 \delta^\mu_\beta \delta^\nu_\alpha$, with c_1, c_2 and c_3 dimensionless constants.

Initial indications are that the Einstein–Aether theories can readily avoid dark energy, but are unable simultaneously to reproduce all of the key features of dark matter in the background expansion and structure formation (Zuntz *et al.*, 2010).

Exercise 14.3.1 Use the trace of the scalar–tensor field equations (14.25) to eliminate R from the Klein–Gordon type evolution equation (14.26). Show that this leads to (where a prime denotes $d/d\varphi$)

$$(2ZF + 3F'^2)\nabla^\alpha \nabla_\alpha \varphi + \tfrac{1}{2}(2ZF + 3F'^2)' \partial^\alpha \varphi \partial_\alpha \varphi$$
$$= 8\pi G F' T^\alpha{}_\alpha - -4F'U + 2U'F . \tag{14.42}$$

14.3.4 Brane-world models: DGP

Modifications to GR within the framework of quantum gravity are typically ultraviolet corrections that must arise at high energies in the very early universe or during collapse to a black hole. A leading candidate for a quantum gravity theory, string theory, is able to remove the infinities of quantum field theory and unify the fundamental interactions, including gravity. But there is a price – the theory is only consistent in nine space dimensions. As discussed in Section 20.3, branes play a fundamental role in the theory. The observable universe is a 4D brane, on which matter and radiation fields are localized, with gravity propagating in the bulk (see Figure 20.1).

Brane-world cosmological models inherit some aspects of string theory, but do not attempt to impose the full machinery of the theory. Instead, simplifications are introduced in order to be able to construct cosmological models that can be used to compute observational predictions (see Maartens and Koyama (2010) for a review). Cosmological data can then be used to constrain the brane-world models, and hopefully provide constraints on string theory, as well as pointers for the further development of string theory.

Most brane-world models modify GR at high energies, and recover GR in the infrared; the premier example is the Randall–Sundrum brane-world (see Section 20.3.4). By contrast, the brane-world model of Dvali–Gabadadze–Porrati (DGP), which was introduced as a particle physics model and then generalized to cosmology by Deffayet, modifies GR at *low* energies, and recovers GR in the ultraviolet. This model produces 'self-acceleration' of the late-time universe due to a weakening of gravity at low energies. Like the Randall–Sundrum model, the DGP model is a 5D model with infinite (but non-warped) extra dimension.

The gravitational action is given by

$$\frac{1}{16\pi G}\left[\frac{1}{r_c}\int_{\text{bulk}} d^5 x \sqrt{-{}^{(5)}g}\,{}^{(5)}R + \int_{\text{brane}} d^4 x \sqrt{-g}\,R\right] . \tag{14.43}$$

The bulk is 5D Minkowski spacetime (in the unperturbed case), with infinite volume. Consequently, there is no normalizable massless spin-2 mode of the 4D graviton in the DGP brane-world – GR is recovered by an ultralight spin-2 graviton mode. In addition, there is a spin-0 mode of the graviton.

Gravity 'leaks' off the 4D brane into the bulk at large scales, $r \gg r_c$, where the first term in the sum (14.43) dominates. On small scales, gravity is effectively bound to the brane and 4D dynamics is recovered to a good approximation, as the second term dominates. The transition from 4D to 5D behaviour is governed by the crossover scale r_c. For a Minkowski brane, the weak-field gravitational potential behaves as $\Phi \propto r^{-1}$ for $r \ll r_c$ and $\propto r^{-2}$ for $r \gg r_c$. On a RW brane, gravity leakage at late times in the cosmological evolution can initiate acceleration – not due to any negative pressure field, but due to the weakening of gravity on the brane.

The energy conservation equation remains the same as in GR, but the Friedmann equation is modified:

$$\dot{\rho} + 3H(\rho + p) = 0, \tag{14.44}$$

$$H^2 + \frac{K}{a^2} - \frac{1}{r_c}\sqrt{H^2 + \frac{K}{a^2}} = \frac{8\pi G}{3}\rho. \tag{14.45}$$

To arrive at (14.45) we have to take a square root which implies a choice of sign. As we shall see, the above choice has the advantage of leading to acceleration but the disadvantage of the presence of a 'ghost' in this background. This is the self-accelerating branch of DGP. We shall discuss the 'normal' DGP model, where the opposite sign of the square root is chosen, in Section 14.3.5.

From (14.45) we infer that at early times, i.e. $Hr_c \gg 1$, the GR Friedmann equation is recovered. By contrast, at late times in an expanding CDM universe, with $\rho \propto a^{-3} \to 0$, we have

$$H \to H_\infty = \frac{1}{r_c}, \tag{14.46}$$

so that expansion accelerates and is asymptotically de Sitter. The above equations imply

$$\dot{H} - \frac{K}{a^2} = -4\pi G\rho \left[1 + \frac{1}{\sqrt{1 + 32\pi G r_c^2 \rho/3}} \right]. \tag{14.47}$$

In order to achieve self-acceleration at late times, we require

$$r_c \gtrsim H_0^{-1}, \tag{14.48}$$

since $H_0 \lesssim H_\infty$. This is confirmed by fitting SNIa observations, as shown in Figure 14.2. The dimensionless cross-over parameter is defined as

$$\Omega_{r_c} = \frac{1}{4(H_0 r_c)^2}, \tag{14.49}$$

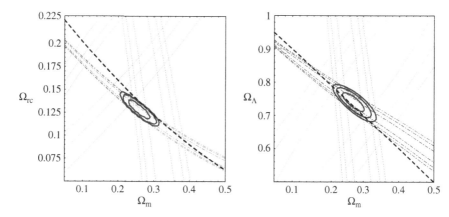

Fig. 14.2   Joint constraints [solid thick (blue)] from SNIa (SNLS) data [solid thin (yellow)], BAO peak at $z = 0.35$ [dotted (green)] and CMB shift parameter (WMAP3) [dot-dashed (red)]. Left plot shows DGP models; right plot shows ΛCDM. Thick dashed (black) line represents flat models. (Reprinted with permission from Maartens and Majerotto (2006). Copyright by the American Physical Society.) A colour version of this figure is available online.

and the ΛCDM relation, $\Omega_{m0} + \Omega_{\Lambda 0} + \Omega_{K0} = 1$, is modified to

$$\Omega_{m0} + 2\sqrt{\Omega_{r_c}}\sqrt{1 - \Omega_{K0}} + \Omega_{K0} = 1. \tag{14.50}$$

ΛCDM and DGP can both account for the SNIa observations, with the fine-tuned values $\Lambda \sim H_0^2$ and $r_c \sim H_0^{-1}$ respectively. When we add further constraints on the expansion history from the BAO and the CMB shift parameter, the DGP flat models are in strong tension with data, whereas ΛCDM models provide a consistent fit. This is evident in Figure 14.2. The open DGP models provide a somewhat better fit to the geometric data – essentially because the lower value of Ω_{m0} favoured by SNIa reduces the distance to last scattering and an open geometry is able to extend that distance.

Observations based on structure formation provide further evidence of the difference between DGP and ΛCDM, since the two models suppress the growth of density perturbations in different ways. The distance-based observations draw only upon the background modified Friedmann equation (14.45) in DGP models – and therefore there are quintessence models in GR that can produce precisely the same expansion history $H(z)$ as DGP. By contrast, structure formation observations require the 5D perturbations in DGP, and one cannot find equivalent quintessence models.

DGP cosmological perturbations are subtle and complicated: although matter is confined to the 4D brane, gravity is fundamentally 5D, and the 5D bulk gravitational field responds to and back-reacts on 4D density perturbations. The evolution of density perturbations requires an analysis based on the 5D nature of gravity. In particular, the 5D gravitational field produces an effective 'dark' anisotropic stress on the 4D universe. If one neglects this stress and other 5D effects, and simply treats the perturbations as 4D perturbations with a modified background Hubble rate – then as a consequence, the 4D Bianchi identity on the brane is violated, i.e. $\nabla^\nu G_{\mu\nu} \neq 0$, and the results are inconsistent. When the 5D effects are incorporated, the 4D Bianchi identity is automatically satisfied. (See Figure 14.3.)

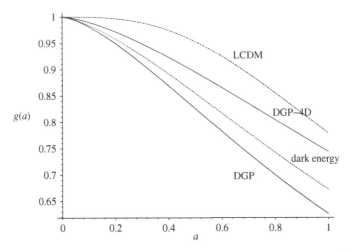

Growth factor $g(a) = \Delta(a)/a$ for ΛCDM (long dashed), DGP (solid, thick), and a dark energy model with the same expansion history as DGP (short dashed). DGP-4D (solid, thin) shows the incorrect result with 5D effects neglected. (From Koyama and Maartens (2006).)

There are three regimes governing structure formation in DGP models:

- On small scales, below the so-called Vainstein radius (which for cosmological purposes is roughly the scale of clusters), the spin-0 scalar degree of freedom becomes strongly coupled, so that the GR limit is recovered.
- On scales relevant for structure formation, i.e. between cluster scales and the Hubble radius, the spin-0 scalar degree of freedom produces a scalar–tensor behaviour. A quasi-static approximation to the 5D perturbations allows us to solve for the density perturbations (see below and Figure 14.3). DGP gravity is like a Brans–Dicke theory with parameter

$$\omega_{BD} = \tfrac{3}{2}(\beta - 1), \tag{14.51}$$

$$\beta = 1 + 2H^2 r_c \left(H^2 + \frac{K}{a^2}\right)^{-1/2} \left[1 + \frac{\dot{H}}{3H^2} + \frac{2K}{3a^2 H^2}\right]. \tag{14.52}$$

At late times in an expanding universe, when $Hr_c \gtrsim 1$, it follows that $\beta < 1$, so that $\omega_{BD} < 0$. (This signals a pathology in DGP which is discussed below.)

- The quasi-static approximation breaks down near and beyond the Hubble radius. On super-horizon scales, 5D gravity effects are dominant. Numerical solutions of the partial differential equation governing the 5D bulk variable have been developed, and the consequent solutions for Φ, Ψ and Δ have been found on super-Hubble scales. The results are illustrated in Figure 14.4.

On sub-Hubble scales relevant for linear structure formation, 5D effects produce a difference between Ψ and Φ:

$$k^2 \Psi = -4\pi G a^2 \left(1 - \frac{1}{3\beta}\right)\rho\Delta, \quad k^2 \Phi = -4\pi G a^2 \left(1 + \frac{1}{3\beta}\right)\rho\Delta, \tag{14.53}$$

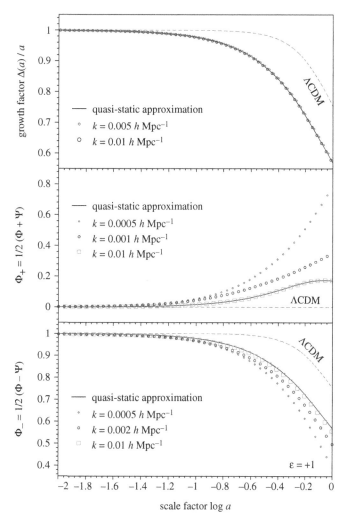

Fig. 14.4   Numerical solutions for DGP density and metric perturbations, showing also the quasistatic solution. In our notation, $\Phi \rightarrow -\Psi, \Psi \rightarrow \Phi$. (Reprinted with permission from Cardoso *et al.* (2008). Copyright by the American Physical Society.)

so that there is an effective dark anisotropic stress on the brane:

$$\Pi_{DGP} = \frac{\rho\Delta}{3\beta^2 k^2} . \tag{14.54}$$

The density perturbations evolve as

$$\ddot{\Delta} + 2H\dot{\Delta} - 4\pi G \left(1 - \frac{1}{3\beta}\right)\rho\Delta = 0 . \tag{14.55}$$

The linear growth factor, $g(a) = \Delta(a)/a$ (i.e. normalized to the flat CDM case, $\Delta \propto a$), is shown in Figure 14.3. This shows the dramatic suppression of growth in DGP relative to ΛCDM – from both the background expansion and the metric perturbations. If we parameterize the growth factor via $f := \mathrm{d}\ln\Delta/\mathrm{d}\ln a = \Omega_\mathrm{m}(a)^\gamma$, (see (12.78)), we can quantify the deviation from a GR model with smooth dark energy:

$$\gamma \approx \begin{cases} 0.55 + 0.05[1 + w(z=1)] & \text{GR, smooth DE} \\ 0.68 & \text{DGP} \end{cases} \tag{14.56}$$

Observational data on the growth factor are not yet precise enough to provide meaningful constraints on the DGP model. Instead, we can look at the large-angle anisotropies of the CMB, i.e. the ISW effect. This requires a treatment of perturbations near and beyond the horizon scale, using the full numerical solutions that are illustrated in Figure 14.4. It is evident from Figure 14.4 that the DGP model is in serious tension with the CMB data on large scales. The problem arises from the large deviation of the ISW potential $(\Phi + \Psi)/2$ in the DGP model from the ΛCDM model. The much stronger decay of the ISW potential leads to an over-strong ISW effect (see (13.29)).

In addition to the severe problems posed by cosmological observations, a problem of theoretical consistency is posed by the fact that the late-time asymptotic de Sitter solution in DGP cosmological models has a ghost. The ghost is signalled by the negative Brans–Dicke parameter in the effective theory that approximates the DGP on cosmological sub-horizon scales:

$$\omega_{BD} < 0. \tag{14.57}$$

The existence of the ghost is confirmed by detailed analysis of the 5D perturbations in the de Sitter limit. There is a ghost mode in the scalar sector of the gravitational field – which is more serious than the ghost in a phantom scalar field. It probably rules out the DGP, since it is hard to see how an ultraviolet completion of the DGP can cure the *infrared* ghost problem. Nevertheless, DGP is a useful and rich toy model for modified gravity, which is very different from the $f(R)$ model. Various attempts are underway to find a generalization of the DGP that cures the ghost problem, but this is proving to be very difficult.

14.3.5 Normal (non-self-accelerating) DGP

The 'normal' (i.e. non-self-accelerating and ghost-free) branch of the DGP arises from a different embedding of the DGP brane in the Minkowski bulk (see Figure 14.5). In the background dynamics, this amounts to a replacement $r_c \to -r_c$ in (14.45) – and there is no longer late-time self-acceleration. It is therefore necessary to include a Λ term in order to accelerate the late universe:

$$H^2 + \frac{K}{a^2} + \frac{1}{r_c}\sqrt{H^2 + \frac{K}{a^2}} = \frac{8\pi G}{3}\rho + \frac{\Lambda}{3}. \tag{14.58}$$

Using the dimensionless crossover parameter defined in (14.49), the densities are related at the present time by

$$\sqrt{1 - \Omega_{K0}} = -\sqrt{\Omega_{r_c}} + \sqrt{\Omega_{r_c} + \Omega_{m0} + \Omega_{\Lambda0}}, \tag{14.59}$$

which can be compared with the self-accelerating DGP relation (14.50).

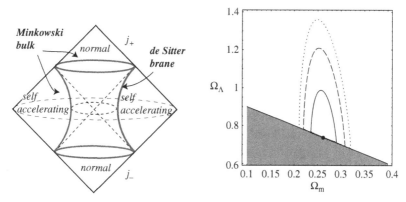

Fig. 14.5 *Left:* Embedding of self-accelerating and normal branches of DGP brane in a Minkowski bulk. (From Charmousis *et al.* (2006).) *Right:* Joint constraints on normal DGP ($K = 0$) from SNIa (SNLS), CMB shift (WMAP3) and BAO (SDSS $z = 0.35$) data. Best-fit is the solid point. Shaded region is unphysical and its upper boundary is flat ΛCDM. (Reprinted with permission from Lazkoz, Maartens and Majerotto (2006). Copyright by the American Physical Society.)

The 'degravitation' feature of normal DGP is that Λ is effectively screened by 5D gravity effects. This follows from rewriting the modified Friedmann equation (14.58) in standard GR form, with

$$\Lambda_{\rm eff} = \Lambda - \frac{3}{r_c}\sqrt{H^2 + \frac{K}{a^2}} < \Lambda. \tag{14.60}$$

Thus 5D gravity in normal DGP can in principle reduce the bare vacuum energy significantly. However, Figure 14.5 shows that best-fit flat models, using geometric data, only admit insignificant screening. The closed models provide a better fit to the data, and can allow a bare vacuum energy term with $\Omega_{\Lambda 0} > 1$. This does not address the fundamental problem of the smallness of $\Omega_{\Lambda 0}$, but it is nevertheless an interesting feature.

We can also define an effective equation of state parameter via

$$\dot{\Lambda}_{\rm eff} + 3H(1 + w_{\rm eff})\Lambda_{\rm eff} = 0. \tag{14.61}$$

At the present time (setting $K = 0$ for simplicity),

$$w_{\rm eff,0} = -1 - \frac{(\Omega_{\rm m0} + \Omega_{\Lambda 0} - 1)\Omega_{\rm m0}}{(1 - \Omega_{\rm m0})(\Omega_{\rm m0} + \Omega_{\Lambda 0} + 1)} < -1, \tag{14.62}$$

where the inequality holds since $\Omega_{\rm m0} < 1$. This reveals another important property of the normal DGP model: effective phantom behaviour of the recent expansion history. This is achieved without any pathological phantom field (similar to what can be done in scalar–tensor theories). Furthermore, there is no 'big rip' singularity in the future associated with this phantom acceleration, unlike the situation that typically arises with phantom fields. The phantom behaviour in the normal DGP model is also not associated with any ghost problem – indeed, the normal DGP branch is free of the ghost that plagues the self-accelerating DGP.

Perturbations in the normal branch have the same structure as those in the self-accelerating branch, with the same regimes – i.e. below the Vainshtein radius (recovering a GR limit), up to the Hubble radius (Brans–Dicke behaviour), and beyond the Hubble radius (strongly 5D behaviour). The quasistatic approximation and the numerical integrations can be simply repeated with the replacement $r_c \to -r_c$ (and the addition of Λ to the background). In the sub-Hubble regime, the effective Brans–Dicke parameter is still given by (14.51) and (14.52), but now we have $\omega_{BD} > 0$ – and this is consistent with the absence of a ghost. Furthermore, a positive Brans–Dicke parameter signals an extra positive contribution to structure formation from the scalar degree of freedom, so that there is *less* suppression of structure formation than in ΛCDM – the reverse of what happens in the self-accelerating DGP. This is confirmed by computations.

The closed normal DGP models fit the background expansion data reasonably well. They may be compared with $f(R)$ models: both types of model lead to less suppression of structure than ΛCDM, but they produce different ISW effects. However, in the limit $r_c \to \infty$, normal DGP tends to ordinary ΛCDM, hence observations which fit ΛCDM will always just provide a lower limit for r_c.

14.4 Constraining effective theories

The concept of low-energy effective theories is extremely useful in physics. One of the most prominent examples is superconductivity. It would be impossible to describe this phenomenon by using full quantum electrodynamics with a typical energy scale of MeV, where the energy scale of superconductivity is milli-eV and less. However, many aspects of superconductivity can be successfully described with the Ginzburg–Landau theory of a complex scalar field. Microscopically, this scalar field is to be identified with a Cooper pair of two electrons, but this is irrelevant for many aspects of superconductivity. Another example is weak interaction and four-Fermi theory. The latter is a good approximation to weak interactions at energy scales far below the Z-boson mass. Most physicists also regard the Standard Model of particle physics as a low-energy effective theory which is valid below some high-energy scale beyond which new degrees of freedom become relevant, be this supersymmetry, grand unified theory or string theory.

Some models of dark energy have unusual Lagrangians that cannot be quantized in the usual way, e.g. because they have non-standard kinetic terms. We then simply call them 'effective low-energy theories' of some unspecified high-energy theory. Similarly, some modifications of GR, such as those we discussed above, should also be seen as effective low-energy theories. Without knowledge of the complete theory, the danger is that unconstrained arbitrariness can be introduced via effective theories. However, some theoretical constraints can be imposed on low-energy effective theories.

The requirements for a physical theory are a matter of debate. A possible list is as follows (see Durrer and Maartens (2008) for further discussion):

1. A fundamental physical theory allows a Lagrangian formulation.
2. Lorentz invariance.
3. No ghosts.

4. No tachyons (i.e. potentials without a minimum).
5. No superluminal motion.

Which of these properties may be lost if we 'integrate out' high-energy excitations and consider only processes which take place at energies below some cutoff scale E_c? We cannot completely ignore all particles with masses above E_c, since in the low-energy quantum theory they can still be produced 'virtually', i.e. for a time shorter than $1/E_c$. This is not relevant for the initial and final states of a scattering process, but plays a role in the interaction.

The Lagrangian formulation will survive if we proceed in a consistent way by simply integrating out the high-energy degrees of freedom.

The high-energy cutoff will be given by some mass scale, i.e. some Lorentz-invariant energy scale of the theory, and therefore the effective low-energy theory should also admit a Lorentz-invariant Lagrangian. Lorentz-invariance is not a high-energy phenomenon which can simply be lost at low energies.

Effective theories with no ghosts or serious tachyon have an energy functional which is bounded from below – and that low-energy property is not removed by integrating out excitations with $E > E_c$.

What about superluminal motion and causality? One can argue that in cosmology we do have a preferred frame, the cosmological frame, hence Lorentz invariance is broken and we can simply demand that all superluminal modes of a field propagate forward in cosmic time. Then no closed signal curves are possible. But there is a problem. Most solutions of a Lagrangian theory do break several or most of the symmetries of the Lagrangian spontaneously. However, when applying a Lorentz transformation to a solution, we produce a new solution that, from the point of view of the Lagrangian, has the same right of existence. If some modes of a field propagate with superluminal speed, this means that their characteristics are spacelike. The condition that the mode has to travel forward in time with respect to a certain frame implies that one has to use the retarded Green's function in this frame. Since spacelike distances have no frame-independent chronology, for spacelike characteristics this is a frame-dependent statement. Depending on the frame of reference, a given mode can represent a normal propagating degree of freedom, or it can satisfy an elliptic equation, a constraint.

14.5 Conclusion

14.5.1 Will evidence for dark energy/ modified gravity persist?

Major efforts are going into explaining the apparent acceleration of the universe; indeed this is a central preoccupation of present-day cosmology. But in accord with the philosophy laid out in the preface, we need to ask: 'Will the apparent acceleration go away as more data are collected and the observations are reinterpreted?' Could it in fact be ephemeral, so that future generations will look back and say, they got it wrong: they misunderstood the data.

The evidence for a late-time cosmic acceleration, or a weakening of Einstein gravity on large scales, continues to mount, as the number of experiments and the quality of data grow. It is possible that some of the evidence may weaken or fall, despite extensive efforts to check that observations are not being misinterpreted or are not subject to strong systematic uncertainties. (See Kirshner (2009) for the case of SNIa.)

For example, if new evolutionary effects are discovered in SNIa, this could undermine the evidence from their luminosities for acceleration/ gravity-weakening. But there is a network of *independent* observations that go into the conclusion about acceleration. For example, if we omit the SNIa data, then the BAO scale and the distance to the last scattering surface of the CMB still provide strong evidence.

It therefore seems unlikely that the evidence will go away, within the framework of a RW geometry, i.e. assuming that the universe at late times is adequately described as a perturbation of a RW background, and that this background is fixed from inflation up to the present day. This assumption is an essential target for ongoing investigation. We need to determine whether there is a *theoretical* misinterpretation of the data, based on a previously unrecognized role of nonlinear inhomogeneities that invalidates a perturbed RW model with fixed background. This alternative is discussed in the following two chapters.

14.5.2 Probing gravity in a RW framework

Assuming that a RW framework is applicable, the revolutionary discovery by observational cosmology confronts theoretical cosmology with a major problem – how to explain the origin of the acceleration. The core of this problem may be 'handed over' to particle physics, since we require, at the most fundamental level, an explanation for why the vacuum energy either has an incredibly small and fine-tuned value, or is exactly zero. Both options violently disagree with naive estimates of the vacuum energy.

If one accepts that the vacuum energy is indeed non-zero, then the dark energy is described by Λ, and the ΛCDM model is the best current model. The cosmological model requires completion via developments in particle physics that will explain the value of the vacuum energy. In many ways, this is the best that we can do currently, since the alternatives to ΛCDM, within and beyond GR, do not resolve the vacuum energy crisis, and furthermore have no convincing theoretical motivation. None of the contenders so far appears any better than ΛCDM, and it is fair to say that at the theoretical level, there is as yet no serious challenger to ΛCDM. One consequence of this is the need to develop better observational tests of ΛCDM, which could in principle rule it out, e.g. by showing, to some acceptable level of statistical confidence, that $w \neq -1$. However, observations are still quite far from the necessary precision for this.

It remains necessary and worthwhile to continue investigating alternative dark energy and modified gravity models, in order better to understand the space of possibilities, the variety of cosmological properties, and the observational strategies needed to distinguish them. The lack of any consistent and compelling theoretical model means that we need to keep exploring alternatives – and also to keep challenging the validity of GR itself on cosmological scales.

We have focused in this chapter on two of the simplest infrared-modified gravity models: $f(R)$ (the simplest scalar–tensor models), and DGP (the simplest brane-world models). In both types of model, the new scalar degree of freedom introduces severe difficulties at theoretical and observational levels. The $f(R)$ models may be ruled out by the presence of singularities that may exclude neutron stars (even if they can match all cosmological observations). And the DGP models are probably ruled out by the appearance of a ghost in the asymptotic de Sitter state – as well as by a combination of geometric and structure-formation data. There is no sign as yet of a serious contender for a consistent and 'not unnatural' modified gravity theory.

Nevertheless, $f(R)$ and DGP models are very important toy models – intensive investigation of them has left an important legacy, in a deeper understanding of:

- the interplay between gravity and expansion history and structure formation;
- the relation between cosmological and local observational constraints;
- the special properties of GR itself;
- the techniques needed to distinguish different candidate models, and the limitations and degeneracies within those techniques;
- the development of tests that can probe the validity of GR itself on cosmological scales, independent of any particular alternative model.

The last point is one of the most important by-products of the investigation of modified gravity models. It involves a careful analysis of the web of consistency relations that link the background expansion to the evolution of perturbations, and opens up the real prospect of testing general GR well beyond the Solar System and its neighbourhood. (For recent reviews and work, see Zhang *et al.* (2007), Reyes *et al.* (2010), Pogosian *et al.* (2010), Daniel *et al.* (2010), Jain and Khoury (2010).)

14.5.3 Dark energy and the far future universe

Will the universe expand forever, cooling indefinitely and so leading to the cessation of all physical phenomena because no free energy is left to sustain any activity of any kind? Or will it rather recontract to a big crunch in the future, with ever-increasing temperatures leading to the destruction of all physical structures in a time-reversed version of the big bang? The outcome depends firstly on the nature of dark energy: will the present acceleration of the universe, driven by a dark energy field, carry on for ever? Is the dark energy a field that might decay away in the future?

If the dark energy is in fact a constant, then in the future as the matter density redshifts to ever smaller values, Λ will dominate because it dominates today, and the universe will undergo an accelerating expansion forever. It will get cooler and cooler, stars will die out and astrophysics will end, life will come to an end, even matter will eventually decay away. This fate has been called a 'heat death' because it signifies the ultimate triumph of entropy, as stated by Thomson (Lord Kelvin) (1862):

> The second great law of thermodynamics involves a certain principle of irreversible action in Nature. It is thus shown that, although mechanical energy is indestructible, there is a universal

tendency to its dissipation, which produces gradual augmentation and diffusion of heat, cessation of motion, and exhaustion of potential energy through the material universe. The result would inevitably be a state of universal rest and death, if the universe were finite and left to obey existing laws.

Suppose, however, that dark energy is time dependent and eventually becomes dynamically irrelevant, decaying away to zero either through a transition to a new lower vacuum state – a potentially catastrophic event – or else smoothly because it eventually decays away at a faster rate than does matter. Then ultimately matter will dominate again, and the universe will decelerate from then on. So the possibility arises of expansion to a maximum radius and subsequent recollapse.

In that case, the outcome depends on the second issue to take into account, namely the sign K of spatial curvature. Assuming the matter always obeys the energy inequalities (which is likely in view of their ever diluting density), recollapse occurs if and only if $K > 0$. Then there will be a heat death in the obvious sense, with everything destroyed by indefinitely increasing temperature, in a time-reversed image of the Big Bang, but not identical in its details. All ordered structures are destroyed, and at even later times the universe either ends up at a final singularity, an end to space and time, mirroring the start of the universe at a final singularity; or quantum gravity effects cause the universe to bounce, leading to an oscillating universe (Section 20.6).

It is not clear how we can ever determine whether a time-varying dark energy will eventually decay away to zero. As the current data are consistent with $w = -1$, perhaps the odds are marginally on the first option: eternal expansion will occur (as favoured by chaotic inflationary models). However, this is certainly not a definite conclusion: it is a topic meriting further investigation, and emphasizes the importance of observationally testing the sign of Ω_{K0} as well as the issue of whether w is constant or not.

A final comment: the discussion above would have to be revisited if it turned out that there is after all no dark energy, but that either modified gravity, or nonlinear effects in GR mimic acceleration.

15

'Acceleration' from large-scale inhomogeneity?

As discussed in the previous chapter, the explanation of dark energy is a central preoccupation of present-day cosmology. Its presence is indicated by the apparent recent speeding up of the expansion of the universe indicated by SNIa observations, which is usually taken to be caused by quintessence or a cosmological constant, and is consistent with other observations such as those of anisotropies and large-scale structure studies. Like dark matter, its existence was discovered, not predicted. The astronomical observations are being refined in many sophisticated ways and used to confirm the acceleration data and test the equation of state of the hypothetical dark energy. Whether its density is constant or varying, its existence is a major problem for theoretical physics. It is therefore crucial to pursue the possibility of other theoretical explanations.

The deduction of the existence of dark energy assumes that the universe has a RW geometry on large scales. However, the interpretation of the observations is ambiguous. They can at least in principle be accounted for without the presence of dark energy, if we allow inhomogeneity. This can contribute in one or both of two ways: locally via backreaction and associated observational effects, as discussed in the next chapter, and globally via large-scale inhomogeneity, considered in this chapter.

Here we consider the possibility that what appears to be the acceleration of an FLRW universe due to dark energy, is in fact rather a manifestation of Hubble-scale inhomogeneity in a universe such as that described by the Lemaître–Tolman–Bondi (LTB) models discussed in the next section, where we are near the centre of a void. There is then no need for dark energy. We show that the redshift–distance relation for distant SNIa is compatible with the proposal that we may be measuring spatial inhomogeneity, rather than acceleration of an FLRW universe. Then we show that the other cosmological data may also be fitted by such models. Finally we consider potentially viable further observational tests of this hypothesis.[1]

A key issue is whether the Copernican Principle holds within the observed region of the universe. It is the foundation of the standard model, and so needs to be subjected to all possible tests, irrespective of the issue of dark energy.

15.1 Lemaître–Tolman–Bondi universes

We assume isotropy of the spacetime, with its matter content well described by zero pressure dust matter. This leads to the simplest inhomogeneous cosmology – the LTB model

---

[1] Szekeres inhomogeneous models have been used in a similar way; see Chapter 19.

(Lemaître, 1933b, Tolman, 1934, Bondi, 1947), which is inhomogeneous in the radial direction only. In its application later in this chapter, we assume the Earth is at or near the centre of isotropy.

This general spherically symmetric metric for dust is given in synchronous comoving coordinates by

$$ds^2 = -dt^2 + X^2(t,r)\,dr^2 + R^2(t,r)\,d\Omega^2 \, , \quad u^\mu = \delta^\mu{}_0, \tag{15.1}$$

where R is the areal radius, since the proper area of a sphere $r = \text{const}$, $t = \text{const}$, is $4\pi R^2$. Solving the EFE (Bondi, 1947) shows that

$$ds^2 = -dt^2 + \frac{[R'(t,r)]^2}{1+2E(r)}\,dr^2 + R^2(t,r)\,d\Omega^2 \, , \tag{15.2}$$

where a prime denotes $\partial/\partial r$ and R obeys a generalized Friedmann equation,

$$\dot{R}(t,r) = \pm\left[\frac{2m(r)}{R(t,r)} + 2E(r) + \frac{1}{3}\Lambda R^2(t,r)\right]^{1/2}. \tag{15.3}$$

Here $E(r)$ is an arbitrary function, often written as $2E = -Kf(r)^2$, where $K = \pm 1$ or 0.[2] The density is given by

$$4\pi G\rho(t,r) = \frac{m'(r)}{R^2(t,r)\,R'(t,r)} \, . \tag{15.4}$$

The differential equation (15.3) with $\Lambda = 0$, like (9.22), can be completely integrated as follows. The solution for $E = 0$ is

$$t - t_B(r) = \pm 2R^{3/2}[18m(r)]^{-1/2}, \tag{15.5}$$

where t_B is an arbitrary function of r. For $E \neq 0$, in terms of a parameter $\eta = \eta(t,r)$,

$$t - t_B(r) = \pm h(\eta)m(r)f^{-3}(r), \quad R = h'(\eta)m(r)f^{-2}(r), \tag{15.6}$$

$$h(\eta) = \{\eta - \sin\eta, \sinh\eta - \eta\} \quad \text{for} \quad \text{sgn}(E) = \{-1, +1\}. \tag{15.7}$$

Putting $h = \eta^3/6$ in (15.6) gives (15.5). Note that, strictly speaking, the three types of solution apply when $RE/m > 0$, $= 0$ and < 0, respectively, since $E = 0$ at a spherical origin in all cases. Solutions of (15.3) with $\Lambda \neq 0$ may, but do not necessarily, require elliptic functions (Stephani et al., 2003).

The LTB model is characterized by three arbitrary functions $E(r)$, $m(r)$ and $t_B(r)$ of the coordinate radius r. $E(r) \geq -1$ has a geometrical role, determining the local 'embedding angle' of spatial slices, and also a dynamical role, determining the local energy per unit mass of the dust particles, and, hence, the type of evolution of R. $m(r)$ is the effective gravitational mass within comoving radius r. $t_B(r)$ is the local time at which $R = 0$, i.e. the local time of the Big Bang – if $t_B \neq \text{const}$, we have a non-simultaneous bang surface. Specification of these three arbitrary functions fully determines the model, and while each of them can be given some type of interpretation for arbitrary choice of the radial coordinate r, there is still a freedom to choose this coordinate, leaving two physically meaningful free functions,

---

[2] Pseudo-spherical and plane versions of these models are discussed in Chapter 19.

e.g. two of $r = r(m)$, $E = E(m)$, and $t_B = t_B(m)$. For more details of the dynamics of these models and its relation to initial data, see Bolejko *et al.* (2010).

A physical limitation on the choices of the arbitrary functions is that if $R' = 0$ we may have a 'shell-crossing singularity', where comoving shells of distinct r collide (Hellaby and Lake, 1985). This equation also holds at an extremum of density if m' and $1 + 2E$ have zeros of the same order.

For particular choices of these initial data, we obtain FLRW dust models:

$$2E(r) = -Kr^2 , \quad R(t,r) = a(t)r , \quad m(r) = \frac{4\pi}{3} \rho(t) R^3 . \tag{15.8}$$

15.1.1 Observer's past light-cone

We now take the area distance R and density ρ as given on the observer's past light-cone as functions of distance, for example being observationally determined as functions of redshift. We wish to express the three arbitrary LTB functions in terms of these observable relations, so characterizing the LTB model that fits the observations, following the argument of Mustapha, Hellaby and Ellis (1999).

Special observational coordinates

As our observations of the sky are essentially based at a single event on cosmological scales, we only need to be able to locate a single light-cone; we do not need a general solution for all null geodesics. On radial null geodesics, $ds^2 = 0 = d\Omega$; so from (15.2) if the past light-cone of the observation event ($t = t_0$, $r = 0$) is given by $t = \hat{t}(r)$, then $\hat{t}$ satisfies

$$d\hat{t} = -\frac{R'(\hat{t}(r),r)}{\sqrt{1+2E}} dr = -\frac{\widehat{R'}}{\sqrt{1+2E}} dr . \tag{15.9}$$

We use a hat to denote a quantity evaluated on the observer's light-cone, $t = \hat{t}(r)$; for example $R(\hat{t}(r),r) := \hat{R}$. Now if we choose r so that, on the past light-cone of (t_0,r),

$$R'(\hat{t}(r),r) = \widehat{R'} = \sqrt{1+2E} , \tag{15.10}$$

then the incoming radial null geodesics are given by

$$\hat{t}(r) = t_0 - r . \tag{15.11}$$

Thus we can choose coordinates so that our particular past light-cone is conveniently given in conformally flat coordinates (note that in general no other past light-cone will be expressed in this way).

With our coordinate choice (15.10), the density (15.4) and the Friedmann equation (15.3) with $\Lambda = 0$ become

$$4\pi \hat{\rho} \hat{R}^2 = \frac{m'}{\sqrt{1+2E}} , \tag{15.12}$$

$$\left[\frac{\partial R(t,r)}{\partial t} \right]_{t=t(r)} = \pm \sqrt{\frac{2m}{\hat{R}} + 2E} . \tag{15.13}$$

Finding the geometry from observables

If we know $\hat{R}(r)$ and $\hat{\rho}(r)$ from observations, where $r = r(z)$, we want to derive the rest from these data. Here we determine the relations in terms of r; in the next section, we turn them into relations in terms of z.

From the total derivative of R on the light-cone, the coordinate conditions give

$$\frac{\mathrm{d}\hat{R}}{\mathrm{d}r} = \widehat{R'} + \widehat{\dot{R}}\,\frac{\mathrm{d}\hat{t}}{\mathrm{d}r}. \tag{15.14}$$

By (15.11) and (15.13) it follows that

$$\frac{\mathrm{d}\hat{R}}{\mathrm{d}r} - \sqrt{1+2E} = -\widehat{\dot{R}} = \mp\sqrt{\frac{2m}{\hat{R}}+2E}. \tag{15.15}$$

We solve for $2E(r)$ by squaring both sides and rearranging:

$$1 + 2E = \left(2\frac{\mathrm{d}\hat{R}}{\mathrm{d}r}\right)^{-2}\left[\left(\frac{\mathrm{d}\hat{R}}{\mathrm{d}r}\right)^2 + 1 - 2\frac{m}{\hat{R}}\right]^2. \tag{15.16}$$

This expression will tell us under what circumstances (or for which regions) the spatial sections are hyperbolic $1 + 2E > 1$, parabolic $1 + 2E = 1$ or elliptic $1 + 2E < 1$, based on data obtained from the light-cone. We now use the expression for the density on the light-cone to find a linear first-order differential equation for $m(r)$. Eliminating $1 + 2E$ between (15.16) and (15.12), we get

$$\frac{\mathrm{d}m}{\mathrm{d}r} + 4\pi G\hat{\rho}\hat{R}\left(\frac{\mathrm{d}\hat{R}}{\mathrm{d}r}\right)^{-1}m = 2\pi G\hat{\rho}\hat{R}^2\left(\frac{\mathrm{d}\hat{R}}{\mathrm{d}r}\right)^{-1}\left[\left(\frac{\mathrm{d}\hat{R}}{\mathrm{d}r}\right)^2 + 1\right]. \tag{15.17}$$

Note the relation,

$$\tau(r) := \hat{t}(r) - t_B(r) = t_0 - r - t_B(r), \tag{15.18}$$

which can be interpreted as proper time from the bang surface to the past light-cone along the particle world lines.

Assuming one knows $\hat{R}(r)$, with m given by (15.17) and E by (15.16), we can solve for $\hat{\eta}$ and then $\tau(r)$ from (15.6)–(15.7). Then $t_B(r)$ follows.

15.1.2 Origin conditions

At the origin of spherical coordinates, $r = 0$, where $R(t,0) = 0$ and $\dot{R}(t,0) = 0$ for all t, we assume that the density is non-zero, that the type of time evolution (hyperbolic, parabolic or elliptic) is not different from its immediate neighbourhood, and that all functions are smooth – i.e. functions of r have zero first derivative there. Thus (15.14) shows that RE/m and $E^{3/2}/m$ must be finite at $r = 0$, and (15.3) shows that $E \to 0$ and hence $m \to 0$ and $E \sim m^{2/3}$ at $r = 0$. Equations (15.14) and (15.15) become

$$\left.\frac{\widehat{\mathrm{d}R}}{\mathrm{d}r}\right|_{r=0} = \widehat{R'}|_{r=0} = \left.\frac{\mathrm{d}\hat{R}}{\mathrm{d}r}\right|_{r=0} = \sqrt{1+2E} = 1, \tag{15.19}$$

and thus $\hat{R} \sim r$ to lowest order near $r = 0$. We can verify that the origin conditions satisfy (15.17) to order r^2 and (15.16) trivially to order r^0.

15.1.3 Redshift–distance formula

We use the fact that in the geometric optics limit, for two light rays emitted on the world line at r_e with time interval $\delta t_e = t^+(r_e) - t^-(r_e)$ and observed on the central world line with time interval $\delta t_o = t^+(0) - t^-(0)$,

$$1 + z = \frac{\delta t_o}{\delta t_e}. \tag{15.20}$$

The incoming radial null geodesics are given by $dt = -R'(t,r)(1 + 2E)^{-1/2}dr$, so for two successive light rays, $-$ and $+$, passing through two nearby comoving world lines r_A and $r_B = r_A + dr$ at times t_A^-, t_B^-, t_A^+ and t_B^+,

$$d(\delta t) = \delta t_B - \delta t_A = dt^+ - dt^- = \frac{\left[-R'(t^+,r) + R'(t^-,r)\right]}{\sqrt{1 + 2E}}\, dr. \tag{15.21}$$

Consequently,

$$d\ln\delta t = -\frac{1}{\sqrt{1 + 2E}}\frac{\partial R'(t,r)}{\partial t}\, dr. \tag{15.22}$$

Then, by integrating along the light-ray and applying this to the log of (15.20), the redshift is given by

$$\ln(1 + z) = \int_0^{r_e} \dot{R}'(t,r)(1 + 2E)^{-1/2}\, dr, \tag{15.23}$$

for the central observer at $r = 0$, receiving signals from an emitter at $r = r_e$.

We need to find the redshift z explicitly in terms of other observables. Differentiating (15.3) with respect to r, and using (15.16), (15.12), (15.15) and (15.23), it follows that

$$\frac{d}{dr}\ln(1 + z) = -\left(\frac{d\hat{R}}{dr}\right)^{-1}\left[\frac{d^2\hat{R}}{dr^2} + 4\pi G\hat{\rho}\hat{R}\right]. \tag{15.24}$$

As mentioned above, observations are in terms of z, rather than the unobservable coordinate r. How this works will be addressed in the next section.

Exercise 15.1.1 Verify that imposing conditions (15.8) on a LTB model does indeed give an FLRW model.

15.2 Observables and source evolution

The particles of the 'dust' are galaxies (or perhaps clusters of galaxies). For simplicity we confine ourselves to one type of cosmic source and only consider bolometric luminosities. We shall assume that the absolute bolometric luminosity L of each source can evolve with time, and that the number density of sources can also evolve. The latter we represent as an evolving mass per source, M, which gives the total local density when multiplied by the source number density.

The two source evolution functions are most naturally expressed as functions of local proper time since the big bang, $L(\tau)$ and $M(\tau)$. However, in a LTB model the time of the bang may vary from point to point, so that the age of objects at redshift z is uncertain both because the bang time is uncertain and because the location of the light-cone is uncertain. The proper time from bang to light-cone will be a function of redshift, $\tau(z)$, and the projections of the evolution functions on the light-cone we write as $\hat{L}$ and $\hat{M}$. Of course, $\tau(z)$ is unknown until we have solved for the LTB model that fits the data. However, for the sake of simplicity, we take $\hat{L}$ and $\hat{M}$ to be given as functions of z, to illustrate how the three quantities – cosmic evolution, cosmic spatial variation, and source evolution – are mixed together in the luminosity and number count observations. A treatment dealing with evolution functions based on τ would involve solving a much more complicated set of differential equations in parallel.

The key point that emerges is that if we have no constraints on source evolution, then we cannot determine the cosmic geometry: on the contrary, what usually happens is that one assumes FLRW geometry and then uses that assumption to determine what the source evolution was. However, if one can find a set of sources where one can believe that source evolution is negligible (i.e. source properties are independent of z), then one can use the observational relations to test spatial homogeneity. SNIa are the sources that have made this possibility a reality.

15.2.1 LTB metric from area distance and number count observations

The area distance gives the true linear extent of the source from the measured angular size. This is by definition the same as the areal radius in the LTB model R, which multiplies the angular displacements to give proper distances tangentially. The projection onto the observer's light-cone gives the observable quantity $\hat{R}$. The luminosity distance is theoretically the same as the area distance (up to redshift factors) (Section 7.4.3), and is measurable provided we know the true absolute luminosity $\hat{L}$ of the source at the time of emission. If the observed apparent luminosity is $\ell(z)$ then

$$\hat{R}^2(z) = \frac{\hat{L}(z)}{\ell(z)} . \tag{15.25}$$

The observational project is to get everything from $\hat{R}(z)$.

Let the observed number density of sources in redshift space be $n(z)$ per steradian per unit redshift interval, so that the number observed in a given redshift interval and solid angle is $n\,d\Omega\,dz$ and over the whole sky this is $4\pi n\,dz$. Thus the total rest mass between z and $z+dz$ is $4\pi\,\hat{M}n\,dz$, where $\hat{M}(z) = M(\tau(z))$ is the mass per source – i.e. the true density over the source number density. This primarily represents the evolution in the number density of sources. Given a local proper density $\rho = \rho(t,r)$, and its value on the light-cone $\hat{\rho}$, the total rest mass between r and $r+dr$ is

$$\hat{\rho}\,\widehat{d^3V} = \hat{\rho}4\pi\,\hat{R}^2\,\widehat{R'}(1+2E)^{-1/2}dr, \tag{15.26}$$

where $\widehat{d^3V}$ is the proper volume on a constant time slice, evaluated on the light-cone. Hence by (15.26) and (15.10),

$$\hat{R}^2\hat{\rho}\,dr = \hat{M}n\,dz . \tag{15.27}$$

Thus we may substitute for $\hat{R}$ and $\hat{\rho}$ from (15.25) and (15.27).

We transform (15.24) to be in terms of redshift z, so finding

$$\frac{\mathrm{d}\hat{R}}{\mathrm{d}z}\frac{\mathrm{d}^2 z}{\mathrm{d}r^2} + (1+z)^{-1}\left[(1+z)^2\frac{\mathrm{d}^2\hat{R}}{\mathrm{d}z^2} + \frac{\mathrm{d}\hat{R}}{\mathrm{d}z}\right]\left(\frac{\mathrm{d}z}{\mathrm{d}r}\right)^2 = -4\pi G\hat{\rho}\hat{R}. \qquad (15.28)$$

Integrating with respect to r and using (15.27) plus the origin conditions $[(\mathrm{d}z/\mathrm{d}r)(\mathrm{d}\hat{R}/\mathrm{d}z)]_0 = [(\mathrm{d}\hat{R}/\mathrm{d}r)]_0 = 1$, gives

$$\frac{\mathrm{d}z}{\mathrm{d}r} = \left[\frac{\mathrm{d}\hat{R}}{\mathrm{d}z}(1+z)\right]^{-1}\left[1 - 4\pi \int_0^z \frac{\hat{M}(\bar{z})n(\bar{z})}{\hat{R}(\bar{z})}(1+\bar{z})\,\mathrm{d}\bar{z}\right], \qquad (15.29)$$

which leads to

$$r(z) = \int_0^z \mathrm{d}\bar{z}\left[\frac{\mathrm{d}\hat{R}}{\mathrm{d}\bar{z}}(1+\bar{z})\right]\left[1 - 4\pi \int_0^{\bar{z}} \frac{\hat{M}(\bar{z})n(\bar{z})}{\hat{R}(\bar{z})}(1+\bar{z})\,\mathrm{d}\bar{z}\right]^{-1}. \qquad (15.30)$$

There are subtleties in the conditions for existence of $r(z)$ and $z(r)$ solutions (15.30) (Mustapha, Hellaby and Ellis, 1999). These exist if and only if: (i) $\hat{M}, n, \hat{R}, (1+z) \geq 0$; (ii) near $z = 0$, $\hat{M}n/\hat{R} \sim z^\sigma$ with $\sigma > -1$; (iii) $\mathrm{d}\hat{R}/\mathrm{d}z$ is finite everywhere; (iv) $z(r)$ is monotonic; (v) a condition on $\mathrm{d}\hat{R}/\mathrm{d}z$ near any maximum in $\hat{R}$. Mustapha *et al.* (1998) show that large enough inhomogeneities can create maxima and minima in $z(r)$ and so make $r(z)$ multi-valued, especially near $\mathrm{d}\hat{R}/\mathrm{d}z = 0$, in which case neither (iii) nor (iv) would be satisfied. However, a multi-valued $r(z)$ manifests itself in a $\hat{R}(z)$ graph that loops. In practice, we do not expect to get a looping $\hat{R}(z)$ from the observational data. The values of ℓ and n at each z are averages over all measured values, and so are single-valued by construction; if $r(z)$ exists, then inverting it should not be a problem.

15.3 Can we fit area distance and number count observations?

The application of these models to the real universe starts by assuming isotropy about the Earth (once our proper motion has been accounted for),[3] and also that the post decoupling universe is well described by zero pressure matter. Assuming that the above existence conditions for $r(z), z(r)$ hold, and that the further existence conditions for $m(r)$ in Exercise 15.3.1 hold, we have (Mustapha, Hellaby and Ellis, 1999):

Theorem 15.1 LTB observations
For any given isotropic observations $\ell(z)$ and $n(z)$, with any given source evolution $\hat{L}(z)$ and $\hat{M}(z)$, LTB metric functions can be found (with $\Lambda = 0$) to fit the observations.

To obtain the LTB mass, energy and bangtime functions (m, E, t_B) from observational data and source evolution we proceed as follows.

- Average the discrete observed data for $\ell(z,\theta,\phi)$, $n(z,\theta,\phi)$ over the sky to obtain $\ell(z)$, $n(z)$, and fit them to some smooth analytic functions (e.g. polynomials). (First correct

---

[3] Later we consider models where we are near the centre.

the data for known distortions and selection effects due to proper motions, absorption, shot noise, image distortion, etc.)

- Choose evolution functions $\hat{L}(z)$, $\hat{M}(z)$.
- Determine $\hat{R}(z)$ from $\hat{L}(z)$ and $\ell(z)$ using (15.25).
- Solve (15.30) for $r(z)$ and hence $z(r)$.
- Solve (15.17) and (15.27) for $m(r)$.
- Determine $E(r)$ from (15.16).
- Solve for $\hat{\eta}$ from (15.6) and (15.7).
- Solve for $\tau(r)$ from (15.6) and (15.7) – then find $L(\tau)$ and $M(\tau)$.
- Determine $t_B(r)$ from (15.18).

In practice, these equations would be solved numerically, and in parallel rather than sequentially; nevertheless the above would determine the numerical procedure within each integration step.

By determining the three arbitrary functions, we have specified the LTB model that fits the given observations and evolution functions. This result simply shows that we can construct an inhomogeneous spherically symmetric exact solution of the field equations that will fit any given source observations combined with any chosen source evolution functions.

We assert, without proof, that if the given observations and source evolution functions are reasonable, then the LTB arbitrary functions will generate a reasonable LTB model. Our definition of 'reasonable' is intentionally rather vague. By reasonable observations we obviously include the actual data, suitably processed to account for selection effects. We also include 'realistic' hypothetical alternatives, but not functions that are wildly different from reality. Reasonable evolution functions are hard to define since the actual ones are not well known, especially at larger z. By a reasonable LTB model, we mainly mean that the density and expansion rate will be within realistic ranges. A less crucial criterion is that there will be no shell crossings too close to the past light-cone. Evolving the model a long time away from the light-cone, either forwards or backwards, may introduce shell crossings because the data are imprecise. In general we do not expect shell crossings on the large scale – i.e. two or more different large-scale flows of galaxies in the same region – nevertheless it is conceivable and in that case the LTB description is inapplicable.

In Section 8.5.5 we discussed how to test for homogeneity via observations on LTB past light-cones. Here we revisit the result.

Theorem 15.2 Conditions for homogeneity

An LTB universe (with $\Lambda = 0$) is FLRW if and only if the area distance and number count relations as functions of z take the FLRW form:

$$\hat{R}(z) = \frac{q_0 z + (1 - q_0)\left(1 - \sqrt{2q_0 z + 1}\right)}{H_0 q_0^2 (1 + z)^2}, \tag{15.31}$$

$$n(z) = \frac{3}{4\pi \hat{M}(z)} \frac{\left[q_0 z + (1 - q_0)(1 - \sqrt{2q_0 z + 1})\right]^2}{H_0 q_0^3 (1 + z)^3 \sqrt{2q_0 z + 1}}. \tag{15.32}$$

This follows by showing from (15.31) and (15.32) that

$$m(z) = H_0^2 q_0 \hat{R}^3(z)(1+z)^3, \quad 2E(z) = (1 - 2q_0) H_0^2 \hat{R}^2(z)(1+z)^2. \tag{15.33}$$

It then follows that $m \propto (2E)^{3/2}$, and consequently the universe has a simultaneous bangtime.

This proof is very model dependent: it relies on a specific theory of gravity (GR) and a specific matter model (dust). What we would really like is a test of spatial homogeneity that is not so model dependent. We shall see below that such tests are indeed possible.

Exercise 15.3.1 Show that necessary and sufficient conditions for existence of solutions $m(r)$ to (15.17) are:
(i) $\hat{M}, n, \hat{R}, \mathrm{d}z/\mathrm{d}r \geq 0$, ensuring $\hat{\rho} \geq 0$;
(ii) $\hat{R}(r) = \hat{R}(z(r))$ has a power-law maximum of the form $\hat{R} \sim (r - r_{\max})^\alpha$ with $1 < \alpha \leq 2$.
(Mustapha, Hellaby and Ellis, 1999).

Exercise 15.3.2 Generalize the LTB observations result to $\Lambda \neq 0$.

Exercise 15.3.3 Fill in the details of the FLRW observations case.

Problem 15.1 Generalize the FLRW observations result to $\Lambda \neq 0$. (While the result will still hold, proving it is complicated by the fact that simple analytic forms for the observational relations are not available in general.)

15.4 Testing background LTB with SNIa and CMB distances

In practice, number count observations are not currently feasible for testing cosmological models, and we need to look for other observations that will constrain the LTB background model. The first acoustic peak in the CMB temperature power spectrum gives the angular extent ℓ_A of the sound horizon at decoupling, and provides a high-redshift distance measure $D_A(z_{\mathrm{dec}})$. This can be used as a standard ruler in an FLRW model (see Section 13.2.4), and it should remain a good approximation in LTB models that are close to FLRW models.

It is useful to recast the background metric in RW-like form:

$$\mathrm{d}s^2 = -\mathrm{d}t^2 + \frac{a_\parallel^2(t,r)}{1 - \kappa(r)r^2}\mathrm{d}r^2 + a_\perp^2(t,r)r^2\mathrm{d}\Omega^2, \tag{15.34}$$

where the radial ($a_\parallel$) and angular ($a_\perp$) scale factors are related by $a_\parallel = (a_\perp r)'$. The curvature $\kappa = \kappa(r)$ is a free function, and the two scale factors define two Hubble rates: $H_\perp(t,r) = (\ln a_\perp)^{\cdot}$ and $H_\parallel(t,r) = (\ln a_\parallel)^{\cdot}$. The Friedmann equation is

$$H_\perp^2(t,r) = H_{\perp 0}^2(r)\Big[\Omega_{m0}(r)a_\perp^{-3}(t,r) + \Omega_{\kappa 0}(r)a_\perp^{-2}(t,r)\Big], \tag{15.35}$$

$$\Omega_{m0}(r) + \Omega_{\kappa 0}(r) = 1. \tag{15.36}$$

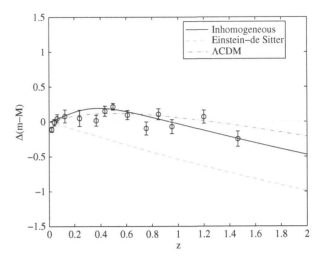

Fig. 15.1   LTB and ΛCDM fits to SNIa data. (Reprinted with permission from Alnes, Amarzguioui and Grøn (2006). Copyright by the American Physical Society.)

Here $\Omega_{m0}(r)$ is a free function, specifying the matter density parameter today. In general, $H_{\perp 0}(r)$ is also free, but if we enforce a uniform bangtime, it is fixed in terms of $\Omega_{m0}(r)$. This single free function is sufficient to fit the SNIa data (see Figure 15.1).

The free function $\Omega_{m0}(r)$ may be parameterized in many different ways; this expands the freedom to fit data – but it also undermines the statistical goodness of fit. It is preferable, but difficult, to find physically motivated choices. One possibility is the Gaussian profile,

$$\Omega_{m0}(r) = \Omega_{\text{out}} + (\Omega_{\text{in}} - \Omega_{\text{out}}) \exp -(r^2/r_0^2), \tag{15.37}$$

where Ω_{in} is the density parameter at the centre, Ω_{out} is the asymptotic density parameter and r_0 determines the void size. A void with much sharper transition from the local to the asymptotic value is given by

$$\Omega_{m0}(r) = \Omega_{\text{out}} + (\Omega_{\text{in}} - \Omega_{\text{out}}) \left[1 + e^{-r_0/\Delta r} \right] \left[1 + e^{(r-r_0)/\Delta r} \right]^{-1}, \tag{15.38}$$

where the transition occurs at r_0, with width Δr.

With one free function we can design LTB models that reproduce any distance modulus. Since the shear vanishes at the centre, the Raychaudhuri equation (6.4) with $\Lambda = 0$, shows that $(\dot{\Theta} + \Theta^2/3)_0 = -4\pi G \rho_0 < 0$ – i.e. $q_0 > 0$ on the central world line. If we choose $\Omega_{m0}(r)$ to reproduce exactly a ΛCDM distance modulus, with $q_0 < 0$, then the LTB model is forced to have a spiked radial density profile which is non-differentiable at the origin. However, this non-differentiability is irrelevant for cosmological modelling: a void model should be constrained directly by *data,* and not matched to a best-fit ΛCDM model. If this is done, a smooth void is perfectly compatible with present SNIa data.

Many specific examples have been given of LTB models that fit the current SNIa data; see e.g. Alnes, Amarzguioui and Grøn (2006), Biswas, Mansouri and Notari (2007), Ishak *et al.* (2008), Yoo, Kai and Nakao (2008), Alexander *et al.* (2009). A typical observationally viable model is one in which we live in a large underdense void, roughly centrally (within ~ 10 Mpc

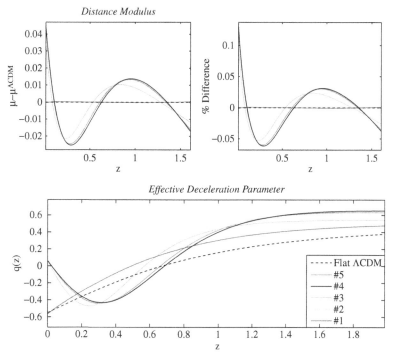

Fig. 15.2 Distance modulus for four void models compared to ΛCDM (top), and the effective deceleration parameter (bottom), after fitting each void with $\Omega_{out} = 1$ to ΛCDM with $\Omega_m^{FLRW} = 0.3$. (From February *et al.* (2009). © RAS.) A colour version of this figure is available online.

of the centre): a location too far off-centre induces too large a dipole in the CMB. If the distance modulus is close to that of ΛCDM, then the comoving size of the void is $O(\text{Gpc})$. February *et al.* (2010) present a parameterization of a void which can reproduce concordance model distances to arbitrary accuracy, but with a smooth density profile everywhere (see Figure 15.2). SNIa data place limits on the size of a void which is roughly independent of its shape. However, the sharpness of the profile at the origin cannot be well constrained due to SNIa data being dominated by peculiar velocities in the local universe.

The background LTB model needs to fit more than the SNIa distance data. All observations that probe the background geometry need to be fitted. If the universe outside the void is approximately a homogeneous Einstein–de Sitter model, the position of the first CMB peak can also be made to match the WMAP data (February *et al.*, 2010). For FLRW models, the BAO scale in the matter power spectrum is a third independent standard ruler that provides powerful constraints on the background model (Section 13.2.2). Since the background is homogeneous, the evolution of the BAO feature is a probe only of the background geometry and not of the density perturbations. *If* the BAO feature evolves in LTB as it does in FLRW, then BAO data could rule out LTB void models together with SNIa and CMB distance data (García-Bellido and Haugbølle, 2008, Zibin, Moss and Scott, 2008). However, the evolution of the BAO feature on an LTB background, especially at low redshifts, could well deviate

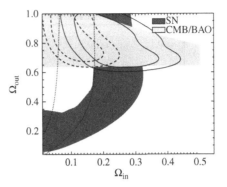

LTB Gaussian model gives a reasonable fit to SNIa and combined CMB/ BAO distance data. (From Sollerman *et al.* (2009). Reproduced by permission of the AAS.)

from the evolution on a homogeneous background. This remains an open problem, which requires further analysis of the perturbations.

Sollerman *et al.* (2009) use SNIa data (SDSS-II Supernova Survey together with other data sets), and then a combined CMB distance and BAO distance measure – assuming that the BAO feature evolves as in FLRW. A model with Gaussian density profile (15.37) is compatible with both constraints, as shown in Figure 15.3.

15.5 Perturbations of LTB

Just as in the case of the FLRW models (see Chapter 13), one can test the background LTB model – Sections 15.3 and 15.4 above – and test the perturbed LTB model, where small-scale inhomogeneities develop dynamically in the large-scale background. The aim is to confront the LTB models with the wealth of data in the CMB anistropies, galaxy distribution and weak lensing.

The equations for perturbations of LTB models have been presented in full generality by Clarkson, Clifton and February (2009). At the time of writing, the application of the general equations to the structure formation problem remains incomplete, given the much greater complexity of the problem than in the FLRW case. We deal with the topic in outline only, and refer the reader to the literature for the details. A good summary is given by Clarkson and Maartens (2010).

15.5.1 1+1+2 formalism

Two different approaches have been used to study perturbations of LTB models. The first is a covariant $1 + 1 + 2$ formalism, building on the work by Clarkson and Barrett (2003) for the case of Schwarzschild black holes. Here the $1 + 3$ covariant approach (Section 10.3) is extended to a $1 + 1 + 2$ covariant formalism by introducing a radial unit vector in addition to the timelike unit vector u^a, and decomposing all covariant quantities with respect to both vectors. The odd and even parity perturbations may be unified by the discovery of a covariant and gauge-invariant transverse-traceless tensor describing gravitational waves, which satisfies a covariant wave equation equivalent to the Regge–Wheeler equation for both even and odd parity perturbations. Clarkson (2007) uses this to obtain a covariant decomposition of the EFE which is particularly suitable for perturbations of spherically symmetric – and general locally rotationally symmetric – spacetimes.

Application of this theory to LTB models is given in Zibin (2008), where the evolution of perturbations is determined by a set of linear transfer functions. If decaying modes are ignored (to be consistent with the standard inflationary paradigm), and the 'silent' approximation is used, where the magnetic part of the Weyl tensor is neglected, then the standard techniques of perturbation theory on homogeneous backgrounds, such as harmonic expansion, can be applied, and results closely paralleling those of familiar cosmological perturbation theory can be obtained. The same approach is used in Dunsby *et al.* (2010). The 'silent' approximation may be a good approximation, but there is no strong motivation for this. Essentially the mode coupling inherent for inhomogeneous backgrounds is switched off, and in general this could remove key physical features. Determination of the observational implications of these equations has yet to be done, but the foundations have been laid.

15.5.2 2+2 covariant formalism

The second approach is based on a $2 + 2$ covariant approach developed for stellar and black hole physics by Gerlach and Sengupta and by Gundlach and Martin-Garcia. It has been adapted and applied to the LTB cosmological case by Clarkson, Clifton and February (2009), who present this theory in a fully general and gauge-invariant form.

In FLRW, linear perturbations split into scalar, vector and tensor modes that decouple from each other, and so evolve independently. Such a split cannot usefully be performed in the same way in a spherically symmetric spacetime, as the background is no longer spatially homogeneous, and modes written in this way couple together. Instead, there exists a decoupling of the perturbations into two independent sectors, called 'polar' (or even) and 'axial' (or odd), which are analogous, but not equivalent, to scalar and vector modes in FLRW. These are based on how the perturbations transform on the sphere: roughly speaking, polar modes are 'curl'-free on S^2 while axial modes are divergence-free. Further decomposition may be made into spherical harmonics, so that all perturbative variables are for a given spherical harmonic index ℓ, and modes decouple for each ℓ – analogously to k-modes evolving independently on an FLRW background. There is a natural gauge – the Regge–Wheeler gauge – in which all perturbation variables are gauge-invariant (rather like the longitudinal gauge in FLRW perturbation theory). Unfortunately, the interpretation of the gauge-invariant variables is not straightforward in a cosmological setting.

Most of the interesting physics happens in the polar sector and the general form of polar perturbations in the Regge–Wheeler gauge is

$$ds^2 = -[1 + (2\eta - \chi - \varphi)Y]dt^2 - \frac{2a_\parallel \varsigma Y}{\sqrt{1 - \kappa r^2}}dt\,dr$$

$$+ [1 + (\chi + \varphi)Y]\frac{a_\parallel^2 dr^2}{(1 - \kappa r^2)} + a_\perp^2 r^2(1 + \varphi Y)d\Omega^2, \tag{15.39}$$

where $\eta(t,r)$, $\chi(t,r)$, $\varphi(t,r)$ and $\varsigma(t,r)$ are gauge-invariant variables. The spherical harmonic Y has an implicit sum over ℓ, m, e.g.

$$\varphi Y := \sum_{\ell=0}^{\infty} \sum_{m=-\ell}^{\ell} \varphi_{\ell m}(t,r) Y_{\ell m}(\theta, \phi). \tag{15.40}$$

φ contains scalar, vector and tensor modes; ς contains vector and tensor modes, while χ is a tensor mode. For $\ell \geq 2$, we have $\eta = 0$. The general form of polar matter perturbations in this gauge is given by

$$u_\mu = \left[1 + \tfrac{1}{2}(2\eta - \chi - \varphi)Y, \frac{\sqrt{1-\kappa r^2}}{a_\parallel} WY + \varsigma Y, VY_{:\theta}, VY_{:\phi} \right], \tag{15.41}$$

$$\rho = \bar{\rho}(1 + \Delta Y), \tag{15.42}$$

where V, W and Δ are gauge-invariant velocity and density perturbations and a colon denotes covariant differentiation on the 2-sphere. In the Regge–Wheeler gauge, the gauge-invariant metric perturbations are master variables and obey a coupled system of PDEs which are decoupled from the matter perturbations. The matter perturbation variables are then determined by the solution to this system. (See Clarkson, Clifton and February (2009) for details.)

Scalar, vector and tensor perturbations interact: at a fixed time, density perturbations in each $r = \mathrm{const}$ sphere grow at different rates and thus can generate vector and tensor modes. We may expect structure to grow more slowly relative to FLRW, since there could be dissipation of potential energy via gravitational radiation. The natural gauge-invariant variables in LTB cosmology do not correspond straightforwardly to the usual FLRW variables, in the limit of spatial homogeneity. Clarkson, Clifton and February (2009) construct new variables that reduce to pure scalar, vector and tensor modes in this limit. Application of this approach to the LTB models, and determination of the observational implications, is ongoing.

15.5.3 CMB in void models

For the CMB anisotropies, except on large angles, there are good arguments that the FLRW results can be used and adapted in a simple way (Zibin, Moss and Scott, 2008, Clifton, Ferreira and Zuntz, 2009, Vonlanthen, Räsänen and Durrer, 2011, Regis and Clarkson, 2010, Biswas, Notari and Valkenburg, 2010, Clarkson and Regis, 2011, Moss, Zibin and Scott, 2011). The physics of decoupling and line-of-sight effects contribute differently to the CMB, and have different dependency on the cosmological model. In general inhomogeneous models, both pre- and post-decoupling effects will play a role, but Hubble-scale void models allow an important simplification for calculating the moderate to high ℓ part of the CMB spectrum. This ℓ range is where the statistical power of the data to constrain models is strongest – on large scales, constraining power is undermined by cosmic variance.

The comoving scale of voids that closely mimic the standard distance modulus is $O(\mathrm{Gpc})$, while the sound horizon, which sets the largest scale seen in the pre-decoupling part of the power spectrum, is about 150 Mpc. Thus, in any causally connected patch prior to decoupling, the density gradient is very small. Furthermore, the comoving radius of decoupling is > 10 Gpc, on which scale the gradient of the void profile is small by assumption. This suggests that before decoupling on small scales we can model the universe in disconnected FLRW shells at different radii, with the shell of interest located at the distance where we see the CMB. This may be calculated using standard FLRW codes, but with the line-of-sight parts corrected for.

For line-of-sight effects, we need to use the full void model. The simple effect is determined by the background dynamics, which modifies the area distance to the CMB. This is the important effect for the small-scale CMB. The more complicated effect is on the largest scales through the ISW effect (see Tomita (2010) for the general formulae in LTB). This requires the solution of the perturbation equations presented above. Indeed the large-angle Sachs–Wolfe effect at decoupling may also be modified.

We note that at least some of the observed CMB dipole can arise because we are a bit off-centre in a large void, and then we would need to re-evaluate the Great Attractor analysis and the observed alignment of the dipole and quadrupole.

CMB anisotropies – excluding the largest scales – can be fitted in void models in different ways. The CMB can be very restrictive on void models (Zibin, Moss and Scott, 2008, Clifton, Ferreira and Zuntz, 2009), although with a varying bangtime the data for H_0, SNIa and CMB can be simultaneously accommodated (Clifton, Ferreira and Zuntz, 2009). Including inhomogeneous radiation in the background, the CMB can be accommodated along with other local observations with a homogeneous bangtime, but with asymptotic curvature at the CMB radius (Regis and Clarkson, 2010). It is an open question exactly what constrains the small-scale CMB places on a generic void solution. It is remarkable that large void models can reproduce the ΛCDM CMB power spectrum so closely. It is often taken for granted that the CMB tells us that the universe is close to flat – these examples show that curvature can in fact be very large, but inhomogeneous.

15.5.4 Large-scale structure and other observations

If LTB models are a viable alternative to ΛCDM, they need to fit not only the SNIa and small to moderate scale CMB anisotropy data, but also the data on the large-scale CMB, the matter distribution (including BAO) and lensing. Additional constraints from ages and primordial element abundances must also be satisfied. Many detailed LTB models with large central voids have been studied, and the resulting observational relations compared with the data, at present assuming that LTB structural growth relations will be similar to those in FLRW models. This may be an acceptable approximation, but it remains an open problem at the time of writing whether the LTB perturbation equations confirm the approximation.

Essentially, one may be able to fit observations that probe different redshifts (e.g. CMB, SNIa, BAO), because they correspond to different values of the radial variable r, and there may be sufficient freedom in the LTB models to fit the data at all the relevant different distances (Alexander *et al.*, 2009, Clifton, Ferreira and Zuntz, 2009). With different observations at different redshifts, corresponding to different distances from the origin, in principle we determine what the universe geometry must be at each radial distance to give the observed data. The real tests come from data that mix observations at different redshifts, e.g. ISW, Sunyaev–Zel'dovich effect, BAO. Then one is seeing how the whole integrates together: are the observations implied by different geometry at different distances all consistent with each other?

Note that the process of fitting the model to the data here is not within an approach of fine-tuning of the model: rather, the approach is to determine what the actual geometry of the universe might be. Whether the resulting model determined in this way is probable or not is a separate and independent issue. Here the issue is – are we using the best ways possible for determining the actual geometry of what is there?

Approximations of the 'silent' type have been used to estimate the BAO and other features of the density perturbations on an LTB background (Zibin (2008), Dunsby *et al.* (2010), Biswas, Notari and Valkenburg (2010), Moss, Zibin and Scott (2011)).

Observations that are more direct, i.e. less dependent on models of structure formation, can lead to interesting results. For example, Regis and Clarkson (2010) consider the lithium problem in the standard FLRW model: a substantial mismatch between the theoretical prediction for ^{7}Li from BBN and the value that we observe today. They find that both the apparent acceleration and the lithium problem can be accounted for as different aspects of cosmic inhomogeneity, without causing problems for other cosmological phenomena such as the CMB; see Figure 15.4.

In summary, the current state of the problem is as follows.

- *CMB*. The details of an inhomogeneous radiation era still have to be investigated. To calculate the large-scale CMB, there are several additional effects which must be taken into account. The most important is the Sachs–Wolfe effect, which requires knowing the perturbation spectrum at the time of decoupling on the largest angular scales. While it might not deviate from FLRW for a central observer, this needs further investigation. In addition, the ISW effect will contribute to the line-of-sight part of the CMB calculation.
- *Structure formation*. The complexity of the perturbation equations is a major stumbling block. Attempts have been made to solve a limited subcase of the equations, but it is

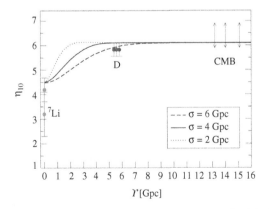

Fig. 15.4   Lithium data are fitted at the centre, deuterium data from more distant objects are fitted further out, and CMB data are fitted at even greater distances. Observations at different distances determine the geometry of the large-scale inhomogeneity, and resolve the lithium abundance discrepancy. (From Regis and Clarkson (2010).)

not clear how close those approximations are to the full solution. This is required for a reliable estimate of corrections to the power spectrum and to the BAO feature.

- At the time of writing, it appears that simple LTB models are ruled out as an explanation of dark energy: a varying bangtime can account for some observables individually, but is not enough to simultaneously explain SNIa observations, the small-angle CMB, the local Hubble rate and the kinetic Sunyaev–Zel'dovich effect (Bull, Clifton and Ferreira, 2011).

15.5.5 Dynamical history

Assuming that we can fit the observations by a large void model, can we find a dynamical history – inflation followed by a HBB era – that can lead to such a model? The void model has the same basic dynamics as the standard model, i.e. evolution along individual world lines governed by the Friedmann equation, but with distance-dependent parameters. Can inflation lead to it? This depends on the initial data, the amount of inflation, and the details of the inflaton fields. With multiple fields, a suitable inflationary potential, and the right choice of initial data, there is sufficient flexibility that this should be possible. Specific models have been proposed where inflation will indeed lead to large voids. Linde, Linde and Mezhlumian (1995) achieve this via a particular measure. Afshordi, Slosar and Wang (2011) consider two-field inflation with a suitable potential ('multi-stream inflation') that provides bifurcating paths from an initial to a final field state, with quantum tunneling possible between them; the result is different numbers of inflationary e-foldings at different places, leading to overdense or underdense spherical bubble formation.

15.6 Observational tests of spatial homogeneity

Assuming that LTB can fit all of the observations, how can we distinguish between it and FLRW with dark energy?

Testing for homogeneity is not entirely straightforward even if the model is actually homogeneous. Ribeiro (1992b), in the course of an attempt to make simple models of fractal cosmologies using LTB models, reminds us of the need to compare data with relativistic models not Newtonian approximations. Taking the Einstein–de Sitter model, and integrating down the geodesics, he plotted the number counts against luminosity distances. At small distances, where a simple interpretation would say the result looks like a uniform density, the calculation is irrelevant because the distances are inside the region where we know things are lumpy, while at greater redshifts the model universe ceases to have a simple power-law relation of density and distance. Thus even Einstein–de Sitter may not seem homogeneous if an inappropriate comparison is used. Kurki-Suonio and Liang (1992) emphasized the ambiguity in reconstruction of density from light-cone observations in LTB models, because one is trying to use one function, $\rho(z)$, to find two metric functions. If the bangtime is inhomogeneous, an overdensity in redshift space may correspond to an underdensity in real space. They conclude that using the FLRW relation between redshift and comoving distance when there are inhomogeneities is 'fundamentally self-inconsistent'.

One must therefore first ask 'do (approximately) homogeneous models look homogeneous?' Of course, they will if the data are handled with appropriate relativistic corrections, and with suitable statistical methods, but to achieve such comparisons in general requires the integration of the null geodesic equations in each cosmological model considered, and this may be difficult. We then need to consider how to distinguish homogeneous and inhomogeneous universes. Ideally, we need a model-independent test of the basic assumption of most present-day cosmology: is an RW geometry the correct metric for the observed universe on large scales? Is the Copernican Principle correct? Four kinds of tests have been proposed, as we now discuss.

15.6.1 CMB-based tests

Some tests use scattered CMB photons to check spatial homogeneity (Goodman, 1995, Caldwell and Stebbins, 2008). If the CMB radiation is anisotropic around distant observers (as will be true in inhomogeneous models), Sunyaev–Zel'dovich scattered photons have a distorted spectrum that reflects the spatial inhomogeneity. However, this test is somewhat model dependent. It also has to take into account other possible causes of spectral distortion.

15.6.2 Direct observational tests: behaviour near origin

The geometry of the light-cone vertex must not have a cusp, as this implies a singularity there. There are general regularity conditions at the centre, given in Section 8.2.1, that must hold also in the special case of spherically symmetric inhomogeneous models (Vanderveld, Flanagan and Wasserman, 2006). If the distance modulus in a LTB void model without Λ behaves for small z as in standard ΛCDM models, it would imply a singularity (Clifton, Ferreira and Land, 2008). Observational tests of this requirement may be possible through sufficient intermediate redshift SNIa.

15.6.3 Direct observational tests: constancy of curvature

There are two geometric effects on distance measurements: the curvature bends null geodesics and the expansion changes radial distances. These are coupled in RW models, as expressed in the relation (13.4):

$$D_L(z) = \frac{(1+z)}{H_0\sqrt{-\Omega_{K0}}} \sin\left(\sqrt{-\Omega_{K0}} \int_0^z \frac{dz'}{H(z')/H_0}\right), \qquad (15.43)$$

but they are decoupled in LTB geometries.

In RW geometries, we can combine the Hubble rate and distance data to find the curvature today, where $D := H_0 D_L/(1+z)$:

$$\Omega_{K0} = \frac{[H(z)D'(z)/H_0]^2 - 1}{D(z)^2}. \qquad (15.44)$$

This relation is independent of all other cosmological parameters, including dark energy – and it is also independent of the theory of gravity. It can be used at a single redshift to determine Ω_{K0}. The exciting result of Clarkson, Bassett and Lu (2008) is that since Ω_{K0} is independent of z, we can differentiate to get the consistency relation,

$$C(z) := H_0^2 + H^2(z)\left[D(z)D''(z) - D'(z)^2\right] + H(z)H'(z)D(z)D'(z) = 0. \qquad (15.45)$$

This is true only for a RW geometry: it is independent of curvature, dark energy, matter content, and theory of gravity. Thus it gives the desired consistency test for spatial homogeneity. In realistic models we should expect $C(z) \sim 10^{-5}$, reflecting perturbations about the FLRW model related to structure formation. Errors may be estimated from a series expansion,

$$C(z) = \left[q_0^{(D_L)} - q_0^{(H)}\right]z + O(z^2), \qquad (15.46)$$

where $q_0^{(D_L)}$ is measured from luminosity distance data and $q_0^{(H)}$ from the Hubble parameter. It is simplest to measure $H(z)$ from BAO data. Carrying out this test is only as difficult as carrying out dark energy measurements of $w(z)$ from Hubble data, which require $H'(z)$ from distance measurements or the second derivative $D_L''(z)$.

This is the simplest direct test of spatial homogeneity, and its implementation should be regarded as a high priority. If it confirms spatial homogeneity, that reinforces the evidence for the standard view in a satisfying way. But if it does not, it has the possibility of undermining the entire project of searching for a physical form of dark energy.

15.6.4 Direct observational tests: redshift drift

Uzan, Clarkson and Ellis (2008) use the time drift of the cosmological redshift as a test of spatial homogeneity. In LTB models, using observational coordinates (Section 8.2.1),

$$\dot{z}(w_0, y) = H_0[1 + z(y)] - H(w_0, y) - 3^{-1/2}\sigma(w_0, y), \qquad (15.47)$$

where σ is the shear scalar [(8.25)] and y is defined on the past light-cone by $\partial_w \ln(B/A^2)|_{w=w_0} = 0$. In RW models,

$$\dot{z} = H_0(1 + z) - H(z), \qquad (15.48)$$

and this leads to a second consistency test. A non-vanishing $\dot{z} - H_0(1 + z) + H(z)$ at any redshift would signal a violation of the Copernican Principle and so determine if our universe is radially inhomogeneous. When combined with distance data, this extra observable allows one to fully reconstruct the geometry of a LTB void, purely from background observations. This is a difficult but practicable test (Dunsby *et al.*, 2010, Clarkson and Maartens, 2010); Figure 15.5 shows the kinds of error bounds that are in principle attainable.

15.6.5 Indirect observational and consistency tests

We discussed consistency tests for FLRW models in Section 13.4. There are a number of these tests that can be regarded as indirect tests of spatial homogeneity, for they probe the homogeneity of these models. We revisit these tests here (see Section 13.4 for references).

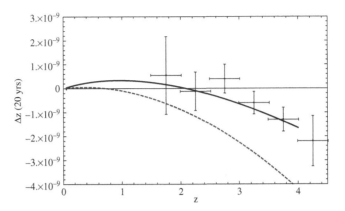

Fig. 15.5 Time drift of redshift for ΛCDM and LTB (mock data). (From Dunsby *et al.* (2010).)

1. *Age consistency at all z.* A crucial observational test for cosmology is that the age of the universe must be greater than the ages of stars. This tension is one area where the standard models are vulnerable to being shown to be inconsistent. Furthermore the age limits must be satisfied at all redshifts. At present this age comparison is acceptable for local objects, assisted by data that Λ is positive. Ages of stellar objects are very difficult to determine, and depend critically on nonlinear astrophysical modelling. With improvements in modelling, continued vigilance is needed on this front. A serious challenge to the standard model from ages could indicate support for a LTB void model with appropriate density parameter and curvature. On the other hand, age tests could also rule out LTB models.

2. *CMB temperature as a function of distance.* The CMB temperature T_γ varies with redshift as $T_\gamma = 2.75(1 + z)$ K. This is a consistency test for both FLRW and LTB: it does not relate to the Copernican issue. If it were violated, it would be a major problem for both FLRW and LTB models.

3. *CMB and number count dipole alignment.* Check that there is a 2% number count dipole parallel to the CMB dipole for all cosmological sources, due to our motion relative to the cosmological rest frame. If this is not true we cannot live in a RW geometry with the CMB coming from the surface of last scattering. In LTB models, the dipoles need not be correlated, since we can be near, but not exactly at, the centre.

4. *Element abundances at high redshift.* Confirm that helium abundances are consistent with a primordial value of 25% at large distances (high redshifts) in all directions. This tests spatial homogeneity at the very early time of nucleosynthesis. It can be applied to the other elements created in primordial nucleosynthesis. Thus the lithium observational fit of LTB models proposed in Regis and Clarkson (2010) (see above) is precisely this kind of test of inhomogeneity.

5. *Physical plausibility.* If the standard analysis of the SNIa data to determine the dark energy equation of state parameter shows there is any redshift range where $w := p/\rho < -1$, this could be a strong indication that a geometric explanation such as an LTB void model is preferable to the Copernican (Robertson–Walker) assumption, for otherwise the matter model indicated by these observations is non-physical (it has a

negative kinetic energy) (see Section 14.2). Although there are ways to avoid negative kinetic energy, they do not seem more plausible than an inhomogeneous geometry. The physically most conservative approach is to assume no unusual dark energy, but rather that an inhomogeneous geometry might be responsible for the observed apparent acceleration.

15.6.6 Improbability

Many dismiss the LTB models on probability grounds: 'It is improbable that the universe is like this, and it is improbable that we are near the centre of such a model.' As discussed in Section 21.4, meaningful statements of this type are hard to make, and in most attempts to do so, the universe is improbable. Additionally, a study by Linde, Linde and Mezhlumian (1995) shows that (for a particular choice of measure) this kind of inhomogeneity actually *is* a probable outcome of inflationary theory, with ourselves being located near the centre. More recently the authors of that paper have decided that the measure used leading to this result is not a probable measure;[4] but the existence of those inflationary models nevertheless shows that one cannot dismiss such models out of hand for probability reasons.

One can shift the improbability around, but the universe in which we live may indeed be improbable. In any case, there is no proof that the universe is probable. That is an unverifiable philosophical assumption. Whatever one's philosophical views may be, they will have to give way to the empirical data. If the tests mentioned above show the universe is spatially inhomogeneous, philosophers and cosmologists alike will have to accept this as a fact. Any theories on probability would have to be adjusted to such an empirical fact.

15.7 Conclusion: status of the Copernican Principle

The Copernican Principle is a foundational principle for the standard models of cosmology. That means it should be queried and tested in all ways possible. It seemed for a long while that it was not testable (Ellis, 2006); but recent work, as discussed above and summarized for example in Shafieloo and Clarkson (2010), Clarkson and Maartens (2010), has shown this is not so.

The acceleration indicated by SNIa and other data could be due to large-scale inhomogeneity. Observational tests of this possibility are as important as pursuing the dark energy (exotic physics) option in a homogeneous universe. What is testable and what is not testable in cosmology is a key issue: theoretical prejudices about the universe's geometry, and our place in it, must bow to such observational tests. It is certainly worth pursuing the present possibility as an alternative to the problematic proposal of dark energy (or of modified gravity). It is a genuinely scientific proposal, for we have shown there are viable ways in which it can be tested.

Overall, the validity or not of the Copernican Principle is a core issue that will not go away. It should be subject to continued examination and testing.

---

[4] A. Linde, private communication

16 'Acceleration' from small-scale inhomogeneity?

The fundamental problem considered in this chapter is, *how do we relate the FLRW model to the non-uniform real world*? The point we emphasize here is that there is a hidden averaging scale in all our descriptions of the universe. There is a hierarchy of different scales of description we can use, with effective equations occurring at each scale. The relation between the descriptions, dynamics, and observations at each of the scales is a key issue. The FLRW model only applies at the largest scales; how does it relate to the inhomogeneities at smaller scales?

A fundamental feature is the non-commutativity of averaging in relation to both dynamics and observations. Through this effect, inhomogeneities in the universe, such as vast walls, filaments, clusters and voids in the distribution of galaxies, can affect both the dynamics and observational properties of the universe. Hence they have the potential to explain at least part, if not all, of the apparent acceleration of the universe indicated by the SNIa data. Note that these are smaller-scale inhomogeneities compared with those considered in the previous chapter, which are of the order of the Hubble scale.

Overall, the question is how to describe the real universe by an (almost) FLRW model, when it is nothing like that on small cosmological scales. What is the meaning of the FLRW metric in relation to the real lumpy universe?

16.1 Different scale descriptions

Any mathematical description of a physical system depends on an implicit *averaging scale* characterizing the nature of the envisaged model (Section 1.4.1), and its tests also have specific scales. The averaging scale, or rather the acceptable range of averaging scales, is usually not explicitly stated but is in fact a key feature underlying the description used, and hence the effective macroscopic dynamical laws investigated. Indeed, different types of physics (particle physics, atomic physics, molecular physics, macroscopic physics, astrophysics) correspond to different assumed averaging scales.

16.1.1 Averaging the matter

When a fluid is described as a continuum, this assumes one is using an averaging scale large enough that the size of individual molecules is negligible (Section 1.4.1). If the averaging scale is close to molecular scale, small changes in the position or size of the averaging volume lead to large changes in the resulting density and velocity of the matter, as individual molecules are included or excluded from the reference volume. In this case the fluid approximation is not applicable; rather one is using a detailed description of the fluid where individual molecules are represented. Note that a model which is very useful on one scale may be quite impractical for use at another scale: one does not calculate the motion of a fluid from the quantum mechanics of its individual atoms.

For a fluid approximation, one assumes a medium-sized averaging scale: not so small that molecular effects matter, but not so large that spatial gradients in the properties of the fluid are significant (Batchelor, 1967): hence there is a range of validity $L_1 < L < L_2$ where the fluid approximation holds. Instead of referring to a density function ρ, one should really refer to a function ρ_L, the density averaged over volumes characterized by scale length L. The key point about the fluid approximation is that, provided this length scale is in the appropriate domain, then its actual value does not matter; i.e. when it is in this range, changing L by a factor of $10, 100, \ldots$ makes no difference: the measured density and average velocity will not change. But if you take L outside this range, this is no longer true.

Similarly, in electromagnetic theory, polarization effects result from a large-scale field being applied to a medium with many microscopic charges. The macroscopic field E differs from the point-to-point microscopic field, which acts on the individual charges because of a fluctuating internal field E_{int}, the total internal field at each point being $D = E + E_{\mathrm{int}}$ (Jackson, 1975). Spatially averaging, one regains the average field because the internal field cancels out: $E = \langle D \rangle$. Indeed this is how the macroscopic field is defined (implying invariance of the background field under averaging: $E = \langle E \rangle$). On a microscopic scale, however, the detailed field D is the effective physical quantity, and so is the field 'measured' by electrons and protons at that scale. Thus, the way different test objects respond to the field crucially depends on their scale (a macroscopic device will measure the averaged field). Furthermore, the two fields obey field equations differing by a polarization tensor. Small-scale and large-scale dynamics are not identical.

During the early stages of structure formation, matter is treated as a pressure-free fluid of baryons and primordial CDM particles, and this is a physically reasonable model. It can be improved via collision-free kinetic theory, but the corrections to the dust model are very small on cosmological scales. When particles begin to condense into galactic structures, things become much more complicated. N-body simulations, based on Newtonian gravity on an expanding FLRW grid, are the best current tool. Here the 'particles' are typically galaxy-sized CDM structures. This is not a fluid model, but neither is it a kinetic gas (there are insufficient galaxies for a consistent kinetic theory model). Baryons in bound structures are added via largely phenomenological prescriptions. But unbound baryons and CDM particles are implicitly neglected. Thus, the model of matter in the late universe is incomplete. Building a physically realistic model is a formidable open question in cosmology. The standard model of cosmology is based on an implicit averaging over matter that leads to an effective fluid model which is pressure-free, and which is assumed to represent the large-scale dynamics across the range of scales from primordial particles to galaxies. This may be a good approximation – but we do not have any plausible argument for this, up to now.

16.1.2 Coarse-graining the gravitational field

Now, exactly the same issue arises with regard to the gravitational field. Applications such as the solar system tests of GR are at Solar System scales. We apply gravitational theory, however, at many other scales: to star clusters, galaxies, and larger-scale structures (clusters, walls, filaments and voids), as well as to black holes (occurring at Solar System and star cluster scales, and possibly at much smaller scales). Averaging effects may then alter the effective gravitational equations at these larger scales. The question then is how do models on two or more different scales relate to each other in Einstein's gravitational theory (Ellis, 1984)? This is a difficult issue both because of the nonlinearity of Einstein's equations, and because of the lack of a fixed background spacetime – one of the core features of Einstein's theory. These cause major problems in defining suitable averaging processes.

In cosmology, we aim to describe the whole observable universe. A range of scales of description are relevant to cosmology. There are levels of approximation in modelling the universe, each with a hidden averaging scale. One can have a description in which every star is represented, or every galaxy (the stars averaged over), or only the largest-scale cosmological structures (even galaxies averaged over, as in the fluid approximation). In using (almost) FLRW models one is assuming a large enough averaging scale both for a fluid approximation to hold and for spatial homogeneity to be valid; this scale should be explicitly indicated (Ellis, 1984) (in the standard models, it is about 100 Mpc at present (Sarkar *et al.*, 2009; Sylos Labini *et al.*, 2009); see Section 1.4.1).

The FLRW models are then the background models for cosmology, and perturbed FLRW models characterize the nature of deviations from the exact FLRW geometry that are expected on smaller scales, while still using a fluid approximation (Chapter 10). The same effect can occur here: the averaged small-scale dynamics can lead to extra terms in the effective equations at cosmological scales (often called 'back reaction' terms). Larger deviations may be characterized by inhomogeneous models that perhaps cannot be adequately represented by perturbed FLRW models; a key issue in what follows is under what circumstances this may happen. And that leads to the question: what kind of well-defined averaging or smoothing process can produce an FLRW model from a genuinely inhomogeneous spacetime model?

One particularly important issue is that of dark energy. A large-scale smoothed-out model of the universe ignores small-scale inhomogeneities, but the averaged effects of those inhomogeneities may alter both observational and dynamical relations at the larger scale, mimicking the effect of a cosmological constant.

16.1.3 Non-commutativity of averaging and dynamics

The key point in considering dynamical effects is that the two processes involved in relating the field equations at different scales do not commute (Ellis, 1984). These processes are:

E: calculating the Einstein tensor $G_{1ab} = R_{1ab} - \frac{1}{2}R_1 g_{1ab}$ from a metric tensor g_{1ab}, and, hence, determining the quantity $\mathcal{E}_{1ab} = G_{1ab} - 8\pi G T_{1ab}$ for g_{1ab}, where T_{1ab} is the matter tensor appropriate to the scale represented by g_{1ab};

A: averaging the metric tensor g_{1ab} to produce a smoothed metric tensor $g_{2ab} : g_{2ab} = \langle g_{1ab} \rangle$ and the matter tensor T_{1ab} to produce a corresponding smoothed matter tensor T_{2ab} : $T_{2ab} = \langle T_{1ab} \rangle$.

In general the averaging process does not commute with taking derivatives: for a function g, usually $\partial_i \langle g \rangle \neq \langle \partial_i g \rangle$. Furthermore the inverse metric g_2^{ab} (nonlinearly dependent on the metric tensor components g_{1ab}) is not the smoothed version of g_1^{ab}. The resulting Christoffel terms Γ^a_{2bc} are therefore not the smoothed version of Γ^a_{1bc}, hence the Ricci tensor components R_{2ab}, nonlinearly dependent on Γ^a_{2bc}, are not the smoothed versions of R_{1ab}. Extra nonlinearities occur in calculating the Einstein tensor $G_{2ab} = R_{2ab} - \frac{1}{2} R_2 g_{2ab}$ from the Ricci tensor R_{2ab}. Thus, if you smooth first and then calculate the field equations, you get a different answer than if you calculate the field equations first and then smooth; symbolically $\mathbf{A}(\mathbf{E}(g_{1ab})) \neq \mathbf{E}(\mathbf{A}(g_{1ab}))$.

Suppose the field equations are true at the first scale; then they will not be true at the second scale:

$$\mathcal{E}_{1ab} := G_{1ab} - 8\pi G T_{1ab} = 0, \quad \mathcal{E}_{2ab} := G_{2ab} - 8\pi G T_{2ab} \neq 0. \tag{16.1}$$

Thus there will be an extra term in the equations at the smoother scale. We can either regard it as an extra term on the left, representing a modified curvature term, or as an extra term on the right, where it is regarded as an extra contribution to the matter tensor:

$$G_{2ab} - \mathcal{E}_{2ab} = 8\pi G T_{2ab}, \quad G_{2ab} = 8\pi G T_{2ab} + \mathcal{E}_{2ab}. \tag{16.2}$$

Which is the more appropriate interpretation depends on the context. In either case, we refer to a *backreaction* from the small-scale inhomogeneity to the smoothed out dynamics.

In the case of gravitational radiation, Isaacson (1967; 1968) showed how to average approximate gravitational wave solutions to the vacuum Einstein equations of GR in situations where the gravitational fields of interest are quite strong. He assumed the wave to be of high frequency, expanded the vacuum field equations in powers of the correspondingly small wavelength, obtaining a gauge-invariant linearized equation for gravitational waves, and solved it in the WKB approximation to show that gravitational waves travel on null geodesics of the curved background geometry. The lowest-order nonlinearities are shown to provide a natural, gauge-invariant, averaged stress tensor for the effective energy localized in the high-frequency gravitational waves. This is the explicit form of the extra term in (16.2) in this case.

Szekeres (1971) developed a polarization formulation for a gravitational field acting in a medium, in analogy to electromagnetic polarization. He showed that the linearized Bianchi identities for an almost flat spacetime may be expressed in a form that is suggestive of Maxwell's equations with magnetic monopoles. Assuming the medium to be molecular in structure, it is shown how, on performing an averaging process on the field quantities, the Bianchi identities must be modified by the inclusion of polarization terms resulting from the induction of quadrupole moments on the individual 'molecules'. A model of a medium whose molecules are harmonic oscillators is discussed and constitutive equations are derived. This results in the form:

$$\mathcal{E}_2^{ab} = \nabla_d \nabla_c Q^{abcd}, \quad Q^{abcd} = Q^{[ab][cd]} = Q^{cdab}, \tag{16.3}$$

i.e. $\mathcal{E}_2^{ab}$ is expressed as the double divergence of an effective quadrupole gravitational polarization tensor Q^{abcd} with suitable symmetries. Gravitational waves are demonstrated to slow down in such a medium.

The problem with such averaging procedures is that they are not covariant. They can be defined in terms of a background unperturbed space, usually either flat spacetime or a RW geometry, and so will be adequate for linearized calculations where the perturbed quantities can be averaged in the background spacetime (although even here the gauge problem arises, see below). But the procedure is inadequate for nonlinear cases, where the integral needs to be done over a generic lumpy (nonlinearly perturbed) spacetime which is not a 'perturbation' of a high-symmetry background. However, it is precisely in these cases that the most interesting effects will occur.

To obtain integrals that are well-defined over a generic spacelike surface or spacetime region (and one interesting issue is which of these one should use in the averaging process) either they have to be for scalars, or one needs the bitensors associated with the world function (Synge, 1971), based on parallel propagation along geodesics, to compare tensors at different points in a normal neighbourhood. The problem then is that the bitensors cannot be used for averaging the metric tensor, for it is the metric tensor itself that defines the parallel propagation used in this process, and so is left invariant by it (since $\nabla_c g_{ab} = 0$). So one has to devise a procedure in which either the field equations are represented only in terms of scalars, possible for example if one takes components relative to a covariantly uniquely defined tetrad (compare Section 17.2), or else bitensors are used to define averages of quantities other than the metric. Zalaletdinov (1997) has taken this issue seriously, and provided the only sustained such attempt based on bitensors. He proposes a macroscopic description of gravitation based on a covariant spacetime averaging procedure. The geometry of the macroscopic spacetime follows from averaging Cartan's structure equations, leading to a definition of correlation tensors. Macroscopic field equations (averaged Einstein equations) can be derived in this framework. It is claimed that use of Einstein's equations with a hydrodynamic stress–energy tensor means neglecting all gravitational field correlations, and a system of macroscopic gravity equations is given when the correlations are taken into consideration. This approach has not won many adherents, but is nevertheless a systematic and coherent attempt to set up the problem generically.

More recently, Korzynski (2010) has given a general definition of coarse-grained quantities for a dust flow, assigning coarse-grained expansion, shear and vorticity to finite-size comoving domains of fluid in a covariant, coordinate-independent manner as relativistic generalizations of simple volume averages of local quantities in a flat space, when the boundary of the domain in question has spherical topology and positive scalar curvature. The time evolution equations for the coarse-grained quantities include additional terms representing backreaction of small-scale inhomogeneities of the flow on the large-scale motion of the fluid.

Finally, a requirement of complete covariance for averages over a spacelike surface or spacetime region demands a covariant definition of the region (Lorentz boosts can stretch the region along the null cone by an arbitrary amount, which almost certainly will change the averages.) In the cosmological case, however, this issue is mitigated because, as emphasized in this book, there are covariantly defined preferred reference frames and associated spacelike surfaces in specific cosmological models.

16.2 Cosmological backreaction

The cosmological application is to understand the nature of the backreaction of perturbations in cosmology. The large-scale solutions used are perturbed FLRW models, approximately spatially homogeneous and isotropic. The real universe is not like this on many scales: it has vast inhomogeneities, walls of clusters of galaxies surrounding voids, forming a soap-bubble-like structure FLRW models are only a smoothed approximation to this complex reality. All the issues discussed above arise.

Varied methods have been used to study this problem, including straightforward perturbation approaches. Isaacson's method of averaging in the gravitational radiation case has been used in the cosmological context (Futamase, 1996). When Zalaletdinov's approach to the averaging problem is applied to cosmology, the Einstein field equations on cosmological scales are modified by appropriate gravitational correlation terms (Coley, Pelavas and Zalaletdinov, 2005; van den Hoogen, 2009); for a spatially homogeneous and isotropic macroscopic spacetime, the correlation tensor is of the form of a positive spatial curvature term. This already shows the effect can potentially be important, because if such averaging were to change the effective value of the spatial curvature from negative to positive, the future evolution of the universe could be drastically changed: only if $K > 0$ is a recollapse to a big crunch in the future possible. Thus the inclusion of backreaction has the potential to lead to a situation where the averaged model seems to recollapse in the future, but the real universe does not.

16.2.1 Fitting problem

Given the difficulty of handling averaging in a covariant and gauge invariant way, the conclusion may be that one should as far as possible use covariant variables, but choose a specific gauge when doing averaging. This involves the

Fitting problem: *How do we determine what is the best FLRW background model for the real lumpy universe?*

(See Ellis and Stoeger (1987); Kolb, Marra and Matarrese (2010).) This decides what the magnitude of the backreaction effects is, because it determines the size of the 'perturbation' away from the background at each point.

A proper fitting procedure underlies detailed backreaction studies. If the wrong background is fitted then it may appear to create a back reaction which vanishes if a better fit is chosen. Thus a gauge choice is inherent in the backreaction issue; this will be apparent in any detailed studies, even if it is not clearly made explicit. It is important to note the following: if you consider an FLRW model $\mathcal{F}_1$ and perturb it to get an almost-FLRW model $\hat{\mathcal{F}}_1$, it is very tempting to assume that $\mathcal{F}_1$ is the best-fit model to $\hat{\mathcal{F}}_1$. Indeed this assumption is effectively hidden in most perturbation studies. But this is not necessarily the case. For example, if you add to $\hat{\mathcal{F}}_1$ a distribution of positive density fluctuations suitably distributed over a spacelike surface, you will have raised the density on that surface: so a process of fitting the background model by averaging densities will no longer result in determining $\mathcal{F}_1$ as the best fit; it will rather yield a model $\mathcal{F}_2$, with a higher average density than $\mathcal{F}_1$. To preserve the best-fit model you started with, you need to add a compensated set of perturbations (e.g. top-hat models) where an under-density surrounds every over-density, so that the average density is unchanged by the perturbation procedure. When this is not done, a change in the average properties will result in a spurious apparent backreaction effect, which is in reality due to a failure to use the correct background model for the perturbed spacetime. A proper fitting procedure will lead to related integral constraints (Traschen, 1985).

An important point needs to be made: *the usual approach to determining a best-fit FLRW model to astronomical observations is indeed a fitting approach as defined here*: one takes an FLRW metric with some arbitrary parameters, and adjusts them to give a best fit to the inhomogeneous real universe by fitting observable relations. This determines $H_0, \Omega_{m0}, \cdots$ for the best-fit background model; one handles inhomogeneities by treating variations from the idealized FLRW model in a statistical way. Most observational studies (e.g. the classic study by Sandage (1961)) implicitly use a light-cone fitting procedure, without making this explicit.

One can also obtain a smoothed out model from a lumpy universe by averaging. If the universe is homogeneous in the large, this will recover an FLRW model with averaged quantities such as densities. These procedures will in general give different results, particularly because the former will be based on a spacelike averaging but the latter on a light-cone averaging (because observations are made on the past light-cone). It is clear that a well-fitted background model should be regained from the realistic lumpy model by a suitable averaging or smoothing procedure. The way to do such averaging (or coarse graining) is still the subject of debate (Buchert and Carfora, 2002; Brown, Robbers and Behrend, 2009; Paranjape and Singh, 2008). In a fitting approach, one starts with an FLRW model with some free parameters, and then tries to see which parameter values give the best fit: so the issue does not arise. The FLRW model was imposed by hand at the start.

16.2.2 Different approaches: key issues

Two key issues arise in all of this, and the approaches taken to these issues shape the subsequent discussion.

Choice of inhomogeneous models

In an *averaging approach*, one starts with a lumpy model: either a perturbed FLRW model or a genuinely inhomogeneous one. The almost-FLRW model we use in the large has to emerge through an averaging (coarse graining) of this more detailed inhomogeneous model. It is not surprising if this works in the perturbed FLRW case, for one has started with an FLRW model in the first place in order to create the lumpy one. The same is true for Swiss cheese models discussed below (Section 16.4.1) provided the matching is done correctly, for again this starts with an FLRW model: averaging will recover that model. The real challenge is to do the process for a fully inhomogeneous model; initial attempts are described in Sections 16.4.5 and 16.4.6 below. It is then not so obvious that an FLRW model will emerge: only a subclass of inhomogeneous models will give that result. Of course the observable part of the real universe corresponds to that subclass, because FLRW models give a good representation of what we see (Section 9.8).

Averaging: spaces and surfaces

Suppose this does work out: then the real spacetime is the lumpy one – the smoothed out FLRW model is a fictitious entity used for convenience to simplify things. The key point now is that: *the dynamics and then the averaging, must be computed in the lumpy model.* Only in this way will the true nature of the effect be tested. The first question is which surfaces to use for the averaging process? Secondly, are the chosen world lines implied in the averaging comoving with the matter or not? This corresponds to choice of the lapse and shift in the ADM approach (Section 3.3.3), which is often used in this context. In the background model, that choice will be obvious; but in using the averaged (smooth) background, which is what is most often done, essential aspects of the problem may be omitted. We want to average in the realistic (lumpy) model. We shall see that this makes a real difference.

16.3 Specific models: almost FLRW

Most approaches in one way or another assume an almost-FLRW context from the beginning.

16.3.1 Buchert formalism

One sustained attack is that by Buchert (2000; 2001; 2008), with averaging defined for observers comoving with the cosmic matter fluid. Here one first thinks of averaging over scalar quantities like the density or the rate of expansion in an inhomogeneous universe model. As long as one works with exact equations for the evolution of those fields in a given foliation of spacetime, such an averaging procedure is covariant. Cosmological parameters like the rate of expansion or the mass density are to be considered as volume-averaged quantities, so the relevant parameters are intrinsically scale-dependent unlike the situation in an FLRW cosmology. Averaging scalar characteristics on a Riemannian spatial domain delivers the effective dynamical sources that an observer would measure, but although the measures are made within the lumpy spacetime, they are going to be interpreted within an FLRW fitting model. This suggests a logical division of the averaging problem into (1) calculating averages in the real manifold, and (2) determining the mapping between averages in the real manifold and values in the FLRW model. The first averaging is straightforward for scalars, and encounters *non-commutativity of averaging and time-evolution*: this is a purely kinematical property that can be expressed, for a scalar field ψ, through the rule

$$\partial_t \langle \psi \rangle - \langle \partial_t \psi \rangle \;=\; \langle \Theta \psi \rangle - \langle \Theta \rangle \langle \psi \rangle \;. \tag{16.4}$$

The fluctuation part on the right-hand side of this rule produces the *kinematical backreaction*. The result of this averaging is the modified Friedmann and Raychaudhuri equations (Buchert, 2008):

$$3\frac{\dot{a}_{\mathcal{D}}^2}{a_{\mathcal{D}}^2} \;=\; \Lambda + 8\pi G \rho_{\mathcal{D}} - \frac{1}{2}\left(Q_{\mathcal{D}} + \langle R \rangle_{\mathcal{D}}\right), \tag{16.5}$$

$$3\frac{\ddot{a}_{\mathcal{D}}}{a_{\mathcal{D}}} \;=\; \Lambda - 4\pi G \langle \rho \rangle_{\mathcal{D}} + Q_{\mathcal{D}}, \tag{16.6}$$

$$Q_{\mathcal{D}} := \frac{2}{3}\langle \Theta^2 - 3\sigma^2 \rangle_{\mathcal{D}} - \frac{2}{3}\langle \Theta \rangle_{\mathcal{D}}^2, \tag{16.7}$$

where $a_{\mathcal{D}}$, $\langle R \rangle_{\mathcal{D}}$ are the volume averaged scale factor and spatial curvature. This shows that averaging in principle allows acceleration terms to arise from the averaging process. The derivation of these equations, and their numerical implementation, has been studied by many others, see for example Behrend, Brown and Robbers (2008).

The second 'averaging' is more adequately thought of as a rescaling of the tensorial geometry. A (Lagrangian) smoothing as opposed to (Eulerian) rescaling of the metric on regional spatial domains has been proposed by Buchert and Carfora (2002; 2008), using a global Ricci deformation flow for the metric initially proposed by Carfora and Piotrkowska (1995). They introduced real-space renormalization group methods, based on properties of the Ricci–Hamilton flow. The smoothing of geometry implies a renormalization of averaged spatial variables, determining the effective cosmological parameters as they appear in the FLRW-fitting model. Two effects that quantify the difference between background and real parameters were identified: *curvature backreaction* and *volume effect* (Buchert and Carfora, 2003). Both are the result of an inherent non-commutativity of averaging and spatial rescaling. In this way we look at the averaging problem in two directions in function space: time-evolution (as a deformation in direction of the extrinsic curvature of the space sections encoding the kinematical variables), and scale-'evolution' (as a deformation in direction of the intrinsic 3-Ricci curvature).

The Buchert method has the advantage of dealing with scalars, so integrals are well-defined, but it does not deal with the full GR dynamics because it does not average the shear and vorticity equations (which in turn would need averaging of the Weyl tensor equations). Consequently closure is obtained by *phenomenological* assumptions about the averaged behaviour – which may or may not be true, in general. This analysis will cover only a subset of possible behaviours. Nevertheless it enables us to discover important aspects of averaging and backreaction.

For example, Barrow and Tsagas (2007) use the Buchert approach to examine the effects of spatial inhomogeneities on irrotational anisotropic cosmologies by looking at the average properties of anisotropic pressure-free models. They recast the averaged scalar equations in Bianchi-type form and close the standard system by introducing a propagation formula for the average shear magnitude. In this case, backreaction effects can modify the familiar Kasner-like singularity and potentially remove Mixmaster-type oscillations; thus they can make a major difference to dynamics near the initial singularity.

Averaging via scalars

What would be most useful is an extension of Buchert's work to averaging a full family of scalars completely representing the spacetime geometry and matter. A spacetime can be completely characterized by scalar invariants (Section 17.2), and this suggests a spacetime averaging scheme based entirely on scalars. Coley (2010) has illustrated such a scheme in a simple, static, spherically symmetric, perfect fluid model, where the averaging scales are clearly identified. However, he does not give an explicit construction for generating the spacetime from a set of invariants; and there are various possible sets of invariants, some of which may be better suited to the job than others (invariants directly interpretable in terms of the fluid flow may be best). Until those issues are fully solved, this averaging proposal is incomplete: it may work for some special classes of models, but not in general. Certainly its full implications have not been worked out. The remaining problem is an aspect of the *equivalence problem* – how to characterize when cosmological models are equivalent to each other – which is discussed further in Section 17.2.

16.3.2 Perturbed FLRW approach

A common approach to studying backreaction effects is based on a 'Newtonianly perturbed FLRW metric' (Ishibashi and Wald, 2006):

$$ds^2 = -(1+2\Psi)dt^2 + a^2(1-2\Psi)\gamma_{ij}dx^i dx^j, \quad a = a(t), \quad \Psi = \Psi(t, \mathbf{x}), \qquad (16.8)$$

which is just the longitudinal gauge (10.55) with $\Pi = 0$ [(10.71)]; Ψ satisfies

$$|\Psi| \ll 1, \quad |\partial_t \Psi|^2 \ll a^{-2}D^i\Psi D_i\Psi, \quad (D^i\Psi D_i\Psi)^2 \ll (D^i D^j \Psi)(D_i D_j \Psi), \qquad (16.9)$$

where D_i is the covariant derivative for γ_{ij}. Note that the matter present is not assumed to be comoving in this frame. This metric gives descriptions of linear deviations from an FLRW model, but does it properly represent the true degrees of inhomogeneity in the real universe? We return to this issue below.

An alternative coordinate system often used is based on an ADM 1+3 foliation of spacetime (Section 3.3.3), taking (3.20) with lapse function N and shift vector N^i such that the 4-velocity n^a of observers is not necessarily comoving with the normals: rather $n^a = N^{-1}(1, n_i)$, $n_a = N(-1, 0, 0, 0)$. Averaging in this system is developed in Larena (2009). The dynamics of the system follow from the standard ADM formalism.

Finally, synchronous coordinates are often used for irrotational pressure-free dust. This is a somewhat restricted situation but easy to calculate; it is the special case of (3.20) with $N = 1$, $N_i = 0$ and comoving matter.

Gauge issue

The gauge problem was described in Section 10.1. The backreaction problem will look very different if described in terms of different gauges (Brown, Behrend and Malik, 2009). While many studies have been carried out quantifying backreaction effects in cosmology, where the smoothed-out effect of the small-scale perturbations causes extra terms in the Friedmann equations for the background metric, none has been done that both clearly takes the gauge issue into account and goes beyond linear order. This is an important issue waiting to be resolved. Gasperini, Marozzi and Veneziano (2010) have made an attempt based on the Buchert formalism , but it is not clear if their method of varying the boundaries of the region of integration gives the correct physical result. One certainly wants to go at least to second order in understanding the effects of nonlinear perturbations. Many of the crucial results at linear order no longer hold, for example scalar, vector and tensor perturbations are no longer independent of each other at second order and then the backreaction in turn affects the perturbations themselves (Martineau and Brandenberger, 2005).

16.4 Inhomogeneous models

An alternative to using these perturbation techniques is looking for suitable exact inhomogeneous solutions of the field equations. One can then study averaging and observational issues in these models.

16.4.1 Swiss cheese models

The 'Swiss cheese' models were originally developed by Einstein and Straus (1945; 1946) to examine the effect of the expansion of the universe on the Solar System (if there were such an effect, we could possibly measure the expansion of the universe by laser ranging within the Solar System). Their matching of a Schwarzschild interior to an FLRW exterior showed that the expansion has no effect on the motion of planets in the Schwarzschild region. Thus one cannot determine the Hubble constant by Solar System observations.

The Swiss cheese model consists of one or more spherically symmetric vacuum regions, each described by the Schwarzschild metric familiar in black hole theory, joined across spherical boundaries to an FLRW dust model. This embodies a natural way to model physical problems, such as describing the boundary between a galaxy and intergalactic space or the relation between bubbles at the end of an inflationary era, by taking two different regions where the behaviour is smooth and joining them at a hypersurface of discontinuity. It does not, however, answer the question as to where the boundary between the regions should be placed – which determines which regions are affected by the universal expansion.

To describe the interface(s), we need first to briefly study junction conditions in GR. Because these conditions are also relevant in modelling domain walls, or cosmologies which change character abruptly at some spacelike slice, or in brane-world models, we introduce them in more generality than the Swiss cheese alone would demand. We return to their other uses in Chapters 19 and 20.

16.4.2 Junction conditions in general

In general a jump discontinuity (step function or Heaviside function) in the metric would, by (2.58), lead to a δ-function in the connection and thence, by (2.39), to products of δ-functions in the curvature, which are not usually considered physically meaningful. So attention is usually restricted to the case where the connection has at worst a step discontinuity and the curvature at worst a δ-function (see e.g. Taub (1980)). A δ-function in the Ricci tensor models a thin shell or surface layer of matter through the Einstein equations (see Section 3.3), for example a domain wall in cosmology (see Section 20.3), while one in the Weyl tensor describes an impulsive gravitational wave.

We consider regions V^+ and V^- with respective metrics g^+ and g^- and bounding hypersurfaces Σ^+ and Σ^-, which are to be identified as a single hypersurface Σ in spacetime. The normal to Σ is taken so that it points from V^- to V^+. In either V^+ or V^-, we can calculate the metric induced on Σ (see Section 2.1). The first requirement at a boundary is that the metrics for Σ calculated from the two sides are the same. This is physically natural as it says that Σ has the same internal geometry no matter which side it is viewed from. Clarke and Dray (1987); Mars and Senovilla (1993) show that then for sufficiently smooth $V^\pm$, there are charts covering the whole spacetime in which the metric coincides with g^+ in V^+ and with g^- in V^-. We use such coordinates in the following discussion.

Junction conditions are hard to use in practice, except when the hypersurface Σ shares a symmetry with the spacetime. This is because we have to specify the surfaces Σ^+ and Σ^- and fix the identification between Σ^+ and Σ^- in such a way that the metrics agree and V^+ and V^- lie on opposite sides of Σ. Hence most of the specific applications have been to cases with spherical, cylindrical or plane symmetry.

In general Σ may be null in some regions and non-null elsewhere. The null case is more awkward because the metric of Σ is degenerate and the vector normal to Σ is also tangent to it. Older literature treated only the cases where Σ has the same character everywhere, and many cosmological applications, such as sharp transitions between cosmological epochs with different behaviours of matter, involve hypersurfaces of fixed character, but there are others in which the character of Σ may vary, e.g. a phase transition occurring suddenly in a limited region which then generates a wave travelling at the speed of light.

Hence we here follow Mars and Senovilla (1993) and treat all cases simultaneously, using a non-zero vector field l^a not lying in Σ, i.e. a rigging as defined in Section 2.1 (in the non-null case, l^a can be taken to be the unit normal n^a to Σ). We define the tensor

$$\mathcal{H}_{ab} = P^c{}_a P^d{}_b \nabla_c l_d, \quad P^a{}_b = \delta^a{}_b - l^a n_b. \tag{16.10}$$

Then if there is a discontinuity, the extrinsic curvature jumps by $[K_{ab}] = n_c n^c [\mathcal{H}_{ab}]$ where $[\mathcal{H}_{ab}] = (\mathcal{H}_{ab})_{|V_+} - (\mathcal{H}_{ab})_{|V_-}$, and the Riemann tensor has a δ-function singularity with coefficient

$$\mathcal{Q}^a{}_{bcd} = 2\left(n^a [\mathcal{H}_{b[c}] n_{d]} - n_b [\mathcal{H}^a{}_{[c}] n_{d]}\right). \tag{16.11}$$

A physical interpretation is that the geodesic deviation along a geodesic which passes through Σ with tangent vector l^a undergoes a sudden change due to the curvature (16.11) (see (2.47)).

It is clear from (16.11) that if surface layers and impulsive waves are to be ruled out, the second condition required for matching is $[\mathcal{H}_{ab}] = 0$. If impulsive waves are allowed, but surface layers are not, we require the Ricci tensor to have no δ-function part and this holds if and only if n^a is non-null and $[\mathcal{H}_{ab}] = 0$, or n^a is null, $n^a[\mathcal{H}_{ab}] = 0$ and $[\mathcal{H}^a{}_a] = 0$.

If $[\mathcal{H}_{ab}] = 0$, the possible (step function) discontinuities in the Riemann tensor satisfy

$$n^a[G_{ab}] = 0 = n_a[C^a{}_{bcd}]P^c{}_e P^d{}_f, \tag{16.12}$$

and appear in specific Riemann tensor components in suitable bases (Mars and Senovilla, 1993). We note in particular that if there are no surface layers, normal components of the Einstein tensor (i.e. from the field equations (3.13), of the energy momentum) are continuous. Thus if the interface is a spacelike hypersurface, representing an abrupt change in evolution of the universe, so n^a is timelike, the energy density and momentum as measured by an observer with velocity n^a must be continuous. Conservation of energy and momentum across the surface will thus be satisfied. When n^a is spacelike, the stresses (for perfect fluids, the pressure) and energy flux normal to the boundary must be continuous.

For the case where the boundary hypersurface Σ is everywhere non-null, discontinuities were first considered by Darmois (1927) and Lichnerowicz (1955). The required conditions for non-singular matching, namely the equality of the metric in the surface and the extrinsic curvature, as calculated from the two sides, were given by Darmois. Israel (1966) similarly discussed shells, and because of the importance of this case to brane-world studies cosmologists often call the conditions the Israel conditions, even if no shell is present. By choosing Gaussian normal coordinates on the two sides of Σ, one can then obtain a coordinate system in which the metric and its first derivative are continuous, which is the form of junction condition given by Lichnerowicz.

The Lichnerowicz form implies the Darmois form is true, and the Darmois form implies that there are coordinates in which the Lichnerowicz formulation is true; in this sense the two formulations are equivalent (Bonnor and Vickers, 1981). Moreover, the Darmois form is equivalent in a similar sense to the conditions of O'Brien and Synge (1952) (the requisite coordinate choice may require a coordinate transformation which is not differentiable at Σ). If the O'Brien–Synge conditions are assumed true in other coordinate systems, they give additional, and physically unnecessary, restrictions. The corresponding restriction of the results above to the case where Σ is null was developed by Taub (1980); Clarke and Dray (1987); Barrabes (1989) and Barrabes and Israel (1991).

Note that if two spacetimes M_1 and M_2 are each divided into two regions, giving V_1^+, V_1^-, V_2^+ and V_2^-, and if V_1^+ is matched with V_2^-, then the same conditions will match V_2^+ with V_1^- (with the opposite sign for the coefficient of any δ-function part). This has been called the complementary matching (Fayos, Senovilla and Torres, 1996).

The conditions stated above concern the gravitational field, and thus the total energy–momentum. However, in non-vacuum spacetimes, the matter content will have its own field equations leading to additional boundary conditions which also have to be imposed.

16.4.3 Swiss cheese matching

This is the best known example in the non-null case, in which the Schwarzschild solution, which models a black hole or the vacuum exterior to a spherically symmetric body, is matched to an exterior dust FLRW metric (Einstein and Straus, 1945; 1946; Schucking, 1954). The complementary matching is the Oppenheimer–Snyder collapsing spherical dust body in a vacuum exterior.

The metric for the Schwarzschild solution can be written

$$\mathrm{d}s^2 = -A\mathrm{d}T^2 + A^{-1}\mathrm{d}R^2 + R^2\mathrm{d}\Omega^2, \quad A = 1 - 2MR^{-1}. \tag{16.13}$$

We attempt to join this metric, as V^-, to (2.65) as V^+. By considering the conditions (16.12) and the properties of the FLRW metric (see Chapter 9) it is clear this can only be done if the FLRW metric has a dust matter content and in it the hypersurface is a sphere, which can be taken to be $r = r_\Sigma = \text{const}$. Physically, these conditions are needed to ensure that a particle at the bounding surface stays in the surface.

We can identify the angular coordinates on the two sides, and take t as the third coordinate in Σ; we have to find $R(t)$ and $T(t)$ for Σ (known only up to the arbitrary choice of an origin for T). The matching of the first fundamental forms leads to $R_\Sigma = a(t)f(r_\Sigma)$ (from the coefficients of the angular coordinates), so the Schwarzschild region expands or contracts with $a(t)$. The equality of the remaining part of the first fundamental form (the tt component) implies that $T(t)$ must obey $A\dot{T}^2 = 1 + \dot{a}^2 f(r_\Sigma)^2/A$, where $\dot{\ }$ means the t-derivative. This together with $R_\Sigma = a(t)f(r_\Sigma)$ gives the equations for Σ^-. The unit (spacelike) normals (from V^- to V^+) on the two sides are $n_a^+ = (0, a, 0, 0)$, and $n_a^- = (-\dot{R}, \dot{T}, 0, 0)$. As the surface Σ is timelike we can take the rigging $l^a = n^a$, so (16.10) is just the extrinsic curvature. Equating the extrinsic curvatures on the two sides of Σ, gives, from the angular coordinates, $R A\dot{T} = aff'(r)$ which implies, as a third equation for the matching, $2M = f^3 a(K + \dot{a}^2)$. After some algebra one can check that the tt component gives nothing extra.

The Einstein field equations for an FLRW model with dust density ρ, using physical units (see Chapter 9), show that the Schwarzschild region has mass $M = 4\pi(\rho a^3)f(r_\Sigma)^3/3$, i.e. the mass a Euclidean region of radius R_Σ and density ρ would have (note that in an FLRW dust model ρa^3 is constant). Given M and ρa^3, the matching conditions (and the FLRW field equations) fix $R(t, r)$; given ρa^3 and an initial value of R they fix M. This means that one cannot match arbitrary masses. The Schwarzschild mass must be the same as the mass that has been removed.

Although we chose coordinates above for the FLRW region which were concentric with the Schwarzschild mass, this is not necessary. One can put any number of non-overlapping Schwarzschild regions into a dust FLRW model – whence the name 'Swiss cheese', by analogy with Emmenthal. Swiss cheese models have been widely used in estimating the effects of localized inhomogeneity in cosmology, in particular on observations.

Mass matching

If the interior and exterior masses were wrongly matched, there would be an excess or deficient gravitational pull from the mass in the interior that would not fit the exterior gravitational field, and the result would be to distort the geometry in the exterior region – which would then no longer be an FLRW model. Another way to explain why the matching of masses is needed is that otherwise we would have fitted the wrong background geometry to the inhomogeneity in the Swiss cheese model – in particular, averaging the masses in the combined model would not give the correct background average (Ellis and Jaklitsch, 1989), and the model could not have arisen from rearranging uniformly distributed masses in an inhomogeneous way (this is the content of the integral constraints of Traschen (1984; 1985)). If there is an overdensity, there has to be a compensating underdensity around it so that the masses match.

Other interiors

Swiss cheese models were originally introduced to deal with matching FLRW exteriors to Schwarzschild ('vacuole') interiors. However, one can use the method for other interiors (see Section 19.2), in particular LTB and Szekeres, and one can use them to match FLRW models with two different sets of parameters, or types of matter. Indeed this is implied in chaotic inflationary models – then proper times have to match across the boundary (the metric must be continuous) even though there is a power-law expansion on one side and an exponential expansion on the other. However, this is possible because the boundaries do not have to be comoving.

16.4.4 Swiss cheese models and backreaction

In a Swiss cheese model of inhomogeneities, there can be no long-range gravitational effects of such inhomogeneities on other masses, because of the matching conditions: the Schwarzschild masses cannot cause large-scale motions of matter in the FLRW region. If the fitting is done properly, matching metrics and masses at the junctions, no strange matter exists and the junctions and the existence of the lumps does not influence the background model FLRW first considered: *there can be no backreaction effects in a Swiss cheese model with the boundary conditions properly satisfied.* However, the observational effects, discussed in Section 16.6, are a different matter. Note in particular that there is no scope for using a Swiss cheese model for the effect of an over-dense or under-dense region on particles outside that region: the non-FLRW region must include the affected particles. Moreover, as we shall see in Chapter 19, they are unstable, and therefore not suitable for describing the histories of objects, but only for such purposes as modelling the effects of lensing. However, they are indeed useful in that context.

16.4.5 Lindquist–Wheeler type models

A rather different way of treating homogeneity was introduced by Lindquist and Wheeler (1957), using a Schwarzschild cell method to model an expanding universe with closed spatial sections (having an S^3 topology). They examined joining many Schwarzschild cells together, with boundaries subject to equations of motion that make the whole expand and recollapse in an approximation to a $K > 0$ FLRW universe.

For simplicity they used a regular lattice, which allows a very limited set of possibilities. They found that for such lattices with N vertices, every vertex can be equidistant from its neighbours only when $N = 5, 8, 16, 120$, or 600. They then derive equations of motion for the expanding universe from junction conditions between the cells. This is quite different conceptually from the Swiss cheese approximation just described, because there are no FLRW domains used in addition to Schwarzschild cells: rather the composite of Schwarzschild cells *is* the approximation to an FLRW universe model. It is locally static but globally expanding.

In the Swiss cheese case one *starts off* with a RW spatially homogeneous and isotropic geometry, and then cuts out 'vacuoles' within which individual masses are imbedded. These masses are thus contained in vacua within a spatially homogenous fluid-filled cosmos. In the Lindquist–Wheeler case, one starts off with the inhomogeneous vacuoles alone, and then glues them together to create an emergent RW geometry when averaged on large scales. There is no fluid filling the spacetime; rather (as in the case of kinetic theory) fluid-like behaviour emerges on large scales when one coarse-grains over the detailed structure. This is a more fundamental approach to the study of the relation between locally static inhomogeneity and a globally expanding universe. However, the method is not strictly realistic (the real universe has matter condensations of different masses, as well as particles that are not condensed), nor is it strictly self-consistent – in that the gravitational fields of the neighbouring particles would in fact deform the field in the neighbourhood of each vertex, thereby resulting in an approximate rather than exact spherically symmetric spacetime region (the real solution will be locally a bit anisotropic about each vertex). Nevertheless it is a plausible approximation, and is a very useful approach to tackling the issue raised here.

The Schwarzschild cell method gives the same relation between radius of the universe and proper time as in a $K > 0$ FLRW universe, except that the connection between maximum radius and mass is different. Thus averaging effects lead to an FLRW counterpart different from an exactly smooth distribution of the same matter – as shown explicitly by Clifton and Ferreira (2009a). What would be interesting would be extension of such models to irregular lattices; but this might be technically difficult. Again the issue will arise of best-fitting of the background model derived in this way.

16.4.6 Wiltshire's models

An interesting new development is the recognition that there are large (but sub-Hubble) voids in the universe (see e.g. Rudnick, Brown and Williams (2007)), and these have important significance for averaging in cosmology. The geometry and dynamics in and out of a void may be quite different. In this case the void and non-void regions would have to be joined with suitable junction conditions so that internal and external geometries agree. This has been examined by Wiltshire (2007a; 2007b; 2008a; 2009), who argues that time runs at different rates in and out of voids in a void-dominated universe because of gravitational redshift effects, and that the difference is cumulative. This can be thought of as the source of an extra redshift contribution over that in an FLRW model, changing the observable relations between cosmological variables. In his model our observable universe is an underdense bubble, with an internally inhomogeneous fractal bubble distribution of bound matter systems, in a spatially flat bulk universe. It is argued that the clocks of the isotropic observers in average galaxies coincide with clocks defined by the true surfaces of matter homogeneity of the bulk universe, rather than the comoving clocks at average spatial positions in the underdense bubble geometry, which are in voids. His model of the average geometry of the universe, depends on two measured parameters, Ω_{m0} and H_0. The observable universe is not accelerating but inferred luminosity distances are larger than naively expected, in accord with the evidence of distant SNIa. Observational consequences are worked out in detail in Leith, Ng and Wiltshire (2008).

This is an interesting series of papers that takes seriously the bound nature of local structures in the universe. However, the details of the models used may or may not be good representations of that structure: the jury is out on this issue.

Exercise 16.4.1 Generalize the Swiss cheese construction to $\Lambda \neq 0$.

16.5 Importance of backreaction effects?

There is no doubt that interesting effects occur (Futamase, 1991; Buchert, Kerscher and Sicka, 2000; Kolb, Matarrese and Riotto, 2006; Buchert, 2008; Li, Seikel and Schwarz, 2008). The question is whether these backreaction effects are significant in cosmology.

On the one hand, they might play a significant role in the inflationary era (Mukhanov, Abramo and Brandenberger, 1997; Geshnizjani and Brandenberger, 2005). On the other, they could possibly help explain the apparent existence either of dark energy and/or of dark matter as effective terms in the macroscopic dynamics at recent times. Various papers suggest the effect may indeed be significant, for example the observed acceleration of the universe could possibly be the result of the backreaction of cosmological perturbations rather than the effect of a negative-pressure dark energy (Wetterich, 2003; Kolb, Matarrese and Riotto, 2006; Kolb, Marra and Matarrese, 2010), if astronomical parameters are restricted in specific ways (Rosenthal and Flanagan, 2008). However, other studies obtain different results (Russ *et al.*, 1997; Buchert, Kerscher and Sicka, 2000; Nambu, 2002; Notari, 2006; Räsänen, 2004; 2008). Gauge effects are problematic (Geshnizjani and Brandenberger, 2002), and many doubt the effect is significant (e.g. Ishibashi and Wald (2006); Baumann *et al.* (2010)). The debate continues (Biswas, Mansouri and Notari, 2007; Li and Schwarz, 2007; Behrend, Brown and Robbers, 2008). We cannot cover all approaches and will just consider some specific aspects of importance.

16.5.1 Validity of coordinates

An assumption made in analyses such as that of Ishibashi and Wald is that one can use the 'Newtonianly perturbed FLRW metric' (16.8) globally. This is probably correct either in a perturbed FLRW domain, or in a locally isolated quasi-static domain, but it is not obvious such coordinates can be simultaneously used in both for any extended length of time. Such coordinates are probably only locally valid in a realistic description of the universe taking both expanding matter-filled domains and locally quasi-static domains into account; but it is the assumption that these coordinates can be used globally that leads to the conclusion that backreaction effects are negligible (Wiltshire, 2008b; Kolb, Marra and Matarrese, 2008). The domain of validity in realistic situations needs to be investigated.

Thus a key issue is, *how large in space and time* is the domain where such quasi-Newtonian coordinates (16.8)–(16.9) can be used in a realistic model of an expanding universe with local virialized structures? One may note here that the timelike reference congruence associated with such coordinates is shear-free, conformally mapping the 3-spaces onto each other. Assuming such a congruence exists is a major assumption, placing strong limits on the Weyl tensor, inter alia excluding occurrence of vector and tensor perturbations. It may be acceptable in the circumstances considered because matter is not comoving in this frame, and metric perturbations are small despite density perturbations being large; but this assumption certainly needs investigation.

Ishibashi and Wald (2006) assert that 'the metric (16.8)–(16.9) appears to very accurately describe our Universe on *all scales*, except in the immediate vicinity of black holes and neutron stars …The basis for this assertion is simply that the FLRW metric appears to provide a very accurate description of all phenomena observed on large scales, whereas Newtonian gravity appears to provide an accurate description of all phenomena observed on small scales.' However, as described above, the crucial issue is to what extent this metric can *simultaneously* provide a good description of both a perturbed expanding universe and a quasi-static domain of local matter condensations. It is locally acceptable in either context on its own, but to describe bound gravitational structures in an expanding universe, it may be that separate coordinate systems of this type should be stitched together in a Swiss cheese kind of construction, as discussed in Section 16.4.1.

There is another important point. The metric (16.8) follows from (10.55) with $\Phi = \Psi$, which is only possible if there are no anisotropic pressure terms in the stress tensor [(10.71)]. But for a perfect fluid, taking the equations to second order, this can only happen if the local velocity vanishes and the flow is therefore along the normals to the time coordinate surfaces. Because the normals are orthogonal to these surfaces, they have zero vorticity and are also shear-free, and shear-free perfect fluid flows are highly restricted (see Section 6.2): inter alia they are self-similar, and so cannot for example represent the Zel'dovich process of pancake collapse. Indeed a warning against assuming that all Newtonian theory solutions are acceptable approximations to GR situations is given by the shear-free theorem which is valid in GR but not in Newtonian gravity; see Section 6.8.2. Therefore, as well as excluding vector and tensor perturbations, these metrics are very restricted in terms of the geometrical situations they can represent; it is not clear they can handle realistic universe models representing nonlinear structure growth, and one should rather use (10.55). But even this sets the magnetic part of the Weyl tensor to zero in the chosen reference frame (Clarkson, Ananda and Larena, 2009). For further discussion see Kolb, Marra and Matarrese (2008).

The particular form of the background metric used in Baumann *et al.* (2010) is the locally Minkowski form,

$$ds^2 \approx -\left[1 - (\dot{H} + H^2)x_G^2\right]dt_G^2 + (1 - \tfrac{1}{2}H^2x_G^2)dx_G^2, \qquad (16.14)$$

where x_G and t_G are local 'physical coordinates' defined from comoving coordinates in a spatially flat FLRW metric by

$$t = t_G - \tfrac{1}{2}x^2, \quad x = a(t)^{-1}x_G\left[1 + \tfrac{1}{4}x_G^2\right]. \tag{16.15}$$

These coordinates are defined only in a local spatial neighbourhood of the world line, but for long time periods. The Hubble expansion is encoded in the Newtonian potential Φ_{FLRW} and background velocity field v_{FLRW}:

$$\Phi_{FLRW} = -\tfrac{1}{2}(\dot{H} + H^2), \quad v_{FLRW} = Hx. \tag{16.16}$$

Adding in perturbation terms as in (10.55) gives a modified form of the metric that can be used to study the backreaction issue.

There are two issues one should note here. Firstly, the velocity relation in (16.16) refers to a coordinate velocity, not a physical velocity, because the time t_G in (16.14) is coordinate time, not physical time, and distance $|x_G|$ is coordinate distance, not physical distance. Thus the linearity of the Hubble law in these coordinates is imposed by hand through coordinate choice (which can always be done) and so is not physically meaningful. It does, however, reduce to the physical velocity in the limit at the origin, and that raises the second issue.

Because the background model (16.14) is written in expanding coordinates, perturbing about it cannot easily represent locally bound systems that do not expand with the universe. One can indeed represent such systems by having a proper motion relative to the expanding background, but then to attain an effectively locally static state such as the Galaxy or the Solar System, an infall velocity will have to compensate for the expansion. Thus the metric (16.14) cannot easily represent locally static systems that are gravitationally bound and are no longer participating in the Hubble expansion. But this is what we want to investigate: how can one stack together quasi-static local spacetime regions in such a way as to attain an overall expanding universe model? What we need in order to analyse the issue properly is a background metric that is static locally but expanding globally; examples are given (in different ways) by the Swiss cheese and Lindquist–Wheeler models.

Problem 16.1 Either show that (16.16) can adequately represent an expanding universe with locally static domains imbedded in it, or show that this is not the case. Explain how this relates to the idea that the Hubble expansion cannot be measured in the Solar System or the Galaxy.

16.5.2 Contrasting views

The contrasting views about these issues are well represented by Baumann *et al.* (2010) and Clarkson, Ananda and Larena (2009). The first uses a perturbed form of (16.14) and uses an effective field theory approach for how the long-wavelength universe behaves, representing it as a viscous fluid coupled to gravity. The expansion is performed in terms of the peculiar velocity v and second-order terms are treated as a stress–energy tensor. Integrating out short-wavelength perturbations renormalizes the homogeneous background and introduces dissipative dynamics into the evolution of long-wavelength perturbations. They find that the backreaction of small-scale nonlinearities is very small, being suppressed by the large hierarchy between the scale of nonlinearities and the horizon scale, and that virialized scales decouple completely from the large-scale dynamics, at all orders in the post-Newtonian expansion. They emphasize that one can linearize in terms of velocities and gravitational potential rather than density fluctuations, which are clearly extremely nonlinear at recent times.

By contrast Clarkson, Ananda and Larena (2009) calculate the backreaction in the longitudinal gauge in a consistent way up to second order in a perturbative expansion about a flat FLRW background, including Λ, but using a Buchert-style averaging scheme. They identify an intrinsic homogeneity scale that arises from the averaging procedure, beyond which a residual offset remains in the expansion rate and deceleration parameter. They give the intrinsic variance that affects the value of the effective Hubble rate and deceleration parameter, leading to a correction of order a few per cent at low redshifts. This is clearly potentially significant in an era of precision cosmology.

The difference in the two results arises from the difference in the approaches used, and in particular how the authors integrate out small scales. Although both employ a similar window function on the background to smooth over small-scale structure in the potential Φ, Clarkson, Ananda and Larena (2009) require that spatial averages are performed in the spacetime itself (consistently up to second order), and not just on the background, whereas Baumann *et al.* (2010) average in the background spacetime. This makes a crucial difference. Requiring that the spatial average remains consistently on a surface in the actual spacetime introduces nontrivial terms $O(\nabla^2\Phi)^2$ when calculating the time derivative of averaged quantities such as the Hubble rate, and these are significant. These terms occur not because of the field equations (which cannot lead to higher than second derivatives) but because of the averaging/smoothing operations (which is what we are investigating).

The key issue is that while the potential may be very small, its derivatives are not; and this has to be the case in order that it represents an inhomogeneous situation where (albeit the velocities are low) $\delta\rho/\rho \simeq 10^{20}$ is typical, so it is certainly not a perturbation. It can of course be described by Newtonian gravity, but the true dynamics that this approximates is GR, where matter causes curvature and so the curvature (coupled to the matter inhomogeneities by the EFE) is way outside the linear approximation regime even though the potential is very small. We need true nonlinear models to examine this properly, such as those discussed above.

At the time of writing, it seems likely that dynamical backreaction effects make a negligible contribution to acceleration on cosmological scales, but they do have to be taken into account in precision cosmology (see Clarkson *et al.* (2011b) for a review).

16.6 Effects on observations

We now turn to looking at observational effects due to local inhomogeneity. In the real universe, as pointed out by Zel'dovich (1964) and Bertotti (1966), observations take place via null geodesics lying in the underdense regions between galaxies. Light rays are focused only by the curvature actually inside the beam, not the matter that would be there in a completely uniform model. The effect on observational relations of introducing inhomogeneities into a given background spacetime is two-fold: it alters redshifts, and it changes area distances.

16.6.1 Redshift effects

The redshift effects can be understood with the following Newtonian analogy. When a void intervenes between the source and the observer, photons drop into a potential well and then climb out, and they exit when the universe is larger than when they went in. If spacetime is static inside the void the redshift changes in and out cancel, but when structure is forming the potential well is changing with time so one gets the Rees–Sciama effect (Rees and Sciama, 1968): a change in redshift due to a change in the potential well as the photon traverses it. A non-zero cosmological constant will also lead to such an effect. When the light source is in the void, the photon has to climb out of the void, giving a contribution to observed redshifts.

16.6.2 Area distance effects

The usual analysis of cosmological observations is based on the equations relating apparent magnitude and redshift in exactly RW spacetimes. In this case (where we ignore the effects of higher-density matter concentrations), the Weyl tensor C_{abcd} vanishes, but the Ricci tensor is non-zero, being given via the Einstein field equations from the matter present. Thus, in the Sachs optical scalar equations, (7.26) and (7.28), $C_{abcd} = 0$ and the relevant solutions are shear-free:

$$\widehat{\sigma}^2 = 0 \quad \Rightarrow \quad \frac{\mathrm{d}\widehat{\Theta}}{\mathrm{d}v} = -R_{ab}k^a k^b - \frac{1}{2}\widehat{\Theta}^2. \tag{16.17}$$

Integration gives the Mattig relations (7.52) for FLRW models.

The situation in a universe with no intergalactic medium (i.e. in which all the matter is concentrated in galaxies) would be the opposite of that above: in the region of spacetime traversed by the geodesics, the Ricci tensor vanishes, so

$$\frac{\mathrm{d}\widehat{\Theta}}{\mathrm{d}v} = -2\widehat{\sigma}^2 - \frac{1}{2}\widehat{\Theta}^2 \, , \qquad \frac{\mathrm{d}\widehat{\sigma}_{ab}}{\mathrm{d}v} = -\widehat{\Theta}\widehat{\sigma}_{ab} - C_{acbd}k^c k^d. \tag{16.18}$$

The Weyl tensor (the tidal gravitational field caused by nearby matter) generates shear that then causes focusing. Thus, the description of the focusing by local inhomogeneities in the vacuum gravitational field ($\widehat{\sigma} \neq 0$, $R_{ab} = 0$, $C_{acbd}k^c k^d \neq 0$) is radically different from the one in a smoothed-out FLRW universe where focusing is caused by a smooth matter distribution only ($\widehat{\sigma} = 0$, $R_{ab} \neq 0$, $C_{acbd}k^c k^d = 0$). Hence the area distance–redshift relation on the small angular scales (i.e. the small solid angle bundles of null geodesics) actually used in observations of individual objects, may be expected to be different from those for large scales (averaging over large solid angles).

Various proposals have been made to deal with this. The most popular is the Dyer–Roeder distance (Dyer and Roeder, 1974; 1975), obtained by assuming that only a fraction $\tilde{\alpha}$ of the total mass density is smoothly distributed, i.e. not bound in galaxies, while a fraction $1 - \tilde{\alpha}$

is bound. For matter not passing through bound 'clumps', one replaces $R_{ab}k^a k^b$ in (16.17) by $\tilde{\alpha} R_{ab} k^a k^b$. The key is the equation for the area distance,

$$(z+1)(\Omega z + 1)\frac{d^2 D_A}{dz^2} + \frac{1}{2}(7\Omega z + \Omega + 6)\frac{dD_A}{dz} + \left[\frac{3}{2}\tilde{\alpha}\Omega + \frac{\hat{\sigma}^2}{(1+z)^5}\right]D_A = 0, \quad (16.19)$$

where $\hat{\sigma}^2$ is the shear induced along the light-ray bundle by the gravitational effect of nearby matter (the equivalent FLRW equation is obtained by setting $\hat{\sigma}^2 = 0$, $\tilde{\alpha} = 1$). The Dyer–Roeder proposal is to ignore the shear term in (16.19) and work out the corresponding area distance (Schneider, Ehlers and Falco, 1992; Demianski *et al.*, 2003). Thus it is the proposal that the main effect of clumpiness is that light-rays by which we observe most distant galaxies pass through less matter than in a corresponding smoothed-out FLRW universe; shear has a negligible effect, because it is only important for near encounters with isolated masses, when it causes gravitational lensing effects.

This is a good approximation when galaxies are embedded in a fairly uniform intergalactic medium of dark matter, but clearly does not take shear effects and caustics properly into account. How good it is will depend on the nature of clustering in the universe and how the averaged distribution impacts along the line of sight (Linder, 1998). If the dark matter is uniform, Dyer–Roeder is good; if dark matter is clustered, it is not so good. But we know that gravitational lensing strong enough to cause significant focusing is a relatively rare effect over the whole sky, suggesting it will be a good approximation.

One can approach the topic in other ways: for example by using stochastic methods (Bertotti, 1966), or detailed examination of geodesics in universes with spherically symmetric lumps, see below. The results of course depend on the statistics of the clumping (see the references in Kainulainen and Marra (2009)). The over- and under-densities lead to a distribution of magnifications, favouring mild demagnifications but with a long tail of magnifications. Holz and Wald (1998) estimated the effects via a Monte Carlo method, assuming that inhomogeneities were correlated only within some fixed radius. Kainulainen and Marra (2009) give a general stochastic method for estimating the probability distribution in models where under-densities occupy more volume than over-densities, which might reasonably represent the observed voids and filaments, and show good agreement with Holz and Wald (1998) and the ray tracing in Holz and Linder (2005). They conclude that lensing by structures on a scale $10^{15} M_\odot/h$ would be important in analysing the current SNIa data. The residuals in the data already indicate some lensing effect (Kronborg *et al.*, 2010).

The influence of a cosmological constant in lensing was mentioned in Section 12.4.2 and may need to be accounted for. Similarly any quintessence present affects distances but otherwise could only be detected by lensing effects if its equation of state were in some regions significantly different from Λ or if it were clumped and so caused a Weyl tensor term.

16.6.3 Averaging over whole sky gives FLRW?

In a fully clustered case ($\tilde{\alpha} = 0$), the Ricci focusing of the averaged model is replaced by Weyl focusing on smaller scales. This Weyl focusing should give FLRW equations when averaged over the whole sky, but how this happens, and whether the inferred FLRW model is the one obtained from the averaged density, is not obvious! It has been suggested (Weinberg, 1976) that energy conservation will imply the correct FLRW all-sky average; this assumes that the areas of a large angular scale bundle of null geodesics are the same in the perturbed and background models, which will not be true when one takes the effect of caustics into account (Ellis, Bassett and Dunsby, 1998). Areas increase slower than in a RW model in the empty spaces between matter, where the Ricci term is zero, and faster in the high-density regions where matter is concentrated, so one might think these effects cancel out. However, the strongly lensed rays soon go through a caustic and emerge highly divergent, so that areas are rapidly increasing again. It is plausible that on average the overall effect is always an increase in area, giving a smaller area distance than in the smooth background model. This suggests that the effect does *not* average out, over the whole sky. In any case specific observations e.g. of SNIa are preferentially made in directions where matter density is lower (universe is transparent) hence they may not represent an all-sky average.

Nevertheless it is possible that for practical purposes this effect is small, and the large-sky average is indeed the same as in an FLRW model (Kibble and Lieu, 2005): as long as the clumps are uncorrelated the average magnification is precisely the same as in a homogeneous universe with equal mean density. This is a remarkable result: completely different focusing mechanisms give the same result on average.

16.6.4 Specific examples

Swiss cheese models and observations

Although Swiss cheese models do not affect the global dynamics of a lumpy universe model, they do affect observations, because they change area distances. Indeed they model precisely the difference between Weyl and Ricci focusing of null geodesics: null geodesics in the empty regions are focused only by shear induced by the Weyl tensor. The null geodesics must be matched across the boundaries between the vacuum and matter-filled regions, and overall focusing calculated. This does indeed lead to interesting observational effects. Detailed examination of geodesics in universes with an FLRW background and spherically symmetric lumps – LTB, Schwarzschild or other models (Dyer, 1976; Kantowski, 1969b; Newman, 1979; Wesson, 1979) – show that the corrections depend on the choice of modelling. For instance Newman's results from a McVittie model differ from the ones based on standard Swiss cheese models. In these exact inhomogeneous solutions, the null geodesic equations can be exactly integrated in each domain and matched; this procedure naturally includes shear effects.

Kantowski (1969b) investigated this effect in a series of papers on standard Swiss cheese models with vacuum interiors, and gave a nice geometrical confirmation of the effect on observational relations. He obtained analytic expressions for distance–redshift relations that have been corrected for the effects of inhomogeneities in the density (Kantowski, 1998; 2001; 2003). The values of the density parameter and cosmological constant inferred from a given set of observations depends on the fractional amount of matter in inhomogeneities and can significantly differ from those obtained by using the Mattig relations for the FLRW universes. As an example, 'a determination of Ω_0 made by applying the homogeneous distance–redshift relation to SN 1997ap at z = 0.83 could be as much as 50% lower than its true value'.

Biswas and Notari (2008) studied an exact Swiss cheese model of the universe, where inhomogeneous LTB patches are embedded in a flat FLRW background. They found a negligible integrated effect on area distances, suppressed by $(LH_0)^3$ (where L is the size of a patch). However they found a Doppler term which is much larger. Marra *et al.* (2007) analysed similar models and found the opposite: that redshift effects are suppressed when the hole is small because of spherical symmetry. However, for the angular diameter distance, strong evolution of the inhomogeneities causes the photon path to deviate from that of the FLRW case, so the inhomogeneities are able to partly mimic the effects of a dark energy component. Marra, Kolb and Matarrese (2008) fitted a phenomenological homogeneous model to describe observables in such a Swiss cheese model. Following a fitting procedure based on light-cone averages, they found that the light-cone average of the density as a function of redshift is affected by inhomogeneities because, as the universe evolves, a photon spends more and more time in the (large) voids than in the (thin) high-density structures. Although the sole source in the Swiss cheese model is matter, the phenomenological homogeneous model behaves as if it has a dark energy component. However they find that the holes must have a present size of about 250 Mpc to be able to mimic the concordance model.

These various papers suffice to show that there may indeed be significant effects on observation from the effects of inhomogeneities in such models. Particularly, a model with genuine vacuum regions, as discussed by Kantowski, will show the largest effects. They will also affect CMB anisotropies; Bolejko (2009) considers the case of Swiss cheese models with Szekeres (Section 19.6) interiors, showing local and uncompensated inhomogeneities can induce temperature fluctuations of amplitude as large as 10^{-3}, and thus can be responsible for the low multipole anomalies observed in the angular CMB power spectrum.

Wiltshire's models

In these models, the observable universe is not accelerating but inferred luminosity distances are larger than naively expected, in accord with the evidence of distant SNIa. Observational consequences are worked out in detail in Leith, Ng and Wiltshire (2008).

Lindquist–Wheeler type models

In the appropriate limits the resulting large-scale dynamics (Clifton and Ferreira, 2009a) approach those of an FLRW universe; the optical properties of such a spacetime, however, do not. Clifton and Ferreira (2009b) show that these differences have consequences for cosmological parameter estimation, and that fitting to recent SNIa observations gives a correction to the inferred value of Ω_Λ of $\sim 10\%$. This broadly concurs with Kantowski's estimates on the basis of Swiss cheese models.

Problem 16.2 Either give a convincing argument as to why the Weyl focusing effects averaged over the whole sky produce the same results as Ricci focusing; or show it is not true (perhaps under some special conditions).

16.7 Combination of effects: altering cosmic concordance?

Averaging processes when there are many local inhomogeneities lead to dynamical back-reaction effects (see Section 16.2), and to optical effects, due both to altered redshifts and area distances. This combination of effects seems to have the potential to significantly influence interpretation of observations such as the SNIa data.

Is this influence of inhomogeneities sufficient to explain fully the apparent acceleration indicated by the SNIa data? Leith, Ng and Wiltshire (2008) make this claim, but the modelling used is controversial, and not yet universally accepted. According to Clifton and Zuntz (2009), 'It is found that intervening voids, between the observer and source, have no noticeable effect, while sources inside voids can be affected considerably. By averaging observable quantities over many randomly generated distributions of voids we find that the presence of these structures has the effect of displacing the average magnitude from its background value, and introducing a dispersion around that average'. A contrasting view, suggesting the effect is much smaller, is in Vanderveld, Flanagan and Wasserman (2007).

CMB secondary anisotropies due to the matter distribution between us and last-scattering can also be modelled using nonlinear inhomogeneities. LTB models have been used by Raine and Thomas (1981); Panek (1992); Arnau, Fullana and Sáez (1994), and Swiss cheese models by Rees and Sciama (1968); Dyer (1976); Meszaros and Molner (1996). The most important outcome is that although in principle such large-scale inhomogeneities will affect the CMB temperature, the quantitative estimates suggest that the variations are at the level of 10^{-6} and therefore hard to distinguish from the variations on the last-scattering surface expected from the seeds of large-scale structure. However, the axes of the CMB and Hubble expansion anisotropies can differ in inhomogeneous models (unlike an FLRW model: the consistency of dipole anisotropies in different data is a test of the latter).

Thus while it is very debatable whether the effect is enough to fully account for the apparent acceleration, it may well be significant (Mattsson, 2010; Amendola *et al.*, 2010). Unlike a large local void, these models respect the cosmological principle, further offering an explanation for the late onset of the perceived acceleration as a consequence of the forming of nonlinear structures. Clearly, this topic needs further careful investigation.

A tentative interim conclusion is that the dynamical effects are certainly there, but are small – not enough to explain fully the apparent cosmic acceleration (Behrend, Brown and Robbers, 2008; Clarkson, Ananda and Larena, 2009). However, when observational effects are added, the total effect may not be negligible. There are claims (Leith, Ng and Wiltshire, 2008; Kolb, Matarrese and Riotto, 2006) that these combined effects may be sufficient to do away with the need for any dark energy or cosmological constant. This seems optimistic: for the present we suggest this is a proposal that needs careful investigation, in particular taking the existence of voids seriously, but, while it may have a measurable effect, it is perhaps not likely by itself to do away with the need for a cosmological constant or some form of 'quintessence'.

This is a nontrivial conclusion: for in an era of precision cosmology, if these effects change the answer by even a few per cent, they must be taken into account in analysing the data. More than that, *they have the potential to upset the cosmic concordance that is a feature of the standard model.* Indeed this effect may show that *if the other energy densities are indeed as measured then the universe does not after all have flat spatial sections.*[1] This in turn raises significant issues for the inflationary explanation of the origin of structure – which is usually stated to necessarily imply a universe with flat spatial sections.

Thus the issue is definitely worth pursuing. One thing we know for sure is that there are indeed significant inhomogeneities in the matter distribution: great walls, filaments, clusters and voids. We must take seriously their effect on cosmological dynamics and observations. In any case, a detailed investigation of backreaction effects helps to improve the fitting of models on regional scales, to give a better interpretation of observational data. A key question is: *what fraction of matter is present outside of bound structures?* If there is enough matter to cause Ricci focusing, it should be represented by Dyer–Roeder distances. According to the Sheth–Tormen elliptical collapse model, 20% of the mass is outside bound haloes (Massey, Kitching and Richard, 2010). This could lead to a significant observational difference for a standard FLRW model.

Finally we note that the effects mentioned here could occur *in addition* to those mentioned in the previous chapter: after all if the Copernican Principle is indeed violated on a large scale (as discussed there), we should still take into account the fact that the universe is also inhomogeneous on smaller scales (as discussed here). Investigation of the combination of these effects has not yet been undertaken. The effects of inhomogeneities on SNIa observations remain poorly understood, because these observations are made on such a small angular scale that the fluid approximation is not applicable and the relevant beam of null geodesics propagates mainly through underdense space (Clarkson *et al.*, 2011a).

Problem 16.3 If there is Weyl focusing it can only work by creating substantial distortion, which should be visible in distant images as a generic defocusing effect across the sky. Can one test for this? What limits does this place on the distribution of matter in voids?

Problem 16.4 Determine the effect on SNIa observations if the Dyer–Roeder distance is used to represent observations made through voids where the density of matter is 50% of the average density of matter in the universe.

16.8 Entropy and coarse-graining

Related to all this is the puzzling question of gravitational entropy. The spontaneous structure growth in the expanding universe due to gravitational attraction appears to be contrary to

---

[1] See the detailed description in Chapter 13.

all the statements about entropy in standard textbooks (Ellis, 1995). This must somehow be related to the nature of the entropy of the gravitational field itself, not just the entropy of matter in a gravitational field.

The key feature regarding the entropy of matter, as clearly explained by Penrose (1989; 2004), is that it is associated with the loss of information that occurs with any coarse-grained description of matter. The most likely macroscopic states will be those that correspond to the largest numbers of microscopic states; that is to the largest volumes of phase space. This is made clear in Boltzmann's definition of entropy: $S = k \ln V_\Gamma$ where k is Boltzmann's constant and V_Γ the volume of phase space with points indistinguishable from each other by means of macroscopic observations of some macro (coarse-grained) variable to some accuracy ε. The dynamics of the system is accompanied by an increase of this entropy as the representative point in phase space moves from less probable to more probable states.

One might therefore expect that a proper definition of gravitational entropy would similarly be related to some kind of coarse-graining of the gravitational field. However, most attempts at definitions of gravitational entropy in the cosmological context (e.g. Pelavas and Coley (2006); Amarzguioui and Grøn (2005)) build on Penrose's proposal (1989, 2004) that it be related to the magnitude of the Weyl tensor, with no introduction of coarse-graining. This is quite puzzling, given the persuasiveness of Penrose's arguments that in the case of matter descriptions, entropy is always related to such coarse-graining. In our view this is one of the most fundamental missing aspects of gravitational theory: a satisfactory relation of gravitational entropy for a general gravitational field in terms of a coarse-grained description of that field, therefore relating to all the issues mentioned in the preceding sections.

A promising start has been made by Hosoya, Buchert and Morita (2004): if we are only concerned with averaging the matter inhomogeneities on an inhomogeneous geometry, one can *deduce* an entropy measure for the distinguishability of the density distribution from its average value directly from the non-commutativity rule:

$$\partial_t \langle \varrho \rangle - \langle \partial_t \varrho \rangle = -V^{-1} \partial_t S\{\varrho || \langle \varrho \rangle\}, \quad S\{\varrho || \langle \varrho \rangle\} := \int \varrho \ln \varrho \langle \varrho \rangle^{-1}, \quad (16.20)$$

where the functional $S\{\varrho || \langle \varrho \rangle\}$ is known in information theory as the Kullback–Leibler relative entropy, and spatial averaging and integration is performed over a domain with volume V. Hosoya, Buchert and Morita (2004) conjecture that this functional is, after a sufficient period of time, always globally increasing. This counter-intuitive statement (in view of canonical considerations, e.g. in isolated Markovian systems) is justified in a self-gravitating system because gravity is long range, the averaging domain is not isolated, and gravity invokes a negative feedback: structural inhomogeneities are amplified due to gravitational instability. We may expect that the information content in the matter inhomogeneities is always increasing.

Given such a definition, the problem is to determine whether increasing *total* entropy (in the gravitational field and in the matter distribution) occurs always, or whether this is true only for special initial conditions. As discussed by Penrose (1989; 2004), it seems plausible that the latter is the case, with the arrow of time in physics arising from boundary conditions at the start and end of the universe: specifically, the Weyl tensor taking a special form at the start of the expansion of the universe but a generic form at the end. The specific details of this proposal have never been clarified, and it is possible that the relation is not due to the Weyl tensor itself, but rather due to a spatial integral of the divergence of the electric part of the Weyl tensor (Ellis and Tavakol, unpublished). A further problem is then relating the arrow of time for structure growth in the universe to that for electromagnetic and gravitational radiation (Ellis and Sciama, 1972). Here again coarse-graining is crucial, for this relates to the kind of multi-scale description of the gravitational field envisaged by Isaacson, as discussed above.

Entropy and the associated arrow of time are fundamental to macroscopic physics. Their foundations in relation to microphysics remain mysterious in the case of general gravitational fields. The entropy of black holes is of course well understood, but this is an extreme case that does not by itself help us understand the relation of entropy to spontaneous structure formation in the expanding universe. Until this is solved, we cannot claim to properly understand the nature of entropy in the cosmological context.

PART 4

ANISOTROPIC AND INHOMOGENEOUS MODELS

17

The space of cosmological models

Although the observations appear to be well fitted by perturbed FLRW models, as described above, more general models need to be considered. One major reason is that the appropriateness of the perturbed FLRW models cannot be said to have been tested unless the consequences of alternatives have been calculated and compared with observation. In particular, there could be drastic changes to the models for the very early universe, since what may now be small and decaying perturbations in the standard picture would have been non-negligible earlier, and could give very different dynamics. Local observations can bound, but could not be sure to detect, such perturbations, so their testable consequences, if any, must arise from effects in the early universe.

We also need to consider the possibility of large-scale anisotropies, for example arising from a cosmic magnetic field aligned on a supergalactic scale, and of large-scale inhomogeneities (advanced as a possible explanation, which we discussed in Chapter 15, of the apparent acceleration seen in the supernova data).

This chapter considers the space of all models and the definition of classes of cosmological models wider than the FLRW models (compare e.g. Ellis (2005)). There are many ways of classifying spacetimes, of which the most common are by symmetry and by Petrov type (see Stephani *et al.* (2003)). In the cosmological case, symmetries are the more relevant and we consider that here. (Some models characterized by other covariant properties are described in Sections 19.6 and 19.7.) These are local properties. We need also to understand global properties, such as causal structure, global topology, asymptotics, singularities and horizons (see Sections 3.3.1, 6.7, 7.9, 8.4.2 and 20.6.1 for discussions of some of these issues); this is in strong contrast to Newtonian theory, where the global spacetime structure is very simple.

We also outline how to check if two apparently different models are locally isometric. Finally, we discuss the space of all cosmological models, its representation and dynamics.

The dynamical and observational properties of the spatially homogeneous models and the more important inhomogeneous models will be considered in the following chapters. The spacetimes with even greater symmetry are so simple that the brief discussion in this chapter suffices.

17.1 Cosmological models with symmetries

In Section 2.7.3, it was shown that an isometry group of dimension r acting on a Riemannian space M of dimension n will define an orbit, of some dimension d, through each point of

M: the orbits are submanifolds of M (so $d \leq n$: in spacetime $d \leq 4$). On an orbit all physical properties defined by the metric are invariant. When $r > d$, there is at each point p an isotropy group of dimension s, and $r = d + s$.

By considering the action of isotropies on the tangent space at p, one can easily show that $r \leq d(d+1)/2$. If $d > 2$, $r \neq d(d+1)/2 - 1$, because the group of isotropies of the tangent space to the orbit at p cannot have dimension one less than the relevant orthogonal group. The case of maximal symmetry, $r = d(d+1)/2$, is the case of a space of *constant curvature* (Section 2.7.7), which can have positive, zero or negative curvature.

In particular $d = n = 4$, $r = 10$ is the maximal symmetry a spacetime can have (de Sitter space, Minkowski space, or anti-de Sitter space respectively for the three curvatures), with $s = 6$. These models apply only if the sole 'matter' content is a cosmological constant: while not realistic as a model of the present-day universe, de Sitter space plays an important role as the approximate metric in (exponential) inflation, while the anti-de Sitter spaces of higher dimension are important in string theory (Section 20.3).

In addition to the isometries of a model, there may be a homothetic motion and one or more conformal symmetries (see (2.61)). Spacetimes with homothety are called self-similar, and since the homothety implies a known dependence on one of the essential coordinates, homothety shares with isometries the property of reducing the number of independent variables in the field equations remaining to be solved.

17.1.1 Isotropy

Cosmological models often have a *perfect fluid matter content* such that $(\rho + p) > 0$. In this case the unit future-pointing timelike eigenvector $\mathbf{u}$ of the Ricci tensor at each point, with eigenvalue ρ, is unique. At each point the isotropy group must act in the three-dimensional tangent space orthogonal to $\mathbf{u}$ (so leaving $\mathbf{u}$ invariant), and thus can have dimension s at most 3. It cannot have dimension 2 since there are no subgroups of dimension 2 of the three-dimensional rotation group. Note that there can be particular orbits where s is larger and d is smaller than is the generic case in the spacetime (e.g. the centre of symmetry of a spherical star has $s = 3$, $d = 0$, whereas the other orbits are spheres with $s = 1$, $d = 2$).

Uniqueness of the timelike Ricci eigenvector, and its consequences, also applies for most other physically acceptable non-zero energy–momentum tensors. The cosmologically relevant exception is when one has only a cosmological constant (which includes the de Sitter family, the spaces of constant curvature), though this is not an exact representation of the real universe at any stage, since there is always some matter or field present.

When the timelike Ricci eigenvector is unique, the only options for the isotropy group are:

$s = 3$: **(Complete) isotropy**, which is the FLRW case: all spatial directions are equivalent; the Weyl tensor vanishes as do all kinematic quantities except Θ.

$s = 1$: **Local rotational symmetry** (LRS). There is one preferred spatial direction, $\mathbf{x}$ say; the Weyl tensor is of type D (or zero) and all kinematic and observable quantities are rotationally symmetric about $\mathbf{x}$. These models fall into three classes, I-III (Ellis, 1967): in class I $\mathbf{u}$ is rotating and the metrics admit a G_4 on timelike hyperplanes; in class II the planes defined by $\mathbf{x}$ and $\mathbf{u}$ are integrable and there is a G_3 acting on the surfaces (spheres, planes or pseudospheres) orthogonal to those planes; in class III the planes are not integrable and the spacetimes are spatially homogeneous.

$s = 0$: **No continuous isotropy**, all neighbouring directions are inequivalent (but there can be discrete isotropies: models which are *locally discretely isotropic*, i.e. have the same discrete isotropy at every point, admit continuous groups of motions (Schmidt, 1969, Mena and MacCallum, 2002)).

17.1.2 Spacetime homogeneous geometries

These models with $d = 4$ are unchanging in space and time. We ignore here the constant curvature spaces and discuss those where the curvature is the same everywhere, but not 'constant'. Then ρ is a constant, so by the energy conservation equation (5.11), if $\rho + p \neq 0$ no invariantly defined timelike congruence can expand: $\Theta = 0$.[1] Thus by (7.17) these spacetimes cannot produce any redshift, in the frame of the matter present, and are not useful as models of the real universe. Nevertheless they have some interesting features.

The *isotropic case* $s = 3$ ($\Rightarrow r = 7$) is the Einstein static universe, the non-expanding FLRW model (Section 6.1.1) that was the first relativistic cosmological model found. Although not a viable cosmology (no redshifts), it laid the foundation for the discovery of the expanding FLRW models.

The *LRS case* $s = 1$ ($\Rightarrow r = 5$) is the Gödel (1949) stationary rotating universe, again with no redshifts. This model was important because it prompted new understanding of the effects of rotation and the nature of time in GR (see Hawking and Ellis (1973), Tipler, Clarke and Ellis (1980), Ellis (1997)). Inter alia, it is a model in which causality is violated (there exist closed timelike lines through each spacetime point) and no cosmic time function whatsoever exists.

The anisotropic models $s = 0$ ($\Rightarrow r = 4$) are all known (Ozsváth, 1965, 1970) but are cosmologically interesting only for the light they shed on effects of rotation and on Mach's principle; see e.g. Ozsváth and Schucking (1962), Rosquist (1980).

17.1.3 Spatially homogeneous geometries

These models with $d = 3$ have played a major role in theoretical cosmology, because they express mathematically the idea of the 'cosmological principle': all points of space at the same time are equivalent to each other (Bondi, 1960). They are discussed further in Chapter 18.

The *isotropic case* $s = 3$ ($\Rightarrow r = 6$) is the family of FLRW models discussed in Chapter 9.

The *LRS case* $s = 1$ ($\Rightarrow r = 4$) contains the Kantowski–Sachs universes (Kompaneets and Chernov, 1965, Kantowski and Sachs, 1966, Collins, 1977) and the LRS orthogonal (Ellis and MacCallum, 1969) and tilted (Farnsworth, 1967, King and Ellis, 1973) Bianchi models.

---

[1] Note that in the spacetimes of constant curvature expanding congruences can be, but need not be, chosen: all timelike vectors at a point are related by spacetime symmetries.

The *anisotropic case* $s = 0$ ($\Rightarrow r = 3$) consists of the Bianchi universes, with a simply transitive group G_3 of isometries acting on spacelike surfaces. They are classified into nine types (Bianchi I to IX) and two major classes: *tilted* or *orthogonal*. Some have an additional homothety, and these special solutions often appear as sources or sinks in the dynamical system for the larger set of models (see Chapter 18).

Caveat: instead of focusing on the nature of the surfaces of homogeneity we could focus on the equivalence of experiences of observers (as discussed in Chapter 13).

17.1.4 Spatially inhomogeneous universes

These models have $d \leq 2$: discussion of their cosmological applications can be found in Chapters 15, 16 and 19, and, more extensively, in Krasiński (1997) and Bolejko *et al.* (2010): for exact solutions in the various classes see also Stephani *et al.* (2003). Models with an additional self-similarity are included in these discussions.

The *LRS cases* ($s = 1 \Rightarrow d = 2, r = 3$) have the metric form

$$ds^2 = -C^2(t,r)\,dt^2 + A^2(t,r)\,dr^2 + B^2(t,r)\,(d\theta^2 + f^2(\theta)\,d\phi^2)\,, \qquad (17.1)$$

where $f(\theta)$ is as in (2.65). These are discussed in Section 19.4 and, for $K = 1$ dust (LTB models), in Sections 15.1 and 19.1. In the dust case, we can set $C(r,t) = 1$ and integrate the field equations analytically. The models may have a centre of symmetry (a timelike world line), and can even allow two such centres, but they cannot be isotropic about a general point (because isotropy everywhere implies spatial homogeneity, see Section 9.1.1).

When $d = 2$, $s = 0$, and there are two commuting Killing vectors (a $G_2 I$), they may act on a timelike surface. The resulting solutions include the stationary axisymmetric solutions important as models of rotating isolated bodies, which we do not discuss further here. They also include static plane and cylindrical solutions used to model domain walls and cosmic strings. We briefly consider those in Section 19.5, along with the models where the $G_2 I$ acts on a spacelike surface. The spatially self-similar models (see Section 19.3) admit a G_2.

Models with exactly two non-commuting Killing vectors (a $G_2 II$) have been relatively little studied. It is known that they cannot admit two-surfaces orthogonal to the group orbits if the fluid flow is orthogonal to the orbits (unless they have an extra symmetry) and that if the fluid is thus orthogonal it is non-rotating (Bugalho, 1987, Van den Bergh, 1988, Aliev and Leznov, 1992). The 'stiff fluid' (Section 5.2.2) is a special case, discussed in detail by Van den Bergh (1992). Most of these solutions have singularities at finite spatial distances or can be regarded as inhomogeneous perturbations of the Bianchi VI_{-1} models. They have not been applied to major issues in cosmology, and we shall not consider them further.

Few models with $d \leq 1$ are known: most have not been seriously applied to significant cosmological issues, and so will not be discussed in Chapter 19. Among them are: Oleson's (1971) perfect fluid solutions of Petrov type N, which in general have no isometries; the solutions obtained by Martín and Senovilla (1986) and Senovilla and Sopuerta (1994) by a generalization of the Kerr–Schild ansatz, which have a G_1; the solutions found by

Martin-Pascual and Senovilla (1988) which belong to the Wainwright (1974) class; and the solutions of Stephani and Wolf (1985), found by assuming the existence of flat three-dimensional slices. Rainer and Schmidt (1995) have discussed the generalization of the last case to surfaces admitting Bianchi groups.

Solutions with no symmetries at all have $r = 0 \Rightarrow s = 0, d = 0$. The real universe, of course, belongs to this class; all the other models can only provide approximations to this actual universe. Remarkably, we know some exact solutions of interest in cosmology which have no symmetries: these are the Szekeres–Szafron models (see Section 19.6), which are in a sense nonlinear perturbations of the FLRW models, and Stephani's models, both those (Stephani, 1987) with a conformally flat 3-space (see Section 19.6) and (Stephani, 1967) the conformally flat metrics (see Section 19.7).

Inhomogeneous universes can also be constructed by combining two or more solutions, either by matching portions of different solutions together or by nonlinearly superposing solutions. The composite solution generally has less symmetry than the parts from which it is made. The main example of the first type is the Swiss cheese model introduced in Section 16.4.1, and analogous models (see Section 19.2).

The McVittie solution (19.4) is an example of nonlinear superposition. Few other such models have been discussed. A number of authors have considered the superposition of a Kerr solution and an FLRW spacetime, aimed at modelling rotating black holes in the cosmos, but in general these require 'null radiation', i.e. a directed lightlike flow of energy, which is not very realistic: see Krasiński (1997), chapter 5. Nolan and Vera (2007) have considered the boundary conditions for a rotating fluid body interfaced to an asymptotically Friedmann background.

17.1.5 Summary of possible symmetric models

Putting this together, we obtain the classification of possible symmetries of cosmological models given in Table 17.1.

Exercise 17.1.1 What further features, other than those mentioned above, make the Einstein static spacetime problematic as a cosmological model?

Exercise 17.1.2 Show that in the case of FLRW models, no causal violations can occur. Does this result extend to perturbed FLRW models? What about Bianchi models? [See Gödel (1952) for some examples.]

Problem 17.1 Investigate whether any of the models mentioned above with $d \leq 1$ have interesting cosmological properties.

Problem 17.2 Solutions of the Einstein equations are of rather diverse nature. What restrictions should one impose to characterize the subspace of solutions that might reasonably be called cosmological?

Table 17.1 Classification of spacetimes by symmetries. s is the dimension of the group of motions and d that of the orbits. There are no isotropies when $d \leq 1$. A Swiss cheese with multiple 'holes' may have no symmetry.

	$d = 4$ (no redshift)	$d = 3$	$d = 2$
$s = 3$ (isotropic)	Einstein static	Friedmann (Chapter 9)	none
$s = 1$ (LRS)	Gödel	Kantowski and Sachs; LRS Bianchi (Chapter 18).	LTB (Sections 15.1, 19.1); Simple Swiss cheese (Sections 16.4.1,19.2); and other LRS cases (Section 19.4).
$s = 0$	Ozsváth/Kerr	Bianchi (Chapter 18).	See Section 19.5. Includes spatially self-similar models, see Section 19.3.

$d = 1$ $d = 0$

Inhomogeneous, no isotropy group ($s = 0$)

Various cases awaiting cosmological application (Section 17.1.4).

The real universe!

Szekeres–Szafron (Section 19.6).

Stephani–Barnes (Section 19.7).

17.2 The equivalence problem in cosmology

In order to consider the space of possible cosmologies unambiguously, one needs to be able to check whether two apparently different models are in fact the same. There is a general procedure for checking the local equivalence of two explicit spacetime metrics, by comparing the complete covariant local classifications based on calculating the Riemann tensor and its derivatives in a canonically chosen frame. The components calculated are known as Cartan invariants. The method has been applied to exact solutions and developed for other problems. A brief outline of it follows: for a fuller survey see Stephani *et al.* (2003), chapter 9.

To relate two apparently different metrics, we need to consider coordinate or basis transformations, and therefore to consider the set of all frames, leading to the frame bundle (which at each point p of M consists of all possible bases of the tangent space $T_p(M)$). If the metrics are equivalent, the frame bundles they define are identical (locally). In the case of spacetimes, this implies that the components of the curvature on the frame bundle will be equal when corresponding points have been correctly identified.

This condition is necessary, but not sufficient. Cartan showed that a sufficient condition is obtained by repeatedly taking derivatives of the curvature until no new functionally independent quantity arises; at that stage the process terminates because then any further derivatives depend on those already known. The relations between the independent invariants and the dependent ones must be the same in (neighbourhoods in) both manifolds for equivalence. Since the number k of functionally independent quantities is at most the dimension m of the manifold, the process necessarily terminates in a finite number of steps. If $k < m$, this is due to the presence of symmetries.

Hence a metric can be locally uniquely characterized by the Riemann tensor and a finite number of its covariant derivatives expressed in a canonical frame (collectively called the Cartan invariants). For practical application one needs to cast the curvature and derivatives at each step of differentiation into a canonical form and only permit those frame changes which preserve the canonical form. Existing computer implementations tend to use null tetrads, but orthonormal tetrads could equally well be used.

It should be noted that for cosmological models, it appears that scalar polynomial invariants in the Riemann tensor and its derivatives would suffice (Coley, Hervik and Pelavas, 2009), but generally these require more computation than the Cartan invariants.

In the cosmological context it is useful to express the method using the 1+3 formalism. Where one knows the full four-dimensional solution, and there is an invariant timelike vector (e.g. a fluid velocity), and its shear uniquely defines a set of eigenvectors, the general procedure can be applied using these vectors as an orthonormal frame. If an invariant timelike vector is determined but the spacelike vectors are not unique, i.e. there is some local isotropy, further discussion is needed. Taking perfect fluid cosmologies in which the kinematic quantities and Weyl tensor are rotationally symmetric leads to LRS solutions, as defined in Section 17.1.1, or Szekeres dust models as in Section 19.6 (Mustapha *et al.*, 2000).

A more difficult problem is to give a complete characterization in terms of initial data on a Cauchy surface, when the solution off the surface is not known explicitly. Although in principle the Cauchy data determine the spacetime (in its domain of dependence) uniquely, as would the characterization by Cartan invariants, relating the two is not easy and so far solved only for Schwarzschild or Kerr data (see e.g. Garcia-Parrado and Valiente Kroon (2008)). The problem should be simpler when the Cauchy surface is defined by an invariant timelike normal vector. We would like to be able to express results using the kinematic and other quantities in the 1+3 approach as presented earlier.

Problem 17.3 Find a general procedure to test equivalence of cosmological models using the 1+3 formalism.

17.3 The space of models and the role of symmetric models

The space of all solutions of Einstein's equations is clearly an infinite-dimensional space, although we do not know an entirely satisfactory way to coordinatize it. One can put a symplectic structure directly on the space of four-dimensional metrics (see e.g. Szczyryba (1976)) but most attempts use some form of 1+3 decomposition (it may be possible to develop an alternative by defining a topology on the space of solutions using values of invariants).

Since we expect cosmologies, or those regions of them which describe the part of a possibly larger universe that we observe, to allow Cauchy surfaces, we can consider the solutions to be described by paths in the space of metrics on spacelike hypersurfaces, with tangent vectors given by the extrinsic curvatures. Because the constraint equations are preserved by the second-order evolution equations, as discussed in Section 6.6, only the latter need be considered once initial conditions have been set. Thus the structure is the one common in physics: second-order governing equations which have a unique solution given the initial 'position' (spatial metric) and 'velocity' (initial extrinsic curvature). Of course, for non-vacuum cosmologies, the geometric variables need to be supplemented by appropriate matter variables and field equations for them.

This description does not take into account either the fact that the same geometry can be represented by different spatial metrics (if they are equivalent under diffeomorphisms of the hypersurface, although it is not always advantageous to identify isometric space-times (see Section 18.5.3)) or the fact that the same four-dimensional geometry may be described by different slicings and hence different paths through the space of hypersurface metrics. On the first aspect, 'geometrodynamics' (Wheeler, 1962) aims to use the space of geometries, 'superspace', rather than that of metrics. Numerous papers have considered ways of making this explicit, and, for example, of recasting the system of equations to make it numerically stable. This representation of the space of spacetimes, with the ADM formalism (Section 3.3.3), is the basis of quantum cosmology (see Section 20.2.1).

The local structure of the infinite-dimensional space is, as in a finite case, described by infinitesimal differentials. One question that has been studied is whether all linearizations about a given metric are in fact tangents to families of metrics approaching the given one, so-called 'linearization stability'. For compact hypersurfaces, metrics with symmetry are known not to be linearization stable: the space of solutions is conical there (see e.g. Marsden (1982)). Only tangents obeying additional conditions, due to Taub (1971), are allowed. However, no restrictions are known for the non-compact non-asymptotically flat spacetimes we might expect as cosmological models, though there may be linearization instability when the neighbouring spacetimes are restricted to those with a common symmetry (Brill and Vishveshwara, 1986). Brill, Reula and Schmidt (1987) showed that linearization stability is always true for any finite region of spacetime with boundary, and Brauer (1991) found that generic spherical inhomogeneities superimposed on FLRW were linearization stable for $K \leq 0$. So while this is a technical worry, e.g. in studying perturbations, it seems not to be of practical importance.

We can thus think of the evolution of cosmological models as taking place in an infinite-dimensional dynamical system on a phase space whose variables are the spatial metric and extrinsic curvature (with the caveat that much of the theory of dynamical systems has only been proved to apply to ordinary, not partial, differential equations). Within this space are subspaces which are invariant sets. Note that defining these sets requires the 'velocity' to be chosen appropriately, as well as the 'position'.

Cosmologies with symmetry form particularly important invariant sets.

Theorem 17.1 *If the metric and extrinsic curvature on a Cauchy surface are invariant under a continuous isometry, then the Cauchy development will also be invariant under that isometry.*

The proof for the general case follows that of Lemma 2.1 in Ellis and King (1974), although the result is stated there only for the spatially homogeneous case. Note that this result is not true in Newtonian cosmology, because its equations are not hyperbolic equations and the evolution can be affected instantaneously by boundary conditions at infinity. An example is given by the FLRW metrics: see Section 9.8.2.

One reason these invariant sets are of interest is that they may form a 'skeleton' in the larger space of less symmetric models, a set of curves which more general models may approximate, so that the higher symmetry cases are behaving like attractors (or, from the time reverse, 'sources' of the flow), or may produce saddle points in the space of solutions. In the direction of expansion these attractors are often models with an additional homothety, self-similar models, and this is also sometimes true in contraction: see Wainwright and Ellis (1997) and Chapter 18. It is conjectured, and has been proved for some cases, that this behaviour extends to the infinite-dimensional dynamical systems describing the evolution of models with less or no symmetry (see Chapter 19).

These properties are clearly shown in the Ehlers–Rindler phase space for FLRW models (see Figure 9.2). More generally, they have been proved to be correct for many of the dynamical systems for particular classes of Bianchi models (where the homogeneity ensures we have ordinary differential equations), which are discussed in the next chapter. They have also been conjectured, and in some cases have been proved, to apply to the infinite-dimensional dynamical systems describing the evolution of inhomogeneous models, with less or no symmetry (see Chapter 19 and Uggla *et al.* (2003)). This is based on the form of the evolution equations in appropriately normalized variables, the attractor (in the time direction towards the big bang) consisting of Kasner and Mixmaster oscillatory models as described in Chapter 18; see Section 19.10.1.

Exercise 17.3.1 Prove Theorem 17.1. (See Collins and Ellis (1979).)

18

Spatially homogeneous anisotropic models

FLRW models are spatially homogeneous, but they are a very restricted subclass of such models because of their isotropy. Why are spatially homogeneous anisotropic models interesting? Basically, because they are tractable solutions of the full non-linear equations since there is only one essential variable, time, so the equations become ordinary differential equations, but they allow investigation of much more general behaviour than the FLRW models. They can represent anisotropic modes, including rotation and global magnetic fields, which could occur in the real universe (indeed, must do so, if the universe is indeed generic, as some claim): here an anisotropic but not necessarily inhomogeneous model is required (see e.g. Thorne (1967)). They allow new classes of singularities, and modification of the BBN–baryon relation in the early universe. They may also be good approximations in regions where there is inhomogeneity but spatial gradients are small, see Section 19.9. They have been explored in various quantum cosmology contexts (see Chapter 20) as well as in GR.

In particular, the tilted cases provide the only tractable cosmological solutions we have which involve rotation: rotation is ubiquitous in the universe, and, because of the vorticity conservation theorems discussed in Chapter 6, this suggests there always was and always will be rotation. Thus it is valuable to have solutions where we can investigate its effects on, for example, the CMB, where we find new classes of anisotropy patterns, and on nucleosynthesis. The models do not provide bounded objects stabilized by rotation such as we see around us, but could provide seeds of rotation for them. The reasons for looking at expanding and rotating spatially homogeneous universes now are as good as when Gödel looked at them in 1952 (and showed they must have non-zero shear).

How could we test if we live in a spatially homogeneous universe if it is anisotropic? (Compare the discussions in Section 9.8 of testing whether we live in an FLRW model, and, in Chapter 15, of the difficulty of testing (statistical) homogeneity.) Direct observational testing is even more difficult than in the FLRW case and there are so far no observations compelling us to assume the background averaged universe is spatially homogeneous but anisotropic. The Postulate of Uniform Thermal Histories (Section 9.8) will also remain a viable option, as there will indeed necessarily be uniform thermal histories in those models. This remains an issue open for investigation.

As stated in Section 17.1, spatially homogeneous anisotropic models have an isometry group G_r, $r \leq 4$. They are either Bianchi models, with a G_3 of one of the nine types I–IX defined in Section 18.3 below acting simply transitively on spacelike hypersurfaces, or, as a special case, the Kantowski–Sachs (K-S) metrics with $K = 1$ and a G_4 containing no simply transitive G_3. The isometry group maps the timelike observer world lines orthogonal to the hypersurfaces into one another, and similarly maps invariantly defined matter world lines (e.g. those of a perfect fluid) intersecting the hypersurfaces into one another. When the normal and matter world lines coincide, we say the model is 'orthogonal': otherwise it is 'tilted'. Tilted models are significantly more complicated than orthogonal models.

Orthogonal models must be irrotational, because the world lines are hypersurface-orthogonal, so rotating models *must* be tilted (compare the discussion in Section 4.6). In the tilted case, whether or not there is rotation, the homogeneous hypersurfaces (and their orthogonal curves) can change character, i.e. there may be a limiting null surface beyond which the hypersurfaces become timelike.

There have been many papers analysing the systems of equations governing the K-S and Bianchi models, finding exact solutions of these equations, and discussing physical properties of the models. These published up to 2003 were discussed in detail in Wainwright and Ellis (1997) and Coley (2003), and for exact solutions Stephani *et al.* (2003), to which the reader is referred for fuller accounts than we can give here. In particular we shall not have space for full discussion of the more complicated energy momenta that have been considered (see e.g. Coley (2003)).

We discuss the simple cases of Kantowski–Sachs and Bianchi I universes, before proceeding to the general dynamics of Bianchi models and the conclusions one can draw concerning cosmological questions.

18.1 Kantowski–Sachs universes: geometry and dynamics

Kantowski (1966) (compare Kantowski and Sachs (1966)) considered metrics of the form

$$ds^2 = -dt^2 + A^2(t)dx^2 + B^2(t)[dy^2 + f^2(y)dz^2], \qquad (18.1)$$

with $K = \pm 1$, where f for various K is as in (2.65). Particular cases, including $K = 0$ metrics, had earlier been discussed by Kompaneets and Chernov (1965), Doroshkevich (1965) and, in his 1965 thesis, Thorne (see Thorne (1967)). The Ricci tensor takes a very simple diagonal form, and the field equations become

$$\frac{2\ddot{B}}{B} + \frac{\dot{B}^2}{B^2} + \frac{K}{B^2} = \Lambda - \kappa_0 p_1,$$

$$\frac{\ddot{B}}{B} + \frac{\ddot{A}}{A} + \frac{\dot{A}}{A}\frac{\dot{B}}{B} = \Lambda - \kappa_0 p_2, \qquad (18.2)$$

$$\frac{2\dot{A}\dot{B}}{AB} + \frac{\dot{B}^2}{B^2} + \frac{K}{B^2} = \Lambda + \kappa_0 \rho.$$

Here p_1 and p_2 are the pressures along the x direction and tangential to the (x, y) surfaces respectively. For a perfect fluid, $p_1 = p_2$. However, these metrics have been used in several discussions of the effects of a cosmic magnetic field, and $p_1 \neq p_2$ when a magnetic field is added. The same applies for other forms of matter with anisotropic stresses.

For $K = -1$ and $K = 0$, these are LRS models of Bianchi types III and I respectively, which we discuss later. For $K = 1$, which we call *the* Kantowski–Sachs form, there is no transitive G_3 subgroup of the G_4 group of isometries. In this case the $t =$ constant,

$x = $ constant, 2-surfaces are spheres, but they all have the same area $(4\pi B^2)$ and so cannot be considered concentric: rather they form a three-dimensional cylinder with spherical cross-section. In particular the models do not include any FLRW geometries, except that a portion of de Sitter space can be put in the form (18.1) (Torrence and Couch, 1988).

Once given equations of state, (18.2) with $K = 1$ gives a two-dimensional dynamical system, which was investigated by Collins (1977) for the case of perfect fluid with $\Lambda = 0$. The models evolve from an initial to a final singularity, with a time of maximal expansion, as do those with $\Lambda < 0$; those with $\Lambda > 0$ approach the de Sitter solution (Weber, 1984; 1985). One can of course, as in the usual $K = 1$ FLRW models, assume that the universe will re-expand after its collapse, but this necessitates providing some model of behaviour of matter at high density which produces such a bounce. The system of equations describes part of the boundary of the LRS Bianchi IX system of Uggla and von Zur-Muhlen (1990), since the group structure is a limit of the LRS Bianchi IX structure in which the Bianchi IX subgroup acts on a sphere rather than a hypersurface (compare Chapter 8 of Wainwright and Ellis (1997)).

Some other cosmological implications have been studied as follows. Only some particular inflationary potentials in the metric (18.1) have been considered (e.g. by Byland and Scialom (1998)). Since in general A/B does not become approximately constant initially isotropic distributions in kinetic theory will not remain so, even approximately, and thus the CMB would show a quadrupolar anisotropy, except in the future of $\Lambda > 0$ models. Nucleosynthesis would be affected by the change of time evolution (Thorne, 1967).

These problems, coupled perhaps with a prejudice against the global topology implied when $K = 1$, explain why the K-S models and their $K \leq 0$ counterparts, despite their simplicity, have not often been applied as models of the actual universe, although their exact solutions have been much explored and they are widely used as testbeds e.g. for quantum effects.

Exercise 18.1.1 Show that the K-S models have no simply transitive group of isometries (i.e. are not Bianchi models) if $K = 1$. Show that the K-S vacuum solution is the Schwarzschild solution for $r < 2M$ (i.e. inside the event horizon).

18.2 Bianchi I universes: geometry and dynamics

These are the simplest generalization of the flat FLRW models to allow for different expansion factors in three orthogonal directions. The metric can be given in the form (Heckmann and Schucking, 1962)

$$\mathrm{d}s^2 = -\,\mathrm{d}t^2 + \ell_1^2(t)\,\mathrm{d}x^2 + \ell_2^2(t)\,\mathrm{d}y^2 + \ell_3^2(t)\,\mathrm{d}z^2\,, \quad u^a = \delta^a{}_0\,. \tag{18.3}$$

The corresponding average expansion scale factor is $\ell(t) = (\ell_1\ell_2\ell_3)^{1/3}$. There are LRS and isotropic (RW) subcases (the latter, for dust, being the Einstein–de Sitter universe).

The Bianchi I metrics are important because other Bianchi models may and often do have phases in their evolution when the terms in spatial curvature in (6.54) become negligible and the dynamics approximates the Bianchi I case. Their solutions for perfect fluids are widely used as backgrounds in which to model other processes, and the metric form is often used with more complicated matter content, because of its simplicity (though note that there has to be no net energy flux and the total spatial stresses must align with the shear).

The space sections $\{t = \text{const}\}$ are flat (in a surface $t = t_0$, all the metric coefficients are constant). The normal to these homogeneous surfaces, which must also be the timelike Ricci eigenvector, is necessarily geodesic and irrotational. Thus these models obey the restrictions

$$\dot{u}^a = \omega^a = 0 , \quad X_a = Z_a = \overline{\nabla}_a p = 0 , \quad {}^3R_{ab} = 0 , \tag{18.4}$$

where we have assumed the matter to be a perfect fluid (X_a and Z_a here are as defined in Section 9.1.1).

We can find a tetrad in the obvious way from the above coordinates ($e_1{}^\mu = \ell_1(t)^{-1}\delta_1{}^\mu$, etc.); then, for a perfect fluid, the tetrad equations of Section 6.5 hold with $\sigma_{ij} = 0$ if $i \neq j$ and

$$\dot{u}^i = \omega^i = \Omega^i = 0 , \quad a^i = n_{ij} = 0 , \tag{18.5}$$

$$\partial_i(\Theta) = \partial_i(\sigma_{jk}) = 0 , \quad \partial_i(\rho) = \partial_i(p) = 0 . \tag{18.6}$$

It follows that the $(0i)$ equations (6.51) are identically satisfied, and that $H_{ab} = 0$ and $\overline{\nabla}_b E^{ab} = 0$. From the Gauss equation (6.22), the shear obeys

$$(\ell^3 \sigma_{ij})\dot{} = 0 \quad \Rightarrow \quad \sigma_{ij} = \frac{\Sigma_{ij}}{\ell^3} , \quad (\Sigma_{ij})\dot{} = 0 , \tag{18.7}$$

which implies

$$\sigma^2 = \frac{\Sigma^2}{\ell^6} , \quad \Sigma^2 := \tfrac{1}{2}\Sigma_{ij}\Sigma^{ij} , \quad (\Sigma^2)\dot{} = 0 . \tag{18.8}$$

All the field equations will then be satisfied if the conservation equation (5.11), the Raychaudhuri equation (6.5), and the Friedmann-like equation (6.23) are satisfied. As in the FLRW case, the last is the first integral of the other two.

Assuming a linear equation of state and using (18.8), equation (6.23) becomes the generalized Friedmann equation,

$$3\frac{\dot{\ell}^2}{\ell^2} = \frac{\Sigma^2}{\ell^6} + \frac{M}{\ell^{3w+3}} + \Lambda . \tag{18.9}$$

We can of course think of the Λ term as a second fluid with $w = -1$, and can similarly add together the effects of several fluids.

The Bianchi I case gives an understanding of some of the critical values of w which arise in many models. When $w > 1$ the matter term dominates as we approach the big-bang singularity and the behaviour is quasi-isotropic, similar to an FLRW universe. Taking $\Lambda = 0$, when $w < 1$, the matter term dominates evolution at large ℓ and drives the metric locally towards isotropy, $\sigma/\Theta = 0$. However, Bianchi I models are atypical in that there is no restoring force term in the $\dot{\sigma}$ equation: one set of such terms will come, in more complicated models, from the spatial curvature, which evolves in general like $1/\ell^2$, suggesting a bifurcation in behaviour at $w = -\tfrac{1}{3}$. (In addition we note from (6.18) that in rotating universes – which are not of course Bianchi I – one can expect a bifurcation at $w = 1/9$.)

Early shear domination

The formula (18.9) shows that no matter how small the shear today, it will if $w < 1$ dominate the very early evolution of a Bianchi I universe, which will then approximate the Kasner vacuum solution (see e.g. Stephani *et al.* (2003), Section 13.3),

$$\mathrm{d}s^2 = -\mathrm{d}t^2 + t^{2p_1}\mathrm{d}x^2 + t^{2p_2}\mathrm{d}y^2 + t^{2p_3}\mathrm{d}z^2, \quad \sum_1^3 p_i = \sum_1^3 (p_i)^2 = 1, \qquad (18.10)$$

where the p_i are constants. Note that the restrictions on the p_i imply the allowed values form a circle in the plane $\sum_1^3 p_i = 1$, the *Kasner circle*.

On writing out the tetrad components of the shear equation (18.7), and using the commutator relations (6.43) to determine the shear components, one finds that the individual length scales are given by

$$\ell_i(t) = \ell(t)\exp(\Sigma_i\, W(t)), \qquad \text{where } W(t) = \int \frac{\mathrm{d}t}{\ell^3(t)} \qquad (18.11)$$

and the constants Σ_α satisfy

$$\Sigma_1 + \Sigma_2 + \Sigma_3 = 0, \qquad \Sigma_1^2 + \Sigma_2^2 + \Sigma_3^2 = 2\Sigma^2\,.$$

These relations can be satisfied by setting

$$\Sigma_j = \tfrac{2}{3}\Sigma \sin \alpha_j\,, \quad \alpha_1 = \alpha\,, \quad \alpha_2 = \alpha + \tfrac{2}{3}\pi\,, \quad \alpha_3 = \alpha + \tfrac{4}{3}\pi\,, \qquad (18.12)$$

where α is a constant. Thus the fluid solutions are given by choosing a value for w and then integrating successively (18.9) and (18.11). For example, in the case of dust ($w = 0$):

$$\ell(t) = (\tfrac{9}{2}Mt^2 + \sqrt{3}\Sigma t)^{1/3}\,, \qquad W(t) = \frac{1}{\sqrt{3}\Sigma}\log\left(\frac{t}{\tfrac{3}{4}Mt + \sqrt{3}\Sigma}\right),$$

so $\ell_i(t) = \ell(t)\left(\dfrac{t^2}{\ell(t)^3}\right)^{\tfrac{2}{3}\sin\alpha_i}$. The generic case is anisotropic; LRS cases occur when (up to a multiple of $2\pi/3$) $\alpha = \pi/6$ (the 'Taub' case) and $\alpha = \pi/2$ in (18.12), and isotropic (Robertson–Walker) cases when $\Sigma = 0$.

At late times these solutions isotropize to give the Einstein–de Sitter model (assuming the matter content is dust at large t), and hence may be a good model of the real universe if Σ is chosen appropriately. However, at early times, the situation is quite different. As $t \to 0$, provided $\Sigma \neq 0$, then $\ell(t) \to (\sqrt{3}\Sigma)^{1/3} t^{1/3}$ and

$$\ell_i(t) \to \ell_{i0}\, t^{\tfrac{1}{3}(1+2\sin\alpha_i)}, \qquad \text{(no sum on } i\text{)},$$

where the ℓ_{i0} are constants. Plotting the functions $1 + 2\sin\alpha_i$, we see that the generic behaviour occurs for $\alpha \neq \pi/2$; in this case two of the powers are positive but one is negative, so as $t \to 0$ the axis with the negative exponent shows a (divergent) expansion, while collapse occurs (divergently) along the two orthogonal directions; the singularity is a *cigar singularity.* Going forward in time, the initial collapse along the preferred axis stops and reverses to become an expansion.

However, when $\alpha = \pi/2$, one exponent is positive but the other two are zero. Hence, going back in time, collapse continues divergently along the preferred direction in these LRS solutions back to the singularity, but in the orthogonal directions it slows down and halts; this is a *pancake singularity.* An important consequence in this special case is that horizons are broken in the preferred direction – communication is possible to arbitrary distance in a cylinder around this axis (Hawking and Ellis, 1973). The limiting vacuum solution (18.10) in this case is in fact a wedge of flat Minkowski space.

To summarize, these models can have arbitrarily small shear at the present day, and so can be arbitrarily close to an Einstein–de Sitter universe since decoupling, but can be quite different early on.

Astrophysical applications

One can work out detailed observational relations in these models. Because the Killing vectors are simple, $\xi_i = \partial/\partial x^i$, the null geodesics can be found explicitly; those along the three preferred axes are particularly simple. Redshift along each of these axes simply scales with the expansion ratio in that direction. Area distances can be found explicitly (Tomita, 1968, MacCallum and Ellis, 1970); an interesting feature, shared with all Bianchi class A models, is that all observations will show an eight-fold discrete isotropy symmetry about the preferred axes (Schmidt, 1969, Mena and MacCallum, 2002).

One can also work out helium production and CMB anisotropy, following the pioneering paper by Thorne (1967). Because the shear can dominate the dynamics at nucleosynthesis or baryosynthesis time, causing a speeding up of the expansion, one can get results quite different from those in the FLRW models. Consequently, one can use the nucleosynthesis observations to limit the shear constant Σ; this still allows extra freedom at the time of baryosynthesis.

The CMB quadrupole anisotropy will directly measure the *cumulative* difference in expansion along the three principal axes since last scattering, and, hence, may also be used to limit the anisotropy parameter Σ. The comparison of these two limits is discussed in Section 18.7.

Bianchi I models have also been investigated in the case of viscous fluid and kinetic theory solutions (Misner, 1967) and with EM fields, and the effects of reheating on the CMB anisotropy and spectrum have been examined, see Rees (1968).

While the t-axis must be a Ricci eigenvector, one can consider forms of matter giving a non-zero π_{ij}. These can then show very different forms of behaviour near the singularity, in particular oscillations of length scales like the 'Mixmaster' cases we discuss below (for example, see LeBlanc (1997) for the magnetic field case, Sandin (2009) for a pair of tilted fluids whose energy fluxes cancel, Rendall (1996) and Heinzle and Uggla (2006) for collisionless particles (compare Misner (1968)), and Calogero and Heinzle (2009) for a general discussion). One way of understanding this similarity is that the extra geometric terms in (e.g.) Bianchi IX can be interpreted as arising from long-wavelength gravitational waves (see e.g. King (1991)).

Exercise 18.2.1 Show how the solutions will be altered by (i) a fluid with simple viscosity: $\pi_{ij} = -\eta\sigma_{ij}$ with constant viscosity coefficient η, (ii) freely propagating massless neutrinos (Misner, 1967).

Exercise 18.2.2 Derive the redshifts down the axes in Bianchi I models. Show that observations in these models have an 8-fold discrete symmetry. Find a formula for area distances in these models. (See MacCallum and Ellis (1970).)

Exercise 18.2.3 Show that in the exceptional (pancake) Bianchi I models, there are no particle horizons in certain directions.

18.3 Bianchi geometries and their field equations

In Bianchi models, the inhomogeneous degrees of freedom have been 'frozen out'. They are thus quite special in geometrical terms; nevertheless they form a rich set of models where one can study the exact dynamics with nonlinear field equations.

18.3.1 Constructing Bianchi models

To write down a metric, we first note that the unit normals **n** to the hypersurfaces of homogeneity are always irrotational and geodesic, and invariant under the Killing vectors of the simply transitive G_3. Hence, taking a Killing vector basis $\{\boldsymbol{\xi}_i\}$, we can choose t so that $\mathbf{n} = \partial_t, n_a = -\nabla_a t$. We can then choose a tetrad of vectors $\mathbf{e}_a$ with $\mathbf{n} = \mathbf{e}_0$ at a point p. These can then be dragged to any other point q by the unique group element moving p to q. Thus the spatial vectors $\mathbf{e}_i$ commute both with $\{\boldsymbol{\xi}_i\}$ and with **n**. They in fact generate a second group of transformations on the spatial surfaces, called the reciprocal group. These transformations are in general not isometries: one can prove that their commutators have the same algebraic structure as the Killing vectors, except that the constants $C^C{}_{AB}$ in (2.62) are replaced by functions of t. The metric will then be

$$ds^2 = -dt^2 + \gamma_{ij}(e^i{}_\alpha\, dx^\alpha)(e^j{}_\beta\, dx^\beta)\,, \tag{18.13}$$

where $e^i{}_\alpha\, dx^\alpha$ are 1-forms inverse to the spatial vector triad $\mathbf{e}_i$. To distinguish between tetrad and coordinate indices in the hypersurfaces, we here label the latter by α, β etc.

The various approaches in the literature differ in whether the time dependence is: wholly in the spatial metric γ_{ij}, this giving the metric approach; wholly in the $\mathbf{e}_i$, giving the orthonormal tetrad approach; or split between the two, giving the automorphism approach. In each approach there are helpful subsequent changes of variable that are often used. One can also make changes of the time coordinate, or Lorentz transform the tetrad so the timelike vector aligns with a tilted matter flow.

The *metric approach* (Taub, 1951, Heckmann and Schucking, 1962) writes the line-element as:

$$ds^2 = -dt^2 + \gamma_{ij}(t)\,(e^i{}_\alpha(x^\gamma)\, dx^\alpha)\,(e^j{}_\beta(x^\gamma)\, dx^\beta)\,, \tag{18.14}$$

where the $e^i{}_\alpha(x^\gamma)$ have the same commutators as the generators of the group of isometries, i.e.

$$[\mathbf{e}_i, \mathbf{e}_j] = C^k{}_{ij}\,\mathbf{e}_k\,, \qquad [\mathbf{e}_0, \mathbf{e}_i] = 0\,, \tag{18.15}$$

the $C^k{}_{ij}$ being the Lie algebra structure constants and satisfying the Jacobi identities (2.63). The EFE (3.13) become ordinary differential equations for $\gamma_{ij}(t)$.

One can modify this to the automorphism approach by factorizing the metric into matrices in the automorphism group of the isometry group (i.e. the set of maps of the isometry group to itself) and residual metric terms: considering the automorphisms as part of the tetrad, this is a form where both the tetrad and the metric have time dependence. The process can be briefly described as follows. Take a transformation

$$\hat{\mathbf{e}}^i = M^i{}_j\mathbf{e}^j.$$

This is an automorphism of the isometry group if the $\{\hat{\mathbf{e}}_i\}$ obey the same commutation relations as the $\{\mathbf{e}_i\}$. The matrices M are time dependent and are chosen so that the new metric coefficients $\hat{g}_{ij}$ take some convenient form, for example, become diagonal. The system of equations then simplifies and the real dynamics is in the remaining components of $\hat{g}_{ij}$, while the components of M become secondary time-dependent variables.

Often this approach is coupled with the parametrization due to Misner (1968), which can be written

$$S^6 := \mathrm{e}^{6\lambda} := \det(g_{ij})\,, \qquad g_{ik} = S^2(\exp 2\boldsymbol{\beta})_{ik}, \tag{18.16}$$

where $\boldsymbol{\beta}$ is a symmetric tracefree matrix function of t. If g_{ik} is diagonal, one may write

$$\beta_{ik} = \mathrm{diag}\left(\beta_1, -\tfrac{1}{2}(\beta_1 - \sqrt{3}\beta_2), -\tfrac{1}{2}(\beta_1 + \sqrt{3}\beta_2)\right). \tag{18.17}$$

One may now take λ, $\Omega = -\lambda$, or S to be a new time variable.

The automorphism idea was present in earlier treatments which grew from Misner's methods for the Bianchi IX Mixmaster case (Ryan and Shepley, 1975) but unfortunately this case is highly misleading in that for Bianchi IX (and no others except Bianchi I) the rotation group is an automorphism group. We shall not give further details here; for those see Jantzen (1979; 1984), Wainwright and Ellis (1997) and Jantzen and Uggla (1998).

The *orthonormal tetrad* approach (Ellis and MacCallum, 1969), which we prefer here because it leads to the widely used expansion-normalized variables (presented in detail in Wainwright and Ellis (1997)), takes an orthonormal tetrad invariant under the group of isometries. This is often with $\mathbf{e}_0 = \mathbf{n}$, but in the tilted case it could also be with $\mathbf{e}_0 = \mathbf{u}$, the matter 4-velocity: we shall discuss only the former choice. In this approach the metric components in the tetrad are spacetime constants, $g_{ab} = \eta_{ab}$, and the time variation is in the commutation functions for the basis vectors, which then determine the time- (and space-) dependence in the basis vectors themselves.

If $\mathbf{e}_0 = \mathbf{n}$ we have an orthonormal basis $\{\mathbf{e}_a, a = 0, 1, 2, 3\}$, such that

$$[\mathbf{e}_a, \mathbf{e}_b] = \gamma^c{}_{ab}(t)\,\mathbf{e}_c\,. \tag{18.18}$$

Note that, unlike the metric approach, here $[\,\mathbf{e}_0, \mathbf{e}_i\,] \neq \mathbf{0}$. The dynamical variables are the commutation functions $\gamma^a{}_{bc}(t)$, together with the matter variables. The EFE (3.13) are first-order equations for these quantities and the matter variables, supplemented by the Jacobi identities for the $\gamma^a{}_{bc}(t)$, which are also first-order equations. Thus the equations needed are just the tetrad equations given in Section 6.5, for the case

$$\dot{u}^i = \omega^i = 0 = \mathbf{e}_i \left(\gamma^a{}_{bc} \right) . \tag{18.19}$$

It is sometimes useful to introduce the Weyl tensor components as auxiliary variables, but this is not necessary.

We note that in some recent papers the expansion-normalized variables are replaced by conformally expansion-normalized quantities defined by conformally transforming the metric by a factor H^{-2} (Röhr and Uggla, 2005).

The spatial commutation functions $\gamma^i{}_{jk}(t)$ can be decomposed into a time-dependent symmetric matrix $n^{ij}(t)$ and vector $a_i(t)$ (see (6.42)), and are equivalent to the structure constants $C^i{}_{jk}$ of the symmetry group at each point.[1] In view of (18.19), the Jacobi identities (2.103) now take the simple form

$$n^{ij} a_j = 0 . \tag{18.20}$$

The tetrad basis can be chosen to diagonalize n_{ij} (i.e. to attain $n_{ij} = \text{diag}\,(n_1, n_2, n_3)$ and $a_i = (a, 0, 0)$), and the Jacobi identities are then simply $n_1 a = 0$. (Some authors have preferred to choose $a_i = (0, 0, a)$.) Consequently we define two major classes of structure constants (and so Lie algebras):

Class A: $a = 0$ (unimodular),

Class B: $a \neq 0$ (non-unimodular).

Following Schucking (see Kundt (2003)), the adaptation of the Bianchi classification of G_3 group types used is as in Table 18.1 (Ellis and MacCallum, 1969). Given a specific group type at one instant, it will be preserved by the evolution equations for the quantities $n_i(t)$ and $a(t)$. This follows from Theorem 17.1.

In some cases, the Bianchi groups allow higher symmetry subcases: isotropic (FLRW) or LRS models. Table 18.2 gives the Bianchi symmetry groups admitted by FLRW and LRS solutions (Grishchuk, 1968, Ellis and MacCallum, 1969), i.e. these are the simply transitive three-dimensional subgroups allowed by the full G_6 of isometries (in the FLRW case) and the G_4 of isometries (in the LRS case). (Remember that the only LRS models not allowing a simply transitive subgroup G_3 are the Kantowski–Sachs models for $k = 1$.)

Tilted models can be described in non-orthogonal bases in various ways (King and Ellis, 1973); those possibilities will not be pursued further here. In the bases described here, before use of the field equations, tilted models differ only in the energy–momentum tensor. In the case of orthogonal perfect fluid models, where the fluid 4-velocity u^a is parallel to the normal vectors n^a, the matter variables will be just the fluid density and pressure (Ellis and MacCallum, 1969). In the case of tilted perfect fluid models, where the fluid 4-velocity

---

[1] That is, they can be brought to the canonical forms of the $C^i{}_{jk}$ by a suitable change of group-invariant basis, except that the final normalization to ± 1 would require unnormalized basis vectors; the required transformation depends on time and spatial position.

Table 18.1 Canonical structure constants for different Bianchi types. The parameter h is $a^2/n_2 n_3$.

Class	Type	n_1	n_2	n_3	a	Notes
A	I	0	0	0	0	Abelian
	II	+ve	0	0	0	
	VI$_0$	0	+ve	−ve	0	
	VII$_0$	0	+ve	+ve	0	
	VIII	−ve	+ve	+ve	0	
	IX	+ve	+ve	+ve	0	
B	V	0	0	0	+ve	
	IV	0	0	+ve	+ve	
	VI$_h$	0	+ve	−ve	+ve	$h < 0$
	III	0	+ve	−ve	$\sqrt{-n_2 n_3}$	same as VI$_{-1}$
	VIII$_h$	0	+ve	+ve	+ve	$h > 0$

Table 18.2 The Bianchi models permitting higher symmetry subcases. The parameter c is zero if and only if the preferred spatial vector is hypersurface-orthogonal.

Isotropic Bianchi models		
FLRW $k = +1$:	Bianchi IX [two commuting groups]	
FLRW $k = 0$:	Bianchi I, Bianchi VII$_0$	
FLRW $k = -1$:	Bianchi V, Bianchi VII$_h$	

	LRS Bianchi models	
Orthogonal	$c = 0$	$c \neq 0$
Taub-NUT 1	[KS $K = 1$: no subgroup]	Bianchi IX
Taub-NUT 3	Bianchi I, VII$_0$ [KS $K = 0$]	Bianchi II
Taub-NUT 2	Bianchi III [KS $K = -1$]	Bianchi VII, III
	(Farnsworth)	
Tilted		
	Bianchi V, VII$_h$	
	(Collins–Ellis)	

is not parallel to the normals, the relation between the $\mathbf{n} = \widetilde{\mathbf{u}}$ and $\mathbf{u}$ decompositions can be calculated using the formulae in Exercise 5.1.1. For this we need the peculiar velocity of the fluid relative to $\mathbf{n}$ (King and Ellis, 1973). Note that a perfect fluid will appear as an imperfect fluid in the $\mathbf{n}$ frame.

As usual, the equations (6.50)–(6.51) are constraints on the initial values. The latter simplify, for the bases discussed in Table 18.1, to

$$0 = q^i - 3a\sigma^{1i} - \sum_j \eta^{ijk} n_j \sigma_{jk}. \tag{18.21}$$

In particular inserting specific values of n_i and a from Table 18.1 in (18.21) restricts the possible energy fluxes q_i for some Bianchi types, e.g. for perfect fluids these equations show which models can have tilt and what direction the fluid velocity can lie in. Conversely, for a given q_i, including $q_i = 0$, the equations constrain σ_{ij}. For example, as already mentioned, Bianchi I models must have $q_i = 0$, and orthogonal cases enforce or allow diagonalization of σ_{ij} (Ellis and MacCallum, 1969). Note that, orthogonal models (18.21), considered as linear equations for σ_{ij}, are of rank 1 rather than 2 for the special case of Bianchi type VI_h, $h = -1/9$, giving rise to models with an additional degree of freedom in σ_{ij}: these cases are denoted $VI^*_{-1/9}$.

For Bianchi universes of class A, but not in general those of class B (MacCallum and Taub, 1972, Sneddon, 1975), a variational (Lagrangian or Hamiltonian) formulation can be given, provided the matter terms admit such a formulation. (The Class B cases with a Lagrangian are those with $n^i_i = 0$, a subclass characterized by Ellis and MacCallum (1969).) This takes the usual form with kinetic and potential terms, the kinetic terms being time derivatives of parameters in the metric. If such a variational form is possible, the kinetic terms define a metric on the space of the time-dependent metric parameters, and its symmetry and other properties can lead to exact solutions, often via choices of time variable (see e.g. Rosquist and Uggla (1991) and references therein).

The potential terms for the orthogonal fluid cases are proportional to the spatial curvature scalar. In the automorphism approach, one can calculate and sketch these potentials in the (β_1, β_2) plane of (18.17), giving an intuitive feel for the behaviour of the evolution (see e.g. MacCallum (1979), Jantzen (1984)). This can be extended to tilted cases by using an additional tilt potential, and the time derivative of an automorphism $M^i{}_j$ gives an additional centrifugal potential (see e.g. Jantzen (1984)). Broadly, each of these provides restoring forces, and may prevent escape to infinity in the (β_1, β_2) plane, leading to evolutions with an infinite sequence of oscillations in (β_1, β_2). For the curvature terms, this was first recognised in studies of Bianchi type IX by Misner (1969a) and Belinski, Lifshitz and Khalatnikov (1971 and earlier) (BLK, in Russian alphabetic order). Such oscillations are generically named 'Mixmaster' after a popular brand of US food mixer (Misner, 1969a).

Since we have a finite-dimensional ODE system, with constraints, one can give the number of freely specifiable constants in the solutions (Siklos, 1976; 1984), which corresponds to the dimension of the dynamical system. This is summarized, for vacuum and tilted barotropic perfect fluid solutions, in Table 18.3. For more complicated forms of matter content extra constants (and hence dimensions) may appear, e.g. if there are two fluids with different 4-velocities.

A number of authors have considered compactified forms of the metrics (the covering groups have topology $\mathbb{R}^3$ except for Bianchi type IX where it is S^3). In particular one can identify those cases which correspond to Thurston's classification of 3-manifolds (see e.g. Fagundes (1985), Barrow and Hervik (2002)). Koike, Tanimoto and Hosoya (1994) showed that for the compactifiable spatial geometries the geometric degrees of freedom can be divided into degrees of freedom of the covering space and degrees of freedom of the Teichmüller deformations of the discrete isometries used in compactification: Kodama (2002), Tanimoto, Moncrief and Yasuno (2003) and papers cited therein give fuller details. Even in the simple Bianchi I case there is a significant moduli space (Hervik, 2000).

One can use the spaces of Bianchi models as finite-dimensional examples in the superspace approach to quantum gravity, so-called 'minisuperspaces' (see Misner (1969b) and

Table 18.3 The number of essential parameters, by Bianchi type, in general solutions for vacuum and for tilted perfect fluids with given equation of state.
The number for a non-tilted fluid, including Λ-term, is one more than for vacuum. Type III is included in VI_h.
h itself is regarded as fixed in a solution, i.e. is not counted as a parameter. The number is reduced if there is extra symmetry or a condition such as $n^i_i = 0$ is imposed.

Energy–momentum			Bianchi type				
	I	II	VI_0 & VII_0	VIII & IX	V	IV, VII_h & VI_h, general	$VI^*_{-1/9}$
Vacuum	1	2	3	4	1	3	4
Perfect fluid	2	5	7	8	5	7	7

Section 20.2.1). Where there is a variational description it provides a symplectic measure which could, if its integral were finite, be used to define probabilities as described in Chapter 21.

The space of dynamical variables composed of the kinematic quantities and commutators is non-compact. It turns out to be very powerful to normalize the variables and compactify this space, which we discuss below in Section 18.4.

Exercise 18.3.1 Find the automorphism groups of the various Bianchi types and choose parameters for them well adapted to the form of the EFE. (See e.g. Harvey (1979), Jantzen (1984).)

18.4 Bianchi universe dynamics

Since the differential equations are ordinary, one can use methods from the theory of dynamical systems to obtain analytic results. These are mathematically powerful and can help us identify true degrees of freedom and simplify the equations, and find exact solutions, as well as give qualitative understanding. There have also been many numerical investigations of these dynamical equations and the resulting solutions.

Shikin (1967) and Collins (1971) introduced the use of 'phase planes' in cases where the dynamical system is two-dimensional (strictly these are not phase planes as the variables are not q and $\dot{q}$ for some q). In this case one can map the qualitative features by finding the equilibrium points or critical points, and the local linear approximations to them, finding any separatrices present, and then completing the set of evolution curves (the differential equations are generally polynomial in the variables used so that continuity of the family of curves is assured, except at critical points). The Poincaré–Bendixson theory for such systems of equations rests on the Jordan curve theorem, which has no suitable analogue in higher dimensions.

Bogoyavlenskii and Novikov (1973) (see also Bogoyavlenskii (1985)) showed how to generalize dynamical systems treatments to higher-dimensional state spaces, by identifying equilibria and homoclinic and heteroclinic orbits (the generalization of separatrices) and finding monotone functions which can be used to show that trajectories must approach certain limits or boundaries. They also initiated the compactification of the state spaces by choice of variables, which facilitates discussion of asymptotic behaviour: the most common current choice is that of the normalized variables we discuss below, though these do not always give a compactification. Subsequent work has used variables based on the automorphism group of the isometry group, Hamiltonian methods, algebraically invariant curves or Darboux polynomials, and other techniques (see Wainwright and Ellis (1997) and below).

Using the expansion-normalized variables we write the EFE as a dynamical system (compare Collins, 1971, Wainwright, 1988) so that one can study the evolution of the various physical and geometrical quantities *relative to the overall rate of expansion of the universe*, as described by the rate of expansion scalar Θ, or equivalently *the Hubble parameter* $H = \frac{1}{3}\Theta$. The main variables used are essentially the commutation functions mentioned above, but rescaled by a common time-dependent factor.

Although other energy-momenta can add oscillatory behaviour, as in the Bianchi I case, studies using these ideas have principally focused on perfect fluids obeying a linear equation of state; for short, we just say a 'fluid' below. (We note that until very recently all the literature used γ rather than w in (5.49).) These studies typically first determine the behaviour in invariant sets, in particular the vacuum behaviour (which will be a limit of the matter-filled behaviour), and the (self-similar) power-law solutions, which are equilibria (fixed points) in the state space of the normalized variables. (Other invariant sets include, for example, the Bianchi II evolutions as limits of Bianchi IX evolutions in which two of the three n_i are zero.) The results on invariant sets and self-similar solutions are then used to provide a skeleton of curves in the way described at the end of Chapter 17.

18.4.1 Reducing the differential equations

To avoid dealing with unnecessarily large systems of equations, the remaining freedom in the choice of orthonormal tetrad needs to be eliminated, first by specifying the variables Ω^i of Section 6.5 implicitly or explicitly (for example by specifying them as functions of the σ_{ij}). One can also simplify other quantities (depending on the particular models studied): for example choice of a shear eigenframe will result in the tensor σ_{ij} being represented by two independent diagonal terms. These measures lead to a reduced set of variables, consisting of H and the remaining commutation functions, which we denote symbolically by

$$\mathbf{x} = (\gamma^a{}_{bc}|_{\text{reduced}}) . \tag{18.22}$$

The physical state of the model is then described by the vector $(H, \mathbf{x})$, together with matter variables. The idea is now to normalize $\mathbf{x}$ with the Hubble parameter H, and similarly normalize matter terms. We denote the resulting variables by a vector $\mathbf{y} \in R^n$. These new variables are *dimensionless*, and will be referred to as *expansion-normalized variables*. The details of this reduction differ for the Class A and B models and the choice of matter terms. Convenient variables may be subject to constraints: for example, in orthogonal

Class B models, using the variables of Hewitt and Wainwright (1993), there is an algebraic constraint of the form

$$g(\mathbf{x}) = 0 , \qquad (18.23)$$

where g is a homogeneous polynomial.

It is clear that each dimensionless state $\mathbf{y}$ determines a 1-parameter family of physical states $(H, \mathbf{x})$. The evolution equations for the $\gamma^a{}_{bc}$ lead to evolution equations for H and $\mathbf{y}$. In deriving the evolution equations for $\mathbf{y}$ from those for $\mathbf{x}$, the *deceleration parameter* q, defined by (6.7), plays an important role, since

$$\dot{H} = -(1+q) H^2 . \qquad (18.24)$$

We can use the scale factor ℓ defined by (4.35) to introduce a *dimensionless time variable* τ according to

$$\ell = \ell_0 e^\tau , \qquad (18.25)$$

where ℓ_0 is the value of the scale factor at some arbitrary reference time. Since ℓ assumes values $0 < \ell < \infty$ in an ever-expanding model, τ assumes all real values, with $\tau \to -\infty$ at the initial singularity and $\tau \to +\infty$ at late times. It follows from equations (4.35) and (18.25) that

$$\frac{dt}{d\tau} = \frac{1}{H} , \qquad (18.26)$$

and the evolution equation (18.24) for H can be written

$$\frac{dH}{d\tau} = -(1+q) H . \qquad (18.27)$$

Since the right-hand sides of the evolution equations for the $\gamma^a{}_{bc}$ are homogeneous of degree 2 in the $\gamma^a{}_{bc}$, the change (18.26) of the time variable results in H cancelling out of the evolution equation for $\mathbf{y}$, yielding an autonomous system of ODEs:

$$\frac{d\mathbf{y}}{d\tau} = \mathbf{f}(\mathbf{y}) , \quad \mathbf{y} \in R^n . \qquad (18.28)$$

Constraints $g(\mathbf{y}) = 0$ are preserved by the DEs. The functions $\mathbf{f}: \mathbb{R}^n \to \mathbb{R}^n$ and $g: \mathbb{R}^n \to \mathbb{R}$ are polynomial functions in $\mathbf{y}$. An essential feature of this process is that the evolution equation for H, namely (18.27), decouples from the remaining equations (18.28). In other words, the DE (18.28) describes the evolution of the Bianchi cosmologies, the transformation to $\mathbf{y}$ essentially scaling away the effects of the overall expansion. An important consequence is that the new variables are bounded near the initial singularity.

Since τ assumes all real values (for models which expand indefinitely), the solutions of (18.28) are defined for all τ and hence define a *flow* $\{\phi_\tau\}$ on $\mathbb{R}^n$. The evolution of the cosmological models can thus be analysed by studying the orbits of this flow in the physical region of state space, which is a subset of $\mathbb{R}^n$ defined by the requirement that the dimensionless energy density Ω be non-negative, i.e.

$$\Omega(\mathbf{y}) = \frac{8\pi G\rho}{3H^2} \geq 0 . \qquad (18.29)$$

18.4.2 Specific systems

To illustrate these ideas we give the systems of equations in two cases, and we consider the resulting behaviour in the next section.

For orthogonal models of Class A, using the forms in Table 18.1, (6.51) shows the shear tensor must be diagonal, and since it is trace-free it can be written, in a manner similar to (18.17), as

$$\sigma_{ik} = \text{diag}\left(-\sigma_+, \tfrac{1}{2}(\sigma_+ - \sqrt{3}\sigma_-), \tfrac{1}{2}(\sigma_+ + \sqrt{3}\sigma_-)\right).$$

The physical state of a Bianchi class A cosmology is thus determined by the vector $\mathbf{x} = (H, \Sigma_+, \Sigma_-, N_1, N_2, N_3)$, where $N_i := n_i/H$, $\Sigma_\pm = \sigma_\pm/H$ (Wainwright and Hsu, 1989). We also use $\Sigma^2 = \sigma^2/3H^2 = \tfrac{1}{6}\Sigma_{ij}\Sigma^{ij}$.

The Einstein field equations are then given by the autonomous system,

$$\Sigma'_\pm = -(2-q)\Sigma_\pm - S_\pm, \quad N'_2 = (q + 2\Sigma_+ + 2\sqrt{3}\Sigma_-)N_2, \qquad (18.30)$$
$$N'_1 = (q - 4\Sigma_+)N_1, \qquad N'_3 = (q + 2\Sigma_+ - 2\sqrt{3}\Sigma_-)N_3,$$

where

$$S_+ = \frac{1}{6}\left[(N_2 - N_3)^2 - N_1(2N_1 - N_2 - N_3)\right]$$

$$S_- = \frac{1}{2\sqrt{3}}(N_3 - N_2)(N_1 - N_2 - N_3),$$

together with the decoupled equation (18.27). Here,

$$\Omega = 1 - \Sigma^2 - K \geq 0, \quad K := \frac{1}{12}\left(\sum_i N_i^2 - 2\sum_{i<j} N_i N_j\right). \qquad (18.31)$$

The equations for orthogonal Class B models were given by Hewitt and Wainwright (1993).

For the general case, the equations can be written as follows (see Hewitt, Bridson and Wainwright (2001), from which the equations in the rest of this section are taken). Normalizing the variables of Section 6.5, introduce

$$\Sigma_{ij} = \frac{\sigma_{ij}}{H}, \quad N_{ij} = \frac{n_{ij}}{H}, \quad A_i = \frac{a_i}{H}, \quad R_i = \frac{\Omega_i}{H}, \quad S_{ij} = \frac{{}^3R_{\langle ij\rangle}}{H^2},$$

$$\Omega = \frac{\rho}{3H^2}, \quad P = \frac{p}{3H^2}, \quad Q_i = \frac{q_i}{H^2}, \quad \Pi_{ij} = \frac{\pi_{ij}}{H^2}, \quad K = -\frac{{}^3R}{6H^2}.$$

The spatial curvatures are algebraic expressions in the n_{ij} and a_i, so there will be no independent evolution equations for them. There are no general evolution equations for P or Π_{ij}, which depend on the nature of the assumed matter: for a tilted barotropic perfect fluid, v^i and the rest frame value of ρ fix the matter variables in the $\mathbf{n}$ frame, so we shall only need evolution equations for Ω and Q_i in that case. Finally there is no general evolution equation for R_i, reflecting the freedom of choice of the frame.

Thus the evolution equations are

$$\Sigma'_{ij} = -(2-q)\Sigma_{ij} + 2\varepsilon^{km}{}_{(i}\Sigma_{j)k}R_m - S_{ij} + \Pi_{ij},$$

$$N'_{ij} = qN_{ij} + 2\Sigma_{(i}{}^k N_{j)k} + 2\varepsilon^{km}{}_{(i}N_{j)k}R_m,$$

$$A'_i = qA_i - \Sigma_i{}^j A_j + \varepsilon_i{}^{km}A_k R_m, \qquad\qquad (18.32)$$

$$\Omega' = (2q-1)\Omega - 3P - \tfrac{1}{3}\Sigma_i{}^j \Pi_j{}^i + \tfrac{2}{3}A_i Q^i$$

$$Q'_i = 2(q-1)Q_i - \Sigma_i{}^j Q_j - \varepsilon_i{}^{km}R_k Q_m + 3A^j\Pi_{ij} + \varepsilon_i{}^{km}N_k{}^j\Pi_{jm}.$$

These are subject to the constraints

$$N_i{}^j A_j = 0, \quad \Omega = 1 - \Sigma^2 - K, \quad Q_i = 3\Sigma_i{}^k A_k - \varepsilon_i{}^{km}\Sigma_k{}^j N_{jm}. \qquad (18.33)$$

For a tilted fluid we write

$$u^a = \frac{1}{\sqrt{1-v_b v^b}}(n^a + v^a),$$

where the spacelike vector $\mathbf{v}$ is orthogonal to the normal vector $\mathbf{n}$. Then (18.33), with $Q_i = 3(w+1)G^{-1}\Omega v_i$, assumes the form

$$3(w+1)G^{-1}\Omega v_i = 3\Sigma_i{}^k A_k - \varepsilon_i{}^{km}\Sigma_k{}^j N_{jk}. \qquad (18.34)$$

where

$$G = 1 + wv^2, \quad v^2 = v_i v^i < 1.$$

The consequent formulae for the kinematic quantities of a general tilted fluid are given in Hewitt, Bridson and Wainwright (2001). In particular we may note that the normalized vorticity is given by

$$W_i = N_i{}^m v_m + \eta_i{}^{mk}v_m A_k + \cosh^2\beta(N^{mk}v_m v_k)v_i, \qquad (18.35)$$

where $u_a n^a = -\cosh\beta$.

For tilted Bianchi II models, we have $A_i = 0$ and $N_{ij} = \mathrm{diag}(N_1, 0, 0)$. Then $v_1 = 0$, assuming $\Omega > 0$ and $w > -1$, and we are free to perform a rotation in the 23-plane to get $v_2 = 0$, $v_3 \neq 0$. The constraints (18.34) now yield

$$\Sigma_{13} = 0 = R_2, \quad \Sigma_{12} \neq 0, \quad 3(w+1)\Omega v_3 = G\Sigma_{12}N_{11}, \qquad (18.36)$$

and the Σ'_{13} equation then implies $R_1 = \Sigma_{23}$, so R_i is uniquely determined in terms of Σ_{ij}.

We now relabel the variables as follows:

$$\Sigma_+ = \tfrac{1}{2}(\Sigma_{22} + \Sigma_{33}), \quad \Sigma_- = \tfrac{1}{2\sqrt{3}}(\Sigma_{22} - \Sigma_{33}), \qquad (18.37)$$

$$N_1 = N_{11}, \quad \Sigma_1 = \tfrac{1}{\sqrt{3}}\Sigma_{23}, \quad \Sigma_3 = \tfrac{1}{\sqrt{3}}\Sigma_{12}. \qquad (18.38)$$

It should be noted that the off-diagonal shear components Σ_1 and Σ_3 determine the angular velocity of the spatial frame.

The set of independent expansion-normalized variables is

$$(\Sigma_+, \Sigma_-, \Sigma_1, \Sigma_3, N_1, v_3),$$

subject to one constraint (18.36), which we now write in the form

$$h(v_3)\Omega = \Sigma_3 N_1, \quad h(v_3) := \frac{\sqrt{3}(w+1)v_3}{G}.$$

The evolution equations for these variables are

$$\Sigma_+' = -(2-q)\Sigma_+ - 3\Sigma_3^2 + \tfrac{1}{3}N_1^2 + \tfrac{1}{2\sqrt{3}}\Sigma_3 N_1 v_3, \tag{18.39}$$

$$\Sigma_-' = -(2-q)\Sigma_- + 2\sqrt{3}\Sigma_1^2 - \sqrt{3}\Sigma_3^2 - \tfrac{1}{2}\Sigma_3 N_1 v_3, \tag{18.40}$$

$$\Sigma_1' = -(2-q+2\sqrt{3}\Sigma_-)\Sigma_1, \tag{18.41}$$

$$\Sigma_3' = -(2-q-3\Sigma_+ - \sqrt{3}\Sigma_-)\Sigma_3, \tag{18.42}$$

$$N_1' = (q-4\Sigma_+)N_1, \tag{18.43}$$

$$v_3' = \frac{v_3(1-v_3^2)}{1-wv_3^2}(3w-1-\Sigma_+ + \sqrt{3}\Sigma_-), \tag{18.44}$$

where

$$q = 2\left(1 - \tfrac{1}{12}N_1^2\right) - \tfrac{1}{2}G^{-1}\Omega\left[3(1-w)(1-v_3^2) + 2(w+1)v_3^2\right].$$

The auxiliary equation for Ω' is

$$\Omega' = G^{-1}[2Gq - (3w+1) - (1-w)v_3^2 - (w+1)(\Sigma_+ - \sqrt{3}\Sigma_-)v_3^2]\Omega. \tag{18.45}$$

The state space is the subset of $\mathbb{R}^6$ defined by the inequality $\Omega \geq 0$, which is equivalent to

$$\Sigma_+^2 + \Sigma_-^2 + \Sigma_1^2 + \Sigma_3^2 + \tfrac{1}{12}N_1^2 \leq 1, \tag{18.46}$$

and the constraint

$$g(\mathbf{x}) = h(v_3)\Omega - \Sigma_3 N_1 = 0.$$

The restriction (18.46), and the fact that $v_3^2 < 1$, implies that the state space is bounded.

18.4.3 Equilibrium points and self-similar cosmologies

Each ordinary orbit in the dimensionless state space corresponds to a one-parameter family of physical universes, which are conformally related by a constant rescaling of the metric. On the other hand, for an equilibrium point $\mathbf{y}^*$ of the DE (18.28) (which satisfies $\mathbf{f}(\mathbf{y}^*) = \mathbf{0}$), the deceleration parameter q is a constant, i.e. $q(\mathbf{y}^*) = q^*$, and we find

$$H(\tau) = H_0\, e^{(1+q^*)\tau}.$$

The parameter H_0 is no longer essential, since it can be set to unity by a translation of τ, $\tau \to \tau + \text{const}$; then (18.26) implies that

$$Ht = \frac{1}{1+q^*}, \tag{18.47}$$

so that the commutation functions are of the form $(\text{const}) \times t^{-1}$. It follows that the resulting cosmological model is self-similar. Thus *each equilibrium point of the DE* (18.28) *corresponds to a unique self-similar cosmological model*. In such a model the physical states at different times differ only by an overall change in the length scale. Such models are expanding, but in such a way that their dimensionless state does not change. They include the flat FLRW model ($\Omega = 1$) and the Milne model ($\Omega = 0$). All vacuum and non-tilted perfect fluid self-similar Bianchi solutions were given by Hsu and Wainwright (1986).

For example, for the Bianchi type II orthogonal models, putting $N_2 = N_3 = 0$ in (18.30), it is easy to work out that the boundaries are $N_1 = 0$ and $\Omega = 0$, i.e. Bianchi I models and vacuum Bianchi II models, and that the LRS Bianchi II models ($\Sigma_- = 0$) are also an invariant set. The equilibrium points are the flat Friedmann solution (a special Bianchi I model), where $\Sigma_+ = \Sigma_- = 0$, the (Bianchi I) Kasner vacua, and an LRS Bianchi II model due to Collins and Stewart at $N_1 = \frac{1}{4}[(1-w)(3w+1)]^{1/2}$, $\Sigma = (3w+1)/8$, $\Omega = \frac{3}{16}(5-w)$ with $-\frac{1}{3} < w < 1$.

18.4.4 More general orbits

The *vacuum boundary*, defined by $\Omega(\mathbf{y}) = 0$, which describes the evolution of vacuum Bianchi models, is always an *invariant set*: this set plays an important role in the qualitative analysis because vacuum models can be asymptotic states for perfect fluid models near the Big Bang or at late times. There are other invariant sets which are also specified by simple restrictions on $\mathbf{y}$ which play a special role: the subsets representing each more specialized Bianchi type (Table 18.1), and the subsets representing the FLRW models and the LRS Bianchi models (according to Table 18.2).

It is desirable that the dimensionless state space D in R^n be a (closed, bounded) compact set. In this case the existence of a monotone function will imply that each orbit has a non-empty future and past limit, and hence there will be a past attractor (the 'α-limit') and a future attractor (the 'ω-limit') in state space. Compactness of the state space in expansion-normalized variables has a direct physical meaning for ever-expanding models: at the Big Bang no physical or geometrical quantity can diverge more rapidly than the appropriate power of H, and at late times no such quantity tends to zero less rapidly than the appropriate power of H. Compactness in expansion-normalized variables happens for many models; it fails, however, where there are recollapsing or bouncing solutions and, for example, for Bianchi type VII_0 and type VIII models. When it fails, following the usual methods may give misleading answers (Goheer, Leach and Dunsby, 2008): the lack of compactness often manifests itself in extreme dominance of the Weyl tensor at late times. To enable a proper understanding of the evolution, a different normalization may be needed.

In discussions of the far future evolution, assuming the matter terms do not dominate like a Λ term will and the model does not recollapse, it is interesting to characterize the behaviour according to whether the Ricci or Weyl terms dominate the curvature, or are in balance: one can distinguish ordinary and extreme Weyl dominance by whether the ratio of Weyl and Ricci invariants tends to infinity or not (Barrow and Hervik, 2002).

The equilibrium points determine the asymptotic behaviour of other more general models. An unstable equilibrium may be arbitrarily closely approached by curves whose initial and final states are at other equilibria. This leads to quasi-equilibrium epochs, which has an important consequence for the observational viability of such models.

We should note that the descriptions derived from these considerations, particularly when infinite sequences of heteroclinic orbits are concerned, in general lack some rigour. A number of the expected evolutions have now been proved (see Weaver (2000), Heinzle and Uggla (2009a), Liebscher *et al.* (2011), Béguin (2010), Reiterer and Trubowitz (2010)), starting with Ringström's proof (2000) for Bianchi type IX, and there is no reason so far to disbelieve any of the arguments (although it should be noted that the present proofs do not extend to the Bianchi VIII or $VI_{-1/9}$ cases, Heinzle and Uggla (2009b)). So there is more work to be done.

18.5 Evolution of particular Bianchi models

A full description of the many cases that have been considered would take far more space than this book allows. Thus we first give here fuller details of two examples, the orthogonal and tilted Bianchi II fluid cases. Then we tabulate some references in which the various specific cases have been discussed and earlier literature cited, and describe briefly some of the most interesting results.

18.5.1 Orthogonal Bianchi II models with $-\frac{1}{3} < w < 1$

We have already given in Section 18.4.3 the boundaries, invariant sets and equilibria for these models. The three invariant sets provide the skeleton for the general solution curves. We can also find three monotone functions: $Z_5 = \Sigma_-^2$ on the non-LRS models, $Z_6 = \Sigma_-/(2 - \Sigma_+)$ on the non-vacuum models, and $Z_7 = N_1^{2m} \Omega^{1-m}/(1 - v\Sigma_+)^2$ on the whole space (except for the Collins–Stewart point P_1^+) where $v = \frac{1}{8}(3w + 1)$ and $m = 3v(1 - w)/8(1 - v^2)$. (Here we use the notation in Wainwright and Ellis (1997).)

The vacuum boundary can be projected into two dimensions with coordinates (Σ_+, Σ_-): the solutions satisfy $\Sigma_- = k(\Sigma_+ - 2)$ for some constant k (these solutions were found explicitly by Taub). The resulting diagram is Figure 18.1.

The LRS invariant set contains points T_1 and Q_1 on the Kasner circle, as well as the flat Friedmann point F and the Collins–Stewart point P_1^+. It is a semicircle with boundaries $N_1 = 0$ and $\Omega = 0$ and the separatrices in these boundaries are easy to find. It can be shown there are no periodic orbits and, from the monotonicity of Z_7, that P_1^+ is a sink, so as F is an unstable point (though it is a sink in the plane $N_1 = 0$), the curve running from it must meet P_1^+. This gives the bold curves in Figure 18.2 which are the skeleton for the remaining curves depicted in the figure. Figures 18.1 and 18.2 were first given by Collins (1971).

As we have now studied all the invariant sets and equilibria (the Bianchi I subset $N_1 = 0$ being covered by Section 18.2) we can combine the results to give the three-dimensional picture of Figure 18.3.

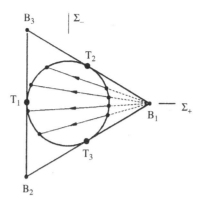

Fig. 18.1 The projections of the Taub orbits into the (Σ_+, Σ_-) plane. The circle here is the Kasner circle and the points marked T_i are the Taub cases given by $\alpha = \pi/6 \bmod 2\pi/3$ in (18.12). (From Wainwright and Ellis (1997), figure 6.6.)

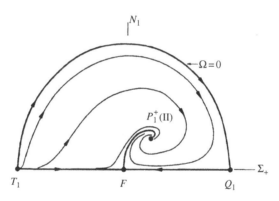

Fig. 18.2 The LRS Bianchi II invariant subset. (From Wainwright and Ellis (1997), figure 6.5.)

18.5.2 Tilted Bianchi II models

Although, on evaluating (18.35), we find that the fluid in tilted Bianchi II models has zero vorticity (so we do not find a critical value $w = 1/9$) these models play a central role in the behaviour of other models.

The configuration space of the tilted Bianchi II models, whose evolution is governed by (18.39)–(18.44), has some important boundary invariant sets (Hewitt, Bridson and Wainwright, 2001). These are the space of orthogonal nonvacuum Bianchi II models $v_3 = 0$ and its three invariant subsets (Bianchi I, LRS Bianchi II and vacuum Bianchi II models) and the extreme tilt models $v_3 = 1$. The vacuum Bianchi II models have been shown in Figure 18.1, but for use in later discussions it is more helpful to sketch them in three dimensions, the third variable being N_1. This gives the picture in Figure 18.4, which is in fact the upper boundary of the region in Figure 18.3.

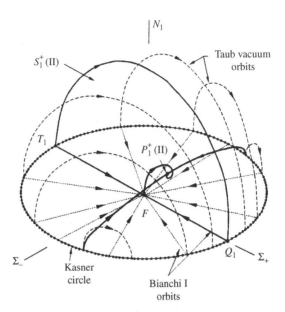

Fig. 18.3   Evolution curves of orthogonal Bianchi II models with $-\frac{1}{3} < w < 1$. (From Wainwright and Ellis (1997), figure 6.7.)

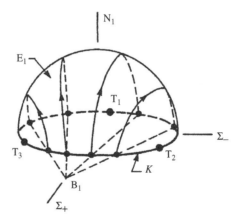

Fig. 18.4   Three-dimensional representation of the evolution of Bianchi II vacua: the upper boundary of Fig. 18.3. (From figure 1 of Ma and Wainwright (1994), reproduced with kind permission from Springer Science+Business Media B.V.)

The nonvacuum equilibrium points in the Bianchi II state space are (Hewitt, Bridson and Wainwright, 2001): the flat FLRW models; the Collins–Stewart solution; the tilted self-similar solution due to Hewitt (1991), valid for $\frac{3}{7} < w < 1$; a line $\mathcal{L}$ of tilted solutions for $w = 5/9$ and varying Σ_1 first found in Hewitt, Bridson and Wainwright (2001); an extreme tilt case; and the Bianchi I $w = 1$ (Jacobs) solutions. ($w > 1$ is not considered as it gives a sound speed greater than light.) The vacuum equilibrium points are the Kasner circle in both its original form and a form with extreme tilt represented in a non Fermi-propagated frame.

The methods for finding the nature of the solution curves are similar to those outlined above (i.e. finding equilibria and their linearizations, using monotone functions on state space or part thereof to determine limit points or sets, and building up a skeleton for the general curves from this information), so we shall just summarize the results.

The future evolution (ω-limit) of the tilted models is one of the self-similar solutions, which one depending on the value of w (Hewitt, Bridson and Wainwright, 2001, Hervik *et al.*, 2010). For $w < -\frac{1}{3}$ it is the flat FLRW model (these models are inflationary and isotropize); for $-\frac{1}{3} < w < \frac{3}{7}$ it is the Collins–Stewart solution; for $\frac{3}{7} < w < \frac{5}{9}$ the Hewitt solution; for $w = \frac{5}{9}$ and varying Σ_1 one moves along $\mathcal{L}$ to the extreme tilt case which applies for $w > \frac{5}{9}$. The bifurcation at $w = -\frac{1}{3}$ is called the spatial curvature bifurcation, since it is where spatial curvature starts to outweigh the matter terms as $t \to \infty$, and the bifurcations at $w = \frac{3}{7}$ and $w = \frac{5}{9}$ are due to the tilt.

The past evolution (α-limit) can be a Jacobs solution if $w = 1$, or some part of the Kasner circle, for particular ranges of the Σ_{ij}, but is in general oscillatory, or of 'Mixmaster' character, following heteroclinic sequences tracking the boundary trajectories that join points on the usual Kasner circle to itself or its extreme tilt copy.

18.5.3 Survey of other cases

Here we first give a table of the most recent papers which cover what is known about the evolution of the various fluid models, and which give references to earlier literature. This table can also be read as the table of references for the following summary of results, which shows both the variety and the interrelations between different types of fluid model. A detailed table of ω limits is given by Hervik *et al.* (2007).

We note that the α-limits (relevant to asymptotics in the very early universe), and ω-limits, which give the far future asymptotics and thus a partial response to questions about isotropization, have been studied extensively, but we lack detailed quantitative studies which could be related to present-day observation and predictions (and also lack a complete set of results for combinations of fluids, including the ΛCDM combination).

All state spaces of orthogonal models except those of Bianchi types I and V contain the orthogonal Bianchi II models as an invariant boundary set (and all, orthogonal or not, contain the flat FLRW model as a limit).

In types VIII and IX there are three distinct such boundary sets, depending on which N_j remains non-zero. These give rise, in their vacuum limits, i.e. when considering the α-limits, to three copies of Figure 18.4 which project into the Kasner circle at angles rotated by $2\pi/3$. Trajectories which approach the big bang may therefore bounce around the Kasner circle, spending most of their time near the Kasner ring but following a series of heteroclinic orbits joining points on it. These may form periodic sequences. This description is equivalent to the original descriptions of the 'Mixmaster': the similar occurrence of oscillations in all Bianchi types except I and V was first noted by Peresetskii (1977). For a recent summary of results about these dynamics, see Sandin and Uggla (2010).

This discussion shows why identifying all points in state space corresponding to the same model is not a good idea (Ellis, 2005). The Kasner ring that serves as a framework for

Type	Tilt	References
Table 18.4	References for studies of tilted Bianchi models	
II	General	Hewitt, Bridson and Wainwright (2001)
VI_0	General	Hervik (2004)
VII_0	Irrotational	Coley and Hervik (2005)
		Lim, Deeley and Wainwright (2006)
	General	Hervik *et al.* (2006)
VIII	General	Hervik and Lim (2006)
IV	General	Coley and Hervik (2005)
V	Irrotational	Hewitt and Wainwright (1992)
	General	Coley and Hervik (2005)
VI_h	Subset	Coley and Hervik (2005)
		Hervik *et al.* (2007)
		Coley and Hervik (2008)
$VI_{-1/9}$	General	Hervik *et al.* (2008)
VII_h	General	Hervik, van den Hoogen and Coley (2005)

evolution of many other Bianchi models contains multiple realizations of the same Kasner model. To identify them as the same point in state space would make the evolution patterns very difficult to follow. It is better to keep them separate, but to learn to identify where multiple realizations of the same model occur (which is just the equivalence problem for cosmological models, discussed in Section 17.2).

An ongoing issue since the discovery of the 'Mixmaster' behaviour of the Type IX models has been whether or not these solutions show chaotic behaviour, rather than just oscillatory, as they approach the initial singularity, and various tests have been applied (see Hobill, Burd and Coley (1994), Hobill in Wainwright and Ellis (1997) and Maciejewski and Szydłowski (1998)). One of the difficulties is the freedom of choice of time coordinate, and another is that the points at which divergences between neighbouring trajectories occur are those where a trajectory meets the Kasner circle, which in the usual approximations are discrete events. A recent review of what has been rigorously proved for this problem (Heinzle and Uggla, 2009b) points out that although it has been proved that the limiting behaviour is an approach to the Mixmaster attractor, it is not certain whether only a subset thereof is involved, nor has it been shown that the Kasner map approximation, which is what is usually discussed as possibly chaotic, is an adequate approximation for this purpose (see also Liebscher *et al.* (2011), Reiterer and Trubowitz (2010)). There may similarly be chaos in other solutions with oscillatory behaviour.

Bogoyavlenskii and Novikov (1973) discuss the form of the solution as it leaves the attractor, showing there are three main types of evolution (compare (Bogoyavlenskii, 1985, section 6.VI)). (The argument is essentially the reverse of the arguments that generic models approach the vacuum boundary.) It could be interesting to study these further as early universe models.

The orthogonal VI_0 and VII_0 models only allow two copies of Figure 18.4. In addition there are trajectories which end at the Taub points on the Kasner circle. For type VI_0 these solutions have a flat Kasner limit, which may be important to the horizon problem in Bianchi VIII models (Heinzle and Ringström, 2009), but for type VII_0 this is not so: there can be unbounded trajectories showing extreme Weyl dominance. Since these are also limit cases of Bianchi VIII and IX, the descriptions of those cases have to incorporate these special trajectories. For the behaviour of compactified type VIII space see Ringström (2003). An inhomogeneously compactified VI_0 model was studied as an 'inhomogenous' Mixmaster case (Weaver, Isenberg and Berger, 1998).

In general models will isotropize in the ω-limit if there is an inflationary matter term ($w < -\frac{1}{3}$), as was first shown for $w = -1$ by Wald (1983): the basic reason is as stated in Section 18.2. However, when there is no such matter, Bianchi models can only isotropize at the ω-limit if they allow as a special case an FLRW model (see Table 18.2): otherwise the anisotropic spatial curvature dominates at late times and the model is driven away from isotropy. We discuss isotropy at intermediate times below (Section 18.6.3). We should also note that under reasonable conditions there may be no infinitely expanded ω-limits in Bianchi type IX: the models always recollapse, as do the $K > 0$ FLRW models, though the Bianchi IX models will again oscillate as they approach a singularity (Lin and Wald, 1991). However, Calogero and Heinzle (2010) have shown that there are Bianchi IX fluid models with negative w obeying the strong energy condition that expand for ever.

The late time behaviour of the models that do allow specialization to Robertson–Walker is to isotropize for types I and V, so only type VI_h requires further discussion. Collins and Hawking (1973b) showed that only a set of measure zero in this class have an isotropic ω-limit, but intermediate isotropization can occur (Wainwright *et al.*, 1998).

The ω-limits of the orthogonal models can show asymptotic self-similarity breaking and Weyl dominance when the state space in expansion-normalized variables is unbounded, see e.g. Nilsson, Hancock and Wainwright (2000) for type VII_0.

Turning to the tilted models, the α-limits are typically vacuum (if $w < 2$) and so their behaviour can often be obtained from the orthogonal cases, though note that new combinations may arise as in the tilted type II case above. Since the tilted type II is a limit of almost all other cases, those cases' limits are also oscillatory although the orthogonal models' limits are not (this was first remarked by Peresetskii (1977)). Most of the work has thus focused on the ω-limits.

There are very few extra possible tilted self-similar attractors in the ω-limit (Apostolopoulos, 2003; 2005, Hervik *et al.*, 2007): Hewitt's type II solution; the rotating Bianchi VI_0 solution of Rosquist and Jantzen; a new, unstable, solution of type $VI_{-1/9}$; and some additional solutions of type VI_h. Just as the orthogonal cases include the orthogonal Bianchi II behaviour as a limit, all tilted cases except Bianchi V include the tilted Bianchi II behaviour as a limit.

However, there is an important additional limit for Bianchi types IV, VI_h and VII_h, namely the (vacuum) plane wave solutions admitting these groups of symmetries. These arise when there is non-extreme Weyl dominance (Barrow and Hervik, 2002) and are the only future attractor for the general type VII$_h$ models. Type VI$_h$ models only approach a plane wave if $h < -1$. However, the waves themselves are unstable to inhomogeneous perturbations (Hervik and Coley, 2005). Remarkably, the state spaces which have plane wave limits also have a 'loophole', a small region of parameter space where the future attractor, the 'Mussel attractor', is a closed loop. Numerical experiment also shows closed trajectories outside the loophole, and there can be bifurcations at which an equilibrium becomes a closed orbit. Type III can show some atypical tilted Bianchi VI$_h$ behaviour. (See Coley and Hervik (2005) for the type IV case, Hervik *et al.* (2007) for the type VI$_h$ case, and Hervik *et al.* (2008) for rotating type VI$_{-1/9}$.)

There is also a special vacuum ω-limit for the VI$_{-1/9}$ class, a particular Robinson–Trautman solution (see Stephani *et al.* (2003), Chapter 28), the Collinson–French solution. This is a Petrov type III vacuum solution and arises in the orthogonal case if $w > 1/9$, while for $-\frac{1}{3} < w < 1/9$ the limit is an orthogonal solution due to Collins. The approach may be oscillatory. In the tilted cases the eventual tilt can be zero, intermediate or extreme.

Tilted type VIII solutions also have an unbounded state space (since they contain a Type VII$_0$ limit) and develop extreme tilt if $w > 0$. There is extreme Weyl dominance if $w > 1/5$, with rapid oscillations and self-similarity breaking. The approach to the vacuum state can be very slow ($\Omega \approx 1/\ln t$) whereas in the orthogonal case it is more like t^{-2w} $(\ln t)^{-(w+1)/2}$.

Type VI$_0$ shows a bifurcation between asymptotically orthogonal and asymptotically extreme tilted at $w = \frac{1}{5}$: there are unstable self-similar solutions which are attractors in a subspace and there is Weyl–Ricci balance at late times. In tilted VII$_0$ the shear and tilt go to zero but the Weyl invariant does not: the shear can oscillate rapidly and lead to Weyl dominance. There is a bifurcation giving extreme tilt as the limit if $w > \frac{1}{3}$.

The Bianchi I models were discussed in Section 18.2 so only the Bianchi V models have not been summarized. Their orthogonal cases are rather similar to Bianchi I, since the surfaces have constant (though negative) curvature, so (18.7) still holds. There is an LRS tilted case which is non-rotating but shows some interesting behaviours, depending on w (Coley *et al.*, 2009). The general tilted cases are more complicated, exhibiting bifurcations for the future behaviour at $w = -\frac{1}{3}, \frac{1}{3}$ and 1 as usual, and also at $\frac{1}{5}$.

It should be noted that all these remarks concern just perfect fluids with linear equations of state. As the remarks at the end of Section 18.2 indicate, more complicated evolutions can arise when intrinsically anisotropic stresses are introduced. For example, a magnetic field in an orthogonal cosmology can mimic the effect of additional spatial curvature, and so introduce additonal ω-limits and oscillation in the α-limit (Collins, 1972, Horwood and Wainwright, 2004). Other cases such as two or more tilted fluids and collisionless particles (Rendall and Tod, 1999) have been considered.

Exercise 18.5.1 Introduce suitable variables for considering the behaviour of the fluid in a comoving (tilted) frame (remember that all rotating Bianchi models are tilted). Show that for $w > \frac{1}{3}$ the tilt generally becomes extreme at late times. (See Coley, Hervik and Lim (2006).)

18.6 Cosmological consequences

18.6.1 Evolution near a big bang

The possible evolutions have been summarized in Section 18.5.3 above. One will typically have oscillatory behaviour in general cases, though whether or not this is chaotic remains unclear, but certain cases have self-similar limits and for $w \geq 1$ there are quasi-isotropic singularities. Most of the interesting dynamics would take place at times expected to be in a quantum gravity regime, so it may be that the issue of how the trajectories leave the α-limit is the right one to pursue.

The oscillations in Bianchi IX were envisaged as a way to remove particle horizons and so resolve the 'horizon problem' (Misner, 1969a). For a variety of reasons this idea did not work (see e.g. the summary in MacCallum (1979)).

We note that these typical simultaneous spacelike singularities are not the only possibilities. As mentioned in Section 6.7, in tilted Class B models, there may be a dramatic change in the nature of the solution, where the surfaces of homogeneity change from being spacelike (at late times) to being timelike (at early times), these regions being separated by a null surface $\mathcal{H}$, the horizon associated with this change of symmetry. At earlier times the solution is no longer spatially homogeneous – it is inhomogeneous and stationary. (This kind of change happens also in the maximally extended Schwarzschild solution at the event horizon.) Associated with the horizon is a 'whimper' singularity where all scalar quantities are finite but components of the matter energy–momentum tensor diverge when measured in a parallelly propagated frame as one approaches the boundary of spacetime (this happens because the parallelly propagated frame gets infinitely rescaled in a finite proper time relative to a family of Killing vectors which in the limit have this singularity as a fixed point). The matter itself originates at an anisotropic big-bang singularity at the origin of the universe in the stationary inhomogeneous region.

Details of how this happens are given in Ellis and King (1974), and phase plane diagrams for the simplest models in which this occurs – tilted LRS Type V models – in Collins and Ellis (1979). These models isotropize at late times, and can be arbitrarily similar to a low-density FLRW model at the present day. Siklos (1981) showed that in general the Kretschmann scalar diverges at the big bang, except for plane wave solutions, that the whimper singularities are unstable, and that except in type $VI_{-1/9}$ the non-scalar singularities are accompanied by horizons.

18.6.2 Occurrence of inflation

An issue of importance is whether these models tend to isotropy at early or late times. Isotropization may occur regardless of a possibly early inflationary phase: we discuss this below. Here we note some of the work on the occurrence and effectiveness of such an inflationary phase in Bianchi models (see also Coley (2003)). Inflation only occurs in Bianchi models if there is not too much anisotropy to begin with (Rothman and Ellis, 1986), and it is not clear that shear and spatial curvature are effectively removed in all inflating cases (Raychaudhuri and Modak, 1988). Hence, some Bianchi models isotropize due to inflation, but not all.

Bianchi I models can isotropize without inflation, and inflation in them can be anisotropic (Gümrükçüoğlu, Himmetoglu and Peloso, 2010), leading to substantial changes in CMB correlations. Aguirregabiria, Labraga and Lazkoz (2002) considered 'assisted inflation', where several scalar fields are used, and found, in Bianchi VI_0 models, that inflation was more likely if there were more non-interacting fields or fewer interacting ones.

Aguirregabiria, Feinstein and Ibañez (1993) considered Bianchi I metrics with an exponential potential for the scalar field proportional to $e^{k\Phi}$ and found they isotropize for $k < \sqrt{2}$ but not for $k > \sqrt{2}$. The same conclusion was reached by Coley and Goliath (2000) for K-S metrics (though these recollapse if inflation fails to isotropize them), and by van den Hoogen and Olasagasti (1999) for a Bianchi IX metric.

For Bianchi class A with a fluid and minimally coupled scalar field, Fay (2004) showed that the solutions only isotropize if scalar field dominated.

In approximately de Sitter inflation, initial velocities are required to determine the 4-velocity of matter at the end of inflation (since the de Sitter solution is locally Lorentz invariant and so has no preferred 4-velocity). Anninos *et al.* (1991) studied the effect of an initial tilted velocity before inflation in an LRS Bianchi V model and found numerically that it did not isotropize on small scales.

18.6.3 Isotropization

Isotropization when just a fluid is present can be studied by use of the evolution curves discussed above (Wainwright *et al.*, 1998). Collins and Hawking (1973a) showed that for ordinary matter, many Bianchi models become anisotropic at very late times, even if they are very nearly isotropic at present. Thus isotropy is unstable in this case. However, Wald (1983) showed that Bianchi models will tend to isotropize at late times if there is a positive cosmological constant present, implying that an inflationary era can cause anisotropies to die away. More detailed discussion of ω-limits was given above (Section 18.5).

Even in the classes of non-inflationary Bianchi models that contain FLRW models as special cases, not all models isotropize at some period of their evolution; and of those that do, most become anisotropic again at late times. Only an inflationary equation of state ($w < -1/3$) will lead to such isotropization for a fairly general class of models; but once inflation has turned off, anisotropic modes will again occur.

However, in many Bianchi types the FLRW models are saddle points in the relevant state spaces, allowing models to be nearly isotropic at intermediate epochs, a behaviour sometimes called 'hesitation dynamics'.

Theorem 18.1 Bianchi Evolution Theorem (1): *Consider a family of Bianchi models that allow intermediate isotropization. Define an ϵ-neighbourhood of an FLRW model as a region in state space where all geometrical and physical quantities are closer than ϵ to their values in an FLRW model. Choose a time scale L. Then no matter how small ϵ and how large L, there is an open set of Bianchi models in the state space such that each model spends longer than L within the corresponding ϵ-neighbourhood of the FLRW model.*

Hence there exist many Bianchi models that are compatible with astronomical observations and therefore viable as models of the real universe. The catch is that the significant deviations from FLRW may occur only at times earlier than those at which quantum gravity effects are normally assumed to dominate, or in the very far future, and hence the difference may be physically unimportant: numerical estimates would be useful.

Another formulation of this idea is

Theorem 18.2 Bianchi Evolution Theorem (2): *In each set of Bianchi models of a type admitting intermediate isotropization, there will be spatially homogeneous models that are linearizations of these Bianchi models about FLRW models. These perturbation modes will occur in any almost-FLRW model that is generic rather than fine-tuned; however, the exact models approximated by these linearizations will be quite unlike FLRW models at very early and very late times.*

The point is that these modes can exist as linearizations of the FLRW model; if they do not occur, then initial data have been chosen to set these modes precisely to zero (rather than being made very small), which requires very special initial conditions. Thus these modes will occur in almost all almost-FLRW cosmologies. Hence, if one believes in generality arguments, they will occur in the real universe. When they occur, they will at early and late times grow until the model is very far from an FLRW geometry (while being arbitrarily close to an FLRW model for a very long time, as in Theorem 18.1).

18.6.4 Light propagation and observations

Although the field equations for Bianchi models are relatively simple, and many exact solutions are known, though not for the most complicated cases, the geodesic equations are not so readily solved. Since there are (at least) three independent Killing vectors $\xi_i, i = 1, 2, 3$, one always has four constants of motion along a geodesic with tangent vector $\mathbf{t}$, namely $\mathbf{t}.\xi_i$ and $\mathbf{t}.\mathbf{t}$, giving first integrals of the second-order geodesic equations, but analytically integrating for a second time may not be possible.

Nilsson *et al.* (1999) considered the general dynamical system with the geodesic equations included, and pointed out in particular that bounds on shear obtained by considering Bianchi I and V models are untypical.

Observational relations for a number of these universes have been examined in detail.

(a) Redshift, area distance, and galaxy observations ((M, z) and (N, z) relations) are considered in MacCallum and Ellis (1970). Anisotropies can occur in all these relations, but many of the models will display discrete isotropies in the sky.

(b) The effect of tilt is to make the universe look inhomogeneous, even though it is spatially homogeneous (King and Ellis, 1973). This will be reflected in particular in a dipole anisotropy in number counts, which will thus occur in rotating universes (Gödel, 1952).[2]

---

[2] They will also occur in FLRW models seen from a reference frame that is not comoving; hence, they should occur in the real universe if the standard interpretation of the CMB anisotropy, as due to our motion relative to an FLRW universe, is correct; see Ellis and Baldwin (1984).

(c) CMB anisotropies will result in anisotropic universe models. For example many Class B Bianchi models will show a hot-spot and associated spiral pattern in the CMB sky (MacCallum and Ellis, 1970, Collins and Hawking, 1973b, Barrow, Juszkiewicz and Sonoda, 1983, 1985, Bajtlik *et al.*, 1986). This enables us to put limits on anisotropy from observed CMB anisotropy limits (Collins and Hawking, 1973b, Bunn, Ferreira and Silk, 1996). If reheating takes place in an anisotropic universe, this will mix anisotropic temperatures from different directions, and hence distort the CMB spectrum (Rees, 1968).

Some of the most detailed work has involved Class B models where the quadrupolar anisotropy found in orthogonal Class A models can be distorted and rotated, leading to hot and cold spots in the sky and contributions to higher multipoles.

Matravers, Madsen and Vogel (1985) showed that Bianchi V models might be of interest in this context, and these and Bianchi VII models have been examined by a number of authors since. Jaffe *et al.* (2006) found that consideration of type VII_h allowed a good fit to some of the suggested CMB anomalies but not a good overall fit. Pontzen (2009) considered the temperature patterns produced by Bianchi VII_h anew and in detail: in particular he noted that certain anisotropies on superhorizon scales could go undetected by nucleosynthesis limits. Taking only models containing an FLRW case, he notes that there are two scales in type VII_h, a spiral scale x and the scalar curvature parameter Ω_K, and derived the CMB patterns. As an example of the type of results obtainable we show Figure 18.5. He also found that in type IX one only got quadrupole terms in the angular distribution, but the E and B modes mixed. Sung and Coles (2009) showed there can also be mixing in the VII_h case and that this case can give localized cold-spots as suggested by the data.

Note that almost-isotropic CMB does not imply, as one might have expected from the almost-EGS theorem (Section 11.1), that a Bianchi metric is close to FLRW. The reason is that one can have cases, e.g. the Bianchi VII_0 example, where the derivative of the shear is large and oscillatory but the shear itself is small (these will be Weyl dominated, as described in Section 18.5.3). One can even have models which at a certain instant have exactly isotropic CMB (Lim, Nilsson and Wainwright, 2001). Pontzen and Challinor (2010) have considered linearization of Bianchi models about FLRW models, highlighting 'the existence of arbitrarily long near-isotropic epochs in models of general Bianchi type'.

18.6.5 Element formation

Element formation will be altered primarily through possible changes in the expansion time scale at the time of nucleosynthesis (Thorne, 1967, Barrow, 1976, Rothman and Matzner, 1984). This enables us to put limits on anisotropy from measured element abundances in particular Bianchi types. This effect could in principle go either way, so a useful conjecture (Matzner, Rothman and Ellis, 1986) is that in fact the effect of anisotropy will always – despite the possible presence of rotation – be to speed up the expansion time scale in Bianchi models.

Matravers, Vogel and Madsen (1984) studied nucleosynthesis in Bianchi V models, showing it could be consistent with observation. More recently Barrow (1997) calculated the

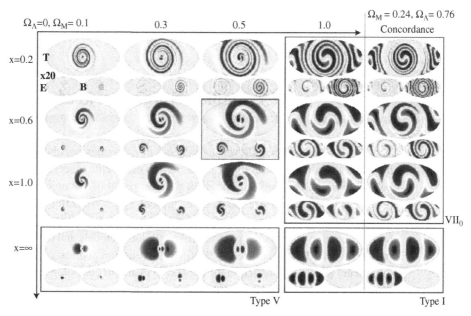

Fig. 18.5 The vector modes of type VII$_h$ and its specializations. For a grid of parameters the temperature pattern (upper panel) and E and B mode polarization exaggerated in scale by a factor 20 (small lower left and right panels respectively) are shown. x varies from top to bottom while Ω_M (and hence Ω_K) varies from left to right. Pontzen assumes $\Omega_\Lambda = 0$ except in the rightmost panels for which the concordance values are used. The primary effect of non-zero Λ is to increase the conformal time to last scattering, causing the final $z = 0$ pattern to be more tightly wound. The model with $(\Omega_M, x) = (0.5, 0.6)$ (shaded panel) is able to mimic known CMB temperature anomalies. (Reprinted with permission from Pontzen (2009). Copyright by the American Physical Society.) A colour version of this figure is available online.

effect of oscillatory behaviour induced by anisotropic stresses in a Bianchi I background. Further large changes in the predictions are unlikely, which may account for the absence of more recent or more detailed work on these issues.

18.6.6 Perturbations, structure formation and the CMB

In addition to the effects of the global geometry on the CMB which largely concern low multipoles, one has to consider whether perturbations will evolve differently and so lead to different small angular scale variations. In Bianchi I models, modes are similar to those of FLRW models but are coupled already at the linear level (see e.g. Dunsby (1993), Gümrükçüoğlu, Contaldi and Peloso (2007)). This can lead to a 'seesaw' effect and resulting statistical patterns dependent on the wave vector $\mathbf{k}$ rather than just its norm $|\mathbf{k}|$, which could be related to the observed CMB anomalies (Pereira, Pitrou and Uzan, 2007). The Jeans instability likewise has directional dependence (Dulaney and Gresham, 2008).

Ghosh, Hajian and Souradeep (2007) using VII_h models argue that these can match some of the anomalies in the CMB but cause unacceptable changes in other parameters unless $\sigma/H \leq 2.77 \times 10^{-10}$.

18.7 The Bianchi degrees of freedom

The above discussion has shown the richness of behaviour possible in these models, and how their evolution can usefully be characterized using state spaces. There may well be such behaviour in the real universe in which we live, possibly dynamically important at early times, even though it is suppressed at the present epoch. Anisotropies might even become important again at late times. The most interesting work to be done in the future is perhaps (a) investigating in further depth the rotating Bianchi models and (b) exploring further the implications for Bianchi models of the CMB and element abundance observations. Bianchi models are also being studied in the context of quantum cosmology (see e.g. references in Damour, Henneaux and Nicolai (2003)), of string cosmology and brane universe models and of loop quantum gravity (Chapter 20; Coley (2003)).

As remarked above, within Bianchi models one can derive limits on the anisotropy from the CMB and from BBN element abundances. Barrow (1976), using those Bianchi types in Table 18.2 which allow FLRW cases, gave limits from BBN which were at the time substantially better than those available from CMB because nucleosynthesis probes to earlier times. However, one has to beware that the limits thus obtained may be misleading if assumed to be correct for other Bianchi models where anisotropic curvature or other effects absent in the Bianchi I case come into play and diminish the instantaneous and/or cumulative distortion.

Indeed, the more recent calculations (Barrow, 1997) gave limits from nucleosynthesis weaker than those from the CMB: this is due to the logarithmic drop in the anisotropy as it evolves in these cases, meaning that the BBN limit is larger than the CMB limit. The same should apply to oscillatory effects due to the geometry, by the analogies mentioned in Section 18.2.

A generic bound from CMB measurements under the assumptions of the almost-EGS theorem is $|\sigma| \lesssim 10^{-4}$, but, as mentioned above, models exist that do not obey those conditions and exceed that bound (Lim, Nilsson and Wainwright, 2001). The more stringent limits sometimes stated for present-day anisotropy, e.g. $|\sigma_0|/\Theta_0 \leq 10^{-6}$ to 10^{-12}, are very model dependent.

Because of the anisotropies that can build up in both directions in time, present-day limits do not imply that either the very early universe (before BBN) or the late universe will also be isotropic. This conclusion applies both to CMB and BBN measurements. In both cases the possibilities are quite model dependent: although very strong limits apply to some Bianchi models, they are much weaker for other types or other matter content. Hence, one should be a little cautious in what one claims in this regard. There needs to be a more careful evaluation of the different possibilities, including quantitative studies of intermediate regimes.

If anisotropies are indeed present, they may well dominate at very early and very late times even if they are very small today; it is therefore important to understand their dynamics and observational effects.

Exercise 18.7.1 Most studies of CMB anisotropies and nucleosynthesis are carried out for the Bianchi types that allow FLRW models as special cases (see Table 18.2). Show that Bianchi models can approximate FLRW models for extended periods even if they do not belong to those types. What kinds of CMB anisotropies can occur in these models?

Problem 18.1 Give a full account of structure formation in perturbed Bianchi models. Consider in particular the orbits leaving the attractor (Bogoyavlenskii, 1985, section 6.VI)).

19

Inhomogeneous models

In the previous chapter, cosmological models which drop the isotropy assumption of FLRW models were considered; here we drop the homogeneity assumption. Of course, perturbed FLRW models also satisfy neither assumption, but they are treated only in perturbation theory. Here we aim to study models in the fully nonlinear theory. Inhomogeneous models have been applied both globally (as shown by the use of LTB models in Chapter 15) and to model localized inhomogeneities and, e.g. their fully nonlinear effects on observation via lensing (as shown by the use of Swiss cheese models in Chapter 16). In the global context, issues such as whether inflation could remove inhomogeneity, or whether hierarchical models could fit the data, can be examined: these are essential to judging the robustness of the assumptions of the standard model.

For example, the evidence cited as support for the standard model can be well fitted by nonstandard models, as we have seen in Chapter 15. Thus one can legitimately ask, *what is the largest family of cosmological models that can fit the observations*? One can then try to devise observational tests to eliminate as many of them as one can.

One may also wonder why we look for exact models of structure formation, when the perturbative theory is so successful? The inflationary paradigm coupled with the perturbation theory of FLRW models has offered the first viable explanation of the observed degree of inhomogeneity in the universe (see Chapter 10). However, the galaxies, clusters and voids we observe now have values of (e.g.) $\delta\rho/\rho$ outside the perturbative regime. Their structure and evolution are often modelled by pseudo-Newtonian N-body calculations (see Section 12.3.7): inhomogeneous solutions of the EFE, especially spherically symmetric models, offer an alternative way to model both collapsed regions (representing galaxies or clusters) and voids. The nonlinear effects of GR may have significant impacts on structure formation: it is helpful to have the best realistic nonlinear GR models we can find in order to study this question. Some such models, especially static ones, are simply descriptions of the present structure: their value is principally as backgrounds in which to study the effects of light propagation in inhomogeneous models: compare Chapter 16. Other work is more concerned with modelling formation and evolution of structures.

Krasiński (1997) gave a very extensive survey of inhomogeneous cosmological models and their physical properties. He fitted the many specific models discussed in the literature into a relatively small number of families, including only those classes which contain an FLRW or Kantowski-Sachs metric, which excludes few models of physical importance. Stephani *et al.* (2003) give the ones with a perfect fluid matter content. (Krasiński placed no restriction on the equation of state, and thus included many metrics outside the scope of Stephani *et al.*) This survey was subsequently complemented by Bolejko *et al.*'s discussion (2010) of the cosmological application of the LTB (Sections 15.1 and 19.1), 'Lemaître' (Section 19.4.1) and Szekeres models (Section 19.6) to structure and evolution. Here we can only summarize the main points from these surveys and the extensive literature.

We first revisit the commonly used Swiss cheese model and LTB models, where simple analytic solutions are possible, and then consider other solutions. Following Section 17.1, we discuss these in order of the amount of their symmetry. (A more detailed classification of possible irrotational perfect fluid models was given by Wainwright (1979, 1981).)

We have grouped applications with the descriptions of the models involved, and we have included brief indications of some attempted applications which did not succeed (this is to help readers avoid repeating these attempts: in Section 17.1.4 we also mention models whose testable predictions have never been considered). This may impede getting a clear overall picture of the value of inhomogeneous models. As well as their lensing applications (Section 12.4 and Chapter 16) and their use as global models (Chapter 15), here are some of the major points among the many that emerge.

- As well as general discussions of nonlinear collapse, detailed fits to a number of actual large-scale structures have been made (Sections 19.1, 19.2, and 19.6): these include for example the Local Group, M87, the Great Attractor, the cluster A2199 and voids with adjacent clusters. The models enable inferences about, e.g. the masses of objects. Nonlinearity can speed up structure formation (Section 19.6).
- In the modelling of voids and clusters, velocity perturbations proved more effective at producing structure than the usual scalar perturbations alone, suggesting that they should be included in general structure formation models (Section 19.1). Remaining incompatibilities between CMB measures and observed density contrasts (Sections 19.1 and 19.3) may be resolved by including radiation (Section 19.4).
- The very early universe may have been significantly different from FLRW models implying that the initial conditions before inflation may be very anisotropic or inhomogeneous (see Sections 19.3, 19.5.1 and 19.10.1). Such inhomogeneous initial conditions can prevent inflation from producing the relatively smooth universe indicated by the CMB (Section 19.10.2). This would imply either that one needs fine tuning before inflation or that the apparent smoothness does not come from inflation.
- The conjecture that universes evolve towards self-similar models, although it is not always correct, has been shown to be a useful guideline characterizing an intriguing part of the dynamics of models of interest, which may help explain some observed structures in an interesting way (Section 19.3).
- Overdensities can grow to underdensities, which does not happen in a linearized picture (Section 19.1).
- Some of the classes considered include exact nonlinear inhomogeneous gravitational waves in an expanding background, whose effects are otherwise only known in the perturbative regime (Section 19.5).
- Models additional to those used in Chapter 15 can be used to fit the SNIa and CMB data with success (Sections 19.6 and 19.7).

19.1 LTB revisited

The LTB models, with metrics given by (15.2) and (15.5)–(15.7) containing arbitrary functions $f(r)$, $m(r)$ and $t_B(r)$ of r, were introduced in Section 15.1, where their application as global models was considered. Here we review other cosmological aspects. Note that we can have LTB models that are FLRW for certain ranges of r, giving concentric shells of differing behaviour.

Spherically symmetric models, especially the evolving LTB models, sometimes with discontinuous density distributions, have for a long time been used to model structures. Lemaître himself (1933a) considered formation of the 'nebulae', Tolman's original paper (1934) predicts the development of condensations and rarefactions, and Sen (1934) argued that inhomogeneous models should form voids, while Bondi's 1947 paper foreshadows black hole formation and notes that an initially expanding central void never recollapses.

The evolution of LTB voids was studied in much more detail by Occhionero and colleagues and Sato and colleagues (see Occhionero, Santangelo and Vittorio (1983), Sato (1984), and papers cited therein). Depending on initial conditions, the void volume may grow or decay, with rarefaction/compression waves in the dust: growth can be asymptotically at the speed of light, so voids may be younger than their surroundings. Collision of the boundaries of growing voids might explain the 'walls' of galaxies. Shell-crossings develop at the edge of expanding voids, but numerical studies (Suto, Sato and Sato, 1984) support the intuitive expectation that non-zero pressure prevents them from forming.

Quantitative modelling of overdensities by LTB models was introduced by Bonnor (1956). Later papers studied galactic scale inhomogeneities (Bonnor, 1972, 1974, Carr and Yahil, 1990), and, on a larger scale, clusters of galaxies (e.g. the Coma cluster, Kantowski (1969a)), variations in the Hubble flow due to the supercluster (Mavrides, 1977), the observed distribution of galaxies, and simple hierarchical models of the universe (Bonnor, 1972, Wesson, 1978, Ribeiro, 1992a). There are studies of evolution of a locally open region in a closed universe (Zel'dovich and Grishchuk, 1984) and of density contrast (Mena and Tavakol, 1999). Meszaros (1991) developed a variation on the usual approach by considering cases with shell-crossings, with the aim of producing 'Great Wall' like structures, rather than the collapse to the centre producing a spherical cluster or galaxy.

Hellaby and Krasiński (2006) considered ways of specifying data for LTB models so as to facilitate production of models fitting observations in various circumstances. For example, Bolejko and Hellaby (2008) used an LTB model for the Shapley concentration and the Great Attractor, and found that 'the peculiar velocity maximum near the SC is $\sim$800 km/s inwards, the density between GA and SC must be about $\sim$0.9 times background, the mass of the GA is probably $4 - 6 \times 10^{15} M_\odot$,' and 'the SC's contribution to the L[ocal] G[roup] motion is negligible'.

Krasiński and Hellaby (2004a) modelled M87, a galaxy believed to contain a black hole, by an LTB metric, showing that models with very different black hole ages were indistinguishable observationally. Their model (2004b) of the galaxy cluster A2199 showed that velocity perturbations produced density variations more effectively than a pure density perturbation. This conclusion, suggesting that a velocity distribution should also be considered in the standard model, was supported by models of the North and South Galactic Pole voids (Bolejko, Hellaby and Krasiński, 2005) where the velocity perturbation was the main factor. Here they also showed that density and velocity perturbations compatible with the CMB observations could not readily produce the observed density contrasts, within this class of models, excluding shell crossings (see, however, Section 19.4.1). Faster expansion produced a larger density contrast.

Sussman (2010b) has recently reformulated the LTB equations using quasi-local integral scalar variables. This provides a covariant interpretation of the parameters of these models, and casts the equations in a dynamical systems form suitable for numerical evolution, which can also be understood in the framework of a gauge invariant and covariant formalism of spherical nonlinear perturbations on an FLRW background. This formulation may be useful in fitting models. It has been applied, using techniques similar to those used for Bianchi models, to LTB models with $\Lambda > 0$ (Sussman and Izquierdo, 2011), generalizing the ΛCDM model. Sussman (2010a) used it to show, among other results, that the inversions of over- to underdensity, or vice versa, can only take overdensities to voids, and only if $K \leq 0$.

Another alternative was introduced by Wainwright and Andrews (2009), who used it to discuss the approach to isotropy and its dependence on Λ, the Weyl to Ricci curvature ratio in the limit, and possible 'hesitation dynamics'.

Exercise 19.1.1 Show how LTB models can be used to characterize the process of gravitational collapse of an overdensity in an FLRW model, demonstrating how it breaks away from the overall cosmic expansion to give a locally collapsing region. Use the model to estimate the relation between turnaround time and initial overdensity. How would this be modified if a cosmological constant is added?

19.2 Swiss cheese revisited

Swiss cheese models were introduced in Section 16.4.1. They show that FLRW models, with metric (2.65), can contain Schwarzschild static vacuum regions, with metric (16.13). We required that there be no surface layer or other discontinuity at the junction: the boundary Σ with the FLRW region was then shown to obey

$$A\dot{T}^2 = 1 + \dot{a}^2 f(r_\Sigma)^2/A, \qquad R_\Sigma = a(t)f(r_\Sigma), \qquad (19.1)$$

where the dot means derivative with respect to the FLRW t, and we must have Schwarzschild mass $M = 4\pi(\rho a^3)f(r_\Sigma)^3/3$. Thus the matching imposes a significant constraint on the models, and raises the issue of the best-fitting background. When two or more Schwarzschild regions are present there is no global spherical symmetry.

The extension to include a cosmological constant is straightforward: just add $-\Lambda r^2/3$ to A, giving the Kottler solution, which can by itself be regarded as a composite of the Schwarzschild and (anti-)de Sitter solutions, but can also be joined to an exterior FLRW dust plus Λ solution. Its main cosmological use has been in trying to clarify the effect of Λ in lensing (see Section 12.4).

As well as their use in studying lensing effects, the Swiss cheese models are suitable for modelling (quasi-) static bound systems, such as voids and galaxy clusters. Applications have included modelling the universe as a patchwork of domains of different curvature $K = 0, \pm 1$ (Harwit, 1992) and the evolution of the boundaries of cosmic voids (Sato, 1984, Hausman, Olson and Roth, 1983, Bonnor and Chamorro, 1990, 1991, Chamorro, 1991).

One can replace either the central portion or the whole of the Schwarzschild region by a spherical source, e.g. a region of an LTB solution. This gives a sort of Swiss cheese model with filled holes. To give the right total mass, any underdense part of an LTB portion has to be compensated by an overdense part (Ribeiro, 1992a), but that may be appropriate in modelling, for example, a void surrounded by galaxies. The cases with dynamic LTB interior regions, giving an exact cosmological model with time-dependent inhomogeneities, can be used to model collapsing systems: for example, Oppenheimer–Snyder collapse in an expanding universe has been studied (Lake, 1980, Hellaby and Lake, 1981, 1983) by taking the interior source to be a different, collapsing, FLRW region.

There is an important difference between the original Swiss cheese model and a matching between LTB and FLRW metrics: in the latter case there is no need for the boundary between the two or more parts to be comoving and the dust can move between them.[1]

Discussion of the Swiss cheese model as a special case of LTB with discontinuous density distribution, and perturbation of this within the LTB class and its generalizations, reveals that the Swiss cheese is unstable (Sato, 1984, Lake and Pim, 1985). As one might guess, if the central mass is too small, the boundary of the void would want to expand faster than the FLRW region does, and if the mass is too large, the vacuole collapses. This suggests that the Swiss cheese model can at best be an approximation to real structures for some (maybe long) period of time but not indefinitely.

The construction in which the mass interior to a sphere in FLRW is conserved but contracted in radius can be used to develop the 'packed Swiss cheese' (Mureika and Dyer, 2004), in which the construction is repeated multiple times both with different centres and inside the already compressed regions. These models, with 3–9×10^4 compressed spheres, were used as models of fractal distributions of mass, but do not match observation well, suggesting that luminosity biasing is important.

One can also change the exterior. Bonnor (2000) considered a Schwarzschild region matched to an LTB exterior, and found that there is then no restriction on the mass of the Schwarzschild part but that, naturally enough, the boundary is infalling if that mass is overdense and moving outward if it is underdense. He points out that such models are inappropriate for the Solar System (because the Galaxy rotates) or the Galaxy itself (because there is a force between it and M31). They could give simple models of the Local Group with, for example, a boundary at ~ 0.7 Mpc and infall velocity ~ 170 km/s, close to the observed values.

A number of authors have considered the possibility of finding generalizations of the Swiss cheese with a stationary interior and expanding exterior, but with nonspherical internal

---

[1] One can also match the spherical Vaidya solution, which approximates a radiating source although spherical gravitational waves do not exist in general relativity, to an FLRW solution (Fayos *et al.*, 1991), with the aim of studying inhomogeneities and collapse. In this case radiation crosses the boundary. This is only possible with a re-interpretation of the FLRW energy–momentum, compare remarks in Section 9.1.4.

or external geometry, or, in some cases, a matter shell at the interface (Lake, 1987). These ways of generalizing the Swiss cheese turn out not to be very promising. Axially symmetric stationary regions in FLRW spacetimes must be static (Nolan and Vera, 2005), and since if static they must have spherical boundaries (Mars, 2001) this leaves only the original Einstein–Straus case. Matching static cylindrical regions to more general anisotropic cosmologies also encounters severe restrictions (Mena, Tavakol and Vera, 2002).

A further alternative is to study whether models which are neither spherically symmetric nor stationary can be joined regularly onto an FLRW model. Bonnor (1976) showed that some Szekeres models can be matched to a Schwarzschild metric across a spherical surface, and (as one might then expect) to a dust FLRW model across a comoving spherical surface: see Section 19.6. Dyer, Landry and Shaver (1993) have shown that one can match FLRW and LRS Kasner (vacuum Bianchi Type I, Section 18.2) models across a flat junction surface. Optical properties of these 'cheese slice' models have been investigated in depth (Landry and Dyer, 1997): they include significant lensing and anisotropic redshifts. Although the models do not match observations, those properties indicate effects that might also be seen in a fully three-dimensionally inhomogeneous situation.

Exercise 19.2.1 Work out the details of the matching in the case with a cosmological constant.

Exercise 19.2.2 Show how appropriate choice of initial data in a LTB model can give an effective Swiss-cheese model with one centre surrounded by a series of successive FLRW and nonFLRW spherical regions. Can you include (i) flat, (ii) vacuum (Schwarzschild) regions in this construction?

19.3 Self-similar models

The definition of self-similarity in the literature is somewhat confusing. Usually it means there is a homothety (Section 2.7.1), i.e. a vector field satisfying

$$\xi_{(a;b)} = 2kg_{ab}, \tag{19.2}$$

where k is a constant. However, this can be generalized in various ways, for example to cases where there are hypersurfaces in which three-dimensional homotheties act (Carter and Henriksen, 1989, Sintes, 1998).

Mathematically, self-similarity reduces the number of independent variables by one and so simplifies the system of equations, but a more convincing physical motivation arises from the following:

Asymptotic self-similarity conjecture:
Expanding models evolve towards self-similar solutions.

The basis for this conjecture lies in the Newtonian theory of blast waves, as treated by Sedov and others. Its applicability in relativity is reviewed by Carr and Coley (2005): it is true for some Newtonian cosmologies, for many classes of Bianchi models, as we have seen in Chapter 18, and for some models of voids in cosmology (Jain and Bertschinger, 1996), and is consistent with cosmic no-hair results.

Apart from Bianchi models with an additional homothety, which are discussed in Chapter 18, self-similar models arise in three main subcases:

(1) generalizations of Bianchi models, with a three-parameter homothety group H_3 on spacelike surfaces S_3;
(2) comparable but truly spatially inhomogeneous models with H_3 on timelike surfaces T_3;
(3) self-similar spherically symmetric models.

We note that all these cases have at least three symmetries, and the first two are special cases of 'G_2 solutions' (Section 19.5), with perfect fluid matter content. We discuss all three only briefly because although their geometry and dynamics have been studied in considerable detail, they have so far been relatively little applied to the major problems of cosmology.

The geometry and dynamics of the first type of self-similar model is very like that of the Bianchi models, and, correspondingly, many details can be found in e.g. Eardley (1974), Luminet (1978), Wu (1981) and Hanquin and Demaret (1984). Hewitt, Wainwright and colleagues (Hewitt and Wainwright, 1990, Hewitt, Wainwright and Goode, 1988, Hewitt, Wainwright and Glaum, 1991) considered the second class, where it is found that the spatial variations can be periodic or monotone; the asymptotic behaviour may be a vacuum or spatially homogeneous model; the periodic cases are unstable to increases in the anisotropy; and the singularities can be acceleration dominated.

The third class, spherically symmetric self-similar models, has been studied by many authors (for a review see Carr and Coley (1999)). Homothety implies that all the metric coefficients in (19.3) are functions of a similarity variable r/t, while in generalized self-similarity they depend on $r/F(t)$ for some function F. For a perfect fluid obeying (5.49) and a spacelike homothety a singularity may arise where the motion reaches the sound speed $\sqrt{w}$: continuation across this surface is only possible in certain cases. The models were classified by their behaviour at large and small r by Carr and Coley, who also studied the asymptotics. After pioneering work of Bogoyavlenskii (1985), the dynamical system for these models was usefully recast by Goliath *et al.* (see Carr *et al.* (2001)).

Applications of this third class include models of self-similar voids and some global models asymptotic to FLRW solutions as $t \to \infty$, but as the self-similarity of the Newtonian blast wave case depends on the effect of pressure, it is more natural to apply the idea in the early universe, for example to set initial conditions for a later LTB phase. However, Carr and Coley (1999) showed one could not fit both the void size and the CMB fluctuations. Other applications include formation of black holes in FLRW models (Carr and Hawking, 1974), other nonlinear FLRW perturbations (Carr and Yahil, 1990) and bubbles in the early universe (Carr and Koutras, 1993). The more convincing uses, such as critical collapse, and the study of cosmic censorship, are on scales small compared with those in cosmology.

Exercise 19.3.1 Which FLRW models are self similar? Are they attractors in the space of FLRW models?

19.4 Models with a G_3 acting on S_2

These models have a metric

$$\mathrm{d}s^2 = -e^{2\nu}\mathrm{d}t^2 + e^{2\lambda}dr^2 + R^2(\mathrm{d}\theta^2 + f^2(\theta)\mathrm{d}\phi^2), \tag{19.3}$$

where $f(\theta) = \sin\theta$, θ or $\sinh\theta$ respectively for the cases with spherical, plane or pseu-dospherical (hyperbolic) symmetry and ν, λ and R are functions of r and t. The precise functional forms in the metric depend on the choice of coordinates and the additional restrictions assumed. The geometry of the time-dependent cases was examined in a covariant way by van Elst and Ellis (1996), and a tetrad analysis was given by Ellis (1967) (for the pressure-free case) and Stewart and Ellis (1968) (for perfect fluids). Note that these metrics are frequently used in the deprecated g-method of Synge (Section 5.2).

As already mentioned, the spherical cases, especially those discussed above (Sections 15.1 and 16.4.1), have been widely applied to issues of cosmological importance. The plane ones have been used mainly for modelling domain walls and the pseudospherical ones hardly at all.

19.4.1 Spherically symmetric models

The spherically symmetric cases are the simplest inhomogeneous models. The perfect fluid cases include stellar models and collapse solutions (see e.g. Misner, Thorne and Wheeler (1973)): Bolejko *et al.* (2010) refer to these as the Lemaître models, since they were discussed in Lemaître (1933b). The dust solutions form the LTB class. A further family that has been extensively studied is the self-similar subclass (Section 19.3).

Lasky and Bolejko (2010) used models with pressure, as discussed in Section 8.5.4, to avoid the shell-crossings of LTB models, at least until after structure formation was complete, and studied the resulting magnitude–redshift relations. They showed that pressure gradients can have significant effects. In particular, radiation can improve the density contrast of the voids obtained, compared with the LTB models discussed above: Bolejko (2006a) concluded radiation was needed to obtain realistic voids.

The shearfree perfect fluid cases belong to the Petrov type D branch of the Stephani–Barnes family (see Section 19.7). Specific shearfree models were first considered by McVittie (1933), Wyman (1946) and Kustaanheimo and Qvist (1948) and all known solutions of this type are included in these papers and Wyman (1976): the analysis of the whole class and the exact solutions in it are described in Stephani *et al.* (2003), Section 16.2.2.

One solution of interest in this class is the McVittie (1933) metric,

$$\mathrm{d}s^2 = \frac{(1+f)^4}{(1+Kr^2/4R^2)^2}e^{g(t)}(r^2\mathrm{d}\Omega^2 + \mathrm{d}r^2) - \frac{(1-f)^2}{(1+f)^2}\mathrm{d}t^2\,, \tag{19.4}$$

$$2f = Me^{-g(t)/2}(1+Kr^2/4R^2)^{1/2}/r, \quad R = \text{constant},$$

where $K = \pm 1$ or 0, defining the spatial curvature of the corresponding Robertson–Walker universe. This gives the Schwarzschild solution if $R \to \infty$, $g \to 0$, and an FLRW solution if $M = 0$; hence it has been interpreted as a mass in a Robertson–Walker universe. It is the unique spherically symmetric metric containing a shearfree perfect fluid with density $\rho = \rho(t)$ and approaching an FLRW solution asymptotically. It has been used in evaluating lensing effects (Section 16.6.4) and Noerdlinger and Petrosian (1971) used it to study the evolution of clusters.

However, it has drawbacks: the global geometry is not what one would want for such an interpretation (see the appendix to Sussman (1988)), and in the $K = 0$ case the surface $r(1 + f)^2 = 2M$ is singular (Nolan, 1999), rather than being a horizon as in the Schwarzschild metric, although other members of the shearfree class can be used to give a regular interior solution for the central region (Nolan, 1993).

A closely related metric form was discussed by Sultana and Dyer (2005): it has the same metric form as (19.4) but with $f_{,t} = 0 = K$, and was obtained by conformal transformation of the Schwarzschild metric (16.13). It contains a null fluid as well as dust, and has global structural drawbacks similar to those of the McVittie solution (Faraoni, 2009): for example, it has no apparent horizon (Sun, 2011).

19.4.2 Plane symmetric models

Exact solutions for domain walls, using plane symmetric models, usually static, have been considered (Vilenkin, 1983, Ipser and Sikivie, 1984, Goetz, 1990, Wang, 1992). Since the sources usually have a boost symmetry in the timelike surface giving the wall, the corresponding solutions have timelike surfaces admitting the (2+1)-dimensional de Sitter group. It may be noted that all these solutions with groups G_3 acting on T_3 are included in the cases considered by Harness (1982).

19.5 G_2 cosmologies

These models all have two commuting Killing vectors, which we shall assume are spacelike. We generally assume also that the orbits of the G_2 are orthogonal to another set of two-dimensional surfaces (block-diagonal metrics). The metrics can then be written as

$$ds^2 = f_{AB}dx^A dx^B + \delta e^{2\gamma}((dx^4)^2 - (dx^3)^2)/f, \tag{19.5}$$

where A, B take values $1, 2$ and the values of f_{AB} can be written as a matrix,

$$\begin{pmatrix} f & -f\omega \\ -f\omega & f\omega^2 + W^2/f \end{pmatrix}. \tag{19.6}$$

Here we assume that the gradient of W has the same character (timelike or spacelike) as the coordinate x^3, so δ parameterizes the nature of the gradient of the determinant of the metric in the surfaces of symmetry.

If $\delta = 1$ then det f_{AB} has the timelike gradient expected for cosmological models and colliding waves: if $\delta = -1$ we may have cylindrical, planar and plane metrics. The distinction here is that in planar metrics there are only two KVs but the orbits are not rolled up into cylinders, and in the plane cases there is a third, rotational, symmetry. So cylindrical models may be locally plane; this has caused some confusion in the literature. Regions of differing δ may be joined across null surfaces. Solutions include colliding waves (see Griffiths (1991)), Bianchi cosmologies with superposed solitonic waves, 'corrugated' cosmologies with spatial irregularities dependent on only one variable, and time-dependent cylindrical metrics.

Cosmic strings, usually static strings,[2] have been modelled by cylindrically symmetric spacetimes, starting with the work of Gott, Hiscock and Linet in 1985. Issues such as the effects on classical and quantum fields in the neighbourhood of the string have been discussed.

Note that special cases of G_2 cosmologies may have additional symmetry: most Bianchi models (all save types VIII and IX) are included, as are the spatially self-similar and other cases with homothety (Section 19.3). In particular the Kasner metrics play an important role as expanding backgrounds on which gravitational waves can be superposed. Centrella and Matzner (1982) found that colliding waves in such a background do focus one another, but that the expansion means that, unlike the case with a flat background (Griffiths, 1991), this need not lead to a singularity.

19.5.1 Gowdy models

The cosmological class is sometimes referred to as the Gowdy models, after Gowdy's early work (Gowdy, 1971, 1974, 1975) but examples appeared elsewhere even earlier (e.g. Belinski, Lifshitz and Khalatnikov (1971)). Gowdy added the restriction that spatial sections should be compact. This led him to spacetimes containing regions with both signs of δ, which can be considered locally to be colliding wave regions, considered with time reversed so that they start from a cosmological singularity, and cylindrically symmetric regions. Space sections of these universes are toroidal (T^3) or have the three-sphere topology S^3 or the hypertorus topology $S^1 \times S^2$. The further structure of these solutions has been described by Chruściel (1990) and Chruściel, Isenberg and Moncrief (1990).

The main interest in recent years has been the highly inhomogeneous asymptotic behaviour (called 'spikes') in the approach of these models to their initial singularities, first found by Berger and Moncrief (1993): some of the apparent discontinuities in the limits of metric variables are 'false spikes', i.e. gauge effects, but there are also 'true spikes'. See Rendall and Weaver (2001) and Andersson, van Elst and Uggla (2004) for more information and fuller references. Recently LeFloch and Rendall (2011) have considered the T^3 case in a low regularity setting allowing discussion of impulsive gravitational waves and matter shock waves.

---

[2] There is some controversy about whether these can correctly represent strings embedded in an expanding universe see e.g. Clarke, Ellis and Vickers (1990). There is an obvious difficulty in reconciling the angular defect formed by a string with a surrounding and pre-existing background which does not have such a defect.

19.5.2 Soliton solutions

There are generating techniques applicable to G_2 models if the matter content has a characteristic propagation speed c, which covers vacuum, electromagnetic, massless neutrino and 'stiff fluid' (or equivalently, massless scalar field with a timelike gradient) cases and combinations thereof: in these cases we can take $W = x^3$. These methods, which are described in Griffiths (1991), Belinski and Verdaguer (2001) and Chapter 10 of Stephani *et al.* (2003), relate solutions with one another, often starting from flat space or vacuum Kasner models. FLRW fluid solutions can be obtained via the same methods in higher dimensions, using dimensional reduction.

Because of the generating techniques, many exact solutions with two commuting KVs are known for the special matter contents. Those of cosmological type are listed in chapter 23 of Stephani *et al.* (2003), together with the known solutions with similar geometry containing perfect fluids other than stiff fluid. Generated solutions of soliton type have received particular attention.

Gravitational solitons, and related solutions, can be defined as localized perturbations of the gravitational field which propagate on a homogeneous background and have no dispersion. (This definition is sometimes broadened to include any solution that has been generated using the Belinski and Zakharov (1978) inverse scattering technique.) They are analogous to classical solitons, which are localized, have a well-defined velocity of propagation, and show persistent structure even in collisions, usually showing only a phase shift, but do not behave like classical solitons in all respects. The interaction of gravitational solitons in a cosmological context could play a role in the process of isotropization in the early universe. For a review see Belinski and Verdaguer (2001).

19.6 The Szekeres–Szafron family

The largest family of solutions described by Krasiński (1997) have in general no Killing vectors. They are the solutions of Szekeres (1975) (for the dust case, which can be re-interpreted as containing a fluid with constant pressure and a compensating cosmological constant) and Szafron (1977) (for the cases where the pressure is non-zero but depends only on time) and their generalizations. Szafron and Collins (1979) proved that they can be characterized as those perfect fluid solutions of Einstein's equations with a geodesic and irrotational fluid flow and with conformally flat comoving slices whose second fundamental form and Ricci tensor possess two equal eigenvalues. Alternative invariant characterizations have been given by Szafron (1977), Wainwright (1977) and Barnes and Rowlingson (1989): these use the properties that the Weyl tensor is of Petrov type D (see Section 2.7.6), and that the fluid flow velocity and the Weyl principal null directions are coplanar.

In coordinates based on those of Goode and Wainwright (1982), the metric of these solutions takes the form

$$ds^2 = -dt^2 + R^2[H^2W^2dr^2 + e^{2\nu}(dx^2 + dy^2)], \qquad (19.7)$$

where R obeys (15.3) with $\Lambda = 0$, $2E = -K$, $K = \pm 1$ or 0, and an arbitrary $m(r)$, and thus has the solutions (15.5) and (15.6), with a bangtime $t_B(r)$. For each r, the evolution of R is like that of an FLRW model obeying (9.22) with $\beta = 0$ but the parameters of the relevant FLRW model depend on r. In these coordinates,

$$H = A(x, y, r) - \beta_+ f_+ + \beta_- f_- \,, \tag{19.8}$$

where $\beta_{\pm}$ are functions of r. The functions $f_{\pm}(t, r)$ each obey the same equation as perturbations of an FLRW metric (Goode and Wainwright, 1982), i.e.

$$\ddot{F} + 2\dot{R}\dot{F}/R - 3m(r)F/R^3 = 0 : \tag{19.9}$$

f_+ is taken to have the form of a growing perturbation, and f_- that of a decaying mode.

Many known solutions are special cases of the Szekeres–Szafron class, as shown in Figs. 2.1 and 2.4 of Krasiński (1997). In particular they include the Kantowski–Sachs solutions and their generalizations, and Ellis's (1967) class II LRS solutions including the LTB family (Sections 15.1 and 19.1), and its plane and hyperbolic counterparts (though the coordinates of (19.7) do not instantly reduce to those of (15.1)).

The solutions split immediately into two subfamilies, one in which $b_{,r} \neq 0$ and the other in which $b_{,r} = 0$, where $e^b = Re^\nu$. The first of these 'has found no useful application in astrophysical cosmology' so far (Bolejko et al., 2010), so we say no more about it.

If $b_{,r} \neq 0$, then $W^{-2} = \varepsilon - Kf^2(r)$, $\varepsilon = \pm 1$ or 0, and

$$e^{-\nu} = [a(r)(x^2 + y^2) + 2b(r)x + 2c(r)y + d(r)]/f(r), \tag{19.10}$$

where f is an arbitrary function of r, $\beta_+ = -Kfm_{,r}/3r$, $\beta_- = ft_{B,r}/6m$ and a, b, c and d obey

$$ad - b^2 - c^2 = \tfrac{1}{4}\varepsilon \,. \tag{19.11}$$

Writing $e^{-\nu} = \sqrt{|g|}\mathcal{E}$ (Hellaby, 1996, Bolejko et al., 2010) provides a useful alternative parameterization. The freedom in choosing r together with (19.11) leaves five physical degrees of freedom.

The surfaces of constant r and t have the intrinsic geometry of spheres, planes or pseudo-spheres, depending on the value of ε, although in general the r dependence ensures the full solution does not have these symmetries (this is an example of 'intrinsic symmetry', Dingle (1933), Szafron and Collins (1979)). The quasi-spherical case $\varepsilon = 1$ has been extensively investigated (whereas the other cases have been little studied):[3] the spheres have a mass dipole whose axis varies with r, while the mass in a comoving volume is constant (Szekeres, 1975, Krasiński, 1997).

The equations governing Szekeres models are the same as those of perturbed FLRW models, so they might be regarded as exact perturbations of FLRW. However, as Krasiński (1997) points out, this interpretation has the following weaknesses: the values of $m(r)$ and $\dot{R}/R$ are taken from the full solution, not a background; the total F has the usual relation

---

[3] Hellaby and Krasiński (2008) studied the other geometries and found quasipseudospherical cases describing a 'snakelike void in a more gently varying inhomogeneous background', and showed that regularity imposed strong conditions on the quasiplanar case but that it could be a boundary between pseudospherical and (quasi-) spherical regions. Krasiński (2008) considered cases with toroidal topology.

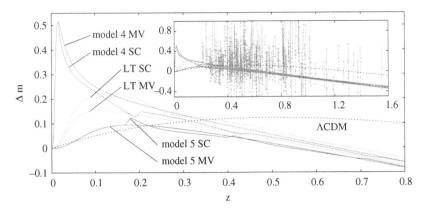

Fig. 19.1 The residual Hubble diagram for four models involving Szekeres regions, all with the same functions of r. SC denotes
Swiss cheese models, and MV ('minimal void') denotes models with a single inhomogeneity. Those labelled model
4 MV and model 4 SC have $\varepsilon = 1$ and model 5 SC and model 5 MV, $\varepsilon = -1$. The corresponding Lemaître–Tolman
models are also presented (LT SC and LT MV). For clarity, the Union supernova data set presented in the inset is for
$z > 0.2$ only. (Reprinted with permission from Bolejko and Célérier (2010). Copyright by the American Physical
Society.) A colour version of this figure is available online.

to the density perturbation only if $A \approx 1$; and only special solutions of (19.9) with the
$\beta_{\pm}$ above are possible, except when $b_{,r} = 0 \neq K$. Note that the decaying mode f_{-} has
amplitude dependent on $t_{B,r}$, the bangtime gradient, while the growing mode has amplitude
dependent on $m_{,r}$ (noted by Silk (1977) for the LTB subcase).

These metrics enable modelling of more complex structures than can be approximated by LTB or spherically symmetric models. Bolejko (2006b, 2007) modelled adjacent
voids and clusters, finding that the evolution depends on the density contrast and
shape rather than the position of the dipole, and that density contrast evolves faster in
larger and more isolated voids and in adjacent high-density regions. Moreover, Bolejko
(2006b) showed that structure formation can be much faster than in corresponding LTB
models.

Nwankwo, Thompson and Ishak (2011) recently considered null geodesics in these
models. Using the coordinates of Bolejko (2006b), they obtain formulae for the luminosity distances and redshifts, which could be employed in arguments such as those in
Section 16.6. Bolejko and Célérier (2010) have studied fitting of the SNIa observations
using an axisymmetric Szekeres model, finding results similar to those of LTB models
discussed in Chapter 15. In particular, they find an inhomogeneity scale of 500 Mpc is
needed to give a good match to the data. Figure 19.1 shows the results for two of their
'minimal-void' models (Szekeres models with a single void region) compared with ΛCDM
and Swiss cheese models.

The generalizations of the Szekeres family add rotation, viscosity, heat flow and/or a
Maxwell field. The most widely used of these have been the solutions with rotation introduced by Stephani (1987) (not to be confused with the Stephani–Barnes family discussed
below). These reduce to members of the Szekeres–Szafron family with $b_{,r} = 0$ when $\omega = 0$,

and have timelike conformally flat surfaces: they include a solution with rotating expanding dust whose nonrotating limit is also nonstatic.

Exercise 19.6.1 Consider the generalization of these models to $\Lambda \neq 0$ (see Barrow and Stein-Schabes (1984)). Identify all cases which are (anti) de Sitter.

19.7 The Stephani–Barnes family

Stephani (1967) considered conformally flat solutions containing an expanding, shearfree and irrotational perfect fluid, and Barnes (1973) found the analogous Petrov type D solutions: the two classes together are known as the Stephani–Barnes family. They are the only perfect fluid solutions with $\sigma = \omega = 0 \neq \Theta$ (whereas the Szekeres–Szafron family have $\omega = 0 = \dot{u}^a$). Krasiński (1989) found a coordinate representation covering all three of the type D subclasses, which allows deformation of one type into another. The conformally flat solutions have no symmetry in general but the type D solutions all have a G_3 acting on a spacelike surface S_2 (and are thus covered by Section 19.4).

The conformally flat cases split into two classes. One has $\Theta = 0$, includes the Einstein static universe (Section 6.1.1), and generalizes the Schwarzschild interior solution. The other, more cosmological, class (the 'Stephani models') has $\Theta \neq 0$ and generalizes the FLRW solutions: the models can thus, like the Szekeres solutions, be considered exact perturbations of FLRW, but with the constant K generalized to $K(t)$ rather than $K(r)$. Their metric reads

$$ds^2 = V^{-2}(dx^2 + dy^2 + dz^2) - (3V_{,4}/V)^2 \Theta^{-2}(t)dt^2,$$

$$\kappa_0 \rho = 3C^2(t), \quad \kappa_0 p = -\kappa_0 \rho, +2CC_{,4}V/V_{,4} \tag{19.12}$$

$$V = V_0(t) + \frac{C^2(t) - \frac{1}{9}\Theta^2(t)}{4V_0(t)} \left\{ [x - x_0(t)]^2 + [y - y_0(t)]^2 + [z - z_0(t)]^2 \right\},$$

where $x_0(t)$, $y_0(t)$, $z_0(t)$, $C(t)$, $\Theta(t)$ and $V_0(t)$ are arbitrary functions. The dust solutions here are FLRW. Note that whereas the Szekeres–Szafron class have spatially constant pressure but varying density, these models have varying pressure and spatially constant density.

A number of cosmological investigations have used these models. For example, Godłowski, Stelmach and Szydłowski (2004) fitted them to the SNIa and CMB data, finding agreement similar to that of ΛCDM but with a higher matter density, and Stelmach and Jakacka (2006) considered angular diameters (relevant to magnitude–redshift diagrams). Clarkson and Barrett (1999) found members of the class as examples of non-FLRW spacetimes with accelerating observers for whom the CMB is isotropic.

If the perfect fluid in these solutions has a barotropic equation of state, they reduce to FLRW or the solutions of Wyman (1946) or Collins and Wainwright (1983). However, one can question whether a barotropic equation of state is appropriate in an inhomogeneous model (including perturbed FLRW models) since it implies isentropy and one might expect entropy to vary in space (see Chapter 5): for an acceptable thermodynamic interpretation see Krasiński, Quevedo and Sussman (1997).

19.8 Silent universes

A novel approach to defining a tractable set of models was introduced by Bruni, Matarrese and Pantano (1995). The 'silent universes' are those in which no information is propagated either by sound waves or gravitational waves. The evolution along distinct world lines is then purely local, i.e. decouples, and is governed by ordinary differential equations. This covers most of the tractable models discussed in this book – all FLRW and Bianchi models, and linearized perturbations of FLRW models, as well as the Szekeres family and its specializations – and it was hoped to obtain some interesting new metrics of cosmological significance.

To ensure the desired conditions, it was assumed that the matter content was irrotational dust and that $H_{ab} = 0$. A series of studies finally led to the result (Apostolopoulos and Carot, 2007) that under these restrictions the only algebraically general spacetimes which are silent are the Bianchi I models discussed in Chapter 18, and that the inhomogeneous models are Petrov type D and are members of the Szekeres family (Section 19.6). Rotating dust models with $H_{ab} = 0$ are similarly restricted (Wylleman and Van den Bergh, 2006). However, silent irrotational dust models with $H_{ab} = 0$ are possible if $\Lambda \neq 0$ (Van den Bergh and Wylleman, 2004), although the ones found so far are not of physical importance. The analogous class with irrotational dust and $E_{ab} = 0$, which would also be silent, reduces to only the FLRW case (Wylleman, 2006).

Dust models, rotating or not, would be silent if both E_{ab} and H_{ab} had zero curl (see (6.34) and (6.36)) but this general class has yet to be investigated: an example with $\omega \neq 0$ and $H_{ab} = 0$ is provided by the Gödel (1949) universe. The only irrotational perfect fluid solution obeying (5.49) is a special model of Bianchi type VI_0 (Wylleman and Van den Bergh, 2006).

Summarizing, the silent universe hypothesis turned out less fruitful, in terms of new models, than initially hoped: however, the concept helps clarify one of the barriers to study of more complex models than those described above. Moreover, inhomogeneous models may be asymptotically silent: we discuss this important characteristic further below.

19.9 General dynamics of inhomogeneous models

A very general study of the dynamics of inhomogeneous cosmologies, aimed at description of the approach to singularities, was developed by Belinski, Lifshitz and Khalatnikov (BLK): see for instance their review (1982). The work assumes that spatial derivatives become in general negligible compared with time derivatives. van Elst, Uggla and Wainwright (2002) refer to this as 'asymptotic silence' by analogy with the silent cosmologies of Section 19.8; in such cases, the past horizon in the limit is just the world line. The same assumption on the magnitude of spatial and time derivatives is used in the long-wavelength approximation scheme (LWAS): see e.g. Comer (1997).

In these approximations, the parameters governing time evolution are position dependent. However, BLK's self-consistency analysis of the temporal behaviour shows that certain spatial terms grow as the singularity is approached, causing a change of evolution parameters. Locally the solution evolves like a spatially homogeneous model (as described in Chapter 18), and in general there is oscillatory behaviour. The set of such solutions is open in the space of all models as it has the maximum number of freely specifiable functions on a spacelike hypersurface.

These studies did not have the standard of rigour of some work in relativity involving theorems on systems of partial differential equations: there were open questions about the relation to linearization stability, domains of dependence, convergence, the relation between neighbouring patches of differing behaviour, and so on (Barrow and Tipler, 1979). However, some of the issues have been resolved e.g. the existence of the required synchronous coordinates (Wald and Yip, 1981): see also Section 19.10.1.

The LWAS scheme, used in various earlier works, was codified and put into Hamilton–Jacobi form by Parry, Salopek and Stewart (1994): they showed how one can expand in successive powers of spatial gradients in a gauge-invariant manner giving a generating function, solved this up to fourth order, and pointed out that one can derive the Zel'dovich approximation by this means (see also Croudace *et al.* (1994)). This was then used (Deruelle and Langlois, 1995) to re-obtain the BLK description.

Work using dynamical systems methods for G_2 models has concentrated on the evolution of vacuum or perfect fluid models with equation of state (5.49) (the latter models not being obtainable by generating techniques, except in the case of 'stiff' fluid, $p = \rho$). van Elst, Uggla and Wainwright (2002) reviewed previous work and rewrote the governing equations, using scale-invariant variables generalizing those used in the studies of Bianchi models, as a first-order symmetric hyperbolic system (which therefore allows stable numerical studies, estimates of asymptotic behaviour, and calculation of propagation of discontinuities). Within the infinite-dimensional phase space thus defined are the finite-dimensional spaces for Bianchi cosmologies other than types VIII and IX.

The ideas of van Elst, Uggla and Wainwright (2002) were generalized to cases without symmetry by Uggla *et al.* (2003), and this formalism with expansion-normalized variables has been used in several investigations mentioned below. A general survey of the method and some of its possible uses was given by Wainwright and Lim (2005).

19.10 Cosmological applications

19.10.1 The Big Bang

A major feature of cosmology is the singularity at the start of the universe predicted by the EFE from the trapped surfaces produced by the thermalized CMB in an approximately FLRW universe, provided the energy conditions are satisfied (see Chapter 6, especially Sections 6.1 and 6.7). The approach to the initial singularity is of interest because it provides the geometric background to early universe studies, with the caveat that very early phases may need to be replaced by a quantum description.

The key issues in this are whether inhomogeneity allows singularity avoidance, what is the generic behaviour at the initial singularity, in particular is it like that of FLRW models, and was the initial singularity in the one universe in which we live of this generic type or not? None of these is wholly decided.

Not all models have initial singularities. Analogously with stellar models, one can consider exact fluid solutions of the EFE where the matter distribution is finite in two directions and infinite in one (an infinite fluid cylinder imbedded in an exterior vacuum domain). These might not cause closed trapped surfaces to occur (null geodesics in some directions may not be refocused), and so can be singularity free. But they are not good models of the real universe. There are similar models where there is matter present everywhere, but its density drops off so fast in some directions that it is insufficient to refocus the past light-cone. The existence of these models has no impact on the conclusion that there must be a start to classical models of the real universe in which the averaged matter density continues at a more or less constant value in all directions until closed trapped surfaces necessarily result, but they do bring out the need for such claims to be carefully stated. See Senovilla (1990), Ruiz and Senovilla (1992) and Senovilla (1998).

Singularities in spatially homogeneous cosmological models may be of various types, as discussed in Chapter 18: we have isotropic (e.g. the FL case), cigar, pancake, oscillatory ('Mixmaster'), and nonscalar ('whimper') singularities. Does inhomogeneity introduce any essentially new possibilities for spacelike singularities? (Timelike singular surfaces can arise in GR (Section 6.7), but no convincing cosmological model of this sort has been obtained so far.)

We do not have a definitive answer. In many cases essentially the same types of singularity can occur, but now in a spatially varying way. There are two particular situations where this has been studied in detail.

Firstly, Eardley, Liang and Sachs (1971) introduced the notions of 'Friedmann-like and 'velocity-dominated' (initial) singularities: in the latter case neither inhomogeneities nor spatial curvature are significant at early enough times. (Hellaby and Lake (1984) later gave arguments, from studying the LTB singularity structure, that the initial singularity should have been Friedmann-like. This case, also called 'isotropic', 'quasi-isotropic' or 'nonchaotic', was reviewed by Tod (2002).) Eardley *et al.* showed that a velocity-dominated singularity can be considered as a three-dimensional manifold with an invariantly and uniquely defined inner metric tensor, extrinsic curvature tensor, and scalar bangtime function, and gave examples from exact solutions for plane symmetric and spherically symmetric expanding dust (LTB) models. These are asymptotically silent universes: the evolution along each world line is independent of that along neighbouring world lines, and hence (like spatially homogeneous models) can be expressed by ODEs along these world lines, even though the solution is spatially inhomogeneous.

Secondly, from the approximations described above, Belinski, Lifshitz and Khalatnikov (1971, 1972) argued that generic inhomogeneous universes locally oscillate like the Bianchi IX ('Mixmaster') case. Although this was the subject of some controversy (Barrow and Tipler, 1979, Belinski, Khalatnikov and Lifshitz, 1982, MacCallum, 1982), it now seems generally to be believed to be the case for fluid dominated models, provided $w < 1$. For instance, Uggla *et al.* (2003) argued that if there is asymptotic silence the attractor behaviour is the same as for spatially homogeneous models, thereby providing conjectures giving a more precise form to the BLK arguments, and Bruni and Sopuerta (2003) showed with an LWAS formulation that it is the magnetic part of the Weyl tensor, H_{ab}, which drives the change to a different Kasner phase in dust examples. If $w \geq 1$ then the BLK arguments predict a non-oscillatory approach to the singularity and this has been verified (Coley and Lim, 2005, Curtis and Garfinkle, 2005) using the methods of van Elst, Uggla and Wainwright (2002).

Recent work of Damour and collaborators (e.g. Damour, Henneaux and Nicolai (2003), Damour and de Buyl (2008)) rediscussed the problem of 'cosmic billiards', and its generalization to higher dimensions and p-form fields, using 'Iwasawa variables'. In a manner analogous to the automorphism group approach to Bianchi models, these write the spatial part of the metric as

$$g_{ij} = \sum_a e^{-2\beta_a} N_i^a N_j^a,$$

where the N_i^a are upper triangular matrices with ones on the diagonal. This enables a rigorous discussion of the oscillatory approach to a singularity. A 'dual' discussion using the expansion-normalized variables has been given by Heinzle, Uggla and Röhr (2009): the two agree in supporting the BLK description.

Andersson *et al.* (2005) concluded that asymptotic silence is true in general G_2 models, and discussed the attractors and spike formation. However, some cases have asymptotic behaviour near the singularity like plane waves (Wainwright, 1983), and others are nonsingular. Moreover the presence of spikes in the Gowdy examples (Section 19.5.1) shows that the BLK approximation can fail if we relax the requirement of matter domination. Consequently the BLK conjecture is now only thought to be true 'almost everywhere' in general, failing at isolated points such as the locations of gravitational wave spikes in the Gowdy models. (An explicit example of this has been constructed by Lim (2008) by applying the generating techniques for G_2 solutions to a Bianchi II metric.)

Examples of asymptotic silence breaking cases were studied by Lim, Uggla and Wainwright (2006): they included some Bianchi models and a Szekeres model, and some solutions with a G_2. These examples illustrate the possibilities that: the event horizon is unbounded in at least one direction and/or does not contract to the world line, or the asymptotic solution is a plane wave, or the singularity is not everywhere spacelike, or the magnetic part of the Weyl tensor remains significant near the singularity.

The BLK arguments aim to show that chaotic oscillatory type behaviour is generic (see the review in Belinski (2009)). The chaotic cosmology programme aimed to show that generic initial conditions could be smoothed out by physical processes to give a smooth RW-like geometry at late times, but was not very successful. The idea was taken up in the inflationary universe scenario. Penrose (1989) powerfully argued that the universe cannot have been truly generic, because of entropy considerations, which imply the initial singularity must have been isotropic. Inflation can be only partially successful for this reason, and also because, as discussed in the next section, if the early universe were anisotropic enough, inflation would never get under way in the first place: anisotropy would dominate instead.

And (see Section 21.4.2) we cannot make definite statements about probabilities as we still do not have a satisfactory measure of probability on the space of cosmological models.

19.10.2 The early universe

As in the Bianchi models discussed above, it is important to study the occurrence and effect of inflation. Numerical and analytic studies implied that inflation would only occur if the scalar field had sufficiently high and uniform value over several Hubble radii (Goldwirth and Piran, 1990, Calzetta and Sakellariadou, 1992). Bounds on how large initial quasi-isotropic inhomogeneities could be before they can prevent inflation were also found in the LWAS (Deruelle and Goldwirth, 1995). The conclusion was borne out in Iguchi and Ishihara (1997), where a massive scalar field in chaotic inflation with a modified long-wavelength approximation was used as initial condition. In the case of exponential potentials in G_2 models (Aguirregabiria, Feinstein and Ibañez, 1993) the result depends on the steepness of the potential. There can be multiple decelerating and inflating phases and the solution may homogenize without isotropizing, for steep potentials. Numerical integrations can be found in Kurki-Suonio, Laguna and Matzner (1993) but the largest grid possible then was 64^3: one example with small-scale and one with large-scale inhomogeneities were studied. The inhomogeneities oscillated and damped, becoming more homogeneous.

The overall conclusion seems to be that inhomogeneity can prevent inflation, and that when inflation occurs it may not be able to reduce the inhomogeneity to the level of linearized perturbations on FLRW, but further work is needed on these important questions.

The presence of growing perturbations in general models of the Szekeres class, and similar results for other cases, suggest that only a set of measure zero among inhomogeneous models (in some reasonable measure, compare Section 21.4.2) can isotropize without inflation. Recently Bolejko and Stoeger (2010) have given arguments that in fact there are sets of initial conditions of non-zero measure which isotropize, using LTB, Szekeres and spherical metrics as examples.

There seem to be rather few attempts to compute (Kurki-Suonio and Centrella, 1991), or to test, the effects of inhomogeneity on spatial variations in element abundances arising from inhomogeneity during the nucleosynthesis era. This is potentially an important consistency test for FLRW models.

19.10.3 The late universe

The BLK work is concerned with behaviour near a singularity and the LWAS work with the early universe. One can also consider asymptotic behaviour in the far future. Rendall (2004) has shown that for $\Lambda > 0$ vacuum models and those with a fluid obeying (5.49), there are in general formal power series for the metric, of the form $e^{2Ht} \sum_n g_{ab}^{(n)} e^{-nHt}$, and discussed the conditions for their validity, making rigorous the methods of Starobinsky (1983).

As we have seen from the individual examples outlined in the previous sections (and those collected in Bolejko *et al.* (2010)), inhomogeneous models can be successfully used to provide exact models of nonlinear behaviour for large-scale structures, both collapsed objects and voids. Some of these become quasi-stationary, just like present-day observed galaxies, clusters and voids. However, observed objects are stabilized by rotational velocities (in spiral galaxies) or random orbital motions (in elliptical galaxies and clusters). The existing models could be claimed to cover the latter case, modelling the effect as that of gas pressure, but there are no satisfactory models of rotating objects.

Problem 19.1 Use new analytic and numerical work to develop a full understanding of whether or when inflation occurs in inhomogeneous models and how successful it is at smoothing out the model.

Problem 19.2 Find tractable models that adequately describe rotation of local systems embedded in an expanding universe.

PART 5

BROADER PERSPECTIVES

20

In approaching the issue of how the universe started, it is common cause that we have to face up to the unsolved problem of quantum gravity: the domain where Einstein's theory of gravity is expected to break down because quantum effects become so dominant that they affect the very nature of space and time. Comparing the gravitational constants of nature with those from quantum theory leads to the Planck length $\ell_P \approx 10^{-33}$cm, which is taken to be the characteristic scale at which quantum gravity dominates. By contrast, most (but not all) variant classical gravitational theories modify GR at low energies (see Chapter 14).

Quantum gravity processes are presumed to have dominated the very earliest times, preceding inflation: the geometry and quantum state that provide the initial data for any inflationary epoch themselves are usually assumed to come from the as yet unknown quantum gravity theory. There are many theories of the quantum origin of the universe, but none has attained dominance. The problem is that we do not have a good theory of quantum gravity (Rovelli, 2004, Weltmann, Murugan and Ellis, 2010), so all these attempts are essentially different proposals for extrapolating known physics into the unknown. A key issue is whether quantum effects can remove the initial singularity and make possible universes without a beginning.

In addition, the weakness of the gravitational force implies that it will be very difficult, though perhaps not impossible, to observationally test theories of quantum gravity.

20.1 Is there a quantum gravity epoch?

Can there be a non-singular start to the inflationary era, thus avoiding the need to contemplate a preceding quantum gravity epoch? In the inflationary epoch the existence of an effective scalar field leads to a violation of the strong energy condition (5.17): therefore at first sight it seems that a bounce may be possible preceding the start of the expanding inflationary era and avoiding the inevitability of a quantum gravity epoch.

However, a series of theorems suggests that inflationary models cannot bounce: they are stated to be future infinite but not past infinite (Guth, 2001). This is an important issue, so it is worth looking at it further. There are two major requirements to get a bounce: $\ddot{a} > 0$ and $\dot{a}(t_*) = 0$ for some t_*. The first condition requires $\rho + 3p < 0$, where ρ and p refer to the total quantities (including a possible Λ).[1] This is a violation of the strong energy

---

[1] If there are semi-classical corrections to GR then ρ and p are the total effective quantities, including gravitational corrections. See Copeland, Lidsey and Mizuno (2006) for examples.

condition (5.17), which follows in the case of inflation. The second condition requires $8\pi G a_*^2 \rho_* = 3K$. If $K \leq 0$, this is possible only if $\rho_* \leq 0$. For a scalar field (see (5.149)) this requires negative potential energies, which appears to be unphysical. Only for $K > 0$ is a bounce possible with $\rho_* > 0$ (Robertson, 1933).

For a bounce in an inflationary universe, it is sensible to consider $K > 0$ inflationary models, which indeed will turn around if inflation occurs for long enough (curvature will eventually always win over a slow-rolling inflaton as we go back into the past) (Ellis *et al.*, 2002a,b). However, the theorems mentioned above do *not* include the case $K > 0$ (Guth, 2001). The scale-free $K = 0$ exponential case clearly is the model underlying many approaches to the problem. But it is highly exceptional – it is of zero measure within the space of all inflationary FLRW models. The fluid flow in these models is singularity free, although the spacetime does have a boundary at finite distance and so is not geodesically complete.

Explicit non-singular models can be constructed, the simplest being the de Sitter universe in the $K = 1$ 'slicing' (Section 9.3.1), which is an exact eternal solution that bounces at a minimum radius a_*. This model has the problem that it does not exit inflation (it corresponds to an exactly constant potential), but variants exist where exit is possible, e.g. a potential that is constant for a long time, but then changes. There are also classical non-singular models that start off in a very special state, asymptotic to the Einstein static universe in the distant past, and avoid the need for a quantum gravity epoch (Ellis and Maartens, 2004). (Quantum versions of the model may tend to stabilize it against quantum fluctuations (Mulryne *et al.*, 2005).)

It seems likely that the options for the start of inflation are (1) avoiding the quantum gravity era, but at the cost of having special ('fine tuned') initial conditions, or (2) having a quantum gravity era preceding the inflationary era. Thus a key issue is whether the start of the universe was very special or generic. We look at this in Section 21.4.1.

20.2 Quantum gravity effects

Lemaître (1931) explored the possibility of a quantum creation of the universe many decades ago. Contemporary efforts to explain the beginning of the universe, and the particular initial conditions that have shaped its evolution, usually adopt some approach to applying quantum theory to the creation of the universe (Gibbons, Shellard and Rankin, 2003).

The attempt to develop a fully adequate quantum cosmology is of course hampered by the lack of a fully adequate theory of quantum gravity, as well as by the problems at the foundation of quantum theory – the measurement problem, collapse of the wave function, etc. (Isham, 1997) – which can be ignored in many laboratory situations, but have to be faced in the cosmological context (Perez, Sahlmann and Sudarsky, 2006). There are many theories of gravity extending and generalizing Einstein's theory to take into account quantum effects, while some attempt to start from completely new foundations. Many of these theories have been specialized to the RW metric form, and we cannot cover here the extensive literature resulting. In making a selection we have included those theories which include applications (often phenomenological) to cosmology, i.e. string theory and loop quantum gravity. The aims of such applications have been principally the avoidance of singularities, and the provision of ways of generating initial perturbations in cosmology.

20.2.1 Quantum gravity theories

Although GR is like modern (quantizable) gauge theories in its use of connections and curvature, it differs fundamentally in its variational principle (3.19): see Section 3.3.2. In attempting to develop full quantum gravity, historically, research tended to try either to force relativity into the mould of quantum theory or vice versa. This has led to treatments that are still in use on the grounds that they should be appropriate limits of some unknown complete theory.

One basic difficulty occurs because quantum theory, including modern gauge field theories, usually treats fields on a fixed background, either Euclidean space, in non-relativistic quantum mechanics, or Minkowski spacetime, in relativistic quantum theory. The formulation is thus not well adapted to considering the situation where the background metric is itself a field variable. For similar reasons even the discussion of quantum field theory in a classical curved background is not simple (see Section 5.7).

A second issue is that in a cosmological context one cannot readily avoid the interpretational difficulties of quantum mechanics. In the 'Copenhagen interpretation' the wave function determines the probabilities with which a macroscopic observer will measure the different possible eigenvalues of the system. However, the observer should also be described by quantum mechanics, and in the case of a wave function for the universe the observer certainly has to be part of what is described, making the wave function hard to interpret. Various solutions to this difficulty have been proposed, one of the more common being the many-worlds hypothesis (DeWitt and Graham, 1973), but none of them seems fully satisfactory.

In quantum cosmology, Section 20.2.2, (and in classical theories of gravity which utilize only three-dimensional quantities, see e.g. Anderson *et al.* (2005)), time is supposed to be an emergent quantity dependent on the three-geometry. This gives rise to further interpretational difficulties (Penrose and Isham, 1986).

If one attempts to treat GR in exactly the way other theories are usually quantized, the attempt fails. For example, in the 'background-field' method one quantizes the difference between the actual field and a background (usually flat space), in the manner of other field theories. This leads to a theory with spin-2 particles, gravitons, moving on the background. One can then use the normal Feynman rules, computing the counterterms which should lead to renormalization (i.e. the systematic removal of divergences), but for gravity the higher terms turn out to be increasingly singular: the theory is not renormalizable (Goroff and Sagnotti, 1985).

A way out of the non-renormalizability is provided by higher-derivative theories of gravity in which quadratic (or, in principle, higher) terms in the curvature are added to the Lagrangian (3.19). The resulting fourth derivative terms affect the short-range behaviour and make the theory renormalizable. However, if a finite number of higher curvature terms is used, they break unitarity, because they introduce a massive spin-2 'ghost' partner to the graviton. Cosmological consequences of corresponding classical theories of gravity have been widely studied.

Another approach is to start from the classical Einstein equations written as evolution equations for a three-dimensional metric as in Section 3.3.3 (or Section 3.3.1). The 3-metric (first fundamental form) and conjugate momentum are then the canonically conjugate variables used in quantization. One would want to consider only inequivalent geometries so that the diffeomorphisms of the three-manifolds are factored out, but such a reduction to the true degrees of freedom is not generally possible. The space of three-geometries is called *superspace*. This then leads to 'quantum cosmology' which we discuss further below (Section 20.2.2).

Other approaches include lattice gauge theory and Regge calculus (two distinct ways of discretizing the underlying manifolds), causal sets, and consistent histories (Brightwell *et al.*, 2003, Isham and Linden, 1995). There have also been extensive investigations of two- and three-dimensional spacetimes as possible models for the full theory.

In recent years the two most prominent proposals for reconciling gravity and quantum theory have taken more radical approaches, going further towards a new theory of which both relativity and the usual quantum theory are only limits: they are (super)string theory, which has developed into M-theory, and loop quantum gravity. Their cosmological applications are discussed in Sections 20.3 and 20.4. The cosmological consequences of a third radical approach, twistor theory (Huggett and Tod, 1994), have yet to be explored at any length, although it has recently been used in potentially testable calculations of quantum scattering amplitudes.

Despite the promising physical and mathematical progress made in both the string and loop quantum gravity approaches (and some others), gravity remains the only observed force for which there is no self-consistent and experimentally tested quantum theory.

The various attempts to apply quantum gravity in cosmology each develop in depth some specific aspect of quantum theory that may be expected to emerge from a successful theory of quantum gravity applied to the universe as a whole. Before turning to string theory and loop quantum gravity, we briefly discuss the Wheeler–DeWitt equation.

20.2.2 Wheeler–DeWitt equation

This is a heuristic approach to quantum cosmology based on the ADM formalism (Section 3.3.3) and the DeWitt approach to quantum gravity (see Bojowald, Kiefer and Vargas Moniz (2010) for a recent review). The core of the approach is the representation of the quantum state of the universe as $\Psi(^3[g])$ on superspace, subject to the '(super)Hamiltonian constraint' of Section 3.3.3, $A^{00} = C_0 = 0$ (and, if the diffeomorphism freedom has not been removed, the 'momentum constraints' $A^{i0} = C_i = 0$), treated as operators. The result is the Wheeler–DeWitt (WDW) equation. Turning $A^{\mu 0}$ into operator form has factor-ordering problems: the method has only been fully implemented for subspaces of metrics, 'minisuperspaces', usually with such high symmetry that Ψ is a function of t alone, which largely avoids the factor ordering issue. However, the imposition of symmetry amounts to simultaneously setting position and momentum to zero for the ignored degrees of freedom, violating the uncertainty principle.

The idea is rather like the Born approach to the atom: it hopes to provide an effective theory of quantum gravity, giving results that will be true whatever the final quantum theory of gravity may turn out to be.

Here we discuss only the minisuperspace of RW models. The state of the universe is thus represented (Hartle, 2003) by a wave function $\Psi(t)$ with a Hamiltonian giving the corresponding Schrödinger equation. One has usual quantum theory with a universal dynamical law,

$$i\frac{\mathrm{d}}{\mathrm{d}t}|\Psi(t)\rangle = \mathcal{H}|\Psi(t)\rangle, \tag{20.1}$$

and an initial quantum state $|\Psi(0)\rangle$. Predictions from probabilities p_α for a set of alternatives are represented by projection operators P_α:

$$p_\alpha = ||P_\alpha|\Psi(t)\rangle||^2. \tag{20.2}$$

The link to geometry is made by specifying that Ψ depends only on 3-geometries: $\Psi = \Psi[h_{ij}]$. However, the gravitational Hamiltonian projects the wave function to zero, so the WDW is

$$\mathcal{H}\Psi[h_{ij}] = 0, \tag{20.3}$$

analogous to the time-independent Schrödinger equation in ordinary quantum mechanics. To solve it one needs boundary conditions for Ψ, i.e. conditions on the spatial geometry.

The usual method of solution is via a path integral, with outcome depending on the conditions assumed. In the Hartle–Hawking no-boundary proposal (Hartle and Hawking, 1983), $\Psi[h_{ij}] = \int_{\mathcal{C}} \mathcal{D}g_{\mu\nu} \exp(-I[g_{\mu\nu}])$, where $\mathcal{D}g_{\mu\nu}$ is a suitable measure on the space of 3-spaces, $I[g_{\mu\nu}]$ is the Euclidean action of a classical solution that is compact and has the S^3 geometry as its boundary, and the sum is over 4-geometries whose only boundary is the 3-surface on which the metric is specified. Euclideanization is used to improve convergence, so the functional integral is to be taken over a complex contour $\mathcal{C}$ (but the return to Lorentzian form can be problematic). This prescription is suggested to lead to a 'start to time' when a classical description becomes valid after evolving from a non-singular Euclidean initial quantum state; thus it is claimed to resolve the issue of the start of the universe.

Variants of this prescription have been given by others. Vilenkin (2003) derives an effective WDW equation in the RW context,

$$\Psi''(a) - [a^2 - a_1^{-2}a^4]\Psi(a) = 0. \tag{20.4}$$

This gives tunnelling probabilities for the wave function 'from nothing' to a closed universe of finite radius a_1, and so represents the birth of an inflationary universe. There are also other variants (Gott and Li, 1998, Vilenkin, 2003).

Various achievements of the WDW approach have been claimed (Page, 2003), including: that a Lorentzian signature spacetime can emerge in the WKB limit of an analytic continuation, and that universe models can inflate to a large size, and can predict a near-critical density and low anisotropies that fit the CMB data. However, there are significant problems. Firstly, various divergences occur (Page, 2003); the path integral is UV divergent and non-renormalizable; in the Hartle–Hawking no-boundary proposal, conformal modes make the Einstein–Hilbert action unbounded below; the sum over all 4-geometries entails a sum over topologies that is not computable. Secondly, there is the problem of time: from (20.1) and (20.3), $\mathrm{d}|\Psi(t)\rangle/\mathrm{d}t = 0$. Hence the solution is static, and so it is difficult to see how it can represent a time-evolving universe. Various attempts derive an effective time evolution for the time-independent function Ψ, for example by regarding other variables such as a in (20.4) as effective time variables. But their success is debatable; Barbour (1999) makes clear how difficult it is to make this work, if one takes the consequences of the WDW equation seriously. There are also major problems of interpretation arising with the concept of a wave function determining probabilities that apply to only a single existing object: what meaning can be given to (20.2) in this context? Does the idea of a wave function of the universe in fact make sense?

Overall, the real problem is the ad hoc nature of the prescriptions that lead to the various solutions of the WDW equation. This approach has now largely been overtaken by developments in quantum gravity theories. We turn to two of those theories, one of which explicitly derives a modified WDW equation.

20.3 String theory and cosmology

String theory is widely believed to be a promising route towards a fundamental theory that unifies the four interactions and includes a theory of quantum gravity. It does so at the price of extra spatial dimensions, and it continues to face huge theoretical challenges. It is not clear that string theory (or any competitor) will indeed lay the basis for quantum gravity. Nevertheless, cosmology is a potential laboratory for testing quantum gravity effects, and we should therefore aim to test the predictions of string theory against cosmological observations. The problem facing this aim is two-fold: firstly, string theory is still very far from being able to make cosmological predictions; and secondly, it is not clear what kinds of generic signatures of quantum gravity one may look for in CMB and other observations. String theory predictions for cosmology are currently based on perturbative string theory, and the lower energy effective theories that result from this. It is not clear how much of the perturbative analysis will survive future developments in non-perturbative string theory. Nevertheless, it is worthwhile to have the perturbative results, which are already highly complicated. It is also useful to develop phenomenological brane-world cosmological models which have qualitative features of string theory but are not derived from the fundamental string equations.

Here we provide a very brief description of string theory and its associated phenomenology, focusing on those aspects relevant to cosmology. For a general treatment of string theory, see Zwiebach (2004). More detailed reviews of string theory and its relation to cosmology are given in Lidsey, Wands and Copeland (2000), Kallosh (2006, 2008), McAllister and Silverstein (2008) and Baumann and McAllister (2009). More phenomenological brane-world cosmology is reviewed in Brax and van de Bruck (2003), Maartens (2004), Maartens and Koyama (2010).

String theory was based on the idea that fundamental states are represented as one-dimensional string-like objects rather than zero-dimensional points – which allows the theory to avoid the divergences in quantum field theory associated with vertices in particle interactions. In particular, it removes the divergences in graviton scattering, which plagued attempts to deal with gravity in quantum field theory. The strings have characteristic length ℓ_s, which is much smaller than characteristic lengths in sub-TeV particle physics, and they are considered to have zero thickness. A multitude of string states exist that in principle should contain as a limit the particle-like states in the Standard Model. These states include a massless spin-2 particle, which plays the role of a graviton at low energies. Thus string theory made the first major progress towards unifying the four interactions in a regular way.

The original string theory was classically straightforward, but its quantization revealed an anomaly that broke Lorentz invariance – unless the spacetime dimension was exactly 1+25. This could be considered a success, in the sense that a definite prediction was made for the number of spacetime dimensions, which until then had been arbitrary. However, the number itself seemed unreasonably large. The original version of the theory incorporated only bosonic states, including the graviton. But it also included unstable tachyonic (imaginary mass) states.

20.3.1 Supersymmetry, p-branes, form fields and moduli

This tachyonic instability in the theory was cured by the inclusion of fermionic states – via supersymmetry. Supersymmetry is a postulated symmetry (independent of string theory) that links fermions (basically the 'matter' states) with bosons (basically the carriers of interactions). Each boson has a fermionic superpartner and vice versa. A feature of supersymmetry is that the zero-point energies of the partners exactly cancel, so that while supersymmetry is unbroken, the total zero-point energy vanishes. This feature was needed to remove the tachyonic instability. The new theory – superstring theory – was found to be consistent only in $1 + 9$ dimensions, an improvement on the original theory. Furthermore, superstring theory retained the original key feature that the string excitation spectrum includes a graviton. Thus superstring theory incorporates a theory of gravity, in fact, supergravity. In addition to the spin-2 graviton, the theory also contains a massless spin-0 scalar, the dilaton. Thus superstring theory includes a low-energy scalar–tensor theory of gravity, which is consistent at the quantum level. (Note that it is customary to use 'string' interchangeably with 'superstring'.)

Superstrings can be either open or closed. Closed strings cannot be transformed to open strings, but open strings can close up, or form closed strings by merging with another open string. Two open strings can also attach at one end to form a new open string. Two closed strings can also merge to produce a new closed string. These properties contribute to the fact that there are different versions of superstring theory. There are five 1+9-dimensional superstring theories. Type I theory describes open and closed strings, Types IIA and IIB describe closed strings, and there are two hybrid theories describing closed strings, Heterotic $E_8 \times E_8$ and Heterotic SO(32).

These apparently distinct theories are related to each other and to the 1+10-dimensional supergravity theory by duality transformations. (Supergravity is only consistent at the quantum level in 1+10 dimensions.) This led to the conjecture that all of these theories arise as different limits of a single theory, which has come to be known as M-theory. The 11th dimension in M-theory is related to the string coupling strength g_s; the size of this dimension grows as the coupling becomes strong. At low energies, M-theory can be approximated by 1+10-dimensional supergravity. In this scenario, the five superstring theories would appear 1+9-dimensional because the 11th dimension would be much smaller than the string scale ℓ_s.

In addition to the duality relations, it was also discovered that p-branes, which are extended objects of spatial dimension p (strings are 1-branes), play a fundamental role in the theory. In the weak coupling limit, $g_s \to 0$, the p-branes with $p > 1$ become infinitely heavy, so that they do not appear in the perturbative theory. But at strong coupling, they are just as important as strings. Of particular importance among p-branes are the Dp-branes, known generically as D-branes, on which open strings can end. This implies that quantum field theories can be associated with D-branes. Roughly speaking, open strings, which describe the non-gravitational sector, are attached at their endpoints to branes, while the closed strings of the gravitational sector can move freely in the bulk. The observable universe may be a D3-brane in string theory.

Apart from the fundamental p-branes, string theory also contains form fields, which are higher-rank antisymmetric analogues of the electromagnetic gauge field. For example, the Kalb–Ramond field is a 2-form field $\mathcal{B}_{AB} = \mathcal{B}_{[AB]}$ which appears in the action via $\mathcal{H}_{ABC}\mathcal{H}^{ABC}$, where $\mathcal{H}_{ABC} = \partial_A \mathcal{B}_{BC} + \partial_C \mathcal{B}_{AB} + \partial_B \mathcal{B}_{CA}$ is its Faraday-like tensor. The electromagnetic field couples to charged point particles in standard quantum field theory, and in a similar way, the form fields can couple to p-branes. Thus one can have anti-branes which carry an opposite charge to their brane partners.

In order to make contact with the 4D universe at lower energies, the six extra spatial dimensions must be compactified (or effectively compactified by 'warping', as we discuss below) on a scale that is small enough not to disturb the well-tested predictions of four-dimensional physics. At each 4D spacetime event, we can think of a six-dimensional 'internal' space being attached, so that the full bulk spacetime has a product topology. Part of the complexity in string theory resides in the enormous variety of compactification spaces for the internal space. Theories with different internal spaces can differ radically, even if they are of the same type. Compact spaces with zero Ricci tensor – known as Calabi–Yau spaces – are of particular importance, since they are associated with the gauge interactions and matter fields in the spacetime.

The huge variety of shapes and sizes in compactification spaces leads to a vast 'landscape' of different possible vacua (see Section 21.5). Moduli fields parameterize these vacua by light scalar fields on the 4D universe that characterize the effects of the internal compactification space (shape, size, etc.) on the 4D universe in string effective theories. On the one hand, it is useful to have many light fields if one is trying to construct inflation within string theory. But on the other hand, there are so many fields that it is difficult to see any predictive power emerging. In typical Calabi–Yau compactifications there can be hundreds of moduli, and it becomes very difficult to find an attractor slow-roll inflation trajectory in this huge moduli field space. Furthermore, the light fields can present serious problems if they decay late enough to affect nucleosynthesis. The moduli fields need to be stabilized in order to avoid this problem.

20.3.2 Higher-dimensional gravity

Extra spatial dimensions revive the original higher-dimensional ideas of Kaluza and Klein in the 1920s, but in a new context of unification and quantum gravity. An important consequence of extra dimensions is that the 4D Planck scale $M_P := M_4$ is no longer the fundamental scale, which becomes M_{4+d}. This can be seen as follows. For an Einstein–Hilbert gravitational action,

$$S_{\text{grav}} = \frac{1}{2\kappa_{4+d}^2} \int d^4x\, d^d y\, \sqrt{-{}^{(4+d)}g}\left[{}^{(4+d)}R - 2\Lambda_{4+d}\right], \tag{20.5}$$

$$^{(4+d)}G_{AB} = -\Lambda_{4+d}\,{}^{(4+d)}g_{AB} + \kappa_{4+d}^2\,{}^{(4+d)}T_{AB}, \tag{20.6}$$

$$\kappa_{4+d}^2 = 8\pi G_{4+d} = \frac{8\pi}{(M_{4+d})^{2+d}}, \tag{20.7}$$

where the coordinates are $X^A = (x^\mu, y^1, \ldots, y^d)$. The static weak field limit of the field equations leads to the $4+d$-dimensional Poisson equation, whose solution is the gravitational potential, $V(r) \propto \kappa_{4+d}^2 r^{-(1+d)}$. If the length scale of the extra dimensions is L, then on scales $r \lesssim L$, the potential is $4+d$-dimensional, $V \sim r^{-(1+d)}$. By contrast, on scales large relative to L, the extra dimensions do not contribute to variations in the potential, and V behaves like a 4D potential, $V \sim L^{-d} r^{-1}$. This means that the usual Planck scale becomes an effective coupling constant, describing gravity on scales much larger than the extra dimensions, and related to the fundamental scale via the volume of the extra dimensions:

$$M_P^2 \sim (M_{4+d})^{2+d} L^d. \tag{20.8}$$

If the extra-dimensional volume is Planck scale, i.e. $L \sim M_P^{-1}$, then $M_{4+d} \sim M_P$. But if the extra-dimensional volume is significantly above Planck scale, then the true fundamental scale M_{4+d} can be much less than the effective scale $M_P \approx 10^{19}$ GeV. In this case, we understand the weakness of gravity as due to 'dilution': gravity 'spreads' into extra dimensions and only a part of it is felt in four dimensions.

A lower limit on M_{4+d} is given by null results in table-top experiments to test for deviations from Newton's law in four dimensions, $V \propto r^{-1}$. These experiments currently probe sub-millimetre scales, so that

$$L \lesssim 10^{-1}\ \text{mm} \sim (10^{-15}\ \text{TeV})^{-1} \Rightarrow M_{4+d} \gtrsim 10^{(32-15d)/(d+2)}\ \text{TeV}. \tag{20.9}$$

The dilution of gravity via extra dimensions not only weakens gravity in 4D, it also extends the range of graviton modes felt in 4D beyond the massless mode of 4D gravity. For simplicity, consider a flat 4D spacetime with one flat extra dimension, compactified through the identification $y \leftrightarrow y + 2\pi n L$, where $n = 0, 1, 2, \ldots$. The perturbative 5D graviton amplitude can be Fourier expanded as $f(x^\mu, y) = \sum_n e^{iny/L} f_n(x^\mu)$, where f_n are the amplitudes of the Kaluza–Klein (KK) modes, i.e. the effective 4D modes of the 5D graviton. To see that these KK modes are massive from the 4D viewpoint, we start from the 5D wave equation that the massless 5D field f satisfies (in a suitable gauge): ${}^{(5)}\Box f = 0$ which is equivalent to $\Box f + \partial_y^2 f = 0$. It follows that the KK modes satisfy a 4D Klein–Gordon equation with an effective 4D mass m_n, $\Box f_n = m_n^2 f_n$, where $m_n = n/L$. The massless mode f_0 is the usual 4D graviton mode. But there is a tower of massive modes, with masses $L^{-1}, 2L^{-1}, \ldots$, which imprint the effect of the 5D gravitational field on the 4D universe. Compactness of the extra dimension leads to discreteness of the spectrum. For an infinite extra dimension, $L \to \infty$, the separation between the modes disappears and the tower forms a continuous spectrum. In this case, the coupling of the KK modes to matter must be very weak in order to avoid exciting the lightest massive modes with $m \gtrsim 0$.

The extra dimensions lead to new scalar and vector degrees of freedom in 4D. In 5D, the spin-2 graviton is represented by a metric perturbation $^{(5)}h_{AB}$ that is transverse traceless: $^{(5)}h^A{}_A = 0 = \partial_B\,^{(5)}h_A{}^B$. In a suitable gauge, $^{(5)}h_{AB}$ contains a 3D transverse traceless perturbation h_{ij}, a 3D transverse vector perturbation Σ_i, and a scalar perturbation β, which each satisfy the 5D wave equation. The other components of $^{(5)}h_{AB}$ are determined via constraints once these wave equations are solved. The five degrees of freedom (polarizations) in the 5D graviton are thus split into:

a 4D spin-2 graviton h_{ij} (two polarizations),

a 4D spin-1 gravi-vector ('gravi-photon') Σ_i (two polarizations), and

a 4D spin-0 gravi-scalar β.

The massive modes of the 5D graviton are represented via massive modes in all three of these fields in 4D. The standard 4D graviton corresponds to the massless zero-mode of h_{ij}. In the general case of d extra dimensions, the number of polarizations of the graviton follows from the irreducible tensor representations of the isometry group as $\frac{1}{2}(d+1)(d+4)$.

20.3.3 Brane-world cosmology

D-branes have open strings attached to them, while closed strings move freely in the bulk spacetime. If our observable universe is a D3-brane, then classically, this is realized via the localization of matter and radiation fields on the brane, with gravity propagating in the bulk (see Figure 20.1).

The Horava–Witten solution in string theory has gauge fields of the E_8 group confined on two 1+9-branes located at the end points of an S^1/Z_2 orbifold, i.e. a circle folded on

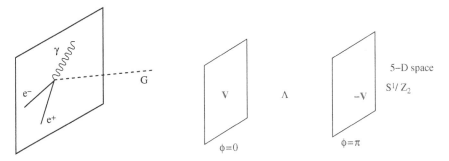

Fig. 20.1   *Left:* Schematic of confinement of matter to the brane, while gravity propagates in the bulk. (From Cavaglià (2003). © World Scientific (2003). Reproduced with permission from World Scientific Publishing Co. Pte. Ltd.). *Right:* The RS 2-brane model (from Cheung (2003)).

itself across a diameter.[2] The six extra dimensions on the branes are compactified on a very small scale close to the fundamental scale, and their effect on the dynamics is felt through moduli fields, i.e. 5D scalar fields. These solutions can be thought of as effectively 5D, with an extra dimension (between the branes) that can be large relative to the fundamental scale. Furthermore, the solution can accommodate the Standard Model of particle states in a low-energy limit.

This solution provides the basis for the Randall–Sundrum (RS) 2-brane models of 5D gravity (see Figure 20.1). The single-brane RS models with infinite extra dimension arise when the orbifold radius tends to infinity.

20.3.4 Randall–Sundrum brane cosmology

The novel features of the RS models compared to previous higher-dimensional models are that they rely on curvature rather than straightforward compactification, and they incorporate the self-gravity of the brane. They have a 'preferred' large extra dimension, with other small extra dimensions treated as ignorable (i.e. stabilized except at energies near the fundamental scale). The large extra dimension is curved or 'warped' rather than flat: the bulk is a portion of anti-de Sitter (AdS_5) spacetime. As in the Horava–Witten solutions, the RS branes are Z_2-symmetric (mirror symmetry), and have a tension, which serves to counter the influence of the negative bulk cosmological constant on the brane.

RS brane-worlds were developed for particle physics on a Minkowski brane, then later generalized to cosmology on a RW brane. They provide phenomenological models that reflect at least some of the features of string theory, and that bring new geometric and particle physics ideas into play. They also provide a framework for exploring holographic ideas that have emerged in string theory. Roughly speaking, holography suggests that higher-dimensional gravitational dynamics may be determined from knowledge of the quantum fields on a lower-dimensional boundary. The AdS/CFT correspondence is an example, in which the classical dynamics of the higher-dimensional gravitational field are equivalent to the quantum dynamics of a conformal field theory (CFT) on the boundary.

What prevents gravity from 'leaking' into the extra dimension at low energies is a negative bulk cosmological constant, $\Lambda_5 = -6/\ell^2$, where ℓ is the curvature radius of AdS_5. The bulk cosmological constant acts to 'squeeze' the gravitational field closer to the brane. We can see this clearly from the metric in Gaussian normal coordinates $X^A = (x^\mu, y)$ based on the brane at $y = 0$,

$$^{(5)}ds^2 = e^{-2|y|/\ell}\eta_{\mu\nu}dx^\mu dx^\nu + dy^2. \tag{20.10}$$

In the bulk, this metric is a solution of the 5D Einstein equations, i.e. $^{(5)}T_{AB} = 0$ in (20.7). The brane is a flat Minkowski spacetime, with self-gravity in the form of brane tension. We shall focus on the RS 1-brane model from now on, perhaps the most simple and geometrically appealing form of a brane-world model, which at the same time provides a framework for the AdS/CFT correspondence.

---

[2] This is different from the projective space P^1 which is the group quotient $S^1/\{-1, 1\}$.

The energy scales are related via $M_5^3 = M_P^2/\ell$. The infinite extra dimension makes a finite contribution to the 5D volume because of the warp factor, and the effective size of the extra dimension probed by the 5D graviton is ℓ. The brane tension is $\lambda = 3M_P^2/(4\pi\ell^2)$, which ensures that there is a zero effective cosmological constant on the brane. The KK modes of spin-1 and spin-0 are pure gauge modes and may be set to zero. The spin-2 modes include a normalizable zero mode, which recovers the GR gravitational potential of a point mass, and a continuous tower of massive modes, that sum to a correction of the 4D potential:

$$V(r) \approx \frac{GM}{r}\left(1 + \frac{2\ell^2}{3r^2}\right). \tag{20.11}$$

Table-top tests of Newton's laws currently find no deviations down to $\mathcal{O}(10^{-1}$ mm$)$, so that $\ell \lesssim 0.1$ mm, and then $\lambda > (1$ TeV$)^4$ and $M_5 > 10^5$ TeV.

The AdS$_5$ bulk, which admits a foliation into Minkowski surfaces, also admits an RW foliation since it is 4-isotropic. The generalization of AdS$_5$ that preserves 4-isotropy and solves the vacuum 5D Einstein equation is Schwarzschild-AdS$_5$, and this bulk therefore also admits an RW foliation. It follows that a RW brane-world, the cosmological generalization of the original RS Minkowski brane-world, is a part of Schwarzschild-AdS$_5$, with the Z_2-symmetric RW brane at the boundary. The junction conditions across the brane determine the modified Friedmann equation as

$$H^2 = \frac{\kappa_4^2}{3}\rho\left(1 + \frac{\rho}{2\lambda}\right) + \frac{m}{a^4} + \frac{1}{3}\Lambda - \frac{K}{a^2}, \tag{20.12}$$

where m is a mass parameter for the bulk black hole, and the term m/a^4 is known as 'dark radiation'. The standard energy conservation equation still holds.

The key modification to the Hubble rate is via the high-energy correction ρ/λ. In order to recover the observational successes of GR, the high-energy regime where significant deviations occur must take place before nucleosynthesis, i.e. cosmological observations impose an upper limit on m and a lower limit, $\lambda > (1$ MeV$)^4$, on λ, which is much weaker than the limit imposed by table-top experiments. Since ρ^2/λ decays as a^{-8} during the radiation era, it will rapidly become negligible after the end of the high-energy regime, $\rho = \lambda$. However, it can have a significant impact on 4D inflation on the brane.

20.3.5 Covariant approach to brane-world dynamics

A broader perspective, with useful insights into the interplay between 4D and 5D effects, can be obtained via the covariant Shiromizu–Maeda–Sasaki (2000) approach, in which the brane and bulk metrics remain general. The Gauss–Codazzi equations are used to project the 5D curvature along the brane, and the Israel–Darmois junction conditions determine the extrinsic curvature of the brane in terms of the energy–momentum tensor on the brane. The induced metric on $y = $ const surfaces (the brane is $y = 0$) is locally given by (where n^A is the unit normal to the brane):

$$g_{AB} = {}^{(5)}g_{AB} - n_A n_B, \quad g_{\mu\nu}(x^\alpha, y)\mathrm{d}x^\mu \mathrm{d}x^\nu = {}^{(5)}\mathrm{d}s^2 - \mathrm{d}y^2, \tag{20.13}$$

The induced field equations on the brane are:

$$G_{\mu\nu} = -\Lambda g_{\mu\nu} + \kappa_4^2 T_{\mu\nu} + 6\frac{\kappa_4^2}{\lambda}\mathcal{S}_{\mu\nu} - \mathcal{E}_{\mu\nu} + 4\frac{\kappa^2}{\lambda}\mathcal{F}_{\mu\nu}, \tag{20.14}$$

$$\kappa_4^2 = \tfrac{1}{6}\lambda\kappa_5^4, \quad \Lambda = \tfrac{1}{2}\left(\Lambda_5 + \kappa_4^2\lambda\right). \tag{20.15}$$

The first correction term relative to GR is quadratic in $T_{\mu\nu}$:

$$\mathcal{S}_{\mu\nu} = \tfrac{1}{12}T T_{\mu\nu} - \tfrac{1}{4}T_{\mu\alpha}T^\alpha{}_\nu + \tfrac{1}{2}g_{\mu\nu}\left[3T_{\alpha\beta}T^{\alpha\beta} - T^2\right]. \tag{20.16}$$

The second correction term is the projected Weyl term,

$$\mathcal{E}_{\mu\nu} = {}^{(5)}C_{ACBD}\,n^C n^D g_\mu{}^A g_\nu{}^B. \tag{20.17}$$

The last correction term is

$$\mathcal{F}_{\mu\nu} = {}^{(5)}T_{AB}g_\mu{}^A g_\nu{}^B + \left[{}^{(5)}T_{AB}n^A n^B - \tfrac{1}{4}{}^{(5)}T\right]g_{\mu\nu}, \tag{20.18}$$

where ${}^{(5)}T_{AB}$ describes any stresses in the bulk apart from the 5D cosmological constant, e.g. a 5D scalar field. (Note that a perfect fluid ${}^{(5)}T_{AB}$ has no physical motivation.)

The conservation equations on the brane are

$$\nabla^\nu T_{\mu\nu} = -2\,{}^{(5)}T_{AB}n^A g^B{}_\mu. \tag{20.19}$$

Thus in general there is exchange of energy–momentum between the bulk and the brane, but we recover the GR conservation when the bulk contains only a cosmological constant. The 4D contracted Bianchi identities ($\nabla^\nu G_{\mu\nu} = 0$), applied to (20.14), lead to

$$\nabla^\mu \mathcal{E}_{\mu\nu} = \frac{6\kappa_4^2}{\lambda}\nabla^\mu \mathcal{S}_{\mu\nu}, \tag{20.20}$$

which shows qualitatively how 1+3 spacetime variations in the matter-radiation on the brane can source KK modes.

The tensor $\mathcal{S}_{\mu\nu}$, which carries local bulk effects onto the brane, may be decomposed relative to a chosen 4-velocity u^μ on the brane as

$$\mathcal{S}_{\mu\nu} = \tfrac{1}{24}\left[2\rho^2 - 3\pi_{\alpha\beta}\pi^{\alpha\beta}\right]u_\mu u_\nu + \tfrac{1}{24}\left[2\rho^2 + 4\rho p + \pi_{\alpha\beta}\pi^{\alpha\beta} - 4q_\alpha q^\alpha\right]h_{\mu\nu}$$
$$- \tfrac{1}{12}(\rho + 3p)\pi_{\mu\nu} - \tfrac{1}{4}\pi_{\alpha\langle\mu}\pi_{\nu\rangle}{}^\alpha + \tfrac{1}{4}q_{\langle\mu}q_{\nu\rangle} + \tfrac{1}{3}\rho q_{(\mu}u_{\nu)} - \tfrac{1}{2}q^\alpha\pi_{\alpha(\mu}u_{\nu)}. \tag{20.21}$$

The tracefree $\mathcal{E}_{\mu\nu}$ carries nonlocal bulk effects onto the brane, and contributes an effective 'dark' radiative energy–momentum on the brane, with energy density $\rho_\mathcal{E}$, pressure $\rho_\mathcal{E}/3$, momentum density $q_\mu^\mathcal{E}$, and anisotropic stress $\pi_{\mu\nu}^\mathcal{E}$:

$$-\frac{1}{\kappa_4^2}\mathcal{E}_{\mu\nu} = \rho_\mathcal{E}\left(u_\mu u_\nu + \tfrac{1}{3}h_{\mu\nu}\right) + q_\mu^\mathcal{E}u_\nu + q_\nu^\mathcal{E}u_\mu + \pi_{\mu\nu}^\mathcal{E}. \tag{20.22}$$

We can think of this as a KK 'fluid'. The brane 'feels' the bulk gravitational field through this effective fluid.

The brane-world corrections can be consolidated into an effective total energy density, pressure, momentum density, and anisotropic stress:

$$\rho_{\text{tot}} = \rho + \frac{1}{4\lambda}\left(2\rho^2 - 3\pi_{\mu\nu}\pi^{\mu\nu}\right) + \rho_{\mathcal{E}}, \tag{20.23}$$

$$p_{\text{tot}} = p + \frac{1}{4\lambda}\left(2\rho^2 + 4\rho p + \pi_{\mu\nu}\pi^{\mu\nu} - 4q_\mu q^\mu\right) + \frac{\rho_{\mathcal{E}}}{3}, \tag{20.24}$$

$$q_\mu^{\text{tot}} = q_\mu + \frac{1}{2\lambda}\left(2\rho q_\mu - 3\pi_{\mu\nu}q^\nu\right) + q_\mu^{\mathcal{E}}, \tag{20.25}$$

$$\pi_{\mu\nu}^{\text{tot}} = \pi_{\mu\nu} + \frac{1}{2\lambda}\left[-(\rho + 3p)\pi_{\mu\nu} - 3\pi_{\alpha\langle\mu}\pi_{\nu\rangle}{}^\alpha + 3q_{\langle\mu}q_{\nu\rangle}\right] + \pi_{\mu\nu}^{\mathcal{E}}. \tag{20.26}$$

Equation (20.20) may be called the 'energy–momentum balance equation for the KK fluid'. In the general nonlinear case, it contains many nonlinear source terms from the brane energy–momentum tensor. If we linearize about an RW background, then (20.20) gives the following KK energy and momentum balance equations:

$$\dot{\rho}_{\mathcal{E}} + \tfrac{4}{3}\Theta\rho_{\mathcal{E}} + \overline{\nabla}^\mu q_\mu^{\mathcal{E}} = 0, \tag{20.27}$$

$$\dot{q}_\mu^{\mathcal{E}} + 4Hq_\mu^{\mathcal{E}} + \tfrac{1}{3}\overline{\nabla}_\mu\rho_{\mathcal{E}} + \tfrac{4}{3}\rho_{\mathcal{E}}A_\mu + \overline{\nabla}^\nu\pi_{\mu\nu}^{\mathcal{E}}$$
$$= \frac{(\rho + p)}{3\lambda}\left[-3\overline{\nabla}_\mu\rho + 2\overline{\nabla}^\nu\pi_{\mu\nu} + Hq_\mu\right]. \tag{20.28}$$

In the RW background, (20.28) is trivially satisfied, while (20.27) has the dark radiation solution $\rho_{\mathcal{E}} = \rho_{\mathcal{E}0}(a_0/a)^4$.

20.3.6   RS brane cosmological perturbations

In general, the four independent equations in (20.27) and (20.28) constrain four of the nine independent components of $\mathcal{E}_{\mu\nu}$ on the brane. What is missing is an evolution equation for $\pi_{\mu\nu}^{\mathcal{E}}$, which has up to five independent components. These five degrees of freedom correspond to the five polarizations of the 5D graviton. Thus in general, the projection of the 5D field equations onto the brane does not lead to a closed system, as expected, since there are bulk degrees of freedom whose impact on the brane cannot be predicted by brane observers. The KK anisotropic stress $\pi_{\mu\nu}^{\mathcal{E}}$ encodes this nonlocality.

Given a solution for $\pi_{\mu\nu}^{\mathcal{E}}$ from the 5D field equations, we can solve the induced equations on the brane. Numerical solutions of the 5D equations to find the 4D cosmological scalar perturbations have been developed (Cardoso *et al.*, 2007). An example of the results for super-Hubble modes is illustrated in Figure 20.2.

The left panel of Figure 20.2 shows the gauge-invariant density perturbation Δ [(10.32)], which evolves from high-energy behaviour to the familiar GR behaviour at low energies. The scalar KK anisotropic stress $\kappa_4^2\delta\pi_{\mathcal{E}}$ steadily decays as the energy falls. Also shown is $\hat{\Omega}_b$, which is the bulk master variable from solving the 5D field equations, evaluated on the brane (where it acts like a source term). The right panel shows the brane metric and curvature perturbations in Newtonian gauge (10.55) – where (Φ, Ψ) should be replaced by $(-\Psi, \Phi)$ in our notation. The GR result for the potentials is recovered at low energy. Notice

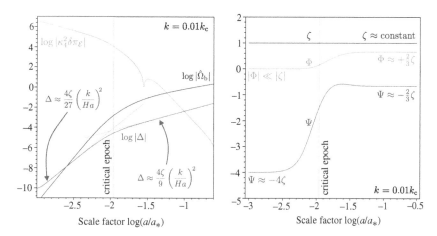

Fig. 20.2   Numerical solution for a super-Hubble mode. (From Cardoso *et al.* (2007).)

that ζ is conserved as in GR [(10.52)], even at high energies. This is due to the fact that energy conservation holds on the brane.

20.3.7 String inflation

There is a 10D solution in type IIB supergravity similar to the RS model,

$$ds_{10}^2 = h(y)^{-1/2}\eta_{\mu\nu}dx^\mu dx^\nu + h(y)^{1/2}\tilde{g}_{mn}dy^m dy^n, \qquad (20.29)$$

where y^m are coordinates on the 6D compact manifold $\mathcal{M}_6$. If there are N coincident D3-branes on the manifold, the 10D spacetime is $AdS_5 \times X_5$, where X_5 is a 5D Einstein manifold. The warp factor is given by $h(r) \sim \left(g_s N \ell_s^2 / r\right)^4$, where r is the distance from the D3-branes in the $\tilde{g}_{mn}$ metric. Near the D3-branes, the geometry is $AdS_5 \times S^5$.

This warped geometry turned out to be very important to construct inflation models in string theory. The idea is to use a mobile D3-brane in this warped 'throat' (see Figure 20.3). For a D3-brane moving in the background very slowly, the D3-brane action is treated as that of a free field. An anti-D3 brane at the tip of the warped throat r_0 has a tension and a five-form charge and perturbs the spacetime and five-form field. Moreover, it has an opposite charge to the D3-brane.

This creates a potential energy that is dependent on the location of the D3-brane:

$$V(r_1) \propto 1 - \frac{1}{N}\left(\frac{r_0}{r_1}\right)^4. \qquad (20.30)$$

The first term is from the potential energy associated with the anti-D3 brane: the force exerted by gravity and the five-form are of the same sign and add. The potential energy is red-shifted due to the warping. The second term is a Coulomb force between the D3 and anti-D3 branes. Again due to warping of the geometry, this force is suppressed. Thus thanks to the warping of the spacetime, the potential for r_1 is very flat and r_1 can act as an inflaton. This model of inflation is known as *D-brane inflation*.

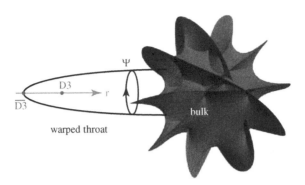

Fig. 20.3 D3-brane inflation in a warped throat geometry. D3-branes are pointlike in the extra dimensions. Circle is base manifold X_5 with angular coordinates Ψ. Brane moves in the radial direction r. At r_{max} the throat attaches to a compact Calabi–Yau space. Anti-D3-branes minimize their energy at the tip of the throat, r_0. (From Baumann (2009).)

Another interesting possibility is that the D3-brane is moving with relativistic speed. In this case, it is possible to realize inflation even if the potential is very steep. This type of inflation is known as *Dirac–Born–Infeld (DBI)* inflation, and the inflaton can develop large non-Gaussianity.

D-brane inflation has attracted significant interest as a concrete example of inflation models in string theory. Detailed studies have shown that it is in general very difficult to keep the flatness of the potential. In fact, r_1 is non-minimally coupled to gravity in the 4D effective theory, due to the fact that the scalar field r_1 is a conformally coupled scalar. This coupling to gravity gives a mass $\mathcal{O}(H^2)$, which spoils the flatness of the potential.

However, there are many other corrections to the inflaton potential and they are sensitive to the stabilization mechanism that is necessary to fix moduli fields in string theory. The stabilization mechanism exploits non-perturbative effects and they are often added in the 4D effective theory. But then it is not clear whether the resultant 4D effective theory is consistent with the 10D equations of motion.

Recently there has been a new development and it has become possible to calculate all significant contributions to the D3-brane potential in the single coherent framework of 10D supergravity. This will provide us with a very interesting bridge between phenomenological brane-world models, where dynamics of higher-dimensional gravity is studied in detail, and string theory approaches, where 4D effective theory is intensively used. It is crucial to identify the higher-dimensional signature of the models in order to test a fundamental theory like string theory.

20.4 Loop quantum gravity and cosmology

Loop quantum gravity (LQG) is one of the alternative attempts to quantize gravity. Unlike string theory, LQG seeks to directly quantize GR itself, in four spacetime dimensions. As part of this, LQG does not pre-suppose a classical spacetime, and aims to quantize the geometry. It does not currently address the issue of matter fields and the unification of the fundamental interactions. (See Ashtekar and Lewandowski (2004), Rovelli (2004), Thiemann (2007) for reviews.) The kinematical structure of LQG has been established, but the dynamical structure is still uncertain, with various proposals under development. Like string theory, LQG is still in its infancy – and either or both of these candidate quantum gravity theories could fail as a result of further discoveries.

Nevertheless, as in the case of string theory, in the absence of an established quantum gravity theory, it remains useful to develop cosmological models that emerge from LQG, or that are at least phenomenologically related to LQG. Given the uncertain status of all current attempts to develop quantum gravity, it is also useful to have competing paradigms.

20.4.1 Basic features of quantum geometry

The quantum theory of geometry is based on formulating the classical theory in the appropriate way before quantizing. Attempts to start from the usual metric formulation of GR do not lead to successful quantization. Trial and error has led to an effective way forward, based on connections, not metrics. GR is formulated as a dynamical theory of connections, with the same phase space as in Yang–Mills gauge theories. The important point is that the spin connection, and not the Levi-Civita connection, is used – and this proves to be crucial for quantization. The internal group in phase space is SU(2). The fundamental variables are the Ashtekar variables:

$$\text{SU(2) (spin) connection } A_a^i \text{ and triad } E_i^a. \tag{20.31}$$

The classical constraints in a Hamiltonian formulation of GR are written in terms of holonomies around loops and fluxes through loops. General covariance of GR leads to a unique representation of the algebra of holonomies and fluxes. A key problem in quantizing lies in the fact that GR, unlike other field theories, has no background field (spacetime is dynamical) – so that the theory is fully constrained in its phase space formulation. LQG adopts the Dirac approach: first the quantum kinematics is constructed for the phase space ignoring the constraints; then quantum operators are found corresponding to the constraints; then the quantum constraints are solved to obtain the physical states and the associated Hilbert space.

The quantum operators corresponding to the diffeomorphism and Gauss constraints can be constructed relatively straightforwardly, but the scalar (Hamiltonian) constraint creates bigger difficulties. Ambiguities in factor ordering produce different possible operators, leading to distinct quantum dynamics. It appears that physical selection criteria will need to be invoked – and then investigated via the consequent predictions.

The quantum kinematics of LQG appears to be well established, but the problem of quantum dynamics remains open, related to the complexities associated with the Hamiltonian constraint. Only in symmetry reduced phase spaces has progress been made, although this begs the key question as to whether symmetry reduction *before* dynamical quantization is consistent with symmetry reduction of the full (and still unknown) general quantum dynamics. Unsurprisingly, it is proving extremely difficult to derive the 'quantum Einstein equations' in the general case.

It is also important to point out that, although the classical smooth spacetime is used in the process of motivating a particular quantization scheme, the quantization itself does *not* rely on assuming a smooth background spacetime. The holonomies and fluxes at the quantum level are constructed via graphs with edges and vertices, without pre-supposing an underlying smooth metric geometry. The quantum operators based on the Ashtekar variables include an area operator, whose eigenvalue spectrum has a minimum,

$$A_{\min} = 4\sqrt{3}\pi\gamma\ell_P^2, \quad \gamma \approx 0.24, \tag{20.32}$$

where γ is the Barbero–Immirzi parameter, whose value is fixed by LQG calculations of black hole entropy.

20.4.2 Loop quantum cosmology

Significant progress has been made in applying LQG to the cosmological RW spacetimes (see Bojowald (2005), Ashtekar (2009a,b) for reviews). This progress is unavoidably based on phenomenology – in the sense that one has to reduce the phase space by rigidly imposing RW symmetries, thus freezing by hand the quantum degrees of freedom in gravity and geometry. However, in the absence of the general quantum dynamics for an arbitrary spacetime, there is no alternative. Phenomenology can often serve as a guide for further developments in the full theory. And the results of the symmetry-reduced LQC are qualitatively in accord with expectations: in particular, as we discuss in the next sub-section, the big bang singularity is removed.

The reduction of LQG to LQC is analogous to the reduction of a quantum field theory to quantum mechanics. In this sense, LQC operates in the same arena as the Wheeler–DeWitt equation discussed above. However, the WDW equation is an ad hoc prescription inspired purely by analogies with quantum mechanics, whereas the LQC analogue arises via a systematic Dirac quantization procedure in the framework of gauge theories of connections. Furthermore, the LQC equation is a difference equation, as befits quantum geometry without a smooth background spacetime. The WDW differential equation is based on a smooth background, and is unable as a consequence to avoid the big bang singularity.

The symmetry-reduced classical Ashtekar variables become

$$A_a^i = C\bar{V}^{-1/3}\bar{e}_a^i, \quad E_i^a = P|\det q|\bar{V}^{-2/3}\bar{e}_i^a, \tag{20.33}$$

$$C = \gamma\bar{V}^{1/3}\dot{a}, \quad |P| = \bar{V}^{2/3}a^2, \tag{20.34}$$

where $\bar{V}$ is the volume of a cubical fiducial cell that must be introduced in the flat RW geometry (physical results are independent of the choice of cell), $\bar{q}_{ab}$ is the fiducial spatial metric and $\bar{e}_i^a$ the associated triad. Note that the expression above for C is only valid in the classical theory. The absolute value of P above reflects the two possible orientations of the triad, to which the classical theory is insensitive.

The Hamiltonian constraint becomes, for a massless scalar field source,

$$0 = C_{\text{grav}} + C_\phi = -6\frac{C^2}{\gamma^2}|P|^{1/2} + 8\pi G\frac{P_\phi^2}{|P|^{3/2}}, \quad P_\phi := \bar{V}a^3\dot{\phi}. \tag{20.35}$$

This constraint gives the classical Friedmann equation, and the Hamilton equation gives the Raychaudhuri equation.

The symmetry-reduced LQG constraints lead to two LQC Dirac operators – the standard momentum operator and an 'instantaneous' volume operator:

$$\hat{P}_\phi \Psi(v,\phi) = -i \partial_\phi \Psi(v,\phi), \tag{20.36}$$

$$|\hat{v}|_{\phi_0} \Psi(v,\phi) = \exp[i\, \hat{Q}^{1/2}(\phi - \phi_0)]|v|\Psi(v,\phi), \tag{20.37}$$

where $|v| \propto \ell_P^{-3}|P|^{3/2}$ is a normalized volume. $\hat{Q}$ is defined by

$$\hat{Q}\Psi(v,\phi) = f_+(v)\Psi(v+4,\phi) + f_0(v)\Psi(v,\phi) + f_-(v)\Psi(v-4,\phi), \tag{20.38}$$

where $f_\pm, f_0$ are complicated functions of v.

The quantum Hamiltonian constraint is then a second-order difference equation,

$$\partial_\phi^2 \Psi(v,\phi) = -\hat{Q}(v)\Psi(v,\phi). \tag{20.39}$$

By contrast, the WDW equation (20.3) in these variables is the second-order differential equation

$$\partial_\phi^2 \Psi(v,\phi) = -12\pi G(v\partial_v)^2 \Psi(v,\phi). \tag{20.40}$$

20.4.3 LQC resolution of the big bang singularity

The quantum difference equation (20.39) is solved numerically by iterating in backward v-steps from an initial semi-classical state peaked at a classical trajectory with $\rho \ll \rho_P$. The result is illustrated in Figure 20.4, which shows how the LQC evolution differs from the classical GR trajectory for

$$\rho \gtrsim 0.02\rho_P, \tag{20.41}$$

until a singularity avoiding bounce is achieved at $\rho \sim 0.4\rho_P$. By contrast, numerical integration of the WDW equation (20.40) shows that the evolution follows the classical trajectory closely, all the way to the big bang singularity.

Note that in quantum gravity there is no obvious definition of time. In the quantum difference equation, the role of time may be played by the monotonically evolving scalar field ϕ.

The bounce can be understood qualitatively as the result of a repulsive LQG 'force' that kicks in near the Planck scale and overcomes the classical attractive force of GR. This effect is geometric, based on the holonomies and curvature effects, and does not arise from high-energy effects in the scalar field.

LQC shows that there is a critical, minimum volume eigenvalue, leading to a critical, maximum energy density given by

$$\rho_{\text{crit}} = (\sqrt{3}/32\pi^2)\gamma^{-3}\rho_P \approx 0.4\rho_P. \tag{20.42}$$

At the onset of inflation, it follows that LQC effects are negligible:

$$(\rho/\rho_{\text{crit}})_{\text{inflation}} \sim 10^{-11}. \tag{20.43}$$

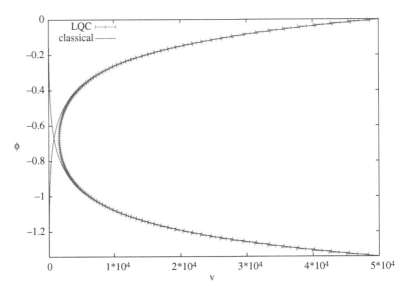

Fig. 20.4 Expectation values and dispersion of the volume operator v in flat RW with massless scalar field ϕ. LQC differs from GR for $\rho \gtrsim 0.02\rho_P$, and the classical singularity is avoided by a bounce at $\rho \sim 0.4\rho_P$. (From Singh (2008).)

It is possible to write an effective classical Hamiltonian which describes the underlying quantum dynamics to an excellent approximation. This then leads to semi-classical modified Friedmann and Raychaudhuri equations that capture the LQC corrections to GR:

$$H^2 = \frac{8\pi G}{3}\rho\left(1 - \frac{\rho}{\rho_{\text{crit}}}\right), \tag{20.44}$$

$$\frac{\ddot{a}}{a} = -\frac{4\pi G}{3}\rho\left(1 - 4\frac{\rho}{\rho_{\text{crit}}}\right) - 4\pi G\, p\left(1 - 2\frac{\rho}{\rho_{\text{crit}}}\right). \tag{20.45}$$

These two equations imply that the GR conservation law is not modified.

20.5 Physics horizon

A basic problem with all these alternatives is that we have no experimental guidance to help us choose. The highest energies we can attain in particle accelerators cannot reach the levels relevant to the very early universe. The uniqueness of cosmology in this regard is that it is the only science contemplating spacetime regions that have experienced such high energies, and with which we are in intimate causal contact despite the huge timescales involved – indeed events at those early times determined much of what we see around us today.

The nuclear reactions underlying nucleosynthesis are well understood, and their cross-sections reasonably well known; the processes of baryogenesis and quark–gluon recombination are partly understood and are on the border of being testable; but physical processes relevant at earlier times are inaccessible to testing by laboratory or accelerator-based experiment. We can define a 'physics horizon' that separates those aspects of physics we can hope to test by high-energy experiments on Earth or in the Solar System, from those where it is reasonable to expect no such test will ever be possible. We have to extrapolate from known physics to the unknown and then test the implications. We cannot experimentally test whether we have got it right.

The physics horizon *limits our knowledge of physics relevant for the very early universe.*

This is independent of the issue of setting of initial conditions for the universe, considered below (Section 20.6.2). The problem arises after the initial conditions have been set and the universe is running according to invariable physical laws. We cannot be confident of the validity of the physics we presuppose then. Since we cannot use known physics to predict the evolution of the very early universe, we end up testing proposals for this physics by exploring their implications in the early universe – the only 'laboratory' where we can test some of our ideas regarding fundamental physics at the highest energies (Yoshimura, 1988). This is particularly true in the case of quantum gravity proposals. The problem is we cannot simultaneously do this and also carry out the aim of physical cosmology, namely predicting the evolution of the early universe from known physical theory.

Our understanding of physics at those times has of necessity to be based on extrapolation of known physics way beyond the circumstances in which it can be tested. The trick is to identify which features are the key to use in that extrapolation: for example, variational principles, broken symmetries and phase changes, duality invariance, entropy limits are candidates. If we confirm our guesses for the relevant physics by their satisfactory implications for the early universe, tested in some suitable way, then this is impressive progress; but if this is the *only* way we can test the proposed physics, the situation is problematic. If the hypothesis solves only the specific issues it was designed to solve in the early universe and nothing else, then in fact it has little explanatory power, rather it is just an alternative (perhaps theoretically preferable) description of the known situation.

One obtains positive observational support for a particular proposal for the relevant physics only if it predicts multiple confirmed outcomes (rather than just one), for example predicting particles that are then confirmed to exist in a laboratory, so that a single hypothesis simultaneously solves several different observational issues. Some of the options may be preferred to others on various theoretical grounds; but one must distinguish this from their having observational support. They lack physical power if they have no other testable consequences.

This issue arises particularly as regards quantum cosmology and the origin of the universe; but it also arises as regards inflation. A particular example is the inflaton field, which has not been identified, much less shown to exist by any laboratory experiment. The consequent arbitrariness of the inflaton potential reflects our inability to experimentally determine the relevant behaviour. What one would like is a laboratory (particle accelerator) test of the inflaton and its potential. However, we may not in fact be able to achieve this.

One key application where this issue becomes significant is the chaotic inflation theory (Section 9.7). As remarked in Section 7.9, its geometric predictions are observationally unverifiable. It would nevertheless be a good physical prediction if it was a more or less inevitable outcome of known and tested underlying physics. However this is not the case. The proposed underlying physics is not experimentally tested: the supposed Coleman–de Luccia tunneling process (Coleman and de Luccia, 1980) that is the core of the creation of new universe bubbles is a hypothetical process that may or may not occur; and the measure that determines probabilities is unknown and the subject of ongoing debate (see Section 21.5).

20.6 Explaining the universe – the question of origins

The core business of physical cosmology is explaining both why the universe has come into existence and evolved to the present very high-symmetry RW geometry on large scales, and how structures come into existence on smaller scales.

20.6.1 Start to the universe

Was there a start to the universe? If so, what was its nature? The issue is unresolved. The major related question is whether the process of expansion only happens once in the life of the universe, or occurs repeatedly. Many theories attempt to address the issue. In effect they attempt either to describe the creation process, or to somehow sidestep the need for one.

Creation theories

These are of two kinds. (1) Simply assume a creation event took place, without investigating it further. This is the traditional position in older texts. (2) Attempt to give a true theory of creation from nothing in terms of quantum field theory processes. Such efforts, however, cannot truly 'solve' the issue of creation, for they rely on some structures or other (e.g. the elaborate framework of quantum field theory and much of the Standard Model of particle physics) somehow pre-existing the origin of the universe, and hence themselves requiring explanation.

No-creation theories

These describe a self-sustaining or self-referential universe which by-passes the issue of creation. One alternative is origination from an eternally pre-existing state, either via a 'phoenix' universe, or via creation from some quite different pre-existing structure. Examples of the phoenix type are: self-repeating universes, e.g. chaotic inflationary models (Linde, 1986, Aguirre and Gratton, 2003); 'pre-big bang' models based on analogues of the dualities of string theory (Gasperini and Veneziano, 1993); cyclic universes (Steinhardt and Turok, 2002, Baum and Frampton, 2007, Penrose, 2006). Examples of pre-existing structures are: emergence from fluctuations in de Sitter spacetime; 'ekpyrotic' universes initiated by a collision between pre-existing branes (Khoury *et al.*, 2001); emerging from an eternal static initial state (Ellis and Maartens, 2004, Mulryne *et al.*, 2005). Another alternative is starting from a state with different properties of time than usual (or with an emergent notion of time), as in: the Hartle–Hawking no-boundary proposal (Hawking, 1987, 1993); the causal violation proposal (Gott and Li, 1998). Any of these may be combined with proposals for an effective ensemble of universes (Tegmark, 2003), realized in spacetime regions that are truly disconnected, or part of a larger entangled quantum entity, or part of a single classical spacetime, but effectively disconnected from each other.

All of these proposals, however, are strongly speculative extrapolations from the known to the unknown. They may or may not be true. One thing is certain: they cannot all be true!

The first option is the standard model, where the entire evolution of the universe is a once-off affair, with all the objects we see, and indeed the universe itself, being transient objects that will burn out like dead fireworks after a firework display. In this case everything that ever happens occurs during one expansion phase of the universe (possibly followed by one collapse phase, which could occur if $K > 0$ and the present dark energy field dies away in the future). This evolution might have a singular start at a spacetime singularity; a beginning where the nature of time changes character; a non-singular bounce from a single previous collapse phase; or a start from a non-singular static initial state.

The major alternative is that many such phases have occurred in the past, and many more will occur in the future, new expansion phases repeatedly arising from the ashes of the old. While the idea of one or more bounces is an old one (Tolman, 1934, Dicke and Peebles, 1979), actual mechanisms that might allow this bounce behaviour are difficult to develop in a fully satisfactory way. A variant is the chaotic inflation idea (Section 9.7) of new expanding universe regions arising from vacuum fluctuations in old expanding regions, leading to a universe that has a fractal-like structure at the largest scales, with many expanding regions with different properties emerging out of each other in a universe that lasts forever.

As discussed above (Section 20.1), it is possible (if the universe has positive spatial curvature) that the quantum gravity domain can be avoided and there was no start to the universe (Ellis and Maartens, 2004); however, this probably requires special initial conditions. If a quantum gravity epoch indeed occurred, we cannot come to a definite conclusion about whether there was a creation event or not because we do not know the nature of quantum gravity, nor how to reliably apply it in the cosmological context where the issue of initial conditions arises.

Eternal existence is also problematic, leading for instance to the idea of Poincaré's eternal return: everything that ever happened will recur an infinite number of times in the future and has already occurred an infinite number of times in the past (Barrow and Tipler, 1984). This is typical of the problems associated with the idea of infinity (see Section 21.5.5). It is not clear which is philosophically preferable: a beginning or eternal existence.

The universe may or may not have a beginning

An initial singularity may or may not have occurred a finite time ago, but a variety of alternatives are conceivable, including singularity avoidance via quantum gravity and eternal universes.

20.6.2 Initial conditions

Even if a beginning does not take place, this does not resolve the underlying issue of what determined why the universe is the way it is. If the proposal is evolution from a previous eternal state then why did that come into existence? And why did the universe expansion start when it did, rather than at some previous time in the pre-existent eternity? Whenever it started, it could have started before!

No physical experiment at all can help here because of the uniqueness of the universe, and the feature that no spacetime exists prior to (in a causal sense) a beginning. So brave attempts to define a 'physics of creation' stretch the meaning of 'physics'. Prior to the start (if there was a start), physics as we know it is not applicable and our ordinary language fails us because time did not exist, so our natural tendency to contemplate what existed or happened 'before the beginning' is highly misleading – there was no 'before' then, indeed there was no 'then' then! Talking as if there was is commonplace, but quite misleading in trying to understand a scientific concept of 'creation' (Grunbaum, 1989).

We run full tilt into the impossibility of testing the causal mechanisms involved, when physics did not exist; this is the 'physics horizon' of Section 20.4 with a vengeance. No experimental test can determine the nature of any mechanisms that may be in operation in circumstances where even the concepts of cause and effect are suspect. This comes particularly to the fore in proposing 'laws of initial conditions for the universe' – for here we are apparently proposing a theory with only one object. Physics laws are by their nature supposed to cover more than one event, and are untestable if they do not do so.

Testable physics cannot explain the initial state and hence specific nature of the universe
Why does the universe have one specific form, when other forms consistent with physical laws seem perfectly possible?

This question cannot be solved by physics alone, unless one can show that only one form of physics is self-consistent; but the variety of proposals made is evidence against that suggestion.

The present state of the universe is very special. Explanation of the present large-scale isotropy and homogeneity of the universe means determining the dynamical evolutionary trajectories relating initial to final conditions, and then essentially either *explaining initial conditions*, where we run into difficulties (Section 20.6.2), or *showing they are irrelevant*: physical processes led to a late time attractor state that is independent of initial conditions (the basic claim of inflation theory). The issue raised is whether the universe started off in a very special geometrical state. We will return to this in the next chapter (Section 21.4).

21

Cosmology in a larger setting

The main part of scientific cosmology today deals with technical issues to do with modelling the origin and evolution of the universe, as discussed in the rest of this book. However cosmology also has wider connotations, as reflected in the broader use of the term in popular use. The link between these two aspects of cosmology resides in two interrelated issues: on the one hand, the relation between cosmology and local physics; and on the other, the foundational question of why the universe is as it is. Both can only be tackled properly by taking philosophical issues seriously. The underlying issue is fundamental: what is the nature of cosmology as a science? How does it relate to issues of testing and verification?

This book does not deal with these issues in depth, as to do so fully would take us too far from our main theme (and our competence), but it does not ignore them either, for to do so would exclude some of the most interesting issues in cosmology. This chapter considers, relatively briefly, how issues in relativistic cosmology relate to these two fundamental themes. It will emphasize two related key topics where these issues come to a head: namely the possible existence of a multiverse, and the question of whether the universe is probable or improbable.

We present theses that can be regarded as reasonable within the current framework of cosmology and physics. Current experiments cannot prove or disprove them; but they are open to debate and potential refutation.

21.1 Local physics and cosmology

The universe is the context for local physics. It provides the environment in which galaxies, stars, and planets develop, thus providing a setting in which local physics and chemistry can function in a way that enables the evolution of life on planets such as the Earth. Thus on the small scale, it provides matter in the form of protons, neutrons, and electrons, combined into atoms of chemical elements (including carbon, nitrogen, oxygen, and iron, as well as hydrogen). On a larger scale, it provides a congenial environment for life, by creating the Galaxy and, within it, the Solar System, with a planet suitable for life at the right distance from the Sun. It does all of this through somehow setting both specific laws of physics, and suitable boundary conditions for those laws. If the cosmological environment were substantially different, local conditions would be different and in most cases we would not be here (Carr and Rees, 1979, Davies, 1982, Barrow and Tipler, 1984, Rees, 1999, 2001) – indeed no biological evolution at all would have taken place. Thus cosmology is of substantial interest to the whole of the scientific endeavour, for it sets the framework for the rest of science, and indeed for the very existence of observers and scientists. It is unique as the ultimate historical/geographical science.

Many of the aspects of how cosmology influences local conditions have been discussed in previous sections, e.g. how elements and large-scale structure are formed. The formation of stars and planetary systems will follow in the resulting galactic environment through standard astrophysical processes. In all these cases, local physics together with suitable boundary conditions (set by the cosmology) determine the outcome (Harwit, 1998). However, there is a different way cosmology influences local conditions: namely through a variety of global to local influences (Ellis and Sciama, 1972, Ellis, 2002). We discuss some of these in turn.

21.1.1 Olbers' paradox and the dark night sky

An old question going back to Halley and Olbers is why the integrated radiation from all sources in the universe is as low as it is; indeed why is it finite? In an infinite static Newtonian universe where stars exist at all times, each shell of stars of thickness Δr about us contributes the same amount of light to the night sky, because the inverse square law for flux of light ($F = F_0/r^2$) is exactly compensated by the growth of numbers of stars with distance, which increases as the square of the distance ($\Delta N = N_0 r^2 \Delta r$). Adding up the light from all such spheres, $F_{tot} = \Sigma (F \Delta N) = F_0 N_0 \Sigma \Delta r$. This shows that the light from all the stars in such a universe should diverge as we sum over all distances: $\Sigma \Delta r \to \infty$. Thus extremely distant matter is more important than local matter, because there is so much of it.

This is incorrect in that further stars would be shielded by nearer stars. In fact in such a universe each line of sight would end up on the surface of a star, so the intensity law for light suggests a better analysis would be that the entire sky should be as bright as the surface of the Sun. But then what about absorption by intervening matter – will this not reduce the result? In the given context the answer is no, for thermal equilibrium would be set up, and all intervening matter would then emit precisely as much radiation as it absorbs. The night sky should be just as bright as the surface of a star.

This is known as *Olbers' Paradox:* Why is the night sky dark? (Bondi, 1960, Harrison, 2000). The resolution is that the universe is not static and has not existed forever. As a consequence of the expansion of the universe, it has a finite age, and hence a finite number of stars are visible to us: we can only see a sphere around us as large as is determined by motion at the speed of light since the origin of the universe. Additionally, the intensity of their light is diminished by their redshift of emission according to the intensity law (7.58), as explained in Section 7.8.2. In fact, most lines of sight intersect the LSS in the early universe, rather than any stars that are closer by. The blackbody radiation emitted from this surface is the radiation that comes to us from most directions in the sky, with intensity redshifted $(1+z)^{-4}$. The temperature of the dark night sky is that of the CMB emitted from the LSS, with its initial temperature of 3000K redshifted to 3K by the expansion of the universe.

A low night sky temperature is thus a result of the particular way the universe is constructed. It is a necessary condition for the existence of life on Earth that this temperature not be too high, because the Earth's biosphere functions by receiving heat from the Sun and disposing of the waste energy to the heat sink of the dark night sky (Penrose, 1989).

This would not work if the sky temperature were greater than about 15°C. Thus one way of explaining why the sky is observed to be dark at night is that if this were not so, for thermodynamic reasons, no observers could exist to observe the sky.

21.1.2 Isolated systems and emergence of classical physics

The nature of physics as we know it, and the chemistry and biology that results, depends on two rather specific further conditions, also related to the cosmological environment.

Firstly, there must be the possibility of existence of local systems, able to evolve on their own according to their own internal logic. Isolated systems should exist in the universe, where local physical and biological processes can take place unimpeded by outside interference. In many universe models, this will never be possible: the background electromagnetic radiation may always be intense and variable; cosmic rays or gravitational waves may always be of high intensity; black holes may always be tidally disrupting local systems. Then local physical systems will be unable to proceed in a predictable way and life will be unable to evolve because of the continuing disruptions interfering with developing complexity.

Thus the concept of locality is fundamental, allowing local systems to function effectively independently of the detailed structure of the rest of the Universe. We need the universe and the galaxies in it to be largely empty, and gravitational waves and tidal forces to be weak enough, so that local systems can function in a largely isolated way (Ellis, 2002). Existence of life requires that physical conditions on planets must be in a quasi-equilibrium state for long enough to allow the delicate balances that enable our existence, through the very slow process of evolution, to be fulfilled. Estimates of non-local influences on local systems in the expanding universe are given by Cox (2007); they are small enough to be unproblematic. It is clear that this non-interference by the universe demands the Weyl tensor C_{abcd} must be suitably small almost everywhere so that tidal forces and gravitational wave effects caused by distant objects in a local domain are small. This presumably implies an almost-RW geometry, because a vanishing Weyl tensor implies an FLRW model, and a small Weyl tensor implies an almost FLRW model in the large (given suitable equations of state) (Stoeger, Maartens and Ellis, 1995). However, although local systems will then be isolated, they will still be influenced in crucial ways by global conditions, particularly requiring a well-defined arrow of time at the macroscopic level, and hence (as just discussed) the initial entropy of the universe must be low, as is required for the second law to hold. Thus the requirement is that the universe should set up conditions that allow local physics to proceed, and then not interfere any further.

Secondly, the emergence of a classical era out of an early quantum state is required. The very early universe would be a domain where quantum physics would dominate, leading to uncertainty and an inability to predict the consequence of any initial situation. For complexity to arise, we need this to evolve to a state where classical physics leads to the properties of regularity and predictability that allow order to emerge (Hartle, 2011, Kiefer and Polarski, 2009). It appears that the process of decoherence will tend to lead to this result, but we do not yet have clarity on precisely what initial conditions will necessarily lead to emergence of a classical era in the history of the universe. Whatever these conditions are, we require them to be fulfilled if complex systems such as life are to emerge in the history of the universe.

21.1.3 Mach's Principle and origin of inertia

Newton carried out a famous bucket experiment, whereby he established that the curvature of the surface of water in a rotating bucket, after the rotation of the bucket had communicated itself to the water, occurred only when the bucket was in rotation relative to distant stars (nowadays, read galaxies); there was no such curvature in a reference at rest relative to distant matter. The question then is whether this relationship between local non-rotating rest frames and distant matter is just a coincidence, or whether it is the result of some causal effect. In his pursuit of unifying explanations, Mach conjectured that this was not a coincidence: he suggested that the local inertial properties of matter are causally determined by the distant distribution of matter in the universe (the most distant matter is most important, just as in Olbers' paradox), so that if the universe were different, inertia would be different. This is Mach's principle (Sciama, 1969), which served as a major impetus for Einstein's ideas about cosmology.

   Like the previous case, it embodies the idea that local physical conditions are determined by the sum properties of very distant matter, i.e. by cosmological boundary conditions. The precise meaning and implications of this idea remain controversial, with some claiming it is already fully incorporated into Einstein's theory of gravity, and others denying this is so (Barbour and Pfister, 1995). The latter viewpoint is supported by realizing that, according to relativistic cosmology ideas, a local inertial frame is at rest relative to distant matter if and only if the vorticity is zero (see Section 4.6); but there exist universes where this is not the case (see Chapter 18). Hence one viewpoint is that Mach's principle holds only in universes with special initial conditions (Raine, 1981).

21.1.4 Arrow of time

The existence and direction of the macroscopic arrow of time in physics – and hence in chemistry, biology, psychology and society – is a considerable puzzle, for the fundamental physical laws are time symmetric[1] and so unable by themselves to explain this feature (Davies, 1974, Ellis and Sciama, 1972, Zeh, 1992). The main current proposed explanation is that the observed arrow of time in local macroscopic physics, and hence in chemistry and biology, is related to differing boundary conditions in the past and future of the universe.

   An argument of this kind is the claim by Penrose (1989) that the existence of the arrow of time is crucially based in the universe having had rather special, low-entropy initial conditions in the past (see also Wald (2005), Carroll (2010)). Thus what appears in ordinary physics as an immutable law of nature (i.e. the second law of thermodynamics with a given arrow of time) may well be the result of specific boundary conditions at the start of the universe. It might not be true in all universes, even if the underlying fundamental physical laws are the same. The existence of the arrow of time, and hence of laws like the second law of thermodynamics, not only underlies the kinds of physics envisaged in inflationary theory and the physics of the hot big bang; it is also probably necessary for evolution of life and

---

[1] Apart from a minor time asymmetry of the weak force which is hard to detect, and so cannot be the cause of the major time asymmetry we perceive in the universe.

the existence of consciousness. This depends on the environmental effect of the universe in the large on local systems. Thus it is another example of how large-scale cosmological features can have important local physical results.

In each of the cases just discussed, proposals have been made as to the possible nature of the deeper underlying unchanging laws, and the relations between the state of the universe and the resultant effective laws in the context of the expanding universe. These proposals are, however, intrinsically untestable, because we cannot change the boundary conditions of the universe and see what happens. But they provide an important explanatory paradigm relating cosmology to local physics. There is an essential difficulty in distinguishing between laws of physics and boundary conditions in the cosmological context of the origin of the universe. Effective physical laws may depend on the boundary conditions of the universe, and may even vary at different spatial and/or temporal locations:

Thesis 1: *Physical laws may depend on the nature of the universe*

21.2 Varying 'constants'

A foundational assumption underlying cosmology is that local physics is the same everywhere; without this, we have no basis to proceed. However, what constitutes local physics changes over time: as new theories supersede older theories, the nature of local physics changes. Some fundamental constants in the current version of local physics may turn out to be dynamical in newer versions.

Any given theory, such as GR, contains constants whose value cannot be predicted by the theory. These are the 'fundamental' constants as opposed to derived constants whose value can be computed within the theory: e.g. in GR, H_0 and $T_{\gamma 0}$ can be calculated, whereas G and c are fundamental. Sometimes a theory with a fundamental constant is superseded by another theory in which that constant turns out to be derived or a dynamical variable. For example, Galilean gravity considered g to be a fundamental constant, whereas Newtonian gravity showed that it was a derived variable, determined by G and by the mass of the Earth, $g = G M_E(r, \theta, \phi)/r^2$.

Any variation in a fundamental constant of a theory would be a signal of the breakdown of the theory. Possible variations must be tested by experiments which take careful account of how the measurement process is defined within that theory. In general, it is best to test for variation in dimensionless quantities, since this avoids the problem of disentangling the role played by units of measurement. (See Uzan (2003) for a review.)

It is easy to set up phenomenological models where some physical constant A is considered to be a function of time $A(t)$ in a generalized RW model; e.g. Dirac proposed that $G \propto t^{-1}$ to explain certain coincidences amongst universal 'constants'. While simple phenomenological models may be useful as an initial step, there is a need eventually to ground this in some proper underlying physical theory, and to work out all the inter-related consequences of this assumption for physics in a consistent way. For example, Dirac's proposal is given a consistent theoretical framework by scalar–tensor theories of gravity.

String theory and GUTs predict that many of the 'constants of nature' within GR and the Standard Model of particle physics are in fact contingent, depending on the nature of the vacuum state, and effectively controlled by scalar fields such as the dilaton. In models of fundamental force unification, coupling constants depend on energy, and in the string theory landscape, they depend on the vacuum (Susskind, 2005).

The possible variation in fundamental constants in time or space has been developed in cosmology specifically in relation to varying-G theories and varying-c theories. It has also received a boost from observations suggesting the fine structure constant α may have varied with time in the past. We shall briefly consider specific cases in turn.

21.2.1 Varying-G theories

In Section 14.3.2 we showed that in scalar–tensor theories, G is proportional to a scalar field rather than a constant (see (14.34)). This proposal can be tested in the Solar System, and also has cosmological implications. As noted above, tests for variation in G are best formulated as tests for variation of a dimensionless quantity, such as $\tilde{G} = Gm^2(\hbar c)^{-1}$, where m is some suitably chosen uniquely determined mass, such as the mass of the proton. Limits on variation of this quantity can be obtained from various sources, e.g. Solar System tests and the geological history of the Earth. An interesting new variant is the proposal that weakened gravity in the early universe, i.e. $G \rightarrow 0$, might be the reason that the entropy of the early universe was so low (Greene *et al.*, 2011).

21.2.2 Varying-c cosmologies

Varying speed of light (VSL) cosmologies have been proposed as potentially solving the problems that inflationary cosmologies seek to solve (see Magueijo (2003) for a review). However, there are some foundational issues to be taken into account by any VSL theory (Ellis and Uzan, 2005). One key point is to distinguish theories where it is the photon velocity v_γ that is supposed to vary, and those where it is a universal causally limiting speed v_{lim} that is supposed to vary. These are the same speed c in standard relativity, but they can differ in VSL theories.

Speed of light and measurement

What is called 'the speed of light' and labelled c is not necessarily the speed of light in physical terms. The key issue relating this to physics is how one measures spatial distances and times, for it is only when we have distance and time units set up that we can start to measure the speed of light. On macroscopic scales, currently the only practical way of determining distance is via radar; other astronomical distance scales are derived from this. Parallax distance measurements for example rely on knowing the physical size of the base used to determine the parallax; and that has to be determined by some method such as radar. One assumes a good clock can be constructed, and then uses the timing of reflected electromagnetic radiation to determine the distance. But then the (physical) speed of light of necessity has to be unity, precisely because all electromagnetic radiation travels at the speed of light, and distances are being determined by use of such radiation. This is reflected in the natural units used for such distance measurements: light seconds and light years. When such units are used, the speed of light is unity by definition - not by definition of how fast light moves, but by the definition used for spatial distances.

In order to be viable, a theory with variable v_γ should be based on some other method of measuring spatial distances than radar.

Speed of light and the metric

In standard relativity theory we are allowed to make basis transformations that are *not* Lorentz transformations; and they can make the metric tensor components anything we want. Sometimes this can be misleading. As an example, suppose that there is an instantaneous 'phase' change in a RW metric:

$$\mathrm{d}s^2 = -c^2\mathrm{d}t^2 + a^2(t)\mathrm{d}\sigma^2, \quad c = c_1 \quad \text{for} \quad t < t_*, \quad c = c_2 \quad \text{for} \quad t > t_*, \quad (21.1)$$

where c_1, c_2 are constants. Then $v_\gamma = (c_1/a)\mathrm{d}\sigma/\mathrm{d}t$, for $t < t_*$, while $v_\gamma = (c_2/a)\mathrm{d}\sigma/\mathrm{d}t$, for $t > t_*$ – giving an apparent change in the speed of light by a factor c_1/c_2. However, in fact the units of measurement have been changed (leading to a consequent change in the metric tensor components), rather than the physical speed of light altering. Indeed one can transform the $t > t_*$ metric to the $t < t_*$ form by the change of coordinates $t \to (c_1/c_2)t$. Then according to the principle of general covariance, the two metric forms are just the same spacetime in different coordinates. The confusion arises by using the same label 't' for what are in fact two different time coordinates used before and after t_*. The two metric forms do not represent different physical speeds of light.

There is a preferred time coordinate that can break this degeneracy in the coordinate speed of light – proper time τ, measured along a (timelike) world line by a perfect clock. In GR, $\tau = \int[-g_{\mu\nu}(x^\alpha)\mathrm{d}x^\mu\mathrm{d}x^\nu]^{1/2}$. Applying this to the fundamental world lines with tangent vector $u^\mu = \delta_0^\mu/c_I$ in (21.1) shows that $\tau = c_1 t$ before t_* and $\tau = c_2 t$ after t_*. The quantity τ is an invariant, and will be the same whatever coordinate system is used. While we can use any coordinates, some are more convenient than others in that they more directly represent the physics of what is going on; we get these preferred coordinates on choosing the time coordinate t as proper time τ along the fundamental world lines at all times. Then $c_1 = c_2 = 1$, and there is no jump in the apparent speed of light. Nothing has changed physically; there has simply been a rescaling in the time coordinate. A VSL theory based on changes in the metric tensor components should explain what replaces τ in the proposed theory.

An integrated whole

There are many other issues that must be considered in a varying speed of light theory (Ellis and Uzan, 2005). They include:

- **Causality and Lorentz group**. The speed of light plays a key role in standard physics because it is the limiting speed $v_{\lim}$ for local relative motion, as indicated by the standard relativistic laws for velocity transformations derived from the Lorentz group. The link to the metric tensor is that Lorentz transformations are precisely those transformations that leave the metric tensor components invariant. A VSL theory should show what replaces this relation.

- **Maxwell's equations.** A VSL theory should not just postulate *ad hoc* changes to the speed of light v_γ – the physics that underlies the variation of v_γ should be made explicit – related to Maxwell's equations or its proposed generalizations. These are the equations that determine the actual speed of light. Possibilities include bimetric theories with one metric determining space and time measurements and another used in Maxwell's equations (Bassett *et al.*, 2000).

The overall message is that we cannot just alter the speed of light in one or two equations and leave the rest of physics unchanged. It plays a central role in modern physics both because it is the invariant limiting speed of the Lorentz group and so is built into any variables that transform under that group, and also because electromagnetism is central to many physical effects. In particular, light is central to measurement. In the standard view, these various roles are tightly integrated together in a coherent package in which the speed of light does not vary. A viable VSL theory needs to propose a similarly integrated viable alternative to the whole package of physical equations and consequent effects (kinematical and dynamical) dependent on c.

21.2.3 Varying-α theories

Some observations of absorption lines in spectra of distant quasars suggest the fine structure constant $\alpha = e^2(\hbar c)^{-1}$ may be varying very slightly with time (Murphy, Webb and Flambaum, 2008). However, these observations are in dispute, and there is no strong theoretical justification for such variation. If such a variation is indeed proven beyond reasonable doubt, this will be an important result requiring a good theoretical explanation, and its cosmological implications will need exploration. Note that one cannot explain it as for example being due to c alone varying, as that has no invariant physical meaning. The source of the variation is c, e, $\hbar$, or some combination of these quantities, according to the units used.

21.3 Anthropic question: fine-tuning for life

One impact of the previous section is its implications for one of the most profound issues in cosmology, namely the anthropic question: *why does the universe have the very special nature required in order that life can exist?* (Davies, 1982, Barrow and Tipler, 1984, Earman, 1987, Fabian, 1989, Davies, 1987, Balashov, 1991, Rees, 1999, 2001, Barrow, 2002). We can imagine many possibilities for how physics could have been: most of them will not allow living beings to exist, so a great deal of 'fine-tuning' is required in order that life be possible. The universe sets the context which allows life to come into being. There are many relationships embedded in physical laws that are not explained by current physics, but are required for life to be possible. In particular various fundamental constants are highly constrained in their values if life as we know it is to exist.

There are three aspects that we consider briefly in turn.

21.3.1 Nature of the laws of physics

The first requirement is *laws of physics that guarantee the kind of regularities that can underlie the existence of life*. These laws as we know them are based on variational and symmetry principles; we do not know if other kinds of laws could produce complexity. As they are, they permit life to exist, but which aspects are key requirements, and which inessential?

- The general nature of quantum theory (e.g. superposition, entanglement, decoherence) and its classical limit?
- The specific nature of quantum field theory and quantum statistics, and specifically quantization that stabilizes matter and allows chemistry to exist through the Pauli exclusion principle?
- The general nature of Yang–Mills gauge theory and its implications in the context of the existence of broken symmetries?
- The basic properties of the known forces (effective existence of four fundamental forces; their unification properties)?
- The specific particles and interactions of the Standard Model of particle physics?

It is not clear precisely which of these are necessary for all possible forms of life, but for example the existence of electromagnetism would certainly seem to be necessary. If there were no electromagnetism, there would be no brain as we know it. It is very difficult to determine if anything else could replace the brain, but within the ambit of physics of the kind we know, nothing else seems a plausible replacement. But electromagnetism is an aspect of the electroweak force; is that unification necessary for the existence of life, or would electromagnetism on its own do? We do not know; nor do we know precisely what else is required.

Given the Standard Model of particle physics and GR, there are tight limits on a number of constants governing the strength of interactions, in order that life can exist (Davies, 1982, Gribbin and Rees, 1989). They include

- The neutron–proton mass differential $m_n - m_p$ must be highly constrained. If m_n were just a little smaller, proton decay could have resulted in no atoms left at all (Davies, 1982).
- Electron–proton charge equality is required to prevent massive electrostatic forces overwhelming the weaker electromagnetic forces that govern chemistry.
- The strong nuclear force must be strong enough that stable nuclei exist (Davies, 1982); indeed complex matter exists only if the strong force properties lie in a tightly constrained domain relative to the electromagnetic force (Tegmark, 2003) (see Figure 21.1).

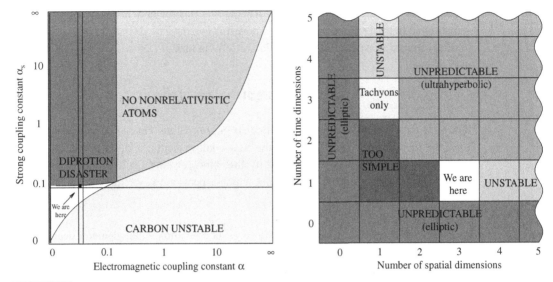

Fig. 21.1   *Left:* Region in the strong/ weak coupling plane that allows complex structures to emerge. *Right:* Numbers of space and time dimensions that allow complex structures to emerge. (From Tegmark (2008). © Elsevier (2008).)

- The chemistry on which the human body depends involves intricate folding and bonding patterns that would be destroyed if the fine structure constant (which controls the nature of chemical bonding) were a little bit different.
- The masses of the light fermions must be very restricted in order that complex molecules can form (Hogan, 2005).
- The number of spatial dimensions must be just three for complexity to exist (Tegmark, 2003, Rees, 2001) (see Figure 21.1).

Hogan (2000) has examined the freedom in the parameters of the Standard Model of particle physics and concluded that five of the 17 free parameters of the Standard Model must lie in a highly constrained domain if complex structures are to exist. This is of course based on the Standard Model of particle physics. It is difficult to determine what the constraints would be in generalizations of the Standard Model. But whatever the nature of fundamental physics, and in particular of particle physics, only a small subset of all possible laws of physics will be compatible with the existence of complexity. Taken together, there are many constraints on the Standard Model resulting from the cosmological context:

- The gravitational force must create large stable structures that can be a habitat for life and its energy source. This requires the ratio of the gravitational to electrical forces to be very small and close to the observed value: $\sim 10^{-36}$ (Rees, 1999).
- The weak force must allow helium production that leaves sufficient hydrogen over; it is related to gravity through a numerical factor of $\sim 10^{-11}$, which cannot be much different. And for this to work, the neutron–proton mass difference must be close to the mass of the electron (Davies, 1982).

- A stellar balance should allow a long lifetime for stars like the Sun, thus enabling the transmutation of light elements into heavy elements. This requires that the nuclear fusion efficiency be close to the observed value ~ 0.007 (Rees, 1999).
- To make heavy elements through nuclear reactions in stars, the beryllium 'bottleneck' must be overcome (Gribbin and Rees, 1989, Susskind, 2005). Production of carbon and oxygen requires the careful setting of two different nuclear energy levels to provide a resonance; if these levels were just a little different, the elements needed for life would not exist (Fabian, 1989). Indeed it was on this basis that Hoyle famously predicted a carbon-12 energy level that has since been experimentally confirmed.
- Something like the existence of neutrinos and the weak interaction with its specific coupling constant are necessary for supernova explosions that spread heavy elements through space, as seeds for planetary formation (Gribbin and Rees, 1989).
- The nuclear force must be weak enough that di-protons do not exist, otherwise no protons will be left over to enable heavier elements to exist (Davies, 1982).
- The neutrino mass must not be too high, or the universe will not last long enough (Davies, 1982).

21.3.2 Specific conditions in the unique universe

Given laws of physics that are suitable in terms of satisfying the requirements of both the previous sections, the universe itself must also be suitable, in terms of its initial or boundary conditions, for life to exist. These are constraints on the contingent parameters describing particular cosmological models. If the laws of physics are basically the same as we now believe them to be, these cosmological requirements include:

- The universe must be sufficiently old ($\sim 15\,\mathrm{Gyr}$) for second generation stars to form and then for planets to have a stable life for long enough that evolution could lead to intelligent life (Gribbin and Rees, 1989); hence we must have $\Omega_{m0} \approx 0.3$ (Rees, 1999).
- Λ must not be too large, or galaxies will not form: $|\Omega_\Lambda| < 1$ (Rees, 1999, Susskind, 2005).
- The amplitude of primordial fluctuations must be large enough for galaxy formation and small enough to avoid collapse into black holes: $\sim 10^{-5}$ (Rees, 1999).

It is easy to emphasize that many of these conditions must be highly restricted (Rees, 1999), but difficult, if not impossible, to state all that is necessary. But for our purposes, it is sufficient to point out that fine-tuning of some of these parameters is demanded. Hence only a small part of parameter space allows life to exist. Tegmark *et al.* (2006) identify 31 dimensionless physical constants required by particle physics and cosmology, and emphasize that both microphysical constraints and selection effects might help elucidate their origin. The overall conclusion is clear:

Thesis 2: *Within the context of our current understanding of physics and biology, life is only possible because both the laws of physics and the boundary conditions for the universe have a very special nature.*

21.3.3 Multiverse proposal

One way to try to handle this scientifically is through some type of multiverse (Rees, 1999, Weinberg, 2000a,b, Susskind, 2005), which tries to show how the biophilic situation is indeed likely to come into existence, even though it is an extremely improbable case. It has been suggested that multiverses explain the parameters of physics and of cosmology, and in particular the very problematic observed value of Λ. The argument is as follows: assume a multiverse exists with many varied regions each having differing properties. Then it is likely that some regions will allow life to exist and others will not (Barrow and Tipler, 1984, Leslie, 1989). Observers can only exist in the highly improbable biophilic outliers where Λ is very small (Hartle, 2004). It is supposed that this will be satisfied somewhere in the multiverse, so we shall indeed observe this condition – and all the necessary requirements for life – to be fulfilled somewhere. A similar argument has been proposed for neutrino (Tegmark, Vilenkin and Pogosan, 2005) and light fermion (Hogan, 2005) masses.

The multiverse proposal: *If there is a large enough ensemble of numerous universes with varying properties, it may be claimed that it becomes virtually certain that some of them will have the right conditions for life to evolve. This could help explain the fine-tuned nature of many parameters whose values are otherwise unconstrained by physics. Then the anthropic principle turns out to be a selection principle.*

A vibrant strand of current cosmology is based on using this multiverse proposal to show that life somewhere in the multiverse is probable. Before looking at this further, we first consider the more fundamental issue: why should we believe that the universe is probable?

21.4 Special or general? Probable or improbable?

21.4.1 Special or general initial conditions?

We decide if a family of universe models is special or generic by considering what subspace it represents within an ensemble of hypothesized universe models (regarded as a possibility space for cosmology). The result obviously depends both on the chosen possibility space, and on the measure imposed on that space; there is a wide variety of choices available in both cases. Nevertheless reasonable results may be obtained from reasonable assumptions, as is illustrated above in our conclusion that anthropic universes are very special within the set of all possible universe models.

The present almost-FLRW state of the universe is very special: whatever measure is used, such models represent a very small part of the space of possibilities because of the very high symmetries of the background FLRW models. However, the issue that is open is whether the initial conditions that led to the present-day special state were special or generic: and both possibilities have been proposed. The universe could have started out in an almost-FLRW state, and stayed that way; or it could have started out in a more generic inhomogeneous state, and then isotropized. Indeed fashions have changed in this regard. The assumption of a geometrically special initial state was encoded in the *cosmological principle,* the assumption that the geometry is RW. This was taken as a founding principle in cosmology until the 1960s, i.e. as an 'explanation' of special initial conditions (see Section 9.8). Then Misner (1968) introduced the 'chaotic cosmology' programme, based on the idea of generic initial conditions being isotropized at later times by physical processes such as viscosity, making initial conditions irrelevant. This concept then became central to the inflationary family of theories (Section 9.7), with the underlying assumption that fine-tuning of initial conditions is unphysical and to be avoided.

Both programmes are, however, only partially successful: one can explain a considerable degree of isotropization and homogenization of the physical universe by either process, but this will not work in all circumstances. Inflation can get rid of much anisotropy (Wald, 1983), but inhomogeneity must be restricted if inflation is to succeed in producing a universe like that we see today (Rothman and Ellis, 1986, Carroll and Tam, 2010). And the success of inflation in solving the horizon problem for HBB models – where exact homogeneity exists to start with – will not necessarily be replicated in anisotropic models. Universes that are initially too anisotropic may never inflate; indeed when perturbations are taken into account, an extremely small fraction of cosmologies will inflate (Carroll and Tam, 2010), and the horizon problem may not be solved in such models if they do. Thus inflation can only be guaranteed to succeed if initial conditions are somewhat restricted; some degree of geometric speciality must have occurred at the start of the observed region of the universe. This conclusion is reinforced by entropy arguments indicating that the universe could not have been in a genuinely generic condition (entropy is maximized by a black hole), and only rather special states lead to ordinary thermodynamics (Penrose, 1989, Wald, 2005, Carroll and Chen, 2005, Greene *et al.*, 2011), which underlies inflationary studies (Penrose, 2004, Carroll, 2010). This special domain might possibly occur within the context of a much larger universe domain where conditions vary randomly, and only isolated regions lead to inflation and eventually domains such as that we see around us; this is essentially a version of the multiverse proposal.

21.4.2 The universe and probability

To explain why the universe is special, we need to consider the further issue: is the universe probable or improbable? Here, probability is intended as a causal explanation, as in the anthropic case mentioned above. Probability only makes sense if there is an actual physically existing ensemble to which probabilities can be applied, not just a hypothetical ensemble (as in the previous section). A hypothetical ensemble only provides an *explanation* of the nature of cosmological existence if it relates to an *actually existing* ensemble.

A fundamental aspect of cosmology is that there is only one observable expanding universe region (McCrea, 1953, 1960, Munitz, 1962, 1986, Ellis, 2006), embodied in a larger unique universe: the single existing physical reality. This essential uniqueness of its object of study sets cosmology apart from all other sciences. We can observe its many aspects in many ways, but they are all aspects of one unique object: the universe itself. Cosmological theory makes statistical predictions of physical outcomes of classes of physical objects in the cosmological context, so probability theory applies to the emergent families of astronomical objects (clusters of galaxies, galaxies, stars, planets for example) in the universe. However, there is no obvious reason why it should apply to the universe itself, unless we live in a multiverse.

When there is just one object in existence, and no other similar object to compare it with, it is not probable or improbable – it simply is. Ordinary probabilities do not apply when there is only one object under consideration. But a concept of probability underlies much of modern argumentation in cosmology. Talk of 'fine-tuning' for example is based on the use of probability (it is a way of saying something is improbable). This assumes both that things could have been different, and that we can meaningfully assign probabilities to the set of unrealized possibilities.

Thesis 3: *The concept of probability is problematic in the context of existence of only one observable object.*

If we base probabilities for cosmology on the concept of 'the wave function of the universe' in quantum cosmology, we face the problem of how to verify that this theory applies to the actual universe. In any case, as we discuss below, there is no well-established measure to use on the space of cosmological models, so the result can be changed by changing the measure used. Additionally, the meaning of the wave function Ψ is not clear in this context. Thus there is no well-established context for determining probability in regard to the universe. But even if there were, there is no guarantee that the single universe that we actually observe is indeed probable.

The concept of probability does not usefully apply to a single object, even though we can make many measurements of that single object to determine its detailed nature: it applies where we can make multiple distinct measurements of the single universe, but not to issues concerning the existence of the universe itself (Ellis, 2006). If we use a Bayesian interpretation, which some suggest can be meaningfully applied to only one object, the results depend on our 'prior knowledge', which in this case can be varied by changing our initial pre-physics assumptions. Bayesian approaches (Garrett and Coles, 1993) are completely dependent on the priors assumed: in the cosmological context, these priors are untestable philosophical assumptions.

21.5 Possible existence of multiverses

If a multiverse existed in physical reality, there would be a physically realized ensemble to apply probability to: probabilities do make sense within the context of a multiverse, because they deny the uniqueness of the universe. They do indeed provide a causal explanation of the anthropic puzzle.

Is there a physically existing ensemble out there, within which probabilities would make sense? This is often assumed to be true. The ensemble is defined by specifying a space of possible models, the measure to use on this space, and the way it is populated by some universe creation mechanism. There are a number of ways in which, theoretically, multiverses could be realized: there might be a fundamental principle that all possible universes occur, not causally related to each other (Lewis, 1986, Tegmark, 2003); they might occur as a realization of different branches of the wave function of the universe envisaged in quantum cosmology (Section 20.2); they might occur through cyclic processes in time, the phoenix universe idea, proposed in various ways (Section 20.6); or they might occur in different spatial places in a greater mega-universe. The pre-eminent proposal today is the latter one: that the emergence of a spatially separated set of universe domains will be an inevitable result of chaotic inflation (Susskind, 2005, Guth, 2007).

21.5.1 Defining multiverses

Possibility space

When considering probability in the context of a space of possible universes, one has to define what constitutes possibilities. This is a deep question which is philosophical, because there is no conceivable way of testing it. It is a choice we make.

What variations in physics do we allow in a multiverse (Ellis, Kirchner and Stoeger, 2004)? For example, are they all based in the same underlying quantum gravity theory? What about the set of allowed logics? Probabilities will depend on what we define as the space of possibilities. The outcome will depend on the choices we make here. Additionally, what geometrical possibilities shall we consider? Only RW models? Will we include Bianchi and LTB and general inhomogeneous models? The outcome will depend on these choices too. One of the proposals of this book is that we should consider all the geometrical possibilities (see Chapters 17–19).

Chaotic inflation resolves these issues by assuming some form of physics applies in the megaverse that allows different local realizations of physics in local domains; and the main proposal for that theory is M-theory, with the possibility space being the landscape of string theory. This idea is usually applied to a very restricted set of geometrical possibilities: usually only RW geometries are considered.

Populating the possibility space

Once the space of possibilities is defined, some hypothetical generating mechanism is needed to populate it with a distribution of models to create a specific multiverse proposal; then we have a suitably defined ensemble of cosmologies in which probability could make sense. This population is defined by a distribution function on the possibility space (Ellis, Kirchner and Stoeger, 2004). The distribution can be assumed a priori, or it can be seen as following from some hypothetical generating mechanism. A popular proposal is Coleman–de Luccia tunneling from a prior space, as in the chaotic inflation proposals.

The issue of measures

But how probabilities work out in this hypothetical ensemble depends on the measure chosen. And we have no agreed measure on the space of possible universes; what seems special or general depends on the choice of such a measure. Even if the ensemble is measurable in a natural way, one may need to apply a weighting to this measure to make its integral over the ensemble converge, and to turn it into a probability measure. A common starting point is a flat a-priori measure, motivated by Laplace's Principle of Indifference (Gibbons and Turok, 2008), later modified to reflect constraints from extra information. Ways to define and refine measures remain an issue of debate. We give some examples of current approaches.

Assigning probabilities on the whole infinite-dimensional space of solutions is so difficult that papers often address only the issue of measures on finite-dimensional subspaces, e.g. the subspace of RW cosmologies with scalar field or barotropic perfect fluid content. Kirschner and Ellis (2003) review earlier proposals and their flaws, for example, that the probability assigned may depend on the $t =$ const slice used to define the measure. They propose probabilities based on applications of Jaynes' principle that the measure should be invariant under a transformation group appropriate to the constraints on the variable, deriving interesting results for the 'flatness problem' from a choice based on considering the time-independent quantities K, Λ and (in the dust case) ρa^3. However, the probabilities thus defined still depend significantly on the assumptions made: for example, if no constraint is put on Λ, the measure is proportional to $\mathrm{d}\Lambda$, while if it is assumed that $\Lambda > 0$, the appropriate measure is proportional to $\mathrm{d}\Lambda/\Lambda$.

Gibbons and Turok (2008) have reconsidered the proposal of Gibbons, Hawking and Stewart (1987) to use the natural symplectic measure on the space of trajectories in finite-dimensional truncations ('minisuperspaces') of the whole space governed by a Hamiltonian and subject to an odd number of holonomic constraints: this counts each classical cosmology only once. The origin of the divergence found in the total measure proposed by Hawking and Page (1988) for scalar field RW models is observationally indistinguishable nearly-flat models. Gibbons and Turok identify such models and arrive at a finite measure. It turns out that this suppresses inflationary models by a factor $\exp(-3N)$ where N is the number of e-folds: they compare this with other proposals favourable to inflation. However, Carroll and Tam (2010) review both proposals, and conclude that the symplectic measure firstly strongly favours flat models, and secondly shows that inflation is suppressed by an even higher number: only 1 in $10^{6.6\times 10^7}$ cosmological histories inflate.

In the context of string theory, one may want to calculate probabilities for the different possible vacua. In (eternal) inflation, one wants to compute probabilities for the parameters of the different 'thermalized' regions (in effect, locally homogeneous regions, i.e. not the full space of inhomogeneous models) within the one universe we are in. A number of papers discuss the two together, i.e. the inflationary multiverse. For example, Garriga *et al.* (2006) used an ergodic assumption on the underlying fields and various choices for weight factors. Linde (2007) noted the dependence of earlier results on cutoffs and proposed an improved volume-weighting avoiding some of the difficulties by making comparisons at the stage when the thermalized volumes enter their stationary phases. This 'stationary measure' was later shown to avoid some problems, in particular the high-temperature CMB background predicted by some alternatives, and to predict the results of some local experiments, such as proton decay (Linde, Vanchurin and Winitzki, 2009). Even when inflation is not eternal and the universe is compact, different measures are possible. Proposals for a proper time measure, scale factor cutoff measure, causal diamond measure, and stationary measure are discussed and compared by Linde and Noorbala (2010).

In a different scenario, the 'no boundary' proposal, Hartle, Hawking and Hertog (2008) considered a probability obtained by weighting the probabilities of the no-boundary wave function by the number of Hubble volumes in the total present volume, proportional to $\exp(3N)$. This favours larger universes and more inflation.

Overall, the measure issue remains unresolved – some measures allow eternal inflation and others non-eternal inflation (Linde and Noorbala, 2010). Some even imply that inflation is improbable by a large factor (Gibbons and Turok, 2008, Carroll and Tam, 2010). The results obtained depend on the measure used.

21.5.2 Testability of multiverse proposals

Multiverse proposals are observationally and experimentally untestable, because we have no causal connections with most of the regions predicted to exist by this theory. They make statements about conditions far beyond the visual horizon, with the term 'infinity' often being used. Extrapolation on such scales is dangerous, and the supposed underlying physics is not yet well established. Any proposed physics underlying a multiverse proposal, such as Coleman–de Luccia tunneling, will be an extrapolation of known physics; but the validity of that major extrapolation to cosmology is untestable. Thus these models are not an inevitable consequence of established physics.

Thesis 4: *Multiverse proposals are unprovable by observation or experiment, but some self-consistency tests are possible.*

Given that the claimed other universes or universe domains in a multiverse are observationally inaccessible, is there any other way of demonstrating their existence? A variety of probability based tests for the multiverse, particularly related to the expected value of Λ, have been proposed (Rees, 2001, Susskind, 2005, Weinberg, 2007), but they cannot be used to prove the existence of a multiverse, for they cannot sensibly be applied to a single object. These tests are only applicable if a multiverse exists, and thus assume the result to be proved:

Thesis 5: *Probability-based arguments cannot demonstrate the existence of multiverses.*

The consistency tests on some multiverse proposals are *necessary* conditions for those specific multiverse proposals, but are hardly by themselves indications that the multiverse proposal is true. The drive to believe this is the case comes from theoretical and philosophical considerations (see e.g. Susskind (2005)) rather than from data. The claim that an ensemble physically exists – as opposed to an explicitly hypothetical ensemble, which can indeed be useful – is problematic as a proposal for scientific explanation, if science is taken to involve testability. Indeed, adopting these explanations invokes theory over testability (Gardner, 2003); but the theories being assumed are not testable. It is therefore a choice made for philosophical reasons. That does not mean it is unreasonable (it can be supported by quite persuasive plausibility arguments); but its scientific status is unclear.

21.5.3 Criteria for scientific theories

Typical criteria for a good scientific theory (Kuhn, 1977), could be the following:

- *Satisfactory structure:* (a) internal consistency, (b) simplicity (Ockham's razor), (c) aesthetic appeal ('beauty' or 'elegance').
- *Intrinsic explanatory power:* (a) logical tightness, (b) scope of the theory – the ability to unify otherwise separate phenomena, (c) probability of the theory within some well-defined measure.
- *Extrinsic explanatory power, or relatedness:* (a) connectedness to the rest of science, (b) extendability – providing a basis for further development.
- *Observational and experimental support:* (a) testability – the ability to make quantitative as well as qualitative predictions that can be tested, (b) confirmation – the extent to which the theory is supported by such tests as have been made.

It is particularly the last that characterizes a scientific theory, in contrast to other types of theories claiming to explain features of the universe and why things happen as they do. It should be noted that these criteria are *philosophical* in nature, in that they themselves cannot be proven to be correct by any experiment. Rather their choice is based on past experience combined with philosophical reflection. Philosophical criteria for satisfactory cosmological theories will in general come into conflict with each other, so that one will have to choose between them to some degree; this choice will shape the resulting theory – in particular in the tension between explanatory power and testability of a theory. These criteria will not all be satisfied to the same degree, and may even lead to opposing conclusions (Ellis, 2006):

Thesis 6: *Conflicts arise in applying criteria for satisfactory cosmological theories.*

We need a meta-theory telling us what criteria to apply in what context: but no such meta-theory exists – and if it did, it would not be testable. It would itself be yet another philosophical presupposition. Because of all the limitations in terms of observations and testing, in the cosmological context we still have to rely heavily on other criteria, such as explanatory power, which is often taken to be more important than observational testing of the theory. The problem arises if there is no observational evidence whatever supporting the theory.

21.5.4 Multiverses and verifiability/falsifiability

Unless the nature of causality and horizons is changed by subsequent developments in physics, multiverse proposals are untestable. They have great explanatory power, for example in terms of explaining the small value of Λ, but in the end that power is too great: as they can explain virtually anything, they are unable to uniquely explain anything in particular. Some versions demand negative spatial curvature (Freivogel *et al.*, 2006), but others are not so constrained (Tegmark, 2003). There are some consistency tests, but they do not uniquely indicate that a multiverse exists. In order to be characterized as science, the testability component seems essential; explanatory power alone cannot replace it. The idea of a multiverse provides a possible route for the explanation of fine-tuning. But it is not uniquely defined, and is not scientifically testable apart from some possible consistency tests. Indeed it explains too much, because (at least in some versions) the multiverse idea can explain *anything*; thus no observation can show it wrong.

Thesis 7: *It can be argued that multiverses cannot be scientifically proven to exist because the usual kinds of scientific proof are not available.*

We emphasize that there is nothing wrong with scientifically based philosophical proposals, indeed they are very useful; but they should be identified for what they are. The issue is discussed further in Carr and Ellis (2008) and Ellis (2009).

Problem 21.1 If you disagree with the criteria for good scientific theories suggested above (Section 21.5.3), propose other ways of defining the issue. Is there for example a more holistic criterion for a theory as a whole, that validates its essential parts, even if they are not testable by themselves? But then what criteria are needed for something to be included as an essential part of a theory?

21.5.5 Infinities in cosmology

One of the reasons that we study cosmology is our fascination with its philosophical aspects; and one that recurs is the idea of infinity in cosmology. The ideas of infinite spatial sections with an infinite number of galaxies, and of an infinite number of universes (or expanding universe domains) in a multiverse, are currently being widely proposed in the context of chaotic inflation, see e.g. Guth (2007). If true, this has strange implications; it plausibly implies that countless identical civilizations to ours are scattered in the infinite expanse of the cosmos, with semi-identical histories to ours replicated an infinite number of times out there (Ellis and Brundrit, 1979).

It has been claimed that this is a necessary outcome of current inflationary theories (Knobe, Ohm and Vilenkin, 2006, Vilenkin, 2006). But the problem lies in the idea that this is currently how things are: that it is the state at the present instant, as is often claimed. The real situation is that physical processes may be such that eventually an infinite number of galaxies, stars, planets, and civilizations will tend to come into existence; but that state is not achieved at any finite time through the supposed physical processes (Ellis and Stoeger, 2009b). And if it were true, it is certainly not verifiable: no observational or experimental process can prove that an infinite number of any entities exist. Thus any such claim is a philosophical rather than scientific statement.

Consequently, any claims of actual existence of physical infinities in the real universe (or an assumed multiverse) should be treated with great caution (Ellis, 2006).

Thesis 8: *The physical existence of infinities is questionable.*

Hilbert expressed a similar view long ago: 'the infinite is nowhere to be found in reality. It neither exists in nature nor provides a legitimate basis for rational thought' (Hilbert, 1964).

Problem 21.2 Consider any way you can of experimentally or observationally testing the idea that there is an infinite number of objects (of any kind) in the universe.

21.6 Why is the universe as it is?

The considerations above, and particularly the fact that it allows life to exist, lead to the conclusion that the physical universe in which we exist is very improbable. Much of current cosmology can be viewed as trying to shift the improbability from one place to another (initial conditions, physical conditions, location in a multiverse, etc.). Thus a key philosophical issue is, *Why do we live in this improbable unverse*? Why does it have this improbable nature? This is the issue of the relation of the universe to ultimate causation (as opposed to the ultimate limits of physical causation, which we considered in Section 20.6). This might be true because:

it happened by pure chance (pure happenstance);

it is the only way things could be (necessity);

an ensemble of universes exists, providing a context in which existence of at least one bio-friendly universe is probable (probability);

it is intended to be that way – in some sense, purpose or design underlies existence (purpose).

In all cases, the final issue of existence remains a mystery: each explanation assumes some kind of existence as a starting point. We shall not enter this discussion here; we simply make a fundamental point relating this to scientific cosmology: the ultimate issue cannot be resolved by the multiverse proposal. If we choose to support the multiverse view on philosophical grounds, ultimate issues remain: Why does this unique larger whole have the properties it does? Why this multiverse rather than any other one? Why is it a multiverse that allows life to exist? Many multiverses will not allow any life at all.

To solve this, we can propose an ensemble of ensembles of universes, with even greater explanatory power and even less prospect of observational verification; and so on. The prospect of an infinite regress looms. Indeed if we declare (as suggested at the start of this book) that 'the universe' is the total of all that physically exists, then when an ensemble of expanding universe domains exists, whether causally connected or not, that ensemble itself should be called 'the universe', for it is then the totality of physically existing entities. All the foundational problems for a single existing universe domain recur for the multiverse – because when properly considered, it is indeed the universe! (Ellis, 2006)

Thesis 9: *Metaphysical uncertainty remains about ultimate causation in cosmology.*

The conclusion of this chapter is that when one probes the far reaches of cosmology, one inevitably starts to engage with philosophical issues. It is best to recognize this fact, and to engage with them in an informed way. If one enters this territory, one cannot get away with the claim that one is just doing physics: an uninformed philosophical attitude is indeed a philosophical stance, albeit it an uninformed one.

Problem 21.3 Reconsider the issue of the relation between philosophy and cosmology in the light of the above discussion, and of Tegmark (2008), White (2007), Ellis (2009), Hogg (2009), Goenner (2010), Balbi (2010).

Conclusion: our picture of the universe

22.1 A coherent view?

We conclude by returning to the question we started with: what is our best current picture of the physical universe, and what are its problems and uncertainties?

There is an agreed basic view of the universe, the standard model of cosmology, in which the universe expands from a hot big bang early phase to a late-time cool, accelerating phase driven by a cosmological constant, with structure formation taking place through gravitational instability around the 'scaffolding' provided by dominant CDM, acting on seed perturbations generated by inflation. This seems to provide a statistically good fit to all the data up to now, with the same set of parameters.

Given the major physical uncertainties concerning this model, we need to continue subjecting it to further observational tests with ever-improving data – and we need to query its uniqueness, and probe the alternative possibilities. A key feature is that uncertainty about both geometry and physics increases with time in the past, and with distance from our world line. Our cosmological claims should make this feature clear. Use of the FLRW models as a starting point of analysis tends to hide this fact. The perturbative inhomogeneity imposed in these models changes the physics equations from PDEs to ODEs and so hides the nature of causality and associated causal domains.

22.1.1 The visible universe

The visible universe is the part of spacetime within our past light-cone since decoupling. The CMB emitted from where our past light-cone intersects the last scattering surface marks the boundary of this domain – the matter emitting this light represents our visual horizon, the furthest away matter that we can detect by electromagnetic radiation of whatever wavelength. Observations give us direct access to this region, and allow us to find the best fit FLRW model – and to test the theory of structure formation within that model. In particular, we can check the consistency between the CMB anisotropies and the large-scale distribution of matter.

The major puzzles here are

- What is the dark matter that we discern through its gravitational effects? Can we succeed in direct detection via experiments on Earth?
- What is the dark energy that we infer through the acceleration of the universe at late times? Is it Λ or dynamical? Is it a modification of GR on large scales, or a mirage due to misinterpretation of observations in a lumpy universe?

Determining the nature of the dark components is a key focus of present-day cosmology. In addition to discovering the CDM particle, we need a much better understanding of the interplay between CDM and baryons in structure formation, involving the complex non-linear magneto-hydrodynamics of baryonic gas. More and more sophisticated simulations need to be combined with analytical progress in understanding the qualitative physics of nonlinear structure formation.

Key developments that will shape our understanding include:

- Wider and deeper galaxy surveys will provide increasingly detailed information about the distribution and peculiar velocities of large-scale structure.
- Weak lensing will give increasingly detailed maps of the total matter, and together with galaxy surveys will provide high-precision tests of GR on cosmological scales.
- The evolution of the BAO feature with redshift – in both radial and transverse directions – will not only provide tight constraints on the expansion history, but will also be a powerful consistency check of the standard model.
- In order to push the frontier back to the reionization era and even to the pre-stellar 'dark ages', deep radio surveys of the HI 21cm signal will provide new advances in our understanding of the origin of galaxies.

22.1.2 The hidden past

The hidden past, as far as observations by electromagnetic radiation of any wavelength are concerned, is everything that lies before the surface of last scattering. The CMB blackbody spectrum provides evidence of the hot big bang era, and its acoustic peaks are relics of the acoustic waves in the plasma before decoupling. Light element abundances in our neighbourhood provide a 'geological' record that probes BBN at early times near our world line.

The major puzzles that remain about the hidden past include the nature of baryogenesis and the fundamental physics underlying inflation. Observational developments that will help to unravel these puzzles include:

- Probing particle physics beyond the Standard Model via higher and higher energy colliders.
- Obtaining detailed CMB polarization data as a test of theories of inflation and alternatives.
- Non-Gaussianity on large scales in the CMB and large-scale structure will also test theories of inflation and probe the primordial perturbations.
- Space-based gravitational wave astronomy as a direct probe of the cosmic gravity wave background, potentially corroborating the evidence from the CMB.
- Neutrino astronomy developed to the point of directly detecting the neutrino background radiation – a very remote possibility, but in principle possible.[1]

---

[1] The possible direct detection of this background is quite different from inferring its existence from the CMB power spectrum.

In principle we can obtain useful information back to the end of inflation, but we can learn much less about pre-inflation, precisely because an inflationary epoch will have wiped out previous data.

The origin of the universe

We cannot test much of the physics important in the very early universe because we cannot attain the required energies in accelerators on Earth. This means we are unable to get certainty on key issues, especially: did the universe have a beginning? Supposing there was a beginning – then major unresolved questions arise:

- Was it special (highly symmetric) or generic (randomly based)?
- What caused it? Testable physics cannot explain the initial state and hence specific nature of the universe.
- Why does the universe appear almost FLRW in the observable domain? Inflationary theory only partly explains this.
- How is the arrow of time built into the structure of the universe, and how does it relate cosmological conditions to local physics?

The ultimate goal is to find a theory of creation of the universe that is fully compatible with quantum theory and GR, or some unification of those theories, and that predicts what we see with a plausible argument for uniqueness.

22.1.3 The inaccessible domains

Observational horizons limit our ability to observationally determine the very large-scale geometry of the universe (unless we live in a 'small universe', where we have already seen right round the universe since last scattering).

Beyond the horizon

There is no observational way to determine what happens outside the visual horizon. The particular issue where the nature of the region outside the horizon is in contention, is the debate

- Does a multiverse exist as a physical entity?

Existence of an effective multiverse is a likely outcome of chaotic inflation, but we have no definitive proof that chaotic inflation occurred.

To the future

Unless we live in a small universe, we do not even in principle have the data necessary to predict to the future from observable data: we can only do so by making untestable assumptions (that no new kinds of data will enter our visual horizon). For want of any better position to adopt, we usually assume things will carry on unchanged. Even in this case, we face uncertainty. What is the future fate of our universe domain: will it expand forever, or recollapse? In the latter case, might it recycle to a new expanding phase? This depends on a possible decay of dark energy in the future; if this does not occur the universe will expand forever. If it does occur, the outcome depends on the sign of the spatial curvature, which we do not yet know. But there is no proof this will happen, and the default assumption must be that the universe will expand forever.

22.2 Testing alternatives: probing the possibilities

Given the above observational restrictions, it is wise to probe the viable alternatives as regards both the physics and geometry.

22.2.1 Alternative physics

Much speculation and research is investigating alternative possible forms of dark energy, and devising observational tests that can test whether acceleration is due to Λ. Equally important is investigating alternative forms of gravity. GR is only soundly tested on Solar System scales, and then in quasi-static situations – testing GR on cosmological scales is opening up a new frontier in cosmology. Furthermore, as we are engaged in investigating the most distant regions of space and the earliest epochs of the universe, it is important to do whatever we can to check whether physics is unchanging with time. In particular, do any of the constants of nature (within our current theories) vary with time?

In summary, further theoretical and observational efforts are needed to:

- Test alternatives to Λ within the standard model.
- Test alternatives to dark energy based on modifications of GR.
- Test our current fundamental theories by probing possible variations in their fundamental constants.

22.2.2 Foundations of homogeneity

The foundation of almost all present-day cosmological analysis is the perturbed RW geometry of the observed region of spacetime. Therefore, it is vital to test this assumption.

- The strongest support for homogeneity is the exact EGS-ETM theorem (isotropic CMB for all observers implies homogeneity) and its partial generalization, the almost-EGS theorem.
- We need to remove the assumptions on derivatives in the almost-EGS result, probably by using observations of matter in addition to the CMB.
- The galaxy distribution on the past light-cone contains crucial information on spacetime geometry, and further work is needed to develop light-cone analysis of galaxy surveys.
- The Copernican assumption underlying these results needs to be tested via Sunyaev–Zel'dovich scattering of the CMB at distant clusters, tests for deviations from RW curvature and other probes.

22.2.3 Fitting, averaging and backreaction

More fundamentally, the homogeneity assumption and its theoretical support make implicit assumptions about how the real universe is smoothed to achieve an RW description.

- Determining the averaging scale, the nature of averaging itself, and the way that the result of averaging is used to fit a given FLRW background constitute a major unresolved problem in cosmology.
- Even if it can be shown that the standard approach of assuming perturbed FLRW is effectively correct, important corrections to precision cosmology may arise from averaging/backreaction effects on the metric and on light propagation.

22.2.4 Alternative homogeneous geometries

Anisotropic modes

There may be non-zero anisotropic (Bianchi) modes, which would tend to dominate dynamics at both early and late stages, while allowing a very long intermediate period very close to a RW geometry. If they occur they may leave traces in the CMB anisotropies and may affect BBN.

Alternative topologies

The topology of the spatial sections of RW geometries can be extremely complex, and no principle is known that will determine what this topology is. An important issue is whether we live in a small universe or not; this should be detectable by searching for identical circles in the CMB sky. These are the only universes where there is no hidden region beyond a visual horizon and where we can in principle predict the future from available initial data.

22.2.5 Alternative geometries: inhomogeneous models

Inhomogeneous models allow us in principle to explore the nonlinear effects of structure within GR, independent of the averaging problem. In addition, they allow us to probe the standard model.

- Large-scale void models, where we are at the centre of an isotropic underdense region, probably cannot pass all observational tests. Further work is needed on how large-scale structure grows in these models, in order to settle the issue of whether they are ruled out.
- Given the central role of dark energy in present-day cosmological studies, it is crucial to test this possibility by all available observational means.

22.3 Limits of cosmology

Cosmological uncertainty mirrors at the largest scales the uncertainty of quantum theory at the smallest scales. It is a key aspect of physical cosmology: we are limited in what we can ever know with reasonable certainty about the geometry of the universe on the largest scales, and the relevant physics at the earliest times. Cosmological theory should acknowledge this uncertainty.

22.3.1 Uniqueness of the universe

The overall problem for cosmology, arising from the uniqueness of the universe, is how does generic theory apply to the specific, when only one specific case exists? What is the relation of chance versus necessity in the origin of the universe? Necessity is represented by physical laws, but they may depend to some degree on the cosmological environment. One has the issue of why specific initial conditions occurred, for both physics and geometry. The place where this makes a practical difference in cosmology is

- Cosmic variance: cosmological theory gives statistical predictions not unique outcomes. How do we handle the relation between the specific one universe that does exist, and the generic theories of the kinds of universe that might exist?

This affects our attitude to such issues as the large-scale power drop-off in the CMB anisotropies, and possible statistical anisotropies between the Northern and Southern hemispheres: do they need an explanation, or are they to be regarded as chance events, not needing an explanation? The difference is substantial.

22.3.2 Significant questions

Finally, there remain fundamental issues which lie beyond the scope of science:

- Uncertainty about ultimate issues will remain. Scientific exploration can tell us much about the universe, but not about its ultimate nature: why things exist, and why they are the way they are?
- How does it come about that the universe provides a viable home for intelligent life?

The anthropic question will remain one of the most interesting issues in terms of relating cosmology to the larger philosophical sphere. Astronomically it relates to the existence of other planetary systems, and theoretically to arguments about why the universe is as it is. Studies of the nature of the full family of plausible cosmological models will inform that argument, but not solve it. Philosophically it leads to speculations about ultimate causation and the nature of existence. Physical cosmology provides the proper context within which to imbed those speculations.

22.3.3 The ephemeral and the long lasting

To return to a point we made at the start of this book: what we would like to do is to clarify which are transient issues, that will eventually go away, and which are foundational issues, that will always remain of concern to the scientific study of cosmology?

We have made choices in this regard in the discussion above, through pointing out the issues we regard as important. Only time will tell how good these choices have been.

Appendix

Some useful formulae

A.1 Constants and units

Some useful constants are given in SI units in Table A.1. It is convenient to use units that simplify the often cumbersome expressions in SI units.

Units with $c = 1$

In units with the speed of light set to unity, length and time, mass and energy, and energy density and pressure, have the same dimensions:

$$c = 1 \;\Rightarrow\; [\text{length}] = [\text{time}] = L, \; [\text{mass}] = [\text{energy}] = M, \; [\rho] = [p] = ML^{-3}, \quad \text{(A.1)}$$

where we have chosen to use length and mass units, but other equivalent choices can be made. Then Newton's constant has dimensions

$$[G] = LM^{-1}. \tag{A.2}$$

The line element has dimension L^2, but coordinates are usually taken to have no dimension (since they are definable even when there is no metric). Thus the metric tensor $g_{\mu\nu}$ is taken to have dimension L^2. Then the Levi-Civita connection components $\Gamma^\mu{}_{\nu\kappa}$, and the curvature components $R^\mu{}_{\nu\kappa\sigma}$ and $R_{\mu\nu}$ are dimensionless. The Ricci scalar and Λ will have dimension L^{-2}, and this will be unaffected by changes of coordinate or tetrad basis. If one uses an orthonormal tetrad, the basis vectors have dimension L and the curvature components have dimension L^{-2}.

It is always possible instead to introduce for each coordinate x^μ a factor k^μ with dimension L, and use coordinates $X^\mu = k^\mu x^\mu$ (no sum on μ), which have dimension L. Then the metric becomes a dimensionless function of $\{X^\mu/k^\mu\}$ (no sum on μ), the connection components have dimension L^{-1} and the curvature components dimension L^{-2}. The drawback is that the extra k^μ factors in general make the expressions rather cumbersome, but in special cases that drawback can be overcome.

In particular, as explained in Section 9.1.3 and below, it is convenient in FLRW cosmology with $K \neq 0$ to choose coordinates adapted to the length scale of the spatial curvature so that in all cases the scale factor $a(t)$ can be considered dimensionless.

$$\lambda_{\text{phys}} = a(t)\lambda_{\text{com}}. \tag{A.3}$$

We can choose how to 'share' the length units; e.g., we could choose $[a] = L$, $[\lambda_{\text{com}}] = 1$. We follow the convention defined by

$$[a] = 1, \; [\lambda_{\text{com}}] = L, \tag{A.4}$$

Table A.1 Some useful constants (to 2 significant figures)		
CONSTANT	SYMBOL	VALUE
light speed	c	3.0×10^8 m/s
Newton constant	G	$6.7 \times 10^{-11}\,\mathrm{m^3/kg/s^2}$
Planck reduced constant	$\hbar$	1.1×10^{-34} Js
Boltzmann constant	k_B	1.4×10^{-23} J/K
Planck mass	M_P	$2.2 \times 10^{-8}\,\mathrm{kg}$
		$1.2 \times 10^{19}\,\mathrm{GeV}/c^2$
Planck time	t_P	5.4×10^{-44} s
electron mass	m_e	$9.2 \times 10^{-31}\,\mathrm{kg}$
		$5.1 \times 10^{-1}\,\mathrm{MeV}/c^2$
proton mass	m_p	$1.7 \times 10^{-27}\,\mathrm{kg}$
		$9.4 \times 10^2\,\mathrm{MeV}/c^2$
Thomson cross-section	σ_T	$6.7 \times 10^{-29}\,\mathrm{m^2}$
		$1.7 \times 10^{-15}\,\mathrm{eV^{-2}\hbar^2 c^2}$
parsec	pc	$3.1 \times 10^{16}\,\mathrm{m}$
year	yr	3.2×10^7 s
solar mass	$M_\odot$	$2.0 \times 10^{30}\,\mathrm{kg}$
solar luminosity	$L_\odot$	3.8×10^{26} W
Hubble constant	H_0	$3.2 \times 10^{-18} h\,\mathrm{s^{-1}}$
		$2.2 \times 10^{-33} h\,\mathrm{eV}/\hbar$
Hubble radius	cH_0^{-1}	$9.1 \times 10^{25} h^{-1}\,\mathrm{m}$
		$3.0 h^{-1}\,\mathrm{Gpc}$
photon temperature	$T_{\gamma 0}$	$2.7\,\mathrm{K}$
		$2.3 \times 10^{-4}\,\mathrm{eV}/k_B$
photon number density	$n_{\gamma 0}$	$4.1 \times 10^8\,\mathrm{m^{-3}}$
equality redshift	$1 + z_{\mathrm{eq}}$	$2.4 \times 10^4 \Omega_{m0} h^2$
decoupling redshift	$1 + z_{\mathrm{dec}}$	1.1×10^3
eV conversions	1 m	$5.1 \times 10^{15}\,\mathrm{GeV^{-1}} c\hbar$
	1 s	$1.5 \times 10^{24}\,\mathrm{GeV^{-1}}\hbar$
	1 kg	$5.6 \times 10^{26}\,\mathrm{GeV} c^{-2}$
	1 J	$6.2 \times 10^9\,\mathrm{GeV}$
	1 K	$8.6 \times 10^{-14}\,\mathrm{GeV} k_B^{-1}$

which implies that

$$[t] = [\tau] = [\chi] = [r] = [\boldsymbol{x}] = L, \quad [K] = L^{-2}, \quad [H] = [k] = L^{-1}, \qquad (A.5)$$

where k is a comoving wavenumber. For non-flat spatial sections, the spatial Ricci scalar defines a curvature scale at the present time,

$$K = \frac{\pm 1}{{}^3 R_0^2}, \quad {}^3 R_0 = \frac{1}{a_0 H_0 \sqrt{|\Omega_{K0}|}}. \qquad (A.6)$$

It is useful to fix the present-day scale factor to unity, $a_0 = 1$.

Units with $\hbar = c = 1$

If we also set Planck's reduced constant to unity, then the Einstein–Hilbert action is dimensionless and we get $[G] = L^2$. It follows from (A.2) that mass and inverse length have the same dimensions:

$$\hbar = c = 1 \;\Rightarrow\; [\text{length}]^{-1} = [\text{mass}] = M. \tag{A.7}$$

This can also be seen via the formulae for Planck mass, length and time,

$$M_P = \sqrt{\frac{\hbar c}{G}}, \;\; \ell_P = \sqrt{\frac{\hbar G}{c^3}}, \;\; t_P = \sqrt{\frac{\hbar G}{c^5}}. \tag{A.8}$$

In these units,

$$[\rho] = M^4, \;\; [\Lambda] = M^2, \;\; [H] = [\varphi] = M, \tag{A.9}$$

where φ is a scalar field.

Units with $k_B = \hbar = c = 1$

If we further set Boltzmann's constant to unity, temperature has the same dimension as mass-energy:

$$k_B = \hbar = c = 1 \;\Rightarrow\; [\text{temperature}] = M. \tag{A.10}$$

In the simplified units, we have

$$\ell_P = t_P = M_P^{-1} = T_P^{-1}, \; G = M_P^{-2}, \tag{A.11}$$

where the Planck temperature is $T_P = M_P c^2 / k_B = 1.4 \times 10^{32}\,\text{K}$.

Then in these units, all physical quantities can be expressed in terms of one dimensionful quantity, for example mass-energy, using some chosen unit, for example eV (see Table A.1 for conversion factors).

A.2 1+3 covariant equations

Metric sign convention: $(-+++)$.
Tensor indices:
general basis: $\quad a, b, \cdots = 0, 1, 2, 3; \; i, j, \cdots = 1, 2, 3$
coordinate basis: $\mu, \nu, \cdots = 0, 1, 2, 3, \; i, j, \cdots = 1, 2, 3.$

Tensor symmetries:

$$S_{(ab)} = \tfrac{1}{2}\{S_{ab} + S_{ba}\}, \;\; S_{[ab]} = \tfrac{1}{2}\{S_{ab} - S_{ba}\}. \tag{A.12}$$

Alternating (volume) tensor:

$$\eta_{abcd} = -\sqrt{-g}\, \delta^0_{[a} \delta^1_{\;b} \delta^2_{\;c} \delta^3_{\;d]}. \tag{A.13}$$

Curvature tensors and covariant derivatives:

$$R^d{}_{abc} = \Gamma^d{}_{ac,b} - \Gamma^d{}_{ab,c} + \Gamma^d{}_{ec}\Gamma^e{}_{ab} - \Gamma^d{}_{eb}\Gamma^e{}_{ac}, \tag{A.14}$$

$$R_{ab} = R^c{}_{acb}, \quad R = R^a{}_a, \tag{A.15}$$

$$\nabla_a g_{bc} = 0, \quad 2\nabla_{[a}\nabla_{b]}W_c = R_{abcd}W^d, \tag{A.16}$$

where $\nabla_a(\cdots)$ is also written as $(\cdots)_{;a}$.

1+3 covariant split relative to u^a ($u^a u_a = -1$):

$$\dot{S}^{a\cdots}{}_{b\cdots} = u^e \nabla_e S^{a\cdots}{}_{b\cdots}, \tag{A.17}$$

$$\overline{\nabla}_c S^{a\cdots}{}_{b\cdots} = h_c{}^f h^a{}_d \cdots h_b{}^e \cdots \nabla_f S^{d\cdots}{}_{e\cdots}, \quad h_{ab} := g_{ab} + u_a u_b. \tag{A.18}$$

Projected symmetric tracefree (PSTF) parts:

$$V_{\langle a \rangle} = h_a{}^b V_b, \quad S_{\langle ab \rangle} = \left\{ h_{(a}{}^c h_{b)}{}^d - \tfrac{1}{3} h_{ab} h^{cd} \right\} S_{cd}. \tag{A.19}$$

Covariant spatial curl:

$$\text{curl } V_a = \eta_{abc}\overline{\nabla}^b V^c, \quad \text{curl } S_{ab} = \eta_{cd(a}\overline{\nabla}^c S_{b)}{}^d \quad \eta_{abc} := \eta_{abcd}u^d. \tag{A.20}$$

For a general spacetime and a general source, the 1+3 covariant equations are:
Evolution:

$$\dot{\rho} + (\rho + p)\Theta + \overline{\nabla}^a q_a = -2\dot{u}^a q_a - \sigma^{ab}\pi_{ab}, \tag{A.21}$$

$$\dot{\Theta} + \tfrac{1}{3}\Theta^2 + 4\pi G(\rho + 3p) - \overline{\nabla}^a \dot{u}_a = -\sigma_{ab}\sigma^{ab} + 2\omega_a\omega^a + \dot{u}_a\dot{u}^a, \tag{A.22}$$

$$\dot{q}_{\langle a \rangle} + \tfrac{4}{3}\Theta q_a + (\rho + p)\dot{u}_a + \overline{\nabla}_a p + \overline{\nabla}^b \pi_{ab} = -\sigma_{ab}q^b + \eta_{abc}\omega^b q^c - \dot{u}^b \pi_{ab}, \tag{A.23}$$

$$\dot{\omega}_{\langle a \rangle} + \tfrac{2}{3}\Theta\omega_a + \tfrac{1}{2}\text{curl}\,\dot{u}_a = \sigma_{ab}\omega^b, \tag{A.24}$$

$$\dot{\sigma}_{\langle ab \rangle} + \tfrac{2}{3}\Theta\sigma_{ab} + E_{ab} - 4\pi G\pi_{ab} - \overline{\nabla}_{\langle a}\dot{u}_{b \rangle} = -\sigma_{c\langle a}\sigma_{b \rangle}{}^c - \omega_{\langle a}\omega_{b \rangle} + \dot{u}_{\langle a}\dot{u}_{b \rangle}, \tag{A.25}$$

$$\dot{E}_{\langle ab \rangle} + \Theta E_{ab} - \text{curl}\,H_{ab} - 4\pi G\left[(\rho + p)\sigma_{ab} + \dot{\pi}_{\langle ab \rangle} + \tfrac{1}{4}\Theta\pi_{ab} + \overline{\nabla}_{\langle a}q_{b \rangle} \right]$$

$$= -8\pi G\dot{u}_{\langle a}q_{b \rangle} + 2\dot{u}^c \eta_{cd(a}H_{b)}{}^d + 3\sigma_{c\langle a}E_{b \rangle}{}^c$$

$$- \omega^c \eta_{cd(a}E_{b)}{}^d - 4\pi G\left(\sigma^c{}_{\langle a}\pi_{b)c} - \omega^c \eta_{cd(a}\pi_{b)}{}^d \right), \tag{A.26}$$

$$\dot{H}_{\langle ab \rangle} + \Theta H_{ab} + \text{curl}\,E_{ab} - 4\pi G\text{curl}\,\pi_{ab} = 3\sigma_{c\langle a}H_{b \rangle}{}^c - \omega^c \eta_{cd(a}H_{b)}{}^d$$

$$- 2\dot{u}^c \eta_{cd(a}E_{b)}{}^c + 4\pi G\left(\sigma^c{}_{(a}\eta_{b)cd}q^d - 3\omega_{\langle a}q_{b \rangle} \right). \tag{A.27}$$

Constraint:

$$\overline{\nabla}^a \omega_a = \dot{u}^a \omega_a, \tag{A.28}$$

$$\overline{\nabla}^b \sigma_{ab} - \text{curl}\,\omega_a - \tfrac{2}{3}\overline{\nabla}_a\Theta + 8\pi G q_a = -2\eta_{abc}\omega^b \dot{u}^c, \tag{A.29}$$

$$\text{curl}\,\sigma_{ab} + \overline{\nabla}_{\langle a}\omega_{b \rangle} - H_{ab} = -2\dot{u}_{\langle a}\omega_{b \rangle}, \tag{A.30}$$

$$\overline{\nabla}^b E_{ab} + \frac{4\pi G}{3}\left(\overline{\nabla}^b \pi_{ab} - 2\overline{\nabla}_a\rho + 2\Theta q_a \right)$$

$$= \eta_{abc}\sigma^b{}_d H^{cd} - 3H_{ab}\omega^b + 4\pi G\left(\sigma_{ab} + 3\eta_{abc}\omega^c \right)q^b, \tag{A.31}$$

$$\overline{\nabla}^b H_{ab} + 4\pi G \left[\operatorname{curl} q_a - 2(\rho + p)\omega_a \right]$$

$$= 3E_{ab}\omega^b - \eta_{abc}\sigma^b{}_d E^{cd} - 4\pi G \left(\eta_{abc}\sigma^b{}_d \pi^{cd} + \pi_{ab}\omega^b \right), \quad (A.32)$$

$$^3R_{\langle ab \rangle} = E_{ab} + 4\pi G \pi_{ab} - \tfrac{1}{3}\Theta \left(\sigma_{ab} + \omega_{ab} \right) + \sigma_{c\langle a}\sigma_{b\rangle}{}^c + \omega_{c\langle a}\omega_{b\rangle}{}^c - 2\sigma_{c[a}\omega_{b]}{}^c, \quad (A.33)$$

$$^3R = 16\pi G \rho - \tfrac{2}{3}\Theta^2 + \sigma_{ab}\sigma^{ab} - \omega_a\omega^a. \quad (A.34)$$

Here ρ, p, q_a, π_{ab} are the total quantities (including a possible Λ term). In (A.33), (A.34), $^3R_{ab} = (R_{ab})_\perp - \Theta\overline{\nabla}_b u_a + \overline{\nabla}_c u_a \overline{\nabla}^c u_b$, and this reduces to the 3-Ricci tensor when $\omega_a = 0$.

A.3 Frequently used acronyms

Frequently used acronyms are listed in Table A.2.

Table A.2 Frequently used acronyms	
Acronym	Meaning
2dF(GRS)	Two-degree Field (Galaxy Redshift Survey)
AGN	Active Galactic Nucleus
BAO	Baryon Acoustic Oscillations
BBN	Big Bang Nucleosynthesis
CCD	Charge Coupled Device
(Λ)CDM	(Lambda) Cold Dark Matter
CMB	Cosmic Microwave Background
COBE	Cosmic Background Explorer (satellite)
DES	Dark Energy Survey
EFE	Einstein's Field Equations
FLRW	Friedmann–Lemaître–Robertson–Walker
GR	General Relativity
GUT	Grand Unified Theory
HBB	Hot Big Bang
ISW	Integrated Sachs–Wolfe (effect)
LG	Local Group (of Galaxies)
LSS	Last Scattering Surface
LTB	Lemaître–Tolman–Bondi
NGT	Newtonian Gravity Theory
QSO	Quasi-Stellar Object (quasar)
RW	Robertson–Walker
SDSS	Sloan Digital Sky Survey
SKA	Square Kilometre Array
SNIa	Supernovae of type Ia
VLBI	Very Long Baseline Interferometry
WMAP	Wilkinson Microwave Anisotropy Probe (satellite)

References

Abbasi, R.U., Abu-Zayyad, T., Amman, J.F., Archbold, G., Belov, K. *et al.* (2008). First observation of the Greisen-Zatsepin-Kuzmin suppression, *Phys. Rev. Lett.* **100**, 101101. arXiv:astro-ph/0703099.

Abbott, L.F. and Schaefer, R.K. (1986). A general, gauge-invariant analysis of the cosmic microwave anisotropy, *Astrophys. J.* **308**, 546.

Abraham, J., Abreu, P., Aglietta, M., Aguirre, C., Allard, D. *et al.* (2008). Observation of the suppression of the flux of cosmic rays above 4×10^{19} eV, *Phys. Rev. Lett.* **101**, 061101. arXiv:0806.4302.

Afshordi, N., Slosar, A. and Wang, Y. (2011). A theory of a spot, *J. Cosmol. Astropart. Phys.* **01**, 019. arXiv:1006.5021.

Agacy, R.L. (1997). *Generalized Kronecker, permanent delta and Young tableaux applications to tensors and spinors: Lanczos-Zund spinor classification and general spinor factorizations*, University of London PhD thesis.

Aguirre, A. and Gratton, S. (2003). Inflation without a beginning: A null boundary proposal, *Phys. Rev. D* **67**, 083515.

Aguirregabiria, J.M., Feinstein, A., and Ibañez, J. (1993). Exponential-potential scalar field universes II: the inhomogeneous models, *Phys. Rev. D* **48**, 4669. arXiv:gr-qc/9309014.

Aguirregabiria, J.M., Labraga, P., and Lazkoz, R. (2002). Assisted inflation in Bianchi VI_0 cosmologies, *Gen. Rel. Grav.* **34**, 341. arXiv:gr-qc/0107009.

Alexander, S., Biswas, T., Notari, A., and Vaid, D. (2009). Local void vs dark energy: Confrontation with WMAP and type Ia supernovae, *J. Cosmol. Astropart. Phys.* **0909**, 025. arXiv:0712.0370.

Aliev, B.N. and Leznov, A.N. (1992). Einstein's vacuum fields with non-Abelian group of motions $G_2 II$, *Class. Quant. Grav.* **9**, 1261.

Allnutt, J.A. (1981). A Petrov type-III perfect fluid solution of Einstein's equations, *Gen. Rel. Grav.* **13**, 1017.

Alnes, H., Amarzguioui, M., and Grøn, Ø. (2006). An inhomogeneous alternative to dark energy?, *Phys. Rev. D* **73**, 083519. arXiv:astro-ph/0512006.

Amarzguioui, M. and Grøn, Ø. (2005). Entropy of gravitationally collapsing matter in FRW universe models, *Phys. Rev. D* **71**, 083011.

Amendola, L., Kainulainen, K., Marra, V., and Quartin, M. (2010). Large-scale inhomogeneities may improve the cosmic concordance of supernovae, *Phys. Rev. Lett.* **105**, 121302. arXiv:1002.1232.

Amendola, L. and Tsujikawa, S. (2010). *Dark Energy: Theory and Observations* (Cambridge University Press, Cambridge).

Anderson, E., Barbour, J., Foster, B.Z., Kelleher, B., and Ó Murchadha, N. (2005). The physical gravitational degrees of freedom, *Class. Quant. Grav.* **22**, 1795.

Andersson, L., van Elst, H., Lim, W.C., and Uggla, C. (2005). Asymptotic silence of generic cosmological singularities, *Phys. Rev. Lett.* **94**, 051101. arXiv:gr-qc/0402051.

Andersson, L., van Elst, H., and Uggla, C. (2004). Gowdy phenomenology in scale-invariant variables, *Class. Quant. Grav.* **21**, S29. Special number 'A spacetime safari: papers in honour of Vincent Moncrief'.

Anninos, P., Matzner, R.A., Rothman, T., and Ryan, Jr., M.P. (1991). How does inflation isotropize the universe?, *Phys. Rev. D* **43**, 3821.

Antoniou, I. and Perivolaropoulos, L. (2010). Searching for a cosmological preferred axis: Union2 data analysis and comparison with other probes, *J. Cosmol. Astropart. Phys.* **12**, 012. arXiv:1007.4347.

Apostolopoulos, P.S. (2003). Self-similar Bianchi models I: Class A models, *Class. Quant. Grav.* **20**, 3371.

Apostolopoulos, P.S. (2005). Self-similar Bianchi models II: Class B models, *Class. Quant. Grav.* **22**, 323.

Apostolopoulos, P.S. and Carot, J. (2007). Uniqueness of Petrov type D spatially inhomogeneous irrotational silent models, *Int. J. Mod. Phys. A* **22**, 1983. arXiv:gr-qc/0605130.

Araujo, M.E. (1999). Exact spherically symmetric dust solution of the field equations in observational coordinates with cosmological data functions, *Phys. Rev. D* **60**, 104020. Erratum: *Phys. Rev. D* **64**, 049002 (2001).

Araujo, M.E., Stoeger, W.R., Arcuri, R.C., and Bedran, M.L. (2008). Solving Einstein field equations in observational coordinates with cosmological data functions: Spherically symmetric universes with cosmological constant, *Phys. Rev. D* **78**, 063513. arXiv:0807.4193.

Arnau, J.V., Fullana, M., and Sáez, D. (1994). Great Attractor-like structures and large scale anisotropy, *Mon. Not. Roy. Astr. Soc.* **268**, L17.

Arnowitt, R., Deser, S., and Misner, C.W. (1962). The dynamics of general relativity, in *Gravitation: An Introduction to Current Research*, ed. L. Witten (Wiley, New York and London), p. 227. Reprinted in *Gen. Rel. Grav.* **40**, 1997 (2008).

Arp, H. and Carosati, D. (2007). M31 and local group QSO's. arXiv:0706.3154.

Ashtekar, A. (2009a). Loop quantum cosmology: an overview, *Gen. Rel. Grav.* **41**, 707. arXiv:0812.0177.

Ashtekar, A. (2009b). Singularity resolution in loop quantum cosmology: A brief overview, *J. Phys.: Conf. Ser.* **189**, 012003. arXiv:0812.4703.

Ashtekar, A. and Lewandowski, J. (2004). Background independent quantum gravity: a status report, *Class. Quant. Grav.* **21**, R53.

Aurich, R., Lustig, S., and Steiner, F. (2005). CMB anisotropy of the Poincaré dodecahedron, *Class. Quant. Grav.* **22**, 2061. arXiv:astro-ph/0412569.

Auslander, L. and MacKenzie, R.E. (1963). *Introduction to Differentiable Manifolds* (McGraw Hill, New York).

Bahcall, N.A., Ostriker, J.P., Perlmutter, S., and Steinhardt, P.J. (1999). The cosmic triangle: revealing the state of the universe, *Science* **284**, 1481. arXiv:astro-ph/9906463.

Bajtlik, S., Juszkiewicz, R., Prozszynski, M., and Amsterdamski, P. (1986). 2.7 K radiation and the isotropy of the universe, *Astrophys. J.* **300**, 463.

Balakin, A.B. and Ni, W.-T. (2010). Non-minimal coupling of photons and axions, *Class. Quant. Grav.* **27**, 055003. arXiv:0911.2946.

Balashov, Y.V. (1991). Resource letter AP-1: The anthropic principle, *Amer. J. Phys.* **54**, 1069.

Balbi, A. (2010). The limits of cosmology. arXiv:1001.4016.

Barbour, J. (1999). *The End of Time: The Next Revolution in our Understanding of the Universe* (Weidenfeld and Nicholson, London).

Barbour, J. and Pfister, H. (1995). *Mach's Principle: from Newton's Bucket to Quantum Gravity* (Birkhauser, Basel and Boston).

Bardeen, J.M. (1980). Gauge-invariant cosmological perturbations, *Phys. Rev. D* **22**, 1882.

Barkana, R. and Loeb, A. (2007). The physics and early history of the intergalactic medium, *Rep. Prog. Phys.* **70**, 627. arXiv:astro-ph/0611541.

Barnes, A. (1973). On shearfree normal flows of a perfect fluid, *Gen. Rel. Grav.* **4**, 105.

Barnes, A. and Rowlingson, R.R. (1989). Irrotational perfect fluids with a purely electric Weyl tensor, *Class. Quant. Grav.* **6**, 949.

Barrabes, C. (1989). Singular hypersurfaces in general relativity: a unified description, *Class. Quant. Grav.* **6**, 581.

Barrabes, C. and Israel, W. (1991). Thin shells in general relativity and cosmology: the lightlike limit, *Phys. Rev. D* **43**, 1129.

Barrow, J.D. (1976). Light elements and the isotropy of the universe, *Mon. Not. Roy. Astr. Soc.* **175**, 359.

Barrow, J.D. (1993). New types of inflationary universe, *Phys. Rev. D* **48**, 1585.

Barrow, J.D. (1997). Cosmological limits on slightly skew stresses, *Phys. Rev. D* **55**, 7451. arXiv: gr-qc/9701038.

Barrow, J.D. (2002). *The Constants of Nature* (Jonathan Cape, London).

Barrow, J.D. and Hervik, S. (2002). The Weyl tensor in spatially homogeneous cosmological models, *Class. Quant. Grav.* **19**, 5173. arXiv:gr-qc/0206061.

Barrow, J.D., Juszkiewicz, R., and Sonoda, D. (1983). The structure of the cosmic microwave background, *Nature* **305**, 397.

Barrow, J.D., Juszkiewicz, R., and Sonoda, D.H. (1985). Universal rotation - How large can it be?, *Mon. Not. Roy. Astr. Soc.* **213**, 917.

Barrow, J.D. and Lip, S.Z.W. (2009). Classical stability of sudden and big rip singularities, *Phys. Rev. D* **80**, 043518.

Barrow, J.D., Maartens, R., and Tsagas, C.G. (2007). Cosmology with inhomogeneous magnetic fields, *Phys. Reports* **449**, 131. arXiv:astro-ph/0611537.

Barrow, J.D. and Stein-Schabes, J. (1984). Inhomogeneous cosmologies with cosmological constant, *Phys. Lett. A* **103**, 315.

Barrow, J.D. and Tipler, F.J. (1979). An analysis of the generic singularity studies by Belinskii, Lifshitz and Khalatnikov, *Phys. Reports* **56**, 371.

Barrow, J.D. and Tipler, F.J. (1984). *The Cosmological Anthropic Principle* (Oxford University Press, Oxford).

Barrow, J.D. and Tsagas, C.G. (2005). New isotropic and anisotropic sudden singularities, *Class. Quantum Grav.* **22**, 1563.

Barrow, J.D. and Tsagas, C.G. (2007). Averaging anisotropic cosmologies, *Class. Quant. Grav.* **24**, 1023. arXiv:gr-qc/0609078.

Bartolo, N., Komatsu, E., Matarrese, S., and Riotto, A. (2004). Non-Gaussianity from inflation: theory and observations, *Phys. Reports* **402**, 103. arXiv:astro-ph/0406398.

Bassett, B.A.C.C. and Hlozek, R. (2010). Baryon acoustic oscillations, in Ruiz-Lapuente (2010). arXiv:0910.5224.

Bassett, B.A.C.C. and Kunz, M. (2004). Cosmic distance-duality as a probe of exotic physics and acceleration, *Phys. Rev. D* **69**, 101305. arXiv:astro-ph/0312443.

Bassett, B.A.C.C., Liberati, S., Molina-Paris, C., and Visser, M. (2000). Geometrodynamics of variable-speed-of-light cosmologies, *Phys. Rev. D* **62**, 10351. arXiv:astro-ph/0001441.

Bassett, B.A.C.C., Tsujikawa, S., and Wands, D. (2006). Inflation dynamics and reheating, *Rev. Mod. Phys.* **78**, 537. arXiv:astro-ph/0507632.

Batchelor, G.K. (1967). *An Introduction to Fluid Dynamics* (Cambridge University Press, Cambridge).

Baum, L. and Frampton, P.H. (2007). Turnaround in cyclic cosmology, *Phys. Rev. Lett.* **98**, 071301. arXiv:hep-th/0610213.

Baumann, D. (2009). TASI lectures on inflation. arXiv:0907.5424.

Baumann, D. and McAllister, L. (2009). Advances in inflation in string theory, *Ann. Rev. Nucl. Part. Sci.* **59**, 67. arXiv:0901.0265.

Baumann, D., Nicolis, A., Senatore, L., and Zaldarriaga, M. (2010). Cosmological non-linearities as an effective fluid. arXiv:1004.2488.

Bean, R. (2010). TASI lectures on cosmic acceleration. arXiv:1003.4468.

Bean, R. and Tangmatitham, M. (2010). Current constraints on the cosmic growth history, *Phys. Rev. D* **81**, 083534. arXiv:1002.4197.

Béguin, F. (2010). Aperiodic oscillatory asymptotic behavior for some Bianchi spacetimes, *Class. Quant. Grav.* **27**, 185005. arXiv:1004.2984.

Behrend, J., Brown, I.A., and Robbers, G. (2008). Cosmological backreaction from perturbations, *J. Cosmol. Astropart. Phys.* **0801**, 013. arXiv:0710.4964.

Bekenstein, J.D. (2010). Alternatives to dark matter: Modified gravity as an alternative to dark matter, in *Particle Dark Matter: Observations, Models and Searches,* ed. G. Bertone (Cambridge University Press, Cambridge), chapter 6, p. 95. arXiv:1001.3876.

Belinski, V.A. (2009). Cosmological singularity, in *The Sun, the Stars, the Universe and General Relativity: International Conference in Honor of Ya.B. Zel'dovich's 95th Anniversary*, ed. R. Ruffini and G. Vereshchagin, volume 1205 of *AIP Conference Proceedings* (American Institute of Physics, Melville, NY), p. 17. arXiv:0910.0374.

Belinski, V.A., Khalatnikov, I.M., and Lifshitz, E.M. (1982). A general solution of the Einstein equations with a time singularity, *Adv. Phys.* **31**, 639.

Belinski, V.A., Lifshitz, E.M., and Khalatnikov, I.M. (1971). The oscillatory regime of approach to an initial singularity in relativistic cosmology, *Sov. Phys. Uspekhi* **13**, 475. See also *Adv. Phys.* **19**, 525 (1970).

Belinski, V.A., Lifshitz, E.M., and Khalatnikov, I.M. (1972). Construction of a general cosmological solution of the Einstein equations with a time singularity, *Sov. Phys. JETP* **35**, 838.

Belinski, V.A. and Verdaguer, E. (2001). *Gravitational Solitons* (Cambridge University Press, Cambridge).

Belinski, V.A. and Zakharov, V.E. (1978). Integration of the Einstein equations by the inverse scattering method and calculation of exact soliton solutions, *Sov. Phys. JETP* **48**, 985.

Bennett, C.L., Banday, A.J., Górski, K.M., Hinshaw, G., Jackson, P. *et al.* (1996). Four-year COBE DMR cosmic microwave background observations: Maps and basic results, *Astrophys. J.* **464**, L1.

Berestetskii, V.B., Lifshitz, E.M., and Pitaevskii, L.P. (1982). *Quantum electrodynamics, 2nd English Edition*, Course of theoretical physics - Pergamon International Library of Science, Technology, Engineering and Social Studies (Pergamon Press, Oxford).

Berger, B. and Moncrief, V. (1993). Numerical investigation of cosmological singularities, *Phys. Rev. D* **48**, 4676.

Bernstein, J. (1988). *Kinetic Theory in the Expanding Universe* (Cambridge University Press, Cambridge and New York).

Bertotti, B. (1966). The luminosity of distant galaxies, *Proc. Roy. Soc. London A* **294**, 195.

Bertschinger, E. (1992). Large scale structure and motions: Linear theory and statistics, in *Current topics in Astrofundamental Physics*, ed. N. Sanchez and A. Zichichi (World Scientific, Singapore).

Bertschinger, E. (2006). On the growth of perturbations as a test of dark energy and gravity, *Astrophys. J.* **648**, 797. arXiv:astro-ph/0604485.

Bertschinger, E., Dekel, A., Faber, S.M., Dressler, A., and Burstein, D. (1990). Potential, velocity and density fields from redshift-distance samples. Application: cosmography within 6000 kilometers per second, *Astrophys. J.* **364**, 370.

Betschart, G., Dunsby, P.K.S., and Marklund, M. (2004). Cosmic magnetic fields from velocity perturbations in the early universe, *Class. Quant. Grav.* **21**, 2115. arXiv:gr-qc/0310085.

Birkinshaw, M. (1999). The Sunyaev-Zel'dovich effect, *Phys. Reports* **310**, 97. arXiv:astro-ph/9808050.

Biswas, T., Mansouri, R., and Notari, A. (2007). Nonlinear structure formation and 'apparent' acceleration: an investigation, *J. Cosmol. Astropart. Phys.* **0712**, 017. arXiv:astro-ph/0606703.

Biswas, T. and Notari, A. (2008). "Swiss-cheese" inhomogeneous cosmology and the dark energy problem, *J. Cosmol. Astropart. Phys.* **0806**, 021. arXiv:astro-ph/0702555.

Biswas, T., Notari, A. and Valkenburg, W. (2010). Testing the void against cosmological data: Fitting CMB, BAO, SN and H0, *J. Cosmol. Astropart. Phys.* **11**, 030. arXiv:1007.3065.

Blake, C. and Wall, J. (2002). A velocity dipole in the distribution of radio galaxies, *Nature* **416**, 150. arXiv:astro-ph/0203385.

Bogoyavlenskii, O.I. (1985). *Methods of the Qualitative Theory of Dynamical Systems in Astrophysics and Gas Dynamics* (Springer-Verlag, Berlin and Heidelberg). Russian original published by Nauka, Moscow, 1980.

Bogoyavlenskii, O.I. and Novikov, S.P. (1973). Singularities of the cosmological model of the Bianchi type IX according to the qualitative theory of differential equations, *Sov. Phys. JETP* **37**, 747.

Bojowald, M. (2005). Loop quantum cosmology, *Living Rev. Relativity* http://relativity.livingreviews.org/Articles/lrr-2008-4/, arXiv:gr-qc/0601085.

Bojowald, M., Kiefer, C., and Vargas Moniz, P. (2010). Quantum cosmology for the 21st century: A debate. arXiv:1005.2471.

Bolejko, K. (2006a). Radiation in the process of the formation of voids, *Mon. Not. Roy. Astr. Soc.* **370**, 924. arXiv:astro-ph/0503356.

Bolejko, K. (2006b). Structure formation in the quasispherical Szekeres model, *Phys. Rev. D* **73**, 123508. arXiv:astro-ph/0604490.

Bolejko, K. (2007). Evolution of cosmic structures in different environments in the quasispherical Szekeres model, *Phys. Rev. D* **75**, 043508. arXiv:astro-ph/0610292.

Bolejko, K. (2009). The Szekeres Swiss cheese model and the CMB observations, *Gen. Rel. Grav.* **41**, 1737. arXiv:0804.1846.

Bolejko, K. and Célérier, M.-N. (2010). Szekeres Swiss-cheese model and supernova observations, *Phys. Rev. D* **82**, 103510. arXiv:1005.2584.

Bolejko, K. and Hellaby, C. (2008). The Great Attractor and the Shapley Concentration, *Gen. Rel. Grav.* **40**, 1771. arXiv:astro-ph/0604402.

Bolejko, K., Hellaby, C., and Krasiński, A. (2005). Formation of voids in the universe within the Lemaître-Tolman model, *Mon. Not. Roy. Astr. Soc.* **362**, 213. arXiv:gr-qc/ 0411126.

Bolejko, K., Krasiński, A., Hellaby, C., and Célérier, M.-N. (2010). *Structures in the Universe by Exact Methods: Formation, Evolution, Interactions* (Cambridge University Press, Cambridge).

Bolejko, K. and Stoeger, W.R. (2010). Conditions for spontaneous homogenization of the universe, *Gen. Rel. Grav.* **42**, 2349. Fifth award in the 2010 Gravity Research Foundation essay competition, arXiv:1005.3009.

Bond, J.R. and Efstathiou, G. (1984). Cosmic background radiation anisotropies in universes dominated by nonbaryonic dark matter, *Astrophys. J. Lett.* **285**, L45.

Bondi, H. (1947). Spherically symmetrical models in general relativity, *Mon. Not. Roy. Astr. Soc.* **107**, 410. Reprinted as *Gen. Rel. Grav.* **31**, 1777–1781 (1999).

Bondi, H. (1960). *Cosmology* (Cambridge University Press, Cambridge).

Bonnor, W.B. (1956). The formation of the nebulae, *Z. Astrophys.* **39**, 143. Reprinted as *Gen. Rel. Grav.* **30** 1113–1132 (1998).

Bonnor, W.B. (1972). A non-uniform relativistic cosmological model, *Mon. Not. Roy. Astr. Soc.* **159**, 261.

Bonnor, W.B. (1974). Evolution of inhomogeneous cosmological models, *Mon. Not. Roy. Astr. Soc.* **167**, 55.

Bonnor, W.B. (1976). Non-radiative solutions of Einstein's equations for dust, *Commun. Math. Phys.* **51**, 191.

Bonnor, W.B. (2000). A generalization of the Einstein-Straus vacuole, *Class. Quant. Grav.* **17**, 2739.

Bonnor, W.B. and Chamorro, A. (1990). Models of voids in the expanding universe, *Astrophys. J.* **361**, 21.

Bonnor, W.B. and Chamorro, A. (1991). Models of voids in the expanding universe II, *Astrophys. J.* **378**, 461.

Bonnor, W.B. and Ellis, G.F.R. (1986). Observational homogeneity of the universe, *Mon. Not. Roy. Astr. Soc.* **218**, 605.

Bonnor, W.B. and Vickers, P.A. (1981). Junction conditions in general relativity, *Gen. Rel. Grav.* **13**, 29.

Bonvin, C. and Durrer, R. (2011). What galaxy surveys really measure, *Phys. Rev. D* **84**, 063505. arXiv:1105.5280.

Bothun, G. (1998). *Modern Cosmological Observations* (Taylor and Francis, London).

Boughn, S. and Crittenden, R. (2004). A correlation between the cosmic microwave background and large-scale structure in the Universe, *Nature* **427**, 45.

Boylan-Kolchin, M., Springel, V., White, S.D.M., Jenkins, A., and Lemson, G. (2009). Resolving cosmic structure formation with the Millennium-II simulation, *Mon. Not. Roy. Astr. Soc.* **398**, 1150. arXiv:0903.3041.

Bozza, V. (2010). Gravitational lensing by black holes. p. 2269 in Jetzer, Mellier and Perlick (2010), arXiv:0911.2187.

Brandenberger, R. (1985). Quantum field theory methods and inflationary universe models, *Rev. Mod. Phys.* **57**, 1.

Brandenberger, R., Laflamme, R., and Mijic, M. (1991). Classical perturbations from decoherence of quantum fluctuations in the inflationary universe, *Physica Scripta* **T36**, 265.

Brauer, U. (1991). Existence of finitely-perturbed Friedmann models via the Cauchy problem, *Class. Quant. Grav.* **8**, 1283.

Brax, P. and van de Bruck, C. (2003). Cosmology and brane worlds: a review, *Class. Quant. Grav.* **20**, R201. arXiv:hep-th/0303095.

Brickell, F. and Clark, R.S. (1970). *Differentiable Manifolds: An Introduction* (Van Nostrand Reinhold, London).

Bridle, S., Balan, S.T., Bethge, M., Gentile, M., Harmeling, S. *et al.* (2010). Results of the GREAT08 challenge: An image analysis competition for cosmological lensing, *Mon. Not. Roy. Astr. Soc.* **405**, 1044. arXiv:0908.0945.

Brightwell, G., Dowker, H.F., Garcia, R.S., Henson, J., and Sorkin, R.D. (2003). "Observables" in causal set cosmology, *Phys. Rev. D* **67**, 084031. arXiv:gr-qc/0210061.

Brill, D., Reula, O., and Schmidt, B. (1987). Local linearization stability, *J. Math. Phys.* **28**, 1844.

Brill, D.R. and Vishveshwara, C.V. (1986). Joint linearization instabilities in general relativity, *J. Math. Phys.* **27**, 1813.

Brown, I.A., Behrend, J., and Malik, K.A. (2009). Gauges and cosmological backreaction, *J. Cosmol. Astropart. Phys.* **0911**, 027. arXiv:0903.3264.

Brown, I.A., Robbers, G., and Behrend, J. (2009). Averaging Robertson-Walker cosmologies, *J. Cosmol. Astropart. Phys.* **0904**, 016. arXiv:0811.4495.

Bruni, M., Crittenden, R., Koyama, K., Maartens, R., Pitrou, C. *et al.* (2011). Disentangling non-Gaussianity, bias and GR effects in the galaxy distribution. arXiv:1106.3999.

Bruni, M., Dunsby, P.K.S., and Ellis, G.F.R. (1992). Cosmological perturbations and the physical meaning of gauge-invariant variables, *Astrophys. J.* **395**, 34.

Bruni, M., Matarrese, S., and Pantano, O. (1995). Dynamics of silent universes, *Astrophys. J.* **445**, 958. arXiv:astro-ph/9406068.

Bruni, M. and Sopuerta, C. (2003). Covariant fluid dynamics: a long wave-length approximation, *Class. Quant. Grav.* **20**, 5275. arXiv:gr-qc/0307059.

Bucher, M., Goldhaber, A.S., and Turok, N. (1995). Open universe from inflation, *Phys. Rev. D* **52**, 3314. arXiv:hep-ph/9411206.

Buchert, T. (2000). On average properties of inhomogeneous fluids in general relativity: Dust cosmologies, *Gen. Rel. Grav.* **32**, 105. arXiv:gr-qc/9906015.

Buchert, T. (2001). On average properties of inhomogeneous fluids in general relativity: Perfect fluid cosmologies, *Gen. Rel. Grav.* **33**, 1381. arXiv:gr-qc/0102049.

Buchert, T. (2008). Dark Energy from structure: a status report, *Gen. Rel. Grav.* **40**, 467. arXiv:0707.2153.

Buchert, T. and Carfora, M. (2002). Regional averaging and scaling in relativistic cosmology, *Class. Quant. Grav.* **19**, 6109.

Buchert, T. and Carfora, M. (2003). Cosmological parameters are dressed, *Phys. Rev. Lett.* **90**, 031101.

Buchert, T. and Carfora, M. (2008). On the curvature of the present-day universe, *Class. Quant. Grav.* **25**, 195001. arXiv:0803.1401.

Buchert, T., Kerscher, M., and Sicka, C. (2000). Backreaction of inhomogeneities on the expansion: the evolution of cosmological parameters, *Phys. Rev. D* **62**, 043525.

Bugalho, M.H. (1987). Orthogonality transitivity and cosmologies with a non-Abelian two-parameter isometry group, *Class. Quant. Grav.* **4**, 1043.

Bull, P., Clifton, T. and Ferreira, P.G. (2011). The kSZ effect as a test of general radial inhomogeneity in LTB cosmology. arXiv:1108.2222.

Bunn, E.F., Ferreira, P.G., and Silk, J. (1996). How anisotropic is our Universe?, *Phys. Rev. Lett.* **77**, 2883. arXiv:astro-ph/9605123.

Burigana, C. and Salvaterra, R. (2003). What can we learn on the thermal history of the Universe from future cosmic microwave background spectrum measurements at long wavelengths?, *Mon. Not. Roy. Astr. Soc.* **342**, 543. arXiv:astro-ph/0301133.

Burstein, D., Faber, S.M., and Dressler, A. (1990). Evidence for the motions of galaxies for a large-scale large amplitude flow toward the Great Attractor, *Astrophys. J.* **354**, 18.

Byland, S. and Scialom, D. (1998). Evolution of the Bianchi type I, Bianchi type III, and the Kantowski-Sachs universe: isotropization and inflation, *Phys. Rev. D* **57**, 6065. arXiv:gr-qc/9802043.

Caillerie, S., Lachièze-Rey, M., Luminet, J.-P., Lehoucq, R., Riazuelo, A. *et al.* (2007). A new analysis of Poincaré dodecahedral space model, *Astron. Astrophys.* **476**, 691. arXiv:0705.0217.

Caldwell, R. and Stebbins, A. (2008). A test of the Copernican principle, *Phys. Rev. Lett.* **100**, 191302. arXiv:0711.3459.

Calogero, S. and Heinzle, J.M. (2009). Dynamics of Bianchi type I solutions of the Einstein equations with anisotropic matter, *Ann. Inst. H. Poincaré* **10**, 225. arXiv:0809.1008.

Calogero, S. and Heinzle, J.M. (2010). On closed cosmological models that satisfy the strong energy condition but do not recollapse, *Phys. Rev. D* **81**, 023520. arXiv:1002.1913.

Calzetta, E. and Sakellariadou, M. (1992). Inflation in inhomogeneous cosmology, *Phys. Rev. D* **45**, 2802.

Capozziello, S. and Francaviglia, M. (2008). Extended theories of gravity and their cosmological and astrophysical applications, *Gen. Rel. Grav.* **40**, 357. arXiv:0706.1146.

Cardoso, A., Hiramatsu, T., Koyama, K., and Seahra, S.S. (2007). Scalar perturbations in braneworld cosmology, *J. Cosmol. Astropart. Phys.* **0707**, 008. arXiv:0705.1685.

Cardoso, A., Koyama, K., Seahra, S.S., and Silva, F.P. (2008). Cosmological perturbations in the DGP braneworld: Numeric solution, *Phys. Rev. D* **77**, 083512. arXiv:0711.2563.

Carfora, M. and Piotrkowska, K. (1995). Renormalization group approach to relativistic cosmology, *Phys. Rev. D* **52**, 4393.

Carr, B.J. and Coley, A.A. (1999). Self-similarity in general relativity, *Class. Quant. Grav.* **16**, R31.

Carr, B.J. and Coley, A.A. (2005). The similarity hypothesis in general relativity, *Gen. Rel. Grav.* **37**, 2165.

Carr, B.J., Coley, A.A., Goliath, M., Nilsson, U.S., and Uggla, C. (2001). The state space and physical interpretation of self-similar spherically symmetric perfect fluid models, *Class. Quant. Grav.* **18**, 303.

Carr, B.J. and Ellis, G.F.R. (2008). Universe or multiverse?, *Astronomy and Geophysics* **49**, 2.29.

Carr, B.J. and Hawking, S.W. (1974). Black holes in the early universe, *Mon. Not. Roy. Astr. Soc.* **168**, 399.

Carr, B.J., Kohri, K., Sendouda, Y., and Yokoyama, J. (2010). New cosmological constraints on primordial black holes, *Phys. Rev. D* **81**, 104019. arXiv:0912.5297.

Carr, B.J. and Koutras, A. (1993). Self-similar perturbations of a Kantowski-Sachs model, *Astrophys. J.* **405**, 34.

Carr, B.J. and Rees, M.J. (1979). The anthropic principle and the structure of the physical world, *Nature* **278**, 605.

Carr, B.J. and Yahil, A. (1990). Self-similar perturbations of a Friedmann universe, *Astrophys. J.* **360**, 330.

Carroll, S.M. (2004). *Spacetime and Geometry: an Introduction to General Relativity* (Addison-Wesley, San Francisco).

Carroll, S.M. (2010). *From Eternity to Here: The Quest for the Ultimate Theory of Time* (Dutton, New York).

Carroll, S.M. and Chen, J. (2005). Does inflation provide natural initial conditions for the universe?, *Gen. Rel. Grav.* **37**, 1671. arXiv:gr-qc/0505037.

Carroll, S.M. and Tam, H. (2010). Unitary evolution and cosmological fine-tuning. arXiv:1007.1417.

Carter, B. and Henriksen, R.N. (1989). A covariant characterisation of kinematic self-similarity, *Ann. de Physique* **14**(colloq. 1), 47.

Cattoën, C. and Visser, M. (2005). Necessary and sufficient conditions for big bangs, bounces, crunches, rips, sudden singularities and extremality events, *Class. Quant. Grav.* **22**, 4913. arXiv:gr-qc/0508045.

Cavaglià, M. (2003). Black hole and brane production in TeV gravity: A review, *Int. J. Mod. Phys. A* **18**, 1843. arXiv:hep-ph/0210296.

Centrella, J. and Matzner, R.A. (1982). Colliding gravitational waves in expanding cosmologies, *Phys. Rev. D* **25**, 930.

Challinor, A. (2000a). Microwave background polarization in cosmological models, *Phys. Rev. D* **62**, 043004. arXiv:astro-ph/9911481.

Challinor, A. (2000b). The covariant perturbative approach to cosmic microwave background anisotropies, *Gen. Rel. Grav.* **32**, 1059. arXiv:astro-ph/9903283.

Challinor, A. and Lasenby, A. (1998). Covariant and gauge-invariant analysis of cosmic microwave background anisotropies from scalar perturbations, *Phys. Rev. D* **58**, 023001. arXiv:astro-ph/9804150.

Challinor, A. and Lasenby, A. (1999). Cosmic microwave background anisotropies in the cold dark matter model, *Astrophys. J.* **513**, 1. arXiv:astro-ph/9804301.

Challinor, A. and Lewis, A. (2011). Linear power spectrum of observed source number counts, *Phys. Rev. D* **84**, 043516. arXiv:1105.5292.

Chamorro, A. (1991). Models of voids in elliptic universes, *Astrophys. J.* **383**, 51.

Chandrasekhar, S. (1960). *Radiative transfer* (Dover, New York).

Charmousis, C., Gregory, R., Kaloper, N., and Padilla, A. (2006). DGP spectroscopy, *J. High Energy Phys.* **10**, 66. arXiv:hep-th/0604086.

Chernoff, D.F. and Tye, S.-H.H. (2007). Cosmic string detection via microlensing of stars. arXiv:0709.1139.

Cheung, K. (2003). Collider phenomenology for models of extra dimensions. arXiv:hep-ph/0305003.

Chruściel, P.T. (1990). On space-times with $U(1) \times U(1)$ symmetric compact Cauchy surfaces, *Ann. Phys. (NY)* **202**, 100.

Chruściel, P.T., Isenberg, J., and Moncrief, V. (1990). Strong cosmic censorship in polarised Gowdy spacetimes, *Class. Quant. Grav.* **7**, 1671.

Chruściel, P.T., Jezierski, J., and MacCallum, M.A.H. (1998). Uniqueness of the Trautman-Bondi mass, *Phys. Rev. D* **58**, 084001. arXiv:gr-qc/9803010.

Clarke, C.J.S. and Dray, T. (1987). Junction conditions for null hypersurfaces, *Class. Quant. Grav.* **4**, 265.

Clarke, C.J.S., Ellis, G.F.R., and Vickers, J.A. (1990). The large-scale bending of cosmic strings, *Class. Quant. Grav.* **7**, 1.

Clarkson, C.A. (2000). *On the observational characteristics of inhomogeneous cosmologies: undermining the cosmological principle*, University of Glasgow PhD thesis arXiv:astro-ph/0008089.

Clarkson, C.A. (2007). A covariant approach for perturbations of rotationally symmetric spacetimes, *Phys. Rev. D* **76**, 104034. arXiv:0708.1398.

Clarkson, C.A., Ananda, K., and Larena, J. (2009). The influence of structure formation on the cosmic expansion, *Phys. Rev. D* **80**, 083525. arXiv:0907.3377.

Clarkson, C.A. and Barrett, R.K. (1999). Does the isotropy of the CMB imply a homogeneous universe? Some generalized EGS theorems, *Class. Quant. Grav.* **16**, 3781.

Clarkson, C.A. and Barrett, R.K. (2003). Covariant perturbations of Schwarzschild black holes, *Class. Quant. Grav.* **20**, 3855. arXiv:gr-qc/0209051.

Clarkson, C.A., Bassett, B.A.C.C., and Lu, T.H-C. (2008). A general test of the Copernican principle, *Phys. Rev. Lett.* **101**, 011301. arXiv:0712.3457.

Clarkson, C.A., Clifton, T., and February, S. (2009). Perturbation theory in Lemaître-Tolman-Bondi cosmology, *J. Cosmol. Astropart. Phys.* **0906**, 025. arXiv:0903.5040.

Clarkson, C.A., Coley, A.A., O'Neill, E.S.D., Sussman, R.A., and Barrett, R.K. (2003). Inhomogeneous cosmologies, the Copernican principle and the Cosmic Microwave Background: More on the EGS theorem, *Gen. Rel. Grav.* **35**, 969. arXiv:gr-qc/0302068.

Clarkson, C.A., Ellis, G.F.R., Faltenbacher, A., Maartens, R., Umeh, O. *et al.* (2011a). (Mis-)Interpreting supernovae observations in a lumpy universe, arXiv:1109.2484.

Clarkson, C.A., Ellis, G.F.R., Larena, J. and Umeh, O. (2011). Does the growth of structure affect our dynamical models of the universe? The averaging, backreaction and fitting problems in cosmology. arXiv:1109.2314.

Clarkson, C.A. and Maartens, R. (2010). Inhomogeneity and the foundations of concordance cosmology, *Class. Quant. Grav.* **27**, 124008. arXiv:1005.2165.

Clarkson, C.A. and Regis, M. (2011). The cosmic microwave background in an inhomogeneous universe. *J. Cosmol. Astropart. Phys.* **02**, 013. arXiv:1007.3443.

Clifton, T. and Ferreira, P.G. (2009a). Archipelagian cosmology: Dynamics and observables in a universe with discretized matter content, *Phys. Rev. D* **80**, 103503. arXiv:0907.4109.

Clifton, T. and Ferreira, P.G. (2009b). Errors in estimating Ω_Λ due to the fluid approximation, *J. Cosmol. Astropart. Phys.* **0910**, 026. arXiv:0908.4488.

Clifton, T., Ferreira, P.G., and Land, K. (2008). Living in a void: Testing the Copernican principle with distant supernovae, *Phys. Rev. Lett.* **101**, 131302. arXiv:0807.1443.

Clifton, T., Ferreira, P.G., and Zuntz, J. (2009). What the small angle CMB really tells us about the curvature of the universe, *J. Cosmol. Astropart. Phys.* **0907**, 029. arXiv:0902.1313.

Clifton, T. and Zuntz, J. (2009). Hubble diagram dispersion from large-scale structure, *Mon. Not. Roy. Astr. Soc.* **400**, 2185. arXiv:0902.0726.

Clowe, D., Gonzalez, A., and Markevitch, M. (2004). Weak-lensing mass reconstruction of the interacting cluster 1E 0657-558: Direct evidence for the existence of dark matter, *Astrophys. J.* **604**, 596.

Cole, S., Percival, W.J., Peacock, J.A., Norberg, P., Baugh, C.M. *et al.* (2005). The 2dF galaxy redshift survey: power-spectrum analysis of the final data set and cosmological implications, *Mon. Not. Roy. Astr. Soc.* **362**, 505. arXiv:astro-ph/0501174.

Coleman, S. and de Luccia, F. (1980). Gravitational effects on and of vacuum decay, *Phys. Rev. D* **21**, 3305.

Coles, P. and Lucchin, F. (2003). *Cosmology: The Origin and Evolution of Cosmic Structure*, Second Edition (Wiley, Chichester).

Coley, A.A. (2003). *Dynamical Systems and Cosmology*, volume 291 of *Astrophysics and Space Science Library* (Kluwer Academic Publishers, Dordrecht, Boston and London).

Coley, A.A. (2010). Averaging in cosmological models using scalars, *Class. Quant. Grav.* **27**, 245017. arXiv:0908.4281.

Coley, A.A. and Goliath, M. (2000). Closed cosmologies with a perfect fluid and a scalar field, *Phys. Rev. D* **62**, 043526. arXiv:gr-qc/0004060.

Coley, A.A. and Hervik, S. (2005). A dynamical systems approach to the tilted Bianchi models of solvable type, *Class. Quant. Grav.* **22**, 579. arXiv:gr-qc/0409100.

Coley, A.A. and Hervik, S. (2008). Bianchi models with vorticity: The type III bifurcation, *Class. Quant. Grav.* **25**, 198001. arXiv:0802.3629.

Coley, A.A., Hervik, S., and Lim, W.C. (2006). Fluid observers and tilting cosmology, *Class. Quant. Grav.* **23**, 3573. gr-qc/0605128.

Coley, A.A., Hervik, S., Lim, W.C., and MacCallum, M.A.H. (2009). Properties of kinematic singularities, *Class. Quant. Grav.* **26**, 215008. arXiv:0907.1620.

Coley, A.A., Hervik, S., and Pelavas, N. (2009). Spacetimes characterized by their scalar curvature invariants, *Class. Quant. Grav.* **26**, 025013.

Coley, A.A. and Lim, W.C. (2005). Asymptotic analysis of spatially inhomogeneous stiff and ultra-stiff cosmologies, *Class. Quant. Grav.* **22**, 3073. arXiv:gr-qc/0506097.

Coley, A.A., Pelavas, N., and Zalaletdinov, R.M. (2005). Cosmological solutions in macroscopic gravity, *Phys. Rev. Lett.* **95**, 151102. arXiv:gr-qc/0504115.

Coley, A.A. and Tupper, B.O.J. (1983). A new look at FRW cosmologies, *Gen. Rel. Grav.* **15**, 977.

Collins, C.B. (1971). More qualitative cosmology, *Commun. Math. Phys.* **23**, 137.

Collins, C.B. (1972). Qualitative magnetic cosmology, *Commun. Math. Phys.* **27**, 37.

Collins, C.B. (1977). Global structure of the 'Kantowki-Sachs' cosmological models, *J. Math. Phys.* **18**, 2116.

Collins, C.B. and Ellis, G.F.R. (1979). Singularities in Bianchi cosmologies, *Phys. Reports* **56**, 65.

Collins, C.B. and Hawking, S.W. (1973a). The rotation and distortion of the Universe, *Mon. Not. Roy. Astr. Soc.* **162**, 307.

Collins, C.B. and Hawking, S.W. (1973b). Why is the Universe isotropic?, *Astrophys. J.* **180**, 317.

Collins, C.B. and Wainwright, J. (1983). On the role of shear in general relativistic cosmological and stellar models, *Phys. Rev. D* **27**, 1209.

Comer, G.L. (1997). 3+1 approach to the long-wavelength iteration scheme, *Class. Quant. Grav.* **14**, 407.

Copeland, E.J., Lidsey, J.E., and Mizuno, S. (2006). Correspondence between loop-inspired and braneworld cosmology, *Phys. Rev. D* **73**, 043503. arXiv:gr-qc/0510022.

Cornish, N.J., Spergel, D.N., and Starkman, G.D. (1998). Circles in the sky: Finding topology with the microwave background radiation, *Class. Quant. Grav.* **15**, 2657. arXiv:gr-qc/9602039.

Cornish, N.J., Spergel, D.N., Starkman, G.D., and Komatsu, E. (2004). Constraining the topology of the universe, *Phys. Rev. Lett.* **92**, 201302. arXiv:astro-ph/0310233.

Courant, R. and Hilbert, D. (1962). *Methods of Mathematical Physics*, volume 2 (Interscience Publishers, New York).

Cox, D.G.P. (2007). How far is infinity?, *Gen. Rel. Grav.* **39**, 87.

Croudace, K.M., Parry, J., Salopek, D.S., and Stewart, J.M. (1994). Applying the Zel'dovich approximation to general relativity, *Astrophys. J.* **423**, 22.

Curtis, J. and Garfinkle, D. (2005). Numerical simulations of stiff fluid gravitational singularities, *Phys. Rev. D* **72**, 064003. arXiv:gr-qc/0506107.

Cutler, C. and Holz, D.E. (2009). Ultra-high precision cosmology from gravitational waves, *Phys. Rev. D* **80**, 104009. arXiv:0906.3752.

Cyburt, R.H., Fields, B.D. and Olive, K. (2008). A bitter pill: the primordial lithium problem worsens, *J. Cosmol. Astropart. Phys.* **0811**, 012. arXiv:0808.2818.

Damour, T. and de Buyl, S. (2008). Describing general cosmological singularities in Iwasawa variables, *Phys. Rev. D* **77**, 043520. arXiv:0710.5692.

Damour, T., Henneaux, M., and Nicolai, H. (2003). Cosmological billiards, *Class. Quant. Grav.* **20**, R145.

Daniel, S.F., Linder, E.V., Smith, T.L., Caldwell, R.R., Cooray, A. *et al.* (2010). Testing general relativity with current cosmological data, *Phys. Rev. D* **81**, 123508. arXiv:1002.1962.

Darmois, G. (1927). *Les équations de la gravitation Einsteinienne*, volume XXV of *Mémorials des sciences mathématiques* (Gauthier-Villars, Paris).

Dautcourt, G. (1969). Small-scale variations in the cosmic microwave background, *Mon. Not. Roy. Astr. Soc.* **144**, 255.

Dautcourt, G. (1983a). The cosmological problem as initial value problem on the observer's past light cone: geometry, *J. Phys. A* **16**, 3507.

Dautcourt, G. (1983b). The cosmological problem as initial value problem on the observer's past light cone: observations, *Astron. Nachr.* **304**, 153.

Dautcourt, G. and Rose, K. (1978). Polarized radiation in relativistic cosmology, *Astron. Nachr.* **299**, 13.

Davies, P.C.W. (1974). *The Physics of Time Asymmetry* (Surrey University Press, London).

Davies, P.C.W. (1982). *The Accidental Universe* (Cambridge University Press, Cambridge).

Davies, P.C.W. (1987). *The Cosmic Blueprint* (Heinemann, London).

Davis, T.M. and Lineweaver, C.H. (2004). Expanding confusion: common misconceptions of cosmological horizons and the superluminal expansion of the universe, *Pub. Astr. Soc. Australia* **21**, 97. arXiv:astro-ph/0310808.

de Groot, S.L., van Leeuwen, W.A., and van Weert, C.G. (1980). *Relativistic Kinetic Theory* (North-Holland, Amsterdam).

de Lapparent, V., Geller, M.J., and Huchra, J.P. (1986). A slice of the universe, *Astrophys. J.* **302**, L1.

de Swardt, B., Dunsby, P.K.S., and Clarkson, C.A. (2010a). Covariant formulation of the lens equation in cosmology. Preprint, University of Cape Town. In draft.

de Swardt, B., Dunsby, P.K.S., and Clarkson, C.A. (2010b). Gravitational lensing in spherically symmetric spacetimes. arXiv:1002.2041.

Demianski, M., de Ritis, R., Marino, A.A., and Piedipalumbo, E. (2003). Approximate angular diameter distance in a locally inhomogeneous universe with a non-zero cosmological constant, *Astron. Astrophys.* **411**, 33. arXiv:astro-ph/0310830.

Deruelle, N. and Goldwirth, D.S. (1995). Conditions for inflation in an initially inhomogeneous universe, *Phys. Rev. D* **51**, 1563. arXiv:gr-qc/9409056.

Deruelle, N. and Langlois, D. (1995). Long wavelength iteration of Einstein's equations near a spacetime singularity, *Phys. Rev. D* **52**, 2007. arXiv:gr-qc/9411040.

DeWitt, B.S. and Graham, R.N. (Eds.) (1973). *The Many-Worlds Interpretation of Quantum Mechanics* (Princeton University Press, Princeton). Includes the original papers of Everett from 1957.

Dicke, R.H. (1963). Experimental relativity, in *Relativity, Groups and Topology*, ed. B. DeWitt and C. DeWitt (Blackie and Son, London), p. 165.

Dicke, R.H. and Peebles, P.J.E. (1979). The big bang cosmology – enigmas and nostrums, in *General Relativity: An Einstein Centenary Survey*, ed. S.W. Hawking and W. Israel (Cambridge University Press, Cambridge), p. 504.

Diemand, J. and Moore, B. (2011). The structure and evolution of cold dark matter halos, *Adv. Sci. Lett.* **4**, 297.

Dingle, H. (1933). On isotropic models of the universe, with special reference to the stability of the homogeneous and static states, *Mon. Not. Roy. Astr. Soc.* **94**, 134.

Dirac, P.A.M. (1938). New basis for cosmology, *Proc. Roy. Soc. A* **165**, 199.

Dirac, P.A.M. (1981). *The Principles of Quantum Mechanics*, volume 27 of *International Series of Monographs on Physics* (Clarendon Press, Oxford). Fourth edition.

Disney, M.J. (1976). Visibility of galaxies, *Nature* **263**, 573.

Dodelson, S. (2003). *Modern Cosmology: Anisotropies and Inhomogeneities in the Universe* (Academic Press, Amsterdam).

Dodelson, S., Gates, E., and Stebbins, A. (1996). Cold + hot dark matter and the cosmic microwave background, *Astrophys. J.* **467**, 10. arXiv:astro-ph/9509147.

Dolag, K., Borgani, S., Schindler, S., Diaferio, A., and Bykov, A.M. (2008). Simulation techniques for cosmological simulations, *Space Science Reviews* **134**, 229. arXiv:0801.1023.

Dominik, M. (2010). Studying planet populations by gravitational microlensing. p. 2075 in Jetzer, Mellier and Perlick (2010).

Doroshkevich, A.G. (1965). Model of a universe with a uniform magnetic field (in Russian), *Astrophysics* **1**, 138.

Dulaney, T.R. and Gresham, M.I. (2008). Direction-dependent Jeans instability in an anisotropic Bianchi type I space-time. arXiv:0805.1078.

Dunkley, J., Komatsu, E., Nolta, M.R., Spergel, D.N., Larson, D. *et al.* (2009). Five-year Wilkinson Microwave Anisotropy Probe observations: Likelihoods and parameters from the WMAP data, *Astrophys. J. Supp.* **180**, 306. arXiv:0803.0586.

Dunsby, P.K.S. (1993). Gauge-invariant perturbations of anisotropic cosmological models, *Phys. Rev. D* **48**, 3562.

Dunsby, P.K.S. (1997). A fully covariant description of cosmic microwave background anisotropies, *Class. Quant. Grav.* **14**, 3391. arXiv:gr-qc/9707022.

Dunsby, P.K.S., Bassett, B.A.C.C., and Ellis, G.F.R. (1997). Covariant analysis of gravitational waves in a cosmological context, *Class. Quant. Grav.* **14**, 1215. arXiv:gr-qc/9811092.

Dunsby, P.K.S., Goheer, N., Osano, B., and Uzan, J.-P. (2010). How close can an inhomogeneous universe mimic the concordance model?, *J. Cosmol. Astropart. Phys.* **1006**, 017. arXiv:1002.2397.

Durrer, R. (2007). Cosmic magnetic fields and the CMB, *New Astronomy Review* **51**, 275. arXiv:astro-ph/0609216.

Durrer, R. (2008). *The Cosmic Microwave Background* (Cambridge University Press, Cambridge).

Durrer, R. and Maartens, R. (2008). Dark energy and dark gravity: theory overview, *Gen. Rel. Grav.* **40**, 301. arXiv:0711.0077.

Dyer, C.C. (1976). The gravitational perturbation of the cosmic background radiation by density concentrations, *Mon. Not. Roy. Astr. Soc.* **175**, 429.

Dyer, C.C., Landry, S., and Shaver, E.G. (1993). Matching of Friedmann-Lemaître-Robertson-Walker and Kasner cosmologies, *Phys. Rev. D* **47**, 1404.

Dyer, C.C. and Roeder, R. (1974). Observations in locally inhomogeneous cosmological models, *Astrophys. J.* **189**, 167.

Dyer, C.C. and Roeder, R. (1975). Apparent magnitudes, redshifts, and inhomogeneities in the universe, *Astrophys. J.* **196**, 671.

Eardley, D.M. (1974). Self-similar spacetimes: geometry and dynamics, *Commun. Math. Phys.* **37**, 287.

Eardley, D.M., Liang, E., and Sachs, R.K. (1971). Velocity-dominated singularities in irrotational dust cosmologies, *J. Math. Phys.* **13**, 99.

Eardley, D.M. and Smarr, L. (1979). Time functions in numerical relativity: marginally bound dust collapse, *Phys. Rev. D* **19**, 2239.

Earman, J. (1987). The SAP also rises: A critical examination of the anthropic principle, *Am. Phil. Qu.* **24**, 307.

Eddington, A.S. (1930). On the instability of Einstein's spherical world, *Mon. Not. Roy. Astr. Soc.* **90**, 668.

Edgar, S.B. (1980). The structure of tetrad formalisms in general relativity: the general case, *Gen. Rel. Grav.* **12**, 347.

Ehlers, J. (1961). Beiträge zur relativistischen Mechanik kontinuerlicher Medien, *Akad. Wiss. Lit. Mainz, Abh. Math.-Nat. Kl.* **11**. English translation by G.F.R. Ellis and P.K.S. Dunsby, in *Gen. Rel. Grav.* **25**, 1225–1266 (1993).

Ehlers, J. (1971). General relativity and kinetic theory, in *General Relativity and Cosmology*, ed. R.K. Sachs, volume XLVII of *Proceedings of the International School of Physics "Enrico Fermi"* (Academic Press, New York and London), p. 1.

Ehlers, J. (1973). Survey of general relativity theory, in *Relativity, Astrophysics and Cosmology*, ed. W. Israel (Kluwer Academic Publishers, Dordrecht).

Ehlers, J., Geren, P., and Sachs, R. K. (1968). Isotropic solutions of the Einstein-Liouville equations, *J. Math. Phys.* **9**, 1344.

Ehlers, J. and Rindler, W. (1989). A phase-space representation of Friedmann-Lemaître universes containing both dust and radiation and the inevitability of a big bang, *Mon. Not. Roy. Astr. Soc.* **238**, 503.

Einstein, A. (1917). Kosmologische Betrachtungen zur allgemeinen relativitätstheorie, *Sitzb. Preuss. Akad. Wiss.* p. 142. English translation in *The Principle of Relativity* by H.A. Lorentz, A. Einstein, H. Minkowski and H. Weyl (Dover, New York), 1923.

Einstein, A. (1956). *The Meaning of Relativity (6th ed.)* (Methuen, London).

Einstein, A. and Straus, E.G. (1945). The influence of the expansion of space on the gravitation fields surrounding the individual stars, *Rev. Mod. Phys.* **17**, 120.

Einstein, A. and Straus, E.G. (1946). Corrections and additional remarks to our paper: The influence of the expansion of space on the gravitation fields surrounding the individual stars, *Rev. Mod. Phys.* **18**, 148.

Eisenhart, L.P. (1924). Space-time continua of perfect fluids in general relativity, *Trans. Am. Math. Soc.* **26**, 205.

Eisenstein, D.J., Seo, H.-J., and White, M. (2007). On the robustness of the acoustic scale in the low-redshift clustering of matter, *Astrophys. J.* **664**, 660. arXiv:astro-ph/0604361.

Eisenstein, D.J., Zehavi, I., Hogg, D.W., Scoccimarro, R., Blanton, M.R. *et al.* (2005). Detection of the Baryon Acoustic Peak in the large-scale correlation function of SDSS luminous red galaxies, *Astrophys. J.* **633**, 560. arXiv:astro-ph/0501171.

Ellis, G.F.R. (1967). Dynamics of pressure-free matter in general relativity, *J. Math. Phys.* **8**, 1171.

Ellis, G.F.R. (1971a). Relativistic cosmology, in *General Relativity and Cosmology*, ed. R.K. Sachs, volume XLVII of *Proceedings of the International School of Physics 'Enrico Fermi'* (Academic Press, New York and London), p. 104. Reprinted as *Gen. Rel. Grav.* **41**, 581–660 (2009).

Ellis, G.F.R. (1971b). Topology and cosmology, *Gen. Rel. Grav.* **2**, 7.

Ellis, G.F.R. (1973). Relativistic cosmology, in *Cargese Lectures in Physics, vol. 6*, ed. E. Schatzman (Gordon and Breach, New York), p. 1.

Ellis, G.F.R. (1975). Cosmology and verifiability, *Q. J. Roy. Astr. Soc.* **16**, 245.

Ellis, G.F.R. (1980). Limits to verification in cosmology, *Ann. N.Y. Acad. Sci.* **336**, 130.

Ellis, G.F.R. (1984). Alternatives to the big bang, *Ann. Rev. Astron. Astrophys.* **22**, 157.

Ellis, G.F.R. (1988). Does inflation necessarily imply $\Omega = 1$?, *Class. Quant. Grav.* **5**, 891.

Ellis, G.F.R. (1989). A history of cosmology 1917-1955, in *Einstein and the History of General Relativity*, ed. D. Howard and J. Stachel, volume 1 of *Einstein Study Series* (Birkhauser Verlag, Boston), p. 367.

Ellis, G.F.R. (1990). The evolution of inhomogeneities in expanding Newtonian cosmologies, *Mon. Not. Roy. Astr. Soc.* **243**, 509.

Ellis, G.F.R. (1993). The physics and geometry of the early universe: changing viewpoints, *Q. J. Roy. Astr. Soc.* **34**, 315. Appeared in first version in Italian in *La cosmologia nella cultura del '900*, Giornale di Astronomia, 17, 6–14 (1991).

Ellis, G.F.R. (1995). Comment on "Entropy and the second law: A pedagogical alternative" by Ralph Baierlein, *Am. J. Phys.* **63**, 472.

Ellis, G.F.R. (1996). Contributions of K. Gödel to relativity and cosmology, in *Gödel 96: logical foundations of mathematics, computer science and physics - Kurt Gödel's legacy*, ed. P. Hajicek, volume 6 of *Lecture Notes in Logic* (Springer Verlag, Berlin and Heidelberg).

Ellis, G.F.R. (1997). Cosmological models from a covariant viewpoint, in *Gravitation and Cosmology*, ed. S. Dhurandhar and T. Padmanabhan, volume 211 of *Astrophysics and Space Science Library* (Springer, Berlin and Heidelberg), p. 53.

Ellis, G.F.R. (2002). Cosmology and local physics, *New Astron. Rev.* **46**, 645. arXiv:gr-qc/0102017.

Ellis, G.F.R. (2005). Dynamical properties of cosmological solutions, *J. Hyper. Diff. Equations* **2**, 381.

Ellis, G.F.R. (2006). Issues in the philosophy of cosmology, in *Handbook in the Philosophy of Science: Philosophy of Physics, Part B*, ed. J. Butterfield and J. Earman (Elsevier, Amsterdam), p. 1183. arXiv:astro-ph/0602280.

Ellis, G.F.R. (2007). Editorial note concerning 'On the definition of distance in general relativity' by I.M.H. Etherington, *Gen. Rel. Grav.* **39**, 1047.

Ellis, G.F.R. (2009). Dark matter and dark energy proposals: maintaining cosmology as a true science?, in *CRAL-IPNL Dark Energy and Dark Matter: Observations, Experiments, and Theories*, ed. E. Pécontal, T. Buchert, P. Di Stefano, and Y. Copin (EAS/EDP Sciences, Les Ulis), p. 325. arXiv:0811.3529.

Ellis, G.F.R. and Baldwin, J.E. (1984). On the expected anisotropy of radio source counts, *Mon. Not. Roy. Astr. Soc.* **206**, 377.

Ellis, G.F.R., Bassett, B.A.C.C., and Dunsby, P.K.S. (1998). Lensing and caustic effects on cosmological distances, *Class. Quant. Grav.* **15**, 2345. arXiv:gr-qc/9801092.

Ellis, G.F.R. and Brundrit, G.B. (1979). Life in the infinite universe, *Q. J. Roy. Astr. Soc.* **20**, 37.

Ellis, G.F.R. and Bruni, M. (1989). Covariant and gauge-invariant approach to cosmological density fluctuations, *Phys. Rev. D* **40**, 1804.

Ellis, G.F.R. and Jaklitsch, M.J. (1989). Integral constraints on perturbations of Robertson-Walker cosmologies, *Astrophys. J.* **346**, 601.

Ellis, G.F.R. and King, A.R. (1974). Was the big-bang a whimper?, *Commun. Math. Phys.* **38**, 119.

Ellis, G.F.R., Kirchner, U., and Stoeger, W.R. (2004). Multiverses and physical cosmology, *Mon. Not. Roy. Astr. Soc.* **347**, 921. arXiv:astro-ph/0305292.

Ellis, G.F.R. and Maartens, R. (2004). The emergent universe: inflationary cosmology with no singularity and no quantum gravity era, *Class. Quant. Grav.* **21**, 223.

Ellis, G.F.R., Maartens, R., and Nel, S.D. (1978). The expansion of the universe, *Mon. Not. Roy. Astr. Soc.* **184**, 439.

Ellis, G.F.R. and MacCallum, M.A.H. (1969). A class of homogeneous cosmological models, *Commun. Math. Phys.* **12**, 108.

Ellis, G.F.R. and Matravers, D.R. (1985). Spatial homogeneity and the size of the universe, in *A Random Walk in Relativity and Cosmology*, ed. N. Dadhich, J.R. Rao, J.V. Narlikar, and C.V. Vishveshwara (Wiley Eastern, Delhi).

Ellis, G.F.R. and Matravers, D.R. (1995). General covariance in general relativity, *Gen. Rel. Grav.* **27**, 777.

Ellis, G.F.R., Matravers, D.R., and Treciokas, R. (1983a). Anisotropic solutions of the Einstein-Boltzmann equations: I. General formalism, *Ann. Phys. (N.Y.)* **150**, 455.

Ellis, G.F.R., Matravers, D.R., and Treciokas, R. (1983b). An exact anisotropic solution of the Einstein-Liouville equations, *Gen. Rel. Grav.* **15**, 931.

Ellis, G.F.R., McEwan, P., Stoeger, W.R., and Dunsby, P. (2002a). Causality in inflationary universes with positive spatial curvature, *Gen. Rel. Grav.* **34**, 1461. arXiv:gr-qc/0109024.

Ellis, G.F.R., Nel, S.D., Maartens, R., Stoeger, W.R., and Whitman, A.P. (1985). Ideal observational cosmology, *Phys. Reports* **124**, 315.

Ellis, G.F.R., Nicolai, H., Durrer, R., and Maartens, R. (Eds.) (2008). Special issue on dark energy, *Gen. Rel. Grav.* **40**.

Ellis, G.F.R., Perry, J.J., and Sievers, A. (1984). Cosmological observations of galaxies: the observational map, *Astron. J.* **89**, 1124.

Ellis, G.F.R. and Rothman, T. (1993). Lost horizons, *Am. J. Phys.* **61**, 883.

Ellis, G.F.R. and Schreiber, G. (1986). Observational and dynamic properties of small universes, *Phys. Lett. A* **115**, 97.

Ellis, G.F.R. and Sciama, D.W. (1972). Global and non-global problems in cosmology, in *General Relativity (Synge Festschrift)*, ed. L. O'Raifeartaigh (Oxford University Press, Oxford), p. 35.

Ellis, G.F.R. and Stoeger, W.R. (1987). The 'fitting problem' in cosmology, *Class. Quant. Grav.* **4**, 1697.

Ellis, G.F.R. and Stoeger, W.R. (1988). Horizons in inflationary universes, *Class. Quant. Grav.* **5**, 207.

Ellis, G.F.R. and Stoeger, W.R. (2009a). A note on infinities in eternal inflation, *Gen. Rel. Grav.* **41**, 1475. arXiv:1001.4590.

Ellis, G.F.R. and Stoeger, W.R. (2009b). The evolution of our local cosmic domain: Effective causal limits, *Mon. Not. Astr. Soc.* **398**, 1527.

Ellis, G.F.R., Stoeger, W.R., McEwan, P., and Dunsby, P. (2002b). Dynamics of inflationary universes with positive spatial curvature, *Gen. Rel. Grav.* **34**, 1445. arXiv:gr-qc/0109023.

Ellis, G.F.R., Treciokas, R., and Matravers, D.R. (1983). Anisotropic solutions of the Einstein-Boltzmann equations: II, *Ann. Phys. (N.Y.)* **150**, 487.

Ellis, G.F.R. and Tsagas, C.G. (2002). Relativistic approach to nonlinear peculiar velocities and the Zel'dovich approximation, *Phys. Rev. D* **66**, 124015. arXiv:astro-ph/0209143.

Ellis, G.F.R. and Uzan, J.-P. (2005). 'c' is the speed of light, isn't it?, *Amer. J. Phys.* **73**, 240. arXiv:gr-qc/0305099.

Ellis, G.F.R. and van Elst, H. (1999a). Cosmological models, in *Theoretical and Observational Cosmology (Cargese Lectures 1998)*, ed. M. Lachièze-Ray, volume 541 of *Nato Series C: Mathematical and Physical Sciences* (Kluwer, Dordrecht), p. 1. arXiv:gr-qc/9812046.

Ellis, G.F.R. and van Elst, H. (1999b). Deviation of geodesics in FLRW spacetime geometries, in *On Einstein's Path: essays in honor of Engelbert Schucking*, ed. A.L. Harvey (Springer-Verlag, New York), p. 203.

Ellis, G.F.R., van Elst, H., and Maartens, R. (2001). General relativistic analysis of peculiar velocities, *Class. Quant. Grav.* **18**, 5115. arXiv:gr-qc/0105083.

Ellis, G.F.R., van Elst, H., Murugan, J. and Uzan, J.-P. (2010). On the trace-free Einstein equations as a viable alternative to general relativity. arXiv:1008.1196.

Etherington, I.M.H. (1933). On the definition of distance in general relativity, *Phil. Mag.* **15**, 761. Reprinted as *Gen. Rel. Grav.* **39** (2007).

Fabian, A.C. (Ed.) (1989). *Origins* (Cambridge University Press, Cambridge).

Fagundes, H.V. (1985). Relativistic cosmologies with closed, locally homogeneous spatial sections, *Phys. Rev. Lett.* **54**, 1200. See also *Gen. Rel. Grav.* **30**, 1437 (1998).

Faraoni, V. (2009). An analysis of the Sultana-Dyer cosmological black hole solution of the Einstein equations, *Phys. Rev. D* **80**, 044013. arXiv:0907.4473.

Farnsworth, D.L. (1967). Some new general relativistic dust metrics possessing isometries, *J. Math. Phys.* **8**, 2315.

Fay, S. (2004). Isotropisation of Bianchi class A models with a minimally coupled scalar field and a perfect fluid, *Class. Quant. Grav.* **21**, 1609. arXiv:gr-qc/0402104.

Fayos, F., Jaen, X., Llanta, E., and Senovilla, J.M.M. (1991). Matching of the Vaidya and Robertson-Walker metric, *Class. Quant. Grav.* **8**, 2057.

Fayos, F., Senovilla, J.M.M., and Torres, R. (1996). General matching of 2 spherically symmetric spacetimes, *Phys. Rev. D* **54**, 4862.

February, S., Larena, J., Smith, M., and Clarkson, C.A. (2010). Rendering dark energy void, *Mon. Not. Roy. Astr. Soc.* **405**, 2231. arXiv:0909.1479.

Feng, J.L. (2010). Dark matter candidates from particle physics and methods of detection, *Ann. Rev. Astron. Astrophys.* **48**, 495. arXiv:1003.0904.

Ferrando, J.J., Morales, J.A., and Portilla, M. (1992). Inhomogeneous space-times admitting isotropic radiation: Vorticity-free case, *Phys. Rev. D* **46**, 578.

Ferreira, P.G. (2007). *The State of the Universe: A Primer in Modern Cosmology* (Phoenix, Los Angeles).

Ferreira, P.G., Juszkiewicz, R., Feldman, H.A., Davis, M., and Jaffe, A.H. (1999). Streaming velocities as a dynamical estimator of Omega, *Astrophys. J. Lett.* **515**, L1. arXiv:astro-ph/9812456.

Ferreira, P.G. and Starkman, G.D. (2009). Einstein's theory of gravity and the problem of missing mass, *Science* **326**, 812. arXiv:0911.1212.

Ferreras, I., Mavromatos, N.E., Sakellariadou, M., and Yusaf, M.F. (2009). Incompatibility of rotation curves with gravitational lensing for TeVeS, *Phys. Rev. D* **80**, 103506. arXiv:0907.1463.

Fixsen, D.J., Cheng, E.S., Gales, J.M., Mather, J.C., Shafer, R.A. *et al.* (1996). The Cosmic Microwave Background spectrum from the full COBE FIRAS data set, *Astrophys. J.* **473**, 576. arXiv:astro-ph/9605054.

Flesch, E. and Arp, H. (1999). Further evidence that some quasars originate in nearby galaxies: NGC3628. arXiv:astro-ph/9907219.

Florides, P.S. and McCrea, W.H. (1959). Observable relations in relativistic cosmology III, *Zs. f. Astrophys.* **48**, 52.

Freedman, W.L. and Madore, B.F. (2010). The Hubble constant, *Ann. Rev. Astron. Astrophys.* **48**, 673.

Freivogel, B., Kleban, M., Rodríguez Martínez, M., and Susskind, L. (2006). Observational consequences of a landscape, *J. High Energy Phys.* **3**, 39. arXiv:hep-th/0505232.

Friedlander, F.G. (1975). *The Wave Equation on a Curved Space-time* (Cambridge University Press, Cambridge).

Frieman, J.A., Turner, M.S., and Huterer, D. (2008). Dark energy and the accelerating universe, *Ann. Rev. Astron. Astrophys.* **46**, 385. arXiv:0803.0982.

Frittelli, S. and Newman, E.T. (1999). Exact universal gravitational lensing equation, *Phys. Rev. D* **59**, 124001. arXiv:gr-qc/9810017.

Fu, L., Semboloni, E., Hoekstra, H., Kilbinger, M., van Waerbeke, L. *et al.* (2008). Very weak lensing in the CFHTLS Wide: cosmology from cosmic shear in the linear regime, *Astron. Astrophys.* **479**, 9. arXiv:0712.0884.

Furlanetto, S.R., Oh, S.P., and Briggs, F.H. (2006). Cosmology at low frequencies: The 21 cm transition and the high-redshift Universe, *Phys. Reports* **433**, 181. arXiv:astro-ph/0608032.

Futamase, T. (1991). A new description for a realistic inhomogeneous universe in general relativity, *Prog. Theor. Phys.* **86**, 389.

Futamase, T. (1996). Averaging of a locally inhomogeneous realistic universe, *Phys. Rev. D* **53**, 681.

Gale, G. (2007). Cosmology: Methodological debates in the 1930s and 1940s. *Stanford Encyclopaedia of Philosophy*. http://plato.stanford.edu/entries/cosmology-30s/.

García-Bellido, J. and Haugbølle, T. (2008). Looking the void in the eyes – the kinematic Sunyaev Zel'dovich effect in Lemaître Tolman Bondi models, *J. Cosmol. Astropart. Phys.* **0809**, 016. arXiv:0807.1326.

Garcia-Parrado, A. and Valiente Kroon, J.A. (2008). Kerr initial data, *Class. Quant. Grav.* **25**, 205018.

Gardner, M. (2003). *Are Universes Thicker than Blackberries? Discourses on Godel, Magic Hexagons, Little Red Riding Hood and Other Mathematical and Pseudoscientific Topics* (W.W. Norton, New York and London).

Garrett, A.J.M. and Coles, P. (1993). Bayesian inductive inference and the anthropic cosmological principle, *Comments Astrophys.* **17**, 23.

Garriga, J., Schwartz-Perlov, D., Vilenkin, A., and Winitzki, S. (2006). Probabilities in the inflationary multiverse, *J. Cosmol. Astropart. Phys.* **0601**, 017. arXiv:hep-th/0509184.

Gasperini, M., Marozzi, G., and Veneziano, G. (2010). A covariant and gauge invariant formulation of the cosmological backreaction, *J. Cosmol. Astropart. Phys.* **1002**, 009. arXiv:0912.3244.

Gasperini, M. and Veneziano, G. (1993). Pre-big-bang in string cosmology, *Astropart. Phys.* **1**, 317.

Gaztañaga, E., Cabré, A., and Hui, L. (2009). Clustering of luminous red galaxies - IV. Baryon acoustic peak in the line-of-sight direction and a direct measurement of H(z), *Mon. Not. Roy. Astr. Soc.* **399**, 1663. arXiv:0807.3551.

Gebbie, T., Dunsby, P.K.S., and Ellis, G.F.R. (2000). 1+3 covariant cosmic microwave background anisotropies II: The almost-Friedmann-Lemaître Model, *Ann. Phys. (N.Y.)* **282**, 321. arXiv:astro-ph/9904408.

Gebbie, T. and Ellis, G.F.R. (2000). 1+3 covariant cosmic microwave background anisotropies I: Algebraic relations for mode and multipole expansions, *Ann. Phys. (N.Y.)* **282**, 285. arXiv:astro-ph/9804316.

Geroch, R. and Lindblom, L. (1990). Dissipative relativistic fluid theories of divergence type, *Phys. Rev. D* **41**, 1855.

Geshnizjani, G. and Brandenberger, R.H. (2002). Backreaction and local cosmological expansion rate, *Phys. Rev. D* **66**, 123507.

Geshnizjani, G. and Brandenberger, R.H. (2005). Backreaction of perturbations in two scalar field inflationary models, *J. Cosmol. Astropart. Phys.* **0504**, 006.

Ghosh, T., Hajian, A., and Souradeep, T. (2007). Unveiling hidden patterns in CMB anisotropy maps, *Phys. Rev. D* **75**, 083007. arXiv:astro-ph/0604279.

Giannantonio, T., Scranton, R., Crittenden, R.G., Nichol, R.C., Boughn, S.P. *et al.* (2008). Combined analysis of the integrated Sachs-Wolfe effect and cosmological implications, *Phys. Rev. D* **77**, 123520. arXiv:0801.4380.

Gibbons, G.W., Hawking, S.W., and Stewart, J.M. (1987). A natural measure on the set of all universes, *Nucl. Phys. B* **281**, 736.

Gibbons, G.W., Shellard, E.P.S., and Rankin, S.J. (Eds.) (2003). *The Future of Theoretical Physics and Cosmology: Celebrating Stephen Hawking's 60th Birthday* (Cambridge University Press, Cambridge).

Gibbons, G.W. and Turok, N. (2008). The measure problem in cosmology, *Phys. Rev. D* **77**, 063518. arXiv:hep-th/0609095.

Giovannini, M. and Kunze, K.E. (2008). Magnetized CMB observables: A dedicated numerical approach, *Phys. Rev. D* **77**, 063003. arXiv:0712.3483.

Gödel, K. (1949). An example of a new type of cosmological solutions of Einstein's field equations of gravitation, *Rev. Mod. Phys.* **21**, 447.

Gödel, K. (1952). Rotating universes in general relativity theory, in *Proc. Int. Cong. Math., Cambridge, Mass., 1950*, ed. L.M. Graves, E. Hille, P.A. Smith, and O. Zariski, volume 1 (American Mathematical Society, Providence, R.I.), p. 175.

Godłowski, W., Stelmach, J., and Szydłowski, M. (2004). Can the Stephani model be an alternative to FRW accelerating models?, *Class. Quant. Grav.* **21**, 3953. arXiv:astro-ph/0403534.

Goenner, H.F. (2010). What kind of science is cosmology?, *Ann. d. Phys.* **522**, 389. arXiv:0910.4333.

Goetz, G. (1990). Gravitational field of plane symmetric thick domain walls, *J. Math. Phys.* **31**, 2683.

Goheer, N., Leach, J.A., and Dunsby, P.K.S. (2008). Compactifying the state space for alternative theories of gravity, *Class. Quant. Grav.* **25**, 035013. arXiv:0710.0819.

Goldwirth, D.S. and Piran, T. (1990). Inhomogeneity and the onset of inflation, *Phys. Rev. Lett.* **64**, 2852.

Gomero, G.I., Teixeira, A.F.F., Rebouças, M.J., and Bernui, A. (2002). Spikes in cosmic crystallography, *Int. J. Mod. Phys. D* **11**, 869. arXiv:gr-qc/9811038.

Goode, S.W. (1989). Analysis of spatially inhomogeneous perturbations of the FRW cosmologies, *Phys. Rev. D* **39**, 2882.

Goode, S.W. and Wainwright, J. (1982). Singularities and evolution of the Szekeres cosmological models, *Phys. Rev. D* **26**, 3315.

Goodman, J. (1995). Geocentrism reexamined, *Phys. Rev. D* **52**, 1821. arXiv:astro-ph/9506068.

Gordon, C., Wands, D., Bassett, B.A.C.C., and Maartens, R. (2001). Adiabatic and entropy perturbations from inflation, *Phys. Rev. D* **63**, 023506. arXiv:astro-ph/0009131.

Goroff, M.H. and Sagnotti, A. (1985). Quantum gravity at two loops, *Phys. Lett. B* **160**, 81.

Gott, J.R. (1985). Gravitational lensing effects of vacuum strings: exact solutions, *Astrophys. J.* **288**, 422.

Gott, J.R. and Li, L.-X. (1998). Can the universe create itself?, *Phys. Rev. D* **58**, 023501. arXiv:astro-ph/9712344.

Gott, III, J.R., Gunn, J.E., Schramm, D.N., and Tinsley, B.M. (1976). Will the universe expand forever?, *Sci. Am.* **234**, 62.

Gowdy, R.H. (1971). Gravitational waves in closed universes, *Phys. Rev. Lett.* **27**, 826.

Gowdy, R.H. (1974). Vacuum spacetimes with two-parameter spacelike isometry groups and compact invariant hypersurfaces: topologies and boundary conditions, *Ann. Phys. (N.Y.)* **83**, 203.

Gowdy, R.H. (1975). Closed gravitational-wave universes: analytic solutions with two-parameter symmetry, *J. Math. Phys.* **16**, 224.

Greene, B., Hinterbichler, K., Judes, S. and Parikh, M.K. (2011). Smooth initial conditions from weak gravity, *Phys. Lett. B* **697**, 178. arXiv:0911.0693.

Greisen, K. (1966). End to the cosmic-ray spectrum?, *Phys. Rev. Lett.* **16**, 748.

Gribbin, J. and Rees, M. (1989). *Cosmic Coincidences* (Bantam Books, New York).

Griffiths, J.B. (1991). *Colliding Plane Waves in General Relativity* (Oxford University Press, Oxford).

Grishchuk, L.P. (1968). Cosmological models and spatial-homogeneity criteria, *Soviet Astronomy - A.J.* **11**, 881.

Grunbaum, A. (1989). The pseudo problem of creation, *Philosophy of Science* **56**, 373.

Gümrükçüoğlu, A.E., Contaldi, C.R., and Peloso, M. (2007). Inflationary perturbations in anisotropic backgrounds and their imprint on the CMB, *J. Cosmol. Astropart. Phys.* **0711**, 005. arXiv:0707.4179.

Gümrükçüoğlu, A.E., Himmetoglu, B., and Peloso, M. (2010). Scalar-scalar, scalar-tensor, and tensor-tensor correlators from anisotropic inflation, *Phys. Rev. D* **81**, 063528. arXiv:1001.4088.

Guth, A. (2001). Eternal inflation, in *Proceedings of 'Cosmic Questions' Meeting* (The New York Academy of Sciences Press, New York). arXiv:astro-ph/0101507.

Guth, A. (2007). Eternal inflation and its implications, *J. Phys. A* **40**, 6811. arXiv:hep-th/0702178.

Guzzo, L., Pierleoni, M., Meneux, B., Branchini, E., Le Fèvre, O. *et al.* (2008). A test of the nature of cosmic acceleration using galaxy redshift distortions, *Nature* **451**, 541. arXiv:0802.1944.

Hamilton, A.J.S. (1998). Linear redshift distortions: a review, in *The evolving universe*, ed. D. Hamilton, volume 231 of *Astrophysics and Space Science Library* (Springer, Berlin and Heidelberg), p. 185. arXiv:astro-ph/9708102.

Hanquin, J.-L. and Demaret, J. (1984). Exact solutions for inhomogeneous generalizations of some vacuum Bianchi models, *Class. Quant. Grav.* **1**, 291.

Hanson, D., Challinor, A., and Lewis, A. (2010). Weak lensing of the CMB. p. 2197 in Jetzer, Mellier and Perlick (2010), arXiv:0911.0612.

Harness, R.S. (1982). Spacetimes homogeneous on a timelike hypersurface, *J. Phys. A* **15**, 135.

Harré, R. (1962). Philosophical aspects of cosmology, *Brit. J. Phil. Sci.* **13**, 104.

Harrison, E.R. (2000). *Cosmology: The Science of the Universe (2nd edition)* (Cambridge University Press, Cambridge).

Hartle, J. (2003). The state of the universe, in Gibbons, Shellard and Rankin (2003).

Hartle, J.B. (2004). Anthropic reasoning and quantum cosmology, in *The New Cosmology*, ed. R.E. Allen, D.V. Nanopoulos, and C.N. Pope, volume 743 of *AIP Conference Proceedings* (American Institute of Physics, Melville, NY). arXiv:gr-qc/0406104.

Hartle, J.B. (2011). The quasiclassical realms of this quantum universe, *Found. Phys.* **41**, 982. arXiv:0806.3776.

Hartle, J.B. and Hawking, S.W. (1983). Wave function of the universe, *Phys. Rev. D* **28**, 2960.

Hartle, J.B., Hawking, S.W., and Hertog, T. (2008). The no-boundary measure of the universe, *Phys. Rev. Lett.* **100**, 201301. arXiv:0711.4630.

Harvey, A.L. (1979). Automorphisms of the Bianchi model Lie groups, *J. Math. Phys.* **20**, 251.

Harwit, M. (1992). Cosmic curvature and condensation, *Astrophys. J.* **392**, 394.

Harwit, M. (1998). *Astrophysical Concepts, 3rd edition* (Springer, New York).

Hasse, W. and Perlick, V. (1999). On spacetime models with an isotropic Hubble law, *Class. Quant. Grav.* **16**, 2559.

Hauser, M.G. and Dwek, E. (2001). The cosmic infrared background: Measurements and implications, *Ann. Rev. Astron. Astrophys.* **39**, 249. arXiv:astro-ph/0105539.

Hausman, M.A., Olson, D.W., and Roth, B.D. (1983). The evolution of voids in the expanding universe, *Astrophys. J.* **270**, 351.

Hawking, S.W. (1966). Perturbations of an expanding universe, *Astrophys. J.* **145**, 544.

Hawking, S.W. (1975). Particle creation by black holes, in *Quantum Gravity: an Oxford Symposium*, ed. C.J. Isham, R. Penrose, and D.W. Sciama (Clarendon Press, Oxford). Also in *Commun. Math. Phys.* **43**, 199 (1975).

Hawking, S.W. (1987). Quantum cosmology, in *300 Years of Gravitation*, ed. S.W. Hawking and W. Israel (Cambridge University Press, Cambridge), p. 631.

Hawking, S.W. (1993). *On the Big Bang and Black Holes* (World Scientific, Singapore).

Hawking, S.W. and Ellis, G.F.R. (1973). *The Large Scale Structure of Space-time* (Cambridge University Press, Cambridge).

Hawking, S.W. and Page, D.N. (1988). How probable is inflation?, *Nucl. Phys. B* **298**, 789.

Hawking, S.W. and Penrose, R. (1970). Singularities of gravitational collapse and cosmology, *Proc. R. Soc. London A* **314**, 529.

Heavens, A. (2009). Weak lensing: Dark matter, dark energy and dark gravity, *Nucl. Phys. B (Proc. Suppl.)* **194**, 76. arXiv:0911.0350.

Heckmann, O. and Schucking, E. (1955). Bemerkungen zur Newtonschen Kosmologie. I, *Zs. f. Astrophys.* **38**, 95.

Heckmann, O. and Schucking, E. (1956). Bemerkungen zur Newtonschen Kosmologie II, *Zs. f. Astrophys.* **40**, 81.

Heckmann, O. and Schucking, E. (1962). Relativistic cosmology, in *Gravitation*, ed. L. Witten (Wiley, New York), p. 438.

Heinzle, J.M. and Ringström, H. (2009). Future asymptotics of vacuum Bianchi type VI_0 solutions, *Class. Quant. Grav.* **26**, 145001.

Heinzle, J.M. and Uggla, C. (2006). Dynamics of the spatially homogeneous Bianchi type I Einstein-Vlasov equations, *Class. Quant. Grav.* **23**, 3463. arXiv:gr-qc/0512031.

Heinzle, J.M. and Uggla, C. (2009a). Mixmaster: Fact and belief, *Class. Quant. Grav.* **26**, 075016. arXiv:0901.0776.

Heinzle, J.M. and Uggla, C. (2009b). A new proof of the Bianchi type IX attractor theorem, *Class. Quant. Grav.* **26**, 075015. arXiv:0901.0806.

Heinzle, J.M., Uggla, C., and Röhr, N. (2009). The cosmological billiard attractor, *Adv. Theor. Math. Phys.* **13**, 293. arXiv:gr-qc/0702141.

Hellaby, C.W. (1996). The null and KS limits of the Szekeres model, *Class. Quant. Grav.* **13**, 2537.

Hellaby, C.W. and Alfedeel, A.H.A. (2009). Solving the observer metric, *Phys. Rev. D* **79**, 043501. arXiv:0811.1676.

Hellaby, C.W. and Krasiński, A. (2006). Alternative methods of describing structure formation in the Lemaître-Tolman model, *Phys. Rev. D* **73**, 023518. arXiv:gr-qc/0510093.

Hellaby, C.W. and Krasiński, A. (2008). Physical and geometrical interpretation of the $\epsilon \leq 0$ Szekeres models, *Phys. Rev. D* **77**, 023529. arXiv:0710.2171.

Hellaby, C.W. and Lake, K. (1981). Local inhomogeneities in a Robertson–Walker background: III. Elementary growth rates in a flat background with a relativistic equation of state, *Astrophys. J.* **251**, 429.

Hellaby, C.W. and Lake, K. (1983). Mass scales in a universe of dust and blackbody radiation, *Astrophys. Lett.* **23**, 81.

Hellaby, C.W. and Lake, K. (1984). The redshift structure of the big bang in inhomogeneous cosmological models. I. Spherical dust solutions, *Astrophys. J.* **282**, 1.

Hellaby, C.W. and Lake, K. (1985). Shell crossings and the Tolman model, *Astrophys. J.* **290**, 381. Erratum: *Astrophys. J.* **300**, 461 (1986).

Heller, M. (1974). On the interpretative paradox in cosmology, *Acta Cosmologica* **2**, 37.

Hervik, S. (2000). The Bianchi type I minisuperspace model, *Class. Quant. Grav.* **17**, 2765. arXiv:gr-qc/0003084.

Hervik, S. (2004). The asymptotic behaviour of tilted Bianchi type VI_0 universes, *Class. Quant. Grav.* **21**, 2301. arXiv:gr-qc/0403040.

Hervik, S. and Coley, A.A. (2005). Inhomogeneous perturbations of plane-wave spacetimes, *Class. Quant. Grav.* **22**, 3391. arXiv:gr-qc/0505108.

Hervik, S. and Lim, W.C. (2006). The late-time behaviour of vortic Bianchi type VIII universes, *Class. Quant. Grav.* **23**, 3017. arXiv:gr-qc/0512070.

Hervik, S., Lim, W.C., Sandin, P., and Uggla, C. (2010). Future asymptotics of tilted Bianchi type II cosmologies, *Class. Quant. Grav.* **27**, 185006. arXiv:1004.3661.

Hervik, S., van den Hoogen, R., and Coley, A.A. (2005). Future asymptotic behaviour of tilted Bianchi models of type IV and VII_h, *Class. Quant. Grav.* **22**, 607. arXiv:gr-qc/0409106.

Hervik, S., van den Hoogen, R.J., Lim, W.C., and Coley, A.A. (2006). The futures of Bianchi type VII_0 cosmologies with vorticity, *Class. Quant. Grav.* **23**, 845. arXiv:gr-qc/0509032.

Hervik, S., van den Hoogen, R.J., Lim, W.C., and Coley, A.A. (2007). Late-time behaviour of the tilted Bianchi type VI_h models, *Class. Quant. Grav.* **24**, 3859. arXiv:gr-qc/0703038.

Hervik, S., van den Hoogen, R.J., Lim, W.C., and Coley, A.A. (2008). Late-time behaviour of the tilted Bianchi type $VI_{-1/9}$ models, *Class. Quant. Grav.* **25**, 015002. arXiv:0706.3184.

Hewitt, C.G. (1991). Algebraic invariant curves in cosmological dynamical systems and exact solutions, *Gen. Rel. Grav.* **23**, 1363.

Hewitt, C.G., Bridson, R., and Wainwright, J. (2001). The asymptotic regimes of tilted Bianchi II cosmologies, *Gen. Rel. Grav.* **33**, 65. arXiv:gr-qc/0008037.

Hewitt, C.G. and Wainwright, J. (1990). Orthogonally transitive G_2 cosmologies, *Class. Quant. Grav.* **7**, 2295.

Hewitt, C.G. and Wainwright, J. (1992). Dynamical systems approach to tilted Bianchi cosmologies: irrotational models of type V, *Phys. Rev. D* **46**, 4242.

Hewitt, C.G. and Wainwright, J. (1993). A dynamical systems approach to Bianchi cosmologies: orthogonal models of class B, *Class. Quant. Grav.* **10**, 99.

Hewitt, C.G., Wainwright, J., and Glaum, M. (1991). Qualitative analysis of a class of inhomogeneous self-similar cosmological models: II, *Class. Quant. Grav.* **8**, 1505.

Hewitt, C.G., Wainwright, J., and Goode, S.W. (1988). Qualitative analysis of a class of inhomogeneous self-similar cosmological models, *Class. Quant. Grav.* **5**, 1313.

Hibler, D.L. (1976). *Construction and some properties of the most general homogeneous Newtonian cosmological models*, University of Texas at Austin PhD thesis. University Microfilms TSZ 77-3913.

Hicks, N.J. (1965). *Notes on Differential Geometry*, volume 3 of *van Nostrand Mathematical Studies* (van Nostrand, Princeton).

Hilbert, D. (1964). On the infinite, in *Philosophy of Mathematics*, ed. P. Benacerraf and H. Putnam (Prentice Hall, Englewood Cliffs, N. J.), p. 134.

Hinshaw, G., Nolta, M.R., Bennett, C.L., Bean, R., Doré, O. *et al.* (2007). Three-year Wilkinson Microwave Anisotropy Probe (WMAP) observations: Temperature analysis, *Astrophys. J. Supp.* **170**, 288. arXiv:astro-ph/0603451.

Hinshaw, G., Weiland, J.L., Hill, R.S., Odegard, N., Larson, D. *et al.* (2009). Five-year Wilkinson Microwave Anisotropy Probe observations: Data processing, sky maps, and basic results, *Astrophys. J. Supp.* **180**, 225. arXiv:0803.0732.

Hiscock, W.A. (1985). Exact gravitational field of a string, *Phys. Rev. D* **31**, 3288.

Hiscock, W.A. and Lindblom, L. (1987). Linear plane waves in dissipative relativistic fluids, *Phys. Rev. D* **35**, 3723.

Hobill, D.W., Burd, A., and Coley, A.A. (Eds.) (1994). *Deterministic Chaos in General Relativity*, volume 332 of *Nato ASI Series B* (Plenum Press, New York).

Hobson, M.P., Efstathiou, G.P., and Lasenby, A.N. (2006). *General Relativity: An Introduction for Physicists* (Cambridge University Press, Cambridge).

Hoekstra, H. (2007). A comparison of weak-lensing masses and X-ray properties of galaxy clusters, *Mon. Not. Roy. Astr. Soc.* **379**, 317. arXiv:0705.0358.

Hogan, C.J. (2000). Why the universe is just so, *Rev. Mod. Phys.* **72**, 1149. arXiv:astro-ph/9909295.

Hogan, C.J. (2005). Quarks, electrons, and atoms in closely related universes, in *Universe or Multiverse?*, ed. B.J. Carr (Cambridge University Press, Cambridge). arXiv:astro-ph/0407086.

Hogan, P. and Ellis, G.F.R. (1989). The asymptotic field of an accelerating point charge, *Ann. Phys. (N.Y.)* **196**, 293.

Hogg, D.W. (2009). Is cosmology just a plausibility argument? arXiv:0910.3374.

Hollands, S. and Wald, R.M. (2005). Conservation of the stress tensor in perturbative interacting quantum field theory in curved spacetimes, *Rev. Mod. Phys.* **17**, 227.

Hollenstein, L., Caprini, C., Crittenden, R., and Maartens, R. (2008). Challenges for creating magnetic fields by cosmic defects, *Phys. Rev. D* **77**, 063517. arXiv:0712.1667.

Holz, D.E. and Linder, E.V. (2005). Safety in numbers: gravitational degradation of the luminosity distance-redshift relation, *Astrophys. J.* **631**, 678. arXiv:astro-ph/0412173.

Holz, D.E. and Wald, R.M. (1998). New method for determining cumulative gravitational lensing effects in inhomogeneous universes, *Phys. Rev. D* **58**, 063501. arXiv:astro-ph/9708036.

Horwood, J.T. and Wainwright, J. (2004). Asymptotic regimes of magnetic Bianchi cosmologies, *Gen. Rel. Grav.* **36**, 799. arXiv:gr-qc/0309083.

Hosoya, A., Buchert, T., and Morita, M. (2004). Information entropy in cosmology, *Phys. Rev. Lett.* **92**, 141302.

Hoyle, F. (1948). A new model for the expanding universe, *Mon. Not. Roy. Astr. Soc.* **108**, 372.

Hoyle, F. (1962). Cosmological tests of gravitational theories, in *Evidence for Gravitational Theories*, ed. C. Møller, volume XX of *Proceedings of the International School of Physics "Enrico Fermi"* (Academic Press, New York and London), p. 147.

Hoyle, F., Burbidge, G., and Narlikar, J.V. (1993). A quasi-steady state cosmological model with creation of matter, *Astrophys. J.* **410**, 437.

Hoyle, F., Burbidge, G., and Narlikar, J.V. (1994). Astrophysical deductions from the quasi-steady state cosmology, *Mon. Not. Roy. Astr. Soc.* **267**, 1007. Erratum: *Mon. Not. Roy. Astr. Soc.* **269**, 1152.

Hsu, L. and Wainwright, J. (1986). Self-similar spatially homogeneous cosmologies: orthogonal perfect fluid and vacuum solutions, *Class. Quant. Grav.* **3**, 1105.

Hu, W., Seljak, U., White, M., and Zaldarriaga, M. (1998). Complete treatment of CMB anisotropies in a FRW universe, *Phys. Rev. D* **57**, 3290. arXiv:astro-ph/9709066.

Hu, W. and Sugiyama, N. (1995). Anisotropies in the cosmic microwave background: an analytic approach, *Astrophys. J.* **444**, 489. arXiv:astro-ph/9407093.

Hu, W. and White, M. (1997). CMB anisotropies: Total angular momentum method, *Phys. Rev. D* **56**, 596. arXiv:astro-ph/9702170.

Hubble, E. (1929). A relation between distance and radial velocity among extra-galactic nebulae, *Proc. Nat. Acad. Sci.* **15**, 168.

Hubble, E. (1936). *The Realm of the Nebulae* (Yale University Press, Yale).

Huggett, S.A. and Tod, K.P. (1994). *An Introduction to Twistor Theory, 2nd edition*, volume 4 of *London Mathematical Society Student Texts* (Cambridge University Press, Cambridge).

Humphreys, N.P., Maartens, R., and Matravers, D.R. (1997). Anisotropic observations in universes with nonlinear inhomogeneity, *Astrophys. J.* **477**, 47. arXiv:astro-ph/9602033.

Huterer, D. (2010). Weak lensing, dark matter and dark energy. p. 2177 in Jetzer, Mellier and Perlick (2010), arXiv:1001.1758.

Iguchi, O. and Ishihara, H. (1997). Onset of inflation in inhomogeneous cosmology, *Phys. Rev. D* **56**, 3216.

Ipser, J.R. and Sikivie, P. (1984). Gravitationally repulsive domain wall, *Phys. Rev. D* **30**, 712.

Isaacson, R.A. (1967). Gravitational radiation in the limit of high frequency. I. The linear approximation and geometrical optics, *Phys. Rev.* **166**, 1263.

Isaacson, R.A. (1968). Gravitational radiation in the limit of high frequency. II. Nonlinear terms and the effective stress tensor, *Phys. Rev.* **166**, 1272.

Ishak, M., Richardson, J., Garred, D., Whittington, D., Nwankwo, A. *et al.* (2008). Dark energy or apparent acceleration due to a relativistic cosmological model more complex than FLRW?, *Phys. Rev. D* **78**, 123531. arXiv:0708.2943.

Ishak, M., Rindler, W., and Dossett, J. (2010). More on lensing, *Mon. Not. Roy. Astr. Soc.* **403**, 2152. arXiv:0810.4956.

Isham, C.J. (1997). *Lectures on Quantum Theory: Mathematical and Structural Foundations* (Imperial College Press, London).

Isham, C.J. and Linden, N. (1995). Continuous histories and the history group in generalised quantum theory, *J. Math. Phys.* **36**, 5392. arXiv:gr-qc/9503063.

Ishibashi, A. and Wald, R.M. (2006). Can the acceleration of our universe be explained by the effects of inhomogeneities?, *Class. Quant. Grav.* **23**, 235. arXiv:gr-qc/0509108.

Israel, W. (1966). Singular hypersurfaces and thin shells in general relativity, *Nuovo Cim. B* **44**, 1. Erratum: *Nuovo Cim. B* **48**, 463 (1967).

Israel, W. and Stewart, J.M. (1979). On transient relativistic thermodynamics and kinetic theory II, *Proc. Roy. Soc. London A* **365**, 43.

Jackson, J.D. (1975). *Classical Electrodynamics* (Wiley, New York).

Jaffe, T.R., Hervik, S., Banday, A.J., and Gorski, K.M. (2006). On the viability of Bianchi type VII$_h$ models with dark energy, *Astrophys. J.* **644**, 701. arXiv:astro-ph/0512433.

Jain, B. and Bertschinger, E. (1996). Selfsimilar evolution of cosmological density fluctuations, *Astrophys. J.* **456**, 43. arXiv:astro-ph/9503025.

Jain, B. and Khoury, J. (2010). Cosmological tests of gravity, *Ann. Phys. (N.Y.)* **325**, 1479. arXiv:1004.3294.

Jain, D. and Dev, A. (2006). Age of high redshift objects - a litmus test for dark energy models, *Phys. Lett. B* **633**, 436. arXiv:astro-ph/0509212.

Jantzen, R.T. (1979). The dynamical degrees of freedom in spatially homogeneous cosmology, *Commun. Math. Phys.* **64**, 211.

Jantzen, R.T. (1984). Spatially homogeneous dynamics: a unified picture, in *Cosmology of the Early Universe*, ed. R. Ruffini and Fang L.-Z. (World Scientific, Singapore), p. 233. Also in *Gamow Cosmology*, (*Proceedings of the International School of Physics 'Enrico Fermi', Course LXXXVI*), (1987) ed. R. Ruffini and F. Melchiorri, pp. 61–147 (North Holland, Amsterdam).

Jantzen, R.T. and Uggla, C. (1998). The kinematical role of automorphisms in the orthonormal frame approach to Bianchi cosmology, *J. Math. Phys.* **40**, 353.

Jarosik, N., Bennett, C.L., Dunkley, J., Gold, B., Greason, M.R. *et al.* (2011). Seven-year Wilkinson Microwave Anisotropy Probe (WMAP) observations: Sky maps, systematic errors, and basic results, *Astrophys. J. Supp.* **192**, 14. arXiv:1001.4744.

Jee, M.J., Rosati, P., Ford, H.C., Dawson, K.S., Lidman, C. *et al.* (2009). Hubble space telescope weak-lensing study of the galaxy cluster XMMU J2235.3-2557 at $z = 1.4$: A surprisingly massive galaxy cluster when the universe is one-third of its current age, *Astrophys. J.* **704**, 672. arXiv:0908.3897.

Jetzer, P., Mellier, Y., and Perlick, V. (Eds.) (2010). Special issue on lensing, *Gen. Rel. Grav.* **42**, 2009.

Jordan, P., Ehlers, J., and Sachs, R.K. (1961). Beiträge zur Theorie der reinen Gravitationsstrahlung, *Akad. Wiss. Lit. Mainz, Abh. Math.-Nat. Kl.* **1**.

Joshi, P. (1996). *Global Aspects in Gravitation and Cosmology*, volume 87 of *International Series of Monographs in Physics* (Oxford University Press, Oxford). Revised paperback edition.

Joyce, M., Labini, F.S., Gabrielli, A., Montuori, M., and Pietronero, L. (2005). Basic properties of galaxy clustering in the light of recent results from the Sloan Digital Sky Survey, *Astron. Astrophys.* **443**, 11.

Kainulainen, K. and Marra, V. (2009). A new stochastic approach to cumulative weak lensing, *Phys. Rev. D* **80**, 123020. arXiv:0909.0822.

Kaiser, N. (1983). Small-angle anisotropy of the microwave background radiation in the adiabatic theory, *Mon. Not. Roy. Astr. Soc.* **202**, 1169.

Kaiser, N. and Squires, G. (1993). Mapping the dark matter with weak gravitational lensing, *Astrophys. J.* **404**, 441.

Kallosh, R. (2006). Towards string cosmology, *Prog. Theor. Phys. Supp.* **163**, 323.

Kallosh, R. (2008). On inflation in string theory, *Lect. Notes Phys.* **738**, 119. arXiv:hep-th/0702059.

Kaluza, T. (1921). Zum Unitätsproblem der Physik, *Sitz. Preuss. Akad. Wiss., Math. Phys. Kl.* p. 966.

Kamionkowski, M., Kosowsky, A., and Stebbins, A. (1997). A probe of primordial gravity waves and vorticity, *Phys. Rev. Lett.* **78**, 2058. arXiv:astro-ph/9609132.

Kamionkowski, M. and Loeb, A. (1997). Getting around cosmic variance, *Phys. Rev. D* **56**, 4511.

Kandus, A. and Tsagas, C.G. (2008). Generalized Ohm's law for relativistic plasmas, *Mon. Not. Roy. Astr. Soc.* **385**, 883. arXiv:0711.3573.

Kantowski, R. (1966). *Some Relativistic Cosmological Models*, University of Texas PhD thesis. Reprinted in *Gen. Rel. Grav.* **30**, 1665-1700 (1998).

Kantowski, R. (1969a). The Coma cluster as a spherical inhomogeneity in relativistic dust, *Astrophys. J.* **155**, 1023.

Kantowski, R. (1969b). Corrections in the luminosity-redshift relations of the homogeneous Friedman models, *Astrophys. J.* **155**, 59.

Kantowski, R. (1998). The effects of inhomogeneities on evaluating the mass parameter Ω_m and the cosmological constant Λ, *Astrophys. J.* **507**, 483. arXiv:astro-ph/9802208.

Kantowski, R. (2001). Distance-redshift in inhomogeneous $\Omega_0 = 1$ Friedmann-Lemaître-Robertson-Walker cosmology, *Astrophys. J.* **561**, 491.

Kantowski, R. (2003). The Lamé equation for distance-redshift in partially filled beam Friedmann-Lemaître-Robertson-Walker cosmology, *Phys. Rev. D* **68**, 123516.

Kantowski, R., Chen, B., and Dai, X. (2010). Gravitational lensing corrections in flat ΛCDM cosmology, *Astrophys. J.* **718**, 913. arXiv:0909.3308.

Kantowski, R. and Sachs, R.K. (1966). Some spatially homogeneous anisotropic relativistic cosmological models, *J. Math. Phys.* **7**, 443.

Kazin, E.A., Blanton, M.R., Scoccimarro, R., McBride, C.K., and Berlind, A.A. (2010). Regarding the line-of-sight baryonic acoustic feature in the Sloan Digital Sky Survey and Baryon Oscillation Spectroscopic Survey luminous red galaxy samples, *Astrophys. J.* **719**, 1032. arXiv:1004.2244.

Kerr, R.P. (1963). Scalar invariants and groups of motions in a four dimensional Einstein space, *J. Math. Mech.* **12**, 33.

Khoury, J., Ovrut, B.A., Steinhardt, P.J., and Turok, N. (2001). Ekpyrotic universe: Colliding branes and the origin of the hot big bang, *Phys. Rev. D* **64**, 123522.

Kibble, T.W.B. (1987). Cosmic strings, in *Cosmology and Particle Physics*, ed. E. Alvarez, R. Dominguez Tenreiro, J.M. Ibañez Cabanell, and M. Quiros, Lectures at GIFT XVIIth International Seminar on Theoretical Physics (World Scientific, Singapore), p. 171.

Kibble, T.W.B. and Lieu, R. (2005). Average magnification effect of clumping of matter, *Astrophys. J.* **632**, 718. arXiv:astro-ph/0412275.

Kiefer, C. and Polarski, D. (2009). Why do cosmological perturbations look classical to us?, *Adv. Sci. Lett.* **2**, 164. arXiv:0810.0087.

Kilbinger, M., Benabed, K., Guy, J., Astier, P., Tereno, I. *et al.* (2009). Dark energy constraints and correlations with systematics from CFHTLS weak lensing, SNLS supernovae IA and WMAP5, *Astron. Astrophys.* **497**, 677. arXiv:0810.5129.

Kim, T.-S., Bolton, J.S., Viel, M., Haehnelt, M.G., and Carswell, R.F. (2007). An improved measurement of the flux distribution of the Lyα forest in QSO absorption spectra: the effect of continuum fitting, metal contamination and noise properties, *Mon. Not. Roy. Astr. Soc.* **382**, 1657. arXiv:0711.1862.

King, A.R. and Ellis, G.F.R. (1973). Tilted homogeneous cosmological models, *Commun. Math. Phys.* **31**, 209.

King, D.H. (1991). Gravity-wave insights to Bianchi IX universes, *Phys. Rev. D* **44**, 2356.

Kirshner, R.P. (2009). Foundations of supernova cosmology, in Ruiz-Lapuente (2010). arXiv:0910.0257.

Kirschner, U. and Ellis, G.F.R. (2003). A probability measure for FLRW models, *Class. Quant. Grav.* **20**, 1199.

Klein, O. (1926). The quantum theory and five-dimensional relativity theory (in German), *Z. Phys.* **37**, 895.

Kling, T.P. and Keith, B. (2005). The Bianchi identity and weak gravitational lensing, *Class. Quant. Grav.* **22**, 2921. arXiv:gr-qc/0506118.

Kling, T.P., Newman, E.T., and Perez, A. (2000). Comparative studies of lensing methods, *Phys. Rev. D* **62**, 024025. arXiv:gr-qc/0003057.

Knobe, J., Ohm, K.D., and Vilenkin, A. (2006). Philosophical implications of inflationary cosmology, *Br. J. Phil. Sci* **57**, 47. arXiv:physics/0302071.

Kobayashi, T., Maartens, R., Shiromizu, T., and Takahashi, K. (2007). Cosmological magnetic fields from nonlinear effects, *Phys. Rev. D* **75**, 103501. arXiv:astro-ph/0701596.

Kodama, H. (2002). Phase space of compact Bianchi models with fluid, *Prog. Theor. Phys.* **107**, 305. arXiv:gr-qc/0109064v1.

Kodama, H. and Sasaki, M. (1984). Cosmological perturbation theory, *Prog. Theor. Phys. Supp.* **78**, 1.

Koike, T., Tanimoto, M., and Hosoya, A. (1994). Compact homogeneous universes, *J. Math. Phys.* **35**, 4855. arXiv:gr-qc/9405052.

Kolb, E.W., Marra, V., and Matarrese, S. (2008). Description of our cosmological spacetime as a perturbed conformal Newtonian metric and implications for the backreaction proposal for the accelerating universe, *Phys. Rev. D* **78**, 103002. arXiv:0807.0401.

Kolb, E.W., Marra, V., and Matarrese, S. (2010). Cosmological background solutions and cosmological backreactions, *Gen. Rel. Grav.* **42**, 1399. arXiv:0901.4566.

Kolb, E.W., Matarrese, S., and Riotto, A. (2006). On cosmic acceleration without dark energy, *New J. Phys.* **8**, 322. arXiv:astro-ph/0506534.

Komatsu, E., Dunkley, J., Nolta, M.R., Bennett, C.L., Gold, B. *et al.* (2009). Five-year Wilkinson Microwave Anisotropy Probe observations: Cosmological interpretation, *Astrophys. J. Supp.* **180**, 330. arXiv:0803.0547.

Komatsu, E., Smith, K.M., Dunkley, J., Bennett, C.L., Gold, B. *et al.* (2011). Seven-year Wilkinson Microwave Anisotropy Probe (WMAP) observations: Cosmological interpretation, *Astrophys. J. Supp.* **192**, 18. arXiv:1004.1856.

Kompaneets, A.S. and Chernov, A.S. (1965). Solution of the gravitation equations for a homogeneous anisotropic model, *Sov. Phys. JETP* **20**, 1303.

Kopeikin, S.M. and Schäfer, G. (1999). Lorentz covariant theory of light propagation in gravitational fields of arbitrary-moving bodies, *Phys. Rev. D* **60**, 124002.

Korzynski, M. (2010). Covariant coarse-graining of inhomogeneous dust flow in general relativity, *Class. Quant. Grav.* **27**, 105015. arXiv:0908.4593.

Kowalski, M., Rubin, D., Aldering, G., Agostinho, R.J., Amadon, A. *et al.* (2008). Improved cosmological constraints from new, old, and combined Supernova data sets, *Astrophys. J.* **686**, 749. arXiv:0804.4142.

Koyama, K. and Maartens, R. (2006). Structure formation in the Dvali Gabadadze Porrati cosmological model, *J. Cosmol. Astropart. Phys.* **0601**, 016. arXiv:astro-ph/0511634.

Krasiński, A. (1989). Shearfree normal cosmological models, *J. Math. Phys.* **30**, 433.

Krasiński, A. (1997). *Inhomogeneous Cosmological Models* (Cambridge University Press, Cambridge).

Krasiński, A. (2008). Geometry and topology of the quasi-plane Szekeres model, *Phys. Rev. D* **78**, 064038. arXiv:0805.0529.

Krasiński, A. and Hellaby, C.W. (2004a). Formation of a galaxy with a central black hole in the Lemaître-Tolman model, *Phys. Rev. D* **69**, 043502. arXiv:gr-qc/0309119.

Krasiński, A. and Hellaby, C.W. (2004b). More examples of structure formation in the Lemaître-Tolman model, *Phys. Rev. D* **69**, 023502. arXiv:gr-qc/0303016.

Krasiński, A., Quevedo, H., and Sussman, R.A. (1997). On the thermodynamical interpretation of perfect fluid solutions of the Einstein equations with no symmetry, *J. Math. Phys.* **38**, 2602.

Kretschmann, E. (1917). Über den physikalischen Sinn der Relativitätspostulate, A. Einsteins neue und seine ursprungliche Relativitätstheorie, *Ann. Physik* **53**, 575.

Krisciunas, K. (2008). Type Ia supernovae and the acceleration of the universe: results from the ESSENCE Supernova survey. arXiv:0809.2612.

Kristian, J. and Sachs, R.K. (1966). Observations in cosmology, *Astrophys. J.* **143**, 379.

Kristiansen, J.R., Elgarøy, Ø., and Dahle, H. (2007). Using the cluster mass function from weak lensing to constrain neutrino masses, *Phys. Rev. D* **75**, 083510.

Kronborg, T., Hardin, D., Guy, J., Astier, P., Balland, C. *et al.* (2010). Gravitational lensing in the supernova legacy survey (SNLS), *Astron. Astrophys.* **514**, A44. arXiv:1002.1249.

Kuhn, T. (1977). Objectivity, value judgment, and theory choice, in *The Essential Tension* (University of Chicago Press, Chicago), p. 3.

Kundt, W. (2003). The Bianchi classification in the Schucking-Behr approach, *Gen. Rel. Grav.* **35**, 475. Notes on a seminar by Schucking, with commentary.

Kurki-Suonio, H. and Centrella, J. (1991). Primordial nucleosynthesis with horizon-scale curvature fluctuations, *Phys. Rev. D* **43**, 1087.

Kurki-Suonio, H., Laguna, P., and Matzner, R.A. (1993). Inhomogeneous inflation: numerical evolution, *Phys. Rev. D* **48**, 3611.

Kurki-Suonio, H. and Liang, E. (1992). Relation of redshift surveys to matter distribution in spherically symmetric dust universes, *Astrophys. J.* **390**, 5.

Kustaanheimo, P. and Qvist, B. (1948). A note on some general solutions of the Einstein field equations in a spherically symmetric world, *Comment. Math. Phys. Helsingf.* **13**, 12.

Lachièze-Rey, M. and Luminet, J.P. (1995). Cosmic topology, *Phys. Reports* **254**, 135. arXiv:gr-qc/9605010.

Lake, K. (1980). Local inhomogeneities in a Robertson-Walker background I. General framework, *Astrophys. J.* **240**, 744.

Lake, K. (1987). Some notes on the propagation of discontinuities in solutions to the Einstein equations, in *Vth Brazilian School of Cosmology and Gravitation*, ed. M. Novello (World Scientific, Singapore).

Lake, K. and Pim, R. (1985). Development of voids in the thin-wall approximation. I. General characteristics of spherical vacuum voids, *Astrophys. J.* **298**, 439.

Landry, S. and Dyer, C.C. (1997). Optical properties of the Einstein-de Sitter-Kasner universe, *Phys. Rev. D* **56**, 3307.

Lang, S. (1962). *Introduction to Differentiable Manifolds* (Interscience, New York).

Langlois, D. and Vernizzi, F. (2005). Conserved nonlinear quantities in cosmology, *Phys. Rev. D* **72**, 103501. arXiv:astro-ph/0509078.

Larena, J. (2009). Spatially averaged cosmology in an arbitrary coordinate system, *Phys. Rev. D* **79**, 084006. arXiv:0902.3159.

Larson, D., Dunkley, J., Hinshaw, G., Komatsu, E., Nolta, M.R. *et al.* (2011). Seven-year Wilkinson Microwave Anisotropy Probe (WMAP) observations: Power spectra and WMAP-derived parameters, *Astrophys. J. Supp.* **192**. arXiv:1001.4635.

Lasky, P.D. and Bolejko, K. (2010). The effect of pressure gradients on luminosity distance-redshift relations, *Class. Quant. Grav.* **27**, 035011. arXiv:1001.1159.

Lazkoz, R., Maartens, R., and Majerotto, E. (2006). Observational constraints on phantomlike braneworld cosmologies, *Phys. Rev. D* **74**, 083510. arXiv:astro-ph/0605701.

LeBlanc, V.G. (1997). Asymptotic states of magnetic Bianchi I cosmologies, *Class. Quant. Grav.* **14**, 2281.

LeFloch, P.G. and Rendall, A.D. (2011). A global foliation of Einstein-Euler spacetimes with Gowdy symmetry on T^3, *Archiv. Rat. Mech. Anal.* DOI 10.1007/5000205. Published online, arXiv:1004.0427.

Leith, B.M., Ng, S.C.C., and Wiltshire, D.L. (2008). Gravitational energy as dark energy: Concordance of cosmological tests, *Astrophys. J.* **672**, L91. arXiv:0709.2535.

Lemaître, G. (1931). The beginning of the world from the point of view of quantum theory, *Nature* **127**, 706.

Lemaître, G. (1933a). La formation des nébuleuses dans l'univers en expansion, *C.R. Acad. Sci. Paris* **196**, 1085.

Lemaître, G. (1933b). L'univers en expansion, *Ann. Soc. Sci. Bruxelles A* **53**, 51. Translation by M.A.H. MacCallum in *Gen. Rel. Grav.*, **29**, 641-680 (1997).

Lena, P., Lebrun, F., and Mignard, F. (2010). *Observational Astrophysics* (Springer, Berlin and Heidelberg).

Lesgourgues, J. and Pastor, S. (2006). Massive neutrinos and cosmology, *Phys. Reports* **429**, 307. arXiv:astro-ph/0603494.

Leslie, J. (1989). *Universes* (Routledge, London and New York).

Levin, J. (2002). Topology and the cosmic microwave background, *Phys. Reports* **365**, 251. arXiv:gr-qc/0108043.

Lewis, A. (2004a). CMB anisotropies from primordial inhomogeneous magnetic fields, *Phys. Rev. D* **70**, 043011. arXiv:astro-ph/0406096.

Lewis, A. (2004b). Observable primordial vector modes, *Phys. Rev. D* **70**, 043518. arXiv:astro-ph/0403583.

Lewis, A. and Challinor, A. (2002). Evolution of cosmological dark matter perturbations, *Phys. Rev. D* **66**, 023531. arXiv:astro-ph/0203507.

Lewis, A. and Challinor, A. (2006). Weak gravitational lensing of the CMB, *Phys. Reports* **429**, 1.

Lewis, A., Challinor, A., and Lasenby, A. (2000). Efficient computation of cosmic microwave background anisotropies in closed Friedmann-Robertson-Walker models, *Astrophys. J.* **538**, 473. arXiv:astro-ph/9911177.

Lewis, D.K. (1986). *On the Plurality of Worlds* (Basil Blackwell, Oxford).

Li, N. and Schwarz, D.J. (2007). On the onset of cosmological backreaction, *Phys. Rev. D* **76**, 083011. arXiv:gr-qc/0702043.

Li, N., Seikel, M., and Schwarz, D.J. (2008). Is dark energy an effect of averaging?, *Fortsch. Phys.* **56**, 465. arXiv:0801.3420.

Liang, E.P.T. (1979). Noncrossing timelike singularities of irrotational dust collapse, *Gen. Rel. Grav.* **10**, 1572.

Lichnerowicz, A. (1955). *Théories relativistes de la gravitation et de l'électromagnétisme* (Masson, Paris).

Lidsey, J.E., Wands, D., and Copeland, E.J. (2000). Superstring cosmology, *Phys. Rep.* **337**, 343. arXiv:hep-th/9909061.

Liebscher, S., Härterich, J., Webster, K. and Georgi, M. (2011). Ancient dynamics in Bianchi models: Approach to periodic cycles, *Commun. Math. Phys.* **305**, 59. arXiv:1004.1989.

Lifshitz, E.M. (1946). On the gravitational stability of the expanding universe, *Acad. Sci. USSR J. Phys.* **10**, 116.

Lifshitz, E.M. and Khalatnikov, I.M. (1963). Investigations in relativistic cosmology, *Adv. Phys.* **12**, 185.

Lim, W.C. (2008). New explicit spike solution – non-local component of the generalized Mixmaster attractor, *Class. Quant. Grav.* **25**, 045014. arXiv:0710.0628.

Lim, W.C., Deeley, R.J., and Wainwright, J. (2006). Tilted Bianchi VII$_0$ cosmologies – the radiation bifurcation, *Class. Quant. Grav.* **23**, 3215. arXiv:gr-qc/0601040.

Lim, W.C., Nilsson, U.S., and Wainwright, J. (2001). Anisotropic universes with isotropic cosmic microwave background radiation, *Class. Quant. Grav.* **18**, 5583. arXiv:gr-qc/9912001.

Lim, W.C., Uggla, C., and Wainwright, J.A. (2006). Asymptotic silence-breaking singularities, *Class. Quant. Grav.* **23**, 2607. arXiv:gr-qc/0511139.

Lin, X.-F. and Wald, R.M. (1991). Proof of the closed universe recollapse conjecture for general Bianchi IX cosmologies, *Phys. Rev. D* **41**, 2444.

Linde, A.D. (1986). Eternally existing self-reproducing chaotic inflationary universe, *Phys. Lett. B* **175**, 395.

Linde, A.D., (2007). Towards a gauge invariant volume-weighted probability measure for eternal inflation, *J. Cosmol. Astropart. Phys.* **0706**, 017. arXiv:0705.1160.

Linde, A.D., Linde, D., and Mezhlumian, A. (1995). Do we live in the center of the world?, *Phys. Lett. B* **345**, 203. arXiv:hep-th/9411111.

Linde, A.D., and Noorbala, M. (2010). Measure problem for eternal and non-eternal inflation, *J. Cosmol. Astropart. Phys.* **1009**, 008. arXiv:1006.2170.

Linde, A.D., Vanchurin, V., and Winitzki, S. (2009). Stationary measure in the multiverse, *J. Cosmol. Astropart. Phys.* **0901**, 031. arXiv:0812.0005.

Linder, E.V. (1998). Averaging inhomogeneous universes: volume, angle, line of sight. arXiv:astro-ph/9801122.

Linder, E.V. (2005). Cosmic growth history and expansion history, *Phys. Rev. D* **72**, 043529. arXiv:astro-ph/0507263.

Linder, E.V. (2010). Constraining models of dark energy. arXiv:1004.4646.

Lindquist, R.W. (1966). Relativistic transport theory, *Ann. Phys. (N.Y.)* **37**, 487.

Lindquist, R.W. and Wheeler, J.A. (1957). Dynamics of a lattice universe by a Schwarzschild shell method, *Rev. Mod. Phys.* **29**, 432.

Linet, B. (1985). The static metrics with cylindrical symmetry describing a model of cosmic strings, *Gen. Rel. Grav.* **17**, 1109.

Lu, T.H.-C. and Hellaby, C. (2007). Obtaining the spacetime metric from cosmological observations, *Class. Quant. Grav.* **24**, 4107. arXiv:0705.1060.

Luminet, J. (1978). Spatially homothetic cosmological models, *Gen. Rel. Grav.* **9**, 673.

Luminet, J., Weeks, J.R., Riazuelo, A., Lehoucq, R., and Uzan, J.P. (2003). Dodecahedral space topology as an explanation for weak wide-angle temperature correlations in the cosmic microwave background, *Nature* **425**, 593.

Luzzi, G., Shimon, M., Lamagna, L., Rephaeli, Y., De Petris, M. *et al* (2009). Redshift dependence of the CMB temperature from S-Z measurements, *Astrophys. J.* **705**, 1122. arXiv:0909.2815.

Lyth, D.H. and Liddle, A.R. (2009). *The Primordial Density Perturbation: Cosmology, Inflation and the Origin of Structure* (Cambridge University Press, Cambridge).

Lyth, D.H. and Mukherjee, M. (1988). Fluid flow description of density irregularities in the universe, *Phys. Rev. D* **38**, 485.

Lyth, D.H. and Woszczyna, A. (1995). Large scale perturbations in the open universe, *Phys. Rev. D* **52**, 3338. arXiv:astro-ph/9501044.

Lyttleton, R. and Bondi, H. (1959). On the physical consequences of a general excess of charge, *Proc. Roy. Soc. London A* **252**, 313.

Ma, C.-P. and Bertschinger, E. (1995). Cosmological perturbation theory in the synchronous and conformal Newtonian gauges, *Astrophys. J.* **455**, 7. arXiv:astro-ph/9401007.

Ma, P.K.-H. and Wainwright, J. (1994). A dynamical systems approach to the oscillatory singularity in Bianchi cosmologies, in Hobill, Burd and Coley (1994), p. 449. A previous version appeared in *Relativity Today*, ed. Z. Perjés (Nova Science, Commack), 1992, the Proceedings of the Third Hungarian Workshop.

Maartens, R. (1980). *Idealised observations in relativistic cosmology*, University of Cape Town PhD thesis.

Maartens, R. (1997). Linearization instability of gravity waves?, *Phys. Rev. D* **55**, 463. arXiv:astro-ph/9609198.

Maartens, R. (1998). Covariant velocity and density perturbations in quasi-Newtonian cosmologies, *Phys. Rev. D* **5812**, 4006. arXiv:astro-ph/9808235.

Maartens, R. (2004). Brane-world gravity, *Living Rev. Relativity* http://relativity.livingreviews.org/Articles/lrr-2004-7/. arXiv:gr-qc/0312059.

Maartens, R. and Bassett, B.A.C.C. (1998). Gravito-electromagnetism, *Class. Quant. Grav.* **15**, 705. arXiv:gr-qc/9704059.

Maartens, R., Ellis, G.F.R., and Siklos, S.T.C. (1997). Local freedom in the gravitational field, *Class. Quant. Grav.* **14**, 1927. arXiv:gr-qc/9611003.

Maartens, R., Ellis, G.F.R., and Stoeger, W.R. (1995a). Improved limits on anisotropy and inhomogeneity from the cosmic background radiation, *Phys. Rev. D* **51**, 5942.

Maartens, R., Ellis, G.F.R., and Stoeger, W.R. (1995b). Limits on anisotropy and inhomogeneity from the cosmic background radiation, *Phys. Rev. D* **51**, 1525. arXiv:astro-ph/9501016.

Maartens, R., Gebbie, T., and Ellis, G.F.R. (1999). Cosmic microwave background anisotropies: Nonlinear dynamics, *Phys. Rev. D* **59**, 083506. arXiv:astro-ph/9808163.

Maartens, R., Humphreys, N.P., Matravers, D.R., and Stoeger, W.R. (1996). Inhomogeneous universes in observational coordinates, *Class. Quant. Grav.* **13**, 253. Erratum: *Class. Quant. Grav.* (1996) **13**, 1689.

Maartens, R. and Koyama, K. (2010). Brane-World gravity, *Living Rev. Relativity* http://relativity.livingreviews.org/Articles/lrr-2010-5/.arXiv:1004.3962.

Maartens, R., Lesame, W.M., and Ellis, G.F.R. (1998). Newtonian-like and anti-Newtonian universes, *Class. Quant. Grav.* **15**, 1005.

Maartens, R. and Majerotto, E. (2006). Observational constraints on self-accelerating cosmology, *Phys. Rev. D* **74**, 023004. arXiv:astro-ph/0603353.

Maartens, R. and Matravers, D.R. (1994). Isotropic and semi-isotropic observations in cosmology, *Class. Quant. Grav.* **11**, 2693.

Maartens, R., Triginer, J., and Matravers, D.R. (1999). Stress effects in structure formation, *Phys. Rev. D* **60**, 103503. arXiv:astro-ph/9901213.

Maartens, R., Tsagas, C.G., and Ungarelli, C. (2001). Magnetized gravitational waves, *Phys. Rev. D* **63**, 123507. arXiv:astro-ph/0101151.

MacCallum, M.A.H. (1973). Cosmological models from the geometric point of view, in *Cargese Lectures in Physics, Vol. 6*, ed. E. Schatzman (Gordon and Breach, New York), p. 61.

MacCallum, M.A.H. (1979). Anisotropic and inhomogeneous relativistic cosmologies, in *General Relativity: An Einstein Centenary Survey*, ed. S.W. Hawking and W. Israel (Cambridge University Press, Cambridge), p. 533.

MacCallum, M.A.H. (1982). Relativistic cosmology for astrophysicists, in *Origin and Evolution of the Galaxies*, ed. V. de Sabbata (World Scientific, Singapore), p. 9. Also, in revised form, in *Origin and Evolution of the Galaxies*, (1983) ed. B.J.T. and J.E. Jones, volume 97 of Nato Advanced Study Institute Series, p. 9 (D.Reidel and Co., Dordrecht).

MacCallum, M.A.H. (1998). Integrability in tetrad formalisms and conservation in cosmology, in *Current Topics in Mathematical Cosmology (Proceedings of the International Seminar)*, ed. M. Rainer and H.-J. Schmidt (World Scientific, Singapore), p. 133.

MacCallum, M.A.H. and Ellis, G.F.R. (1970). A class of homogeneous cosmological models II: observations, *Commun. Math. Phys.* **19**, 31.

MacCallum, M.A.H. and Taub, A.H. (1972). Variational principles and spatially homogeneous universes, including rotation, *Commun. Math. Phys.* **25**, 173.

Maciejewski, A.J. and Szydłowski, M. (1998). On the integrability of Bianchi cosmological models, *J. Phys. A* **31**, 2031. arXiv:gr-qc/9702045.

Madsen, M.S. and Ellis, G.F.R. (1988). Evolution of Ω in inflationary universes, *Mon. Not. Roy. Astr. Soc.* **234**, 217.

Magueijo, J. (2003). New varying speed of light theories, *Rep. Prog. Phys* **66**, 2025. arXiv:astro-ph/0305457.

Malik, K.A. and Wands, D. (2009). Cosmological perturbations, *Phys. Rep.* **475**, 1. arXiv:0809.4944.

Manasse, F.K. and Misner, C.W. (1963). Fermi normal coordinates and some basic concepts in differential geometry, *J. Math. Phys.* **4**, 735.

Mantz, A., Allen, S.W., Ebeling, H., and Rapetti, D. (2008). New constraints on dark energy from the observed growth of the most X-ray luminous galaxy clusters, *Mon. Not. Roy. Astr. Soc.* **387**, 1179. arXiv:0709.4294.

Marra, V., Kolb, E.W., and Matarrese, S. (2008). Light-cone averages in a Swiss-cheese universe, *Phys. Rev. D* **77**, 023003. arXiv:0710.5505.

Marra, V., Kolb, E.W., Matarrese, S., and Riotto, A. (2007). On cosmological observables in a Swiss-cheese universe, *Phys. Rev. D* **76**, 123004. arXiv:0708.3622.

Mars, M. (2001). On the uniqueness of the Einstein-Straus model, *Class. Quant. Grav.* **18**, 3645.

Mars, M. and Senovilla, J.M.M. (1993). Geometry of general hypersurfaces in spacetime – junction conditions, *Class. Quant. Grav.* **10**, 1865.

Marsden, J.E. (1982). Spaces of solutions of relativistic field theories with constraints, in *Differential Geometric Methods in Mathematical Physics*, volume 905 of *Springer Lecture Notes in Mathematics* (Springer, Berlin and Heidelberg), p. 29.

Martin, J. and Brandenberger, R.H. (2001). Trans-Planckian problem of inflationary cosmology, *Phys. Rev. D* **63**, 123501. arXiv:hep-th/0005209.

Martín, J. and Senovilla, J.M.M. (1986). Petrov type D perfect-fluid solutions in generalized Kerr-Schild form, *J. Math. Phys.* **27**, 2209. Erratum: *J. Math. Phys.*, 27, 2209.

Martin-Pascual, F. and Senovilla, J.M.M. (1988). Petrov types D and II perfect fluid solutions in generalized Kerr-Schild form, *J. Math. Phys.* **29**, 937.

Martineau, P. and Brandenberger, R.H. (2005). The effects of gravitational backreaction on cosmological perturbations, *Phys. Rev. D* **72**, 023507. arXiv:astro-ph/0505236.

Massey, R., Kitching, T., and Richard, J. (2010). The dark matter of gravitational lensing, *Rep. Prog. Phys.* **73**, 086901. arXiv:1001.1739.

Massey, R., Rhodes, J., Ellis, R., Scoville, N., Leauthaud, A. *et al.* (2007). Dark matter maps reveal cosmic scaffolding, *Nature* **445**, 286. arXiv:astro-ph/0701594.

Mather, J.C., Cheng, E.S., Eplee, Jr., R.E. Isaacman, R.B., Meyer, S.S. *et al.* (1990). A preliminary measurement of the cosmic microwave background spectrum by the cosmic background explorer (COBE) satellite, *Astrophys. J.* **354**, L37.

Matravers, D.R. and Ellis, G.F.R. (1989). Evolution of anisotropies in Friedmann cosmologies, *Class. Quant. Grav.* **6**, 369.

Matravers, D.R., Madsen, M.S., and Vogel, D.L. (1985). The microwave background and (m, z) relations in a tilted cosmological model, *Astrophys. Sp. Sci.* **112**, 193.

Matravers, D.R., Vogel, D.L., and Madsen, M.S. (1984). Helium formation in a Bianchi V universe with tilt, *Class. Quant. Grav.* **1**, 407.

Mattsson, T. (2010). Dark energy as a mirage, *Gen. Rel. Grav.* **42**, 567. arXiv:0711.4264.

Matzner, R.A., Rothman, T., and Ellis, G.F.R. (1986). Conjecture on isotope production in the Bianchi cosmologies, *Phys. Rev. D* **34**, 2926.

Mavrides, S. (1977). Anomalous Hubble expansion and inhomogeneous cosmological models, *Mon. Not. Roy. Astr. Soc.* **177**, 709.

McAllister, L. and Silverstein, E. (2008). String cosmology: a review, *Gen. Rel. Grav.* **40**, 565. arXiv:0710.2951.

McClure, M.L. and Hellaby, C.W. (2008). Determining the metric of the cosmos: stability, accuracy, and consistency, *Phys. Rev. D* **78**, 044005. arXiv:0709.0875.

McCrea, W.H. (1935). Observable relations in relativistic cosmology. I, *Zs. f. Astrophys.* **9**, 290.

McCrea, W.H. (1939). Observable relations in relativistic cosmology. II, *Zs. f. Astrophys.* **18**, 98. Reprinted as *Gen. Rel. Grav.* **30**, 315 (1998).

McCrea, W.H. (1953). Cosmology, *Rep. Prog. Phys.* **16**, 321.

McCrea, W.H. (1960). The interpretation of cosmology, *Nature* **186**, 1035.

McCrea, W.H. (1970). A philosophy for big-bang cosmology, *Nature* **228**, 21.

McCrea, W.H. and Milne, E.A. (1934). Newtonian universes and the curvature of space, *Q. J. Math.* **5**, 73.

McGaugh, S.S., de Blok, W.J.G., Schombert, J.M., Kuzio de Naray, R., and Kim, J.H. (2007). The rotation velocity attributable to dark matter at intermediate radii in disk galaxies, *Astrophys. J.* **659**, 149. arXiv:astro-ph/0612410.

McLenaghan, R.G. and Sasse, F.D. (1996). Non-existence of Petrov type III space-times on which Weyl's neutrino or Maxwell's equations satisfy Huygens' principle, *Ann. Inst. H. Poincaré* **65**, 253.

McVittie, G.C. (1933). The mass-particle in an expanding universe, *Mon. Not. Roy. Astr. Soc.* **93**, 325.

Mena, F.C. and MacCallum, M.A.H. (2002). Locally discretely isotropic space-times, in *Proceedings of the Ninth Marcel Grossmann Meeting on General Relativity*, ed. V.G. Gurzadyan, R.T. Jantzen, and R. Ruffini (World Scientific, Singapore), p. 1976.

Mena, F.C. and Tavakol, R.K. (1999). Evolution of the density contrast in inhomogeneous dust models, *Class. Quant. Grav.* **16**, 435.

Mena, F.C., Tavakol, R.K., and Vera, R. (2002). Generalization of the Einstein-Strauss model to anisotropic settings, *Phys. Rev. D* **66**, 044004.

Meszaros, A. (1991). On shell crossing in the Tolman metric, *Mon. Not. Roy. Astr. Soc.* **253**, 619.

Meszaros, A. and Molner, Z. (1996). On the alternative origin of the dipole anisotropy of microwave background due to the Rees-Sciama effect, *Astrophys. J.* **470**, 49.

Meszaros, P. (1974). The behaviour of point masses in an expanding cosmological substratum, *Astron. Astrophys.* **37**, 225.

Milne, E.A. (1934). A Newtonian expanding universe, *Q. J. Math.* **5**, 64.

Milne, E.A. (1935). *Relativity Gravitation and World-Structure* (Clarendon Press, Oxford).

Miralda-Escudé, J. (1996). The mass distribution in clusters of galaxies from weak and strong lensing, in *Astrophysical Applications of Gravitational Lensing*, ed. C.S. Kochanek and J.N. Hewitt, volume 173 of *IAU Symposium* (Kluwer Academic Publishers, Dordrecht), p. 131. arXiv:astro-ph/9509077.

Misner, C.W. (1967). Neutrino viscosity and the isotropy of primordial blackbody radiation, *Phys. Rev. Lett.* **19**, 533.

Misner, C.W. (1968). The isotropy of the universe, *Astrophys. J.* **151**, 431.

Misner, C.W. (1969a). Mixmaster universe, *Phys. Rev. Lett.* **22**, 1071.

Misner, C.W. (1969b). Quantum cosmology I, *Phys. Rev.* **186**, 1319.

Misner, C.W., Thorne, K.S., and Wheeler, J.A. (1973). *Gravitation* (W.H. Freeman and Co., San Francisco).

Mollerach, S. and Roulet, E. (2002). *Gravitational Lensing and Microlensing* (World Scientific, Singapore).

Moniez, M. (2010). Microlensing as a probe of the galactic structure; 20 years of microlensing optical depth studies. p. 2047 in Jetzer, Mellier and Perlick (2010), arXiv:1001.2707.

Morganson, E., Marshall, P., Treu, T., Schrabback, T. and Blandford, R.D. (2010). Direct observation of cosmic strings via their strong gravitational lensing effect: II. Results from the HST/ACS image archive, *Mon. Not. Roy. Astr. Soc.* **406**, 2452. arXiv:0908.0602.

Moss, A., Zibin, J.P. and Scott, D. (2011). Precision cosmology defeats void models for acceleration, *Phys. Rev. D* **83**, 103515. arXiv:1007.3725.

Mota, B., Rebouças, M.J., and Tavakol, R. (2010). Circles-in-the-sky searches and observable cosmic topology in a flat universe, *Phys. Rev. D* **81**, 103516. arXiv:1002.0834.

Mukhanov, V.F. (2005). *Cosmology* (Cambridge University Press, Cambridge).

Mukhanov, V.F., Abramo, L.R., and Brandenberger, R.H. (1997). Backreaction problem for cosmological perturbations, *Phys. Rev. Lett.* **78**, 1624.

Mukhanov, V.F., Feldman, H.A., and Brandenberger, R.H. (1992). Theory of cosmological perturbations, *Phys. Reports* **215**, 203.

Mulryne, D.J., Tavakol, R., Lidsey, J.E., and Ellis, G.F.R. (2005). An emergent universe from a loop, *Phys. Rev. D* **71**, 123512. See also M. Bojowald. (2005). Original Questions, *Nature* **436**, 920-921 (2005).

Munitz, M.K. (1962). The logic of cosmology, *Br. J. Phil. Sci.* **13**, 104.

Munitz, M.K. (1986). *Cosmic Understanding: Philosophy and Science of the Universe* (Princeton University Press, Princeton).

Mureika, J.R. and Dyer, C.C. (2004). Multifractal analysis of packed Swiss cheese cosmologies, *Gen. Rel. Grav.* **36**, 151. arXiv:gr-qc/0505083.

Murphy, M.T., Webb, J.K., and Flambaum, V.V. (2008). Revision of VLT/UVES constraints on a varying fine-structure constant, *Mon. Not. Roy. Astr. Soc.* **384**, 1053.

Mustapha, N., Bassett, B.A.C.C., Hellaby, C.W., and Ellis, G.F.R. (1998). The distortion of the area distance-redshift relation in inhomogeneous isotropic universes, *Class. Quant. Grav.* **15**, 2363.

Mustapha, N., Ellis, G.F.R., van Elst, H., and Marklund, M. (2000). Partially locally rotationally symmetric perfect fluid cosmologies I, *Class. Quant. Grav.* **17**, 3135.

Mustapha, N., Hellaby, C.W., and Ellis, G.F.R. (1999). Large-scale inhomogeneity versus source evolution: can we distinguish them observationally?, *Mon. Not. Roy. Astr. Soc.* **292**, 817. arXiv:gr-qc/9808079.

Nakahara, M. (1990). *Geometry, Topology and Physics* (Institute of Physics Publishing, Bristol).

Nambu, Y. (2002). Backreaction and the effective Einstein equation for the universe with ideal fluid cosmological perturbations, *Phys. Rev. D* **65**, 104013.

Narlikar, J.V. (1963). Newtonian solutions with shear and rotation, *Mon. Not. Roy. Astr. Soc.* **126**, 203.

Nel, S.D. (1980). *Observational Space-times*, University of Cape Town PhD thesis.

Newman, R.P.A.C. (1979). *Singular perturbations of the empty Robertson-Walker cosmologies*, University of Kent PhD thesis.

Nicastro, F., Mathur, S., Elvis, M., Drake, J., Fang, T. *et al.* (2005). A mass measurement for the missing baryons in the warm-hot intergalactic medium via the X-ray forest, *Nature* **433**, 495.

Nilsson, U.S., Hancock, M.J., and Wainwright, J. (2000). Non-tilted Bianchi VII$_0$ models - the radiation fluid, *Class. Quant. Grav.* **17**, 3119. arXiv:gr-qc/9912019.

Nilsson, U.S., Uggla, C., Wainwright, J., and Lim, W.C. (1999). An almost isotropic cosmic microwave temperature does not imply an almost isotropic universe, *Astrophys. J. Lett.* **522**, L1. arXiv:astro-ph/9904252.

Noerdlinger, P.D. and Petrosian, V. (1971). The effect of cosmological expansion on self-gravitating ensembles of particles, *Astrophys. J.* **168**, 1.

Nolan, B. (1993). Sources for McVittie mass particle in an expanding universe, *J. Math. Phys.* **34**, 178.

Nolan, B. (1999). A point mass in an isotropic universe: II. Global properties, *Class. Quant. Grav.* **16**, 1227.

Nolan, B.C. and Vera, R. (2005). Axially symmetric equilibrium regions of Friedmann-Lemaître-Robertson-Walker universes, *Class. Quant. Grav.* **22**, 4031. arXiv:gr-qc/0505093.

Nolan, B.C. and Vera, R. (2007). On global models for finite rotating objects in equilibrium in cosmological backgrounds, in *Beyond General Relativity, Proceedings of the 2004 Spanish Relativity Meeting 2004 (ERE 2004)*, ed. N. Alonso-Alberca, E. Álvarez, T. Ortín, and M.A. Vázquez-Mozo, volume 119 of *Colección de Estudios* (Ediciones de la Universidad Autónoma de Madrid, Madrid). arXiv:gr-qc/0501074.

North, J.D. (1965). *The Measure of the Universe* (Oxford University Press, Oxford). Re-issued by Dover, New York, 1990 with new preface.

Norton, J. (1998). The cosmological woes of Newtonian theory, in *The Expanding Worlds of General Relativity*, ed. H. Goenner, J. Renn, J. Ritter, and T. Sauer, volume 7 of *Einstein Studies* (Birkhäuser, Basel and Boston), p. 271.

Notari, A. (2006). Late time failure of the Friedmann equation, *Mod. Phys. Lett. A* **21**, 2997. arXiv:astro-ph/0503715.

Nwankwo, A., Thompson, J. and Ishak, M. (2011). Luminosity distance and redshift in the Szekeres inhomogeneous cosmological models, *J. Cosmol. Astropart. Phys.* **1105**, 028. arXiv:1005.2989.

O'Brien, S. and Synge, J.L. (1952). Jump conditions at discontinuities in general relativity, *Commun. Dublin Inst. Adv. Studies A* **9**, 1. Reprinted 1973.

Occhionero, F., Santangelo, P., and Vittorio, N. (1983). Holes in cosmology, *Astron. Astrophys.* **117**, 365.

Oleson, M. (1971). A class of type [4] perfect fluid space-times, *J. Math. Phys.* **12**, 666.

Olive, K.A. (2010). The violent Universe: the Big Bang. arXiv:1005.3955.

Ozsváth, I. (1965). New homogeneous solutions of Einstein's field equations with incoherent matter obtained by a spinor technique, *J. Math. Phys.* **6**, 590.

Ozsváth, I. (1970). Dust-filled universes of class II and class III, *J. Math. Phys.* **11**, 2871.

Ozsváth, I. and Schucking, E. (1962). An anti-Mach metric, in *Recent Developments in General Relativity* (Pergamon Press and PWN Warsaw, Oxford), p. 339.

Padmanabhan, T. (1993). *Structure Formation in the Universe* (Cambridge University Press, Cambridge).

Padmanabhan, T., Seshadri, T.R., and Singh, T.P. (1989). Making inflation work – damping of density perturbations due to Planck energy cutoff, *Phys. Rev. D* **39**, 2100.

Page, D.N. (2003). Quantum cosmology, in Gibbons, Shellard and Rankin (2003).

Page, L., Hinshaw, G., Komatsu, E., Nolta, M.R., Spergel, D.N. *et al.* (2007). Three-year Wilkinson Microwave Anisotropy Probe (WMAP) observations: Polarization analysis, *Astrophys. J. Supp.* **170**, 335. arXiv:astro-ph/0603450.

Panek, M. (1992). Cosmic background radiation anisotropies from cosmic structures – Models based on the Tolman solution, *Astrophys. J.* **388**, 225.

Paoletti, D., Finelli, F., and Paci, F. (2009). The scalar, vector and tensor contributions of a stochastic background of magnetic fields to cosmic microwave background anisotropies, *Mon. Not. Roy. Astr. Soc.* **396**, 523. arXiv:0811.0230.

Paraficz, D. and Hjorth, J. (2010). The Hubble constant inferred from 18 time-delay lenses, *Astrophys. J.* **712**, 1378. arXiv:1002.2570.

Paranjape, A. and Singh, T.P. (2008). Explicit cosmological coarse graining via spatial averaging, *Gen. Rel. Grav.* **40**, 139. arXiv:astro-ph/0609481.

Pareja, M.J. and MacCallum, M.A.H. (2006). Local freedom in the gravitational field revisited, *Class. Quant. Grav.* **23**, 5039. arXiv:gr-qc/0605075.

Parry, J., Salopek, D.S., and Stewart, J.M. (1994). Solving the Hamilton-Jacobi equation for general relativity, *Phys. Rev. D* **49**, 2872. arXiv:gr-qc/9310020.

Peacock, J. (1999). *Cosmological Physics* (Cambridge University Press, Cambridge).

Peebles, P.J.E. (1968). Recombination of the primeval plasma, *Astrophys. J.* **153**, 1.

Peebles, P.J.E. (1971). *Physical Cosmology* (Princeton University Press, Princeton).

Peebles, P.J.E. (1980). *The Large-scale Structure of the Universe* (Princeton University Press, Princeton).

Peebles, P.J.E. (1982). Large-scale background temperature and mass fluctuations due to scale-invariant primeval perturbations, *Astrophys. J. Lett.* **263**, L1.

Peebles, P.J.E. and Yu, J.T. (1970). Primeval adiabatic perturbation in an expanding universe, *Astrophys. J.* **162**, 815.

Peiris, H.V. and Easther, R. (2006). Recovering the inflationary potential and primordial power spectrum with a slow roll prior: methodology and application to WMAP three year data, *J. Cosmol. Astropart. Phys.* **0607**, 002. arXiv:astro-ph/0603587.

Pelavas, N. and Coley, A.A. (2006). Gravitational entropy in cosmological models, *Int. J. Theor. Phys.* **45**, 1258. arXiv:gr-qc/0410008.

Penrose, R. (1965). Gravitational collapse and space-time singularities, *Phys. Rev. Lett.* **14**, 579.

Penrose, R. (1968). Structure of space-time, in *Battelle rencontres 1967: Lectures in Mathematics and Physics*, ed. C.M. DeWitt and J.A. Wheeler (W.A. Benjamin, New York), p. 121.

Penrose, R. (1989). Difficulties with inflationary cosmology, in *14th Texas Symposium on Relativistic Astrophysics*, ed. E.J. Fergus (Proc. New York Academy of Science, New York).

Penrose, R. (1999). *The Emperor's New Mind: Concerning Computers, Minds, and the Laws of Physics* (Oxford University Press, Oxford). Second edition.

Penrose, R. (2004). *The Road to Reality: A Complete Guide to the Laws of the Universe* (Jonathan Cape, London).

Penrose, R. (2006). Before the big bang: An outrageous new perspective and its implications. Proc. of EPAC 2006 (Edinburgh, Scotland), online at CERN, Geneva. http://accelconf.web.cern.ch/AccelConf/e06/.

Penrose, R. and Isham, C.J. (1986). *Quantum Concepts in Space and Time* (Oxford University Press, Oxford).

Penrose, R. and Rindler, W. (1984). *Spinors and Space-time I: Two-spinor Calculus and Relativistic Fields* (Cambridge University Press, Cambridge).

Penrose, R. and Rindler, W. (1985). *Spinors and Space-time II: Spinor and Twistor Methods in Space-time Geometry* (Cambridge University Press, Cambridge).

Percival, W.J., Nichol, R.C., Eisenstein, D.J., Frieman, J.A., Fukugita, M. *et al.* (2007). The shape of the Sloan Digital Sky Survey Data Release 5 galaxy power spectrum, *Astrophys. J.* **657**, 645. arXiv:astro-ph/0608636.

Percival, W.J., Reid, B.A., Eisenstein, D.J., Bahcall, N.A., Budavari, T. *et al* (2010). Baryon acoustic oscillations in the Sloan Digital Sky Survey Data Release 7 galaxy sample, *Mon. Not. Roy. Astr. Soc.* **401**, 2148. arXiv:0907.1660.

Pereira, T.S., Pitrou, C., and Uzan, J.-P. (2007). Theory of cosmological perturbations in an anisotropic universe, *J. Cosmol. Astropart. Phys.* **0709**, 006. arXiv:0707.0736.

Peresetskii, A.A. (1977). Singularity of homogeneous Einstein metrics, *Math. Notes* **X**, 39.

Perez, A., Sahlmann, H., and Sudarsky, D. (2006). On the quantum origin of the seeds of cosmic structure, *Class. Quant. Grav.* **23**, 2317. arXiv:gr-qc/0508100.

Perivolaropoulos, L. (2010). Consistency of ΛCDM with geometric and dynamical probes, *J. Phys.: Conf. Series* **222**, 012024. arXiv:1002.3030.

Perlick, V. (2004). Gravitational lensing from a spacetime perspective, *Living Rev. Relativity* http://relativity.livingreviews.org/Articles/lrr-2004-9/.

Peter, P. and Uzan, J.-P. (2009). *Primordial Cosmology* (Oxford University Press, Oxford).

Petters, A.O., Levine, H., and Wambsganss, J. (2001). *Singularity Theory and Gravitational Lensing*, volume 21 of *Progress in Mathematical Physics* (Birkhäuser, Boston, U.S.A.).

Petters, A.O. and Werner, M. (2010). Mathematics of gravitational lensing: multiple imaging and magnification. p. 2011 in Jetzer, Mellier and Perlick (2010), arXiv:0912.0490.

Pettini, M. (1999). Element abundances at high redshifts, in *Chemical Evolution from Zero to High Redshift*, ed. J.R. Walsh and M.R. Rosa (Springer-Verlag, Berlin). arXiv:astro-ph/9902173.

Pettini, M., Lipman, K., and Hunstead, R.W. (2005). Element abundances at high redshifts: The N/O ratio in a primeval galaxy. arXiv:astro-ph/9502077.

Pirani, F.A.E. (1956). On the physical significance of the Riemann tensor, *Acta. Phys. Polon.* **15**, 389.

Pitrou, C. (2009). The radiative transfer at second order: a full treatment of the Boltzmann equation with polarization, *Class. Quant. Grav.* **26**, 065006. arXiv:0809.3036.

Pitrou, C. and Uzan, J.-P. (2007). Quantization of perturbations during inflation in the 1+3 covariant formalism, *Phys. Rev. D* **75**, 087302. arXiv:gr-qc/0701121.

Pogosian, L., Silvestri, A., Koyama, K., and Zhao, G.-B. (2010). How to optimally parametrize deviations from general relativity in the evolution of cosmological perturbations, *Phys. Rev. D* **81**(10), 104023. arXiv:1002.2382.

Polnarev, A.G. (1985). Polarization and anisotropy induced in the microwave background by cosmological gravitational waves, *Sov. Astron.* **29**, 607.

Pontzen, A. (2009). Rogue's gallery: the full freedom of Bianchi CMB anomalies, *Phys. Rev. D* **79**, 103518. arXiv:0901.2122.

Pontzen, A. and Challinor, A. (2010). Linearization of homogeneous, nearly-isotropic cosmological models. arXiv:1009.3935.

Pullen, A.R. and Hirata, C.M. (2010). Non-detection of a statistically anisotropic power spectrum in large-scale structure, *J. Cosmol. Astropart. Phys.* **1005**, 027. arXiv:1003.0673.

Pyne, T. and Birkinshaw, M. (1996). Beyond the thin lens approximation, *Astrophys. J.* **458**, 46Å 56.

Raine, D.J. (1981). *The Isotropic Universe* (Adam Hilger, Bristol).

Raine, D.J. and Thomas, E.G. (1981). Large-scale inhomogeneity in the universe and the anisotropy of the microwave background, *Mon. Not. Roy. Astr. Soc.* **195**, 649.

Rainer, M. and Schmidt, H.J. (1995). Inhomogeneous cosmological models with homogeneous inner hypersurface geometry, *Gen. Rel. Grav.* **27**, 1265.

Räsänen, S. (2004). Dark energy from backreaction, *J. Cosmol. Astropart. Phys.* **0402**, 003. arXiv:astro-ph/0311257.

Räsänen, S. (2008). Evaluating backreaction with the peak model of structure formation, *J. Cosmol. Astropart. Phys.* **0804**, 026. arXiv:0801.2692.

Räsänen, S. (2009). Relation between the isotropy of the CMB and the geometry of the universe, *Phys. Rev. D* **79**, 123522. arXiv:0903.3013.

Raychaudhuri, A.K. (1955). Relativistic cosmology I, *Phys. Rev.* **98**, 1123.

Raychaudhuri, A.K. and Modak, B. (1988). Cosmological inflation with arbitrary initial conditions, *Class. Quant. Grav.* **5**, 225.

Rees, M.J. (1968). Polarization and spectrum of the primeval radiation in an anisotropic universe, *Astrophys. J. Lett.* **153**, L1.

Rees, M.J. (1999). *Just Six Numbers: The Deep Forces that Shape the Universe* (Weidenfeld and Nicholson, London).

Rees, M.J. (2001). *Our Cosmic Habitat* (Princeton University Press, Princeton).

Rees, M.J. and Sciama, D.W. (1968). Large scale density inhomogeneities in the Universe, *Nature* **217**, 511.

Regis, M. and Clarkson, C.A. (2010). Do primordial Lithium abundances imply there's no dark energy? arXiv:1003.1043.

Reiterer, M. and Trubowitz, E. (2010). The BKL conjectures for spatially homogeneous spacetimes. arXiv:1005.4908.

Rendall, A.D. (1996). The initial singularity in solutions of the Einstein-Vlasov system of Bianchi type I, *Class. Quant. Grav.* **16**, 1705.

Rendall, A.D. (2004). Asymptotics of solutions of the Einstein equations with positive cosmological constant, *Ann. Inst. H. Poincaré* **5**, 1041. arXiv:gr-qc/0312020.

Rendall, A.D. and Tod, K.P. (1999). Dynamics of spatially homogeneous solutions of the Einstein-Vlasov equations which are locally rotationally symmetric, *J. Math. Phys.* **37**, 438.

Rendall, A.D. and Weaver, M. (2001). Manufacture of Gowdy spacetimes with spikes, *Class. Quant. Grav.* **18**, 2959.

Reyes, R., Mandelbaum, R., Seljak, U., Baldauf, T., Gunn, J.E. *et al.* (2010). Confirmation of general relativity on large scales from weak lensing and galaxy velocities, *Nature* **464**, 256. arXiv:1003.2185.

Riazuelo, A., Weeks, J., Uzan, J.-P., Lehoucq, R., and Luminet, J.-P. (2004). Cosmic microwave background anisotropies in multi-connected flat spaces, *Phys. Rev. D* **69**, 103518. arXiv:astro-ph/0311314.

Ribeiro, M.B. (1992a). On modelling a relativistic hierarchical (fractal) cosmology by Tolman's spacetime. I. Theory, *Astrophys. J.* **388**, 1.

Ribeiro, M.B. (1992b). On modelling a relativistic hierarchical (fractal) cosmology by Tolman's spacetime. II. Analysis of the Einstein-de Sitter model, *Astrophys. J.* **395**, 29.

Ricci, G. and Levi-Civita, T. (1901). Méthodes de calcul differentiel absolu et leurs applications, *Math. Ann.* **54**, 125.

Rindler, W. (1956). Visual horizons in world models, *Mon. Not. Roy. Astr. Soc.* **116**, 662.

Rindler, W. (1977). *Essential Relativity (2nd edition)* (Springer, New York).

Rindler, W. (2001). *Relativity: Special, General, and Cosmological* (Oxford University Press, Oxford).

Ringström, H. (2000). *On the asymptotics of Bianchi class A spacetimes*, Royal Institute of Technology, Stockholm, PhD thesis.

Ringström, H. (2003). Future asymptotic expansions of Bianchi VIII vacuum metrics, *Class. Quant. Grav.* **20**, 1943. arXiv:gr-qc/0301101.

Robertson, H.P. (1933). Relativistic cosmology, *Rev. Mod. Phys.* **5**, 62.

Röhr, N. and Uggla, C. (2005). Conformal regularization of Einstein's field equations, *Class. Quant. Grav.* **22**, 3775.

Rosenthal, E. and Flanagan, É.É. (2008). Cosmological backreaction and spatially averaged spatial curvature. arXiv:0809.2107.

Rosquist, K. (1980). Global rotation, *Gen. Rel. Grav.* **12**, 649.

Rosquist, K. and Uggla, C. (1991). Killing tensors in two-dimensional space-times with applications to cosmology, *J. Math. Phys.* **32**, 3412.

Rothman, T. and Ellis, G.F.R. (1986). Can inflation occur in anisotropic cosmologies?, *Phys. Lett. B* **180**, 19.

Rothman, T. and Matzner, R.A. (1984). Nucleosynthesis in anisotropic cosmologies revisited, *Phys. Rev. D* **30**, 1649.

Roukema, B.F., Bulinski, Z., Szaniewska, A., and Gaudin, N.E. (2008). The optimal phase of the generalised Poincaré dodecahedral space hypothesis implied by the spatial cross-correlation function of the WMAP sky maps, *Astron. Astrophys.* **486**, 55. arXiv:0801.0006.

Rovelli, C. (2004). *Quantum Gravity* (Cambridge University Press, Cambridge).

Rudnick, L., Brown, S., and Williams, L.R. (2007). Extragalactic radio sources and the WMAP cold spot, *Astrophys. J.* **671**, 40. arXiv:0704.0908.

Ruiz, E. and Senovilla, J.M.M. (1992). General class of inhomogeneous perfect-fluid solutions, *Phys. Rev. D* **45**, 1995.

Ruiz-Lapuente, P. (2010). *Dark Energy: Observational and Theoretical Approaches* (Cambridge University Press, New York, NY).

Russ, H., Soffel, M.H., Kasai, M., and Börner, G. (1997). Age of the universe: influence of the inhomogeneities on the global expansion factor, *Phys. Rev. D* **56**, 2044.

Ryan, M.P. and Shepley, L.C. (1975). *Homogeneous Relativistic Cosmologies* (Princeton University Press, Princeton).

Rykoff, E.S., Evrard, A.E., McKay, T.A., Becker, M.R., Johnston, D.E. *et al.* (2008). The L_X-M relation of clusters of galaxies, *Mon. Not. Roy. Astr. Soc.* **387**, L28. arXiv:0802.1069.

Sachs, R.K. and Wolfe, A.M. (1967). Perturbations of a cosmological model and angular variations of the microwave background, *Astrophys. J.* **147**, 73.

Sakagami, M. (1988). Evolution from pure states into mixed states in de-Sitter space, *Prog. Theor. Phys.* **79**, 442.

Salam, A. (1990). *Unification of the Fundamental Forces* (Cambridge University Press, Cambridge).

Sandage, A. (1961). The ability of the 200-inch telescope to discriminate between selected world-models, *Astrophys. J.* **133**, 355.

Sandin, P. (2009). Tilted two-fluid Bianchi type I models, *Gen. Rel. Grav.* **41**, 2707. arXiv:0901.0800.

Sandin, P. and Uggla, C. (2010). Perfect fluids and generic spacelike singularities, *Class. Quant. Grav.* **27**, 025013. arXiv:0908.0298.

Sarkar, P., Yadav, J., Pandey, B. and Bharadwaj, S. (2009). The scale of homogeneity of the galaxy distribution in SDSS DR6, *Mon. Not. Roy. Astr. Soc.* **399**, L128. arXiv:0906.3431.

Sasaki, M. (1993). Cosmological gravitational lens equation – its validity and limitation, *Prog. Theor. Phys.* **90**, 753.

Sato, H. (1984). Voids in the expanding universe, in *General Relativity and Gravitation: Proceedings of the 10th International Conference on General Relativity and Gravitation*, ed. B. Bertotti, F. de Felice, and A. Pascolini (D. Reidel and Co., Dordrecht), p. 289.

Schmidt, B.G. (1969). Discrete isotropies in a class of cosmological models., *Commun. Math. Phys.* **15**, 329.

Schneider, P., Ehlers, J., and Falco, E.E. (1992). *Gravitational Lenses* (Springer-Verlag, New York).

Schneider, P., Kochanek, C., and Wambsganss, J. (2006). *Gravitational Lensing: Strong, Weak and Micro*, volume 33 of *Saas-Fee Advanced Course* ed. G. Meylan, P. Jetzer and P. North (Springer, Berlin and Heidelberg).

Schouten, E. (1954). *Ricci Calculus: An Introduction to Tensor Analysis and its Geometrical Applications (2nd edition)*, volume X of *Die Grundlehren der Mathematischen Wissenschaften* (Springer, Berlin).

Schrödinger, E. (1956). *Expanding Universe* (Cambridge University Press, Cambridge).

Schucking, E. (1954). The Schwarzschild line element and the expansion of the universe (in German), *Zs. f. Phys.* **137**, 595.

Schutz, B.F. (1980). *Geometrical Methods of Mathematical Physics* (Cambridge University Press, Cambridge).

Schutz, B.F. (2009). *A First Course in General Relativity* (*2nd edition*) (Cambridge University Press, Cambridge).

Schwab, J., Bolton, A.S., and Rappaport, S.A. (2010). Galaxy-scale strong lensing tests of gravity and geometric cosmology: Constraints and systematic limitations, *Astrophys. J.* **708**, 750. arXiv:0907.4992.

Sciama, D.W. (1969). *The Physical Foundations of General Relativity* (Doubleday, New York).

Seitz, S., Schneider, P., and Ehlers, J. (1994). Light-propagation in arbitrary spacetimes and the gravitational lens approximation, *Class. Quant. Grav.* **11**, 2345.

Sekino, Y., Shenker, S., and Susskind, L. (2010). On the topological phases of eternal inflation, *Phys. Rev. D* **81**, 123515. arXiv:1003.1347.

Seljak, U. and Zaldarriaga, M. (1996). A line-of-sight integration approach to cosmic microwave background anisotropies, *Astrophys. J.* **469**, 437. arXiv:astro-ph/9603033.

Seljak, U. and Zaldarriaga, M. (1997). Signature of gravity waves in the polarization of the microwave background, *Phys. Rev. Lett.* **78**, 2054. arXiv:astro-ph/9609169.

Sen, N.R. (1934). On the stability of cosmological models, *Zs. f. Astrophys.* **9**, 215. Reprinted in *Gen. Rel. Grav.* **29**, 1477–1488 (1997).

Senovilla, J.M.M. (1990). New class of inhomogeneous cosmological perfect-fluid solutions without big-bang singularity, *Phys. Rev. Lett.* **64**, 2219.

Senovilla, J.M.M. (1998). Singularity theorems and their consequences, *Gen. Rel. Grav.* **30**, 701.

Senovilla, J.M.M. and Sopuerta, C.F. (1994). New G_1 and G_2 inhomogeneous cosmological models from the generalized Kerr-Schild transformation, *Class. Quant. Grav.* **11**, 2073.

Senovilla, J.M.M., Sopuerta, C.F., and Szekeres, P. (1998). Theorems on shear-free perfect fluids with their Newtonian analogues, *Gen. Rel. Grav.* **30**, 389. arXiv:gr-qc/9702035.

Sethi, S.K., Nath, B.B., and Subramanian, K. (2008). Primordial magnetic fields and formation of molecular hydrogen, *Mon. Not. Roy. Astr. Soc.* **387**, 1589. arXiv:0804.3473.

Shafieloo, A. and Clarkson, C.A. (2010). Model independent tests of the standard cosmological model, *Phys. Rev. D* **81**, 083537. arXiv:0911.4858.

Shan, H-Y., Qin, B., Fort, B., Tao, C., Wu, X.-P. *et al.* (2010). Offset between dark matter and ordinary matter: evidence from a sample of 38 lensing clusters of galaxies, *Mon. Not. Roy. Astr. Soc.* **406**, 1134. arXiv:1004.1475.

Shapiro Key, J., Cornish, N.J., Spergel, D.N., and Starkman, G.D. (2007). Extending the WMAP bound on the size of the universe, *Phys. Rev. D* **75**, 084034. arXiv:astro-ph/0604616.

Shikin, I.S. (1967). A uniform axisymmetrical cosmological model in the ultrarelativistic case, *Dokl. Akad. Nauk SSSR* **176**, 1048.

Shiromizu, T., Maeda, K.-I. and Sasaki, M. (2000). The Einstein equations on the 3-brane world, *Phys. Rev. D* **62**, 024012. arXiv:gr-qc/9910076.

Shoji, M., Jeong, D., and Komatsu, E. (2009). Extracting angular diameter distance and expansion rate of the Universe from two-dimensional galaxy power spectrum at high redshifts: Baryon Acoustic Oscillation fitting versus full modeling, *Astrophys. J.* **693**, 1404. arXiv:0805.4238.

Sigurdson, K. and Furlanetto, S.R. (2006). Measuring the primordial deuterium abundance during the cosmic dark ages, *Phys. Rev. Lett.* **97**, 091301. arXiv:astro-ph/0505173.

Siklos, S.T.C. (1976). *Singularities, Invariants and Cosmology*, University of Cambridge PhD thesis.

Siklos, S.T.C. (1981). Non-scalar singularities in spatially homogeneous cosmologies, *Gen. Rel. Grav.* **13**, 433.

Siklos, S.T.C. (1984). Einstein's equations and some cosmological solutions, in *Relativistic Astrophysics and Cosmology* (*Proceedings of the XIVth GIFT International Seminar on Theoretical Physics*), ed. X. Fustero and E. Verdaguer (World Scientific, Singapore), p. 201.

Silk, J. (1967). Fluctuations in the primordial fireball, *Nature* **215**, 1155.

Silk, J. (1968). Cosmic black-body radiation and galaxy formation, *Astrophys. J.* **151**, 459.

Silk, J. (1977). Large-scale inhomogeneity of the universe: spherically symmetric models, *Astron. Astrophys.* **59**, 53.

Silk, J. (2008). *The Infinite Cosmos: Questions from the Frontiers of Cosmology* (Oxford University Press, New York).

Singh, P. (2008). Transcending big bang in loop quantum cosmology: Recent advances, *J. Phys.: Conf. Ser.* **140**, 012005. arXiv:0901.1301.

Sintes, A.M. (1998). Kinematic self-similar locally rotationally symmetric models, *Class. Quant. Grav.* **15**, 3689.

Skordis, C. (2009). The tensor-vector-scalar theory and its cosmology, *Class. Quant. Grav.* **26**, 143001. arXiv:0903.3602.

Smarr, L. and York, Jr., J.W. (1978). Kinematical conditions in the construction of spacetime, *Phys. Rev. D* **17**, 2529.

Smith, M.S., Kawano, L.H., and Malaney, R.A. (1993). Experimental, computational, and observational analysis of primordial nucleosynthesis, *Astrophys. J. Suppl.* **85**, 219.

Smith, T.L. (2009). Testing gravity on kiloparsec scales with strong gravitational lenses. arXiv:0907.4829.

Sneddon, G.E. (1975). Hamiltonian cosmology: a further investigation, *J. Phys. A* **9**, 229.

Sollerman, J., Mörtsell, E., Davis, T.M., Bassett, B.A.C.C. *et al.* (2009). First-year Sloan Digital Sky Survey-II (SDSS-II) Supernova results: Constraints on non-standard cosmological models, *Astrophys. J.* **703**, 1374. arXiv:0908.4276.

Song, Y.-S., Peiris, H., and Hu, W. (2007). Cosmological constraints on f(R) acceleration models, *Phys. Rev. D* **76**, 063517. arXiv:0706.2399.

Sopuerta, C.F. (1998). Covariant study of a conjecture on shear-free barotropic perfect fluids, *Class. Quant. Grav.* **15**, 1043.

Sotiriou, T.P. and Faraoni, V. (2010). f(R) theories of gravity, *Rev. Mod. Phys.* **82**, 451. arXiv:0805.1726.

Springel, V., White, S.D.M., Jenkins, A., Frenk, C.S., Yoshida, N. *et al.* (2005). Simulations of the formation, evolution and clustering of galaxies and quasars, *Nature* **435**, 629. arXiv:astro-ph/0504097.

Stabell, R. and Refsdal, S. (1966). Classification of general relativistic world models, *Mon. Not. Roy. Astr. Soc.* **132**, 379.

Starobinsky, A.A. (1983). Isotropization of arbitrary cosmological expansion given an effective cosmological constant, *JETP Lett.* **37**, 66.

Starobinsky, A.A. (1985). Multicomponent de Sitter (inflationary) stages and the generation of perturbations, *JETP Lett.* **42**, 152.

Steigman, G. (2006). Primordial nucleosynthesis:. Successes and challenges, *Int. J. Mod. Phys. E* **15**, 1. arXiv:astro-ph/0511534.

Steigman, G. (2010). Primordial nucleosynthesis after WMAP, in *Chemical Abundances in the Universe: Connecting First Stars to Planets*, ed. K. Cunha, M. Spite, and B. Barbuy, volume 265 of *IAU Symposia* (Cambridge University Press, Cambridge), p. 15.

Steinhardt, P.J. and Turok, N. (2002). Cosmic evolution in a cyclic universe, *Phys. Rev. D* **65**, 126003. arXiv:hep-th/0111030.

Stelmach, J. and Jakacka, I. (2006). Angular sizes in spherically symmetric Stephani cosmological models, *Class. Quant. Grav.* **23**, 6621.

Stephani, H. (1967). Konform flache gravitationsfelder, *Commun. Math. Phys.* **5**, 337.

Stephani, H. (1987). Some perfect fluid solutions of Einstein's field equations without symmetries, *Class. Quant. Grav.* **4**, 125.

Stephani, H. (2004). *Relativity: An Introduction to Special and General Relativity* (Cambridge University Press, Cambridge). Third edition, translated by J.M. Stewart.

Stephani, H., Kramer, D., MacCallum, M.A.H., Hoenselaers, C.A., and Herlt, E. (2003). *Exact Solutions of Einstein's Field Equations, Second edition* (Cambridge University Press, Cambridge). Corrected paperback reprint, 2009.

Stephani, H. and Wolf, T. (1985). Perfect fluid and vacuum solutions of Einstein's field equations with flat 3-dimensional slices, in *Galaxies, Axisymmetric Systems and Relativity. Essays Presented to W.B. Bonnor on his 65th Birthday.*, ed. M.A.H. MacCallum (Cambridge University Press, Cambridge), p. 275.

Stewart, J.M. (1971). *Non-equilibrium Relativistic Kinetic Theory*, volume 10 of *Lecture Notes in Physics* (Springer, Berlin and Heidelberg).

Stewart, J.M. (1990). Perturbations of Friedmann-Robertson-Walker cosmological models, *Class. Quant. Grav.* **7**, 1169.

Stewart, J.M. (1994). *Advanced General Relativity* (Cambridge University Press, Cambridge). Paperback edition.

Stewart, J.M. and Ellis, G.F.R. (1968). On solutions of Einstein's equations for a fluid which exhibit local rotational symmetry, *J. Math. Phys.* **9**, 1072.

Stewart, J.M. and Walker, M. (1974). Perturbations of space-times in general relativity, *Proc. Roy. Soc. London A* **341**, 49.

Stoeger, W.R., Maartens, R., and Ellis, G.F.R. (1995). Proving almost-homogeneity of the universe: an almost Ehlers-Geren-Sachs theorem, *Astrophys. J.* **443**, 1.

Subramanian, K. (2010). Magnetic fields in the early Universe, *Astron. Nachr.* **331**, 110. arXiv:0911.4771.

Sultana, J. and Dyer, C.C. (2005). Cosmological black holes: A black hole in the Einstein-de Sitter universe, *Gen. Rel. Grav.* **37**, 1347.

Sun, C.-Y. (2011). Does the apparent horizon exist in the Sultana–Dyer space-time?, *Commun. Math. Phys.* **55**, 597. arXiv:1004.1760.

Sung, R. and Coles, P. (2009). Polarized spots in anisotropic open universes, *Class. Quant. Grav.* **26**, 172001. arXiv:0905.2307.

Sunyaev, R.A. and Zel'dovich, Y.B. (1970). Small-scale fluctuations of relic radiation, *Astrophys. Space Sci.* **7**, 3.

Susskind, L. (2005). *The Cosmic Landscape: String Theory and the Illusion of Intelligent Design* (Little Brown, New York).

Sussman, R.A. (1988). On spherically symmetric shear-free perfect fluid configurations (neutral and charged). III. Global view, *J. Math. Phys.* **29**, 1177.

Sussman, R.A. (2010a). Evolution of radial profiles in regular Lemaître-Tolman-Bondi dust models, *Class. Quant. Grav.* **27**, 175001. arXiv:1005.0717.

Sussman, R.A. (2010b). A new approach for doing theoretical and numeric work with Lemaître-Tolman-Bondi dust models. arXiv:1001.0904.

Sussman, R.A. and Izquierdo, G. (2011). A dynamical systems study of the inhomogeneous ΛCDM model, *Class. Quant. Grav.* **28**, 045006. arXiv:1004.0773.

Suto, Y., Sato, K., and Sato, H. (1984). Nonlinear evolution of negative density perturbations in a radiation-dominated universe, *Prog. Theor. Phys.* **72**, 1137.

Suyu, S.H., Marshall, P.J., Auger, M.W., Hilbert, S., Blandford, R.D. *et al.* (2010). Dissecting the gravitational lens B1608+656. II. Precision measurements of the Hubble constant, spatial curvature, and the dark energy equation of state, *Astrophys. J.* **711**, 201.

Sylos Labini, F., Vasilyev, N.L., Pietronero, L., and Baryshev, Y.V. (2009). Absence of self-averaging and of homogeneity in the large-scale galaxy distribution, *Europhys. Lett.* **86**, 49001. arXiv:0805.1132.

Synge, J.L. (1937). Relativistic hydrodynamics, *Proc. Lond. Math. Soc.* **43**, 376.

Synge, J.L. (1971). *Relativity: The General Theory* (North-Holland, Amsterdam). Fourth printing.

Synge, J.L. and Schild, A. (1949). *Tensor Calculus* (University of Toronto Press, Toronto). Reprinted 1961.

Szafron, D.A. (1977). Inhomogeneous cosmologies: new exact solutions and their evolution, *J. Math. Phys.* **18**, 1673.

Szafron, D.A. and Collins, C.B. (1979). A new approach to inhomogeneous cosmologies: intrinsic symmetries II. Conformally flat slices and an invariant classification, *J. Math. Phys.* **20**, 2354.

Szczyryba, W. (1976). A symplectic structure of the set of Einstein metrics: a canonical formalism for general relativity, *Commun. Math. Phys.* **51**, 163.

Szekeres, P. (1971). Linearized gravitational theory in macroscopic media, *Ann. Phys. (N.Y.)* **64**, 599.

Szekeres, P. (1975). A class of inhomogeneous cosmological models, *Commun. Math. Phys.* **41**, 55.

Tanimoto, M., Moncrief, V., and Yasuno, K. (2003). Perturbations of spatially closed Bianchi III spacetimes, *Class. Quant. Grav.* **20**, 1879. arXiv:gr-qc/0210078.

Taub, A.H. (1951). Empty space-times admitting a three-parameter group of motions, *Ann. Math.* **53**, 472. Reprinted as *Gen. Rel. Grav.* **36**, 2689-2719 (2004).

Taub, A.H. (1971). Variational principles in general relativity, in *Relativistic Fluid dynamics (C.I.M.E., Bressanone, 1970)*, ed. C. Cattaneo (Edizioni Cremonese, Rome), p. 206.

Taub, A.H. (1980). Space-times with distribution-valued curvature tensors, *J. Math. Phys.* **21**, 1423.

Tegmark, M. (2003). Parallel universes, in *Science and Ultimate Reality: From Quantum to Cosmos*, ed. J.D. Barrow, P.C.W. Davies, and C. Harper (Cambridge University Press, Cambridge). arXiv:astro-ph/0302131.

Tegmark, M. (2008). The mathematical universe, *Found. Phys.* **38**, 101. arXiv:0704.0646.

Tegmark, M., Aguirre, A., Rees, M.J., and Wilczek, F. (2006). Dimensionless constants, cosmology and other dark matters: Theory, *Phys. Rev. D* **73**, 023505. arXiv:astro-ph/0511774.

Tegmark, M., Blanton, M.R., Strauss, M.A., Hoyle, F., Schlegel, D. *et al.* (2004). The three-dimensional power spectrum of galaxies from the Sloan Digital Sky Survey, *Astrophys. J.* **606**, 702. arXiv:astro-ph/0310725.

Tegmark, M., Vilenkin, A., and Pogosan, L. (2005). Anthropic predictions for neutrino masses, *Phys. Rev. D* **71**, 103523. arXiv:astro-ph/0304536.

Thiemann, T. (2007). *Modern Canonical Quantum General Relativity* (Cambridge University Press, Cambridge).

Thomson (Lord Kelvin), W. (1862). On the age of the sun's heat, *Macmillan's Magazine* **5**, 388.

Thorne, K.S. (1967). Primordial element formation, primordial magnetic fields and the isotropy of the universe, *Astrophys. J.* **148**, 51.

Thorne, K.S. (1980). Multipole expansions of gravitational radiation, *Rev. Mod. Phys.* **52**, 299.

Thorne, K.S. (1981). Relativistic radiative transfer – Moment formalisms, *Mon. Not. Roy. Astr. Soc.* **194**, 439.

Thurston, W.P. (1997). *Three-dimensional Geometry and Topology* (Princeton University Press, Princeton).

Tipler, F.J., Clarke, C.J.S., and Ellis, G.F.R. (1980). Singularities and horizons: a review article, in *General Relativity and Gravitation: One Hundred Years after the Birth of Albert Einstein*, ed. A. Held, volume 2 (Plenum, New York), p. 97.

Titov, O. (2009). Systematic effects in the radio source proper motion, in *19th European VLBI for Geodesy and Astrometry Working Meeting*, ed. G. Bourda, P. Charlot, and A. Collioud (Université Bordeaux 1 - CNRS, Bordeaux), p. 14. arXiv:0906.4840.

Tod, K.P. (2002). Isotropic cosmological singularities, in *The Conformal Structure of Space-time*, ed. J. Frauendiener and H. Friedrich, volume 604 of *Lecture Notes in Phys.* (Springer, Berlin), p. 123.

Tolman, R. (1934). Effect of inhomogeneity on cosmological models, *Proc. Nat. Acad. Sci.* **20**, 169. Reprinted in *Gen. Rel. Grav.*, **29**, 935-943 (1997).

Tolman, R. and Ward, M. (1932). On the behaviour of non-static models of the universe when the cosmological constant is omitted, *Phys. Rev.* **39**, 835.

Tomita, K. (1968). Theoretical relations between observable quantities in an anisotropic and homogeneous universe, *Prog. Theor. Phys.* **40**, 264.

Tomita, K. (1978). Inhomogeneous cosmological models containing space-like and time-like singularities alternately, *Prog. Theor. Phys.* **59**, 1150.

Tomita, K. (1982). Tensor spherical and pseudo-spherical harmonics in four-dimensional spaces, *Prog. Theor. Phys.* **68**, 310.

Tomita, K. (2010). Gauge-invariant treatment of the integrated Sachs-Wolfe effect on general spherically symmetric spacetimes, *Phys. Rev. D* **81**, 063509. arXiv:0912.4773.

Torrence, R.J. and Couch, W.E. (1988). Note on Kantowski-Sachs spacetimes, *Gen. Rel. Grav.* **20**, 603.

Traschen, J. (1984). Causal cosmological perturbations and implications for the Sachs-Wolfe effect, *Phys. Rev. D* **29**, 1563.

Traschen, J. (1985). Constraints on stress-energy perturbations in general relativity, *Phys. Rev. D* **31**, 283.

Trautman, A. (1965). Foundations and current problems of general relativity, in *Lectures on General Relativity, Vol.1, Brandeis 1964.* (Prentice-Hall, Englewood Cliffs, New Jersey).

Treciokas, R. (1972). *Relativistic kinetic theory*, University of Cambridge PhD thesis.

Treciokas, R. and Ellis, G.F.R. (1971). Isotropic solutions of the Einstein-Boltzmann equations, *Commun. Math. Phys.* **23**, 1.

Trümper, M. (1962). Beitrage zur Theorie der Gravitations-Strahlungsfelder, *Akad. Wiss. Lit. Mainz, Abh. Mat.-Nat. Kl.* **12**.

Tsagas, C.G., Challinor, A., and Maartens, R. (2008). Relativistic cosmology and large-scale structure, *Phys. Reports* **465**, 61. arXiv:0705.4397.

Turok, N., Pen, U., and Seljak, U. (1998). Scalar, vector, and tensor contributions to CMB anisotropies from cosmic defects, *Phys. Rev. D* **58**, 023506. arXiv:astro-ph/9706250.

Uggla, C., van Elst, H., Wainwright, J., and Ellis, G.F.R. (2003). Past attractor in inhomogeneous cosmology, *Phys. Rev. D* **68**, 103502. arXiv:gr-qc/0304002.

Uggla, C. and von Zur-Muhlen, H. (1990). Compactified and reduced dynamics for locally rotationally symmetric Bianchi type IX perfect fluid models, *Class. Quant. Grav.* **7**, 1365.

Uzan, J.-P. (1998). Dynamics of relativistic interacting gases: from a kinetic to a fluid description, *Class. Quant. Grav.* **15**, 1063. arXiv:gr-qc/9801108.

Uzan, J.-P. (2003). The fundamental constants and their variation: observational status and theoretical motivations, *Rev. Mod. Phys.* **75**, 403. arXiv:hep-ph/0205340.

Uzan, J.-P., Clarkson, C.A., and Ellis, G.F.R. (2008). Time drift of cosmological redshifts as a test of the Copernican principle, *Phys. Rev. Lett* **100**, 191303. arXiv:0801.0068.

Uzan, J.-P., Lehoucq, R., and Luminet, J.-P. (2000). New developments in the search for the topology of the universe, *Nucl. Phys. B: Proc. Suppl.* **80**. In *Proceedings of the XIXth Texas meeting*, ed. E. Aubourg, T. Montmerle, J. Paul and P. Peter. CD-Rom version, article-no: 04/25, arXiv:gr-qc/0005128.

Van den Bergh, N. (1988). Perfect-fluid models admitting a non-Abelian and maximal two-parameter group of isometries, *Class. Quant. Grav.* **5**, 861.

Van den Bergh, N. (1992). A qualitative discussion of the Wils inhomogeneous stiff fluid cosmologies, *Class. Quant. Grav.* **9**, 2297.

Van den Bergh, N. and Wylleman, L. (2004). Silent universes with a cosmological constant, *Class. Quant. Grav.* **21**, 2291.

van den Hoogen, J. (2009). A complete cosmological solution to the averaged Einstein field equations as found in macroscopic gravity, *J. Math. Phys.* **50**, 082503. arXiv:0909.0070.

van den Hoogen, R. and Olasagasti, I. (1999). Isotropization of scalar field Bianchi type-IX models with an exponential potential, *Phys. Rev. D* **59**, 107302.

van der Walt, P.J. and Bishop, N.T. (2010). Observational cosmology using characteristic numerical relativity, *Phys. Rev. D* **82**, 084001. arXiv:1007.3189.

van Elst, H. (1996). *Extensions and applications of 1+3 decomposition methods in general relativistic cosmological modelling*, Queen Mary and Westfield College, London, PhD thesis.

van Elst, H. and Ellis, G.F.R. (1996). The covariant approach to LRS perfect fluid spacetime geometries, *Class. Quant. Grav.* **13**, 1099.

van Elst, H. and Ellis, G.F.R. (1998). Quasi-Newtonian dust cosmologies, *Class. Quant. Grav.* **15**, 3545. arXiv:gr-qc/9805087.

van Elst, H. and Ellis, G.F.R. (1999). Causal propagation of geometrical fields in relativistic cosmology, *Phys. Rev. D* **59**, 024013.

van Elst, H. and Uggla, C. (1997). General relativistic 1+3 orthonormal frame approach, *Class. Quant. Grav.* **14**, 2673.

van Elst, H., Uggla, C., and Wainwright, J. (2002). Dynamical systems approach to G_2 cosmology, *Class. Quant. Grav.* **19**, 51.

van Oirschot, P., Kwan, J. and Lewis, G.F. (2010). Through the looking glass: Why the 'cosmic horizon' is not a horizon, *Mon. Not. Roy. Astr. Soc.* **404**, 1633. arXiv:1001.4795.

Vanderveld, R.A., Flanagan, É.É., and Wasserman, I. (2006). Mimicking dark energy with Lemaître-Tolman-Bondi models: Weak central singularities and critical points, *Phys. Rev. D* **74**, 023506. arXiv:astro-ph/0602476.

Vanderveld, R.A., Flanagan, É.É., and Wasserman, I. (2007). Systematic corrections to the measured cosmological constant as a result of local inhomogeneity, *Phys. Rev. D* **76**, 083504. arXiv:0706.1931.

Vaudrevange, P.M. and Kofman, L. (2007). Trans-Planckian issue in the Milne Universe. arXiv:0706.0980.

Velden, T. (1997). Dynamics of pressure-free matter in general relativity. Diplomarbeit, University of Bielefeld.

Vilenkin, A. (1983). Gravitational field of vacuum domain walls, *Phys. Lett. B* **133**, 177.

Vilenkin, A. (2003). Quantum cosmology and eternal inflation, in Gibbons, Shellard and Rankin (2003).

Vilenkin, A. (2006). *Many Worlds in One. The Search for Other Universes* (Hill and Wang, New York).

Vonlanthen, M., Räsänen, S., and Durrer, R. (2010). Model-independent cosmological constraints from the CMB, *J. Cosmol. Astropart. Phys.* **8**, 23. arXiv:1003.0810.

Wagoner, R.V., Fowler, W.A., and Hoyle, F. (1967). On synthesis of elements at very high temperatures, *Astrophys. J.* **148**, 3.

Wainwright, J. (1974). Algebraically special fluid space-times with hypersurface-orthogonal shearfree rays, *Int. J. Theor. Phys.* **10**, 39.

Wainwright, J. (1977). Characterization of the Szekeres inhomogeneous cosmologies as algebraically special space-times, *J. Math. Phys.* **18**, 672.

Wainwright, J. (1979). A classification scheme for non-rotating inhomogeneous cosmologies, *J. Phys. A* **12**, 2015.

Wainwright, J. (1981). Exact spatially inhomogeneous cosmologies, *J. Phys. A* **14**, 1131.

Wainwright, J. (1983). A spatially homogeneous cosmological model with plane-wave singularity, *Phys. Lett. A* **99**, 301.

Wainwright, J. (1988). On the asymptotic states of orthogonal spatially homogeneous cosmologies, in *Relativity Today. Proceedings of the Second Hungarian Relativity Workshop*, ed. Z. Perjes (World Scientific, Singapore), p. 237.

Wainwright, J. and Andrews, S. (2009). The dynamics of Lemaître–Tolman cosmologies, *Class. Quant. Grav.* **26**, 085017.

Wainwright, J., Coley, A.A., Ellis, G.F.R., and Hancock, M. (1998). On the isotropy of the universe: do Bianchi VII$_h$ cosmologies isotropize?, *Class. Quant. Grav.* **15**, 331.

Wainwright, J. and Ellis, G.F.R. (1997). *Dynamical Systems in Cosmology* (Cambridge University Press, Cambridge).

Wainwright, J. and Hsu, L. (1989). A dynamical systems approach to Bianchi cosmologies: orthogonal models of class A, *Class. Quant. Grav.* **6**, 1409.

Wainwright, J. and Lim, W.C. (2005). Cosmological models from a dynamical systems perspective, *J. Hyper. Diff. Equat.* **2**, 437. arXiv:gr-qc/0409082.

Wald, R.M. (1983). Asymptotic behavior of homogeneous cosmological models in the presence of a positive cosmological constant, *Phys. Rev. D* **28**, 2118.

Wald, R.M. (1984). *General Relativity* (University of Chicago Press, Chicago).

Wald, R.M. (1993). Correlations beyond the cosmological horizon, in *The Origin of Structure in the Universe*, ed. E. Gunzig and P. Nardone, volume 393 of *NATO ASI Series C* (Kluwer, Dordrecht), p. 217.

Wald, R.M. (1994). *Quantum Field Theory in Curved Space-time and Black Hole Thermodynamics* (University of Chicago Press, Chicago).

Wald, R.M. (2005). The arrow of time and the initial conditions for the universe. arXiv:gr-qc/0507094.

Wald, R.M. and Yip, P. (1981). On the existence of simultaneous coordinates in spacetimes with spacelike singularities, *J. Math. Phys.* **22**, 2659.

Walker, A.G. (1944). Completely symmetric spaces, *J. Lond. Math. Soc.* **19**, 219.

Walsh, D., Carswell, R.F., and Weymann, R. (1979). 0957 + 561 A, B – twin quasistellar objects or gravitational lens, *Nature* **279**, 381.

Wambsganss, J. (2001). Gravitational lensing in astronomy, *Living Rev. Relativity* http://relativity.livingreviews.org/Articles/lrr-1998-12/.

Wands, D., Malik, K., Lyth, D.H., and Liddle, A. (2000). A new approach to the evolution of cosmological perturbations on large scales, *Phys. Rev. D* **62**, 043527. astro-ph/0003278.

Wang, A.-Z. (1992). Planar domain walls emitting and absorbing electromagnetic radiation, *Phys. Rev. D* **44**, 1705.

Weaver, M. (2000). Dynamics of magnetic Bianchi VI_0 cosmologies, *Class. Quant. Grav.* **17**, 421. arXiv:gr-qc/9909043.

Weaver, M., Isenberg, J., and Berger, B.K. (1998). Mixmaster behaviour in inhomogeneous cosmological spacetimes, *Phys. Rev. Lett.* **80**, 2984. arXiv:gr-qc/9712055.

Weber, E. (1984). Kantowski-Sachs cosmological models approaching isotropy, *J. Math. Phys.* **25**, 3279.

Weber, E. (1985). Kantowski-Sachs cosmological models as big-bang models, *J. Math. Phys.* **26**, 1308.

Weeks, J. and Gundermann, J. (2007). Dodecahedral topology fails to explain quadrupole-octupole alignment, *Class. Quant. Grav.* **24**, 1863.

Weeks, J., Luminet, J.-P., Riazuelo, A., and Lehoucq, R. (2004). Well-proportioned universes suppress CMB quadrupole, *Mon. Not. Roy. Astr. Soc.* **352**, 258. arXiv:astro-ph/0312312.

Weinberg, S. (1972). *Gravitation and Cosmology; Principles and Applications of the General Theory of Relativity* (Wiley, New York).

Weinberg, S. (1976). Apparent luminosities in a locally inhomogeneous universe, *Astrophys. J.* **208**, L1.

Weinberg, S. (1989). The cosmological constant problem, *Rev. Mod. Phys.* **61**, 1.

Weinberg, S. (2000a). The cosmological constant problem. arXiv:astro-ph/0005265.

Weinberg, S. (2000b). A priori distribution of the cosmological constant, *Phys. Rev. D* **61**, 103505. arXiv:astro-ph/0002387.

Weinberg, S. (2007). Living in the multiverse, in *Universe or Multiverse?*, ed. B.J. Carr (Cambridge University Press, Cambridge). arXiv:hep-th/0511037.

Weinberg, S. (2008). *Cosmology* (Oxford University Press, New York).

Weltmann, A., Murugan, J., and Ellis, G.F.R. (2010). *Foundations of Space and Time: Reflections on Quantum Gravity* (Cambridge University Press, Cambridge).

Wesson, P.S. (1978). General-relativistic hierarchical cosmology: an exact model, *Astrophys. Space Sci.* **54**, 489.

Wesson, P.S. (1979). Observable relations in an inhomogeneous self-similar cosmology, *Astrophys. J.* **228**, 647.

Wetterich, C. (2003). Can structure formation influence the cosmological evolution?, *Phys. Rev. D* **67**, 043513.

Wheeler, J.A. (1962). *Geometrodynamics* (Academic Press, New York).

White, S.D.M. (2007). Fundamentalist physics: why dark energy is bad for astronomy, *Rep. Prog. Phys.* **70**, 883. arXiv:0704.2291.

Will, C.M. (2006). The confrontation between general relativity and experiment, *Living Rev. Relativity* http://relativity.livingreviews.org/Articles/lrr-2006-3/.

Wilson, M.L. (1983). On the anisotropy of the cosmological background matter and radiation distribution. II – The radiation anisotropy in models with negative spatial curvature, *Astrophys. J.* **273**, 2.

Wiltshire, D.L. (2007a). Cosmic clocks, cosmic variance and cosmic averages, *New. J. Phys.* **9**, 377. arXiv:gr-qc/0702082.

Wiltshire, D.L. (2007b). Exact solution to the averaging problem in cosmology, *Phys. Rev. Lett.* **99**, 251101. arXiv:0709.0732.

Wiltshire, D.L. (2008a). Cosmological equivalence principle and the weak-field limit, *Phys. Rev. D* **78**, 084032. arXiv:0809.1183.

Wiltshire, D.L. (2008b). Dark energy without dark energy, in *Dark Matter in Astroparticle and Particle Physics: Proceedings of the 6th International Heidelberg Conference*, ed. H.V. Klapdor-Kleingrothaus and G.F. Lewis (World Scientific, Singapore), p. 565. arXiv:0712.3984.

Wiltshire, D.L. (2009). Average observational quantities in the timescape cosmology, *Phys. Rev. D* **80**, 123512. arXiv:0909.0749.

Wolf, J.A. (1972). *Spaces of Constant Curvature* (*2nd edition*) (J.A. Wolf, Berkeley).

Wright, E.L. (2007). Constraints on dark energy from supernovae, Gamma ray bursts, acoustic oscillations, nucleosynthesis and large scale structure and the Hubble constant, *Astrophys. J.* **664**, 633. arXiv:astro-ph/0701584.

Wu, Z.-C. (1981). Self-similar cosmological models, *Gen. Rel. Grav.* **13**, 625.

Wylleman, L. (2006). Anti-Newtonian universes do not exist, *Class. Quant. Grav.* **23**, 2727.

Wylleman, L. and Van den Bergh, N. (2006). Complete classification of purely magnetic, non-rotating and non-accelerating perfect fluids, *Phys. Rev. D* **74**, 084001.

Wyman, M. (1946). Equations of state for radially symmetric distributions of matter, *Phys. Rev.* **70**, 396.

Wyman, M. (1976). Jeffery-Williams lecture 1976: Nonstatic radially symmetric distributions of matter, *Can. Math. Bull.* **19**(3), 343.

Yamamoto, K. and Suto, Y. (1999). Two-point correlation function of high-redshift objects: an explicit formulation on a light-cone hypersurface, *Astrophys. J.* **517**, 1. arXiv:astro-ph/9812486.

Yamazaki, D.G., Ichiki, K., Kajino, T., and Mathews, G.J. (2010). New constraints on the primordial magnetic field, *Phys. Rev. D* **81**, 023008. arXiv:1001.2012.

Yoo, C-M., Kai, T., and Nakao, K-I. (2008). Solving the inverse problem with inhomogeneous universes, *Prog. Theor. Phys.* **120**, 937. arXiv:0807.0932.

Yoo, J. (2010). General relativistic description of the observed galaxy power spectrum: Do we understand what we measure?, *Phys. Rev. D* **82**, 083508. arXiv:1009.3021.

Yoo, J., Fitzpatrick, A.L., and Zaldarriaga, M. (2009). New perspective on galaxy clustering as a cosmological probe: General relativistic effects, *Phys. Rev. D* **80**, 083514. arXiv:0907.0707.

Yoshimura, M. (1988). The universe as a laboratory for high energy physics, in *Cosmology and Particle Physics*, ed. L.-Z. Fang and A. Zee (Gordon and Breach, New York), p. 293.

Zalaletdinov, R.M. (1997). Averaging problem in general relativity, macroscopic gravity and using Einstein's equations in cosmology, *Bull. Astron. Soc. India* **25**, 401. arXiv:gr-qc/9703016.

Zalaletdinov, R.M., Tavakol, R.K., and Ellis, G.F.R. (1996). On general and restricted covariance in general relativity, *Gen. Rel. Grav.* **28**, 1251.

Zaldarriaga, M., Seljak, U., and Bertschinger, E. (1998). Integral solution for the microwave background anisotropies in nonflat universes, *Astrophys. J.* **494**, 491. arXiv:astro-ph/9704265.

Zatsepin, G.T. and Kuz'min, V.A. (1966). Upper limit of the spectrum of cosmic rays, *JETP Lett.* **4**, 78.

Zeh, H.D. (1992). *The Physical Basis of the Direction of Time* (Springer Verlag, Berlin).

Zel'dovich, Y.B. (1964). Observations in a universe homogeneous in the mean, *Sov. Astr.* **8**, 13.

Zel'dovich, Y.B. (1970). The hypothesis of cosmological magnetic inhomogeneity, *Sov. Astr.* **13**, 608.

Zel'dovich, Y.B. and Grishchuk, L.P. (1984). Structure and future of the 'new' universe, *Mon. Not. Roy. Astr. Soc.* **207**, 23P.

Zel'dovich, Y.B., Kurt, V.G., and Sunyaev, R.A. (1968). Recombination of hydrogen in the hot model of the Universe, *Zh. Eksp. Teor. Fiz.* **55**, 278.

Zel'dovich, Y.B. and Sunyaev, R.A. (1969). The interaction of matter and radiation in a hot-model universe, *Astrophys. Space Sci.* **4**, 301.

Zhang, P., Liguori, M., Bean, R., and Dodelson, S. (2007). Probing gravity at cosmological scales by measurements which test the relationship between gravitational lensing and matter overdensity, *Phys. Rev. Lett.* **99**, 141302. arXiv:0704.1932.

Zibin, J.P. (2008). Scalar perturbations on Lemaître-Tolman-Bondi spacetimes, *Phys. Rev. D* **78**, 043504. arXiv:0804.1787.

Zibin, J.P., Moss, A., and Scott, D. (2008). Can we avoid dark energy?, *Phys. Rev. Lett.* **101**(25), 251303. 0809.3761.

Zibin, J.P. and Scott, D. (2008). Gauging the cosmic microwave background, *Phys. Rev. D* **78**, 123529. arXiv:0808.2047.

Zuntz, J., Zlosnik, T.G., Bourliot, F., Ferreira, P.G., and Starkman, G.D. (2010). Vector field models of modified gravity and the dark sector, *Phys. Rev. D* **81**, 104015. arXiv:1002.0849.

Zwiebach, B. (2004). *A First Course in String Theory* (Cambridge University Press, Cambridge).

Index

Printed in the United States
by Baker & Taylor Publisher Services